HAZARDOUS WASTE OPERATIONS AND EMERGENCY RESPONSE MANUAL AND DESK REFERENCE

HAZARDOUS WASTE OPERATIONS AND EMERGENCY RESPONSE MANUAL AND DESK REFERENCE

Christian L. Hackman, C.E.M.

E. Ellsworth Hackman, III, Ph.D., P.E.

Matthew E. Hackman, P.E., CHMM,
LSP (MA), LEP (CT)

McGRAW-HILL

New York Chicago San Francisco Lisbon London Madrid
Mexico City Milan New Delhi San Juan Seoul
Singapore Sydney Toronto

Library of Congress Cataloging-in-Publication Data

Hackman, Christian L.
  Hazardous waste operations and emergency response manual / Christian L. Hackman,
E. Ellsworth Hackman, III, Matthew E. Hackman.
        p. cm.
  Includes index.
  ISBN 0-07-137881-2 (alk. paper)
      1. Hazardous wastes—Safety measures—Handbooks, manuals, etc.   2. Hazardous
wastes—Health aspects—Handbooks, manuals, etc.   3. Waste spills—Safety
measures—Handbooks, manuals, etc.   4. Hazardous substances—Accidents—Handbooks,
manuals, etc.   5. Hazardous waste sites—Safety measures—Handbooks, manuals, etc.   6.
Hazardous waste management industry—Employees—Training of—United
States—Handbooks, manuals, etc.   I. Hackman, E. Ellsworth.   II. Hackman, Matthew E.
III. Title.

TD1050.S24 H33 2001
628.4'2—dc21

2001054375

McGraw-Hill

A Division of The McGraw-Hill Companies

1 2 3 4 5 6 7 8 9 0   AGM/AGM   0 9 8 7 6 5 4 3 2 1

P/N 137887-1
PART OF

ISBN 0-07-137881-2

*The sponsoring editor for this book was Kenneth P. McCombs, the editing
supervisor was Steven Melvin, and the production supervisor was Sherri
Souffrance. It was set in Times Roman by Pro-Image.*

Printed and bound by Quebecor/Martinsburg.

McGraw-Hill books are available at special quantity discounts to use as pre-
miums and sales promotions, or for use in corporate training programs. For
more information, please write to the Director of Special Sales, McGraw-
Hill, 2 Penn Plaza, New York, NY 10121-2298. Or contact your local book-
store.

This book is printed on recycled, acid-free paper containing a minimum of 50% recycled, de-inked fiber.

CONTENTS

Section 2. Hazardous Waste Defined 2.1

Section 3. Material Hazards 3.1

Section 4. Chemical Incompatibility   **4.1**

Section 5. Toxicology   **5.1**

Section 6. Personal Protective Equipment (PPE)   **6.1**

Section 9. Material Safety Data Sheets (MSDSs) and International Chemical Safety Cards (ICSCs)   9.1

Section 10. Confined Spaces   10.1

Section 11. Other Workplace Hazards

Section 13. Work at Hazardous Waste Generators and at Treatment, Storage, and Disposal Facilities (TSDFs)   13.1

Section 14. Superfund Sites and Brownfields: Site Investigation, Control, and Remediation

14.1

Section 15. Emergency Response   **15.1**

Section 16. Health and Safety System   **16.1**

CONTENTS OF THE COMPANION CD

The role of the Companion CD is threefold, to provide:

1) a selection of downloadable and printable resources for trainers, managers, students, and others engaged in the environmental, health and safety fields

2) the latest 'NIOSH (2002) Pocket Guide to Chemical Hazards and Other Databases', including the 365 page 'Emergency Response Guidebook'

3) Appendixes B through S of the Book.

The CD is written in Hypertext Markup Language (.html) for ease of 'navigation'. It also contains files in Microsoft WORD® (.doc) and Adobe Portable Document (.pdf) formats. An Adobe 5 Reader® is included. As the CD requires your computer to have a web browser, a copy of Netscape Communicator® version 4.77 is also included. Although the CD is a 'stand alone' resource, an Internet connection is recommended as there are many 'links' included on the CD to other valuable Internet resources.

Any prospective HAZWOPER personnel, having the Book as a study guide, taught by a competent trainer using the CD's printable teaching resources, and understanding the 'Key Facts' and the 'Glossary' terms, will be able to pass an examination for 40-Hour HAZWOPER Certification. This is also basic training for all emergency responders, and in fact, anyone who might contact hazardous materials.

The 34 selected downloadable and printable resources include all the major regulation references, sample HAZWOPER exams, checklists, hazardous waste reporting and medical exam forms, equipment inspection logs, and more.

The NIOSH Pocket Guide in CD form, for use with wireless laptops, is rapidly becoming a necessity for support of all field work and emergency response. The Appendices provide the in-depth detail needed in a reference work.

The CD has 'Autorun' capability. Insert it in your CD ROM. Allow about one minute for Autorun to complete its startup. **Then choose 'Start Databases'** and make your selection from the 34 topics as they are listed below. Be sure to read the CD's 'Read Me' file for easiest use of the CD. Choose 'Important Information' to see the 'Read Me' file. Many files are searchable. When you are linked to the Internet, you can immediately access many more files and databases.

TOPIC

'Ice Breaker' Exam and Answers
Sample HAZWOPER Worker Exam
Sample HAZWOPER Worker Exam Key
Sample HAZWOPER Supervisor Exam
Sample HAZWOPER Supervisor Exam Key
Key Facts from Sections 1–16 of this Book

NIOSH DATABASES

Dear Friends,

It is my pleasure to offer an introduction to this important and useful book. The success of my key initiative, Livable Delaware, depends upon the reclamation and redevelopment of Brownfields and other "wasted" land resources within our communities.

Building a Livable Delaware is based upon preventing continued unnecessary expansion into green fields and taking advantage of the infrastructure already available within our towns and cities. Land that was productive in a former life but is now a blight, or even a hazard, is a key target for new, or renewed, development. Bringing that land back into being of use and benefit to the community is essential to my Livable Delaware effort.

"Hazardous Waste Operations and Emergency Response, Manual and Desk Reference" provides the information necessary for those involved in Brownfields and Superfund cleanups and redevelopment. It provides valuable information to my partners in creating a Livable Delaware and livable towns and cities all over the country.

Sincerely,

Ruth Ann Minner
Governor

Dear Readers:

In my Navy career as a Safety and Environmental Senior Manager I have experienced numerous training efforts and read many books and training manuals on HAZWOPER subjects. Most of them fall far short of the mark. Invariably the authors of many of those materials rely on information from other sources, not from their own experiences, and much of the time outdated. What I always hope for, but rarely find, are materials prepared by authors with extensive work and training experience in each of the topics presented.

Having reviewed a draft of "Hazardous Waste Operations and Emergency Response, Manual and Desk Reference", I find the authors approach to the topics and coverage of the subjects to be highly commendable. The writing is clear and intelligent. The organization of the text provides easy access to answers to many relevant operational questions. The authors have spent considerable effort clearing up many misconceptions about HAZWOPER work that are found in other sources.

During my tenure at the Philadelphia Naval Shipyard as a Production Superintendent, I was responsible to assure training and certification for one hundred and twenty Gas Free Engineering Technicians to inspect approximately six hundred confined spaces daily for the safe entry and working conditions for various shipyard and shipboard personnel. Over a period of two decades, due to improvements to the Navy's Gas Free Engineering program, we were able to achieve a Zero Defects status for the confined spaces program.

One of my responsibilities was for both land-based and shipboard confined space inspections. Due to the space limitations imposed by naval architecture, there are more rigorous testing and inspection requirements to certify shipboard confined space entry as compared with most HAZWOPER projects. I find that the authors have successfully captured the vital elements of the Navy program.

In this book I have found what I have been searching for as a HAZWOPER training manual and desk reference. I intend to keep it close at hand for frequent reference.

Sincerely,

Angelo P. Marchiano
United States Navy

FOREWORD

HAZARDOUS MATERIALS ARE BEING FOUND EVERYWHERE; WHO MUST RESPOND?

Hazardous materials, and the hazardous wastes they can produce, are being found almost everywhere. Horror stories of unexpected backyard, undeveloped properties, and farmland hazardous waste discoveries are becoming daily TV and headline news. Highway, railroad, and industrial site accidents requiring emergency response to control spills and emissions vie for position as the top story. In the wake of the September 11, 2001 terrorist attacks we have been warned of possible terrorist incidents involving hazardous materials.

Who must respond to such discoveries and incidents? The answer is clear. Only those who are properly trained and certified. OSHA has stepped up to the plate with the Hazardous Waste Operations and Emergency Response (HAZWOPER) 40-hour certification. This standard includes any personnel drawn from five distinct groups of employers. They are personnel who might be exposed to hazardous substances, including hazardous wastes, and who are engaged in any of the following operations:

1. Work involving hazardous wastes at treatment, storage, and disposal facilities;
2. Cleanup operations required by a governmental body, involving hazardous substances, that are conducted at uncontrolled hazardous waste sites;
3. Corrective actions involving cleanup operations at sites covered by the Resource Conservation and Recovery Act (RCRA);
4. Voluntary cleanup operations at sites recognized by a governmental body as uncontrolled hazardous waste sites;
5. Emergency response operations due to releases of, or substantial threats of release of, hazardous substances.

In this book the duties and responsibilities of the personnel and the typical operations within these five distinct groups, are covered as follows. In Section 13, we describe operations at treatment, storage, and disposal facilities (group 1 above). In Section 14, we describe the work at Superfund sites and brownfields (groups 2, 3, and 4 above). In Section 15, we describe the work of emergency response (group 5 above). The first 12 sections of this book provide the background and knowledge base for such work. The last section, Section 16, presents the concept of the health and safety system; what every U.S. organization needs to keep its prime asset, its employees, fit for long-term health and productivity.

In the opinion of the authors, most employers and employees in the United States will benefit from an understanding of the topics involved in HAZWOPER certification, whether or not they are currently working with hazardous materials or their wastes. If not yet, that time will come.

In many cases, it would be wise to certify employees to prevent costly delays in the future. The authors have received many anxious calls from employers. Due to the discovery of unforeseen environmental contaminants, we have had to rush to sites to provide the required 40-hour HAZWOPER training. Work had been stopped, personnel were just marking time, and nonproductive expenses were mounting rapidly. Workers waited for training and certification to be able to enter a site where hazardous wastes had been discovered.

This book has also been designed to provide both experienced and new trainers and educators in a variety of organizations with HAZWOPER manual and desk reference material for use as content in course or seminar preparation.

A MUCH LARGER PART OF THE POPULATION NEEDS HAZWOPER UNDERSTANDING

Individuals ranging from executives to laborers must know how to protect themselves from a myriad of hazards while planning and executing solutions to health, environmental, and remediation problems. Students in environmental engineering and technology need to understand the workplace challenges they are bound to face sooner or later. Chemical, medical, and industrial plant and laboratory personnel need to know "What must I do if an almost unthinkable hazardous waste threat occurs at my site?"

Legislators, regulators, lawyers, real estate owners, and managers need to know more about specific hazards that are being encountered in the workplace, at brownfields, and at redevelopment sites. Knowing more about what experienced engineers, remediation workers, chemists, and safety experts have learned and have been able to put into practice can only help their daily decision-making.

Violations of the HAZWOPER standard have resulted in crushing OSHA citations. They have been followed by U.S. Department of Justice prosecutions. One case involved multimillion-dollar fines and a nearly two-decade prison term for a company owner for crimes that left an employee with permanent brain damage. OSHA is proving its commitment to enforcing the protection of the health and safety of America's workers.

THE AUTHORS' BACKGROUND AND EXPERIENCE

The authors of this manual and desk reference have almost 100 years of combined experience. They have worked in:

- Preparation of environmental legislation and regulations
- Hazardous materials management
- Site environmental, safety, and health inspections
- Hands-on cleanup and disposal of wastes
- Site investigation and remediation
- Training for HAZWOPER 24-, 40-, and 48-hour certification
- Training of mutual aid groups and other organizations in Emergency Response; from First Responders' through Incident Commanders' levels
- Training for certified hazardous materials managers (CHMMs)
- Waste treatment, storage, and disposal facilities
- Emergency response
- Development and implementation of site-specific health and safety plans
- Environmental technology and pollution control and the related chemistry and chemical engineering
- Worldwide forensic investigations for Fortune 100 corporations

This manual and desk reference highlights the authors' recommendations for the training of workers who operate in the presence of hazardous materials and who must meet the 40-hour OSHA HAZWOPER training requirements.

The authors have worked with propellants and explosives, toxic gases, liquids and solids, flammables, corrosives, solvents, and a host of other hazardous materials. Some of these materials are present in our factories, laboratories, and even offices. We all encounter transporters of hazardous materials when traveling on our highways and railroads, in the air, and on waterways. Incidents involving hazardous materials can intrude in the lives of all of us and our families at one time or another.

WHY ARE TRAINING AND CERTIFICATION NECESSARY?

Workers sometimes ask, "Can't we just learn on the job?" No. What you learn from job experiences is invaluable. However, it is not the broader or more complete training you need. You need training so you will know how to respond to situations, problems, materials, and emergencies that you have not yet faced on the job. That is why OSHA has specific, minimum, HAZWOPER training requirements. Then certification proves that workers not only have been trained, but also have learned.

WHAT IS NEEDED FOR CERTIFICATION

You need to understand the Key Facts in this book, and the explanation of terms and phrases in the Glossary and Acronyms, Appendix A. As you absorb this material, you also need answers to your questions by a qualified trainer in a 40-hour HAZWOPER course. That, followed by achieving a score of at least 70 on an examination of the material and 24 hours of supervised on-site work experience, will qualify you to receive 40-hour HAZWOPER certification. That covers not only the training of workers, but other employees and contractors.

OSHA's requirements are stated in 29 CFR 1910.120. The complete text of the standard is reproduced in Appendix D of the Companion CD. The Companion CD serves to aid trainers in preparing course handout materials containing specific citations.

RESPONDING TO OSHA'S REQUIREMENTS

To respond to OSHA's requirements and the regulations that have become laws, we offer the following 16 Sections, Appendix, and the Companion CD. Our goal is to provide a training manual, assistance for educators and trainers, and a desk reference for a wide readership in one package. We need to emphasize "desk reference." There is far more content here than is required understanding for the OSHA 40-hour certificate. However, we have found all of this material to be important for our job understanding and safety at one time or another.

We want to answer logical questions you might ask when planning training, after an on-the-job occurrence, or when planning brownfields remediation or preparing for emergency response. Or you might have heard news accounts, attended courses, read other books or regulations, or pondered legislation in that growing field of hazardous waste operations and emergency response, all of which have triggered questons in your mind.

The following is a brief introduction to the book's contents. No matter whether you are just beginning or your background is deep in work with hazardous wastes, we trust that you will find the book an informative, helpful, and easy read. Note that the book sections are not exhaustive for their subjects. They focus on the HAZWOPER standard and the training it mandates, plus the authors' explanatory and recommended reference material.

SUMMARY OF THE BOOK'S CONTENTS AND THE AUTHORS' OBJECTIVES

Section 1. Regulators, Legislation, and the HAZWOPER Working Environment. Here we will show you who makes the rules and laws and what you need to know and do to be in compliance. We introduce the kinds of work that require HAZWOPER certification. They are detailed later in Sections 13 through 15.

Section 2. Hazardous Waste Defined. You are familiar with many kinds of hazards. You know what you consider wastes. You need to know how OSHA and the other regulators define hazardous waste.

Section 3. Material Hazards. You need to appreciate the kinds of hazards you should know about when working with, or encountering, waste materials. Materials that attack body tissues, such as acids and alkalis; heat-, friction-, and shock-sensitive materials; and explosives, and more are covered.

Section 4. Chemical Incompatibility. There are materials that just cannot be mixed with each other without a hazardous outcome. One or both of the materials may not be hazardous, but the mixture is just plain incompatible. You may already know not to pour water into concentrated acid. A large reaction heat will cause eruption and hazardous splashing.

Here is another rule of thumb. Never store or mix oxidizers with fuels. The mixture can easily become an explosive. An example is ammonium nitrate and fuel oil (ANFO), which is what was said was used in the terrorist Oklahoma City bombing. You need to know how to recognize which materials are oxidizers and which are fuels. Most readers probably need to learn more about what makes materials incompatible.

Section 5. Toxicology. Determining how materials will or might affect your health, depending upon the amount and length of exposure (or dose) you receive, is a major goal of toxicology studies. You will learn about exposure limits explained as PELs, RELs, or TLVs®. The permitted dosages, formerly considered safe for many toxic materials, have been lowered quite a bit in the past few years as more evidence about their adverse effects on our health has been collected and evaluated.

Section 6. Personal Protective Equipment (PPE). We all need to know what is available to wear for our personal protection. Then we need to know how to wear it properly to be protected when we are near, or in contact with, hazardous materials or wastes. Examples of PPE include respiratory protection (masks of many types), hearing, eye, head, foot, and hand protection, and chemical protective clothing. Become aware of the extra physical demands and heat stress incurred by PPE wearers. Those necessary, but extra, burdens can reduce job performance expectations.

Section 7. Decontamination. Here we stress the processes for safely removing hazardous material from your PPE, body, and equipment. Decontamination can become quite involved in heavily contaminated work areas. Work practices must be designed and implemented to avoid tracking contaminated material off a site. All contaminated solids and liquids must be packaged and collected for approved disposal.

Section 8. Labels, Placards, and Other Identification. Easily understandable signage and readily available documents are required on hazardous material and waste transport vehicles and at sites. They warn all viewers and readers of specific types of material hazards. The agencies that regulate standards for signage are gradually approaching multicountry agreements. More signs will appear having graphics with limited text support.

Section 9. Material Safety Data Sheets (MSDS). How do you know whether or not a specific material you are working with is hazardous? Many materials have chemical properties you should know about. Every material has physical properties. Sulfuric acid (battery acid) will chemically attack your skin. It also has the physical property (specific gravity) of being about 1.8 times as heavy as water.

MSDSs are designed to tell you about all properties of materials that might affect how you can safely work with, store, clean up, and dispose of those materials. MSDSs are required to be provided by manufacturers for products that have, or might be thought to have, toxic or otherwise hazardous effects following improper storage, use, or disposal. This can be carried to extremes. There are even MSDSs for some kinds of packaged water! International trade is increasing yearly. Here we will also give a description and our appraisal of the international equivalent of the MSDS, the International Chemical Safety Card.

Section 10. Confined Spaces. The previous nine sections will have prepared readers to deal with hazardous materials and hazardous wastes. Now you need to know about one of the most common and most hazardous working conditions: working in a permit-required confined space. This means a space not designed for continuous human occupancy, large enough for a person to enter, with limited means of access or egress and unfavorable natural ventilation. The space might contain atmospheric, physical, or chemical hazards. Spaces like this are more common then you might imagine. Fatal accidents occur yearly to untrained and unprotected workers.

Section 11. Other Workplace Hazards. Be aware that in HAZWOPER activities there are hazards other than those due to hazardous waste and work in confined spaces. Know how to avoid falls, heat stress, heat stroke, electrocution, entrapment, engulfment, harmful radiation exposure, and other unprotected exposures. Learn the recovery measures to use when the unavoidable occurs.

Section 12. Sampling and Monitoring. This testing is done to determine the nature and extent of hazardous materials and wastes threats at a site. You can get some answers you need right in the field. Others may require laboratory analysis. The analytical results help you to determine how you should respond to any hazard detected. During and following cleanup of a contaminated site, monitoring may be continued for months or many years to ensure and record cleanup effectiveness.

Section 13. Work at Hazardous Waste Generators and at Treatment, Storage, and Disposal Facilities (TSDFs). This section deals with the special characteristics of two kinds of HAZWOPER work. Work at a TSDF involves continuous work with hazardous wastes. Some TSDFs work routinely with a rather narrow range of wastes. Others work with a broader range of materials requiring much more extensive worker training.

On the other hand, the processes and materials used at a production facility working with hazardous materials, and the hazardous wastes generated, are rather fixed in comparison with the work at a TSDF and the work described in the two following sections. This can lead to a narrowing of the scope of materials handled. But in some cases the degree of hazard in daily work can be quite high due to the extremely hazardous materials being processed or used as processing aids.

Section 14. Superfund Sites and Brownfields: Site Investigation, Control, and Re-mediation. Personnel working in these specialities generally move from site to site. There can be large variability in materials hazards they face. Further, variations in topography, remediation equipment required, weather conditions, and more are to be expected. These workers need environmental site assessement information before starting work.

How do you safely determine the best course of action to clean up a site known or believed to be contaminated with hazardous waste? How do you keep hazards on the site from migrating off the site? How do you keep the public from entering the site and becoming contaminated? How do you conduct site remediation safely? These questions and more are answered in Section 14.

Section 15. Emergency Response. A tanker truck overturns on an interstate and hazard-ous liquid spills on the highway. We all expect some dedicated team to arrive at the site within minutes. We trust that they will stop release of hazardous vapors and liquids from the area, extinguish fires, prevent explosions, and begin decontamination and cleanup swiftly. We also expect that they will keep bystanders and blocked traffic safe from any related threats. We expect their cleanup efforts will allow routine highway traffic in a few hours at most. Who performs such miracles? Emergency responders. You might not think of one of their initial stumbling blocks at the site. Environmental site assessment information, the basic planning information available at all other sites, probably will not be available.

This work, compared with the types in Sections 13 and 14, has the highest variability in its daily challenges. Logically, members of all fire and police departments must be trained in at least a basic understanding of response work and the duties and responsibilities they are expected to fulfill. Special emergency response teams need exhaustive training and special organization. Preplanned response and command organization details among local fire, police, and special response teams are essential.

Section 15 does not present all of the information needed to become a specialist in responder work. It does give sufficient information about the roles of the various responder specialists to meet 40-hour HAZWOPER requirements. Further, emergency responders at specified levels must also have 40-hour HAZWOPER certification before starting their specialist training. Incident commanders are an example.

Section 16. Health and Safety System. The previous 15 sections have covered the special information you need to know for safe work with hazardous wastes. A major requirement of any facility where hazardous wastes might be expected is to have in operation an overall health and safety system. That system, consisting of several required programs, must be written and available to be read by employees and OSHA representatives. Within such a system you need an approved and site-specific health and safety plan (HASP) for any site where work with hazardous wastes occurs. The HASP is the written document that is designed to foresee and address any conditions that might arise at a site covered by HAZWOPER. Then it must give readers detailed instructions on how they must react to those conditions to maintain site safety and health. The HASP also informs workers of the required routine medical exams. It tells how to get emergency aid and more.

THE APPENDICES

You will find a number of appendices to the book. The idea is to place longer, more detailed, but important reference material where it would not interrupt the flow of text. Appendix A, "Glossary and Acronyms," must not be overlooked. It presents many definitions and explanations that HAZWOPER personnel need to know. The Companion CD contains the remaining 18 Appendices in Microsoft® Word® format for ease of reading and local reproduction.

"README" COMMENTS ABOUT THE COMPANION CD

To facilitate training and provide ready desk reference material we offer the Companion CD. It includes the 2002 version of the NIOSH Pocket Guide with its Emergency Response Guidebook. The CD contains highly valuable tables and searchable files. It also provides summary text and regulations for handouts and sample exams. These are aids in Microsoft® Word® format. Competent trainers can customize the training material to prepare their own site-specific training program. Other readers can use these sample exams to test their comprehension of the material.

REFRESHER OR NEW MATERIAL FOR YOU?

If you are using this material as part of refresher training, we trust you will have been over a good bit of this before. But we guarantee there will be much new material here for you to appreciate. If you are seeing this information for the first time, you might ask, "Isn't this an overpowering amount to learn?" Remember that this book serves as a desk reference in addition to providing training material.

The material that needs to be taught for the 40-hour HAZWOPER certificate is emphasized in the Objectives and Summary in each section and in the Key Facts statements within the sections. The purpose of the extra details is to help trainers to provide new information in OSHA-required 'refresher' training. Eight hours of refresher training, you might recall, are required yearly.

OSHA'S MISSION AND PARTNERS

The mission of OSHA is to save lives, prevent injuries, and protect the health of America's workers. To accomplish the laudable goal of steadily reducing accident and injury frequencies, federal and state governments must work in partnership with workers and employers. The 100 million working men and women and their 6.5 million employers in this country are covered by the Occupational Safety and Health Act of 1970 and its revisions to date.

Nearly every working man and woman in the nation, whether company-employed or an outside contractor, comes under the jurisdiction of OSHA. However, miners, transportation workers, many public employees, and the self-employed are not included.

OSHA state partners have over 2000 inspectors, in addition there are complaint and discrimination investigators, engineers, physicians, educators, standards writers, and other technical and support personnel. They are assigned to 10 regions and more than 200 offices throughout the country. This staff establishes protective standards and enforces those standards. It reaches out to employers and employees through technical assistance and consultation programs.

HAZCOMM COMPARED TO HAZWOPER

If you work where there is the possibility that you might come in contact with hazardous materials at your job, you have probably already taken a class in the Hazard Communication Standard (HAZCOMM). That class told you how to recognize potential hazards in the workplace, where to find additional information about materials you work with, and your rights and responsibilities. HAZCOMM training, defensive in nature, basically trains you in the fundamental information you can get from the labels on materials containers, and from MSDSs.

HAZWOPER training is required where personnel routinely work with hazardous waste and where there is the reasonable possibility for worker exposure to related safety and health hazards. HAZWOPER training extends much more broadly and deeply in filling in all the information gaps that have become evident following HAZCOMM training. When HAZWOPER training has been successfully completed and routinely put into practice, it delivers workers capable of performing hazardous, and many times difficult, tasks SAFELY.

NOTES, CAUTIONS, AND WARNINGS

Note

Notes present information that is not necessarily safety-oriented. They are intended to capture the special attention of the operator.

CAUTION

CAUTIONS ARE PRESENTED TO CALL THE READER'S ATTENTION TO ACTIONS THAT MAY CAUSE DAMAGE TO MACHINERY OR EQUIPMENT. THEY ARE ALSO USED TO CALL ATTENTION TO ACTIONS THAT COULD HAVE LEGAL OR REGULATORY IMPLICATIONS.

WARNING

WARNINGS ARE PRESENTED TO ALERT READERS TO ACTIONS OR CONDITIONS THAT CAN CAUSE HARM TO WORKERS. WARNINGS EXPLAIN PROTECTIVE ACTIONS OR RESPONSES TO DANGEROUS CONDITIONS.

CAUTION

REMEMBER THAT NOTHING IN THIS MANUAL AND DESK REFERENCE IS INTENDED TO CONTRADICT OR NEGATE ANY PART OF A COMPLETE REGULATORY TEXT. THIS MANUAL AND REFERENCE MATERIAL IS PRESENTED TO AMPLIFY REGULATIONS AND TO EASE WORKERS' UNDERSTANDING OF THE REQUIREMENTS. FURTHER, IT IS ESSENTIAL THAT EVERY WORKER BE TRAINED TO WORK MORE SAFELY WITH THE UNIQUE AND SPECIFIC HAZARDS PRESENTED BY HIS OR HER WORK SITE, INSTALLATION, OR FACILITY.

ACKNOWLEDGMENTS

The authors wish to express their appreciation to a number of individuals for the special expertise they have shared with us and the assistance they have provided us during the preparation of this book.

We extend warm thanks to The Honorable Ruth Ann Minner, Governor of the State of Delaware, for her sincere and continuing support of improvements in Superfund and brownfields remediation projects, typically the workplace of 40-hour HAZWOPER certified personnel. We salute her for her testimony to Congress, as a member of the National Governors Association's Committee on Natural Resources, representing all 50 state governors, wherein she provided recommendations for innovative government, industry, and community cooperation for revitalization projects.

Marianne Cloeren, M.D., Master of Public Health, Fellow American College of Physicians, Delegate of the American College of Occupational & Environmental Medicine, and Principal of Cloeren Occupational Health Associates provided expert advice, valuable content, and many helpful suggestions in the preparation of Section 5, "Toxicology."

Angelo Marchiano, United States Navy, provided us with Section 10, "Confined Spaces," background and applications from what we believe is the most successful confined space testing and permit entry program in the world. As Group Superintendent and manager of the Navy's Gas Free Engineering Program at the venerable Philadelphia Naval Shipyard, Mr. Marchiano was a pioneer of modern confined spaces safety management and technology development. His innovations were adopted first by the Philadelphia Naval Shipyard and then the U.S. Navy worldwide.

We are grateful for the expert help and assistance of Mr. Henry S. Chan, CIH, Manager of the CD version of the *NIOSH Pocket Guide to Chemical Hazards and Other Databases*, and of Dr. Paul Schulte, Director of the Education and Information Division of NIOSH, Cincinnati, Ohio. We appreciate the opportunity to include the 2002 version of the *NIOSH Pocket Guide to Chemical Hazards and Other Databases* as part of our companion CD. We have relied upon previous hardcopy versions of the Pocket Guide for years. The CD allows emergency responders on-site laptop access to invaluable information in support of incident response.

We extend our sincere thanks to David L. Volz, HAZMAT Team Member, Los Alamos National Laboratory, Environmental Safety and Health Division. His assistance with Section 14 on the subjects of safe management of hazardous waste containers, and particularly bulging and dangerously corroded drums, is much appreciated. The drum-handling guidelines that have been developed based upon his expert advice are destined to save HAZWOPER personnel lives, prevent injuries, and preserve environmental integrity.

For continuing valuable editorial assistance, we thank Edna O. Hackman, B.S., Temple University. Her careful reading of text followed by comments and suggestions for improved clarity was greatly appreciated.

We thank our principal contact at McGraw-Hill, Kenneth P. McCombs, Senior Acquisitions Editor of the Professional Book Group in New York City, for the numerous creative discussions he has held with us and the recommendations he has provided for book content over the past year. They were extremely helpful in formulating our approaches to text, graphics, appendices, and the Companion CD.

The authors would also like to thank all of the equipment suppliers who were so generous with their technical information and support. We have first-hand experience with these folks and feel confortable recommending their products to the reader. We extend special thanks to the following for their generous time, product technical support, and encouragement:

Karmen Lawson from Kappler (www.kappler.com)
Ed Bickrest from Delloz Fall Protection (Miller) (www.christiandelloz.com)
Ray Chromer from Scott Safety (www.scotthealthsafety.com)
Ken Bostwick from Action Training Systems, Inc. (www.action-training.com)
Alan Runk and Tony Jones from Sibak Industries, Co., Inc. (www.sibak.com)
Barry Ford from Conbar Environmental Products (www.conbar.com)
A. Joseph Irons from Easy Lift Equipment Co., Inc. (www.easylifteqpt.com)
Chris Wrenn from RAE Systems, Inc. (www.raesystems.com)
Paul Nyfield from EMSL Analytical, Inc. (www.emsl.com)
Melanie Balastracci from North Safety Products (www.northsafety.com)
John Shein from Niton Corporation (www.niton.com)
Janet Pierson from Safe Expectations, International (www.safeexpectations.com)
Debbie Bucar from MSA (www.msasafetyworks.com)
Stephanie Lantgen from Best Gloves (www.bestglove.com)
Brett Hanson and Dave Clos from customhardhats.com (www.customhardhats.com)
Michael Esris from WESCO Industrial Products (www.wescomfg.com)
Lesa Hawkey from Kelley Supply, Inc. (www.kelleysupply.com)
Lea Miller and Bryan Oneschuk from U.S. Filter, Wilmington, DE
Marian R. Young from WIK, Associates, Inc. (www.wik.net)
Carl DeCaspers from New Pig® Corp. (www.newpig.com)

We have attempted not to forget anyone. Our sincerest apologies if we have accidentally omitted anyone.

We want to hear from any reader having questions or comments about the book's contents, and anyone having suggestions for future editions. Please feel free to contact the authors at www.nstengineers.com

SECTION 1
REGULATORS, LEGISLATION, AND THE HAZWOPER WORKING ENVIRONMENT

OBJECTIVES

Our objective is to explain regulatory affairs as they impact upon those involved in hazardous waste operations. You, as a HAZWOPER worker, trainer, or supervisor, must be aware of the applicable regulatory framework and how it affects you as well as your company or organization.

In addition to the roles and responsibilities of the three principal federal regulators, we summarize the functions and contributions of other federal agencies and private organizations. Learn the information and services that they will provide to make your workplace safer. We trust you will also learn how to find and understand the regulations that govern how all of us must operate when dealing with hazardous waste. HAZWOPER is OSHA-regulated. However, it is important to know the special roles played by EPA and DOT in managing hazardous waste.

We list the minimum requirements for HAZWOPER training. We also list a number of other topics that we feel anyone working in this area will be well-served in learning. The topics in both lists are covered in the sections in this book. The painful and injurious accidents we hear about weekly can only be reduced by greater worker knowledge of workplace hazards and how to avoid them. We aim to help in that reduction.

Although it is not required by the HAZWOPER standards, the authors recommend that management of every candidate organization should investigate the benefits of participation in OSHA's Voluntary Protection Program (VPP).

1.1   THREE PRINCIPAL FEDERAL REGULATORS

O—π   The three principal regulators involved in hazardous waste operations are OSHA, EPA, and DOT.

Throughout this book and its companion CD, we will be referring to the three principal federal regulators, each charged by law with regulating certain kinds of operations with hazardous materials and their wastes. They are the Occupational Safety and Health Administration (OSHA), the Environmental Protection Agency (EPA), and the Department of Transportation (DOT).

OSHA's mission is to protect and improve the safety of the worker. EPA concerns itself with control of polluting and hazardous materials that have the potential to be released, or have been released into the environment. DOT regulates the transportation of hazardous materials to ensure workers' and public safety. The regulations that affect hazardous material workers are found in the United States Code of Federal Regulations (CFR).

1.2   SOURCES OF REGULATIONS AND LAWS

We will be referring to various regulations and laws (Acts) throughout this book. An Act is a bill that has passed both houses of Congress and has been signed by the President. A bill also may be passed over the President's veto. If the bill is not acted upon during the President's veto time limit, it becomes a law by default.

he don't like that

1.2.1   Regulations (Standards), Titles, and Codes

It is important that you understand how these regulations (sometimes called standards) were created. You should know their relationship to laws having to do with hazardous materials. It will be helpful to know how to read the titles. They tell you where to find the regulation or act. Trainers should bear in mind that it is not important for trainees to memorize the laws or regulations individually. What is important is that the trainee be familiar with what each regulation or law covers, which agency did the writing, and where to find it. All of the three major regulators have posted their regulations on the Internet. All are completely searchable.

The complete texts of OSHA's regulations (Standards) are found at http://www.osha-slc.gov/OshStd_toc/OSHA_Std_toc.html.

The complete texts of EPA's regulations are found at http://www.epa.gov.

The complete texts of DOT's regulations are found at http://hazmat.dot.gov/rules.htm.

The United States Code, the official record of all United States laws, is found at and can be searched from http://www4.law.cornell.edu/uscode/.

A regulation is a rule that can be enforced, with monetary fines and other penalties for noncompliance, like a law.

1.2.2   How Laws and Regulations Work Together

Laws and regulations are a major tool in protecting the safety and health of Americans and the environment. Congress passes laws that govern the United States. To put those laws into effect, Congress authorizes certain government agencies, including OSHA, EPA, and DOT to create and enforce regulations. Here is a basic description of how laws and regulations come into being. We explain where to find them, with an emphasis on hazardous waste laws and regulations.

1.2.3   Creating a Law

Step 1:   A member of Congress proposes a bill, which is a document that, if approved, will become law.

Step 2:   If both houses of Congress approve a bill, it goes to the President, who has the option to either approve it or veto it. If approved, the new law is called an act, and the text of the act is known as a public statute.

Step 3:   Once an act is passed, the House of Representatives standardizes the text of the law and publishes it in the United States Code (USC). The U.S. Code is the official record of all federal laws.

Note

OSHA was created and empowered under the Occupational Safety and Health Act of 1970 (29 USC 61 and following).

1.2.4   Putting the Law to Work

Once the law has been enacted, how is it put into practice? Laws often do not include all the details. The U.S. Code would not tell you, for example, what you must know and do to become HAZWOPER-certified. In order to make the laws work on a day-to-day basis, Congress authorizes certain government agencies, including OSHA, EPA, and DOT, to create and enforce regulations.

Regulations set specific rules about what is legal and what is not. For example, a regulation issued by OSHA does tell what a worker must know and do to become HAZWOPER certified. Once the regulation is in effect, OSHA then works to help Americans comply with the law. OSHA's compliance personnel work to enforce the regulations.

1.2.5   Creating a Regulation

First, an authorized agency, such as OSHA, EPA, or DOT, decides that a regulation may be needed. The agency researches the situation. If it deems necessary, the agency proposes a regulation. The proposal is listed in the Federal Register so that members of the public can consider it and send their comments to the agency. The agency considers all the comments, revises the regulation as it feels is appropriate, and issues a final rule. At each stage in the process the agency publishes a notice in the Federal Register. These notices include the original proposal, requests for public comment, notices about meetings where the proposal will be discussed (open to the public), and ultimately the text of the final regulation.

Twice a year, each agency publishes a comprehensive report that describes all the regulations it is working on or has recently finished. These are published in the Federal Register, usually in April and October, as the Unified Agenda of Federal and Regulatory and Deregulatory Actions.

Once a regulation is completed and has been printed in the Federal Register as a final rule, it is codified by being published in the Code of Federal Regulations (CFR).

⊙━π   The CFR is the official record of all regulations created by the federal government.

The CFR is divided into 50 volumes, called titles, each of which focuses on a particular area. Almost all worker safety regulations appear in Title 29. Environmental regulations appear in Title 40 and transportation regulations in Title 49. The CFR is revised yearly, with one fourth of the volumes updated every three months.

1.2.6   An Example of How to Read and Understand a Regulation Title

29 CFR 1910.120

Title 29 of the Code of Federal Regulations (CFR) is concerned with Labor (that is where OSHA regulations are found)

Part 1910 Occupational Safety and Health Standards (General Industry)

(Subpart H), Hazardous Materials

Section .120 Hazardous Waste Operations and Emergency Response

From this example, we see that all OSHA regulations will be found in Title 29 of the Code of Federal Regulations. A Part number will follow this. Part number 1910 is where General Industry standards are found, and 1926 is where Construction Industry standards are found. Further, there will be a Subpart number, and sometimes a section letter, subsection number, sub-subsection letter and a sub-sub-subsection number, with each section of the title becoming more specific: for example, 29 CFR 1910.120(a)(2)(iii)(B).

This may seem very confusing. However, once you become accustomed to reading regulations, you will find that the system is not that difficult to understand.

1.2.7 An Example of a Specific Citation

29 CFR 1926.102(a)(4)

Title 29 CFR = OSHA

Part 1926 = Safety and Health Regulations for Construction

(Subpart E) = Personal Protective and Life Saving Equipment

Section .102 = Eye and Face Protection

Subsection (a)(4) = "Face and eye protection equipment shall be kept clean and in good repair. The use of this type equipment with structural or optical defects shall be prohibited."

This citation prohibits the use of safety glasses, goggles, and face shields that are scratched enough to impair the worker's vision.

You will find the HAZWOPER regulations, or standards, in 29 CFR 1910.120 and 29 CFR 1926.65.

1.3 THE OCCUPATIONAL SAFETY AND HEALTH ADMINISTRATION (OSHA)(29 CFR)

United States **OSHA** Occupational Safety & Health Administration U.S. Department of Labor

1.3.1 OSHA's Mission

The Occupational Safety and Health Administration (OSHA) is the federal agency charged with protecting and improving workers' health and safety on the job. The mission of OSHA is to save lives, prevent injuries, and protect the health of America's workers. To accomplish this, federal and state governments must work in partnership with the more than 100 million working men and women and their six and a half million employers who are covered by the Occupational Safety and Health Act of 1970. OSHA has established categories of safety and health violations: willful, serious, repeated, and other-than-serious.

1.3.2 Safety and Health Violations Defined

OSHA is primarily charged with the safety and health of the workforce; not the population in general or the environment.

OSHA divides major violations of the regulations into two categories:

- *Willful violations* are those committed with an intentional disregard of, or plain indifference to, the requirements of the Occupational Safety and Health Act and regulations.
- *Serious violations* involve actions on the job such that there exists a substantial probability that death or serious physical harm could result. They result in citations because they are activities or tasks that the employer knew, or should have known, involved hazard(s) for which there could have been protective measures.

OSHA is empowered by the U.S. Code to issue citations to anyone it believes is not following OSHA's regulations. Citations call for penalties and remedial activities by the recipient. The recipient is an employer as defined in the regulations. OSHA takes unsettled disputes with employers to federal court. Federal courts make the final determination. See Section 6.2 for an example of how serious OSHA is about violations.

1.3.3 Potential Misunderstandings between Site Workers about Safety and Health Responsibilities

CAUTION

EACH EMPLOYER MUST REMEMBER THAT PROTECTION OF THE WORKER IS THE EMPLOYER'S OBLIGATION!

The authors, as HAZWOPER personnel, on occasion work on sites where construction as well as our remediation work are being performed simultaneously. Sometimes we are working in the same area as other contractors (e.g., utilities installers) and both parties are working for the same general contractor. The construction trades workers see us using a Health and Safety Plan (HASP; see Section 16), performing ambient monitoring and related activities. The construction workers assume that we are there performing monitoring services covering all site workers. We continually have to remind them that this is not the case! Unless we are explicitly hired to perform *job-site health and safety services,* we are only monitoring and collecting environmental data for ourselves, as required by our own HASP.

Furthermore, we are checking our breathing zones—say at ground surface—not theirs, possibly down in a trench or in a confined space on the site.

We absolutely may not give them our HASP for their use. Nor can they copy it and attach their own cover page. This can cause frustration, confusion, and even work stoppage. Readers who find themselves in this situation must realize that there are significant professional liability insurance implications. Construction workers injured at hazardous waste sites often try to sue engineers or environmental professionals since they can't sue their employers under Workers' Compensation.

Insurers constantly remind us of the dangers of potential litigation. Consequently, we include a specific clause in our contract with the general contractor that explains our duties and responsibilities and the limitations of our environmental and safety work. This is especially important since we often deal with urban soils contaminated with lead at concentrations that would invoke the OSHA lead-in-construction standard.

1.3.4 Minimum Content Required of HAZWOPER Training

In order to be qualified for operations with hazardous wastes, according to OSHA regulations, you must, at a minimum, be trained in the following nine areas:

1. Containers and drums, selection, and handling
2. Emergency response

3. Flammable and combustible liquids

4. Hazardous materials, hazardous wastes, and hazardous substances

5. Materials handling and storage

6. Personal Protective Equipment (PPE)

7. Storage areas

8. Training required

9. Waste disposal

Note that in addition to this training, to get 40-hour HAZWOPER certification, an individual must experience 24 hours of supervised on-site work.

1.3.5   The Authors' Recommended Additional Training Content

Through years of work experience and training of workers and trainers, the authors have found that more than the minimum is advisable. Further clarification and definition of operations in the real world of hazardous wastes is appropriate. We have added detailed discussions on:

• Chemical incompatibilities

• Decontamination

• Toxicity and exposure levels

• Placarding and labeling

• Material safety data sheets

• Confined spaces

• Sampling and monitoring

• Typical working environments

• Health and safety plans

We believe that these recommended additions have been vindicated by OSHA's adoption of the Voluntary Protection Program (VPP). Its goal is to boost workers' health and safety far above that provided by the minimum standards. It seems reasonable that to achieve the benefits of the VPP, additional training will be required.

1.3.6   The OSHA Voluntary Protection Program (VPP)

An organization that is active in the VPP has a genuine commitment to the safety and health of its workforce.

"Voluntary" is the operative word here. Companies join with OSHA in this initiative voluntarily. Once they do, they get what many managers may consider a big benefit. None of those routine, unexpected visits by OSHA inspectors. OSHA's VPP on-site reviews, by the cooperating company, ensure that company safety and health programs provide superior worker protection. The cooperative companies provide OSHA with valuable specific input on safety and health matters.

For instance, when an OSHA compliance officer, on an unexpected visit to a non-VPP site, issues a citation for lack of personal protective equipment (PPE), the company must pay a fine. Then the site owner must provide the appropriate PPE. When a worksite participates in the VPP, slip-ups like the wrong or missing PPE just don't occur.

OSHA studies show companies participating in the VPP experience substantially lower-than-average worker injury rates. They are only about half the expected rate for non-VPP sites. OSHA also claims that participants experience decreased costs in workers' compensation and lost work time. Further, they often experience increased productivity and improved employee morale.

Proactive safety and health work is just good sense. The authors experienced two "gee-whiz" case histories in one year. While performing other work at two plants we noticed some apparent OSHA violations. We proposed inexpensive plant walk-through inspections that we would follow up with recommendations for safety and health improvements. The proposals were turned down. "We'll get to that," we were told. Apparently not. Within a year we saw both companies listed as cited by OSHA, one for $80,000 and one for $500,000.

> **A lack of commitment to the safety and health of the workforce can lead to fines, injuries, higher compensation costs, and death.**

A horrible incident, which the authors believe might well have been avoided by this kind of cooperation with OSHA, resulted in the total destruction of a facility processing hazardous material. Several workers died.

Companies can apply for VPP membership and then work toward "Merit," or better yet, "Star" rating. The voluntary aspect, and the potential benefits of cooperation, remind us of the ISO 9000 quality and ISO 14000 environmental management programs. Both those programs have been beneficial to the business success of many companies who saw the handwriting on the wall; strong quality and environmental management initiatives were essential.

1.3.7 How to Apply for VPP Participation

To apply for VPP membership, a facility must be prepared to provide:

- Management leadership and employee and union involvement
- Work-site analysis, including how you inspect for hazards, how employees report them, and how accidents are investigated
- Hazard prevention and control
- Safety and health training and the details of medical service available at your facility

The management, employee, and union involvement factors rank high in importance to OSHA. The documentation required of the company applying to participate will ultimately be extensive. However, many companies already have information that OSHA needs, but in the company's own format. That is acceptable to OSHA. What OSHA wants to see most of all is your company and workforce joined in commitment to continuous improvement in workers' safety and health.

1.3.8 Company Statement of Commitment

Union Statement. If your site is unionized, the authorized collective bargaining agent(s) must sign a statement that supports, or that at least indicates no objection to, the site's participation in the VPP. The statement should be submitted with your application and must be on file before OSHA will schedule an on-site visit. Expressions of the commitment of non-union employees are welcomed but not required.

OSHA is rather sticky about requiring that the management representative for your site sign the following verbatim Statement of Commitment or attach a letter that provides the same assurances.

Management Statement. We agree that all employees, including newly hired employees and contract employees when they reach the site, will have the VPP explained to them, including employee rights under the program and under the Act; all hazards discovered through employee notification, self-inspections, an OSHA on-site review, accident investigations, process hazard reviews, annual evaluations, or any other means or report, investigation, or analysis will be corrected in a timely manner, with interim protection provided as necessary; if employees are given health and safety duties as part of our safety and health program, we will ensure that those employees will be protected from discriminatory actions resulting from their carrying out such duties, just as section 11(c) of the OSH Act of 1970 protects employees for the exercise of rights under the Act; and employees will have access to the results of self-inspections and accident investigations upon request. (In construction, this requirement may be met through joint labor-management committee access to these results.)

We agree to provide the following information for OSHA review on-site:

- Written safety and health programs
- All documentation enumerated under III,I.2.d of the current VPP Federal Register Notice
- Any agreements between management and collective bargaining agent(s) concerning the functions of any joint labor-management safety and health committee and its organization and any other employee involvement in the safety and health program

We will retain these records until OSHA communicates its decision regarding initial VPP participation.

We will likewise retain comparable records for the period of VPP participation to be covered by each subsequent evaluation until OSHA communicates its decision regarding continued approval.

We agree to make available for evaluation purposes any data necessary to evaluate the achievement of goals not listed above.

We will provide OSHA each year by February 15 our injury incidence and lost workday case numbers and rates, hours worked, estimated average employment for the past full calendar year, and a copy of the most recent annual evaluation of the site's safety and health program.

In addition, we will send our combined injury incidence and lost workday case numbers and rates, hours worked, and estimated average employment for the past full calendar year for all contractors' employees who worked at least 500 hours in any one quarter on our site during the year.

For construction sites only, injuries of all employees at the site, no matter who the employer is, will be recorded together. Rates will be calculated based on information for the site as a whole as long as we participate in VPP.

We understand that we may withdraw our participation at any time or for any reason should we so desire.

Signature

Manager of the applicant worksite

(You may add the signatures of any others
you wish.)

Note

This document, and supporting information, must be sent to your Regional OSHA office. You may discuss your potential participation with OSHA by calling OSHA's Division of Voluntary Programs at (202) 693-2213.

1.3.9   OSHA Regional Office Directory

OSHA has 10 regional offices to serve industry and the workforce in the United States.

Region I: Connecticut, Massachusetts, Maine, New Hampshire, Rhode Island, Vermont
John F. Kennedy Federal Building, Room E340
Boston, MA 02203
(617) 565-9860

Region II: New Jersey, New York, Puerto Rico, Virgin Islands
201 Varick St., Room 670
New York, NY 10014
(212) 337-2378

Region III: District of Columbia, Delaware, Maryland, Pennsylvania, Virginia, West Virginia
3535 Market St.
Gateway Building, Suite 2100
Philadelphia, PA 19104
(215) 596-1201

Region IV: Alabama, Florida, Georgia, Kentucky, Mississippi, North Carolina, South Carolina, Tennessee
1375 Peachtree St., NE
Suite 587
Atlanta, GA 30367
(404) 347-3573

Region V: Illinois, Indiana, Michigan, Minnesota, Ohio, Wisconsin
230 South Dearborn St., Room 3244
Chicago, IL 60604
(312) 353-2220

Region VI: Arkansas, Louisiana, New Mexico, Oklahoma, Texas
525 Griffin St., Room 602
Dallas, TX 75202
(214) 767-4731

Region VII: Iowa, Kansas, Missouri, Nebraska
City Center Square
1100 Main St., Suite 800
Kansas City, MO 64105
(816) 426-5861

Region VIII: Colorado, Montana, North Dakota, South Dakota, Utah, Wyoming
1999 Broadway, Suite 1690
Denver, CO 80202-5716
(303) 844-1600

Region IX: Arizona, California, Guam, Hawaii, Nevada
71 Stevenson St.
San Francisco, CA 94105
(415) 975-4310
(800) 475-4019

Region X: Alaska, Idaho, Oregon, Washington
1111 Third Ave., Suite 715
Seattle, WA 98101-3212
(206) 553-5930

1.4 THE U.S. ENVIRONMENTAL PROTECTION AGENCY (EPA) (40 CFR)

⊕EPA United States
Environmental Protection Agency

" ... to protect human health and to safeguard the natural environment..."

⊙─π **The Environmental Protection Agency is mainly concerned with protecting the environment and human health by regulating activities that will clean up our soil, water, and air and keep them clean.**

The Environmental Protection Agency is the federal agency charged with protecting the environment. The EPA was established in 1970 by Executive Order of President Nixon. Hazardous waste operations are guided by three major laws: RCRA, CERCLA, and SARA. Note that these are laws, not regulations. The regulations that EPA has produced under RCRA are listed in the parts of 40 CFR.

⊙─π **The three major laws, also called acts, bearing upon HAZWOPER work are RCRA, CERCLA, and SARA.**

1.4.1 Resource Conservation and Recovery Act (RCRA)
(42 USC 6901 *et seq.* [1976])

⊙─π **The Resource Conservation and Recovery Act (RCRA) "cradle-to-grave" legislation was enacted to assure the American public that hazardous materials and hazardous wastes would be managed safely from the time they were generated to the time they were disposed of safely.**

This statute spells out how any operation that generates hazardous waste must manage it, from its creation to its final destination (cradle-to-grave legislation). Although RCRA also addresses other Solid Wastes (garbage and trash), Underground Storage Tanks, and Medical Waste, among others, we will only be addressing RCRA Subtitle C, Managing Hazardous Waste.

RCRA Subtitle C, Managing Hazardous Waste. The following nine statements summarize the goals of Subtitle C.

1. The hazardous waste management program, Subtitle C, is intended to ensure that hazardous waste is managed safely from the moment it is generated to the moment of final disposal.

2. The Subtitle C program includes procedures to facilitate the proper identification and classification of hazardous waste.

3. While waste recycling and recovery are major components of RCRA's goals, they must be implemented consistently with proper hazardous waste management. As a result, RCRA contains provisions to ensure safe hazardous waste recycling and facilitate the management of commonly recycled waste streams.

4. The program also includes standards for those facilities that generate (i.e., produce), transport, treat, store, or dispose of hazardous waste. These include requirements for general facility management and specific hazardous waste management units. The provisions for Treatment, Storage, and Disposal Facilities (TSDFs) include additional precautions to protect ground water and air resources.

5. The hazardous waste management program includes safeguards to protect human health and the environment from hazardous waste that is disposed on the land or burned. These safeguards are known as Land Disposal Restrictions (LDRs).

6. Because EPA wants to limit hazardous waste treatment, storage, or disposal to those facilities that can adequately protect human health and the environment, RCRA requires such facility owners and operators to obtain a hazardous waste permit from the Agency.

7. Since hazardous waste management may result in spills or releases into the environment, RCRA Subtitle C also contains provisions governing corrective action, or the cleanup of contaminated air, ground water, and soil.

8. The Statute also grants EPA broad enforcement authority to require all hazardous waste management facilities to comply with the regulations.

9. The Subtitle C program also contains provisions to allow EPA to authorize state governments to implement and enforce the hazardous waste regulatory program.

CFR Guide to Hazardous and Solid Waste Regulations.   To review the details of RCRA regulations, consult the following citations in Title 40 of the Code of Federal Regulations (40 CFR):

- Part 240—Guidelines for the thermal processing of solid wastes
- Part 241—Guidelines for the land disposal of solid wastes
- Part 243—Guidelines for the storage and collection of residential, commercial, and institutional solid waste.
- Part 256—Guidelines for development and implementation of state solid waste management plans
- Part 257—Criteria for classification of solid waste disposal facilities and practices
- Part 258—Criteria for MSW (Municipal Solid Waste) landfills
- Part 260—Hazardous waste management system: general
- Part 261—Identification and listing of hazardous waste
- Part 262—Standards applicable to generators of hazardous waste
- Part 263—Standards applicable to transporters of hazardous waste
- Part 264—Standards for owners and operators of hazardous waste treatment, storage, and disposal facilities
- Part 265—Interim status standards for owners and operators of hazardous waste Treatment, Storage and Disposal Facilities (TSDFs)
- Part 266—Standards for the management of specific hazardous wastes and specific types of hazardous waste management facilities
- Part 268—Land Disposal Regulations (LDRs)
- Part 270—EPA-administered permit programs: the Hazardous Waste Permit Program
- Part 271—Requirements for authorization of state hazardous waste programs
- Part 272—Approved state hazardous waste management programs
- Part 273—Standards for universal waste management
- Part 279—Standards for the management of used oil
- Part 280—Technical standards and corrective action requirements for owners and operators of Underground Storage Tanks (USTs)
- Part 281—Approval of state USTs
- Part 282—Approved UST programs

RCRA's Guiding Principle.   To encourage environmentally sound methods for managing household, municipal, commercial, and industrial waste, RCRA requires the EPA to identify and regulate hazardous waste. That includes developing programs to ensure that hazardous waste is handled safely from cradle to grave: from generation through transportation, treatment, storage, and ultimate disposal.
    Remember: the goals of RCRA are to:

- Protect human health and the environment
- Reduce waste and conserve energy and natural resources
- Reduce or eliminate the generation of hazardous waste as expeditiously as possible

Do not think for a minute that your improper actions at the worksite go unnoticed by the EPA. On April 3, 2001, a California man was sentenced to 18 months in prison for violating the Clean Water Act, at the plant where he worked, by discharging toxics into the community sewer system.

He violated the facility's CWA discharge permit by intentionally releasing higher-than-permitted levels of cyanide, nickel, and corrosives into the sewer system. He also attempted to conceal his illegal activity by trying to circumvent a monitoring device placed in the sewer by the city.

1.4.2 "Superfund," the Comprehensive Environmental Response, Compensation, and Liability Act (CERCLA)
(42 USC 9601 *et seq.* [1980])

O—π **The Comprehensive Environmental Response, Compensation, and Liability Act (CERCLA) established Superfund. A Superfund site is one that poses a major threat to the environment. The cleanup of these sites is administered by the EPA.**

CERCLA, commonly known as "Superfund," was enacted by Congress on December 11, 1980. This law created a tax on the chemical and petroleum industries and provided broad federal authority to respond directly to releases or threatened releases of hazardous substances that may endanger public health or the environment. A trust fund was established for cleaning up abandoned or uncontrolled hazardous waste sites. CERCLA:

- Established prohibitions and requirements concerning closed and abandoned hazardous waste sites
- Provided for liability of persons responsible for releases of hazardous waste at these sites
- Established a trust fund to provide for cleanup when no responsible party could be identified.

CERCLA authorizes two kinds of response actions:

1. *Short-term removals:* These actions may be taken to address releases or threatened releases requiring prompt response.
2. *Long-term remedial response actions:* These actions permanently and significantly reduce the dangers associated with releases, or threats of releases, of hazardous substances that are serious but not immediately life-threatening.

Superfund responses are conducted only at sites listed on EPA's National Priorities List (NPL).

CERCLA also enabled the revision of the National Contingency Plan (NCP). The NCP provided the guidelines and procedures needed to respond to releases and threatened releases of hazardous substances, pollutants, or contaminants. The NCP also established the NPL.

1.4.3 Superfund Amendments and Reauthorization Act (SARA)
(42 USC 9601 *et seq.* [1986])

O—π **The Superfund Amendments and Reauthorization Act (SARA) ensures that community members will be able to find out about hazardous materials and wastes in their community.**

CERCLA was amended in 1986. SARA reauthorized CERCLA to continue cleanup activities around the country. Several site-specific amendments, definitions clarifications, and technical requirements were added to the legislation, including additional enforcement authorities.

Emergency Planning and Community Right-To-Know (RTK) Act.   Title III of SARA authorized the Emergency Planning and Community Right-to-Know (RTK) Act (EPCRA). EPCRA set up local emergency planning commissions (LEPCs). This allowed for better communication between the emergency response agencies and the community. RTK's most important contribution was the part of the law that required factories and other facilities to tell what feedstocks they used in their processes, how much was stored on-site, and what products and wastes were produced in what quantity. All this information was required to be divulged at the request of the residents. The main points of EPCRA follow.

Notification of Extremely Hazardous Substances.   EPCRA section 302 requires facilities to notify the state emergency response commission (SERC) and the LEPC of the presence of any extremely hazardous substance if it has the substance in excess of the specified threshold planning quantity (threshold quantity, or TQ). (Do not confuse this with the RQ reportable quantity.) The list of such substances is in 40 CFR Part 355, Appendices A and B. Section 302 also direct the facility to appoint an emergency response coordinator.

Notification during Releases.   EPCRA section 304 requires facilities to notify the SERC and the LEPC in the event of a release exceeding the reportable quantity (RQ) of a CERCLA hazardous substance or an EPCRA extremely hazardous substance (EHS). Reportable Quantities are listed in Appendix J of the Companion CD.

Emergency Planning.   EPCRA sections 311 and 312 require facilities to notify SERC, LEPC, and the local fire department of all hazardous chemicals used and stored at their facility, for which OSHA requires material safety data sheets (MSDSs). The facility must submit either the MSDSs or a list of the substances for which MSDSs are maintained. If a list is submitted, hazardous chemical inventory forms (also known as Tier I and II forms) must also be submitted. Tier I forms must provide information about hazardous chemicals grouped by hazard category. Tier II forms must provide information about each specific hazardous chemical. This information helps the local government respond in the event of a spill or release of the chemical (40 CFR 370).

Toxic Release Inventory (TRI).   EPCRA section 313 requires manufacturing facilities included in SIC (Standard Industrial Classification) codes 20 through 39 that have 10 or more employees and that manufacture, process, or use specified chemicals in amounts greater than TQs to submit an annual toxic chemical release report to EPA. This report, commonly known as the Form R, covers releases and transfers of toxic chemicals to various facilities and environmental media and allows EPA to compile the national Toxic Release Inventory (TRI) database (40 CFR 372).

TRI incorporates a *de minimis* rule that refers to the amount of an EHS present in a mixture below which the concentration is too small to be considered hazardous. For purposes of TQ determinations, the *de minimis* value is 1%. This means that if any EHS is present in a mixture or solution at concentrations below 1% by weight, it does not need to be accounted for in a TQ determination.

EPA's EPCRA Hotline is (800) 535-0202, and the TRI website is http://www.epa.gov/tri/.

SARA Instructs OSHA to Establish HAZWOPER and Its Training Requirements.   SARA also required that any operation at a hazardous waste site must be carried out in a safe and healthy way.

○──π   **SARA instructed OSHA to write the specifics of the regulations that became HAZWOPER.**

SARA requires the following training standards for employees who will be involved in cleanup operations occurring under authority of these laws:

- General site workers with potential exposure to hazardous substances shall receive 40 hours of off-site instruction (*NOTE: this book meets and exceeds those requirements*) and an additional 24 hours of supervised field experience. The three days of supervised on-site experience are not required for those who have had equivalent training in the past. Those workers with unique or special exposures need additional training.
- Supervisors must receive 8 hours of additional training.
- Any emergency response worker who might be exposed to hazardous substances must also be trained.
- SARA directs OSHA to issue specific regulations for site health and safety.
- Any state or local government worker who is not covered by an OSHA state plan is covered under the OSHA regulations by authority of EPA. As of 2001, only 26 states had OSHA-approved plans.

O—ᴨ **Anyone not HAZWOPER-trained cannot work on a hazardous waste operation, or even visit the site, within the controlled area.**

1.4.4 Environmental Laws that Establish EPA's Authority

- 1938 Federal Food, Drug, and Cosmetic Act
- 1947 Federal Insecticide, Fungicide, and Rodenticide Act
- 1948 Federal Water Pollution Control Act (also known as the Clean Water Act)
- 1955 Clean Air Act
- 1958 Federal Food, Drug and Cosmetic Act, Delaney Clause (21 USC 409)
- 1965 Shoreline Erosion Protection Act
- 1965 Solid Waste Disposal Act
- 1966 Freedom of Information Act (5 USC 552)
- 1970 Clean Air Act (42 USC 7401 et seq.)
- 1970 National Environmental Policy Act (42 USC 4321–4347)
- 1970 Occupational Safety and Health Act (29 USC 61 et seq.)
- 1970 Pollution Prevention Packaging Act
- 1970 Resource Recovery Act
- 1971 Lead-Based Paint Poisoning Prevention Act
- 1972 Coastal Zone Management Act
- 1972 Federal Water Pollution Control Act Amendments of 1972 (PL 92-500)
- 1972 Federal Insecticide, Fungicide and Rodenticide Act (7 USC 135 et seq.)
- 1972 Marine Protection, Research, and Sanctuaries Act (33 USC 1401 et seq.; PL 92-532)
- 1972 Ocean Dumping Act
- 1973 Endangered Species Act (7 USC 136; 16 USC 460 et seq.)
- 1974 Safe Drinking Water Act (43 USC 300f et seq.)
- 1974 Shoreline Erosion Control Demonstration Act
- 1975 Hazardous Materials Transportation Act
- 1976 Resource Conservation and Recovery Act (42 USC 321 et seq.)
- 1976 Toxic Substances Control Act (15 USC 2601 et seq.)
- 1977 Surface Mining Control and Reclamation Act

- 1977 Clean Water Act (33 USC 121 et seq.)
- 1978 Uranium Mill-Tailings Radiation Control Act
- 1980 Asbestos School Hazard Detection and Control Act
- 1980 Comprehensive Environmental Response, Compensation, and Liability Act (Superfund) (42 USC 9601 et seq.)
- 1982 Asbestos School Hazard Abatement Act (PL 98-377, Title II of TSCA)
- 1982 Nuclear Waste Policy Act
- 1986 Asbestos Hazard Emergency Response Act (PL 99-519, Title II of TSCA)
- 1986 Emergency Planning and Community Right to Know Act (42 USC 11011 et seq. [Title III of Superfund Amendments Reauthorization Act of 1986 (SARA)])
- 1986 Superfund Amendments and Reauthorization Act (SARA) (42 USC 9601 et seq.)
- 1988 Asbestos Information Act of 1988 (42 USC 7401, 7412, 7414, 7416 [amends Clean Air Act])
- 1988 Indoor Radon Abatement Act (PL 100-551, Title III of TSCA)
- 1988 Lead Contamination Control Act (PL 100-572 [amends Safe Drinking Water Act])
- 1988 Medical Waste Tracking Act
- 1988 Ocean Dumping Ban Act (PL 100-688 [amendment to Marine Protection, Research and Sanctuaries Act])
- 1988 Shore Protection Act
- 1990 Clean Air Act Amendments of 1990 (42 USC 7401 et seq.)
- 1990 National Environmental Education Act (PL 101-619)
- 1990 The Oil Pollution Act (33 USC 2702-2761; PL 101-380—amends section 311 of Clean Water Act)
- 1990 The Pollution Prevention Act (42 USC 13101 and 13102 et seq.)
- 1992 Residential Lead-Based Paint Hazard Reduction Act (PL 102-550, Title IV of TSCA)
- 1996 Food Quality Protection Act (PL 104-170; 110 Stat. 1489 (amended FIFRA and FFDCA))
- 1999 Chemical Safety Information, Site Security and Fuels Regulatory Relief Act (PL 106-40; 42 USC 7412(r)—amends Section 112(r) of Clean Air Act)

As we stated before, however, this is not meant as a comprehensive regulatory reference. These laws are listed simply to reinforce the fact that there are many laws and regulations that affect our hazardous waste work in one way or another. That prompts many companies to have full-time regulatory affairs employees within their Environmental, Safety and Health departments.

1.4.5   EPA Regional Office Directory

○━🗝   **EPA, like OSHA, has 10 regional offices to serve you.**

Region 1:   Connecticut, Maine, Massachusetts, New Hampshire, Rhode Island, and Vermont
EPA New England
1 Congress St., Suite 1100
Boston, MA 02114-2023
(888) 372-7341
http://www.epa.gov/region01/

Region 2:   New Jersey, New York, Puerto Rico, and the U.S. Virgin Islands
290 Broadway
New York, NY 10007-1866
(212) 637-3000
http://www.epa.gov/Region2/

Region 3:   Delaware, Maryland, Pennsylvania, Virginia, West Virginia, and the District of
Columbia
1650 Arch St.
Philadelphia, PA 19103-2029
(800) 438-2474
http://www.epa.gov/region03/

Region 4:   Alabama, Florida, Georgia, Kentucky, Mississippi, North Carolina, South Carolina, and Tennessee
Atlanta Federal Center
61 Forsyth St., SW
Atlanta, GA 30303-3104
(404) 562-9900
http://www.epa.gov/region4/reg4.html

Region 5:   Illinois, Indiana, Michigan, Minnesota, Ohio, and Wisconsin
77 West Jackson Blvd.
Chicago, IL 60604
(312) 353-2000
http://www.epa.gov/Region5/

Region 6:   Arkansas, Louisiana, New Mexico, Oklahoma, and Texas
1445 Ross Ave., Suite 1200
Dallas, TX 75202
(214) 665-2200
http://www.epa.gov/earth1r6/index.htm

Region 7:   Iowa, Kansas, Missouri, and Nebraska
901 N. 5th St.
Kansas City, KS 66101
(800) 223-0425
http://www.epa.gov/rgytgrnj/

Region 8:   Colorado, Montana, North Dakota, South Dakota, Utah, and Wyoming
999-18th St., Suite 300
Denver, CO 80202-2466
(800) 227-8917 or (303) 312-6312
http://www.epa.gov/unix0008/

Region 9:   Arizona, California, Hawaii, Nevada, and the territories of Guam and American
Samoa
75 Hawthorne St.
San Francisco, CA 94105
(415) 744-1500
http://www.epa.gov/region09/

Region 10:   Alaska, Idaho, Oregon, and Washington
1200 6th Ave.
Seattle, WA 98101
(206) 553-1200
http://www.epa.gov/r10earth/

1.5   THE DEPARTMENT OF TRANSPORTATION (DOT) (49CFR)

DOT regulates the transportation of hazardous materials: what materials can and cannot be shipped, what size, type, and amount of containers are allowed, and how and where they may be transported.

1.5.1   The Hazardous Materials Transportation Act (HMTA)

The Department of Transportation is responsible for transportation by air, rail, sea, and road. DOT's Office of Hazardous Materials Safety (OHMS) is part of DOT's Research and Special Programs Administration (RSPA), established by 49 U.S.C. 5101 *et seq.,* as amended, the Hazardous Materials Transportation Act (HMTA). OHMS regulates the transport of hazardous materials in the United States of America according to 49 CFR 100–180. Enforcement of the HMTA is shared by each of the following administrations under the Secretary of the DOT:

- Research and Special Programs Administration (RSPA): responsible for container manufacturers, reconditioners, and retesters and shares authority over shippers of hazardous materials
- Federal Highway Administration (FHA): enforces all regulations pertaining to motor carriers
- Federal Railroad Administration (FRA): enforces all regulations pertaining to rail carriers
- Federal Aviation Administration (FAA): enforces all regulations pertaining to air carriers
- U.S. Coast Guard (USCG): enforces all regulations pertaining to shipments by water

1.5.2   Hazardous Materials Tables

DOT's Hazardous Materials Tables contain everything you need to know about packaging, labeling, and shipping of hazardous materials, including hazardous waste.

The DOT regulations most used in HAZWOPER are the Hazardous Materials Tables in Appendix A to 49 CFR 172.101. If, as a HAZWOPER employee, you are required to ship hazardous waste, all the DOT information you will require for manifesting, marking, packing, labeling, and placarding the shipment will be found in the Hazardous Materials Tables. They are reproduced for reference in the Companion CD as part of the NIOSH Pocket Guide.
    DOT's OHMS website is http://hazmat.dot.gov/.

1.5.3   Overview of DOT Hazardous Materials Regulations

An Overview of the Department of Transportation (DOT) Hazardous Materials Regulations is provided as Appendix B to this book, in the Companion CD.

1.5.4   DOT's HAZMAT Safety Field Offices

DOT has five regional Safety Field Offices to serve you.

Central: Illinois, Indiana, Iowa, Kentucky, Michigan, Minnesota, Missouri, Nebraska, North Dakota, Ohio, South Dakota, Wisconsin
Enforcement, Initiatives and Training
2350 East Devon Ave., Suite 136

Des Plaines, IL 60018
(847) 294-8580

Eastern: Connecticut, District of Columbia, Delaware, Maine, Maryland, Massachusetts, New Hampshire, New Jersey, New York, Pennsylvania, Rhode Island, Vermont, Virginia, West Virginia
Enforcement, Initiatives and Training
820 Bear Tavern Rd., Suite 306
West Trenton, NJ 08628
(609) 989-2256

Southern: Alabama, Georgia, Florida, Mississippi, North Carolina, Puerto Rico, Tennessee, South Carolina
Enforcement, Initiatives and Training
1701 Columbia Ave., Suite 520
College Park, GA 30337
(404) 305-6120

Southwest: Arkansas, Colorado, Kansas, Louisiana, New Mexico, Oklahoma, and Texas
Enforcement, Initiatives and Training
2320 LaBranch St., Suite 2100
Houston, TX 77004
(713) 718-3950

Western: Alaska, Arizona, California, Hawaii, Idaho, Montana, Nevada, Oregon, Utah, Washington, Wyoming
Enforcement, Initiatives and Training
3200 Inland Empire Blvd., Suite 230
Ontario, CA 91764
(909) 483-5624

Special Investigations: National (cylinder manufacturers, explosives, and radioactive materials)
DHM-40
400 7th St., S.W.
Washington, DC 20590
(202) 366-4700

CAUTION

THERE WILL ALSO BE STATE REGULATORS, AND POSSIBLY OFFICIALS IN LOCAL MUNICIPALITIES THAT HAVE JURISDICTION OVER A WORK SITE. ALTHOUGH IT WOULD BE IMPOSSIBLE TO COVER EACH INDIVIDUAL STATE AND LOCAL REGULATION HERE, UNDERSTAND THAT YOU SHOULD FAMILIARIZE YOURSELF WITH THESE REGULATIONS AS THEY MAY BE MORE STRINGENT THEN THE FEDERAL REGULATIONS.

1.6   OTHER FEDERAL AGENCIES AND PRIVATE ORGANIZATIONS THAT OFFER GUIDANCE AND ASSISTANCE

O—π   **OSHA, EPA, and DOT often rely on other organizations to aid them with technical information.**

A number of other federal agencies and private organizations have a direct impact on the HAZWOPER worker. The federal government and corporate America rely on these for research, clarification, guidance, training, standards, and preparedness information. Sometimes private sector or municipal expertise in a certain area exceeds that available in the federal government. Some of these agencies and organizations set their own standards. It has happened that they are then used as a basis for federal regulatory compliance.

1.6.1 The National Institute of Occupational Safety and Health (NIOSH)

> **NIOSH is the governments' prime research organization for occupational safety and health. All of its information is advisory unless added to OSHA, EPA or DOT's regulations.**

NIOSH is part of the U.S. Department of Health and Human Services, Public Health Service, Centers for Disease Control and Prevention. NIOSH was originally funded to be the research arm of OSHA. Their job is research and outreach versus OSHA's outreach and enforcement. NIOSH has no enforcement mandate. All of their information is advisory.

The NIOSH website is http://www.cdc.gov/niosh/homepage.html.

1.6.2 Federal Emergency Management Agency (FEMA)

> **FEMA is a prime resource for any type of disaster or other large-scale emergency.**

FEMA's job is emergency preparedness for all types of emergencies: floods, hurricanes, earthquakes, fires, terrorist attacks and major environmental releases of hazardous materials among others. FEMA coordinates with State Emergency Management Agencies working in either a leading or supporting role in case of emergency. FEMA has a tremendous amount of human and material assets available on short notice.

The FEMA website is http://www.fema.gov/.

1.6.3 National Response Center (NRC)

> **The U.S. Coast Guard's National Response Center (NRC) is the sole federal point of contact for reporting oil and chemical spills. It is the home of the National Response Team (NRT).**

The U.S. Coast Guard's National Response Center is the sole federal point of contact for reporting oil and chemical spills. If you have a spill to report, call 1-800-424-8842. The NRC is the home of the National Response Team (NRT). The NRT can respond to oil or hazardous materials spills nationwide on an almost instantaneous basis. It will also act as liaison between all federal agencies for the incident. The NRC is staffed by Coast Guard personnel who maintain a 24 hours a day, 365 days a year telephone watch. NRC watch personnel enter telephonic reports of pollution incidents into the Incident Reporting Information System (IRIS) and immediately relay each report to the predesignated Federal On-Scene Coordinator (FOSC). Federal Law requires that you report oil and hazardous material spills to the NRC.

Note

One call to the NRC fulfills the requirement to report releases of hazardous substances under CERCLA and several other regulatory programs. That includes those under CWA section 311, RCRA, and the U.S. Department of Transportation's Hazardous Materials Transportation Act.

The NRC website is http://www.nrc.uscg.mil/index.htm.

1.6.4 The National Transportation Safety Board (NTSB)

O—π **The NTSB investigates large-scale transportation accidents of hazardous materials and gives recommendations to prevent future accidents.**

The NTSB is an independent federal agency that investigates every civil aviation accident in the United States and significant accidents in the other modes of transportation. NTSB conducts special investigations and safety studies, and issues safety recommendations to prevent future accidents. Safety Board investigators are on call 24 hours a day, 365 days a year.

The NTSB website is http://www.ntsb.gov/.

1.6.5 The American Conference of Governmental Industrial Hygienists (ACGIH)

O—π **The ACGIH provides its own exposure data, called TLVs®. Although it has no regulatory effect, this information is often relied upon by safety professionals.**

This research body conducts independent scientific research and publishes the results. The ACGIH is best known for its research into the effects of materials on exposed humans. It publishes recommended exposure levels called Threshold Limit Values (TLVs®). Although not legally enforceable limits, many Certified Industrial Hygienists (CIHs) rely on this information when designing safe work practices for workers.

The ACGIH website is http://www.acgih.org/.

1.6.6 The National Fire Protection Association (NFPA)

O—π **The NFPA is a private organization most widely known for its 704M® labeling system for materials. Its fire codes and emergency response procedures have been adopted nearly nationwide.**

This association produces more than 300 fire codes and standards. Initially it predominantly supported professional and volunteer firefighters, providing information about the varied hazards and dangers firefighters might encounter in their job.

The NFPA is most widely known among HAZWOPER workers for the NFPA 704M® material label. This labeling system was instituted to allow firefighters a quick hazard reference when surveying a fire scene. This is the most widely recognized labeling system in the United States.

The NFPA website is http://www.nfpa.org/.

1.6.7 The American Society for Testing and Materials (ASTM)

O—π **The ASTM writes technical standards that are used nationwide. ASTM's Standard Practice for Environmental Site Assessments and its Transaction Screen Process can provide valuable site information before HAZWOPER work begins.**

This society writes technical standards for industry. By producing products that meet these standards, American manufacturers and other organizations can assure their customers that they are producing a product that meets an accepted standard. Of the thousands of standards and test methods produced, two are of special interest to the HAZWOPER worker:

1. E1527-00 Standard Practice for Environmental Site Assessments: Phase 1 Environmental Site Assessment Process
2. E1528-00 Standard Practice for Environmental Site Assessments

The results and information produced by these practices is invaluable to HAZWOPER cleanup workers, as described in Section 14 of this book.

The ASTM website is http://www.astm.org/.

1.6.8 The American National Standards Institute (ANSI)

> **ANSI writes performance standards for nearly every type of material in use in America today. HAZWOPER workers benefit from their standards for every type of PPE used. ANSI standards are incorporated by reference in OSHA regulations.**

The institute writes standards for the performance of materials in various configurations. HAZWOPER workers will benefit particularly from ANSI's standards for the materials and construction of eye, foot, head, hand, and Chemical Protective Clothing. These standards all have been adopted by OSHA.

The ANSI website is http://www.ansi.org/.

1.6.9 The American Society of Safety Engineers (ASSE)

ASSE is the world's oldest and largest professional safety organization. Its approximately 33,000 members manage, supervise, and consult on safety, health, and environmental issues in industry, insurance, government, and education. ASSE has 12 divisions and 148 chapters in the United States and abroad. The ASSE is a valuable resource for professional help with safety issues.

The ASSE website is http://www.asse.org/.

1.6.10 The National Academy of Sciences, the National Academy of Engineering, the Institute of Medicine, and the National Research Council (Collectively Known as the National Academies)

The academies are the advisors to the Nation on Science, Engineering, and Medicine.

The National Academies website is http://www.nas.edu/.

1.6.11 The National Institute of Environmental Health Sciences (NIEHS)

The NIEHS is one of the National Institutes of Health of the U.S. Department of Health and Human Services. Its sister organization is the National Toxicology Program (NTP).

The NIEHS website is http://www.niehs.nih.gov/home.htm.

1.6.12 The U.S. Fire Administration (USFA)

> **The USFA is part of FEMA. The *USFA Hazardous Materials Guide for First Responders* is one of the prime resources for that sector of HAZWOPER workers.**

The USFA is part of FEMA. It is best known to HAZWOPER emergency response workers for their publication *USFA Hazardous Materials Guide for First Responders*. The guide is available in print or as a searchable version at the USFA website.

The USFA website is http://www.usfa.fema.gov/hazmat/.

1.6.13 National Institute of Standards and Technology (NIST)

The NIST is one of largest public research laboratories in the United States. It prepares standards for, among other things, the measurement of chemical constituents in the hazardous material samples we take in HAZWOPER field work. It also has branches such as:

- Building and Fire Research Laboratory
- Chemical Science and Technology Laboratory
- Computer Security Division
- Information Technology Laboratory
- Manufacturing Engineering Laboratory
- Materials Science and Engineering Laboratory
- NIST Computer Security Resource Clearinghouse
- NIST/MSEL Center for Theoretical and Computational Materials Science
- Physics Laboratory
- Software Diagnostics and Conformance Testing Division

The NIST website is http://www.nist.gov/.

1.6.14 The EPA's Office of Research and Development

This is the research arm of the EPA. It provides the latest information on all the EPA's R&D projects and grants. It includes the National Exposure Research Laboratory (NERL), National Health and Environmental Effects Research Lab (NHEERL), National Center for Environmental Assessment (NCEA), National Risk Management Research Lab (NRMRL).
The EPA's Office of Research and Development website is http://www.epa.gov/ORD/.

1.7 ADDITIONAL RESOURCES

The authors recommend the following videos for aiding readers to see "real life" examples of what they have learned in Section 1.

- OSHA At Work.
- Protecting Workers: How OSHA Conducts Inspections
- Partnering With OSHA: New Ways of Working

These videos are available from OSHA. (Note: Although these videos are "dated," they still contain pertinent information for today's workplace.)

1.8 SUMMARY

Section I has explained briefly the mission of the three principal regulating agencies. The ones you will most likely become involved with during hazardous waste operations are EPA, OSHA, and DOT. The differences between the regulations enforced by those agencies and the laws, also called acts, of the federal government were described.

We traced the birth of HAZWOPER and its training requirements. The EPA, established in 1970, concerns itself with control of polluting and hazardous materials. That is, when hazardous materials have the potential to be released, or have been released, into the environment. EPA is also charged with improving the environment. The Resource Conservation and Recovery Act (RCRA) of 1976 established cradle-to-grave legislation with regulation by EPA. Congress enacted RCRA so that hazardous materials and hazardous wastes would be managed safely. That is, from the time they were generated to the time they were disposed of safely.

Congress, responding to public sentiment, realized more was needed. It enacted the Comprehensive Environmental Response, Compensation, and Liability Act (CERCLA) of 1980, commonly known as Superfund. A Superfund site is one that poses a major threat to the environment. The cleanup of these sites is administered by the EPA. In 1986, Congress enacted the Superfund Amendments and Reauthorization Act (SARA), the third of the three major legislative acts that bear upon our work. SARA reauthorized CERCLA to continue cleanup activities around the country.

Of interest to our readers is that SARA also instructed OSHA to establish HAZWOPER and its training requirements. As we have learned, OSHA was formed in 1970 and charged with the overall mission of protecting and improving the safety and health of the American worker. OSHA, as instructed, has specified minimum requirements for HAZWOPER training, experience, and certification. The keys within our text have emphasized the minimum information that HAZWOPER trainees must know to pass this portion of a reasonable basic certification exam on relevant regulations and legislation.

The regulations that affect hazardous material workers are found in the United States Code of Federal Regulations (CFR). We have listed the specific applicable locations.

DOT, the third agency (after OSHA and EPA) of importance in our work, regulates the transportation of hazardous materials to ensure workers' and public safety. DOT's Hazardous Materials Tables contain everything you need to know about packaging, labeling, and shipping of hazardous materials, including hazardous waste.

Relevant information you can obtain from other federal agencies and private organizations has also been described. Over the years the authors have found those other agencies to be invaluable sources of information that assists HAZWOPER work.

SECTION 2
HAZARDOUS WASTE DEFINED

WARNING

FOR THE PURPOSE OF PERSONAL PROTECTION, THERE IS ONLY ONE MATERIAL THAT SHOULD ENTER YOUR BODY AT A HAZARDOUS WASTE SITE: CLEAN AIR. ANYTHING ELSE MAY DO YOU HARM SO CONSIDER ANYTHING ELSE HAZARDOUS UNTIL YOU ARE SURE IT IS NOT!

OBJECTIVES

If you were asked to define hazardous waste, how would you respond? You might say that it is something that is contaminated, you want to dispose of it, and you believe it can harm you. You would be correct . . . as far as that goes. What about something that will not harm you, but might harm the environment? How about something hazardous you have that is not really waste (someone could use it) but that you do not need anymore?

Defining hazardous waste is not so easy, is it?

In this Section the authors will give the definitions from the three principal regulatory agencies. OSHA considers hazardous substances to be anything that will, or has the capacity to, harm humans who may come into contact with such a substance.

The EPA considers there to be a difference between what it calls hazardous material and hazardous waste. The EPA has the most complete definition. For instance, if there is a usable portion of the hazardous material in the original container and it is the owner's intention to use or recycle the material, EPA says that it is a hazardous material. If there is a usable amount of the material left in the original container but the owner wants to dispose of it, it becomes hazardous waste.

The DOT considers all hazardous substances, whether new product in the original container, cleaned-up hazardous waste product, marine pollutants, or elevated temperature materials that will be shipped in commerce, hazardous materials.

Note

There is really no difference between a "substance" and a "material." The different regulators simply defined "matter" with different words.

2.1 THE OSHA DEFINITION

OSHA defines "hazardous waste" as the waste form of a "hazardous substance"—that is, a substance that will, or may, result in adverse effects on the health or safety of employees.

"Hazardous substance," according to OSHA, means any substance designated or listed under 1–4 below, exposure to which results, or may result, in adverse effects on the health or safety of employees:

1. Any substance defined under section 101(14) of CERCLA.

2. "Any biologic agent and other disease-causing agent which after release into the environment and upon exposure, ingestion, inhalation, or assimilation into any person, either directly from the environment or indirectly by ingestion through food chains, will or may reasonably be anticipated to cause death, disease, behavioral abnormalities, cancer, genetic mutation, physiological malfunctions (including malfunctions in reproduction) or physical deformations in such persons or their offspring."

3. Any substance listed by the U.S. Department of Transportation as a hazardous material under 49 CFR 172.101 and appendices (the Hazardous Materials Table).

4. 'Hazardous waste' is also a hazardous substance, with the following additions:
   - A waste or combination of wastes as defined in 40 CFR 261.3
   - Those substances defined as hazardous wastes in 49 CFR 171.8

Note

HAZWOPER training is OSHA-regulated. OSHA depends upon EPA and DOT for help with definitions. This is reasonable because there are some shared responsibilities between these three agencies.

2.2   EPA DEFINITIONS

Hazardous waste is a legal term that describes certain toxic, ignitable, corrosive, or reactive wastes generated in manufacturing, industrial, or other processes.

To be considered a "hazardous waste," the waste must be a "solid waste." The EPA defines "solid waste" as "solids, semi-solids, liquids, and contained gaseous materials."

The hazardous waste regulations can be found in Title 40 of the Code of Federal Regulations (CFR) Part 261–299. According to the *EPA Methodology for Identifying Hazardous Waste,* you must first determine if your waste is a "solid waste" according to EPA. EPA defines "solid waste" as garbage, refuse, sludge, or other discarded material, including solids, semi-solids, liquids, and contained gaseous materials. Notice that uncompressed gases and vapors are not included, but liquids are included with solids. Further, notice that contained gaseous materials will include as solid waste a nitrogen or oxygen cylinder if you are considering disposing of either one as a waste.

If your waste is a solid waste, you must then determine if it is a hazardous waste. It is your responsibility as a generator (the person or organization for whom it is a waste) either to test your waste or to use your knowledge of the waste to make a determination about its properties. Once you know what is in your waste, you can then determine if EPA considers it to be hazardous.

All EPA hazardous wastes have a four-character Hazardous Waste Code beginning with the letters F, K, P, U, or D and followed by three digits—for example D001, F002, or K145.

Note

Appendix I, on the Companion CD, provides the details of all these wastes.

2.2.1 Is the Waste a Listed (F, K, P, or U) or Characteristic (D) Hazardous Waste?

○━π   **F, K, P, and U listed hazardous wastes are hazardous regardless of the concentration of their hazardous constituents.**

○━π   **D, or Characteristic, wastes are treated differently than F, K, P, and U wastes. Because of D wastes' specified characteristics, they are considered nonhazardous until proven hazardous.**

EPA has developed five lists, or categories (F, K, P, U, and D), of hazardous wastes. They are published in 40 CFR 261.30. With the exception of the D Characteristic wastes, if your waste is currently listed, it is hazardous regardless of the concentrations of hazardous constituents in the waste.

The F List Waste Codes

○━π   **"Nonspecific source wastes" are designated with F waste codes. They are material-specific rather than industry-specific. They are wastes that can come from a wide variety of industries.**

Nonspecific source wastes (40 CFR 261.31) are material-specific wastes, such as solvent wastes, electroplating wastes, or metal heat-treating wastes, commonly produced by a wide variety (nonspecific sources) of manufacturing and industrial processes (designated with F waste codes). *Examples:* Wastewater treatment sludges from electroplating operations (F006), process wastes such as distillation residues, heavy ends, tars, and reactor cleanout wastes (F024).

The K List Waste Codes

○━π   **"Specific source wastes" from industries such as wood preserving, petroleum refining, and organic chemical manufacturing are designated with K waste codes.**

Specific source wastes (40 CFR 261.32) are wastes from specifically identified industries such as wood preserving, petroleum refining, and organic chemical manufacturing (designated with K waste codes). *Examples:* Wastewater treatment sludge from the production of chrome yellow and orange pigments (K002), tar storage tank residues from coal tar refining (K147).

The P and U Waste Codes

○━π   **Discarded commercial chemical products that are toxic are designated with P and U waste codes, with P wastes being highly toxic.**

Discarded commercial chemical products (40 CFR 261.33), which are off-specification products, container residuals, spill residue runoff, or active ingredients that have spilled or are unused and intended to be discarded and are toxic. These materials are designated with P and U waste codes. P listed wastes are considered highly toxic. If the intent is to use the material or recycle it, it is not considered a hazardous waste. *Examples:* Aldicarb (P070), parathion (P089), and vinyl chloride (U043).

The Delisting Procedure for F, K, P, and U Listed Wastes.   If your waste is of the F, K, P, or U type, it is automatically designated as RCRA-regulated hazardous waste, based on its listing, regardless of the nonhazardous characteristics it might have. That is why there is a delisting procedure. If you can show that your F, P, K, or U waste consistently has neither the characteristic for which that class of wastes was originally listed nor any other hazardous characteristic, you can petition EPA for your specific waste stream to be delisted.

The waste lists were originally prepared based on EPA's surveys and subsequent determinations. EPA found that the wastes, as a class, typically have some characteristic that causes them to be hazardous. Then EPA decided not to run specific studies of the characteristics for each waste stream in the many plants throughout an industry. Rather, EPA has determined that that class of waste (F, K, P, or U listed) is to be considered hazardous until proven otherwise (something like "guilty until proven innocent").

This is as opposed to the ruling on D Characteristic wastes. EPA presumes all of them not to be RCRA-regulated hazardous wastes—that is, unless examination shows a waste to have a hazardous characteristic (something like "innocent until proven guilty").

The D List Characteristic Wastes

○━ㅍ **There are four types of characteristic hazardous waste. They all have D waste codes. The four hazardous characteristics are ignitability, corrosivity, reactivity, and toxicity.**

EPA identified four characteristics, or traits, of hazardous waste: ignitability, corrosivity, reactivity, and toxicity (designated with D waste codes). Your waste is considered hazardous if it exhibits any of these characteristics (40 CFR 261.20-24). These properties are measurable by standardized and available testing methods that can be found in a manual entitled *Test Methods for Evaluating Solid Waste, Physical/Chemical Methods* (*SW-846*). (www.epa.gov/epaoswer/hazwaste/test/main.htm).

Determining D List Characteristics of Hazardous Waste (40 CFR 261.20)

1. "A solid waste, as defined in Sec. 261.2, which is not excluded from regulation as a hazardous waste under Sec. 261.4(b), is a hazardous waste if it exhibits any of the characteristics identified in this subpart. It is the generators' responsibility to determine whether their waste exhibits one or more of the characteristics identified in this subpart."

2. "A hazardous waste which is identified by a characteristic in this subpart is assigned every EPA Hazardous Waste Number that is applicable as set forth in this subpart. This number must be used in complying with the notification requirements of section 3010 of the Act and all applicable record keeping and reporting requirements under parts 262 through 265, 268, and 270 of 40 CFR."

The Characteristic of Ignitability (40 CFR 261.21)

○━ㅍ **Hazardous waste with the characteristic of ignitability is labeled with the EPA hazardous waste code D001.**

A solid waste exhibits the characteristic of ignitability if a representative sample of the waste has any of the following properties:

- It is a liquid, other than an aqueous solution, containing less than 24% alcohol by volume.
- It has a flash point of less than 60°C (140°F), as determined by a Pensky-Martens Closed Cup Tester, that is, by using the test method specified in ASTM Standard D-93-79 or D-93-80, or a Setaflash Closed Cup Tester, using the test method specified in ASTM Standard D-3278-78. It may also be determined by an equivalent test method approved by the Administrator under procedures set forth in sections 260.20 and 260.21.
- It is not a liquid and is capable, under standard temperature and pressure, of causing fire through friction, absorption of moisture or spontaneous chemical changes. When ignited, it burns so vigorously and persistently that it creates a hazard.
- It is an ignitable compressed gas as defined in 49 CFR 173.300 and as determined by the test methods described in that regulation.
- It is an oxidizer as defined in 49 CFR 173.151.

The Characteristic of Corrosivity (40 CFR 261.22)

⊙—π **Hazardous waste with the characteristic of corrosivity is labeled with the EPA hazardous waste code D002.**

A solid waste exhibits the characteristic of corrosivity if a representative sample of the waste has any of the following properties:

- An aqueous liquid is corrosive if it has a pH less than or equal to 2 or greater than or equal to 12.5, that is, as determined by a pH meter using Method 9040 in *Test Methods for Evaluating Solid Waste, Physical/Chemical Methods,* EPA Publication SW-846.
- A liquid is corrosive if it corrodes steel (SAE 1020) at a rate greater than 6.35 mm (0.250 in.) per year at a test temperature of 55°C (130°F). This is as determined by the test method specified in NACE (National Association of Corrosion Engineers) Standard TM-01-69. That test was standardized in SW-846.

The Characteristic of Reactivity (40 CFR 261.23)

⊙—π **Hazardous waste with the characteristic of reactivity is labeled with the EPA hazardous waste code D003.**

A solid waste exhibits the characteristic of reactivity if a representative sample of the waste has any of the following properties:

- It is normally unstable and readily undergoes violent change without detonating.
- It reacts violently with water.
- It forms potentially explosive mixtures with water.
- When mixed with water, it generates toxic gases, vapors, or fumes in a quantity sufficient to present a danger to human health or the environment.
- It is a cyanide- or sulfide-bearing waste that, when exposed to pH conditions between 2 and 12.5, can generate toxic gases, vapors, or fumes in a quantity sufficient to present a danger to human health or the environment.
- It is capable of detonation or explosive reaction if it is subjected to a strong initiating source or if heated under confinement.
- It is readily capable of detonation or explosive decomposition or reaction at standard temperature and pressure.
- It is a forbidden explosive as defined in 49 CFR 173.51, or a Class A explosive as defined in 49 CFR 173.53 or a Class B explosive as defined in 49 CFR 173.88.

The Characteristic of Toxicity (40 CFR 261.24)

⊙—π **Hazardous waste with the characteristic of toxicity is labeled with an EPA hazardous waste code ranging from D004 to D043.**

Here the EPA hazardous waste code corresponds to the toxic contaminant causing it to be hazardous. Refer to Appendix I, in the Companion CD, for details.

Toxicity Characteristic Leaching Procedure (TCLP)

⊙—π **The TCLP is the EPA test method that is used to decide if waste will be characterized as toxic. If waste is found to be nontoxic (passes the TCLP test) it may be disposed of in a sanitary landfill.**

Is the TCLP Sample a Hazardous Waste? A solid waste exhibits the characteristic of toxicity (and therefore is a hazardous waste) if a representative sample of the waste fails the

Toxicity Characteristic Leaching Procedure (TCLP). Extracts (leachates) produced by the TCLP test fail if they contain higher than the regulatory level concentrations of the materials listed in Table 2.1. Materials examples are metals such as cadmium, lead, and mercury and organic chemicals such as benzene, carbon tetrachloride, DDT, and vinyl chloride.

If the individual o-, m-, and p-cresol concentrations cannot be differentiated, the (total) cresol $(D026)_1$ concentration is used. The regulatory level of total cresol is 200 mg/L. If the quantitation limit (least concentration measurable in the test) is greater than the regulatory level, then the quantitation limit$_2$ becomes the regulatory level.

TABLE 2.1   Maximum Concentrations (specified in 40 CFR 261.24) of Contaminants for Rating a Waste Extract as Having the Toxicity Characteristic

Contaminant	Regulatory Level (mg/L)
Arsenic	5.0
Barium	100.0
Benzene	0.5
Cadmium	1.0
Carbon tetrachloride	0.5
Chlordane	0.03
Chlorobenzene	100.0
Chloroform	6.0
Chromium	5.0
o-cresol	200.0_1
m-cresol	200.0_1
p-cresol	200.0_1
Cresol	200.0_1
2,4-D	10.0
1,4-dichlorobenzene	7.5
1,2-dichloroethane	0.5
1,1-dichloroethylene	0.7
2,4-dinitrotoluene	0.13_2
Endrin	0.02
Heptachlor (and its hydroxide)	0.008
Hexachlorobenzene	0.13_2
Hexachloro-1,3-butadiene	0.5
Hexachloroethane	3.0
Lead	5.0
Lindane	0.4
Mercury	0.2
Methoxychlor	10.0
Methyl ethyl ketone	200.0
Nitrobenzene	2.0
Pentachlorophenol	100.0
Pyridine	5.0_2
Selenium	1.0
Silver	5.0
Tetrachloroethylene	0.7
Toxaphene	0.5
Trichloroethylene	0.5
2,4,5-trichlorophenol	400.0
2,4,6-trichlorophenol	2.0
2,4,5-TP (silvex)	1.0
Vinyl chloride	0.2

Note

Urban fill soils often have high concentrations of lead (Waste Code D008) due to the long-term use of leaded gasoline and oxidized lead from lead-based paint. This lead contamination buildup has taken place over decades. But the results are astounding. In Boston, Massachusetts, the average urban lead background concentration is 570 mg/kg!

The test method used for the Toxicity Characteristic Leaching Procedure is test Method 1311 in EPA Publication SW-846 (see above, The D List Characteristic Wastes). Where the waste contains less than 0.5% filterable solids, the waste itself, after filtering using the methodology outlined in Method 1311, is considered to be the extract.

The TCLP (pronounced "tee clip") is designed to simulate the leaching a waste will undergo if disposed of in a sanitary landfill. The TCLP test procedure is described in more detail in Section 12.5.2 of this book. This extract, the "TCLP extract," is then analyzed to determine if any of the thresholds established for the 40 toxicity characteristic (TC) constituents listed in Table 2.1 have been exceeded.

Under the toxicity characteristic, a representative solid waste sample exhibits the characteristic of toxicity if the TCLP extract from the sample contains high levels of any of the contaminants listed in Table 2.1, that is, at a concentration greater than, or equal to, the regulatory level given in that table.

Land Disposal Restrictions (LDR). Land disposal is defined as disposal to landfills, waste piles, and surface impoundments (ponds). The next question is: does the sample meet the standards for Land Disposal? If the treatment standards established for the constituents listed in 40 CFR section 268.41 have been met for the Land Disposal Restrictions (LDR) program, the answer is yes.

Under the Land Disposal Restrictions program, a restricted waste identified in 40 CFR 268.41 may be land disposed only if the proper TCLP extract of the waste does not exceed the Regulatory Levels shown in Table 2.1 for any hazardous contaminant.

"If the TCLP extract contains LDR contaminants in an amount exceeding the concentrations specified in 40 CFR 268.41 (Waste Codes D004 through D011), the treatment standard for that waste has not been met, and further treatment is necessary prior to land disposal."

2.2.2   Is the Waste a Mixture?

If your waste is a mixture of nonhazardous solid wastes and listed hazardous wastes, it is considered hazardous (40 CFR 261.3). A few wastes are listed only because they are ignitable or reactive. In these cases, if the resulting mixture no longer is ignitable or reactive, the mixture is not considered a listed waste. *Examples:* Spent solvents (F003), such as methanol and acetone, are listed hazardous wastes and are ignitable. If these solvents are mixed with a nonignitable nonhazardous waste, the mixture will still be considered hazardous unless the mixture can be proven to be not ignitable. However, it is illegal to dilute a hazardous waste and then dispose of the waste as non-hazardous!

2.2.3   Is the Waste a "Derived-from" Hazardous Waste?

Any solid waste generated (derived) from the management (treatment, storage, or disposal) of a listed hazardous waste is considered hazardous waste (40 CFR 261.3). *Example:* Any ash or residue left from the incineration process at a hazardous waste incinerator is considered hazardous waste. That includes sludge, spill residue, ash, emission control dust, or leachate.

2.2.4 Is the Waste Contaminated Media?

Environmental media (ground water, soil, or sediment) sometimes comes in contact with listed hazardous waste. If these media become contaminated with (and therefore contain) hazardous waste, they must be managed as a hazardous waste. *Example:* If a tank leaks listed hazardous waste into ground water, the ground water must be managed as a listed hazardous waste.

2.2.5 Is the Waste Contaminated Debris?

Manufactured objects, plant or animal matter, and natural geological material that exceeds 60 mm (2.36 in.) particle size and that is intended for disposal are considered "debris" (40 CFR 268.2). Debris is not considered a solid waste, but if a hazardous waste is mixed with debris, it becomes hazardous waste. *Example:* Rags, personal protective equipment, or wood pallets that are contaminated with a hazardous waste must be managed as a hazardous waste until they are decontaminated and no longer contain the listed waste.

2.2.6 Exclusions

Some wastes are excluded from the EPA definition of hazardous waste. That does not mean they are not covered under some other federal, state, or local regulation as far as disposal is concerned. Further, it does not mean that these wastes are not harmful to humans.

Be sure to note that some wastes are excluded from the definition of solid waste and therefore from the hazardous waste regulations. The following are excluded from the definition of solid waste: domestic sewage, irrigation return flows, and *in situ* mining wastes. Household wastes, agricultural wastes used as fertilizers, and cement kiln dust are further examples of wastes that are excluded from the definition of hazardous waste.

CAUTION

THESE WASTES ARE NOT SUBJECT TO THE FEDERAL HAZARDOUS WASTE REGULATIONS BUT MAY BE SUBJECT TO OTHER FEDERAL REGULATIONS OR STATE OR LOCAL WASTE PROGRAMS.

2.2.7 "Universal Waste" Rule

The EPA introduced the "Universal Waste" rule to encourage resource conservation while at the same time ensuring adequate protection of human health and the environment. The rule's intent is to facilitate compliance with the regulations governing universal waste by making the regulations easy to understand. Their third intended purpose is to divert universal wastes from the municipal waste stream. The EPA acknowledges that there are certain wastes that are generated in such a quantity throughout the nation (universally) that, if recycled, they would benefit the generators and the nation as a whole. The EPA will undoubtedly add to this list in the future. Currently, the rule covers the following wastes:

• Batteries that can be recycled, such as lead-acid car batteries
• Mercury-containing thermostats
• Fluorescent, high-intensity discharge, neon, incandescent, and other mercury-containing lamps
• Pesticides

By recycling these materials instead of disposing of them as hazardous waste, the generator benefits in a number of ways:

- The generator can store these wastes for up to one year instead of the 90 or 180 days for other hazardous wastes. In the case where the generator needs more material to make recycling feasible, they may exceed the one-year term.
- The generator can ship the wastes to the recycler under a normal bill of lading. A Uniform Hazardous Waste Manifest is not required.
- The generator will not have to pay for hazardous waste disposal costs.
- The generator, after ensuring that the recycler recycles the materials, no longer has the long-term liability of CERCLA's "cradle-to-grave" rule.

As the generator, you must understand that the universal waste rule only applies to intact materials. Generators cannot crush lamps and ship the debris to the recycler. EPA considers that process as waste treatment, and the resultant waste must be considered hazardous. The exception to this rule is that batteries need not contain their electrolyte (acid).

2.2.8   Rebuttable Presumption (Waste Oils)

The EPA does allow for a rebuttable presumption for used oil. Used oil containing more than 1000 ppm total halogens (such as the chlorofluorocarbon refrigerants) is presumed to be a hazardous waste. This is because it is assumed that the oil has been mixed with halogenated hazardous waste listed in Subpart D of Part 261. Companies may rebut this presumption by demonstrating that their used oil does not contain hazardous waste. A company might show, for example, that the used oil does not contain significant concentrations of halogenated hazardous constituents. Rather, the company might prove that the halogen is chlorine from saltwater contamination. Therefore the waste oil would not be hazardous and could be recycled in a number of ways, such as fuel blending.

- The rebuttable presumption does not apply to metalworking oils/fluids containing chlorinated paraffins if they are processed, through a contracting agreement, to reclaim metalworking oils/fluids. The presumption does apply to metalworking oils/fluids if such oils/fluids are recycled in any other manner, or disposed.
- The rebuttable presumption does not apply to used oils contaminated with chlorofluorocarbons (CFCs) removed from refrigeration units where the CFCs are destined for reclamation.
- The rebuttable presumption does apply to used oils contaminated with CFCs that have been mixed with used oil from sources other than refrigeration units.

2.9.9   "Generators" Defined

Whether or not a manifest is required for a shipment of hazardous waste is determined by the status of the generator. All shipments of hazardous waste, regardless of the size of the individual shipment, from Large Quantity Generators (LQG) or Small Quantity Generators (SQG), *must* be accompanied by a manifest. See Section 13.2 for more complete definitions.

A LQG is a facility that generates more than 1000 kg (2200 lb) of hazardous waste per month and greater than 1 kg (2.2 lb) of acutely hazardous waste per month.

A SQG is a facility that generates more than 100 kg (220 lb) of hazardous waste or 1 kg (2.2 lb) of acutely hazardous waste (P list waste) per month.

O━━π   **A Conditionally Exempt Small Quantity Generator (CESQG) is exempt from all manifest requirements. These are defined as generators that produce, monthly, 100 kg or less of hazardous waste; 1 kg or less of acute hazardous waste; or 100 kg or less of contaminated soil, waste, or debris resulting from cleanup of an acute hazardous waste spill.**

CAUTION

IF YOU QUALIFY AS A CESQG, THE AUTHORS STRONGLY RECOM-MEND THAT YOU KEEP ACCURATE RECORDS TO PROVE THAT YOU NEVER EXCEEDED YOUR MONTHLY ALLOTMENT OF HAZARDOUS WASTE.

2.3   THE DOT DEFINITION

2.3.1   Hazardous Materials, Hazardous Substances, and Reportable Quantities

O━━π   **The DOT defines a hazardous material as any material listed in the Hazardous Materials Table. That includes hazardous substances, hazardous wastes, marine pollutants, and elevated temperature materials in one package that equals or exceeds the reportable quantity (RQ).**

O━━π   **The RQ of a material is the minimum quantity, in pounds, that must be reported to the NRC and EPA in the event of a spill.**

Note

Appendix J on the Companion CD to this book lists the Reportable Quantities of Hazardous Substances.

The Hazardous Materials Tables tell the people who package, offer for transport, load, and transport hazardous materials everything the DOT wants them to know about all the hazardous materials listed there. This following information is included:

Symbol

Proper shipping name

Hazard Class and Division of the material

Identification number (UN/NA number)

Packing group

Label codes

Special provisions packaging

   Exceptions
   Non-bulk
   Bulk

Quantity limitations

   Passenger aircraft/rail—limit/no limit/forbidden
   Cargo aircraft only—limit/no limit/forbidden

Vessel stowage

   Location
   Other

Penalties for noncompliance in the transportation and disposal of hazardous waste can be severe.

> A Michigan contractor was sentenced to two years in prison and ordered to pay $44,000.00 in restitution to the U.S. government for failure to follow regulations related to the handling of harmful substances. The defendant pleaded guilty to violating the Resource Conservation and Recovery Act by illegally causing the transportation of paint and solvent wastes to four rural Michigan sites, where they were abandoned. One site was near an elementary school. The materials were ignitable and contained xylene, ethyl benzene (a known carcinogen), toluene, and lead.

"Hazardous material" means a substance or material that has been determined by the Secretary of Transportation to be capable of posing an unreasonable risk to health, safety, and property when transported in commerce. The term includes hazardous substances, hazardous wastes, marine pollutants, and elevated temperature materials. These are materials designated as hazardous under the provisions of 49 CFR 172.101 and materials that meet the defining criteria for hazard classes and divisions in 49 CFR 173.

"Hazardous substance" for DOT means a material, including its mixtures and solutions, that meets the following criteria:

1. The material is listed in Appendix A to 49 CFR 172.101.

2. The material is in a quantity, in one package, that equals or exceeds the reportable quantity (RQ); that is the minimum quantity, in pounds, that must be reported to EPA in the event of a spill.

3. The material, when in a mixture or solution:
   - For radionuclides, conforms to 49 CFR 172.101
   - For other than radionuclides, is in a concentration by weight that equals or exceeds the concentration corresponding to the RQ of the material, as shown in Table 2.2.

Example: Ammonium hydroxide has an RQ of 1,000 lb. If you spill 1,000 pounds or more, the spill must be reported. Ammonium hydroxide commonly exists as about a 5% solution in water as household ammonia. If you had this household ammonia diluted with water to less than 2%, no matter how much was spilled, it would not be a reportable quantity, not even if there were more than 1000 lb of ammonium hydroxide in the total solution, just as long as the concentration was less than 2%.

CAUTION

THE EXAMPLE USING AMMONIUM HYDROXIDE IS NOT MEANT TO SUGGEST THAT DILUTING A CONTAMINANT TO BELOW REGULATORY LEVELS AND THEN WILLFULLY SPILLING THE TOTAL MIXTURE IS LEGAL DISPOSAL. IT IS NOT LEGAL TO DILUTE HAZARDOUS

TABLE 2.2 Concentration by Weight That Equals or Exceeds the Concentration Corresponding to the RQ of the Substance

RQ lb (kg)	Percent	Concentration by weight, ppm
5000 (2270)	10	100,000
1000 (454)	2	20,000
100 (45.4)	0.2	2,000
10 (4.54)	0.02	200
1 (0.454)	0.002	20

MATERIAL TO BELOW REGULATORY CONCENTRATIONS AND THEN IMPROPERLY DISPOSE OF THE MIXTURE.

2.3.2 Dot Exclusions

The DOT excludes petroleum and natural or synthetic gas and related liquids from its definition of hazardous substances, even though they are often flammable and must be placarded as such.

"Hazardous substance" does not include petroleum, including crude oil or any fraction thereof, which is not otherwise specifically listed or designated as a hazardous substance in Appendix A to 49CFR 172.101. Further, "hazardous substance" does not include natural gas, natural gas liquids (propane, LPG, butane in lighters), liquefied natural gas (LNG), or synthetic gas usable for fuel (or mixtures of natural gas and such synthetic gas).

Table 2.3 gives examples of hazardous waste generated by selected industries.

TABLE 2.3 Typical Hazardous Wastes Generated by Selected Industries

Waste Generators	Waste Type
Chemical manufacturers	Strong acids and bases Reactive wastes Ignitable wastes Discarded commercial chemical products
Vehicle maintenance shops	Paint wastes Ignitable wastes Spent solvents Acids and bases
Printing industry	Photography waste with heavy metals Heavy metal solutions Waste inks Spent solvents
Paper industry	Ignitable wastes Corrosive wastes Ink wastes, including solvents and metals
Construction industry	Ignitable wastes Paint wastes Spent solvents Strong acids and bases Cleaning agents and Heavy metal dusts and sludges
Cosmetic manufacturing	Ignitable wastes Solvents Strong acids and bases
Furniture and wood manufacturing and refinishing	Ignitable wastes Spent solvents Paint wastes
Metal manufacturing	Paint wastes containing heavy metals Strong acids and bases Cyanide wastes Sludges containing heavy metals

2.3.3   The Hazardous Waste Manifest System

Note

The application of the Uniform Hazardous Waste Manifest System will be covered in greater detail in Section 13 of this book.

○━━π   **A Uniform Hazardous Waste Manifest, EPA Form 8700-22, must accompany all shipments of hazardous waste. This is a requirement of both the EPA and the DOT.**

○━━π   **OSHA's duty is the protection of the health and safety of workers employed in the shipping and transport tasks involved with hazardous wastes.**

Now that we have reviewed the different methodologies, examples, and wording of the three major regulators' definitions, there is another way that hazardous waste can easily be identified. Particularly where transportation is involved, hazardous waste means any material that is subject to the Hazardous Waste Manifest Requirements of the EPA, specified in 40 CFR part 262.

A hazardous waste manifest is the descriptive, identifying documentation that MUST accompany any shipment of hazardous waste. That covers any mode of transport, anywhere in the United States. The manifest must stay with the waste until it reaches its final disposal site. Figure 2.1 is a copy of a blank Uniform Hazardous Waste Manifest.

2.3.4   Manifest System Components

The Hazardous Waste Manifest System is a set of forms, reports, and procedures. They are designed to track hazardous waste from the time it leaves the generator facility, where it was produced or accumulated, until it reaches the off-site waste management facility. That facility will store, treat, or dispose of the hazardous waste. The system requires waste generators to verify that their wastes have been properly delivered. The system ensures that no waste has been lost or unaccounted for in the process.

2.3.5   How the Manifest Accompanies the Waste from Cradle-to-Grave

The key component of this system is the Uniform Hazardous Waste Manifest, shown in Figure 2.1. All generators first fill out this form. They are the entities who transport, or offer for transport, hazardous waste for off-site treatment, recycling, storage, or disposal. For years, the manifest has been a paper document containing multiple copies of a single form. It contains one section each for the generator, the transporter, and the facility (TSDF). It also contains information on the type and quantity of the waste being transported, special handling instructions for the waste, and signature lines for all parties involved in the disposal process. Both Department of Transportation and EPA require the manifest. Each party that handles the waste signs the manifest and retains a copy for itself. This ensures critical accountability in the transportation and disposal processes. Once the waste reaches its destination, the receiving facility returns a signed copy of the manifest to the generator. This confirms that the waste has been received by the facility designated by the generator on the form.

2.3.6   An Electronic Uniform Hazardous Waste Manifest System?

In January 2001, the EPA proposed to improve the Uniform Hazardous Waste Manifest system by automating procedures and standardizing the manifest form. EPA believes that waste handlers could realize savings from $24–37 million a year. States could possibly save up to 25% in manifest-related costs while ensuring the continuous, safe management of

Please print or type   (Form designed for use on elite (12 - pitch) typewriter)   Form Approved. OMB No. 2050 - 0039 Expires 9 - 30 - 91

UNIFORM HAZARDOUS WASTE MANIFEST	1 Generator's US EPA ID No.	Manifest Document No.	2. Page 1 of	Information in the shaded areas is not required by Federal law

3. Generator's Name and Mailing Address	A. State Manifest Document Number
	B. State Generator's ID
4. Generator's Phone ( )	

5. Transporter 1 Company Name	6.	US EPA ID Number	C. State Transporter's ID
			D. Transporter's Phone
7. Transporter 2 Company Name	8.	US EPA ID Number	E. State Transporter's ID
			F. Transporter's Phone
9. Designated Facility Name and Site Address	10.	US EPA ID Number	G. State Facility's ID
			H. Facility's Phone

	11. US DOT Description (Including Proper Shipping Name, Hazard Class, and ID Number)	12. Containers No.	Type	13. Total Quantity	14. Unit Wt/Vol	I. Waste No.
G E N E R A T O R	a.					
	b.					
	c.					
	d.					

J. Additional Descriptions for Materials Listed Above	K. Handling Codes for Wastes Listed Above

15. Special Handling Instructions and Additional Information

16. **GENERATOR'S CERTIFICATION:** I hereby declare that the contents of this consignment are fully and accurately described above by proper shipping name and are classified, packed, marked, and labeled, and are in all respects in proper condition for transport by highway according to applicable international and national government regulations.

If I am a large quantity generator, I certify that I have a program in place to reduce the volume and toxicity of waste generated to the degree I have determined to be economically practicable and that I have selected the practicable method of treatment, storage, or disposal currently available to me which minimizes the present and future threat to human health and the environment; **OR**, if I am a small quantity generator, I have made a good faith effort to minimize my waste generation and select the best waste management method that is available to me and that I can afford.

Printed/Typed Name	Signature	Month Day Year

T R A N S P O R T E R	17. Transporter 1 Acknowledgement of Receipt of Materials		
	Printed/Typed Name	Signature	Month Day Year
	18. Transporter 2 Acknowledgement of Receipt of Materials		
	Printed/Typed Name	Signature	Month Day Year

F A C I L I T Y	19. Discrepancy Indication Space		
	20. Facility Owner or Operator: Certification of receipt of hazardous materials covered by this manifest except as noted in item 19.		
	Printed/Typed Name	Signature	Month Day Year

EPA Form 8700 - 22 (Rev. 9 - 88) Previous editions are obsolete.

FIGURE 2.1 The Uniform Hazardous Waste Manifest. See Appendix F of the Companion CD for a printable version of the Manifest.

hazardous waste. EPA proposes to allow waste handlers the option of using an electronic manifest to track their waste shipments. EPA plans to standardize further the manifest form and procedures.

The automation options should significantly reduce the manifest paperwork burden and should also improve the tracking of hazardous waste shipments. Although waste handlers still may opt to use a paper manifest, the Agency expects that automation will provide better and more timely data on waste shipments for both waste handlers and regulators.

The new standard format announced in the proposal could be used in all states. These proposed changes would eliminate the different manifest forms that are currently required by many authorized states. Each form would have a unique, preprinted manifest tracking number. Under this option, for example, waste handlers with multistate operations could register and use their own manifest forms everywhere they do business.

This proposal affects approximately 92,000 businesses in 45 economic sectors that conduct hazardous waste manifest-related activities. These include about 90,000 small and large quantity waste generators, 500 waste transporters, and 2,000 treatment, storage, and disposal facilities in 24 states. Simply automating the manifest form could save up to $27 million a year.

2.3.7   The Manifest System Audit Trail

The audit trail is as follows. See Figure 2.2 for a diagram of the manifest trails.

1. Generators who transport, or offer for transportation, hazardous waste for off-site management prepare a Uniform Hazardous Waste Manifest form. They follow the instructions included on the form and amplified in the appendix to 40 CFR 262.

2. Generators sign the manifest certification. Rubber stamps are not allowed. They obtain the handwritten signature of the transporter and date of acceptance on the manifest. The generator retains one copy and gives the transporter the remaining copies of the manifest. Generators may be required by the state to send a copy of the manifest (signed by both the generator and the transporter) to the state EPA. Generators must keep a copy of each signed manifest for three years. The authors recommend that the generator keep the returned, completed manifests indefinitely.

3. Transporters arriving at the treatment, storage, and disposal facility (TSDF) designated on the manifest, give the manifest to a representative of that TSDF. The representative signs and dates the manifest. The transporter keeps a signed copy of this manifest on file.

4. Any discrepancies between the waste description on the manifest and the actual waste received by the TSDF are noted in the "Discrepancy Indication Space" on the manifest. Such discrepancies include differences in size, number, weight, volume, type of containers, or type of waste.

5. The TSDF sends a copy of the manifest, signed by the TSDF's representative, to the generator; thereby closing the loop in the manifest cycle. This enables the generator to verify that the waste has been disposed of properly. The TSDF also retains a copy of the manifest on file.

6. If a generator does not receive a copy of the manifest signed by the designated facility owner or operator within 45 days of the date that the waste was accepted by the initial transporter (60 days for small quantity generators), the generator must file an exception report.

Most states also require the TSDF to send the state a copy of the manifest signed by the TSDF. The requirement to submit copies to states is not based on federal requirements. It is based on the needs of state programs that use manifest data for purposes of compliance monitoring, capacity planning, and program management. You should contact your authorized state program for the RCRA regulations that apply in your state.

2.3.8   Rail, Water, and Export Shipments

Domestic shipments made by rail or water are subject to streamlined manifesting requirements. Hazardous waste shipments exported from the United States are subject to additional requirements. The full text of these regulations is at 40 CFR 261 and 262.

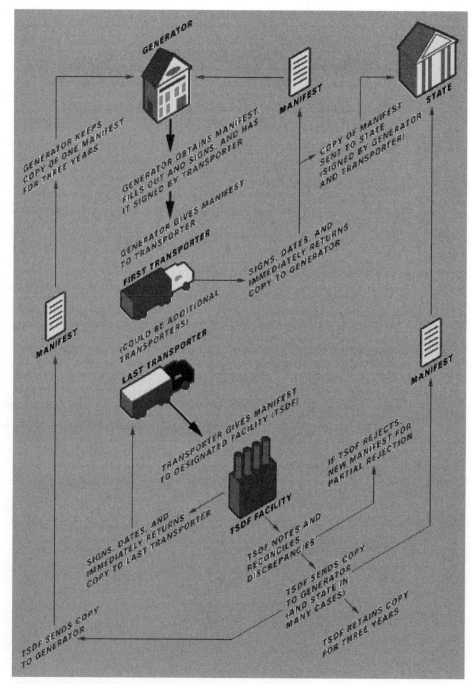

FIGURE 2.2 Diagram of the automated manifest process.

2.4   TRAINING AIDS AND ADDITIONAL RESOURCES

The authors recommend the following video for aiding readers to see real life examples of what they have learned in this Section:

- *Introduction to Hazardous Waste,* available from Safe Expectations (formerly BNA Communications, Inc.) (www.safeexpectations.com).

2.5   SUMMARY

In this Section we have seen how the different regulators define hazardous materials, substances, and waste. OSHA, EPA, and DOT all have somewhat different ideas as to the exact definition, but they agree on one thing: If the hazardous substance or hazardous material is a waste, it is a hazardous waste.

OSHA's definition is the most important for HAZWOPER personnel to remember. If a material can harm you, the worker, it is hazardous. OSHA includes in its regulations all of the materials rated as hazardous in the EPA and DOT regulations.

The EPA and DOT must deal with thousands of industries manufacturing, transporting, and disposing of tens of thousands of waste chemicals and materials. To make it possible for those industries to continue to produce the products we use in our daily lives, while unavoidably generating hazardous wastes, EPA and DOT have established specific hazardous waste definitions and accompanying regulations. The EPA established the four characteristic hazardous waste codes, which defined the hazardous characteristics as ignitability, corrosivity, reactivity, and toxicity. EPA also requires the TCLP test to determine whether a waste can be sent to a sanitary landfill. It also developed the concept of reportable quantities (RQs), which governs the reporting of spills.

As you can see, the EPA is mainly concerned with hazardous waste or materials that are released into the environment. It not only regulates how and where hazardous waste may be disposed of, but how and where hazardous materials can be stored, used, and recycled.

The DOT developed the Hazardous Waste Manifest System and the Hazardous Materials Tables. The DOT is concerned chiefly with how hazardous materials, including wastes, are shipped safely. It is concerned with the health of the transporter as the material is shipped, and with the environment if there is an accident during shipment.

All of the three principal regulators are tied together to form a chain that, if kept intact, will allow us to work more safely with hazardous materials and their wastes.

SECTION 3
MATERIAL HAZARDS

OBJECTIVES

Many hazardous materials will be discussed in this section. However, this is not meant to be an all-inclusive list of material hazards. Many more material hazards may exist at your facility or site! Our objective is to heighten the sensitivity of HAZWOPER personnel to typical members of four broad categories of hazardous materials, and then for personnel to be prepared for protection from those hazards using the information in the following sections of this book.

Section 2 described hazardous waste as defined by the regulatory agencies OSHA, EPA, and DOT. Simply stated, in their terms, waste is hazardous because it contains one or more material hazards and meets the other regulatory requirements summarized in Section 2. You must be cautioned that those definitions do not include all materials that might harm you. Section 2 explained a few regulatory exemptions.

In Section 3 we present those four categories of material hazards that may be present at any site. They are: toxic, fire and explosion, corrosive and chemical, and radiation. We treat toxic hazards first. Our experience leads us to believe that they are the most prevalent. It becomes more and more difficult to go through a week without encountering, or hearing that others have encountered, some objectionable gas or vapor. We all wonder, "Is it toxic?" The same goes for what we eat and drink. In HAZWOPER work such concerns are magnified. Section 5 will deal with the answers we can get from the science of toxicology.

It seems reasonable for fire and explosion material hazards to be second in importance for us to investigate. Easily ignitable fuels, gasoline, diesel, and natural gas, are with us in our daily lives—in fact, with us more than at any other time in history!

We can also say that our third category, corrosive and chemical hazards, particularly man-made chemical hazards, are with us more than ever before. The good news is that we are more adept at detecting and controlling them. Detection, presented in Section 12, and personal protection, presented in Section 6, are of prime importance to HAZWOPER personnel. Section 4 will dig deeper into the topic of the hazards of incompatibility between chemicals. Before studying that topic, you can be sure you are unaware of some dangerous incompatibilities between relatively common materials. The hazards of practically all materials, including individual chemicals and chemical mixtures, are explained in MSDSs. Section 9 tells how you learn of those details.

Our last category, radiation hazards, including radioactive materials hazards, may seem of relatively little concern. Don't be fooled. The proliferation of nuclear power sites, the wastes they generate, and the challenges of radioactive hazardous wastes management will be increasing throughout our lives.

Sections 10 and 11 explain the major nonmaterial hazards you might encounter at any worksite or facility. Sections 13 through 16 tell how HAZWOPER personnel manage material and non-material hazards in daily work.

3.1 TOXIC HAZARDS

○━━π **Although poisons are only one of seven different subcategories of toxic hazard, in HAZWOPER work you will often find the terms "*toxic*" and "*poison*" used as synonyms.**

Toxic materials can cause local or systemic detrimental effects in your body and in other organisms. In Section 5 you will learn more about the effects of toxic materials and how exposures are rated. Exposure to such materials does not always result in death, although that is often the most immediate concern. Poisons and asphyxiants are the fastest-acting hazards. However, each type of toxic material can pose a serious threat to some significant population of workers.

Since there are so many different toxic materials with which you may come in contact in HAZWOPER duties, always rely on the instructions in the MSDS for materials you might contact at the worksite. Be sure you understand the Section 9 explanation of MSDSs.

We classify toxic hazards in seven broad categories:

1. Systemic poisons and biohazards
2. Asphyxiants
3. Carcinogens
4. Mutagens (affecting genes) and teratogens (affecting offspring)
5. Irritants
6. Allergens
7. Sensitizers

3.1.1 Systemic Poisons and Biohazards

○━━π **Materials that will harm the body if inhaled or ingested, or if they enter the body via puncture wounds (injection) or skin absorption, are called poisons.**

Systemic poisons are materials that are toxic to specific organs or organ systems following exposure by any of the four major routes of exposure: inhalation, ingestion, absorption, and injection. These toxic hazards can be grouped in categories based on the organ or system they affect.

Central Nervous System (CNS) Depressants. CNS depressants slow down normal brain functions. They cause symptoms such as headaches, nausea, impaired coordination, lethargy, and confusion. Examples include beverage alcohol, toluene, and methanol.

Convulsants. Convulsants are materials that act in the brainstem or spinal cord to produce convulsions and possible coma. Examples include the organo-phosphate insecticides parathion and malathion, the wartime nerve gases, and a number of pharmaceuticals.

Hemolytic Agents. Hemolytic agents are materials that affect blood and blood components. Examples include arsine, benzene, carbon monoxide, and lead. The poisonous gas carbon monoxide is generated when almost any carbon-containing fuel is not completely burned to form the much less dangerous carbon dioxide, among other products. Every year charcoal fires in enclosed spaces cause deaths and hospitalizations of many people, due mostly to deadly carbon monoxide concentrations.

Hepatotoxins. Hepatotoxins are materials that damage the liver. Examples include carbon tetrachloride, nitro-hydrocarbons, chlordane, and chloroform.

Nephrotoxins.   Nephrotoxins are materials that damage the kidneys. Examples include carbon disulfide, chloroform, mercury, and parathion.

Neurotoxins.   Neurotoxins are materials that can destroy portions of the nervous system. Examples include cyanides, organo-mercury compounds, mercury, lead, PCBs, and methanol.

Gold mining, jewelry manufacturing, and some other operations involving metals use cyanides in processing. Those process wastes, as sludges, are poisonous materials. Possibly even more dangerous is the fact that the cyanides in those wastes, if not destroyed first, can release the deadly poisonous gas hydrogen cyanide if acid is accidentally added to the sludge. Cyanide's physiological affect is due to its inhibiting tissue respiration by combination with the iron in blood and to its paralyzing of the CNS.

A university professor conducting research using the liquid dimethyl mercury died within a half hour after receiving skin absorption of about a 25 cent piece-sized spot of the chemical on her palm. She was even wearing laboratory gloves, but she did not realize that the particular kind of glove was permeable to dimethyl mercury. (You will learn about the factors to be considered when selecting personal protective equipment, such as gloves, in Section 6.)

Reproductive System Toxins.   Reproductive system toxins affect the male or female reproductive system. Examples include steroids, cadmium chloride, and lead.

As you can see, a number of these systemic poisons fall into more than one category. These materials are frequently lethal and can cause death rapidly. They can enter the body and attack life-sustaining systems. Commonly, entry can be through breathing, eating, swallowing, and by puncture wounds. Rarely is the mode of entry skin absorption.

Biohazards.   Biohazards fill a special niche for HAZWOPER workers. A biohazard describes bacteria, viruses, fungi, or other infectious materials. Biohazard waste is usually packaged in red or orange plastic bags or red or yellow plastic containers. These are usually sent directly to incinerators for destruction. However, the HAZWOPER worker may encounter any one of these biohazards on the site:

• Known etiologic (infectious) agents such as bacteria, fungi, viruses, protozoa, and parasites

• Animals or materials that contain or may potentially contain known infectious agents or yet unknown agents

• Human and animal tissues and body fluids

Although the chance of encountering extremely hazardous viruses such as Ebola or HIV is slight, the HAZWOPER workers must know to avoid any container marked with the biohazard symbol as described above until they are properly trained and properly protected. Please see the further discussion of biological hazards in Section 11.10.

3.1.2 Asphyxiants

○━━╥ **Asphyxiants are gaseous or vapor contaminants that make it impossible for us to absorb sufficient oxygen to support life.**

○━━╥ **Asphyxiants are divided into two categories, simple asphyxiants and chemical asphyxiants.**

Simple asphyxiants dilute or displace atmospheric oxygen, lowering the concentration of oxygen in air. They are physiologically inert. When they are present in high concentration, they displace the oxygen in the atmosphere. Victims exposed to them will suffer from lack of oxygen. Normal air contains about 20.9% oxygen. Simple asphyxiants can create an oxygen-deficient atmosphere (<19.5% oxygen), resulting in headaches, unconsciousness, and

(handwritten: dry ice is frozen CO₂)

(handwritten: Argon)

eventually death. Examples are nitrogen (N_2), carbon dioxide (CO_2), helium (He), and methane (CH_4).

Any gas or vapor that displaces oxygen in the breathing atmosphere is an asphyxiant. A common example in this category is a nitrogen atmosphere. Although nitrogen is almost four fifths of the air we breathe, the remaining one fifth that is oxygen is critical to our health and life.

Any gas or vapor that reduces the concentration of oxygen to less than 19.5% in a breathing atmosphere is a potential asphyxiant.

Every year someone working with nitrogen dies from asphyxiation. It seems so harmless. It is odorless, completely unreactive, and it is not really toxic. The problem is that if a person enters and breathes a nitrogen atmosphere, respiration stops almost immediately and the person collapses and, within a few minutes, dies. Nitrogen is used to blanket flammable liquids in storage and purge lines in flammable liquids processing. Nitrogen is nonflammable and excludes the oxygen that is necessary for the flammable liquid to catch fire. Liquid nitrogen, at about 300°F below zero, is used to quick-freeze foods.

Chemical Asphyxiants interfere with the body's ability to utilize oxygen. They prevent the uptake of oxygen into the blood or prevent normal oxygen transfer from the blood to the body. Examples include carbon monoxide (CO), hydrogen sulfide H_2S, nitrobenzene ($C_6H_5NO_2$), and hydrogen cyanide (HCN).

(handwritten: MTBE — Methyltertiary Butal Ether)

(handwritten: LEAD)

3.1.3 Carcinogens

(handwritten: = if they say its a carcinogen it is known.)

(handwritten: KNOWN —)

A carcinogen is a material that has been proven to cause some type of cancer.

Carcinogens are materials that are proven to cause some form of cancer. There are also materials that are suspected carcinogens. You want to be protected from both. Airborne fibers of asbestos, dusts of beryllium metal, and tobacco smoke are all known to cause lung cancer, for instance. Arsenic is another example. It is an OSHA cancer hazard, an NTP human carcinogen, and an IARC Group I human carcinogen.

It is important to note that different agencies and groups examine their data on carcinogenicity of different materials and come to different conclusions. The International Agency for Research on Cancer (IARC) is the most widely accepted group when it comes to carcinogen data. However, you will find on some MSDSs that a material is a suspected carcinogen by the IARC but not considered a carcinogen by OSHA, NIOSH, the National Toxicology Program (NTP) or the ATSDR (Agency for Toxic Substances and Disease Registry).

(handwritten: SUSPECT — is one)

(handwritten: fernoldhyde)

The authors agree that if any accepted source (OSHA, NIOSH, ACGIH, IARC, ATSDR, NTP, etc.) determines a material to be a carcinogen, or a suspected carcinogen, then managers should treat it as such and protect their workers. Accepted sources include:

(handwritten: Benzene in — gasoline — solvents.)

- If the material is regulated by OSHA as a carcinogen
- If it is listed under the category "known to be carcinogens" in the *Annual Report on Carcinogens* published by the NTP
- If it is listed under Group 1 ("carcinogenic to humans") by the International Agency for Research on Cancer Monographs (IARC)
- If it is listed in either Group 2A or 2B by IARC or under the category "reasonably anticipated to be carcinogens" by the NTP

3.1.4 Mutagens and Teratogens

Mutagens are agents that cause mutations to cells in the body. In laboratory animal tests they are also found to be carcinogens and teratogens.

(handwritten: modifies the genetic structure.)

A ***mutagen*** is an agent that can cause an increase in the rate of mutation: Examples include X-rays, ultraviolet irradiation, and various chemicals. Mutagens induce DNA damage that either kills cells or, when misrepaired, produces abnormal sequences that will be passed on to daughter cells. These actions, in turn, induce birth defects by injuring developing organs or by initiating mosaicism that disorganizes growth and differentiation.

We are now more aware and more concerned about the presence of mutagens in the environment. Mutagens are ubiquitous; some are naturally present in plants, many are produced by combustion of organic materials (including cooking), and others are products of industry. Given daily exposures to mutagens, it is natural to ask which agents are responsible for birth defects and what reproductive risks may arise from specific exposures. As you will see, the effects we can document in humans are not those commonly anticipated by the public.

Mutagens, on the other hand, produce less predictable patterns of anomalies, even under experimental conditions, because they are indiscriminate in the types of cells they damage. However, experiments with ionizing radiation, the simplest mutagen to study, indicate that the developing brain is particularly ill prepared to deal with mutagenic insults and is consistently affected. Hence, microcephaly is the only consistent finding in nonlethal mutagen exposures during pregnancy, but is clearly not an effect exclusive to mutagen exposures. As with any developmental toxin, increasing the dosage of experimental mutagen exposure during pregnancy also leads to miscarriage.

Mutagens can also affect germ cells, and there is general concern about environmental exposures leading to birth defects through mutations in egg or sperm cells. Experimental studies, however, show that significant exposures to mutagens are not likely to be detected by increased rates of birth defects. Germ cells are more likely to be killed by DNA damage than to be mutated. This loss of cells will lead to infertility. Mutations in germ cells could introduce new genetic conditions into families, but only those conditions that are dominant and whose phenotypes are viable will be seen.

Experimental data indicate that most dominant mutations induced in germ cells at high levels of exposure are lethal. Conceptions carrying dominant lethal genes are likely to abort early or even fail to implant. Thus, the perceived effect is again infertility. Recessive mutations induced by mutagens will not be detected until they become homozygous (double-gened) in future generations, and many changes in sequence in noncoding regions of the genome will simply remain silent. Each of these effects has been well documented in animal studies.

Providers of prenatal care rarely encounter pregnancies complicated by high dose exposures to ionizing radiation or chemical mutagens such as may occur with cancer treatment. On the other hand, they are routinely asked questions about effects of diagnostic X-rays and exposures to known carcinogens such as tobacco smoke.

O—π   Teratogens are agents that cause malformations in embryos (birth defects).

A *teratogen* is an agent capable of causing malformations in embryos. Thalidomide, a drug given to pregnant mothers to help with morning sickness from 1956 to 1961, is one example. Teratogens can be recognized from the patterns of birth defects they produce. They interact with cellular receptors that define developmental fields. Teratogens can be chemicals, reproductive toxins, maternal disorders, maternal infections, or medications. To appreciate the wide variety of teratogens and their effects, review Table 3.1. A chemical teratogen that has become notorious because of its toxicity at very low concentrations in air is dioxin (2,3,7,8 tetra chloro dibenzo p-dioxin).

3.1.5 Irritants

O—π   Irritants cause irritation and inflammation to exposed skin and mucous membranes. Although irritants are usually thought of as nontoxic, a severe reaction to an irritant can be fatal.

Irritants are much less toxic than those in the previous four categories. They attack exposed tissue such as skin and eyes. Breathing brings the irritants into nasal passages and

Women can declare trying to get "pregnant" so they don't have to be around these

TABLE 3.1   Known Human Teratogens[a]

	Effect
Drugs	
Thalidomide	Limb reduction defects, ear anomalies
Diethylstilbestrol	Vaginal adenosis/adenocarcinoma, cervical erosion and ridges
Warfarin	Nasal hypoplasia, stippled epiphyses, CNS defects
Hydantoin	Dysmorphic facial features, hypoplastic nails, growth and developmental retardation
Trimethadione	Developmental retardation, dysmorphic facial features
Aminopterin and methotrexate	Pregnancy loss, hydrocephalus, low birth weight, dysmorphic facial features
Streptomycin	Hearing loss
Tetracycline	Stained teeth, enamel hypoplasia
Valproic acid	Neural tube defects, dysmorphic facial features
Isotretinoin	Pregnancy loss, hydrocephalus, other CNS defects, small or absent thymus, microtia/anotia, conotruncal heart defects
Antithyroid drugs	Hypothyroidism, goiter
Androgens and high doses of nor-progesterones	Masculinization of external female genitalia
Penicillamine	Cutis laxa
ACE Inhibitors	Renal dysgenesis, oligohydramnios sequence, skull ossification defects
Carbamazepine	Neural tube defects
Cocaine	Pregnancy loss, placental abruption, growth retardation, microcephaly
Lithium	Ebstein anomaly
Chemicals	
Methylmercury	Cerebral atrophy, spasticity, mental retardation, rapid death
Lead	Pregnancy loss, CNS damage
Polychlorobiphenyls (PCBs)	Low birth weight, skin rashes and discoloration
Reproductive toxins	
Cigarette smoking	Pregnancy loss, low birth weight
Hyperthermia	Neural tube defects
Chronic alcoholism	Growth and developmental retardation, microcephaly, craniofacial dysmorphism
Therapeutic radiation	Growth and developmental retardation, microcephaly

[a]Courtesy of "Teratogen Update," *Genetic Drift Newsletter,* vol. 12, Fall 1995, Carol Clericuzia, M.D., Editor, Department of Pediatrics, The University of New Mexico, Albuquerque, NM.

throat and lung tissues. Common tear gas is an example. Many other materials (chemical raw materials, wastes, and products) cause effects similar to those of tear gas. The vapor of the liquid benzoyl peroxide causes tearing of the eyes. Ozone reactions with vehicle exhausts cause complex gases and vapors that are eye irritants especially.

Sulfur dioxide from burning sulfur causes irritation and inflammation of the breathing passages. Vapors from sodium sulfite and sodium hydrosulfite solutions cause some people to experience asthma-like attacks from the lung irritation and edema.

As with all materials, some people are more sensitive to irritants than others. A professional insulator may be able to work with fiberglass insulation all day with no apparent affect. Another person only has to rub an arm on some exposed fiberglass with the result of a terrible, itching rash.

3.1.6 Allergens

Allergens are substances capable of producing immediate hypersensitivity reactions known as allergies.

Medical science seems unable to explain why one person is allergic to a given *allergen,* such as ragweed pollen or dusts, and another person is not affected. Both persons can be healthy and have no other unusual health complaints. One person can lose an allergy with aging, and another can develop an allergy with aging.

Most of us are now aware that some people have an allergic reaction to such common foods as peanut butter (or any food containing peanuts) and wheat flour-containing foods. Upon exposure, these people can suffer severe allergic reactions including asthma-type breathing difficulties (anaphylactic reactions to foods) and even death.

Allergens, which usually have a protein component, cause the body organs and tissues of an allergic person to release histamine. Its release causes all the sneezing, itching eyes, skin rashes, lethargy, and other symptoms. It has been estimated that enough histamine is stored in body tissues and can be produced by the body that if it were all released at one time the result would be fatal. The antihistamine drugs, which have been in use for more than half a century, have given major relief to untold millions.

If a worker experiences an allergic breathing reaction to a material at a site, a particulate respirator should provide the relief needed if an attending physician agrees. If a site contains poison ivy or oak, proper suiting, footwear, gloves and head protection, might be required. Some people, however, are not allergic even to those troublesome plants.

3.1.7 Sensitizers

A sensitizer is a material that causes people to develop an allergic reaction, usually, but not necessarily, after repeated exposure.

A *sensitizer* is defined by OSHA as "a chemical that causes a substantial proportion of exposed people or animals to develop an allergic reaction in normal tissue after repeated exposure to the chemical." The condition of being sensitized to a chemical is a chemically induced immune system disorder called chemical hypersensitivity. Perfume is an excellent example. We now see companies not allowing employees to wear fragrances because of the possibility of another employee's chemical hypersensitivity. Latex gloves are another example.

Certain chemicals have no immediate health effect. However, if you are exposed to them several times, they can make you allergic or sensitive to other chemicals. When your exposure to the substance that you are hypersensitive to causes a rash, it is called allergic contact dermatitis. A classic example is formaldehyde.

Some other people suffer from multiple chemical sensitivities (MCS). Exactly how this occurs, and why it affects some people and not others, are unknown. What is known is that these people can be affected by acute (short-term) or chronic (long-term) exposures to the material. Their sensitivities, once started, can lead to other sensitivities to chemicals that they have only been exposed to one time. Further, since some of the symptoms mimic the symptoms of more common ailments such as the flu, diagnosing MCS can be extremely difficult.

Symptoms of Sensitization.   Symptoms of chemical hypersensitivity cover a wide range. The following symptoms are characteristic:

• Skin—sores, rashes, itching, burning, cracking

• Eyes—redness, pain, burning, vision problems

• Ears—dizziness or balance problems, ringing (tinnitus)

- Nose—running, stopped up, congested, burning, nosebleeds
- Throat—dry, burning, phlegm, hoarseness, voice problems
- Chest—pain, burning, shortness of breath, trouble breathing, asthma, excess mucus, cough
- Gastrointestinal—nausea, upset stomach, pain, cramps, vomiting, diarrhea
- Genital (women)—abnormal periods, irregular periods, excess bleeding
- Urinary—blood in urine
- Muscle pain, aches, joint pain
- Headache
- Nausea
- Chest pain
- Diarrhea
- Swelling hands, feet, or other parts of the body
- Poor concentration
- Blurred vision
- Sleep disturbances
- Depression
- High irritability
- Extreme fatigue
- Memory lapses
- Organic brain disorders
- Fainting or near-fainting
- Spleen pain and liver discomfort

There is a spreading phenomenon associated with chemical hypersensitivity. In one study of 6800 chemically hypersensitive people, 95% had problems when exposed to: petroleum products, phenol, formaldehyde, pesticides, detergents, enzymes, synthetic fragrances, and flavors. Table 3.2 lists some occupational sensitizers.

TABLE 3.2   Occupational Sensitizers[a]

Airway	Skin
Beryllium,[b] cobalt,[b] colophony (rosin), diphenylmethane-4,4'-diisocyanate (MDI), glutaraldehyde, hexane-1,6-diisocyanate (HDI), phthalic anhydride, platinum,[b] toluene diisocyanate (TDI), trimellitic anhydride, chromium,[b] ethylene diamine, formaldehyde, maleic anhydride, methyl methacrylate, nickel,[b] piperazine	Chromium,[b] cobalt,[b] colophony (rosin), ethylene diamine, formaldehyde, glutaraldehyde, mercury,[b] nickel,[b] p-phenylenediamine, platinum,[b] benzofuran,[b] benzoyl peroxide, beryllium,[b] butyl acrylate, copper,[b] dibutyl phthalate, dichloropropane, ethylene oxide, hydrazine,[b] hydroquinone, iodine,[b] maleic anhydride, methyl methacrylate, polyvinyl chloride, resorcinol, toluene diisocyanates, turpentine

[a] *Source:* The National Foundation for the Chemically Hypersensitive.
[b] Evaluation does not necessarily apply to all individual chemicals within the group.

3.2   FIRE HAZARDS

////, FIRE LINE - DO NOT CROSS \\\\\ FIRE LINE - DO NOT CROSS /////, FIRE LINE - DO NOT CROSS \\\\\

3.2.1   Flammability, Combustion, and Deflagration

Note

Unfortunately, you will find these terms used interchangeably when referring to materials that can catch fire and burn, producing a flame. Flammability denotes the ability to burn. Combustion and deflagration are used to describe the act of burning. We speak of combustion and deflagration "rates." *Deflagration* is a term much less often used.

However, there is a definition difference among flammable, combustible, and ignitable when referring to liquids. OSHA, DOT, and NFPA agree that a flammable liquid has a flash point below 100°F, while a combustible liquid has a flash point at or above 100°F. EPA defines ignitable as a liquid with a flashpoint of less than 140°F. These are important differences when classifying hazardous material for purposes of consolidation, storage, or shipment.

3.2.2   The Three Factors Required for a Fire

   Normally, three factors are needed for a fire: oxygen, heat, and fuel.

Three factors must be present for a common combustion reaction, or fire, to occur in the atmosphere:

1. Enough oxidizer to sustain combustion. The most common oxidizer is the oxygen in the atmosphere. Less common oxidizers are chlorine and fluorine gases.
2. Enough heat, spark, or other ignition source to raise the material to its ignition temperature.
3. A fuel: a combustible or flammable, or any material that can burn. The concentrations of the fuel and the oxygen must be in the proper percentages to allow ignition and maintain the burning process.

Heat is either supplied by the ignition source and maintained by the combustion or supplied from an external source. The relationship of these three components that produce fire is illustrated by the fire triangle shown in Figure 3.1.

Most fires can be extinguished by removing one of these components. For example, water applied to a fire removes the heat, thereby extinguishing the fire. When a material heats up enough to self-ignite, spontaneous combustion occurs, either as a fire or explosion. This has been known to occur in stored coal piles heated by sunlight. The interior temperature of the coal pile can become hot enough (heat) to ignite in the presence of air.

WARNING

DISCUSSIONS OF COMBUSTION, FLAMMABILITY, AND FIRE HAZ-ARDS IN THIS BOOK WILL FOCUS ON WHAT HAPPENS IN ATMO-SPHERIC AIR. WHEN OXYGEN CONTENT IN THE ATMOSPHERE RISES ABOVE THE NORMAL BREATHING ATMOSPHERE OF 19.5–23.5%, THE PROBABILITY OF FIRE INCREASES GREATLY! SO DOES THE SPEED OF FLAME PROGRESSION. THE APOLLO 1

FIGURE 3.1 The fire triangle.

**CREW OF THREE BURNED TO DEATH ON THE LAUNCH PAD DUR-
ING PREFLIGHT TESTING. A FLASH FIRE ERUPTED WHEN AN
ELECTRICAL SPARK IGNITED IN THEIR CAPSULE IN A 100% OX-
YGEN ATMOSPHERE WITH PLENTY OF MATERIALS ACTING AS
FUELS.**

3.2.3 Uncommon Combustion Reactions

○—π **Some material combinations do not require an ignition source to result in a
fire reaction.**

The descriptions in Section 3.2.1 generally apply to the vast number of common combustion reactions that occur under atmospheric conditions. However, two other types of
combustion have become more common in recent decades: pyrophoric and hypergolic reactions.

Pyrophoric materials are those fine particles that can catch fire when exposed to air
without an ignition source. This is due to the very large, active surface area of the fine
particles for a given weight.

Hypergolic reactions are combustion reactions that occur when a fuel material and an
oxidizer material are mixed without an ignition source. Some liquid rocket engines have
depended upon hypergolic reactions of two liquids. No ignition source is required then. The
heat of reaction is so great that it provides ignition. An example is the mixing in the combustion chamber of red fuming nitric acid (strong oxidizer) and an alcohol (rich fuel).

Another hypergolic reaction occurs when mixing sodium hypochlorite (dry "pool chlorine" crystals) with brake fluid (a nonflammable oil). The reaction can be spectacular. Can
you imagine what would happen if you accidentally mixed pool chlorine with gasoline?

Hypergolic reactions are an extreme example of chemical incompatibility that is described
in greater detail in Section 4.

3.2.4 Flammability

○—π **Flammability is the ability of a material to generate enough combustible vapors
to ignite and produce a flame.**

Note

If your site has a wide variety of flammable liquids, you must understand that
OSHA has very strict definitions of each material. They are defined under Flammable and combustible liquids, 29 CFR 1910.106.

29 CFR 1910.106(a)(14)(iii)

For a liquid that is a mixture of compounds that have different volatilities and flashpoints, its flashpoint shall be determined by using the procedure specified in paragraph (a)(14) (i) or (ii) of this section on the liquid in the form it is shipped. If the flashpoint, as determined by this test, is 100 deg. F. (37.8 deg. C.) or higher, an additional flashpoint determination shall be run on a sample of the liquid evaporated to 90 percent of its original volume, and the lower value of the two tests shall be considered the flashpoint of the material.

(a)(14)(iv)

Organic peroxides, which undergo auto accelerating thermal decomposition, are excluded from any of the flashpoint determination methods specified in this subparagraph.

..1910.106(a)(18)

"Combustible liquid" means any liquid having a flashpoint at or above 100 deg. F. (37.8 deg. C.) Combustible liquids shall be divided into two classes as follows:

(a)(18)(i)

"Class II liquids" shall include those with flashpoints at or above 100 deg. F. (37.8 deg. C.) and below 140 deg. F. (60 deg. C.), except any mixture having components with flashpoints of 200 deg. F. (93.3 deg. C.) or higher, the volume of which make up 99 percent or more of the total volume of the mixture.

..1910.106(a)(18)(ii)

"Class III liquids" shall include those with flashpoints at or above 140 deg. F. (60 deg. C.) Class III liquids are subdivided into two subclasses:

(a)(18)(ii)(a)

"Class IIIA liquids" shall include those with flashpoints at or above 140 deg. F. (60 deg. C.) and below 200 deg. F. (93.3 deg. C.), except any mixture having components with flashpoints of 200 deg. F. (93.3 deg. C.), or higher, the total volume of which make up 99 percent or more of the total volume of the mixture.

(a)(18)(ii)(b)

"Class IIIB liquids" shall include those with flashpoints at or above 200 deg. F. (93.3 deg. C.). This section does not cover Class IIIB liquids. Where the term "Class III liquids" is used in this section, it shall mean only Class IIIA liquids.

(a)(18)(iii)

When a combustible liquid is heated for use to within 30 deg. F. (16.7 deg. C.) of its flashpoint, it shall be handled in accordance with the requirements for the next lower class of liquids.

(a)(19)

"Flammable liquid" means any liquid having a flashpoint below 100 deg. F. (37.8 deg. C.), except any mixture having components with flashpoints of 100 deg. F. (37.8 deg. C.) or higher, the total of which make up 99 percent or more of the total volume of the mixture.

Flammable liquids shall be known as Class I liquids. Class I liquids are divided into three classes as follows:

..1910.106(a)(19)(i)

Class IA shall include liquids having flashpoints below 73 deg. F. (22.8 deg. C.) and having a boiling point below 100 deg. F. (37.8 deg. C.).

(a)(19)(ii)

Class IB shall include liquids having flashpoints below 73 deg. F. (22.8 deg. C.) and having a boiling point at or above 100 deg. F. (37.8 deg. C.).

(a)(19)(iii)

Class IC shall include liquids having flashpoints at or above 73 deg. F. (22.8 deg. C.) and below 100 deg. F. (37.8 deg. C.).

○━π **The flashpoint (Fl.P.) of a material is the temperature at which that material will generate sufficient vapors to ignite.**

Flammability is the ability of a material to generate enough concentration of combustible vapors under normal atmospheric conditions to be ignited and produce a flame. It is necessary to have a proper fuel-to-air ratio (expressed as the percentage of fuel in air) to allow combustion and/or explosion. The flashpoint of a material is the temperature at which that material will generate sufficient vapors to ignite.

As you can see, the flammability of materials is closely related to the surrounding temperature.

○━π **The flammable range or explosive range is the range of fuel concentrations in air for each material that will support combustion or ignite.**

○━π **The lowest concentration of fuel in the flammable or explosive range is the lower flammable limit (LFL) or lower explosive limit (LEL).**

Concentrations less than the LFL/LEL are not flammable or explosive because there is too little fuel—that is, the mixture is too "lean."

Note

LEL is the lowest concentration of vapors or gases that will ignite in a normal atmosphere. LFL is the lowest concentration of vapors that will ignite and continue to burn. The concentrations are so close to each other that the terms are used as synonyms, although, as you can see, there is a difference.

○━π **The highest fuel concentration in air that is flammable is the upper flammable limit (UFL) or upper explosive limit (UEL).**

Concentrations greater than the UFL/UEL are not flammable or explosive because there is too much fuel—that is, the mixture is too "rich."

For example, the LFL/LEL for benzene is 1.3% (13,000 ppm), and the UFL/UEL is 7.1% (71,000 ppm), and thus the flammable range is 1.3% to 7.1%.

WARNING

ATMOSPHERES THAT ARE TOO RICH WITH FUEL MUST NEVER BE CONSIDERED SAFE FOR ENTRY. THE SIMPLE ACT OF OPEN-

ING THE HATCH INTO SUCH A SPACE CAN LET SUFFICIENT AIR IN TO DILUTE THE FUEL AIR MIXTURE TO AN EXTREMELY EXPLOSIVE ATMOSPHERE!

3.2.5 Examples of Flammable Material

⚷ **Flammable materials can be gases, liquids, or solids.**

Some examples of flammable materials are:

Common Flammable Liquids

- Ketones (methyl ethyl ketone MEK)
- Amines (methyl amine)
- Ethers (ethyl ether—the anesthetic)
- Aliphatic hydrocarbons (methane, ethane, propane)
- Aromatic hydrocarbons (benzene, toluene, xylene)
- Alcohols—methyl (wood alcohol), ethyl (grain alcohol)
- Nitroaliphatics (nitromethane ("top fuel" dragster fuel))

Common Flammable Solids

- Phosphorus
- Magnesium dust
- Zirconium dust
- Titanium dust
- Aluminum dust
- Zinc dust

Common Flammable Gases

- Hydrogen
- Carbon monoxide
- Hydrogen sulfide
- Methane (natural gas)
- Acetylene (welding and burning gas)
- MAPP gas (welding and burning gas)

Water-Reactive Metals Producing Flammable Hydrogen Gas

- Potassium
- Sodium
- Lithium
- Calcium

Pyrophoric Materials

O—π   **Pyrophoric materials can spontaneously ignite in the presence of air.**

WARNING
PYROPHORIC SUBSTANCES IGNITE SPONTANEOUSLY UPON CONTACT WITH AIR! THEY DO NOT REQUIRE AN IGNITION SOURCE!

- Organometallic compounds (tetraethyl lead, former antiknock compound for gasoline)
- *Cadmium
- *Chromium
- *Cobalt
- Boron
- *Iron
- *Magnesium
- *Manganese
- *Nickel
- *Titanium
- **Yellow phosphorus
- *Zinc
- *Zirconium

WARNINGS
*NOTE THAT THESE METALS, SUCH AS IRON, ARE EXPLOSIVELY COMBUSTIBLE WHEN FINELY DIVIDED. HANDLE FINE METAL POWDERS WITH GREAT CAUTION. AVOID POURING THEM THROUGH AIR FROM HEIGHTS.
**YELLOW PHOSPHORUS MUST BE STORED UNDERWATER.

3.2.6 Combustibles

Combustibles are differentiated from flammables by the temperature required, under normal atmospheric conditions, to start to burn. Combustible liquids are described in detail above. So-called "normal combustibles" such as wood and paper will not burn until a more elevated temperature is reached (approximately 450°F).

3.3 EXPLOSIVE AND PROPELLANT HAZARDS

3.3.1 Explosives

Explosives Defined

O—π   **Explosives contain both the fuel and oxidizer combined. All they need is an initiation source.**

An explosive is a material that contains both fuel and oxidizer combined. Nitroglycerin is an example. The "nitro" component acts as the oxidizer, and the "glycerine" component is the fuel. An explosive requires only initiation to cause an extremely rapid reaction. Initiation is some action that produces the necessary heat in the required fuel-oxidizer-heat combination. That produces large amounts of gases and heat. Due to the heat, the gases and vapors produced, such as nitrogen oxides, carbon dioxide, and steam, rapidly expand, causing a shock wave. The primary hazard of an explosion is the great amount of energy it releases in that shockwave in a small fraction of a second. Explosions are rated according to the speed of their shockwave. Unlike a fire, which can often be controlled easily, an explosion, once initiated, cannot be stopped. Table 3.3 shows the relative power of different commercial explosives.

High Explosives.   *High explosives* (primary and secondary) are capable of detonating and are used in military ordinance, blasting for earth moving, and mining. They have a very high rate of reaction. The high-pressure gases produced cause a detonation wave that moves faster than the speed of sound (1400–9000 m/s).

Primary explosives, such as nitroglycerin, can detonate with little or no stimulus. They are extremely dangerous to handle.

The hazards of primary explosives prompted the development of *secondary explosives,* such as dynamite, detonating cord (det cord), RDX, and plastic explosives. They require a strong initiation shock from a detonator, such as a blasting cap. They are relatively insensitive to shock, heat, or friction. They can be broken apart and cut, and the plastics can be molded, with very little danger of explosion. Composition Four (C4) is a composite explosive containing approximately 91% RDX and 9% nonexplosive plasticizer. C4 is effective in temperatures between −70 and +170°F but loses its plasticity in colder temperatures.

Low Explosives.   *Low explosives* are initiated by spark, flame, heat, or impact. They then burn very rapidly, producing gases that expand very rapidly producing a powerful pressure wave. They have a lower reaction rate than high explosives. The overall effect ranges from rapid combustion to a low-order detonation (generally less than 2000 m/s). The original gunpowder (black powder) is a common example.

3.3.2   Solid Propellants

Just as in the case of explosives, these materials are fuels and oxidizers combined in one mixed (composite) material. They are generally shock-insensitive. They are designed to burn at rates of only a few tenths of an inch per second from the exposed surface. They only require a flame ignition source. Everyone is familiar with the TV pictures of the Space Shuttle solid propellant boosters burning with a dense white cloud at liftoff. The propellant is chiefly ammonium perchlorate, aluminum powder, and polymer.

Actually, although not designed to explode, the ammonium perchlorate ingredient itself can explode if uncontrolled heating takes place within a large mass of the stored material. That is also true of ammonium nitrate, the fertilizer ingredient that was used in the Oklahoma City bombing. Careful storage of these, and any similar materials, to prevent any accidental heating is absolutely essential. Once a solid propellant starts burning, it is almost impossible to extinguish it. It burns readily underwater!

3.3.3   Gas and Vapor Explosions

A gas or vapor explosion is a very rapid, violent release of energy. If combustion is extremely rapid, large amounts of stored energy, heat, and gaseous byproducts are released. A major factor contributing to the force of the explosion is the confinement of a flammable material. When vapors or gases cannot freely dissipate, they enter the combustion reaction more

TABLE 3.3 Relative Power of Different Commercial Explosives

Explosive Name	Principal Use	Velocity of Detonation (V.O.D.) m/s f/s	Relative Effectiveness (TNT = 1)	Intensity of Toxic Fumes	Water Resistance
TNT	Demolition charge and composition explosives	6,900 m/s 22,600 f/s	1.00	Dangerous	Excellent
PETN	Detonation cord, blasting caps and demolition charges	8,300 m/s 27,200 f/s	1.66	Slight	Excellent
PETN	Demolition charges as M118 block or M186 roll	7,100 m/s 23,000 f/s	1.14	Slight	Excellent
RDX	Detonation cord, blasting caps and demolition charges	8,350 m/s 27,400 f/s	1.60	Dangerous	Excellent
RDX	Demolition charges as M118 block or M186 roll	7,100 m/s 23,000 f/s	1.14	Dangerous	Excellent
Tetryl	Booster charge and composition explosives	7,100 mm/s 23,300 f/s	1.25	Dangerous	Excellent
Tetrytol 75/25	Demolition charge, M2 block	7,000 m/s 23,000 f/s	1.20	Dangerous	Excellent
Amatol 80/20	Bursting charge	4,900 mm/s 16,000 f/s	1.17	Dangerous	Very Poor
Pentolite 50/50	Booster charge and bursting charge	7,450 m/s 24,400 f/s		Dangerous	Excellent
Composition A3	Booster charge and bursting charge	8,100 m/s 26,500 f/s		Dangerous	Good
Composition B	Bursting charge	7,800 m/s 25,600 f/s	1.35	Dangerous	Excellent
Composition C3	Demolition charge, M3 block	7,625 m/s 25,000 f/s	1.34	Dangerous	Good
Composition C4	Demolition charge, M5A1 block and M112 block	8,040 m/s 26,400 f/s	1.34	Slight	Excellent
Black powder	Time blasting fuse Historical small arms	400 m/s 1,300 f/s	0.55	Dangerous	Poor
Ammonium nitrate	Demolition charge	3,400 m/s 11,000 f/s	0.42	Dangerous	None
Military dynamite, M1	Demolition charge	6,100 m/s 20,000 f/s	0.92	Dangerous	Good
Nitroglycerine	Commercial dynamites	7,700 m/s 25,200 f/s	1.50	Dangerous	Good
Straight dynamite (commercial 40%)	Demolition charge	4,600 m/s 15,000 f/s	0.65	Dangerous	Good if fired within 24 hours
Straight dynamite (commercial 50%)	Demolition charge	5,500 m/s 18,000 f/s	0.79	Dangerous	Good if fired within 24 hours
Straight dynamite (commercial 60%)	Demolition charge	5,800 m/s 19,000 f/s	0.83	Dangerous	Good if fired within 24 hours

TABLE 3.3 Relative Power of Different Commercial Explosives (*Continued*)

Explosive Name	Principal Use	Velocity of Detonation (V.O.D.) m/s f/s	Relative Effectiveness (TNT = 1)	Intensity of Toxic Fumes	Water Resistance
Ammonia dynamite (commercial 40%)	Demolition charge	2,700 m/s 8,900 f/s	0.41	Dangerous	Poor
Ammonia dynamite (commercial 50%)	Demolition charge	3,400 m/s 11,000 f/s	0.46	Dangerous	Poor
Ammonia dynamite (commercial 60%)	Demolition charge	3,700 m/s 12,000 f/s	0.53	Dangerous	Poor
Gelatine dynamite (commercial 40%)	Demolition charge	2,400 m/s 7,900 f/s	0.42	Slight	Poor
Gelatine dynamite (commercial 50%)	Demolition charge	2,700 m/s 8,900 f/s	0.47	Slight	Poor
Gelatine dynamite (commercial 60%)	Demolition charge	4,900 m/s 16,000 f/s	0.76	Slight	Poor
Ammonia gelatine dynamite (commercial 40%)	Demolition charge	4,900 m/s 16,000 f/s		Slight	Excellent
Ammonia gelatine dynamite (commercial 60%)	Demolition charge	5,700 m/s 18,700 f/s		Slight	Excellent

Courtesy of the USGS.

rapidly. The rate of the explosion or combustion reaction always increases with increased pressure.

Poorly ventilated buildings, sewers, drums, and bulk liquid containers are examples of confined spaces where potentially explosive atmospheres can accumulate. Natural gas (methane) accounts for many gas explosions and fires yearly. Methane is also a major component of swamp gas and the gases from landfills.

3.3.4 Boiling Liquid Expanding Vapor Explosions (BLEVEs)

BLEVEs are a fairly common occurrence when a container holding flammable liquid is involved in a fire.

BLEVEs (see Figure 3.2), are a fairly common occurrence when a container holding flammable liquid is involved in a fire. The heat from the fire raises the pressure inside the container until it bursts and releases tremendous amounts of flammable vapor, which then ignites in the atmosphere. These are the spectacular explosions that are associated with petroleum refinery fires and rail cars.

3.4 CORROSIVE HAZARDS

Corrosives include acids (low pH), bases (high pH), and halogens. They can cause serious burns to the unprotected HAZWOPER worker.

FIGURE 3.2   A railcar BLEVE. (*Courtesy of USEPA.*)

3.4.1 Corrosion Defined

Corrosion, in the industrial sense, is the oxidative destruction of metals (such as rusting of iron), which can occur in the presence of air, moisture, and a variety of chemicals. Considered more broadly, upon contact, a corrosive material also may destroy body tissues, plastics, and other materials.

Technically, corrosivity is the ability of a corrosive material to increase either the hydrogen ion or the hydronium ion concentration of or on another material. A corrosive agent is a reactive compound or solution that produces a destructive chemical change in the material upon which it is acting. Common corrosives are the acids, bases and halogens. Skin irritation and burns are typical results when the body contacts an acidic or basic material. The halogen most widely used, and therefore most frequently affecting humans, is chlorine.

The corrosiveness of acids and bases can be rated as to their ability to dissociate in solution, forming a high concentration of ions. If moisture could be excluded completely, pure acids and pure bases would not ionize very much and would not be very corrosive. The ability to ionize is largely dependent on the presence of water. The hazard to your skin and eyes is partly due to the moisture present in those tissues.

3.4.2 Corrosives Effects on Humans

On contact with corrosives, your skin will feel warm, then hot, and then painful as though it is being burned. With a long enough contact, the skin will die, and gradually the damaged layer will peel off. Corrosive splashed in your eye may feel like sand or grit. If not immediately flushed out with copious quantities of water, your eye will start to feel like it is being burned, and then tissue will be destroyed. A lot of wash water is needed immediately, until the corrosive is diluted to the point that it is not a hazard. That is why emergency eyewash and shower stations are so important where corrosives are being handled.

When diluting acids or bases with water, always slowly add the acid or base to the water, not the water to them. If you do it the other way around, a violent reaction will occur and possibly burn you and contaminate the work area.

3.4.3 pH, Measurement of Acid and Base Strength

Those corrosive solutions that have the greatest concentration of hydrogen ions (H+) are the strongest acids, while those that have the greatest concentration of hydroxyl ions (OH−) are the strongest bases. The accepted measure of acidity and basicity is pH (*potenz Hydrogen*), German for "power of hydrogen." Because of the way pH is defined, strong acids have a low pH (high concentration of H+, low OH− in solution) while strong bases have a high pH (low H+ concentration, high concentration of OH− in solution). The pH scale ranges from 0 to 14:

<Increasing acidity						Neutral					Increasing basicity>		
1	2	3	4	5	6	**7**	8	9	10	11	12	13	14

Measurements of pH can be quickly and easily performed on-site, providing immediate information on the corrosive hazard.

3.4.4 Common Corrosives

Common corrosives include:

Acids

Acetic acid (common household vinegar is a weak [about 5%] solution of acetic acid)
Hydrochloric acid (used in cleaning concrete and pools as muriatic acid)
Hydrofluoric acid (etches glass)
Nitric acid (widely used in the chemical industry)
Sulfuric acid (battery acid is a several percent solution)

Bases (Alkali or Caustics)

Potassium hydroxide
Sodium hydroxide (caustic soda, an ingredient in Draino®)

Halogens

Bromine (increasingly being used in water purification)
Chlorine (widely used in water purification)
Fluorine (the most reactive element; about 1 ppm added to drinking water to reduce the incidence of dental cavities)

3.5 CHEMICAL REACTIVITY HAZARDS

3.5.1 Chemical Reactions

Chemical reactions are usually classified in two broad categories: exothermic reactions generate heat, endothermic reactions absorb heat from their surroundings.

A chemical reaction is the decomposition, molecular rearrangement, polymerization, or other reaction of a single compound. It is also the interaction of two or more compounds, resulting in formation of other chemical compounds. *Exothermic* chemical reactions, which give off heat, can be the most dangerous. They tend to occur spontaneously. An example of an exothermic reaction is mixing calcium hypochlorite (pool "chlorine") with brake fluid.

Brake fluid is almost nonflammable, but mixing it with calcium hypochlorite, a strong oxidizer, provides great amounts of heat.

Endothermic chemical reactions absorb heat from their surroundings. Generally they are the most difficult to start and are not spontaneous. An example of an endothermic reaction is the addition of sodium chloride (table salt) to water. Endothermic reactions are rarely hazardous.

3.5.2 Reactivity Hazards

○━┳ **Materials are said to react with each other when two or more are mixed together and chemical changes occur. The reaction may generate or absorb heat and give off toxic, corrosive, or flammable gases or vapors, or the mixture may form new liquid or solid materials.**

Reactive materials are those that can undergo a chemical reaction under certain specified conditions. *Incompatibility* is the term widely used in HAZWOPER work when two materials react when they are simply mixed together. This is explained further in Section 4.

Many times this kind of reaction occurs in the presence of water or air. Water poured into acid produces an immediate heat release that can cause an eruption and splashing. Sodium or potassium metal dropped into water vigorously attacks and decomposes the water into hydrogen and oxygen, with enough heat of reaction to ignite or explode the hydrogen.

Sulfur chlorides, on the other hand, react in contact with water to form corrosive hydrochloric acid, toxic sulfur dioxide, and flammable and toxic hydrogen sulfide.

Pyrophoric metal powders will ignite in air due to the extreme reaction heat produced by very fast oxidation. This can occur at or below normal room temperature in the absence of any added heat, shock, or friction.

Another reactive hazard is hazardous polymerization. Polymerization is the chemical reaction in which a series of simple molecules, known as monomers, combine to form large, chain-like, and sometimes branched molecules, known as polymers. Polymerization can be a mild exothermic reaction. But in hazardous polymerization a large amount of heat is given off and is contained. This can lead to a pressure burst or even an explosion. An example is mixing part A and part B of epoxy paint while consolidating waste paint in a drum and then sealing it.

peroxide formers — Ethyl Ether
THF

3.6 RADIOACTIVITY HAZARDS

○━┳ **According to OSHA, nuclear radiation exposure must be kept "as low as reasonably achievable" (ALARA).**

HAZWOPER personnel are primarily concerned with two broad types of radiation: electromagnetic and nuclear. Electromagnetic radiation is the way power is broadcast through the air. Your cell phone is the transmitter and receiver for electromagnetic radiation that transmits to, and receives from, microwave towers in your area. Another type of electromagnetic radiation is used to cook dinner in your microwave oven. Obviously, we do not want to be in close proximity to a transmission device (antenna, radar dish, and the like) while it is transmitting (radiating).

Nuclear radiation comes from either naturally occurring or man-made sources. Naturally occurring sources include radon and uranium. Man-made sources are the man-made elements listed near the end of the periodic table of the elements. All man-made elements are radioactive. Radioactive materials are very useful in a number of ways, such as power generation, radiation therapy for cancer, glow-in-the-dark paint, and element analysis in hazardous waste utilizing X-ray fluorescence.

We consider radioactive sources a nuisance when we are done using them and they become waste. Some radioactive elements have half-lives of many millions of years. (Half-life is the time required for one half of a given quantity of a substance to disintegrate or decay.) Uranium 238 has a half-life of 4.468×10^9 years! As these wastes continue to decay, they give off energy that can be very harmful to humans and other living organisms. Nuclear radioactive material is the type we will concentrate on as it is the only type of which there can be waste.

3.6.1 Radioactivity Defined

Radioactivity is defined as the spontaneous disintegration or decay of the nucleus of an atom by emission of particles, usually accompanied by electromagnetic radiation.

In 1896, Antoine Henri Becquerel discovered natural radioactivity. He was working with compounds containing the element uranium. To his surprise, he found that photographic plates covered to keep out light became fogged, or partially exposed, when these uranium compounds were anywhere near the plates. This fogging suggested that some kind of ray had passed through the plate coverings. Several materials other than uranium were also found to emit these penetrating rays. Materials that emit this kind of radiation are said to be radioactive and to undergo radioactive decay.

In 1899, Ernest Rutherford discovered that uranium compounds produce three different kinds of radiation. He separated the radiations according to their penetrating abilities and named them alpha (α), beta (β), and gamma (γ) radiation, after the first three letters of the Greek alphabet.

3.6.2 Researchers, and the Price One Paid

Marie and Pierre Curie conducted research on radioactive substances. They found that the uranium ore, or pitchblende, contained much more radioactivity than could be explained solely by the uranium content.

The Curies began a search for the source of the radioactivity and discovered two highly radioactive elements, radium and polonium. The Curies won the 1903 Nobel Prize for physics for their discovery (she was the first woman to win the prestigious honor). They shared the award with the aforementioned Becquerel.

After the death of her husband in 1906, Madame Curie (Figure 3.3) continued her work on radioactive elements and won the 1911 Nobel Prize for chemistry for isolating radium and studying its chemical properties. In 1914, she helped found the Radium Institute in Paris.

In 1934, at the age of 67, Madame Curie died of leukemia, generally accepted to have been caused by exposure to the high levels of radiation involved in her research. After her death the Radium Institute was renamed the Institut Curie in her honor.

Environmental measurements of radiation are commonly measured in picocuries (pCi), named after Mme. Curie. One pCi is equal to the decay of about two radioactive atoms per minute.

3.6.3 Types of Radioactivity

> **Alpha radiation is the weakest radiation; it poses the least threat to humans and can be stopped by a single sheet of paper.**

Alpha Radiation (α). *Alpha particles* are energetic, positively charged particles (helium nuclei) that rapidly lose energy when passing through matter. They are commonly emitted in the radioactive decay of the heaviest radioactive elements, such as uranium and radium, as well as by some man-made elements. Alpha particles lose energy rapidly in matter and do not penetrate very far; however, they can cause damage over their short path through tissue. These particles are usually completely absorbed by the outer dead layer of the human skin, and so alpha-emitting radioisotopes are not a hazard outside the body. However, they

FIGURE 3.3   Mme. Marie Curie. (*Courtesy of Institut Curie.*)

can be very harmful if they are ingested or inhaled. Alpha particles can be stopped completely by a sheet of paper.

🔑 **Beta particles travel appreciable distances in air but can be reduced or stopped by a layer of clothing or by a few millimeters of a substance such as aluminum.**

Beta Radiation (β).   *Beta particles* are fast-moving, positively or negatively charged particles emitted from the nucleus during radioactive decay. Humans are exposed to beta particles from man-made and natural sources such as tritium, carbon-14, and strontium-90. Beta particles are more penetrating than alpha particles but are less damaging over equally traveled distances. Some beta particles are capable of penetrating the skin and causing radiation damage; however, as with alpha emitters, beta emitters are generally more hazardous when they are inhaled or ingested. Beta particles travel appreciable distances in air but can be reduced or stopped by a layer of clothing or by a few millimeters of a substance such as aluminum.

🔑 **Gamma radiation is the most harmful to humans. To shield from gamma radiation requires several feet of concrete or several inches of lead!**

Gamma Radiation (γ).   Like visible light and X-rays, *gamma rays* are weightless packets of energy called photons. Gamma rays often accompany the emission of alpha or beta particles from a nucleus. They have neither a charge nor a mass and are very penetrating. One source of gamma rays in the environment is naturally occurring potassium-40. Man-made sources include plutonium-239 and cesium-137. Several feet of concrete or a few inches of lead may be required to stop the more energetic gamma rays.

WARNING
GAMMA RAYS CAN EASILY PASS COMPLETELY THROUGH THE

**HUMAN BODY OR BE ABSORBED BY TISSUE, THUS CONSTITUT-
ING A RADIATION HAZARD FOR THE ENTIRE BODY.**

X-Rays.   *X-rays* are invisible, highly penetrating electromagnetic radiation of much shorter wavelength (higher frequency) than visible light. Hard X-rays have higher frequencies and are more penetrating; soft X-rays have lower frequencies and are less penetrating.

X-rays were discovered in 1895 by the German physicist Wilhelm Conrad Roentgen; for this discovery he received the first Nobel Prize in physics in 1901. They are produced by the impact of high-energy electrons on a metal anode in a highly evacuated glass bulb.

X-ray exposure is measured by dosimetry. A badge containing unexposed film is worn in the area where the dose is to be measured. At the end of the exposure period, the film is developed to measure the extent of the exposure. Our exposure to radiation is measured in millirems (mRems) (thousandths of a Roentgen equivalent in man).

☛ **Worker exposure to radiation is measured in millirems (mRems).**

Diagnostic radiation usually involves exposures that result in absorbed doses of radiation measured in millirads, or thousandths of rads (radiation absorbed dose). A "rad" is a unit of absorbed radiation that is now being replaced by another unit called a "gray" (Gy). One Gy is equal to 100 rads.

X-rays are generally lower in energy and therefore less penetrating than gamma rays. Thousands of X-ray machines are used daily in medicine, dentistry, and industry for examinations, inspections, and process controls. X-rays are also used for cancer therapy to destroy malignant cells. Because of their many uses, X-rays are the single largest source of man-made radiation exposure. A few millimeters of lead can stop medical X-rays.

3.6.4 Naturally Occurring Radiation

Humans are primarily exposed to natural radiation from the sun, cosmic rays, and naturally occurring radioactive elements found in the earth's crust. Cosmic rays from space include energetic protons, electrons, gamma rays, and X-rays. Natural radioactivity is exhibited by several elements, including radium and uranium. The radiation produced is of the three types previously discussed: alpha, beta, and gamma rays. The rate of disintegration (decay) of a radioactive substance is commonly designated by its half-life, which is the time required for one half of a given quantity of the substance to disintegrate or decay. Radioactive decay is a natural, spontaneous process in which an atom of one element decays to form another element by losing atomic particles (protons, neutrons, or electrons).

Radon, which emanates from the ground, is another important, and sometimes dangerous, source of natural radiation. See Figure 3.4. Radon is a gas produced by the radioactive decay of the element radium. This produces alpha radiation. Radon itself is radioactive because it also decays, losing an alpha particle and forming the element polonium. Radon can be expected to be most concentrated in cellars.

Radon has been linked to certain cancers. Depending on where you live or work in the United States, you will be exposed to varying levels of radon. See Figure 3.5. Because it emits alpha particles, it is not considered dangerous until it is trapped and thus concentrated, such as in a house without good ventilation.

3.6.5 Man-Made Radiation

Radiation is used on an ever-increasing scale in medicine, dentistry, and industry. Main users of man-made radiation include medical facilities such as hospitals and pharmaceutical facilities; research and teaching institutions; nuclear reactors and their supporting facilities such

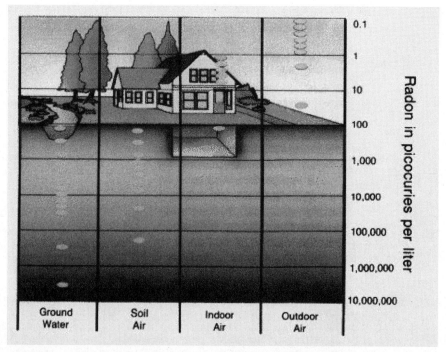

FIGURE 3.4   Environmental concentrations of radon. (*Courtesy of the USGS.*)

as uranium mills and fuel preparation plants; and federal facilities involved in nuclear weapons production as part of their normal operation.

Many of these facilities generate some radioactive waste, and some release a controlled amount of radiation into the environment. Radioactive materials are also used in common consumer products such as luminous-dial wristwatches and compasses, ceramic glazes, night sights for small arms, and smoke detectors.

Spent uranium is used in some military ammunition projectiles because of its superior penetrating power. It is far more dense than lead. However, it is a source of residual radiation.

Another source of man-made radiation is electromagnetic radiation, commonly referred to as electric and magnetic fields (EMF). EMF emanates from radio and TV broadcasting, computer display terminals, cell phone communications, power transmission lines, and all wireless transmissions. As wavelength decreases and frequency increases, the radiation does more damage to humans. In order of decreasing wavelength and increasing frequency, the various types of electromagnetic radiation are:

- Radio waves
- Microwaves
- Infrared radiation
- Visible light
- Ultraviolet radiation
- X-rays
- Gamma radiation

3.6.6   Health Effects of Radiation Exposure

Depending on the level of exposure, radiation can pose a health risk. It can adversely affect individuals directly exposed, as well as their descendants. Radiation can affect cells of the

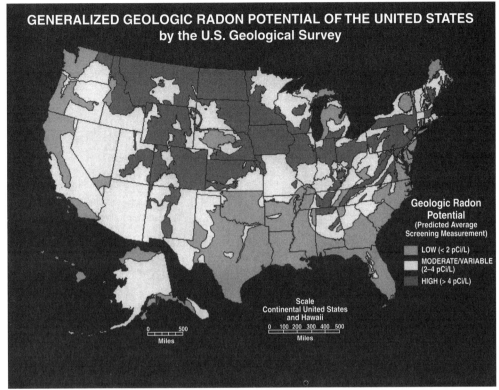

FIGURE 3.5 Generalized geologic radon potential for the United States. (*Courtesy of the USGS.*)

body, increasing the risk of cancer or harmful genetic mutations that can be passed on to future generations; or, if the dosage is large enough to cause massive tissue damage, it may lead to death within a few weeks of exposure. Neither OSHA nor NIOSH considers EMF a proven health hazard. They have done extensive research over the past decade. Certain studies have shown increased cancer occurrence in workers exposed to high levels of EMF, but most were inconclusive, according to NIOSH.

See Section 5 for further information on health effects.

Diagnostic radiation usually involves exposures that result in absorbed doses of radiation measured in millirads, or thousandths of rads. Experimental data and epidemiology place the threshold for birth defects at levels of exposure during pregnancy greater than 100 rads (1 Gy). The threshold for microcephaly is lower, somewhere between 10 and 100 rads (0.1–1 Gy). Diagnostic exposures, even multiple studies, rarely result in doses that exceed 1 rad (0.01 Gy). On the other hand, fetal exposures as low as 1 rad (0.01 Gy) have been shown to increase risks for leukemia in childhood significantly. Hence the recommendations to limit exposures during pregnancy.

3.6.7 Protection from Radiation

The three factors in protecting us from radiation overexposure are time, distance, and shielding. (See Figure 3.6.)

The shorter the amount of time we are exposed, the fewer rays we receive. The farther away we are, the weaker the rays are and therefore the less damage can be done to us.

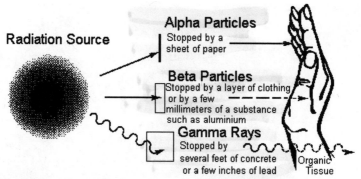

FIGURE 3.6 The penetrating powers of alpha and beta particles and gamma rays. (*Courtesy of USEPA.*)

Radiation particles are subject to Newton's inverse square law:

$$\text{Exposure} = 1/d^2$$

Radiation exposure will drop off by the inverse square law. As you increase your distance from a radioactive source, your exposure will decline by the square of the distance. For example, if the radiation exposure is 100 mRem/hr at 1 ft from a source, the exposure will be 0.01 mRem/hr at 100 ft.

$$100 \text{ mRem/hr} = 1/1\text{ft}^2$$

$$\text{Exposure} = 1/100^2$$

$$\text{Exposure} = 1/10,000$$

$$100 \text{ mRem/hr}/10,000$$

$$\text{Exposure} = 0.01 \text{ mRem/hr}$$

The denser the blocking material, the fewer rays can travel through. Lead is considered one of the best shields. Concrete is also good, but it takes several feet of concrete to block the same amount of radiation as a few inches of lead.

3.7 TRAINING AIDS AND ADDITIONAL RESOURCES

The authors recommend the following videos for aiding readers to see real-life examples of what they have learned in this Section.

- *Introduction to Chemical Safety*
- *Corrosives*
- *Oxidizers*
- *Gases*
- *Introduction to Reactive and Explosive Materials*
- *Explosives*

These videos are available from Safe Expectations (formerly BNA Communications, Inc.) (www.safeexpectations.com)

- *Chlorine—Handle with Care*
- *Hydrogen Sulfide—A Matter of Life and Death*
- *Lead—Treat It with Respect*
- *Asbestos Awareness—Are You at Risk?*

These videos are available from Coastal Safety and Health (www.coastal.com)

3.8 SUMMARY

We have discussed the four broad groups that material hazards fall into: toxic, fire and explosion, chemical, and radiation. We want to reiterate that we have not described nor listed all hazardous materials. We do believe, however, that we have covered the four broad groups that include all hazardous materials.

Under *toxic material hazards* we have shown that although the terms poison and toxin are used interchangeably, they have different meanings. Poisons are just one of seven kinds of toxins. Now you should be able to define poisons, asphyxiants, carcinogens, mutagens and teratogens, irritants, allergens, and sensitizers. You should also be familiar with examples of each category, their effect on our health, and how to avoid contact with them.

Next we discussed *material hazards that lead to fire and explosion.* You should know that common fires require heat, fuel, and oxygen. You should also remember that some materials are especially hazardous. They supply both the oxygen and the fuel. All they need for combustion is the initiating heat. Organic peroxides, explosives, and propellants are examples. Routine firefighting procedures will not be effective when these catch fire. You have been introduced to flammability and explosive limits in air. There are explosive and flammable ranges of fuel vapors in air. Outside of those ranges, combustion can't take place because the fuel-air mixture is either too rich or too lean.

Under *chemical material hazards* we discussed corrosivity and defined acids, bases, and halogens. Acids, with their low pH, and bases, with their high pH, are both hazardous to humans on contact. Acids and bases are also extremely reactive with some other rather common materials. This brought our discussion to reactivity hazards, what happens when two or more incompatible materials are brought into contact with each other. The reactivity hazard can be the reaction itself. The hazard of an exothermic (heat-producing) reaction can be the heat itself. The hazard can also be products of the reaction such as a toxic or explosive gases or vapors. Because of the unique hazards presented, chemical incompatibility will be investigated at greater length in Section 4.

Finally we discussed *radiation hazards,* with a focus on radioactive materials hazards. Here we learned that radiation can occur from natural sources or can be man-made. All man-made elements are radioactive. There are hazards associated with electromagnetic radiation as well as nuclear radiation. Of the types of rays emitted by radioactive sources (alpha, beta, gamma, and X-rays), we emphasized that gamma rays are the most harmful to the human body.

SECTION 4
CHEMICAL INCOMPATIBILITY

OBJECTIVES

Section 4 is intended to heighten reader awareness of the fact that there are materials, containing chemicals, that cannot be mixed together without creating some kind of hazard. We have already introduced those kinds of hazards in Section 3.5. They are due to chemical reactions. In HAZWOPER terminology, we say they are due to "chemical incompatibility." Section 4 gives further explanation of compatibility and incompatibility.

You will be cautioned about hazardous materials cleanup tasks. Those are tasks in which materials are being collected and consolidated for disposal. Most efforts are directed toward minimizing disposal containers. Accidents have occurred when mixing incompatibles in disposal containers. Containers have overheated, become pressurized, and ruptured. You will learn the special value of following the instructions in MSDSs, placards, and warning labels. Sections 8 and 9 deal with these information sources in detail.

We want to alert HAZWOPER personnel to incompatibility signs they can see whenever they come upon unfamiliar storage containers. Bulging or corrosion of containers is a frequent early warning sign of hazardous incompatibility. Remote handling of such containers may be required. You will be given special warnings about crystal formation on containers. The crystals may be shock-sensitive.

Another objective of Section 4 is to reinforce understanding of the fire and explosion hazards of mixing fuels and oxidizers. This has already been introduced in Sections 3.2 and 3.3. We want HAZWOPER personnel to have almost a sixth sense about materials that are oxidizers and those that are fuels. Never mix or store them together.

4.1 COMPATIBILITY VERSUS INCOMPATIBILITY

O—π **Beware of incompatibility when mixing or even storing two or more "chemicals" or materials containing chemicals.**

If two or more materials, or chemicals, remain in contact indefinitely *without reaction*, they are compatible. If two or more materials in contact begin to react immediately, or even over a period of months or years, they are incompatible.

O—π **The words "chemical" and "material" in regulations and hazardous materials information often refer to the same substance.**

O—π **We frequently speak of "chemical incompatibility" in HAZWOPER activities. Chemists and engineers usually call that incompatibility a "reaction."**

O—π **"Chemical reaction" and "incompatibility" are terms describing practically the same effects.**

HAZWOPER personnel should understand that there are far more examples of materials or chemicals that can be mixed together without reaction (that are compatible) than there are examples of incompatibility. However, never let down your guard when you encounter unlabeled wastes, or materials with which you are unfamiliar.

One of the HAZWOPER activities is consolidation, which means putting identical, or at least compatible, materials together in a large container to save shipping and disposal costs. For instance, you receive 500 partially used gallon cans of different colors of latex paint. They have some hazardous constituents and thus must be treated as hazardous. Do you ship out 500 cans, each with its own label? No, you consolidate the paint into several 55-gallon (208-L) drums and recycle the cans as scrap metal. But beware of incompatible materials. One gallon of an acid or base added to the water-based paint consolidation drum will yield a large and potentially dangerous mess.

Throughout this book we will remind readers to consult materials' MSDSs in Section 9. What do you do if the materials are not labeled? We will go into further detail on this situation in Sections 12 and 13. However, if you do not have MSDSs or know MSDS equivalent information about two or more materials you are working with, treat those materials as incompatibles until you are sure they are compatible.

4.2   EARLY WARNING SIGNS OF INCOMPATIBILITY

⊙━━π   **Warning signs of incompatibility of mixtures include temperature rise or fall, color change, viscosity change, and formation of vapors, foams, sludges, or crystals. Exterior signs are container corrosion, bulging, or rupture.**

Frequently you can observe early warning signs of reaction or incompatibility. They sometimes occur immediately when you mix materials or pour them together in a container. Reactions most often cause a rise in temperature (an exothermic reaction). Infrequently a fall in temperature occurs (an endothermic reaction). Color or viscosity changes can occur. The mixture may thin out and exude vapors or gases, which may be toxic or flammable. The mixture may thicken and even solidify. Formation of foams and sludges is fairly common. Crystal formation is usually a longer-term result of reaction. However, crystal formation can be the most hazardous. Picric acid has been used in medicinals for burn treatment. However, when picric acid attacks a number of metals, it forms metal picrate crystals, a friction-sensitive explosive. In that case, immersion of the container in water can be an acceptable approach. Water tends to dissolve and deactivate the crystals.

4.2.1   Government Tests of Drum Failure

⊙━━π   **Container bulging, although not conclusive evidence that the container is under pressure, does mean that it has been overpressurized at some time.**

Corrosion of containers should always raise a red flag. For metal drums it may be rusting of steel or oxidation of other metals due to atmospheric moisture or high humidity. Or it may signal corrosive action from the inside. Bulging or discoloration of containers is a definite warning sign. It could mean imminent bursting.

Following over 100 bulging drum incidents, the Los Alamos National Laboratory performed failure analysis on several types of drums. Because there is currently no economical way of checking internal drum pressure, each type had to be tested to failure. Drums of different types were pressurized from zero pounds per square inch gauge (psig) to failure, at 5.0 psig intervals. All test drums were in new condition. The test results in the three following subsections should prompt HAZWOPER caution when approaching a bulging drum.

Fifty-Five-Gallon Metal Open-Head Drums (for Solids)

1. Drums appear to vent immediately adjacent to the nut-and-bolt fastener on the ring.
2. One hundred percent of drums tested vented at pressures at or below 32 psig.
3. "Pinging" (a banging sound as the head bulges) was noticeable between 15 and 20 psig.
4. The 55-gallon metal open-head drums appeared to bulge at only the top and bottom ends.
5. Body seams (top to bottom) experienced no visible distortion or apparent weakening.

Fifty-Five-Gallon Metal Closed-Head Drums (for Liquids)

1. Ninety-five percent of the drums tested failed explosively.
2. Of the catastrophic failures, 68% failed at the bottom end, making the entire drum a projectile. (You may remember this effect from the movie *Backdraft*.)
3. One hundred percent of the drums tested failed at the top or bottom ends.
4. When filled with liquid (½ or ¾ full), bottom failures appeared to be increasingly violent with increasing water levels up to ¾ full.
5. Approximately 5.0 psig before failure, a significant amount of distortion of the drum's chime was apparent. There are two chimes per drum. They are what might be called the top and bottom rims of a liquid-filled drum.
6. The 55-gallon metal open-head drums appeared to bulge at only the top and bottom ends.
7. Body seams (top to bottom) experienced no visible distortion or apparent weakening.
8. Pinging was noticeable between 15 and 20 psig and increased dramatically immediately before drum failure.
9. Probability analysis predicted that 99% of the failures would occur above 48.7 psig.

Fifty-Five-Gallon Plastic Drums (Plastic and Plastic-Lined Metal—for Corrosive Liquids and Solids)

1. Four of five failures occurred through the sides of the drums at no particular or identifying location. One failed at 30 psig out of the top end of the drum.
2. These are significant observations because they show the potential for seamless high-density polyethylene drums to fail out of the sides. Deformation was observed at the tops, bottoms, and sides of the drum.
3. The two 55-gallon open-head high-density polyethylene drums failed explosively at 23 and 24 psig, ejecting the entire top off the drum. One 55-gallon plastic-lined metal drum self vented at 50 psig with a top bulge characteristic of the curve for the closed-head top deformation.

The range of failure pressures for the plastic drums was 23–68 psig, and for the metal drums was 13–125 psig (125 psig was the maximum test pressure). You can imagine the injury that could result, not only from the physical explosion of the drum, but from the instantaneous, explosive release of the drum's contents!

4.2.2 Warnings on Crystal Formation on Containers

The formation of crystals on containers, especially around openings, should be treated with great caution. Some crystallized materials are so sensitive that the friction created by twisting a cap can cause a reaction.

Be wary of crystal formation on containers. There are some chemicals, such as organic peroxides, that form shock-sensitive crystals as they decompose or are mixed with incompatible materials.

An example of such a change is a crate of old dynamite. Dynamite was originally diatomaceous earth, or other absorbent, and nitroglycerin. As it ages, the nitroglycerin can desorb from the absorbent and form an oily liquid that may then crystallize. As we all know, nitroglycerin is an extremely shock-sensitive, explosive material. Dynamite can be handled quite safely. It requires a detonator (blasting cap) to explode.

Another example is picric acid, used as a dye and for the formulation of some medicines. Picric acid is a flammable solid when wet with 10% water. It is a high-powered explosive when dehydrated. It is not-shock sensitive, but in contact with metals such as lead, iron, and copper it can form shock-sensitive metal picrates. Picric acid can be detonated by extreme heat, a blasting cap, or an electric charge. Dehydrated picric acid forms dry, orange-colored crystals. We caution workers who see crystals appearing around openings of containers. Metal picrates are extremely shock-sensitive and will detonate with the slightest movement or vibration.

WARNING

NEVER OPEN OR MOVE ANY CONTAINER THAT HAS CRYSTALS FORMED AROUND THE CAP OR TOP OR FROM LEAKAGE ON THE SIDE. IF THESE ARE SHOCK-SENSITIVE CRYSTALS, A FIRE OR EXPLOSION MAY RESULT. OBTAIN FURTHER GUIDANCE IN THIS SITUATION.

4.3 HAZARDOUS CONDITIONS RESULTING FROM INCOMPATIBILITY

Hazardous conditions that can result from incompatibility include smoke or toxic vapor generation, combustion, explosion, container distortion, and container heating.

⊙━⚲ **Mixing of incompatible materials can produce a number of types of hazardous conditions. Examples are smoke, toxic vapor, or explosive gas generation. Also combustion, explosion, container distortion, and container heating can occur to such an extent that a handler can suffer burns.**

Problems can occur rapidly when incompatibility causes a large amount of heat or vapor generation, or both, in a sealed container.

Container failure or destructive pressure bursts can occur rapidly. However, incompatibilities in the open atmosphere can also be extremely serious. Formation of toxic vapor plumes is one example.

In the authors' work with TSDFs we experienced the critical importance of segregating acid wastes from the extremely toxic cyanide wastes. Waste tankers unloading acids were fitted with special connections that only fit the acid storage tanks. The cyanide tankers had hose fittings that only mated with the cyanide storage tank fittings. *Note that acid contact with cyanides releases deadly hydrogen cyanide gas!*

Many operations on waste or accident sites involve mixing (known as consolidation) or unavoidable contact between various hazardous materials. It is important to know ahead of time if such materials are compatible. Table 4.1 lists a sampling of hazards and the incompatibilities that produce them. Table 4.2 lists a sampling of common chemicals and the materials with which they are incompatible.

TABLE 4.1 Hazards Caused by Incompatible Materials

Hazard	Incompatibility
Heat generation	Acid and water
Fire and explosion	Hydrogen sulfide and calcium hypochlorite; dusts in air
Toxic gas or vapor production	Sulfuric acid and aluminum
Flammable gas or vapor production	Acids and metals
Formation of a substance with greater toxicity than the reactants (separately)	Chlorine and ammonia
Formation of shock or friction-sensitive compounds	Peroxides and organics or liquid oxygen and petroleum products
Solubilization of toxic substance	Hydrochloric acid and chromium
Dispersal of toxic dusts and mists	Sodium or potassium cyanide and water or acid vapor
Violent polymerization and heat release	Ammonia and acrylonitrile

Courtesy of NIOSH.

4.4 INCOMPATIBILITY, MILD OR VIGOROUS?

O—π   **Incompatibilities, or reactions, can be mild or vigorous. Vigorous incompatibilities cause immediate hazards. Mild incompatibilities cause time-delayed hazards. Some materials have a limited lifetime. They decompose spontaneously.**

There are reactions that are not necessarily incompatibility problems. Be aware that there are materials that decompose, apparently spontaneously, either slowly or rapidly. Readers all know of the dated labels on their prescriptions. Some of these reactions may be due to impurities in the material or slow reactions with container or packaging materials. There are materials that decompose when stored under high-temperature and high-humidity conditions. Information about such instability is required on a material's MSDS.

Incompatibility, however, does not necessarily indicate a lasting hazard. For example, acids and bases (both corrosive) react to form salts and water. That solution might not be corrosive and might or might not cause hazards. Note that the resultant material might not be hazardous. However, there can be a considerable heat of reaction during the mixing of the acid with the base.

You will see liquid oxygen (LOX) trucked throughout the world as "refrigerated liquid," and with its "oxidizer" placard prominently displayed on the tanker. Should enough of that LOX spill on asphalt-containing pavement, a shock-sensitive explosive compound will be formed that can be detonated on impact. The authors know of a situation where some LOX had spilled on an asphalt runway. A technician dropped a wrench on the spot, and the asphalt/LOX (fuel/oxidizer) mixture exploded. That explosion propelled the wrench at the technician's head, killing him instantly!

WARNING

ANY PAVEMENT REMOVAL, OR CLEANUP, OF SUCH AN AREA MUST BE CONDUCTED WITH PERSONNEL PROTECTION FROM EXPLOSION.

Simply adding water to some materials is hazardous. Water poured on sodium or potassium metal, or either metal added to water, causes an immediate decomposition of the water, release of hydrogen, then fire or explosion. Water poured onto a pool or into a container of concentrated sulfuric acid causes an immediate strong heat release as ionization takes place.

TABLE 4.2   Common Incompatible Materials

Material	Incompatible Materials
Acetic acid	Aldehydes, bases, carbonates, hydroxides, metals, oxidizers, peroxides, phosphates, xylene
Acetylene	Halogens (chlorine, fluorine, etc.), mercury, potassium, oxidizers, silver
Acetone	Acids, amines, oxidizers, plastics
Alkali and alkaline earth metals	Acids, chromium, ethylene, halogens, hydrogen, mercury, nitrogen, oxidizers, plastics, sodium chloride, sulfur
Ammonia	Acids, aldehydes, amides, halogens, heavy metals, oxidizers, plastics, sulfur
Ammonium nitrate	Acids, alkalis, chloride salts, combustible materials, metals, organic materials, phosphorous, reducing agents, urea
Aniline	Acids, aluminum, dibenzoyl peroxide, oxidizers, plastics
Azides	Acids, heavy metals, oxidizers
Bromine	Acetaldehyde, alcohols, alkalis, amines, combustible materials, ethylene, fluorine, hydrogen, ketones (acetone, carbonyls, etc.), metals, sulfur
Calcium oxide	Acids, ethanol, fluorine, organic materials
Carbon (activated)	Alkali metals, calcium hypochlorite, halogens, oxidizers
Carbon tetrachloride	Benzoyl peroxide, ethylene, fluorine, metals, oxygen, plastics, silanes
Chlorates	Powdered metals, sulfur, finely divided organic or combustible materials
Chromic acid	Acetone, alcohols, alkalis, ammonia, bases
Chromium trioxide	Benzene, combustible materials, hydrocarbons, metals, organic materials, phosphorous, plastics
Chlorine	Alcohols, ammonia, benzene, combustible materials, flammable compounds (hydrazine), hydrocarbons (acetylene, ethylene, etc.), hydrogen peroxide, iodine, metals, nitrogen, oxygen, sodium hydroxide
Chlorine dioxide	Hydrogen, mercury, organic materials, phosphorous, potassium hydroxide, sulfur
Copper	Calcium, hydrocarbons, oxidizers
Hydroperoxide	Reducing agents
Cyanides	Acids, alkalis, aluminum, iodine, oxidizers, strong bases
Flammable liquids	Ammonium nitrate, chromic acid, hydrogen peroxide, nitric acid, sodium peroxide, halogens
Fluorine	Alcohols, aldehydes, ammonia, combustible materials, halocarbons, halogens, hydrocarbons, ketones, metals, organic acids
Hydrocarbons (such as butane, propane, benzene, turpentine)	Acids, bases, oxidizers, plastics
Hydrofluoric acid	Metals, organic materials, plastics, silica (glass), (anhydrous) sodium
Hydrogen peroxide	Acetaldehyde, acetic acid, acetone, alcohols, carboxylic acid, combustible materials, metals, nitric acid, organic compounds, phosphorous, sulfuric acid, sodium, aniline
Hydrogen sulfide	Acetaldehyde, metals, oxidizers, sodium
Hypochlorites	Acids, activated carbon
Iodine	Acetaldehyde, acetylene, ammonia, metals, sodium
Mercury	Acetylene, aluminum, amines, ammonia, calcium, fulminic acid, lithium, oxidizers, sodium
Nitrates	Acids, nitrites, metals, sulfur, sulfuric acid
Nitric acid	Acetic acid, acetonitrile, alcohols, amines, (concentrated) ammonia, aniline, bases, benzene, cumene, formic acid, ketones, metals, organic materials, plastics, sodium, toluene
Oxalic acid	Oxidizers, silver, sodium chlorite
Oxygen	Acetaldehyde, secondary alcohols, alkalis, ammonia, carbon monoxide, combustible materials, ethers, flammable materials, hydrocarbons, metals, phosphorous, polymers
Perchloric acid	Acetic acid, alcohols, aniline, combustible materials, dehydrating agents, ethyl benzene, hydrochloric acid, iodides, ketones, organic material, oxidizers, pyridine
Peroxides, organic	Acids (organic or mineral)
Phosphorus (white)	Oxygen (pure and in air), alkalis
Potassium	Acetylene, acids, alcohols, halogens, hydrazine, mercury, oxidizers, selenium, sulfur

TABLE 4.2 Common Incompatible Materials (*Continued*)

Material	Incompatible Materials
Potassium chlorate	Acids, ammonia, combustible materials, fluorine, hydrocarbons, metals, organic materials, sugars
Potassium perchlorate (also see chlorates)	Alcohols, combustible materials, fluorine, hydrazine, metals, organic matter, reducing agents, sulfuric acid
Potassium Permanganate	Benzaldehyde, ethylene glycol, glycerol, sulfuric acid
Silver	Acetylene, ammonia, oxidizers, ozonides, peroxyformic acid
Sodium	Acids, hydrazine, metals, oxidizers, water
Sodium nitrate	Acetic anhydride, acids, metals, organic matter, peroxyformic acid, reducing agents
Sodium peroxide	Acetic acid, benzene, hydrogen sulfide, metals, oxidizers, peroxyformic acid, phosphorous, reducers, sugars, water
Sulfides	Acids
Sulfuric acid	Potassium chlorate, potassium perchlorate, potassium permanganate

Courtesy of NIOSH.

Acid and solution are violently expelled. Water added to sodium hydroxide (a strong base) will also result in large heat release.

4.5 COMMON FUELS AND OXIDIZERS

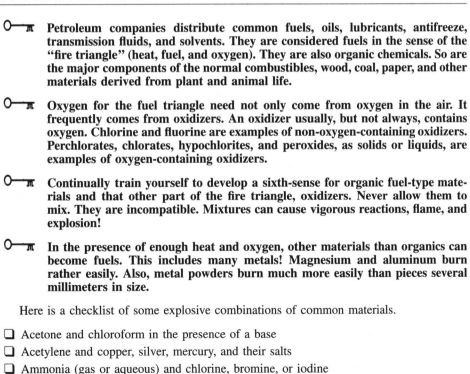

Petroleum companies distribute common fuels, oils, lubricants, antifreeze, transmission fluids, and solvents. They are considered fuels in the sense of the "fire triangle" (heat, fuel, and oxygen). They are also organic chemicals. So are the major components of the normal combustibles, wood, coal, paper, and other materials derived from plant and animal life.

Oxygen for the fuel triangle need not only come from oxygen in the air. It frequently comes from oxidizers. An oxidizer usually, but not always, contains oxygen. Chlorine and fluorine are examples of non-oxygen-containing oxidizers. Perchlorates, chlorates, hypochlorites, and peroxides, as solids or liquids, are examples of oxygen-containing oxidizers.

Continually train yourself to develop a sixth-sense for organic fuel-type materials and that other part of the fire triangle, oxidizers. Never allow them to mix. They are incompatible. Mixtures can cause vigorous reactions, flame, and explosion!

In the presence of enough heat and oxygen, other materials than organics can become fuels. This includes many metals! Magnesium and aluminum burn rather easily. Also, metal powders burn much more easily than pieces several millimeters in size.

Here is a checklist of some explosive combinations of common materials.

❑ Acetone and chloroform in the presence of a base
❑ Acetylene and copper, silver, mercury, and their salts
❑ Ammonia (gas or aqueous) and chlorine, bromine, or iodine
❑ Carbon disulfide and sodium azide
❑ Chlorine and alcohol

❑ Carbon tetrachloride or chloroform and powdered aluminum or magnesium
❑ Diethyl ether and chlorine
❑ Dimethyl sulfoxide and an acetyl halide
❑ Dimethyl sulfoxide and chromic oxide
❑ Ethanol and calcium hypochlorite
❑ Ethanol and silver nitrate
❑ Nitric acid and acetic anhydride or acetic acid
❑ Picric acid and heavy metal salt, such as lead, mercury or silver
❑ Silver oxide and ammonia and ethanol
❑ Sodium and chlorinated hydrocarbons
❑ Sodium hypochlorite and amines

Here is a checklist of some hypergolic reactants. These exothermic reactions release sufficient heat to cause a fire. No ignition source is necessary.

❑ Perchloric acid and magnesium powder
❑ Acetone and 85% nitric acid
❑ Nitric acid and phenol
❑ Concentrated nitric acid and triethylamine
❑ Red fuming nitric acid and aromatic amines
❑ Divinyl ether and 96% nitric acid and 5% sulfuric acid
❑ Ninety percent solution of potassium permanganate in red fuming nitric acid and alcohols
❑ Solid potassium permanganate and ketones, esters, alcohols

4.6 PREFERRED STORAGE, CLEANUP, AND DISPOSAL PRACTICES

☛ **Keep chemicals packaged and separated as far as practical from each other.**

☛ **Never arrange or store chemicals alphabetically without regard to incompatibility. That could place dangerous incompatibles too close together.**

We have reviewed industrial, academic, government, and medical institution plans for segregating incompatible materials. The U.S. Coast Guard has one of the more elaborate storage schemes, based on 24 segregated groups. However, there are roughly 10 or so categories of materials that find widespread usage. Keep the following checklist of groups segregated from one another. Further information on chemical storage is in Section 15.

❑ Flammables (including solvents)
❑ Oxidizers (including peroxides, peroxide-forming materials, perchlorates, and light-sensitive materials)
❑ Reducers
❑ Acids
❑ Bases or alkalis
❑ Water reactives
❑ Poisons, or extremely toxic materials

❑ Pyrophorics
❑ Gas cylinders
❑ Radioactive materials

Flammables should be segregated where, if they catch fire, they can be treated with the proper extinguishing agent without harming the other materials.

The *oxidizers* category should be kept cool and in shade or darkened storage. Unless you are a manufacturing employee with specific knowledge of alternative safe storage of specific oxidizers, you should not allow piles of oxidizers to accumulate. Limit them to as-delivered, drum-size containers, or less according to labeling. Explosions of oxidizers have occurred where it is believed that explosion could have been prevented if the mass of contained oxidizer had been limited.

As explained in Section 3.3, there are materials (some propellants and explosives) that are considered oxidizers. They have both fuel and oxidizer components in their molecules. Ammonium perchlorate and ammonium nitrate are prime examples. The ammonium part of the molecule can act as a fuel. The perchlorate and nitrate parts can act as oxidizers.

There are believed to be 'critical diameters' that are characteristic of stored oxidizers such as ammonium perchlorate and ammonium nitrate. That is, you can consider a stored pile of material as being roughly spherical in shape and thus having a diameter you can measure. The larger that diameter, the more difficult it is for any heat that forms in the center of the pile to be dissipated. Any heat present can cause decomposition. Any decomposition causes more heat. The effects can start to accelerate. Fire or explosion can be the final result.

WARNING

EXPLOSIONS OF OXIDIZERS HAVE OCCURRED WHERE IT IS BE-LIEVED THAT THE EXPLOSION COULD HAVE BEEN PREVENTED IF THE MASS OF CONTAINED OXIDIZER HAD BEEN LIMITED. THERE ARE BELIEVED TO BE CRITICAL DIAMETERS CHARAC-TERISTIC OF STORED OXIDIZERS SUCH AS AMMONIUM PER-CHLORATE AND AMMONIUM NITRATE. NEVER STORE OXIDIZ-ERS NEAR FUELS.

Reducers (or reducing agents), such as metal hydrides, must be kept tightly sealed and stored away from oxidizers to avoid the flammable and explosive oxidation-reduction reactions. Reducers are potent fuels.

Acids should be kept tightly sealed and in containers constructed of the proper material. Many common acids (acetic, nitric, and hydrochloric, for instance) will be stored in special stainless steel drums or plastic or glass bottles. Other acids, such as hydrofluoric (HF), must never be stored in glass. HF etches glass. Acids must never be stored where they can contact fuel, oxidizers, bases, or water.

Many *bases and alkalis* are stored in the form of dry powders. This allows them to be contained in boxes with plastic liners. Special care must be exercised to keep these containers dry. They must never be stored where they can contact fuel, oxidizers, acids, or water.

Water reactives need to be protected from the potential for being drenched with fire-extinguishing water.

Poisons and other toxics need to be stored separately from all other materials. This is necessary in case of fire or other emergency. When these materials are involved in a fire, the responders will often let them burn versus subjecting the emergency personnel to the toxic gases and vapors they produce. Always handle these containers with proper PPE. Some materials are so toxic that skin contact with a trace amount left on a "clean" container could be deadly.

Pyrophorics need to be protected from air. You need to experience pyrophorics (finely divided metals or metal powders) being poured through the air and spontaneously catching fire to really appreciate this hazard!

Gas cylinders need to be protected from heat and from mechanical damage.

- Oxygen and other oxidizers must be stored separately from flammable fuel gases such as acetylene, propane, and MAPP gas.
- All cylinders must be stored in the upright position with their valve cap in place. They must also be securely chained or otherwise held in place. These cylinders can contain gas pressurized to 3500 psi. You may have heard that a gas cylinder can become a rocket! It is true. The authors know of a number of such incidents. Should the valve end of a gas cylinder become broken off, allowing free gas discharge, the cylinder will be propelled by the same forces that propel rockets. They have been known to fly hundreds of yards and crash through walls!
- Toxic and poison gases may need to be stored separately in a well-ventilated area.

Radioactive materials deserve special mention. As we will continue to stress, this book is not intended as sufficient preparation to work with radioactive materials. They are extremely well-regulated materials that most of us will never come into contact with. The storage watchwords here are distance and shielding. We reiterate: when you see the radioactive symbol, back off until you have more information and training.

Other sources of information on chemical incompatibility include:

- The National Fire Protection Association's publication 491M, Hazardous Chemical Reactions
- The National Research Council's *Prudent Practices for Handling Hazardous Chemicals in Laboratories*

Note

The *NIOSH Pocket Guide to Chemical Hazards* in the Companion CD contains a listing of the incompatibilities of nearly 700 chemicals.

A storage space for incompatible chemical substances should be designed with the worst-case scenario in mind. Ask yourself, if this material were to accidentally mix with others in this space, what could happen?

Here is a checklist of some storage safety factors.

- ❑ Label all containers stating their contents.
- ❑ Provide MSDSs for each substance stored in the area.
- ❑ Separate incompatible chemicals by means of fire-resistant material. Provide a fire resistance of at least two hours. Concrete block walls and specially designed chemical lockers are recommended.
- ❑ Construct storage areas to contain any spillage. Calculate the volume of the maximum quantity of liquid material normally stored. Plan to contain 110% of that quantity. Floor drains must lead to an isolated, dedicated sump. Don't allow floor drains to lead to sanitary sewers, storm sewers, or adjacent storage areas.
- ❑ Equip storage areas with suitable shelving constructed of nonporous and noncombustible material.
- ❑ Properly ventilate the area. Include forced ventilation from floor to ceiling, with exhaust above roof level. Provide sufficient exhaust so that vapors cannot accumulate inside the area. Place the exhaust outlet so as not to impact on other buildings or workers.
- ❑ Clearly mark the area with a sign indicating "Hazardous Material Storage." Indicate the maximum allowable amount of stored materials.
- ❑ Keep aisles clear of pallets, debris, electrical cords, hoses, and the like.
- ❑ Store PPE for use in the area in such a fashion that it cannot be contaminated. Have it readily available to personnel without exposing them to the hazards of the storage area.

❑ Equip the area with washing facilities for personal hygiene.

❑ Equip the area with fire extinguishers of a quantity and type suitable for the expected fire risk. Provide any automatic fire-extinguishing system with fire fighting agents that will not contribute to the fire, health, or safety risk.

❑ Provide smoke detectors.

❑ Provide a communication system with the main office or emergency system.

❑ Firmly secure shelf assemblies to walls.

❑ Provide antiroll lips on all shelves.

❑ Provide spill control and clean-up materials.

❑ Place an approved eyewash station and safety shower in the immediate area with a dedicated sump.

MSDSs usually describe proper clean-up and disposal procedures. See Section 9. We describe spill cleanup in more detail in Section 13.

We want to emphasize that the materials used to clean up a spill must be compatible with the spilled material. There are many specialized products available for cleaning up specific hazardous materials. Sawdust may be ideal for absorbing spilled motor oil, but using it to clean up spilled sulfuric acid will likely result in a fire. Treat the media and equipment used for the cleanup as hazardous until you are sure they are not, or are safely decontaminated.

4.7   IMPORTANCE OF HAVING MSDSs AND REQUIRED LABELING

Obviously this Section cannot describe all kinds of incompatibilities. Be sure always to read and to understand the MSDS for materials with which you will be working. Even those materials that you might be contacting. Known incompatibilities are required to be listed in the MSDS. See Section 9 for MSDS details.

MSDSs supplied with materials are required by regulations to describe incompatibilities. Owners of materials in drums or other containers are required to label them as to contents and potential hazards.

Refer to Section 8 for more details on labels, placards, and other Identification.

WARNING
WHEN YOU DO NOT KNOW ABOUT THE INCOMPATIBILITIES OF MATERIALS YOU ARE DEALING WITH, PROCEED CAUTIOUSLY. WHEN YOU HAVE NO MSDS INFORMATION ABOUT A CONTAINER'S CONTENTS, THE PROPER RESPONSES ARE TO MAINTAIN A SAFE DISTANCE FROM THE CONTAINER, DON PERSONAL PROTECTIVE EQUIPMENT, AND HAVE THE CONTAINER SEGREGATED USING MECHANICAL HANDLING EQUIPMENT.

Learn to recognize and identify as many common incompatibilities as possible. Especially in emergency response, MSDSs may not be available.

4.8   DISPOSAL COSTS—P² IS THE PREFERRED WAY

With hazardous waste disposal costs increasing, pollution prevention (P^2) is the only economical way to proceed. We must continue to find new and innovative ways to reduce our hazardous waste streams. These ways include reducing, recycling and reusing along with finding less hazardous or nonhazardous replacements. We will discuss this further in Section 13.

That said, what can we do right now with the waste we cannot reuse or recycle economically? Here is a checklist of just a few things to keep in mind.

❑ Do not mix hazardous waste with nonregulated waste.

❑ The average cost of disposal for a 55-gallon drum of hazardous waste is in the $400–$600 range. Minimize the number of drums being used without mixing incompatibles. Mixing hazardous waste with nonregulated waste greatly increases costs.

❑ Hazardous waste disposal is calculated in pounds.

❑ Keep the load as light as possible. If you can use boxes or fiberpack drums, do so. This will reduce the overall weight of the shipment. Large tri-wall boxes, properly lined, are excellent for consolidating smaller containers or large solids loads. Formerly these were packaged in steel drums.

❑ Rather than shipping a large number of partially full containers of compatible materials, consolidate the material into a fewer number of full drums.

❑ Always focus on recycling possibilities.

❑ Consolidate and possibly reuse. Consider mixing the same type of paints together. That is, water-based with water-based, and oil-based with oil-based. Painters call this boxing. Mix containers of same-color paints together so that the resultant mixture is uniform in color. We have found that when various colors are mixed together with a little white (the most common color), the resultant mixture is some shade of gray. Consider using this mixture to paint buildings or equipment. If usable, it beats shipping it out as waste at $500 a drum!

4.8.1 Lab Packs and Highly Hazardous Waste

Small-quantity wastes, such as from laboratories, are disposed of in lab packs. A lab pack is simply a drum containing individual containers, usually bottles, of compatible chemicals. Vermiculite is added to cushion the contents and to help absorb leakages. Lab packs are limited to 15 gallons of total waste.

The approximate cost of disposal of a normal lab pack in 2001 was about $400. Add about $23 for the drum and around $80 for such additional costs as labor, insurance, permits, and fees. Since a 55-gallon lab pack can only contain 15 gallons of waste, the per-gallon cost is almost $50.

If we look at highly hazardous waste (the EPA P waste codes), the costs skyrocket. For example, reactives such as shock-sensitive fulminates and picrates can have disposal costs in the neighborhood of $100 per pound. The authors know of one facility that paid over $6000 to dispose of a single liter of methyl ethyl ketone peroxide!

We reiterate, always use your best efforts to segregate nonregulated wastes from hazardous wastes. As you can see, this is especially important when dealing with highly hazardous wastes.

4.9 TRAINING AIDS AND ADDITIONAL RESOURCES

The authors recommend the following videos for aiding readers to see 'real-life' examples of what they have learned in this Section.

• *Packaging Hazardous Waste*
• *Lab Packing*

These videos are available from Safe Expectations (formerly BNA Communications, Inc.) (www.safeexpectations.com).

- *Bulging Drums—What Every Responder Should Know* is available by contacting Michael Larranaga, Los Alamos National Laboratory, at 505-665-9396.

4.10 SUMMARY

As you can see, there a large number of materials that cannot come into contact with each other without some sort of reaction occurring. We call that chemical incompatibility or unwanted chemical reaction. Chemical reactions are used every day (under controlled conditions) to make products we use and need. Problems arise when reactive (incompatible) materials are mixed accidentally.

Mixing of incompatible materials can produce a number of types of hazardous conditions. Examples are smoke, toxic vapor or explosive gas generation, combustion, explosion, container distortion and rupture, and container heating such that a handler can suffer burns.

You must study and observe to develop a feel, or sixth sense, for common incompatibilities. It is particularly important that you recognize oxidizers and fuels. You do not need to memorize long lists of chemical names. Your best ally is the MSDS. Always consult the MSDSs before mixing or storing chemicals together. If no MSDSs are available, assume a worst-case scenario and treat the materials as incompatible—that is, until you know for sure they are compatible. Acids, bases, oxidizers, reducing agents, fuels, and poisons should never be mixed or stored together.

As HAZWOPER personnel, you are being prepared to work safely around all types of hazards. A typical assignment will be to clean up someone else's mess. The MSDSs you need may not be available. Labels, placards, signs, and markings may be partially or totally defaced. Incompatible materials may have been stored, or mixed, together in containers that are failing. In these situations, approach materials and containers with caution. Use mechanical or remote handling to keep your distance from suspect materials or containers. We will describe remote opening, sampling and handling of suspect containers in further detail in Sections 12, 13, and 14.

We trust that this Section has demonstrated that HAZWOPER workers can never let their guard down. There is no such thing as routine HAZWOPER work!

SECTION 5
TOXICOLOGY

ACKNOWLEDGMENT

The authors wish to express their appreciation to Marianne Cloeren, MD, MPH, FACP, and Principal of Cloeren Occupational Health Associates (www.cloeren-OHA.com) for her constructive review and helpful additions to this section. Dr. Cloeren is a Board member of the Maryland College of Occupational and Environmental Medicine and a Member and Delegate of the American College of Occupational and Environmental Medicine. Dr. Cloeren held a three-year Fellowship in Occupational Medicine at Johns Hopkins and is certified in Internal Medicine and Occupational and Environmental Medicine.

OBJECTIVES

Section 5 is meant to give readers involved in any aspect of HAZWOPER work an understanding of the science of toxicology and the practical benefits it provides. Those benefits include lifesaving protective information about the effects of toxic substances on humans. We want HAZWOPER workers to be convinced that when toxicology studies prove that a material has toxic effects, workers must protect themselves from those effects.

Workers need to understand allowable concentrations and dosages and the ways of expressing exposures. We explain PELs, TWAs, RELs, TLV®s, STELs, and IDLHs. Workers must know how to access toxicity information: data that tell the limits of exposure they can withstand without adverse health effects. The Companion CD is a handy resource.

More people die from job-related toxic effects than from all job-related accidents. These are usually people who are not knowledgeable about the toxic hazards and who are not outfitted with proper PPE for the job.

Another of our objectives is to impress upon readers that there are toxins that are odorless, colorless, tasteless, painless, and may have no apparently hazardous immediate effects. However, even short exposures can result in death. On the other hand, in some cases it takes 20 to 30 years for cancer or other diseases to develop after exposure. This is called the latency period.

HAZWOPER workers must understand the four routes of exposure to toxins: inhalation, injection, absorption, and ingestion. Inhalation is the most common route. Injection is the most direct entry route. Absorption through body tissues (including skin and eyes) has become less common due to more routine use of PPE. Skin absorption of PCBs can produce chloracne, a serious form of dermatitis. However, absorption exposure can be deadly. We present the example of methyl mercury. Ingestion (eating or swallowing) is the exposure route most preventable by following common safety restrictions against eating in contaminated areas and by following proper decon procedures.

Acute exposures usually involve single doses, at high levels, of relatively short duration. Workers can easily appreciate the acute exposure hazards of toxic gases and vapors, but we want them to develop greater appreciation of the more sinister hazards of chronic exposures. Chronic exposures usually involve frequently repeated doses, sometimes at relatively low levels, over longer periods of time. Asbestos inhalation hazards are a prime example.

We point out that newly discovered and widely used chemical compounds may be important factors in the increase of some rates of disease. Some of them are synthetic organic compounds. Others, such as asbestos and lead, have been with us throughout history. They are natural minerals and elements. Toxic synthetic and natural materials are widespread in the environment because early on, before their toxicity was fully understood, they were found to provide tremendous benefits. They were materials that provided exceptional fire safety, crop protection, disease fighting, structural, and other benefits.

Later the health and environmental hazards of many materials became better understood. Unfortunately, indiscriminate dumping and burying of hazardous materials had become common. A frequent HAZWOPER task now is to be a specialist who will help correct those wrongs with special understanding of toxic hazards and thus with special protection and skill.

5.1 TOXICOLOGY DEFINED

Toxicology is the scientific study of the properties of toxic substances and of their chemical, biological, and radiological health effects on all kinds of organisms.

In Section 3.1 we classified toxic substances into a number of hazards categories* depending upon their mode of action on humans. In Section 5 we will delve deeper into those harmful actions we all want to avoid. The following categories were listed:

- Systemic poisons and biohazards
- Asphyxiants
- Carcinogens
- Mutagens (affecting genes) and teratogens (affecting offspring)
- Irritants
- Allergens
- Sensitizers

Organisms include humans, animals, and plants. While the subject of toxicology is quite complex, understanding some basic concepts is necessary to assist us in making logical decisions about the protection from toxic injuries to ourselves, those we train, and those we supervise.

Toxicology studies on the effects of materials on laboratory animals do not always correctly predict the effects on humans. However, they have proven to supply valuable human health information that has saved many lives. Toxicology studies with laboratory animals, performed by caring researchers, are absolutely essential in the prevention of worker illness, injury, and death.

Unfortunately, the words "toxins," "toxic substances," and "poisons" are all used in regulations and literature to describe the same kinds of materials. Actually, poisons are only one of several kinds of toxins. We restrict the word

* You may have noticed that many chemicals can be assigned to more than one of these groups. However, we do believe that all toxic substances can be assigned to at least one of these groups.

"poisons" to mean materials that have predominantly a harmful internal, systemic, and chemical effect on the body.

5.2 IMPORTANCE OF TOXICOLOGY FINDINGS TO HAZWOPER PERSONNEL

O—π **More people die from job-related toxic effects than from all job-related accidents.**

The study of toxins and toxic effects rates its own section because of the frequency with which workers are affected by toxic substances. Did you know that it has been estimated by the National Institutes of Health that about 140 workers die from job-related toxic effects every day? This is more than eight times the number of people who die from job-related accidents. HAZWOPER personnel, probably more than any others in the workforce, must be knowledgeable about toxins and toxic effects—especially how to avoid being injured by them.

5.3 ANCIENT RECORDINGS OF TOXIC EFFECTS

"All materials are poisons. The only difference between a poison and a 'remedy' is the proper dose."

—*Paracelsus (1493–1541).*

Observations of the effect of occupational exposure to substances on humans were reported as early as 63 B.C. Pliny the Elder, a Roman statesman, noted that slaves who worked in asbestos mines developed coughs and died gasping for breath. He also noted that lead-smelter workers lived for only a few years after they began work at the furnaces (see Section 5.5.3 below). Today, due to the science of toxicology, we know what probably caused many people from both groups to die at an early age.

Some of the toxic effects we will describe have been at least partially known for a long time. Some are recent discoveries. Unfortunately, the toxic effects of many materials we live with are still unproven and undiscovered.

5.4 ENVIRONMENT AND LIFE EXPECTANCY

According to David P. Rall, Director of the National Institute of Environmental Health Sciences and Director and National Toxicology Program from 1971 to 1990, life expectancy in the United States has increased significantly over the last century. Our lifespans today extend well into the 70s.

A major contributor to this increase is a better available diet. (This is true only if more people take advantage of it.) We also have better housing, greatly improved sanitation, and a better understanding of toxic hazards and their remediation. HAZWOPER personnel, take note of the last two factors. The public depends upon your expertise being put into action.

An individual's life expectancy after the age of 50 has remained fairly constant since 1900. However, the pattern of deaths has changed. We are dying of different causes. Death rates for such pulmonary diseases as tuberculosis and pneumonia have decreased. Deaths from cardiovascular disease have increased in the past, but appear to be leveling off or decreasing. However, cancer death rates have risen steadily and still seem to be increasing.

As a result of these different patterns of life and death, we have to contend with different diseases today. The medical profession and researchers are working hard to deal successfully with chronic long-term diseases, such as cancer, heart disease, kidney disease, and neurological diseases. Scientists do not fully understand the causes of these diseases. Many factors appear to be involved.

○━π **Newly discovered and widely used chemical compounds may be important factors in the increase of some rates of disease.**

One of the prominent factors in the causation of many of these diseases may be chemical compounds, especially some of those synthesized, produced, and widely used in the 20th century. We are finding that sometimes the byproducts or impurities present in the materials we use are the bad actors, not the materials themselves. Sometimes waste treatment causes problems. The infamous dioxins are found in the exhaust from the incineration of common household and industrial wastes.

5.5   TOXIC MATERIALS AND HUMAN DISEASE

The following statements about a sampling of common elements, metals, materials, and chemical compounds explain some of the vast differences in how toxic substances act on humans. Sometimes one human and not another is affected. Some substances can be highly beneficial in one concentration and toxic in another. Some substances may interact with each other in the body to cause ill effects.

WARNING

YOU MUST UNDERSTAND THAT THE SYMPTOMS OF EXPOSURE TO TOXIC MATERIALS OFTEN MIMIC OTHER AILMENTS OR DISEASES. A HIGH EXPOSURE TO LEAD, FOR INSTANCE, MAY CAUSE THE SAME SHORT-TERM SYMPTOMS AS FLU. BECAUSE OF THIS, PHYSICIANS CAN MISDIAGNOSE THE AILMENT. THEREFORE, IT IS IMPERATIVE FOR YOU, AS A HAZWOPER TRAINER, SUPERVISOR, OR WORKER, TO BE KNOWLEDGABLE ABOUT YOUR OWN, AND OTHERS', EXPOSURES TO TOXINS.

FURTHER, ALL INDIVIDUALS MUST INFORM THEIR PHYSICIANS OF ANY EXPOSURE THEY KNOW, OR SUSPECT, IS HARMFUL. THAT INFORMATION ASSISTS MEDICAL PERSONNEL IN THE PROPER TESTING AND DIAGNOSIS OF ANY AILMENT. FOR EXAMPLE, OVEREXPOSURE TO LEAD CAN BE DIAGNOSED WITH A SIMPLE BLOOD LEAD LEVEL (BLL) TEST.

○━π **Early information about your exposure to a toxin must be reported to a physician.**

5.5.1   Iodine (Symbol I)

One hundred years ago it was common in the middle United States (away from the coastal areas) for people to get insufficient iodine from the foods they ate. This can cause a small gland in the neck, called the thyroid, to grow very large as it tries to compensate for lack of essential iodine. This condition, called goiter, became rare after public health officials decided that iodine (as potassium iodide) should be added to salt. (Read the fine print on your salt box!) However, we also know that iodine solutions used for external antiseptic purposes, are poisonous if ingested.

O—π **The dosage, or exposure, to some materials can spell the difference between health benefit and toxicity.**

5.5.2 Fluorine (Symbol F)

In the 1930s, dentists in areas of Texas and Oklahoma noticed that children who lived in areas where the drinking water contained naturally occurring salts of fluorine (fluorides) had none or fewer cavities. Today, nearly half of all Americans drink water that is either naturally fluoridated or treated with fluorides. Fluoride addition, at about the 1 ppm level, has lowered the incidence of dental cavities as much as 65%. At a somewhat higher level, tooth discoloration occurred. At much higher concentrations, fluorides are toxins.

5.5.3 Lead (Symbol Pb)

O—π **HAZMAT workers must be alert to potential lead contamination in pre-1978 homes, industrial buildings, and urban fill.**

Lead and its chemical compounds, have been used for thousands of years. In fact the word "plumbing" comes from the Latin word for lead, *plumbum*. For most of those millennia, however, little public attention was directed toward any of its potential toxic effects. That is, except for the efforts of Gaius Plinius Secundus, known as Pliny the Elder, and a few others. Pliny, a Roman statesman, author, scientist, and military leader, wrote a 37-volume encyclopedia entitled *Natural History*. He observed the mining industries, most notably lead mining and smelting, and noted that the slaves who worked at the lead smelters developed loss of control of their limbs and died after a relatively short exposure.

Lead has been used for centuries. Its use by the Egyptians and Romans has been recorded in paints, facial decoration, piping, food and beverage containers, and wine (for sweetening).

In the 20th century, lead has been used in lead-based paint, leaded gasoline (as the antiknock compound tetraethyl lead), and lead pipes. You still hear of plumbing systems being removed because of the lead solder used to make connections in copper piping. Lead compounds have been used in food and drink containers, glazes for pottery, and even (illegally) as a sweetener in paprika (lead acetate or "sugar of lead"). Farmers gave up leaded-white or light-colored paints on their barns when they found that farm animals licking the paint became sick. They switched to red iron-oxide pigments. That is why for many years all farms had red barns.

Although lead has been banned from house and barn paints since 1978, it can still be found in lead carbonate painted walls, woodwork, furniture, and plumbing. Lead exists as lead halides and other lead compounds in the soil all over the country due to the use of leaded gasoline for much of the 20th century. We have used unleaded gasoline for years now, but many countries have not.

O—π **Children, pregnant women, seniors, and people who are in ill health suffer more from toxic lead exposures than young adults.**

Infants and children are attracted to lead-based paint dust due to its perceived sweet taste. Children eat non-food items, a behavior called pica. That leads to lead ingestion in houses or other areas where lead paint is found. Lead dust, fumes, and lead-contaminated water can also introduce lead into the body. Lead can damage the brain, kidneys, liver, reproductive system, and nervous system and can cause anemia and nerve deafness. Many children may have been misdiagnosed as having Attention Deficit Disorder (ADD) when in reality they had hearing loss due to lead poisoning. Severe lead poisoning can produce headaches, cramps, convulsions, and even death. Even small amounts can cause learning problems and changes in behavior.

Lead poisoning research and lead-abatement efforts tend to center on children because their nervous systems are still growing. However, even healthy adults are not immune to lead's toxic effects. Anemia is a common result of lead poisoning in healthy adults. This is because the red blood cells will attach more readily to lead than to iron. Adults can have higher blood lead levels (BLLs), up to 40 μg/dl versus 20 μg/dl (micrograms per deciliter of blood) for children, without adverse effect. However, both groups will begin to show toxic effects at higher levels. Pregnant women and children both absorb 50% of the lead they take into their bodies. OSHA has gone so far as to recommend that those adults wishing to have children maintain their BLL below 30 μg/dl. Lead is a heavy metal, regulated (for disposal purposes) by RCRA.

5.5.4   Mercury (Symbol Hg)

Mercury, sometimes called quicksilver, is a silvery heavy metal also regulated by RCRA. It is the only metal that is a liquid at room temperature. It is extremely poisonous. Its widest use has possibly been in thermometers, thermostats, fluorescent lamps, and barometers. Very small amounts can damage the kidneys, liver and brain. Years ago, workers in hat factories were poisoned by breathing the fumes from mercury used to finish the hats. Remember the Mad Hatter in Lewis Carroll's classic *Alice in Wonderland*?

Today, mercury exposure usually results from eating contaminated fish and other foods that contain small amounts of mercury compounds. The fish pick it up from their water environment, into which mercury wastes have found their way. One source of that mercury is the wastes from the mercury cell process for production of chlorine and sodium hydroxide. The body cannot rid itself quickly of mercury. With ingestion, it gradually builds up inside the body's fat tissue. This is one type of bioaccumulation. If not treated, mercury poisoning can eventually cause pain, numbness, weak muscles, loss of vision, paralysis, and even death.

5.5.5   Dusts of All Kinds

○──π   Avoid breathing all kinds of dusts. Even nuisance dusts can cause disruptive sneezing and coughing. Some dusts, including asbestos fibers and silica dusts, are serious toxins.

HAZMAT workers must always keep the following in mind: the only thing you should breathe regularly is pure air of comfortable temperature and humidity. When breathing anything else, you face risks. Some dusts are considered more of a nuisance than a toxin. You react to them by sneezing or coughing, which tends to rid your body of the contaminant. However, asthma and skin irritation can still result over time.

On the other hand, some airborne dusts or particles can be very dangerous. These include fibers from asbestos, cotton, and hemp and dusts from such compounds as silica, cadmium, graphite, coal, iron, and clay. Coal dust, for example, can damage sensitive areas of the lung, turning healthy tissue into scar tissue. This condition is called pneumoconiosis, or black lung.

Silica dust causes silicosis. Breathing cotton or hemp dust causes brown lung, or byssinosis. Asbestos fibers cause asbestosis (white lung), lung cancer, pleural thickening (scarring of the lining of the lungs), and, in rare cases, mesothelioma (a cancer of the lung lining). Aside from cancers and lung disease, inhaled dusts and fibers can cause chest pains, shortness of breath (often progressing to bronchitis), emphysema, and/or early death. Proper ventilation and the use of personal protective equipment (PPE, as described in Section 6), in the form of respirators, can greatly reduce the risk of these diseases and ailments.

5.5.6   Uranium (Symbol U)

○──π   Radioactive materials are the only toxins that can harm you from a distance, usually without any warning pain.

Uranium is the best-known member of the family of elements that are extremely dangerous because they are radioactive. Uranium comes in three main forms. *Natural uranium* is found in the soil and in small quantities in our food and water. It is only slightly radioactive, and there is no good evidence that exposure to natural uranium found normally in the environment causes any health problems. An increased cancer rate in miners working in uranium mines is thought to be due mainly to their exposure at the same time to radon gas. Radon gas is a product of several steps of the radioactive decay of uranium.

When uranium is taken from the ground and processed to make it more radioactive for power plants and weapons, it is called *enriched uranium*. This form of uranium is much more radioactive and dangerous than natural uranium. It is highly regulated by RCRA, the Nuclear Regulatory Commission, and the DOE, among other agencies.

The third common form of uranium, *depleted uranium,* is a byproduct of the enrichment process. Depleted uranium is used for weapons and some industrial applications. It is much less radioactive than natural uranium.

There are several kinds of radiation. Some radioactive materials give off forms of radiation that can penetrate through body tissue at a distance. Other radioactive materials give off particles that can not travel far. They need to be inhaled or ingested to cause harm. Uranium is one of the latter. However, the decay process of radioactive materials can change them into other elements with other radioactive properties. Therefore, you also have to consider the properties of the decay products as well as the original radioactive material.

Small doses over a long period can also be harmful. Some cases of lung cancer are due to unrecognized exposure to radon in homes over years. In addition to its radioactive properties, uranium and other radioactive elements can cause toxicity related to their chemical activity in the body. Uranium, like lead and other heavy metals, can damage the kidneys at high doses.

Those working on hazardous waste sites are more likely than most people to come in contact with uranium and other hazardous radioactive wastes. People who work with medical X-rays or radioactive compounds are also at risk. Their practice is to wear lead-containing garments and follow recommended safety guidelines to protect themselves from unnecessary exposure.

HAZWOPER personnel must be able to recognize immediately the labels and placards that identify containers of radioactive materials such as uranium. You will require specialized, additional training to work with, and around, radioactive material. Pay close attention to the decontamination measures discussed in Section 7.

5.5.7   PCBs (Polychlorinated Biphenyls) and Yusho Poisoning

In 1968, more than 1000 people in western Japan became seriously ill. They suffered from fatigue, headache, coughs, numbness in the arms and legs, and unusual skin sores. Pregnant women later delivered babies with birth defects. These people had eaten food that had been cooked in contaminated rice oil. PCBs (polychlorinated biphenyls), used as heat transfer agents in the rice oil manufacturing process, had accidentally leaked from the process heating coils into the rice oil. Health experts refer to the resulting illnesses as "Yusho," which means "oil disease."

For years PCBs were used in many applications because of their outstanding stability and resistance to flammability. They were also a much safer choice as electrical insulators (called dielectrics) than the hydrocarbon oils they replaced. The dielectric and nonflammable properties of PCBs allowed for the manufacture of more compact, higher-power transformers. When scientists discovered that low levels of PCBs could kill fish and other wildlife and physicians discovered Yusho, PCB use was dramatically reduced.

However, by that time, PCBs were already leaking into the environment from waste disposal sites and other sources. One of the authors worked on PCB waste treatment recommendations for the EPA. It was found that PCBs were some of the most difficult organic compounds to destroy. Today, small amounts of these compounds can still be found in our air, water, soil, and some of the foods we eat.

It took decades for the toxic hazards of PCBs to be fully realized. In fact, during a World War II shortage of natural chicle for chewing gum, PCBs were suggested as a substitute. Fortunately, for some reason that suggestion was not acted upon! Waste PCBs were used in rural areas as road-oiling materials on dirt roads to allay dust.

Although PCBs have been virtually banned, our body burden of that family of chemicals remains fairly constant. Furthermore, enormous amounts of PCBs are in the environment. PCBs enter the food chain and into human diets primarily through consumption of certain freshwater fish. Infants as well as adults can be exposed to these chemicals. PCBs can be passed to babies through breast milk.

PCBs tend to be stored in the fatty parts of the body. They can be released upon starvation conditions and enter the blood stream. Numerous studies have been conducted over the years on the health effects of stored small amounts of PCBs in the blood, but the results have conflicted. The Japanese patients who had Yusho had damage to their nervous systems, eyes, and livers. It is now thought that many of the toxic effects of Yusho were due to other contaminants in the oil. Thus, it is still "up in the air" as to what kind of long-term toxic effects to expect from PCB exposure. What is known is that skin contact can produce chloracne, a serious form of dermatitis. And it is also known that PCBs, once in the body, remain stored in fatty tissue for a long time.

PCBs became widely used in the mid-20th century because of their outstanding ability to act as safe transformer fluids. No substitute has quite matched their ability to prevent dangerous transformer fires during electrical overloads. That wonderful stability in transformers also cautions us about an environmental hazard. PCBs are only very slowly decomposed by any environmental reactions. We can expect them to be around for a long while.

WARNING

HAZWOPER REMEDIATION WORKERS SHOULD BEWARE OF PCB CONTAMINATION WHEN DISCOVERING ANY DISCARDED TRANS-FORMERS THAT CONTAIN DIELECTRIC (INSULATING) FLUID. FOR THAT MATTER, ANY WASTE OIL FROM AN UNKNOWN SOURCE SHOULD BE SUSPECT UNTIL IT IS ANALYZED FOR PCBS.

5.5.8 Tobacco Smoke

A well-established example of common exposure to toxins is tobacco smoking, which involves the inhaling of carcinogens in the form of tars and other materials. That practice is a major cause of lung cancer. Among the wide range of toxins in tobacco smoke is carbon monoxide (CO). Hemoglobin has a 200 times greater affinity for CO than it does for oxygen. This leads to a lower blood oxygen level in smokers. Another problem with smoking is that toxins in the environment can get onto cigarettes (usually from the hands) and can get into the body by being inhaled during smoking.

According to the U.S. Department of Health and Human Services report *The Health Benefits of Smoking Cessation: A Report of the Surgeon General,* an estimated 470,000 premature smoking-related deaths occur per year in the United Stataes. These premature deaths represent more than those from alcohol-related causes, infectious diseases, exposures to other toxic agents, firearms/gunshot wounds, high-risk sexual behavior, and motor vehicle fatalities combined.

Lung cancer is the most common cause of cancer-related death in the United States. Cigarette smoking is the major risk factor for lung cancer. After 10 years of not smoking, a patient has a 30–50% decrease in the risk of developing lung cancer. After 15–20 years, this risk returns toward the incidence for nonsmokers. After 5 years of smoking cessation, the incidence of oral cancer is reduced 50%.

The risk of cardiovascular disease is reduced by 50% within the first year after the patient stops smoking and continues to decline toward that of a nonsmoker. After several years of abstinence, the risk of cardiovascular disease may return to that of a nonsmoker.

The good news is that the American Cancer Society reports that lifetime smokers who stop smoking for just 10 years have the same life expectancy as nonsmokers. That sounds like a good reason to kick the habit!

5.5.9 Asbestos, a Mineral

As previously stated, another cause of disease is the fibrous mineral asbestos. Asbestos is a group of six different minerals (actinolite, amosite, anthophyllite, chrysotile, crocidolite, and tremolite) that occur naturally in the environment. The most common type (approximately 95% of all asbestos used in the United States) is chrysotile or white asbestos. The other two common types are crocidolite (blue) and amosite (brown).

Asbestos, which does not burn, has been used for over 2000 years as an outstanding insulating material. It was routinely used in insulation for furnace, boiler, heating duct, steam and other hot fluid and gas piping. Asbestos-containing materials (ACMs) were also used as brake linings, fire blankets, and fireproof curtains and in kitchens as hot pads, mats, and oven mitts. However, our greatest chance of exposure comes from poorly maintained, friable, asbestos-containing building materials (ACBMs). (Friable means able to release fibers and to be crumbled with hand pressure when dry.)

Prior to the early 1970s, asbestos was used extensively in sprayed-on or troweled-on fireproofing and decorative and acoustical material. Other widespread uses were in vinyl asbestos floor tile (VAT) and sheet goods, roofing felt and shingles, and siding. It has been called the miracle mineral because it is able to protect property from fire almost universally.

The downside of its use is that breathing its dust causes several diseases. One example is mesothelioma, an unusual tumor of the linings of the chest and abdominal cavity. Others are asbestosis (a chronic fibrous disease of the lung), lung cancer, and emphysema. The possibility of gastrointestinal cancer is also increased.

⊙━π   **HAZMAT workers must be alert to potential asbestos contamination in pre-1970s buildings and discarded building materials.**

WARNING

WE STILL USE ASBESTOS IN THE UNITED STATES IN SOME PRODUCTS. BY THE END OF THE 20ᵀᴴ CENTURY, WE WERE STILL IMPORTING NEARLY 34,000 TONS PER YEAR. ASBESTOS IS STILL OPEN-PIT MINED IN LARGE QUANTITIES IN CANADA. ASBESTOS PRESENTS A CONTINUING HEALTH PROBLEM. HAZWOPER PERSONNEL ENGAGED IN REMEDIATION WORK SHOULD BE PREPARED FOR THE HAZARD OF ASBESTOS DUST. THIS IS ESPECIALLY TRUE WHEN WORKING DURING DEMOLITION OF PRE-1970s BUILDINGS. AVOID CREATING DUST WITH ANY SUSPECT INSULATION MATERIAL, FLOOR TILES, SIDING, FIREPROOFING, AND THE LIKE UNTIL IT IS PROVEN NOT TO BE ASBESTOS-CONTAINING. IF ASBESTOS IS PRESENT, SPECIAL OPERATING PROCEDURES AND PPE (SECTION 6) ARE REQUIRED.

5.5.10 Fungicides, Insecticides, and Rodenticides

A number of chemicals have been discovered to cause sterility, particularly in men. Two pesticides, kepone (also called chlordecone) and dibromochloropropane, have been cited. In addition, the experiences of workers at a (now-closed) kepone-producing plant in Hopewell, Virginia, have provided dramatic evidence that chemicals can cause neurological diseases. Kepone inhaled by workers who were not protected from exposure resulted in the gradual onset of nervous tremors, twitching and flickering eyes. In addition, the neighborhood surrounding the plant was found to be contaminated. Organophosphate and carbamate pesticides are widely used and very toxic to the nervous system.

5.5.11 Synthetic Organic Chemicals

O—π **All organic compounds are not found in nature. Many in the public consider organic to mean healthful, wholesome, and naturally occurring. Many organic compounds (chemicals) are man-made (synthetic).**

O—π **Chemists divide all chemicals into two large groups: organic and inorganic. Organic chemicals are characteristically those containing carbon and hydrogen. Many also contain nitrogen, sulfur, and other elements. They were originally discovered as products of living systems. Now there are large numbers of new synthetic organic chemicals produced yearly.**

We all are familiar with organic farming and organic gardening. Their organic produce is considered more healthful, due particularly to the lack of synthetic organic chemicals use in their growth. Typical synthetics are insecticides, fungicides, and growth stimulants.

O—π **Thousands of new man-made chemicals are coming on the scene yearly. We know relatively little about their toxic effects.**

Pesticides and PCBs are just two examples of the problems created by the increasing production and widespread use of synthetic organic chemicals. "Synthetic" means manufactured by chemical reaction from other starting materials. Generally it also means not found in nature.

However, chemists have learned to synthesize many natural compounds. The synthetics can be cheaper than the natural materials. Synthetic vanilla is about half the price of the natural product. Plastics, paints, polymers, pesticides, and synthetic fibers are all made from synthetic organic chemicals.

Organic compounds include almost all of the carbon-containing compounds. Among the relatively few common inorganics are carbon monoxide and dioxide and the carbonates. Calcium carbonate is a common antacid.

5.6 CANCER AND TOXIC MATERIALS

O—π **In some cases it takes 20 to 30 years for cancer or other diseases to develop after exposure to a toxic material. This is called the latency period.**

There is increasing evidence that many uncontrolled chronic diseases are caused in part by exposure to toxic materials in the environment. Studies and discussions of chronic diseases frequently focus more on the man-made or synthetic chemicals (and their wastes and by-products) than on naturally occurring chemicals. The chronic disease usually cited is the universally feared cancer.

Cancer is unquestionably an important disease. Approximately 20% of all people in the United States die of cancer. We concentrate on cancer because we know much more about it than any other chronic disease. We have better records about it, and it is better diagnosed and followed than many other important chronic diseases.

Basically, there are two ways to determine the causes of cancer. You can study people or study laboratory animals. Both techniques provide valuable information. Also, both techniques have strengths and weaknesses. It is important to understand those strengths and weaknesses.

5.7 EPIDEMIOLOGY, RELATING DISEASES TO CAUSES

   Epidemiology is the study of the occurrence of disease in people.

Epidemiology is the most common method of determining a cause of cancer. This technique is easily accepted and certainly the most easily understood. It associates exposure to a specific material or chemical compound with a large number of people either dead or ill suffering from an identified disease. This seems like it should be easy to do. But even the link between smoking and disease took a long time to prove by epidemiologic means. Investigators examine groups of people. The groups must be large enough for statistical methods to be applied to the study data. Enough must be known about the exposures of the groups and the other factors, such as smoking, that can cause the same type of illness being studied.

Some epidemiological studies, conducted without the aid of statistical analysis, are very old. We have already cited the 2000-year-old observations of lead and asbestos workers' diseases. The report, in 1775, by Sir Percival Pott in England that chimney sweeps had a very high incidence of scrotal cancer may have been the first epidemiological study in more recent times. Chimneysweeps were poorly fed in order to keep them thin enough to climb inside flues. They often worked practically naked. It is easy to understand how they were directly exposed.

However, there can be many problems with epidemiology. Usually people do not know for certain the materials they have been exposed to or the amount of their exposure. Further, people are most often exposed to mixtures of chemicals and compounds. That makes the individual constituent determination more difficult. To produce a good epidemiological study, it is critical to know the type of material exposure and its quantity—that is, not only the exposure of the workers producing a product, but also the exposure of consumers. Another problem is that people find it very difficult to remember the important exposures of 5, 10, or 15 years ago as well as the exposures of today or yesterday. And someone with a disease may be more likely to recall an exposure years ago than someone who is well (this is called recall bias).

Epidemiology suffers from the fact that it cannot detect small differences in the rates of common diseases. Although a common disease may affect a very large number of people, relatively small differences cannot be picked up by epidemiology. For example, it would be very difficult to determine epidemiologically if a chemical or other material increased the rate of a common disease.

Perhaps the most significant problem is that epidemiology is entirely an after-the-fact science. Diseases can be related to causes only after people have experienced exposures of sufficient intensity and duration to produce illness and death. This delay is less important when the effect, such as an adverse drug reaction, happens very rapidly. However, chronic diseases, particularly cancer and asbestosis, have a latent, or silent, period of decades. Most cancers do not develop until 20 or 30 years after exposure has occurred.

It may be necessary for an entire generation of people to be exposed to a suspect material before epidemiological studies can prove that the material does cause cancer or some other chronic disease.

Epidemiological studies are an imperfect tool. But they are the best tool we have for learning about the effects of exposures in people. Therefore, most health regulations today are based on epidemiological studies.

5.8 LABORATORY ANIMAL STUDIES

Mice and other laboratory animals are used in tests to determine the toxicity of materials. Animal tests are fairly good, but not perfect, predictors of the toxic effects to be expected in humans.

Use of laboratory animal studies as a method to identify carcinogens is less than 100 years old. These studies have been much improved in recent decades. Today they are better constructed, performed, and evaluated. A good animal study is very complex. It is not the simple observation of especially-bred rats in a box. The critical question to ask about a laboratory animal study is its predictability. Do results of laboratory animal studies predict for those effects in a human population?

It is much easier to study the effects of a specific chemical compound with a known formula, like C_6H_6 for benzene, than those of a complex material such as tobacco smoke. Tobacco smoke can consist of many individual chemical compounds that could be suspected as hazardous, and it can differ from brand to brand.

The International Agency for Research on Cancer (IARC) is part of the World Health Organization. It assembles expert groups of independent scientists from all over the world. Their task is to determine whether there is adequate evidence to state that a chemical is carcinogenic (causes cancer). This agency publishes a list of a few dozen chemicals that are carcinogenic in man (Group 1 chemicals). Nearly all of the chemicals on this list also are carcinogenic in laboratory animals. This list strongly suggests that chemicals that are carcinogenic in man will have been found to be carcinogenic in animals.

Another important issue focuses on the amount of a chemical needed to cause cancer. Are the amounts or doses of chemicals that cause cancer in man the same as those that cause cancer in the laboratory animals? A study committee of the National Academy of Sciences National Research Council carefully reviewed this problem. The committee compared the available evidence for amounts of chemicals needed to cause human cancer with laboratory data about cancer. This was in the animal species most sensitive to a particular chemical. Three out of the six situations studied showed a direct relationship. This relationship exists between the amount of chemical exposure and cancer in animals and in man.

A National Institute of Health researcher has said that it is very difficult to encourage a mouse or other lab animal to smoke cigarettes. On the other hand, it is not at all difficult to persuade people to smoke. Still, there seems to be a very close relationship between the number of cigarettes smoked and the production of cancer in both people and animals.

In other tests, animals have been much more susceptible to cancer than humans. Animal tests are not perfect predictors of the human response, but they can have great value. Vinyl chloride was discovered to cause cancer in animal studies before like effects were observed in factory workers in Louisville, Kentucky. Aflatoxins were observed to cause cancer first in trout, then in laboratory animals, and finally in human populations by use of epidemiological studies in Africa.

Another human and animal toxic material example is bischloromethyl ether, a chemical byproduct of a manufacturing process. An epidemic of lung cancer had been noticed in a small factory. Laboratory animal tests pinpointed bischloromethyl ether as the cause. The manufacturing process was closed and the incidence of lung cancer among the factory workers decreased.

5.9   FACTORS CONSIDERED IN DESIGNING ANIMAL TESTS

There are problems in understanding and using laboratory animal tests. Much of the public, and some scientists, are reluctant to put faith in laboratory animal tests. These people feel there is too big a gulf between humans and mice. There is empirical evidence from a variety of studies that animals do predict for man, not only for cancer but also for other forms of disease and quantities of toxicity.

Mammalian species, from tiny mice to enormous elephants, are very similar in their biological functions, physiology, and biochemistry. The stomachs of swine respond to toxins much more as the stomachs of people do than do the stomachs of rats or mice. Such extensive similarities have permitted us to learn much of what we know about human physiology and biochemistry by the convenient and effective study of laboratory animals.

5.10 WHY LARGE DOSES ARE USED IN ANIMAL TESTS

Many people who are unfamiliar with this type of research question the need to administer large doses of chemicals to animals in laboratory tests. The amount of chemical used is related primarily to what is referred to as the statistical power of the animal tests. It is extremely difficult to detect anything less than a very strong carcinogen, unless researchers use either an enormous number of animals, hundreds of thousands, or larger doses and fewer animals. Negative test results from small numbers of animals simply mean that a test chemical is probably not an extremely powerful carcinogen.

For many generations a special strain of lab mice has been bred, and in-bred, only for lab use. Animals are like humans in one respect that can confuse and complicate test results: nearly all strains have some incidence of cancer in the general population, whether being exposed to test materials or not. First an animal test is designed and a test chemical is administered to the animals. Then the scientists must decide which of the resulting cancers were caused by the chemical and which might have been present in the normal course of the animal's life. This decision poses severe statistical problems. The chance of detecting a twofold incidence of cancer using 50 rats is just 2 in 1000. That is an almost infinitesimal chance of detecting a cancer-causing agent. When the number of rats is increased to 100, there still is only a 2 in 100 chance of detecting a doubling of the background rate of cancer due to a test chemical. It is unreasonable to use thousands upon thousands of animals.

The only other alternative is to use larger doses of the chemical being studied. These doses must not be life-threatening in themselves. They must not cause other toxicities in the animal population. They are simply larger doses. There are other reasons for using large doses. Animals excrete chemicals faster than humans do. However, the main reason large doses are necessary and valid is the statistical requirements cited above.

We do this kind of exaggerated testing routinely in research. To find out how well materials will age, we conduct accelerated aging studies. For example, we know that heat causes loss of desirable properties in many consumer materials over time. So we expose various formulations of the materials to elevated temperatures for periods far shorter than their expected lifetime. Then we choose the formulation that best withstands those tests.

Many people in the general population have expressed the belief that all chemicals cause cancer. This is not true. In the initial large-scale National Cancer Institute study, scientists selected 120 compounds to study that were suspected of being carcinogenic. Of these 120 compounds, less than 10% were found to cause cancer.

In addition to using animals for studying cancer, scientists at the National Institute of Environmental Health Sciences and other institutions are working on animal tests for other chronic diseases. However, these tests are not as well developed as those for cancer. It will take a number of years before the results of these tests will have validity similar to that of the cancer tests.

5.11 FACTORS INFLUENCING TOXICITY

Many factors affect the reaction of an organism to a toxic substance. The specific response that is brought about by a given dose of a substance can vary greatly. Responses depend upon the species being tested and variations that occur among individuals of the same species. In this book we are focusing on the human organism while gaining valuable insights from animal studies.

5.11.1 Personal Characteristics That Influence Substance Interactions

Age and Maturity.   Infants and children are often more sensitive to toxic action than are younger adults. The elderly have diminished physiological capabilities to deal with toxic insult. These two extreme age groups appear to be more susceptible than others to toxic effects at relatively lower doses.

Gender and Hormonal Status.   Some substances may be more toxic to one gender than the other. Certain substances can affect the reproductive system of either the male or female. Additionally, since women have a larger percentage of body fat than men, they may accumulate more fat-soluble substances. DDT and PCBs, for example, are more fat-soluble than many other toxic substances. Other variations in response have also been shown to be related to physiological differences between males and females. For example, pregnant women can often suffer far greater toxic insult. This is due to their rapidly changing hormonal status and their requirement to supply necessary materials such as calcium to their growing baby.

Genetic Makeup.   Genetic factors influence individual responses to toxins. If the necessary physiological processes are diminished or defective, the natural body defenses are impaired. The decoding of the human genome and the progress of understanding the functions of its components will lead to revolutionary new treatments for toxic insults to humans.

Physical Health.   Persons in poor health or who do not exercise regularly or those with poor eating habits or diets are generally more susceptible to toxic damage. This is largely due to the body's decreased ability to cleanse itself of toxins. But the damage can also be due to the body absorbing the toxin in place of a missing nutrient. An example of this is the absorption of lead into the bodies of exposed individuals whose diet is deficient in potassium and calcium and high in saturated fat. This should be a wakeup call to all HAZ-WOPER personnel—yet another reason to maintain a healthy diet.

5.12   ACUTE AND CHRONIC EXPOSURES; DURATION, FREQUENCY, AND DOSAGE LEVEL

Type and severity of toxic effects vary depending on the duration of dosage and how often the dose is received.

Acute exposures usually involve single doses at high levels of relatively short duration.

Acute exposures are usually single doses at high levels of relatively short duration. Short-term effects have a relatively quick onset, usually minutes to possibly a few days after brief or acute exposures to relatively high concentrations of substances. The effect may be local or systemic (affecting the body as a whole). Local effects occur at the point of contact between the toxin and the body. This point is usually the lungs, skin, or eyes.

Chronic exposures usually involve frequently repeated doses, sometimes at relatively low levels, over longer periods of time.

Chronic exposures are frequent doses, sometimes at relatively low levels over longer periods of time. Long-term (chronic) effects are those with a long period of time between initial exposure and injury. This is called the latency period. These effects may occur after apparent recovery from acute exposure, or they may occur as a result of repeated or chronic exposures to low concentrations of substances over a period of years. In the case of cancer, it can take decades for effects to occur. In the case of asbestos exposure, the latency period is from 15 to 30 years.

Chronic exposures can also lead to hypersensitivity to certain materials. For instance, continued dermal exposures to some solvents and oils can appear to have no effect for long periods of time but can lead to hypersensitivity (allergic reaction) to those same materials. Without warning, you might start to develop rashes upon contact, when that had never happened before. Here is further evidence that you should never allow the materials you work with to contact your skin or respiratory tract. Wear your PPE!

WARNING

OUR EXPERIENCE SHOWS THAT LATENCY CAN GIVE HAZWOPER WORKERS A FALSE SENSE OF SECURITY. IF THEY DO NOT SUFFER IMMEDIATE SYMPTOMS (FROM SECONDS TO A FEW DAYS), THEY OFTEN FIGURE THEY ARE OUT OF THE WOODS. HOWEVER, A DELAYED REACTION IS ALWAYS POSSIBLE. IT CAN OCCUR WITHIN A COUPLE OF HOURS, DAYS, WEEKS, OR YEARS. THIS CAN BE A DANGEROUS TRAP! THIS OFTEN CAUSES WORKERS TO DISREGARD PPE REQUIREMENTS. KNOW SYMPTOMS OF EXPOSURE, BUT ALSO KNOW THE LONG-TERM HEALTH EFFECTS!

5.12.1 Laboratory and Clinical Exposure Testing

In laboratory toxicology studies, exposure of subject animals is achieved by giving carefully administered dosages of test materials. To test the effects of inhalation exposure, various concentrations of a material are added to the breathing atmosphere of the subject. To test the effects of absorption exposure, various concentrations of a material are placed on a patch of measured area on the skin of the subject. To test the effects of ingestion or injection, various weights of the material are used. Then the weights given, divided by the weight of the subject, are recorded.

In clinical studies of pharmaceuticals, researchers administer dosages (exposures) to both animals and humans. Then they measure the responses to those exposures. Read those detailed instructions that come with any oral prescriptions you take. You will see that the drug company researchers have used ingestion tests on both animals and humans. Then they determined what dose an average-weight adult should take, and how frequently.

5.13 THE 'DOSE-RESPONSE' RELATIONSHIP

○—🔑 **The 'dose-response' relationship means that the higher the dose over the same time period, the greater the response.**

If a dose is administered slowly so that the rate of elimination or the rate of detoxification keeps pace with intake, it is possible that no toxic response will occur. However, the same dose could produce a toxic response when administered rapidly.

In general, a given amount of a toxic agent administered over a given period of time will cause a given type and intensity of response. The higher the dose over the same period will cause a greater response. This 'dose-response' relationship is a fundamental concept in toxicology. It is the basis for measurement of the relative harmfulness of a substance.

A 'dose-response' relationship is a consistent mathematical agreement between the number of individuals responding to a given dose over an exposure period. The biological response to dosage must be consistent and explainable to a satisfactory degree. Table 5.1 shows a range of toxic materials. The extremely toxic substances are a thousand times more toxic than the slightly toxic substances. LD_{50} is the dose that is lethal for 50% of the animals in the study. (Expressed as amount of material per kilogram of body weight.)

Please understand that many materials may have a 'dose-response' relationship on the overall population while your body has a different response. In numerous studies people exposed only once to asbestos, for instance, died of asbestosis, while others were exposed all their lives and died of other causes in old age. People have lived to their 90s as lifetime smokers, while others have died of lung cancer in their 30s. We all vary in our dose-response relationships. That is why statistical analysis is so important. It shows us that we can best report probabilities, not precise outcomes.

TABLE 5.1 Toxicity Ratings and Examples of Toxic Substance Doses

Toxicity Rating	Oral Acute LD_{50} for Rats
Extremely toxic	1 mg/kg or less (dioxin, tributyltin)
Highly toxic	1–50 mg/kg (strychnine)
Moderately toxic	50–500 mg/kg (DDT)
Slightly toxic	0.5–5 g/kg (morphine)
Borderline toxic	5–15 g/kg (ethyl alcohol)

5.14 DOSAGE AND EXPOSURE TERMS

Dosage is the act of causing an exposure. Dosages are administered in laboratory tests to tell us what health effects can be expected from exposures during daily activities.

- *Inhalation (Inh)* is rated as the volume of substance inhaled per unit volume of air. It is usually stated as μl of vapor or gas per liter of air. This is the same as parts-per-million-by volume (ppmv) of the substance in air. Particulates and liquids are usually given as milligrams of material per cubic meter of air (mg/m^3) or micrograms per cubic meter of air ($\mu g/m^3$). Fiber concentrations, such as those of asbestos, are expressed in terms of fibers-per-cubic-centimeter (f/cc) of inhaled air.
- *Absorption (Abs)* is the quantity absorbed per unit of skin (dermal) surface, usually expressed as milligrams per square centimeter (mg/cm^2).
- *Ingestion or injection* is the quantity ingested (eaten or swallowed) or injected per unit weight of the test subject per day. This is usually expressed as milligrams per kilogram of body weight (mg/kg). These dose terms are abbreviated in some references as Ing or Inj, respectively.

Where dosage results in the death of the subjects, two measurements are used.

An LD_{50} dose of a toxin, by any route other than inhalation, is expected to kill one-half of the experimental animals given that dose.

- *Lethal dose fifty* (LD_{50}): A calculated dose of a substance, which is expected to cause the death of 50% of an entire defined experimental animal population. It is determined from the exposure to the animal by any route other than inhalation.

An LC_{50} dose of a toxin is expected to kill one-half of the experimental animals that breathe that dose.

- *Lethal concentration fifty* (LC_{50}): A calculated concentration of a substance inhaled in air, exposure to which for a specified length of time is expected to cause the death of 50% of an entire, defined, experimental animal population.

A TD_{50} is a numerical description of carcinogenic potency.

- *Tumor development fifty* (TD_{50}): A numerical description of carcinogenic potency. A tumor often occurs in control animals. TD_{50} is defined as a daily dose-rate in mg/kg of body weight, one that is administered chronically for the standard lifespan of the species. The result is that that dose will halve the probability of remaining tumor-free throughout that period.

5.15   ROUTES OF EXPOSURE AND TARGET ORGANS

O—π **The four ways toxins can enter your body and harm your health are by breathing (inhalation), getting them on your skin or eyes (absorption), swallowing (ingestion), or receiving a puncture wound from a toxin-covered sharp object (injection).**

Biological results can be different for the same dose, depending on whether the substance is inhaled, ingested, absorbed (after application to the skin), or injected. Natural barriers impede the intake and distribution of material once it is in the body. These barriers can lessen the toxic effects of the same dose of a substance given another way. The effectiveness of these barriers is partially dependent upon the route of entry of the substance.

O—π **Inhalation (breathing) is the most frequent entry route of toxins into the body.**

5.15.1 Inhalation

Most substances enter the body through inhalation in the form of vapors, gases, mists, or particulates. Once inhaled, substances are either exhaled or deposited in the respiratory tract. If deposited, damage can occur through direct contact with tissue. Then the substance may diffuse into the blood through the lung-blood interface.

Upon contact with tissue in the upper respiratory tract or lungs, substances may cause health effects ranging from simple irritation to severe tissue destruction. Substances absorbed into the blood are transported to organs, tissue, or bones and teeth, where they may accumulate. Health effects can then occur in those areas that are sensitive to the substance.

5.15.2 Exposed Body Tissue Absorption

Skin (dermal) contact can have minor effects such as redness or mild dermatitis. More severe effects are destruction of skin tissue and other debilitating conditions. Many substances can also cross the skin barrier and be absorbed into the blood system. Once absorbed, they may produce damage to internal organs and systems.

The eyes are particularly sensitive to many substances, and even a short toxic exposure can cause severe effects. Substances can also be absorbed through the eyes and transported to other parts of the body.

Contact lenses can absorb toxins and then slowly release them into the eyes. This is a major reason why contact lenses should not be worn when you are working around hazardous materials.

5.15.3 Ingestion

Substances that inadvertently get into the mouth and are swallowed do not generally harm the gastrointestinal tract itself unless they are irritating or corrosive.

Substances that are insoluble in the fluids of the gastrointestinal tract (stomach small, and large intestines) are generally excreted. Soluble substances are absorbed through the lining of the gastrointestinal tract. They are then transported by the blood to internal sites where they can cause damage.

5.15.4   Injection

⊙━𝓽   **Injection is the most direct entry route of toxins into the body.**

Substances may enter the body if the skin is penetrated or punctured by contaminated objects. This is the most direct route of exposure. The contaminant is then introduced directly into the bloodstream. Prime injection hazards on the job site are broken glass, nails from broken pallets, and sharp, rusty metal objects such as deteriorated drums, and broken glass.

WARNING

WHEN DEALING WITH MEDICAL WASTE, HAZWOPER WORKERS MUST BE ESPECIALLY CAREFUL. THIS WASTE IS OFTEN CONTAMINATED WITH BLOOD-BORNE PATHOGENS. THESE ARE MICROORGANISMS THAT ARE PRESENT IN HUMAN BLOOD AND CAN CAUSE DISEASE IN HUMANS. OSHA'S REGULATION COVERING THE PROTECTION OF WORKERS FROM BLOOD-BORNE PATHOGENS IS 29 CFR 1910.1030.

Systemic effects are those that occur if the toxic substance has been absorbed into the body from its initial contact point and then transported to other parts of the body. This can cause adverse effects in susceptible organs. Many toxic substances can cause both local and systemic effects.

5.16   FATES OF TOXIC MATERIALS AFTER ENTERING THE BODY

⊙━𝓽   **Toxins entering the body can be metabolized, stored, or excreted.**

Once a substance has entered the body, it has three fates: it can be metabolized, stored, or excreted. All three of these can vary drastically from person to person, depending on age, physical health, sex, ethnic origin, and medical history.

- *Metabolism:* This is the chemical process by which absorbed matter is broken down into simpler substances. This is what happens as the human body processes food. If a non-beneficial substance or even a large dose of a beneficial substance is metabolized, it may cause harm to the body. It may be broken down into more toxic or less toxic constituents, thereby causing more or less harm to the rest of the system.

- *Storage or bioaccumulation:* Bioaccumulation is the chemical storage buildup in the body of foreign substances. Some chemical elements, such as mercury, are stored in the body's fatty tissue. Other elements, such as lead, accumulate in the bones and teeth. Your blood is made in the marrow of your bones. Toxic bioaccumulated elements, such as lead and mercury, can have serious health effects on humans.

  Other materials, such as asbestos fibers, can be physically stored in the body. The fibers can become trapped by piercing the tissue in the air sacs of the lungs, called alveoli, and stay there permanently. These stored fibers can lead to the body's own defense mechanisms creating scar tissue that affects the oxygen transfer to the blood and CO_2 transfer from the blood. This scarring is asbestosis.
  Otherwise beneficial substances such as fat-soluble vitamin A can be stored in the body and, if taken in very large doses, can be harmful or even fatal.

- *Excretion* is the body's natural ability to eliminate waste through urination and defecation. It is rare for someone to overdose on water-soluble vitamin C. This is because it is excreted soon after it is taken and the body absorbs only what it needs.

5.17   SPECIFIC INHALATION EFFECTS ON THE RESPIRATORY TRACT

○━ℼ **Toxins in breathing air can produce irritation, cell death and edema (fluid buildup) of the lungs, fibrosis or scarring of the lungs, allergic reactions, and cancer.**

The respiratory system is the only organ system with vital functional elements in constant, direct contact with the environment. The lungs also have the largest exposed surface area of any organ, approximately 70 to 100 m². Compare this to 2 m² for the skin and 10 m² for the digestive system.

Many substances used or produced in industry can cause acute or chronic diseases of the respiratory tract when they are inhaled. These substances can be classified according to the way in which they affect the respiratory tract.

Note

The effects described below have been observed for the population in general. This does not take into account an individual's hypersensitivity to certain materials.

- *Asphyxiants, simple and chemical,* are gases or vapors that cause unconsciousness or death due to insufficient oxygen being delivered to the body. *Simple asphyxiation* is suffocation, which occurs if there is not enough (at least 19.5%) oxygen in the atmosphere you are breathing. *Chemical asphyxiation* occurs when there is a chemical in the breathing atmosphere that interferes with, or stops, the proper absorption of oxygen by the blood or body tissues. See Section 3.1.2 for a detailed description.

WARNING

OSHA HAS INVESTIGATED A NUMBER OF DEATHS INVOLVING NITROGEN ASPHYXIATION. IN THOSE CASES, WORKERS ATTACHED BREATHING AIR LINES TO FACTORY AIR CONNECTIONS. THESE AIR LINES ARE USED FOR PNEUMATIC TOOL OPERATION. WHILE THE FACTORY AIR SYSTEM WAS BEING REPAIRED THE LINES HAD BEEN RENDERED INERT WITH NITROGEN SO THAT HOT WORK COULD BE PERFORMED. DEATH FROM NITROGEN ASPHYXIATION CAN OCCUR WITHIN SECONDS!

NEVER ATTACH BREATHING AIR LINES TO ANY OTHER SOURCE THAN A BREATHING AIR COMPRESSOR OR CASCADE SYSTEM!

Nitrous oxide ("laughing gas"), the breathing of which some people find entertaining, has been used as a medical analgesic for over 100 years. When it is not administered by an anesthetist, however, anyone breathing it takes the risk of asphyxiation due to lack of oxygen. Further, the nitrous oxide used as an analgesic is of a tested medical grade. The grade of "recreational" nitrous oxide is unknown.

- *Toxic or chemical asphyxiants* are gases that prevent body tissues from getting enough oxygen. Examples are carbon monoxide and hydrogen cyanide. Carbon monoxide binds to hemoglobin 200 times more readily than oxygen. Hydrogen cyanide prevents the transfer of oxygen from blood to tissues by inhibiting the necessary transfer enzymes.
- *Irritants* are substances that irritate the air passages. Constriction of the airways occurs and may lead to edema (liquid in the lungs) and infection. Examples are hydrogen fluoride,

chlorine, hydrogen chloride, and ammonia. You get some appreciation for these kinds of irritants when you sniff household bleach or ammonia.

- *Necrosis producers* are substances that result in cell death and edema. Examples are ozone and nitrogen oxides.

- *Fibrosis producers* are substances that produce scarring of the lungs (pulmonary fibrosis), which, if massive, blocks oxygen transfer to the blood, decreases lung capacity, and can lead to death. Examples are silicates (sand used in glass manufacturing), asbestos dust, and beryllium dust. Although these substances act in different ways to produce the fibrosis, the injury is equally harmful.

- *Allergens* are substances that induce an allergic response characterized by asthma or other allergic responses. Examples are ragweed, isocyanates (Crazy Glue®), and sulfur dioxide.

- *Carcinogens* are substances that are known to cause cancer. Examples are cigarette smoke, benzene, and asbestos.

O—π   Toxins entering your system can affect a variety of organs and systems.

5.18   TOXIC EFFECTS ON THE LIVER DUE TO INHALATION, INGESTION, ABSORPTION, OR INJECTION

O—π   Substances that poison the liver are called hepatotoxins.

The liver filters the body's blood. Substances that poison the liver are called *hepatotoxins*. Effects on the liver from contaminants can be necrosis, or cell death (for example, from chlorinated solvents); cirrhosis (for example, from alcohol); and carcinomas (from carcinogens).

5.19   TOXIC EFFECTS ON THE KIDNEYS DUE TO INHALATION, INGESTION, ABSORPTION, OR INJECTION

O—π   Substances that are toxic to the kidneys are called nephrotoxins.

The kidneys filter the body's liquid waste and make urine. Substances that are toxic to the kidneys are called *nephrotoxins*. The kidneys are susceptible to toxic agents for several reasons:

- The kidneys constitute 1% of the body's weight but receive 20–25% of the blood flow. Therefore, large amounts of circulating toxins reach the kidneys quickly.

- Changes in kidney pH may increase passive diffusion and thus cellular concentrations of toxins.

- Active secretion processes may concentrate toxins.

5.20   TOXIC EFFECTS ON THE BLOOD DUE TO INHALATION, INGESTION, ABSORPTION, OR INJECTION

The blood system can be damaged by substances that affect blood cell production. This can occur in the bone marrow where blood and its components are produced. These blood components include the red blood cells, the white blood cells (which are responsible for defending the body from infection and cancer), and the platelets (which help the blood clot). Some compounds affect the oxygen carrying capabilities of red blood cells. The notable example is carbon monoxide, which combines with hemoglobin to form carboxyhemoglobin. As stated previously, hemoglobin has an affinity for carbon monoxide 200 times greater than it has for oxygen.

5.21 TOXIC EFFECTS ON EXPOSED BODY TISSUES DUE TO ABSORPTION

In our discussion of exposed body tissues, we focus attention on the skin and the eyes. The skin, in terms of weight, is the largest single organ of the body. (The lungs have the largest surface area of any organ.) Providing a barrier between the environment and the organs, except for the lungs and eyes, the skin is a defense against many substances.

The skin consists of the epidermis (outer layer) and the dermis (inner layer). In the dermis are sweat glands and ducts, sebaceous glands, connective tissue, fat, hair follicles, and blood vessels. Hair follicles and sweat glands penetrate both the epidermis and dermis. Substances can penetrate through the sweat glands, sebaceous glands, or hair follicles. An extreme example is the organic solvent dimethylsulfoxide (DMSO). Within minutes of DMSO contacting your skin, you sense a garlic-like odor, caused by DMSO, in your breath.

WARNING

SUBSTANCES THAT ARE ABSORBED AT AN EXTREMELY HIGH RATE, SUCH AS DMSO, CAN CAUSE ANOTHER HAZARD. DMSO, USED AS A HIGHLY EFFECTIVE INDUSTRIAL SOLVENT, IS NOT CONSIDERED TOXIC. HOWEVER, UPON SKIN CONTACT, IT CAN CARRY IMPURITIES OR TOXINS THAT ARE WITH IT ON THE SKIN INTO THE BLOODSTREAM RAPIDLY.

An example of a chemical that is extremely hazardous when deposited as a liquid on the skin is 2,4-dichlorophenol (2,4-DCP), which is the chief raw material in the production of 2,4-dichloro phenoxy acetic acid (2,4-D). 2,4-D is one of the most widely used herbicides in the world.

A chemical plant worker was steam-cleaning a process pump that was shut down. The heat melted some solid 2,4-DCP. The molten chemical sprayed on the worker. He ran to a locker room shower and collapsed. About an hour later, the man, 29, was pronounced dead from acute 2,4-DCP poisoning. Prior to this death in 1998, the extreme toxicity of this chemical when absorbed through the skin was apparently largely unknown. Since that time warnings have been sent throughout industry and the scientific communities. As a result, reports have turned up on four others during the last two decades who have died quickly after 2,4-DCP exposure.

What can we learn from tragedies such as this? Keep foreign materials off of your skin and out of your eyes! Wear the proper personal protective equipment (see Section 6). Keep a safe distance from operations where splashing or dumping of materials might cause them to contact your skin or eyes.

The eyes are affected by many of the same substances that affect skin, but the eyes are much more sensitive. Many materials, such as particulates, gases, and vapors, can damage the eyes by direct contact. Smog, caused by ozone reactions with motor vehicle exhausts and other combustion gases, produces a stinging effect on the eyes before the skin appears to be affected.

5.22 TOXIC EFFECTS ON THE REPRODUCTIVE SYSTEM DUE TO INHALATION, INGESTION, ABSORPTION, OR INJECTION

Years of occupational health research have shown that toxins can interfere with the reproductive capabilities of both sexes. Toxins can cause sterility, infertility, birth defects, spontaneous abortions, abnormal sperm, and low sperm count and can affect hormonal activity in humans. Examples include lead dust and dioxin. There are two kinds of reproductive toxins.

- *Teratogens* are any agent or process that interferes with the normal development of the embryo or fetus, causing physical abnormalities.
- *Mutagens* are any chemical or physical agent that increases a mutation rate of the genetic code.

5.23 *MEASUREMENTS OF HUMAN EXPOSURE TO TOXIC SUBSTANCES*

◎—π **NIOSH provides research and testing data to assist OSHA in making toxic materials exposure regulations for worker safety.**

There are three primary organizations that develop exposure guidelines: the Occupational Safety and Health Administration (OSHA), the National Institute for Occupational Safety and Health (NIOSH), and the American Conference of Governmental Industrial Hygienists (ACGIH). Other organizations also develop exposure guidelines, such as the Workplace Environmental Exposure Level (WEEL) Guides series published by the American Industrial Hygiene Association (AIHA). Other countries and organizations also publish guidelines such as the DFG MAK in Germany. However, for the purposes of HAZWOPER training, we will concentrate on the first three.

◎—π **The only federal legally enforceable exposure levels for U.S. workers are the PELs published by OSHA.**

Note

All three organizations develop their guidelines independently. It is not uncommon to look up a substance in each organization's publications and find different levels listed. The ONLY LEGALLY ENFORCEABLE FEDERAL exposure levels are the Permissible Exposure Limits (PELs) published by OSHA.

Exposure levels can change from year-to-year as more research is completed and more data is collected. The allowable exposure levels usually go down! We strongly recommend that you follow the lowest published allowable exposure level!

5.23.1 OSHA's PELs, TWAs, NIOSH's RELs, and ACGIH's TLVs®

◎—π **OSHA has set Permissible Exposure Limits (PELs) for thousands of toxins. These limits, usually expressed as 8-hour time weighted averages (TWAs), limit the average exposure of workers during an 8-hour day and a 40-hour week.**

OSHA's Permissible Exposure Limits (PELs) are legally enforceable. They are found in 29 CFR 1910.1000 (and following) Subpart Z, Toxic and Hazardous Substances. This is also referred to as The Z List and is included as Appendix G on the Companion CD.

◎—π **NIOSH publishes Recommended Exposure Limits (RELs).**

NIOSH, a branch of the CDC (Centers for Disease Control and Prevention) publishes Recommended Exposure Limits (RELs). NIOSH was formed at the same time as OSHA to act as a research organization. It is charged with making recommendations for new standards and revising old ones. NIOSH developed RELs, which are used to develop new OSHA standards. RELs are defined in the NIOSH Recommendations for Occupational Health Standards. They are the TWA concentrations for up to a 10-hour workday during a 40-hour workweek. OSHA has not adopted many RELs as PELs. Therefore, they hold the same advisory status as the exposure guidelines of the ACGIH and other groups.

Note

Many of these exposure limits can be found in the NIOSH Pocket Guide to Chemical Hazards, available on the Companion CD with the permission of the National Institute of Occupational Safety and Health. This is an extremely valuable resource, and we encourage all interested parties to take advantage of it.

WARNING

ALTHOUGH THE NIOSH POCKET GUIDE IS AN EXTREMELY VALUABLE TOOL WITH COMPREHENSIVE INFORMATION, IT DOES NOT CONTAIN ALL THE CHEMICALS AND COMPOUNDS THAT MIGHT BE ENCOUNTERED AT THE WORK SITE. IT IS NOT INTENDED TO TAKE THE PLACE OF A COMPLETE INVESTIGATION OF THE INDIVIDUAL CONTAMINANTS AT THE SITE. THE OSHA Z-LISTS (APPENDIX G ON THE COMPANION CD) ARE COMPREHENSIVE LISTS OF ALL THE REGULATED AIR CONTAMINANTS.

ACGIH publishes Threshold Limit Values (TLVs®).

ACGIH publishes Threshold Limit Values (TLVs®). These values are intended for use in the practice of industrial hygiene as guidelines or recommendations in the control of potential health hazards and for no other use. These limits are not fine lines between safe and dangerous concentration. They are not a relative index of toxicity. They should not be used by anyone untrained in the discipline of industrial hygiene.

5.24   MEASUREMENTS BASED ON DURATION OF EXPOSURE

5.24.1   Time Weighted Averages (TWAs)

A time weighted average exposure is determined by averaging the concentrations of the exposure with each concentration weighted based on the duration of the exposure. For example, for a person exposed to 0.1 mg/m³ for 6 hours and 0.2 mg/m³ for 2 hours, the 8-hour TWA is $(0.1 \times 6 + 0.2 \times 2)/8$, which equals 0.125 mg/m³.

A TWA can be the average concentration over any period of time. Most TWAs are the average concentration of a substance most workers can be exposed to during a 40-hour week and a normal 8-hour workday without showing any toxic effects. NIOSH TWA recommendations, on the other hand, can also be based on exposures up to 10 hours.

OSHA Short Term Exposure Limits (STELs) put a cap on exposure of 15 or 30 minutes on some materials. The STEL can be 10 times the PEL but is only used in emergency situations, and the worker is allowed no further exposure to that material for that workday.

5.24.2   Short Term Exposure Limits (STELs)

The STEL is also referred to as an excursion limit. Because the excursions allowed by the TWA could involve very high concentrations (but still within the allowable average) and cause an adverse effect, some limits were set to these excursions. In 1976, ACGIH added STELs to its Threshold Limit Values (TLVs®). The STEL is a 15-minute TWA exposure. Because the excursions are calculated into the 8-hour TWA, the exposure must be limited to avoid exceeding the daily TWA.

WARNING

MANAGEMENT MUST PROVIDE ADEQUATE ENGINEERING CONTROLS AND PROPER PERSONAL PROTECTIVE EQUIPMENT. MANAGEMENT MAY NOT USE STELs AND EMPLOYEE ROTATION AS A SUBSTITUTE.

5.24.3 Ceiling Limits to Exposure

O━π OSHA ceiling limits are exposures that may not be exceeded even instantaneously.

Ceiling limits exist for substances for which overexposure could result in serious health hazards. A ceiling limit is used where a TWA (with its allowable excursions) allows a person to be exposed to an overly dangerous amount of a substance. ACGIH and OSHA state that a ceiling value should not be exceeded even instantaneously. They each denote a ceiling value by a "C" preceding the exposure limit.

5.24.4 Immediately Dangerous to Life or Health (IDLH) Exposures

O━π OSHA's IDLH (Immediately Dangerous to Life and Health) limits are maximum concentrations of toxins that may be breathed in the event of respirator failure.

According to the NIOSH Pocket Guide, IDLH is defined in terms of a "maximum concentration," one from which, in the event of respiratory protective equipment failure, one could escape within 30 minutes without a respirator. Furthermore, the escape must be effected without experiencing any escape-impairing (e.g., severe eye irritation) or irreversible health effects.

5.24.5 "Skin" Notation

While the above exposure limitations are based on exposure to airborne concentrations of substances, remember the absorption routes of exposure in the workplace. Some substances have the potential to contribute to overall exposure by direct contact with the skin, mucous membranes, or eyes. These substances are given a "skin" notation in the NIOSH Pocket Guide.

5.26 *TRAINING AIDS AND ADDITIONAL RESOURCES*

The National Toxicology Program (http://ntp-server.niehs.nih.gov/) has a wealth of information on workplace and environmental toxicological studies. The Chemical Industry Institute of Toxicology (CIIT) (http://www.ciit.org/) is useful for specific areas of toxicology study. The University of Pittsburgh's Medical, Clinical and Occupational Toxicology Resource Home Page is also an excellent resource (http://www.pitt.edu/~martint/welcome.htm).

For aiding readers to see real-life examples of what they have learned in this section, the authors recommend the video

- *Carcinogens and Poisons,* available from Safe Expectations (formerly BNA Communications, Inc.) (www.safeexpectations.com).

5.27 SUMMARY

Anyone involved in HAZWOPER activities, including field workers, supervisors, managers, planners, regulators, legislators, educators, and medical personnel, must know how serious toxic effects can be in the workplace. Far more people die from toxic effects than from other job-related accidents.

One can not be fooled by the fact that exposure to small amounts of any given material, or short exposures to it, do not immediately appear to result in health effects. We have learned that there can be long latency periods.

Workers require routine medical exams, with full disclosure of contacts with hazardous materials. New chemical compounds are being brought into use in the United States every year by the thousands. We can not be completely sure of their safe use. That can only come from many years of study and use. What we can do is protect ourselves from exposure.

Epidemiology and laboratory studies using test animals have been found to be valuable guides to uncovering toxic substances.

Workers must be alert to the four ways toxins can enter the body: inhalation, absorption through exposed tissues, ingestion, and injection. Inhalation is the most common entry route.

An awareness of the types of effects that common toxins can cause is valuable protection. Workers are then attuned to effects and prepared to search for causes. All involved personnel must know how time weighted averages of exposures are calculated. OSHA's permissible exposure limits (PELs) are legally enforceable limits and they are usually expressed as TWAs. OSHA also has established short term exposure limits (STELs), ceiling limits to exposure, and immediately dangerous to life or health (IDLH) exposure limits.

Staying within these limits and using the proper personal protective equipment (Section 6) are vital to good health and survival.

SECTION 6
PERSONAL PROTECTIVE EQUIPMENT (PPE)

OBJECTIVES

The major objective of Section 6 is to convince all readers of the critical importance of wearing the proper PPE for the task. You may wonder, "why *not* wear what is needed?" The authors have probably heard all of the reasons at one time or another. "I do not need it for the short time on the task." (We know of a death occurring because an engineer believed he could run in and out of a test nitrogen atmosphere while holding his breath. He was found dead later.) "It makes me get overheated and tired from working with all the extra weight." (Probably true. This complaint is accepted as reasonable. It must be addressed by providing suitable breaks together with a consistent supply of caffeine-free and non-alcoholic drinks.) "It is uncomfortable and dirty." (Supervisors must ensure that fit is correct and the PPE is clean.) There are many more excuses. None are sufficient to risk your health and life.

Section 5 should have convinced you that inhalation is the most prevalent route for toxins to enter your body. Thus, another essential objective of this Section is to present to readers the various kinds of respirators and the tasks for which they are required. You must understand the limitations and proper application of each type of respirator. Also, you must know when you will need the ultimate protection. The fully encapsulating suit (FES), boots, and gloves together with an atmosphere supplying respirator offer the maximum protection to the wearer.

Personal protective equipment is the LAST line of defense against health and safety hazards. As a manager, you have the responsibility to ensure that engineering and administrative controls, when operational and used, are the first and second lines of defense. In the event that the engineering and administrative controls in place do not provide adequate health and safety protection, then PPE must be used.

Also, as a HAZWOPER trainer, it is your duty to ensure that all workers are adequately trained. This Section and the Companion CD meet and exceed the OSHA requirements for PPE training material.

As a worker, you have the duty to check that engineering controls are operating at your worksite. You also must comply with your duties under administrative control. Then you must be prepared to don the proper PPE and work safely when that last line of defense is required.

Examples of engineering controls include using forced air ventilation (blowers) to remove hazardous atmospheres from the workplace and enclosing processes so that workers are never in contact with hazardous materials.

Administrative controls, for example, include setting a mandatory 5 mph speed limit for forktrucks in warehouses. Work practices might be changed in exceptionally hot environments, such as giving workers frequent breaks and offering fluids to avoid heat-related injuries or illnesses. Also, work shifts can be altered so that HAZWOPER personnel who must wear fully encapsulating suits or other cumbersome PPE can work during the cool of the night instead of the heat of the day.

6.1   THE OSHA PPE REQUIREMENTS

29 CFR 132(a) states:

> Protective equipment, including personal protective equipment for eyes, face, head, and extremities, protective clothing, respiratory devices, and protective shields and barriers, shall be provided, used, and maintained in a sanitary and reliable condition *wherever it is necessary* by reason of hazards of processes or environment, chemical hazards, radiological hazards, or mechanical irritants (irritation caused by solids such as dusts and fibers such as fiberglass), encountered in a manner capable of causing injury or impairment in the function of any part of the body through absorption, inhalation or physical contact. [Emphasis added]

○━ㅠ   **Personal protective equipment is the last line of defense in protecting worker health and safety.**

Supervisors and other managers take note: the availability of adequate PPE including respiratory protection equipment does not absolve you of your first responsibility: to eliminate hazards. Remember that PPE is the last choice for worker protection. PPE should only be used when you have exhausted all other resources, especially engineering and administrative controls.

○━ㅠ   **All required and practical engineering controls must be applied to the worksite to preserve worker health and safety.**

Once the hazards of a workplace have been identified and eliminated, or reduced to the extent feasible; supervisors must determine the suitability of the PPE presently available. If necessary, they must select new or additional equipment, which ensures a level of protection greater than the minimum required to protect the employees from the hazards. Care must be taken to recognize the possibility of multiple and simultaneous exposure to a variety of hazards. Adequate protection against the *highest* level of each of the hazards must be provided to workers before their entry to a contaminated workplace.

○━ㅠ   **Administrative controls are the workplace health and safety rules, regulations, warning signs, and worker training that management maintains, upgrades, reinforces, and enforces.**

6.2   HOW IMPORTANT IS PPE, AND HOW SERIOUS IS OSHA ABOUT VIOLATIONS?

Consider the following excerpt from a Department of Justice press release:

IDAHO MAN GIVEN LONGEST-EVER SENTENCE FOR ENVIRONMENTAL CRIME
Company Owner to Serve 17 Years, Pay $6 Million
for Crimes That Left Employee with Brain Damage

WASHINGTON, D.C.—In the longest sentence ever imposed for an environmental crime, a federal judge has ordered an Idaho man to serve 17 years in prison for his crimes that left a 20-year-old employee with permanent brain damage from cyanide poisoning, the Department of Justice announced. [The company owner] also was ordered to pay $6 million in restitution to the victim and his family.

In May 1999, a jury in Pocatello, Idaho, found that [the company owner] had ordered employees of [his fertilizer manufacturing company] to enter and clean out a 25,000-gallon storage tank containing cyanide without taking required precautions to protect his employees. Occupational Safety and Health Administration inspectors repeatedly had warned [the owner] about the dangers of cyanide and explained the precautions he must take before sending his employees into the tank, such as giving workers protective gear.

[An employee] was overcome by hydrogen cyanide gas while cleaning the tank and sustained permanent brain damage as a result of cyanide poisoning.

. . .

The jury convicted [the owner] of three counts of violating the Federal Resource Conservation and Recovery Act. In addition to the RCRA charge that [he] knowingly endangered his employees, [the owner] was convicted under RCRA of illegally disposing of hazardous cyanide waste on two separate occasions. . . . [He was] ordered . . . to pay $400,000 to clean up the site.

. . .

. . . Over a period of two days in August 1996, [the owner] directed his employees—wearing only jeans and T-shirts—to enter an 11-foot-high, 36-foot-long storage tank and clean out cyanide waste. . . . [The owner] did not first test the material inside the tank for its toxicity, nor did he determine the amount of toxic gases present. After the first day of working inside the tank, several employees met with [the owner] and told him that working in the tank was causing them to have sore throats, which is an early symptom of exposure to hydrogen cyanide gas.

The employees asked [the owner] to test the air in the tank for toxic gases and bring them protective gear—which is required by OSHA and which was available to [the owner] free of charge in this case. [The owner] did not provide the protective gear, and he ordered the employees to go back into the tank, falsely assuring them that he would get them the equipment they sought. Later that morning, [an employee] collapsed inside the tank. And could not be rescued for nearly an hour because [the owner] also had not given employees the required rescue equipment.

[After the trial's conclusion, the OSHA Administrator said,] "The sentence handed down today sends a clear message that employers who flagrantly disregard workers' safety will be held accountable."

All HAZWOPER workers must wear all assigned PPE while in the work area. Otherwise, management faces the probability of an OSHA citation, fine, or jail time. The worker will face company disciplinary action from reprimand to removal!

CAUTION

REFER BACK TO SECTION 1 DEALING WITH OSHA'S VPP PROGRAM. THIS OWNER, ACCORDING TO THE REPORT, NOT ONLY DID NOT PARTICIPATE IN THE VPP, HE DID NOT HEED REPEATED ADVICE AND WARNINGS FROM OSHA.

6.3 RESPIRATORY PROTECTION (29 CFR 1910.134)

Note

There are numerous variations of respiratory protective equipment. Many already are, or are becoming, obsolete. In this Section we will concentrate only on the respiratory protection and the types of respirators that we expect to be in common use in HAZWOPER work during the first decade of the 21st Century.

WARNING

RESPIRATORY PROTECTION MUST ONLY BE USED AFTER ALL OTHER MEANS OF ELIMINATING THE RESPIRATORY HAZARD HAVE BEEN EXHAUSTED. ENGINEERING CONTROLS, SUCH AS VENTILATION AND REMOVAL OF THE SOURCE OF THE HAZARD,

**MUST FIRST BE APPLIED BEFORE YOU CAN RELY ON RESPIRA-
TORS.**

6.3.1   Workers' Requests for Respiratory Protection Must Be Honored

⊶   **There are three times when the employer MUST provide respiratory protection: when the employee requests it, when the work site contains respirable
hazardous materials at or above the PEL, and when atmospheric monitoring
has not taken place and there is a reasonable likelihood that the atmosphere is
contaminated.**

OSHA gives employees, including contract workers, very wide latitude when it comes to
respiratory protection. The reason is that the greatest amount of toxic substances that get
into the human body at the work site get there by inhalation. OSHA requires that workers
be supplied with respirators under these conditions:

1. When the worker requests it
2. When the level of hazardous materials in the air is at or above the PEL
3. When atmospheric monitoring has not taken place and there is a reasonable likelihood
   that a hazard may exist

Note

Supervision has the option of providing the requested respiratory equipment or
reassigning the employee to other work.

The first concern of workers and supervisors is that whatever choice of respiratory protection has been made, proper fitting must be ensured. This means that the equipment fitted
on the worker has been demonstrated to provide the designated protection.

HAZWOPER workers must know, for their own respiratory protection, how to select the
proper equipment. This means that they must know when to use a self-contained breathing
apparatus, an air-line respirator, or one of the many air-purifying respirators. This decision
must be based on a proper hazard assessment using the results of approved monitoring and
sampling techniques. That sampling must accurately report the amount of hazardous contamination in the atmosphere of the workplace. (See Section 12 for an in-depth discussion of
sampling and monitoring.)

WARNING

**THE OSHA RESPIRATORY PROTECTION STANDARD, 29 CFR
1910.134, STATES THAT YOU, THE WORKER, MUST BE SUPPLIED
WITH A RESPIRATOR UNDER CERTAIN CONDITIONS. IT SHALL BE
ADEQUATE FOR THE MATERIAL THAT MUST BE FILTERED FROM
THE AIR IN THE WORKPLACE ATMOSPHERE. YOU MUST BE IS-
SUED PROPER RESPIRATORY PROTECTION IF:**
1. **YOU REQUEST IT, OR**
2. **AIR SAMPLES SHOW THE PRESENCE OF A CONTAMINANT AT
   A CONCENTRATION THAT IS AT OR ABOVE THE PEL, OR**
3. **WHEN ATMOSPHERIC MONITORING HAS NOT TAKEN PLACE
   AND THERE IS A REASONABLE LIKELIHOOD THAT A HAZARD
   MIGHT EXIST.**

⊶   **Even when supervision finds a work site atmosphere to be healthy and safe for
breathing, a worker may demand respiratory protection. Supervision has the
option of providing the PPE or reassigning the worker.**

CAUTION

OSHA REQUIRES THAT MANAGEMENT PROVIDE A WRITTEN RESPI-
RATORY PROTECTION PROGRAM. THIS DOCUMENT MUST BE MADE
AVAILABLE TO ALL EMPLOYEES. ALL EMPLOYEES WHO ARE RE-
QUIRED TO WEAR RESPIRATORS MUST BE FULLY TRAINED ON EACH
TYPE OF EQUIPMENT THAT THEY MAY BE REQUIRED TO USE.

Workers must be trained in the use of each type of respiratory equipment they may be required to use in their work.

6.3.2 The Respiratory System

The human respiratory system is an excellent filter. However, we do not want to use it for that purpose when it comes to hazardous materials.

The lungs' function is to oxygenate the blood. They add oxygen, O_2, to the blood at the same time that they remove carbon dioxide, CO_2, from the blood. You will suffer, and even die, if there is too little oxygen in the atmosphere you breathe.

The lungs' function is to bring air in contact with the body's blood so that oxygen can be added and carbon dioxide removed. The lungs are two spongy organs that fill your chest cavity. They lie against your ribs and extend from the bottom of your chest to a point slightly above the collarbone. The back surface of each lung is curved to allow room for the heart. Space is also provided for the esophagus, trachea, nerves, and blood vessels of the chest.

Atmospheric Oxygen Contents and Their Effect on the Body. The normal atmosphere consists of approximately 78% nitrogen, 21% oxygen, 0.96% inert gases, and 0.04% carbon dioxide. An atmosphere containing toxic hazardous materials, even at very low concentrations, could be a hazard to the lungs and body. A concentration of inert gas such as nitrogen, large enough to decrease the percentage of oxygen in the air to below 19.5%, can lead to asphyxiation.

OSHA requires that the breathing atmosphere at a work site shall be between 19.5% and 23.5% oxygen.

The normal atmosphere contains approximately 21% oxygen. The physiological effects of reduced oxygen begin to be evident as the percentage drops below 19.5%. Without regard to hazardous materials, the atmosphere must contain a minimum of 19.5% oxygen to permit use of an air-purifying respirator. Further, due to the increased fire hazard with atmospheres of greater than 23.5% oxygen (**oxygen enriched**), OSHA states that the working atmosphere

TABLE 6.1 The Likely Effects of an Oxygen-Deficient Atmosphere

Percent of oxygen in air	Effect
17%	Faster, deeper breathing
15%	Dizziness, buzzing in ears, rapid heartbeat
13%	May lose consciousness with prolonged exposure
9%	Fainting, unconsciousness
7%	Life endangered
4–6%	Convulsions, death

Source: Courtesy of NIOSH.

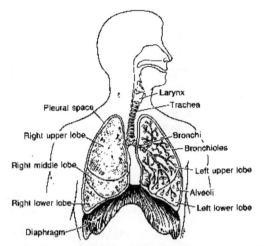

FIGURE 6.1   Diagram of lung components. (*Courtesy of the Long Island Jewish Medical Center.*)

must contain between 19.5% and 23.5% oxygen. Atmospheres with oxygen concentrations below 19.5% can have adverse physiological effects and are considered oxygen deficient. Table 6.1 illustrates the likely effects of an **oxygen deficient** atmosphere.

A Simplified Description of Your Breathing System.   See Figure 6.1. The trachea, or windpipe, carries air from outside your body, through your mouth or nose, into your lungs. The trachea divides into two main branches known as the left and right main stem bronchi, which lead to the left and right lobes of your lungs.

These main stem bronchi are the actual passages for air moving to and from each lung. The bronchi continue to divide into smaller tubes called bronchioles that eventually end in little air sacs called alveoli, which resemble a microscopic bunch of hollow grapes. It is at the level of the alveoli, the tiny air sacs where the oxygen you breathe in is absorbed into the blood. The carbon dioxide that your body has produced is desorbed and exhaled.

⊙━π **The alveoli are the tiny air sacs in your lungs where O_2 and CO_2 transfer takes place.**

6.3.3   Categories of Respiratory Protection

⊙━π **Respirators can be divided into two broad categories: air purifying respirators (APRs) and atmosphere supplying respirators (ASRs).**

There are two broad categories of respiratory protection: air purifying respirators (APRs) and atmosphere supplying respirators (ASRs).

Note

ASRs are treated in detail in Section 6.4.

While ASRs are capable of protecting the wearer against a broad range of atmospheres, they can be cumbersome. They interfere somewhat with mobility and job performance. They are also more expensive to buy, operate, and maintain. If the ASR is a self-contained breathing apparatus (SCBA), its air cylinders must be refilled or exchanged periodically. This limits the wearer's work period.

The vast majority of respirators used in industry are APRs. They afford the wearer much greater operating flexibility than ASRs and are less expensive. However, before you choose an APR for a job, you must be certain that an ASR is not required. Next, you must be able to choose the correct filter cartridge for the selected APR.

○──π **The APR MUST only be considered for use when you are certain that the atmosphere contains between 19.5% and 23.5% oxygen. ASRs MUST be used in all oxygen deficient atmospheres.**

6.3.4 Air-Purifying Respirators (APRs)

Note

Our discussion of air-purifying respirators will focus on those respirators that can prove their efficiency with a fit test. We will mention the nuisance dust mask. However, it is not suitable for a majority of HAZWOPER work situations.

The simplest kind of APR consists of a mask with straps that attach to the back of your head and hold the mask to your face, covering your nose and mouth. This is called a *half-face respirator*. One or two cartridges (constructed of fabric, metal or plastic) are attached to the mask. When you inhale, a flapper valve (*inhalation valve*) opens in the mask at the cartridge, allowing filtered air into the mask and then into your lungs. When you exhale, that valve closes and the *exhalation valve* opens. Your exhaled air is discharged to the atmosphere. Since you create a vacuum in the face-piece with the power of your lungs, this is called a *negative pressure respirator*.

Full-face respirators cover from below the chin to your forehead. In the same breathing configuration, this full-face respirator is also a negative pressure respirator.

APRs are also available in *positive pressure respirators*. The powered air purifying respirator (PAPR) is a positive pressure respirator. PAPRs have a battery-powered fan that sucks air through the filter cartridge and then blows filtered air onto your face. The most common PAPRs are full-face respirators, but half-face PAPRs are available.

Negative pressure respirators provide less protection than positive pressure respirators. That is because as you breathe in you create a vacuum. Then there is a chance of breathing in atmospheric contaminants through any gaps in the face seal. PAPRs, on the other hand, afford greater protection because they will tend to leak from the inside out.

There are two kinds of APR cartridges (filters): particulate cartridges and gas and vapor cartridges. You must not make the wrong choice when protection is required!

○──π **A particulate filter cartridge only removes dusts, fumes, mists, and other particulates from your breathing air. It does not remove hazardous gases or vapors.**

1. *Particulate filter cartridges* use a fabric medium inside or a covering on the outside of the cartridge to trap particulates. This prevents dusts, fumes, mists, and fibers from entering the respirator and then getting into your lungs.

WARNING

PARTICULATE FILTER CARTRIDGES ALONE DO NOT REMOVE GASES AND VAPORS!

○──π **A chemical vapor or gas cartridge only removes gases or vapors for which it is approved. It is not approved for particulate removal.**

2. *Chemical vapor or gas cartridges* are packed with granular chemical adsorbents, principally activated carbon. The common gas mask that has been in use for about 100 years is of this type. In some cases, special chemical reactants are added to the carbon to remove specific gases, vapors, fumes, and mists. The reactants can be impregnated on the carbon.

3. *Combination particulate filter and chemical cartridges* are required when protection from both particulate and chemical hazards is required in the workplace. These cartridges possess both the filtering ability of the particulate cartridge and the chemical adsorption of the chemical cartridge.

WARNING

CHEMICAL VAPOR AND GAS CARTRIDGES ALONE DO NOT RE-MOVE ALL PARTICULATES!

○━━π   **There are combination cartridges that filter out particulates and gases and vapors.**

How Chemical Vapor or Gas Cartridges Work.   There are chemical cartridges manufactured both with and without added purifying reactants. Chemical cartridges are used on respirators to help remove, and lower worker exposures to, harmful gases and vapors in the workplace. There are a number of types of chemical adsorbing cartridges, including organic vapor and mist, ammonia, formaldehyde, mercury vapor, and acid gases (such as hydrogen chloride and sulfur dioxide). No one cartridge protects the wearer from all gases, vapors, and mists.

It is important to understand how chemical cartridges purify the air. You will then have a better appreciation of the need to care for cartridges, and when to change them. All chemical cartridges consist of a container filled with a sorbent. The term *sorbent* is used to indicate that both weaker absorbent and stronger adsorbent (see Appendix A, Glossary) forces will be at work to trap gases and vapors in the cartridge. The container, called a cartridge, filter, filter cartridge, or canister (either metal or plastic) is usually cylindrical. A cartridge is usually several inches in diameter but only one or two inches in height. The top and bottom are perforated to allow air to pass through the sorbent in the cartridge. A canister is usually worn on the belt and usually holds 10 to 20 times as much sorbent material as is found in a regular cartridge or filter.

A chemical cartridge sorbent is a granular porous material with a large active surface area per unit weight. The greater the active surface area, the longer the cartridge service life. The target gas or vapor toxin molecules, upon striking the sorbent active surface, are trapped from the air stream and held tightly to the sorbent surfaces. The most common sorbent is activated carbon, which is a specially treated and purified form of granular carbon. It is characterized by high adsorptivity for many gases and vapors.

Some Cartridge Usage Considerations to Keep in Mind.   Where only activated carbon is used, the toxin molecules are held to the carbon surfaces by physical forces. Although stronger than absorption forces (such as water held by a sponge) these are not the even stronger chemical bonds formed in a chemical reaction. Therefore, the process can be reversed to some extent. This reversal is called *desorption.* Desorption occurs naturally over time. It occurs faster at higher temperatures. When the adsorbed, or trapped, molecules break away from the activated carbon, they will move in the direction of breathing air flow. The cartridge is spent when the contaminants are able to pass through the cartridge sorbent into the wearer's breathing air. This is called *breakthrough.*

Some desorption and movement of toxin through the cartridge bed can occur naturally during periods of respirator storage after use. This can move toxins closer to the inhaling end of the cartridge, making it unfit for later use.

There is one more consideration to keep in mind. Use of the respirator and its cartridge in clean, but moist, air can cause the active carbon surfaces to become saturated with water molecules. Then the cartridge will be unfit for trapping toxin molecules. Should this happen, and you notice an odor, or taste, passing through the cartridge, leave the area immediately, discard the cartridge(s), and then replace the cartridge(s) with new ones.

TABLE 6.2 Chemical Cartridge Types and Removal Mechanisms

Chemical cartridge type	Removal mechanism	Examples of impregnant
Organic vapors	Adsorption	Not used
Ammonia/methylamine	Chemisorption	Nickel chloride, cobalt salts, copper salts, acids
Acid gases	Chemisorption	Carbonate salts, phosphate salts, potassium hydroxide, copper oxide
Formaldehyde	Chemisorption	Copper oxide + metal sulfates, salts of sulfamic acids
Mercury vapor	Chemisorption	Iodine, sulfur
Hydrogen fluoride	Chemisorption	Carbonate salts, phosphate salts, potassium hydroxide, copper oxide

Types of Chemical Vapor or Gas Cartridges.   To make the cartridges more selective for certain chemicals, sorbents can be impregnated with chemical reagents. Impregnated, activated carbon removes specific gas and vapor molecules by *chemisorption*. Chemisorption is the formation of bonds between molecules of the impregnant and the chemical contaminant. These bonds are much stronger than the attractive forces of physical adsorption. The binding is usually irreversible. Table 6.2 shows several types of chemical cartridges, the mechanism of removal, and the impregnants used for removal of specific gases or vapors. Where no impregnant is used, the adsorbent does all the work of trapping the toxins.

Unfortunately, little information has been published about the effect of desorption or migration on cartridge service life. The two safest approaches when the service life estimate is longer than the use period are:

1. Never reuse an organic vapor chemical cartridge; dispose of it safely and properly after the period or shift in which it is used

2. Conduct desorption studies in a laboratory mimicking the conditions of use/reuse at your work site. Use these data when establishing the change schedule.

The American National Standards Institute (ANSI) Z88.2-1992, *American National Standard for Practices for Respiratory Protection,* recommends desorption studies unless cartridges are changed daily or are equipped with End of Service Life Indicators (ESLIs).

The following two subsections will focus on you, the person being qualified for respirator use.

6.3.5 Medical Qualifications for Your Use of Respirators

O—π **HAZWOPER personnel who enter atmospheres at or above the PEL for any atmospheric hazard for 30 or more days a year, or who are required to wear an APR, MUST have a respirator physical.**

The Strains Caused By Respirator Use.   The wearing of all respirators, APRs and ASRs, causes a strain on the wearer's cardiovascular system. Negative pressure air-purifying respirators place the entire burden on the wearer's heart and lungs. All wearers of negative pressure APRs MUST pass a respirator physical.

CAUTION

THE AUTHORS STRONGLY RECOMMEND THAT ALL WEARERS OF ALL TYPES OF RESPIRATORS PASS A RESPIRATOR PHYSICAL.

Positive pressure APRs (PAPRs) and ASRs offer the wearer breathing assistance in the form of air pressure. Your physical examination, administered by a health professional, will determine whether or not you are physically capable of working with a respirator.

Diseases such as asthma, chronic bronchitis, emphysema, asbestosis, hypertension, and heart conditions are among the conditions that may cause a physician to recommend that you not wear a respirator. This is a decision that MUST be made by a physician. There are also known cases of claustrophobia that, while admittedly rare, also will render you unable to wear a respirator.

The Required Pulmonary Function Test

○——π **Due to the extra strain placed on your cardiopulmonary system by the breathing restrictions of a negative pressure APR, users are required to have a respirator physical.**

Your respirator physical must include a pulmonary function test (PFT), also called a spirometry test. This will indicate whether you have sufficient lung capacity for negative pressure APR use. This measurement is obtained by first having you inhale as deeply as possible, then exhale forcefully through a mouthpiece to the fullest extent for a count of at least 10 seconds, and finally inhale deeply and rapidly.

Your nose is pinched tightly shut during this test to avoid leakage, which would lead to incorrect results and an incorrect diagnosis. You are usually required to perform at least three of these tests. A respiratory therapist coaches you through each maneuver and reviews the results to ensure that there are no technical defects.

The physical must include the OSHA mandatory 10-minute questionnaire (see Appendix H on the Companion CD).

If the respirator is a negative pressure respirator, the health professional may find a medical condition that could place the employee's health at increased risk. In that case, if a respirator must be used, the employer must provide a PAPR or ASR if the health professional's medical evaluation finds that the employee can use such a respirator.

6.3.6 Proper Fitting of Respirators

Respirator Fit Check

○——π **Inspect for defects and conduct a fit check each time you don a respirator.**

Only after you have passed your respirator physical can you proceed to the fit test. The fit test requires that you wear the respirator. But first you must perform the positive-negative fit check.

Note

Once qualified to wear a respirator, you must inspect and conduct a fit check each time you don the respirator. You must inspect for missing or worn parts. Look for cracks in the rubber, broken or excessively worn straps, damaged facepiece, if applicable, and seating of the cartridges.

○——π **To conduct a positive fit check, hold your hand over the exhaust valve and breathe out. The respirator facepiece should lift away from your face.**

First, you will be properly fit with a respirator of the correct type and size by an individual trained in application and fit check. Then you perform a fit check. While holding your hand over the exhaust valve, you breathe out. The respirator facepiece should lift away from your face and stay there (positive fit check). If it does not lift away, there is not a good seal.

O—π   **To conduct a negative fit check, hold your hand(s) over the filter(s) inlet(s) on an APR or block the air line of an ASR and breathe in. The respirator should suck down on your face and stay there.**

Second, while holding your hand(s) over the filter(s) inlet(s) on an APR, or blocking the air line of an ASR, you inhale. The respirator should suck down on your face and stay there (negative fit check). If the respirator facepiece draws back off your face while you are holding your breath, there is not a good seal.

If positive-negative fit check fails, readjust the respirator's straps and try again. If you cannot successfully perform a positive-negative fit check, another make, model, or size of respirator must be selected.

If the fit check is successful, you can proceed to the fit test.

O—π   **A positive-negative fit check is required each time you don a respirator.**

Respirator Fit Test.   You must, at a minimum, pass a qualitative fit test for the make, model, and size of respirator you are going to use. You must be provided with instructions for the qualitative tests. The OSHA regulations, in 29 CFR 1910.134(e)(5)(i), state: "Every respirator wearer shall receive fitting instructions including demonstrations and practice in how the respirator is worn, how to adjust it, and how to determine if it fits properly."

OSHA says that a worker whose job requires a respirator must be fit tested yearly. The fit test may be either qualitative or quantitative and determines whether the respirator provides a good seal. The qualitative test uses the wearer's sense of smell. The quantitative test actually measures the leakage rate of the mask. Although not required, the quantitative fit test offers a more objective test result.

O—π   **You must be fit tested yearly if your job requires you to wear a respirator.**

Who Cannot Wear Respirators.   Respirators cannot be worn when physical conditions prevent a good face seal. Such conditions may be growth of beard, sideburns, facial scarring, a skull cap that projects under the facepiece, or temple pieces on glasses. Individually made spectacle kits, incorporating the worker's prescription, are available for all full-face respirators. They must be supplied if your vision requires glasses. Also, the absence of one or both dentures can seriously affect the fit of a facepiece. Further, persons with religious proscriptions against shaving or removing a prayer cap may not be able to wear a respirator. Also, some people are naturally claustrophobic to the extent that they cannot function for more than a few minutes wearing a full-face mask. They cannot be approved to wear that type of mask.

WARNING

UNDER NO CIRCUMSTANCES MAY CONTACT LENSES BE WORN ON A HAZARDOUS WASTE SITE OR DURING EMERGENCY RESPONSE. CHEMICALS CAN MIGRATE BETWEEN THE LENS AND THE EYE AND CAUSE PERMANENT DAMAGE, INCLUDING BLINDNESS. ALSO, IF ARC WELDING OPERATIONS ARE TAKING PLACE AT THE SITE, THE ULTRAVIOLET RADIATION PRODUCED HAS BEEN KNOWN TO "WELD" THE CONTACT LENSES TO THE WEARER'S CORNEA! THE AUTHORS ARE AWARE OF SEVERAL OCCASIONS WHERE THIS HAS OCCURRED!

O—π   **The physical conditions that prevent a good facepiece seal and that prohibit you from wearing your respirator include excess facial hair, facial scarring, a skullcap, interfering eyeglasses, absence of dentures that are normally worn, and the gain or loss of more than 30 pounds.**

Qualitative Fit Test (QLFT) for Gas or Vapor Protection.   This is a subjective test of respirator *fit* only. It is not intended to check cartridge efficiency. For this qualitative fit test, the test technician can vaporize saccharin, banana oil (isoamyl acetate, IAA, also known as isopentyl acetate), and/or irritant smoke, stannic (tin) chloride, to determine the facepiece seal. The tester will make up a solution of the test chemical. This solution, along with water, will be placed in unmarked jars. The subject will sniff each jar to ensure that the subject can detect the odor of the test chemical. The tester will have you, as the subject, stand in a containment. Usually this is an inverted clear plastic bag suspended from the ceiling. Then you will talk, exercise, and move your head around, simulating actual work. Appendix A to 29 CFR 1910.134: *Fit Testing Procedures (Mandatory)*, OSHA-Accepted Fit Test Protocols, Section 14, states that the following activities must be performed:

(1) **Normal breathing.** In a normal standing position, without talking, the subject shall breathe normally.

(2) **Deep breathing.** In a normal standing position, the subject shall breathe slowly and deeply, taking caution so as not to hyperventilate.

(3) **Turning head side to side.** Standing in place, the subjects shall slowly turn their heads from side to side between the extreme positions on each side. The head shall be held at each extreme momentarily so the subject can inhale at each side.

(4) **Moving head up and down.** Standing in place, the subjects shall slowly move their heads up and down. The subject shall be instructed to inhale in the up position (i.e., when looking toward the ceiling).

(5) **Talking.** The subject shall talk out loud slowly and loud enough so as to be heard clearly by the test conductor. The subject can read from a prepared text such as the Rainbow Passage, count backward from 100, or recite a memorized poem or song.

Rainbow Passage

When the sunlight strikes raindrops in the air, they act like a prism and form a rainbow. The rainbow is a division of white light into many beautiful colors. These take the shape of a long round arch, with its path high above, and its two ends apparently beyond the horizon. There is, according to legend, a boiling pot of gold at one end. People look, but no one ever finds it. When a man looks for something beyond reach, his friends say he is looking for the pot of gold at the end of the rainbow.

(6) **Grimace.** The test subject shall grimace by smiling or frowning. (This applies only to QNFT [Quantitative Fit Test] testing; it is not performed for QLFT)

(7) **Bending over.** The test subject shall bend at the waist as if he/she were to touch his/her toes. Jogging in place shall be substituted for this exercise in those test environments such as shroud type QNFT or QLFT units that do not permit bending over at the waist.

(8) **Normal breathing.** Same as exercise (1).

If you, the test subject, can smell anything at all, the seal is not complete. Another make, model, or size of respirator must be selected and the entire fitting process repeated.

For decades the U.S. Army used tear gas for two purposes. The first was to demonstrate whether a mask was poorly fitted. A solider would enter into a room with an atmosphere laced with tear gas. If the fit was poor, the soldier reacted rapidly.

The second purpose was to demonstrate how well the mask provided protection. While in the room, a soldier wearing a properly fitted mask would be ordered to remove the mask and then answer questions. The stinging eyes and choking that ensued would prove to the soldier that his correctly fitted mask had been effective.

Quantitative Fit Test (QNFT).   The OSHA-approved quantitative fit test uses nonhazardous aerosols. Examples are corn oil, polyethylene glycol 400, and di-2-ethyl hexyl sebacate. The respirator is connected to a particle analyzer. Use an ambient-aerosol, condensation-nuclei-counter such as the respirator manufacturer recommends. The counter measures the concentration of particles outside and inside the respirator. The ratio of concentration outside to the concentration inside is the protection factor (PF) for that respirator.

Once you have successfully passed the fit tests, you are ready to wear your respirator on the job for which it is suited.

O—π **OSHA allows either a QLFT or a QNFT for the yearly fit test.**

6.3.7 Respirator and Cartridge Approval and Selection

O—π **The only approved respirator/cartridge combinations permitted are those approved by NIOSH.**

In the past, both the National Institute for Occupational Safety and Health (NIOSH) and the Mine Safety Hazard Administration (MSHA) approved respirators and cartridges. According to the latest regulations, NIOSH alone will approve the respirator cartridge. Also, approval "Tested and Certified" (TC) numbers for products certified under 42 CFR Part 84 (NIOSH) will carry the designation, 84A: for instance, TC-84A-000. The entire respirator assembly (facepiece, harness, air-purifying elements (cartridges), blower, air tanks, regulators, and valves) must be approved for protection against the contaminant at the concentration that is present in the work area.

Maximum Use Concentrations and Maximum Use Limits. For any APR you plan to use, the concentration of contaminants in the worksite must not exceed the manufacturer's maximum use concentration (MUC) for that type and size cartridge or canister fitted on the APR. It is assumed that if the APR is used above the MUC, the wearer might be exposed to potentially damaging health effects despite wearing the respirator. These MUC designations are listed on the side of the cartridge or canister. The MUC is assigned by the manufacturer for each NIOSH-approved cartridge or canister.

The maximum use limit (MUL) is the maximum amount of protection provided by a specific respirator. The MUL is calculated by multiplying the respirator's assigned protection factor (APF) by the permissible exposure level (PEL) for the contaminant.

$$APF \times PEL = MUL$$

O—π **You may not wear an APR at a contaminated site if the approved MUC is exceeded. Further, the MUL may not be exceeded. Also, the MUL may never be greater than the IDLH for that contaminant.**

The MUL may never be greater than the Immediately Dangerous to Life or Health (IDLH) for any gas, vapor, or mist. (IDLH is defined in Section 5.24.5.)

CAUTION

ALL RESPIRATORS AND RESPIRATOR CARTRIDGES MUST BE TESTED AND CERTIFIED (T&C) BY NIOSH. THIS IS THE LAW. RESPIRATOR CARTRIDGES MUST BE PERMANENTLY MARKED WITH A T&C NUMBER AND THE ATMOSPHERES FOR WHICH THEY ARE SUITABLE.

Respirator Selection. ANSI Z-88.2 states that the selection of the properly approved respirator depends upon a number of factors. Here is a checklist:

❑ The nature of the hazard

❑ The characteristics of the hazardous operation or process

❑ The location of the hazardous area with respect to a safe area having respirable air

❑ The period of time for which respiratory protection may be needed

❑ The activity of workers in the hazardous area

❏ The physical characteristics, functional capabilities, and limitations of respirators of various types

❏ The respirator/protection factors and respirator fit

The Joint NIOSH/OSHA Standards Completion Respirator Committee agrees that all these criteria must be considered in the selection of a respirator.

6.3.8   Common Types of Air-Purifying Respirators

○━ᴫ **APRs purify the air you breathe. They are not a source of oxygen. You can only wear them in an atmosphere that contains 19.5% to 23.5% O_2.**

Respiratory protection must be used when the concentration of a substance in the ambient atmosphere exceeds the permissible exposure limit (PEL). Several exposure limits exist, as you will remember from Section 5.23. They are the OSHA PELs, NIOSH recommended exposure limits (RELs), and American Congress of Governmental Industrial Hygienists (ACGIH) threshold limit values (TLVs®). Remember that the OSHA PELs are the only federal legally enforceable limits. Figures 6.2(a) through (e) show examples of APRs ranging from the nuisance dust mask to the PAPR with an APF of 100.

Note

Due to the latest 42 CFR 84 (NIOSH) Respirator Standard, familiar terms such as Organic Vapors and Mists and HEPA assigned to cartridges will have to be reclassified as they will no longer meet the more stringent standard.

All air-purifying respirators have a similar appearance within their class. Canisters are much larger than cartridges (see Figures 6.2(c) and (d) for comparison). They are usually belt- or harness-mounted and have a greater filtering capacity than cartridges.

By definition, APRs do just that, purify the air. Using a myriad of specially designed filter cartridges, the APR can be fitted to filter-out rated concentrations of gases and vapors or particulates, including dusts, fumes, and mists. Through the use of *combination filters,* both types of material are removed from breathing air.

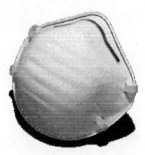

Assigned Protection Factor = 0

FIGURE 6.2 Example of air-purifying respirators (APRs). (*a*) Generic disposable dust mask APR; service: for light duty, nuisance (nonhazardous), dusts such as sawdust.

Assigned Protection Factor: 10

FIGURE 6.2 (*b*) MSA Advantage® half-mask APR; service: for organic vapors and mists and higher particulate concentrations than the disposable dust mask. (*Courtesy of Mine Safety Appliances Co.* [www.epartner. msanet.com/].)

Assigned Protection Factor: 50

FIGURE 6.2 (*c*) MSA Ultra Elite® twin cartridge full-face respirator; service: for organic vapors and mists and higher particulate concentrations than the half-face mask. (*Courtesy of Mine Safety Appliances Co.* [www.epartner.msanet.com/].)

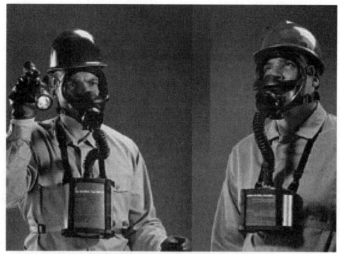

Assigned Protection Factor: 50

FIGURE 6.2 (*d*) MSA canister-type full-face APR (gas masks) with P-100 pre-filter; service: for organic vapors, gases, particulates, and mists. (*Courtesy of Mine Safety Appliances Co.* [www.epartner.msanet.com/].)

Protection Factor: 100

FIGURE 6.2 (*e*) Tight-fitting full-face PAPR MSA OptimAir® 6A PAPR with Ultravue™ facepiece; service: for higher concentrations of more hazardous particles. Can include P100 or combination filter and belt-mounted eight-hour battery pack with full-face respirator. For use when it is necessary to wear other chemical protective clothing. Note that there are also half-face PAPRs but they are in much less common use. (*Courtesy of Mine Safety Appliances Co.* [www.epartner.msanet.com/].)

WARNING

**THE MOST SERIOUS MISTAKE THAT CAN BE MADE IN THE SE-
LECTION OF AN APR CAN BE FATAL. THIS IS TO ASSUME THAT
THE OXYGEN CONTENT IN THE WORK ATMOSPHERE IS IN THE
19.5%–23.5% BREATHABLE RANGE. IT MAY NOT BE. ALWAYS
CHECK FOR A BREATHABLE OXYGEN CONCENTRATION FIRST!
THE APR DOES NOT SUPPLY OXYGEN!**

**6.3.9 Ten Widely Accepted Workplace Requirements for the Use of Air-
Purifying Respirators (APRs)**

Based on years of wearing APRs in hazardous atmospheres, the authors believe that these
10 requirements pretty well cover what must be done and known prior to wearing one in
hazardous atmosphere duty. Here is a checklist.

☐ All personnel planning to don an APR must have passed their respirator physical. Men
   must be clean-shaven.

☐ The oxygen content in the work atmosphere is between 19.5% and 23.5%.

☐ The identity and concentration of the hazardous materials are known.

☐ The ambient concentration of every contaminant is below the IDLH concentration.

☐ The ambient concentration of every contaminant is below the maximum use level (MUL)
   for the particular respirator.

☐ The ambient concentration of every contaminant is below the maximum use concentration
   (MUC) for the particular respirator cartridge chosen.

☐ The respirator assembly and cartridges are NIOSH-approved for protection against the
   specific concentration of every contaminant present in the work area.

☐ There is periodic monitoring of the work area.

☐ The respirator assembly (make, size, and type) has been successfully fit tested on the
   user within the last year.

☐ The contaminant has adequate warning properties of odor, taste, or irritation.

6.3.10 Identification of Particulate, Gas, or Vapor Breathing Hazards

**Atmospheric hazards in the work area must be identified and quantified before
appropriate respiratory protection can be selected.**

It is absolutely imperative that the contaminant(s) in the working atmosphere be known
so that:

1. The toxic effects of inhaling the contaminant can be determined.

2. Appropriate particulate filters or chemical cartridges or canisters can be chosen.

3. It can be determined whether adequate warning properties exist for the contaminant.

4. The appropriate facepiece can be selected. A full-face mask is required if the material
   causes eye irritation.

WARNING

**IN SITUATIONS WHERE NO MONITORING HAS TAKEN PLACE BUT
THERE IS REASON TO SUSPECT A CONTAMINANT, OSHA RE-
QUIRES USE OF SUPPLIED AIR RESPIRATORS UNTIL THE TYPE
AND LEVEL OF CONTAMINATION IS KNOWN.**

The maximum concentration depends on the contaminant and the respirator. For APRs:

1. The concentration must not exceed IDLH.
2. The MUC for the particular cartridge must not be exceeded.
3. The MUL of the respirator must not be exceeded.
4. The expected service life (cartridge or canister efficiency) must be determined.

Cartridge Efficiency.   Cartridge and canister filter media vary in their ability to remove hazardous materials from breathing air. For example, the adsorbent activated carbon in an organic vapor cartridge does a much better job of removing chlorobenzene than vinyl chloride. Cartridge efficiencies need to be considered when selecting and using APRs. Cartridges are rated by OSHA and NIOSH for their assigned protection factor (see Table 6.3). This tells the wearers whether they will be protected when entering an atmosphere containing a known concentration of a specific hazardous material or a specified class of materials.

Assigned Protection Factor (APF)

○─π **The assigned protection factor (APF) for a respirator is equal to the concentration of the contaminant outside the facepiece, divided by the concentration inside the facepiece as measured by laboratory instrumentation.**

Assigned protection factors (APFs) are assigned to each type and configuration of respirator and hazardous gas or vapor. An APF is measured in a laboratory with an instrument capable of measuring concentrations of particles, gases, or vapors inside and outside of a respirator. An APF is defined by the following calculation:

$$APF = \frac{CO}{CI}$$

where CO = the measured concentration outside the respirator
CI = the measured concentration inside the respirator

An APF is used to determine the maximum use limitation (MUL) and maximum use concentration (MUC) of a correctly fit-tested respirator. The MUL and MUC values are the highest concentrations, not exceeding IDLH concentration, of a specific contaminant in which a respirator may be worn:

○─π **The maximum use limit (MUL) for any APR respirator is equal to the APF times the PEL.**

$$MUL = APF \times PEL$$

The manufacturer assigns an MUC for each cartridge.

For example, say a contaminant has a PEL of 10 ppm, then using Table 6.3 the MUL for an air-purifying half-mask respirator is 100 ppm; the MUL for a full-face APR or demand SCBA is 500 ppm. If the ambient concentration is greater than 1000 ppm, either a PAPR or an alternate Atmosphere Supplying Respirator must be worn.

○─π **Half-face APRs have an OSHA APF of 10.**

○─π **Full-face APRs have an OSHA APF of 50.**

○─π **Full-face powered air-purifying respirators (PAPRs) have an OSHA APF of 100.**

○─π **Full-face air-line respirators have an OSHA APF of 1000.**

TABLE 6.3   OSHA and NIOSH Assigned Protection Factors for Various Respirators

Respirator class and type	OSHA APF	NIOSH APF
Air Purifying		
Half-mask	10	10
Full-facepiece	50	50
Powered Air Purifying (PAPR)		
Half-mask	50	50
Full-facepiece	100	50
Loose-fitting facepiece	25	25
Hood or helmet	25	25
Supplied Air		
Half-mask—demand	10	10
Half-mask—continuous	50	50
Half-mask—pressure demand	1000	1000
Full-facepiece—demand	50	50
Full-facepiece—continuous flow	250	50
Full-facepiece—pressure demand	1000	2000
Loose-fitting facepiece	25	25
Hood or helmet	25	25
Self-Contained Breathing Apparatus (SCBA)		
Demand	50	50
Pressure demand	greater than 1,000	10,000

O—π **Full-face self-contained breathing apparatus (SCBAs) respirators have an OSHA APF of greater than 1000.**

Note

You will note some wide variances in APFs issued from OSHA and NIOSH. This is due in part to the differences in the test methods of the two agencies. As we stressed before, we will use OSHA's values in this, an OSHA-oriented manual and reference. There is no reason why you cannot choose the lower of the two APFs for an additional safety buffer. OSHA APFs are the only federal legally enforceable ones. A quantitative fit test will give the PF for the specific respirator worn by a specific person under specific test conditions.

6.3.11 Particle-Filtering Cartridges and Canisters

O—π **Particle-filtering cartridges or canisters will not protect against any gases or many vapors and mists.**

Particles (or particulates) can occur as dusts, fibers, grit, smoke, fumes, or mists. Particle sizes can range from visible to the unaided eye to microscopic. Particle toxicological effects can be severe or nonexistent. The hazard posed by a particulate can be determined by its PEL. A nuisance particulate will generally have a PEL of 10 mg/m^3 or greater. Toxic particulates such as platinum soluble salts have PEL values of 0.007 $\mu g/m^3$, or about one million times lower than nuisance particulates. A particulate filter's life is as long as you can breathe through it without excessive effort. Actually, as a filter gets clogged, it becomes more efficient in blocking particulate hazardous materials. However, it will deliver a lower airflow rate for the same breathing effort.

WARNING

IF BREATHING STARTS TO BECOME DIFFICULT, LEAVE THE AREA AND CHANGE YOUR CARTRIDGES. ALSO, BE AWARE THAT DIFFICULTY IN BREATHING COULD BE A SIGN OF TOXIC HAZARDOUS MATERIALS IN THE ATMOSPHERE PENETRATING THE RESPIRATOR CARTRIDGE. FURTHERMORE, IT COULD BE A SIGN OF PHYSICAL ILLNESS SUCH AS HEART ATTACK.

Aerosols.   *Aerosol* is a term used to describe fine particulates (solid or liquid) readily suspended in air. For our purposes, aerosols can be classified in two ways: by their physical form and origin and by their physiological effect on the body.

Physical classification examples include:

1. Mechanical dispersions, such as paint spray from an airless paint sprayer
2. Compressor-powered, such as a pinhole leak in a high-pressure hydraulic line or compressed liquefied materials that provide vapor sprays from the common aerosol can
3. Condensation dispersions (mist, fog, and smog)
4. Thermal dispersions (fumes and smoke)

Physiological classification examples include:

1. Nuisance
2. Inert pulmonary reaction
3. Allergy-producing
4. Chemical irritation
5. Systemic poison

Since 1998, all nonpowered, air-purifying, particulate respirators have been required to meet a revised standard for testing and certification, commonly referred to as 42 CFR 84. 42 CFR (Public Health) Title 84 (Approval of Respiratory Protective Devices) categorizes filters according to filter efficiency along with resistance to filter efficiency degradation. According to 42 CFR 84, the P series filters are the only series certified for use without any time or use restrictions other than those normally associated with particulate filters. That is, when it becomes harder for the wearer to breathe through the filter, it has become clogged and must be replaced. The familiar color codes of filters will be redesigned to reflect the new designations. However, the P100 filter (old high efficiency particulate air [HEPA]) will continue to be identified by the familiar magenta color.

ANSI has approved the ANSI/AIHA Z88.7: "Color Coding of Air-Purifying Respiratory Cannisters, Cartridges, and Filters" to become a National Standard. OSHA will most likely adopt the Standard in 2002.

Particle-Filtering Efficiencies

○─π **The highest particle-filtering efficiency cartridge is the P100, which is still commonly referred to as the HEPA filter.**

○─π **A P100 cartridge will filter out 99.97% of all particles 0.3 μm and larger.**

Filter efficiency is rated at three levels: 95%, 99%, and 99.97%, that is, having those efficiencies for retaining particles with diameters of 0.3 μm or greater. The highest particulate filter rating of any APR cartridge is the true HEPA filter (now called N100, P100, or R100), which will remove 99.97% of all particles greater than 0.3 μm.

TABLE 6.4 The Nine Classifications of
Particulate Filters

Filter type	Filter efficiency		
	95%	99%	99.97%
N	N95	N99	N100
R	R95	R99	R100
P	P95	P99	P100

Note

According to 42 CFR 84, all filter tests will use "the most penetrating aerosol size," which is 0.3 μm "aerodynamic, mass median, diameter." This may seem illogical but studies have shown that particles smaller than 0.3 μm do not penetrate filters as readily as 0.3 μm particles. Respirators certified under 42 CFR Part 84 will filter all other sizes of particles at or greater than the certified efficiency level.

Resistance to filter efficiency degradation is also rated on three levels:

- *N*, not resistant to oil
- *R*, resistant to oil; but time use restrictions may apply
- *P*, oil proof; some time restrictions may also apply

Consequently, under the new standard there are nine classifications of particulate filters. See Table 6.4.

Under 42 CFR 84, prior to selecting any particular respirator, the end user must know or determine the specific particulate present and its concentration within the work area, and whether or not oil aerosols are present.

NIOSH no longer approves particulate respirators for specific uses or applications. Therefore, the end user is required to be knowledgeable regarding potential exposure to particulate hazardous materials. However, NIOSH has stated that in those instances where a HEPA filter has been required under the OSHA Substance Specific Standard, under 42 CFR 84, a filter efficiency of 100% (99.97%) is now required.

6.3.12 Gas- and Vapor-Filtering Cartridges and Canisters

A gas- or vapor-removing filter must be chosen for protection against a specific type of contaminant. Gas and vapor cartridges (Figure 6.3) are available in different styles just as are particle filters. The smallest are cartridges that contain 50-200 cm^3 of adsorbent and attach directly to the facepiece of a half- or full-face mask, sometimes in pairs and sometimes singly. Chin-mounted filters have a volume of 250–500 cm^3 of adsorbent and are usually attached to a full-face mask. Belt- and harness-mounted canisters can have over 500 cm^3 of adsorbent and are usually attached to a full-face mask.

No air-purifying respirator is permitted for use in an IDLH atmosphere.

The difference in applications is the maximum use limitation (MUL) for which the cartridge or canister can be used in accordance with its NIOSH approval. For instance, cartridges are approved for use in atmospheres up to 1,000 ppm (0.1%) organic vapors, chin-style canisters up to 5,000 ppm (0.5%), and larger belt- or harness-mounted canisters up to 20,000 ppm (2.0%). Remember, no air-purifying respirator is permitted in an IDLH atmosphere.

FIGURE 6.3   MSA GME® (left) and GME-P100® Super (right) (combination particulate and gas, vapor, and mist) cartridges for MSA Comfo® respirators. (*Courtesy of Mine Safety Appliances, Co.* [www.epartner.msanet.com/].)

Service Life and End-of-Service Life Indicators (ESLIs).   Service life of a gas or vapor cartridge or canister is dependent on several factors, including the breathing rate of the wearer, moisture content in the atmosphere, contaminant concentration, and absorption/adsorption efficiency of the filter media.

A higher breathing rate brings a larger amount of gas or vapor contaminant in contact with the filter media in a given period of time. This in turn increases the rate of filter saturation and shortens service life.

As concentration goes up, the mass flow rate (pounds of contaminant per hour) through the filter increases. This brings more contaminant in contact with the adsorbent in a given period of time. Refer also to Section 6.3.4 for an explanation of absorption and adsorption efficiency.

To be in compliance with the revised OSHA Respiratory Standard, employers who provide air-purifying respirators to their employees must either provide respirators that are equipped with an end-of-service life indicator (ESLI) or develop a respirator cartridge change-out schedule that will ensure cartridges are changed before the end of their service life.

In the past, OSHA required that all APRs be equipped with cartridges with an ESLI certified by NIOSH for contaminants with poor warning properties. With the new respirator standard, **ALL** APRS must have ESLIs or use a change-out schedule.

If there is no ESLI appropriate for conditions in the employer's workplace, the employer must implement a change-out schedule for canisters and cartridges. That schedule must be based on objective information or data that will ensure that canisters and cartridges are changed before the end of their service life. The employer must describe in the written respiratory protection program the information and data relied upon and the basis for the canister and cartridge change schedule and the basis for reliance on the data.

Some hazards have no odor, so we cannot tell when breakthrough has occurred. A prime example is mercury fumes. Cartridges for this toxic materials must be equipped with ESLIs. These filters must be worn with the ESLI visible to the wearer, for instance on the belt, or else with the buddy system, so the buddies can monitor each other's filter effectiveness.

Figure 6.4 shows OSHA's three ways of estimating a cartridge's service life.

> **Breakthrough, when the wearer detects an odor or sensation, is no longer a legal way of determining a cartridge's efficiency, as the worker has already been exposed.**

3 VALID WAYS FOR YOU TO ESTIMATE

A CARTRIDGE'S SERVICE LIFE:

1. Conduct Experimental Tests

 Can save money by providing a more accurate service life value instead of relying on conservative assumptions made by other methods

 Most reliable method, especially for multiple contaminants

 Can be used to validate an existing change schedule

 Will likely take time and money to perform the tests

2. Use the Manufacturer's Recommendation

 Can result in a more accurate estimate for your particular brand of respirator

 Relies on the manufacturer's broad knowledge and expertise

 May not be possible if the manufacturer is unable to provide a recommendation

 May not accqunt for all workplace and user factors adequately

3. Use a Math Model

 Inexpensive and takes little time

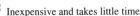

 Requires no math calculations if you use the Advisor Genius

 Not as accurate as experimental testing. May result in a service life estimate that is shorter than it needs to be due to conservative assumptions

 Generally limited to single contaminant situations

FIGURE 6.4   The three OSHA-approved methods for estimating an APR's cartridge service life. (*Courtesy of OSHA.*)

OSHA's Advisor Genius is an excellent tool for estimating cartridge life for single contaminants. It is available on-line at http://www.osha-slc.gov/SLTC/respiratory_advisor/advisor_genius_wood/advisor_genius.html.

Gaseous and Vapor Hazardous Materials.   Gases and vapors (see the Glossary in Appendix A for the definition of the difference) are filtered to some degree at various distances on their trip through the respiratory tract. They can be absorbed by the airways on route to the alveoli. Not all will be absorbed, however. Some gases and vapors finally reach the alveoli, where they can be directly absorbed into the bloodstream. Of course, some fraction of the hazardous materials will be exhaled.

Gaseous hazardous materials can be classified as *chemical* and *physiological* hazards. Refer to Sections 3.1 and 3.4 and Section 5 for further details.

Chemical hazardous materials: corrosives, such as chlorine, and toxics, such as hydrogen sulfide and carbon monoxide

Physiological hazardous materials: Irritants, asphyxiants, anesthetics, and central nervous system (CNS) depressants.

Warning Properties.   A warning property is a signal to a respirator wearer that a cartridge or canister in use is beginning to lose its effectiveness. This is called breakthrough. The warning property might be detected as an odor, taste, irritation, or other physiological effect.

○──ᴫ **When a respirator wearer detects an odor, taste, irritation, or other physiological change, it must be treated as a warning sign of filter breakthrough. Immediately leave the area and replace the cartridges.**

In order for an APR to be chosen as adequate respiratory protection, the contaminant(s) must have adequate warning properties. Further, there must be a suitable, NIOSH-approved cartridge for the contaminant.

Most substances have warning properties at some concentration. A warning property detected only at dangerous levels, i.e., greater than PEL, is not considered adequate.

An odor, taste, or irritation detected at extremely low concentrations is also not adequate because the warning is being given all the time or long before the filter will begin to lose its effectiveness.

WARNING

SOME SUBSTANCES CAUSE RAPID OLFACTORY FATIGUE. THAT IS, THE SENSE OF SMELL IS NO LONGER EFFECTIVE. FOR THOSE GASES OR VAPORS, ODOR IS NOT AN ADEQUATE WARNING PROPERTY. THE BEST EXAMPLE OF THIS IS HYDROGEN SULFIDE, H_2S. IT HAS THE ODOR OF ROTTEN EGGS. HOWEVER, IT QUICKLY FATIGUES THE SENSE OF SMELL. THEN H_2S CAN BE INHALED AT IDLH LEVELS WHILE THE UNSUSPECTING WORKER SMELLS NOTHING.

Most people are familiar with a similar sulfide odor, that of the mercaptans added to natural gas to warn of gas leaks. The purest natural gas (nearly 100% methane) is odorless.

○──ᴫ **When wearing an APR in a hazardous atmosphere that does not have the necessary warning properties, always use a filter cartridge with ESLIs (end-of-service-life indicators).**

Quality Inspection Checklist for Air-Purifying Respirators.   The following is a detailed fitness-for-use checklist for APRs:

1. Disposable respirator inspection: check for service readiness (and perform this action) as required:
   - ❏ Holes in the filter or filter deformation (replace respirator)
   - ❏ Straps, for loss of elasticity and deterioration (replace respirator)
   - ❏ Metal nose clip for deterioration or excessive deformation (replace respirator)

2. Re-useable air-purifying respirators (Half-mask or Full facepiece): check for service readiness (and perform this action) as required:

   A. Rubber facepiece:
- ❏ Excessive dirt (clean thoroughly)
- ❏ Cracks, tears, or holes (replace facepiece)
- ❏ Distortion (replace facepiece)
- ❏ Cracked or loose-fitting lenses (contact manufacturer to see if replacement is possible; otherwise, replace facepiece)

   B. Head straps or nets:
- ❏ Breaks or tears (replace head straps or nets)
- ❏ Loss of elasticity (replace head straps or nets)
- ❏ Broken buckles, snaps, or attachments (replace)
- ❏ Excessively worn serrations on the headpiece harness that could cause a loose fit (replace head piece straps)

   C. Inhalation valve(s) and exhalation valve:
- ❏ Detergent or disinfectant residue, dust particles or dirt on valve or valve seat that may allow blowby (clean thoroughly)
- ❏ Cracks, tears, or distortion of the valve material or valve seat (contact manufacturer for replacement)
- ❏ Missing or defective valve cover (contact manufacturer for replacement)

   D. Filter(s) cartridge(s) or canisters:
- ❏ Proper filter for the hazard (check label)
- ❏ NIOSH approval designation (beware of counterfeit cartridges)
- ❏ Missing or worn gaskets (contact manufacturer for replacement)
- ❏ Worn threads on filter or facepiece (replace filter or facepiece, whichever is worn)
- ❏ Cracks, dents, or holes in filter housing (replace filter)
- ❏ Change of color on the ESLI (indicating end of service life, replace filter)
- ❏ Canister has passed expiration date (replace canister)

3. Powered air-purifying respirators (PAPRs): check for service readiness (and perform this action) as required:

- ❏ Check all items under A through D above (perform actions indicated)
- ❏ Battery in good condition, fully charged (if not, recharge)
- ❏ Blower in good condition (if not, replace blower)
- ❏ Visible cracks or holes in breathing tube (if so equipped) or tube leaks air in water test (replace tube)

Table 6.5 is a generic log format for recording respirator quality inspections. The form should be stored with the respirator. When the respirator requires cleaning, repair, or replacement, take the respirator out of service until required maintenance is completed.

APR Cost.   Although cost of PPE should be the last consideration, it is a consideration nonetheless. We have already discussed that it is easier on workers to wear PAPRs than half-face or full-face APRs because the fan makes breathing easier. The following are the authors' estimates of the costs of various APRs through the year 2010.

Half-face APRs cost between $15 and $50 and their cartridges cost between $10 and $30 per set. This is one reason disposable APRs are gaining popularity. There is no maintenance, and the cost difference is relatively minor.

Full-face negative pressure APRs cost from $80 to approximately $500. There are no disposable full-face respirators on the market yet. Their cartridges cost about the same as half-face cartridges. Canisters range between $35 and $150.

The range in costs for cartridges and canisters is based on the type of material they are designed to filter from the air.

PAPRs range from around $500 to $800 per respirator and their cartridges cost between $15 and $30 each.

TABLE 6.5   Generic Respirator Inspection Log

Location:

Respirator type(s):	_____ chemical cartridge	_____ disposable/single use
	_____ canister	_____ quarter mask
	_____ powered air-purifying (PAPR)	_____ half mask
	_____ airline	_____ full face
	_____ self-contained (SCBA)	_____ helmet/hood
Date:	Inspected by:	Inspector's Job Title:
Comments (items to check are listed above under Quality Inspection Checklist for Air-Purifying Respirators and below in Section 6.3.13 under Inspection of SCBA Equipment)		

These are all retail prices, and there are substantial discounts for buying in quantity. These prices are simply to give the reader an idea of the cost of air-purifying respirators and the need to care for them properly.

6.3.13 Atmosphere-Supplying Respirators (ASRs)

There is a great deal of confusion regarding atmosphere-supplying respirators. Many references describe the term as synonymous with supplied air respirators. OSHA, however, makes a clear distinction. Because this is an OSHA-oriented reference, we will use its definitions.

Atmosphere-supplying respirators (ASRs) are any respirators that supply the user with breathing air from a source independent of the ambient atmosphere. This includes:

1. Supplied-air respirators (SARs), also called airline respirators

2. Self-contained breathing apparatus (SCBA) units

Supplied-air respirators (SARs) or air-line respirators (also called Type C respirators) are atmosphere-supplying respirators for which the source of breathing air is not designed to be carried by the user. The supply of the breathing air comes from a hose connected to a remote air source (breathing air compressor or cylinders).

Self-contained breathing apparatus (SCBA) means an atmosphere-supplying respirator for which the breathing air source is designed to be carried by the user in the form of breathing air cylinders. Note that SCBA and SCUBA are entirely different systems. Self-contained underwater breathing apparatus (SCUBA) supplies breathing air to a mouthpiece. The SCBA supplies air to the facepiece of the respirator.

In this book we will distinguish the two types of ASRs by using their OSHA-preferred names, *air-line respirators* and *SCBAs*.

Advantages and Disadvantages of Working with ASRs.   As we mentioned previously, both types of ASRs afford the highest level of respiratory protection available. One advantage of the air-line respirator over the SCBA is that workers can perform their jobs for extended periods of time without the need to replace air cylinders. Air-line respirators are available in half-face and in full-face configuration for added eye and skin protection.

The SCBA, with its full-face configuration, can be cumbersome. The cylinders can impede movement in a tightly configured space. As with other full-face respirators, some facepiece configurations can partially obscure vision.

With air-line respirators there is the danger of tangling hoses with other workers. There is also the danger of the air supply hose becoming trapped under mobile equipment, tangled in machinery, or pinched or severed on sharp or pointed objects. Further, OSHA limits hose length to 300 ft from air source to mask. For these reasons, among others, airline respirators are not allowed in IDLH atmospheres unless accompanied by an escape cylinder with a minimum of 15 minutes of breathing air.

Air-line respirators have an OSHA length restriction of 300 ft of air supply hose.

The IDLH limitation does not apply to the wearer of the SCBA. However, SCBA equipment is fairly heavy, adding to worker fatigue. The air supply cylinders must be recharged frequently. SCBAs also are fairly expensive and require frequent and more costly maintenance.

Conditions Requiring ASRs

You must use an ASR when the oxygen content of the atmosphere is less than 19.5% or greater than 23.5% (an IDLH atmosphere), when air contaminants are greater than IDLH, MUL, or MUC levels, when no atmospheric monitoring has taken place and you suspect hazardous air contaminants, and in emergencies.

Air-line respirators, when used in an IDLH atmosphere or in emergencies, must be accompanied by a 15-minute escape air cylinder.

Due to the greater freedom of movement of SCBAs, the authors recommend SCBA use under these four conditions:

1. When the oxygen content is less than 19.5% or greater than 23.5% and wearers require extensive mobility
2. When measured hazardous material concentrations are greater than their IDLH, MUL, or MUC valves and 15-minute escape cylinders are unavailable
3. When no monitoring has taken place and you suspect presence of a hazardous material that does not provide adequate warning properties
4. In emergencies

Three types of atmosphere-supplying respirators are shown in Figures 6.5(a) through (c). As with all APRs, all ASRs must come with NIOSH Tested and Certified (TC) certification as shown in Figure 6.6.

Air-Delivery Modes.   HAZWOPER work frequently requires use of respiratory protective equipment.

There are three modes of ASR operation: 'negative pressure' demand mode, 'positive pressure' pressure demand mode, and 'positive pressure' continuous flow mode (escape or bypass mode for SCBAs).

In *demand mode* the pressure inside the facepiece in relation to the immediate environment is positive during exhalation and negative during inhalation. This offers the lowest level of ASR protection due to the potential for influx of atmospheric contaminants upon inhalation. Exhaled air is vented to the atmosphere and not rebreathed.

(a)

FIGURE 6.5 Atmosphere-supplying regulators. (*a*) SCOTT Ska-Pak 5-, 10-, and 15-minute escape SCBA; service: for use in conjunction with APRs, for disconnect from air-line respirators, and in emergency escape situations. (*Courtesy Scott Health and Safety* [www.scotthealthsafety.com].)

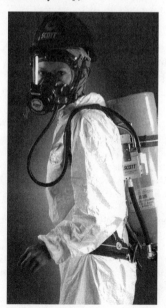

(b)

FIGURE 6.5 (*b*) Scott Air Pak 30-, 45-, or 60-minute full-facepiece SCBA; service: for the highest level of respiratory protection and mobility. (*Courtesy Scott Health and Safety* [www.scotthealthsafety.com]).

(c)

FIGURE 6.5 (*c*) Scott EZ® air-line respirator (the regulator is at chest level. It is shown disconnected from facepiece. The wearer is connecting quick-disconnect coupling to the airline); service: for the highest level of respiratory protection and longer uninterrupted service. (*Courtesy Scott Health and Safety* [www.scotthealthsafety.com]).

In *pressure demand* mode an ASR's exhalation valve maintains a positive pressure in the facepiece of about 1.5 in. of water column (w.c.). It opens only when pressure exceeds 1.5 in., w.c. Pressure demand mode maintains positive pressure in the facepiece at all times. This exhausts the air supply sooner than demand mode but offers 20 times or greater respiratory protection.

In *continuous-flow* mode an ASR supplies respirable air at a constant flow at all times, rather than only on demand. In place of a demand or pressure demand regulator, an airflow control valve or orifice partially controls the airflow. This means that, by design, the control valve cannot be closed completely. Continuous-flow mode is used routinely on airline respirators for increased protection. Further, air supply for air-line respirators is not a consideration as long as the breathing air compressor continues in operation or the cascade or supply cylinders are not depleted.

Continuous-flow mode is only used in emergencies on an SCBA. For the SCBA it is called *escape* or *bypass* mode because it is used only for escape and it bypasses the regulator. Escape mode on SCBAs is used when the wearer detects odors, taste, eye or skin irritation, or some other physiological change. This is a sign that there is a leak in the SCBA. The increased airflow dilutes any contamination. Continuous flow rapidly depletes the air supply.

Air-Line Respirators (Also Called Type C Respirators). Air-line respirators are almost always run in continuous-flow mode. The system provides continuous airflow of at least 115 L (4 ft³) of air per minute for tight-fitting facepieces. It must provide at least 170 L (6 ft³) of air per minute for loose-fitting types such as abrasive blasting helmets (Type CE respirators) or Level A suits. The wearer does not have to inhale and provide a vacuum. Exhaled air and some fresh air are constantly exhausting out the exhalation valve. For these reasons

FIGURE 6.6 MSHA/NIOSH Tested and Certified (TC) approval for a pressure demand SAR.

the air-line respirator in pressure-demand mode offers an APF of 1,000. There is also no pressing need to conserve air. As long as the continuous supply (cascade) system has air or the breathing air compressor is operating, you have all the air you need. The air-line respirator is lighter weight than the SCBA and thus puts much less physical strain on the worker. It also allows workers to remain in the hazardous atmosphere longer.

If the air-line respirator is being used with 8–50 ft of hose, the inlet pressure must be 10–15 psig. If the maximum length of hose, 300 ft, is used, the inlet pressure must be 35–40 psig.

> **Air supplied to either an air-line respirator or an SCBA must meet the requirements for Grade D breathing air.**

Air-line respirators may be fed from an air compressor or from breathing air cylinders. The respirable air must meet the requirements of at least Grade D Air as defined by the Compressed Gas Association, Inc. (CGA). This requirement is defined in *CGA's "Commodity Specification for Air, G-7.1."*

CAUTION

OLDER REFERENCES MIGHT CITE BOTH ANSI Z86.1-1973 AND CGA'S G-7.1. OSHA NOW REQUIRES THAT THE STRICTER CGA G-7.1 STANDARD BE MET FOR GRADE D AIR.

Grade D air must meet the following minimum requirements:

- Hydrocarbon (condensed) content of 5 mg/m^3 of air or less
- Oxygen (O_2) content of 19.5% to 23.5%
- Carbon monoxide (CO) content of 10 ppm or less
- Carbon dioxide (CO_2) content of 1000 ppm or less

The production of this Grade D air falls under CGA standard G-7, *"Compressed Air for Human Respiration."* This standard specifies the types of compressors and other equipment allowed for use in Grade D air production.

When an air-line respirator is to be used, supervision and the wearer must be assured that the air-line is attached to a supply source of breathing quality air. This cannot be air from plant utility air. That air is not of breathing air quality.

When air-line respirators are fed from breathing air cylinders, there is no requirement for testing or monitoring the quality of the air.

WARNING

NEVER ATTACH AN AIR-LINE RESPIRATOR SUPPLY HOSE TO A SUPPLY SOURCE THAT IS, OR MIGHT BE, PLANT UTILITY AIR. THAT AIR IS PURE ENOUGH FOR OTHER PLANT OPERATION PURPOSES, SUCH AS PNEUMATIC TOOL OPERATION OR SPRAYPAINTING, BUT IS NOT OF BREATHING QUALITY.

When air-line respirators are fed from a compressor, there are several additional safety rules you must obey.

Employers must ensure proper construction and work site placement of compressors used to supply breathing air to respirators. Here is a checklist to help readers to meet these requirements.

- ❏ Prevent entry of contaminated air into the air supply system. (This is accomplished by keeping the compressor air inlet at least 50 ft upwind of the work area, cordoned off so that internal combustion engines cannot run in that area.)
- ❏ Minimize moisture content so that the dewpoint at 1 atm is at least 10°F (5.56°C) below the ambient temperature.
- ❏ Have suitable in-line air-purifying sorbent beds and filters to further ensure breathing air quality. Sorbent beds and filters shall be maintained and replaced or refurbished periodically following the manufacturer's instructions.
- ❏ Sorbent beds and filters must have a tag containing the most recent sorbent bed change date and the signature of the person authorized by the employer to perform the change. The tag shall be maintained at the compressor.
- ❏ For compressors that are oil-lubricated, the employer shall ensure that carbon monoxide (CO) levels in the breathing air do not exceed 10 ppm. (Toxic carbon monoxide is a potential breathing air contaminant produced by decomposition of lubricating oil at high compressor operating temperatures.) CO contamination warning is accomplished by the use of a CO alarm. This alarm must be audible or in the form of a beacon so that all air-line users can be alerted while working. (We recommend both forms of warning.)
- ❏ For oil-lubricated compressors, the employer must use a high-temperature or carbon monoxide alarm, or both, to monitor carbon monoxide levels. If only high-temperature alarms are used, the air supply must be sampled and analyzed at intervals sufficient to prevent carbon monoxide in the breathing air from exceeding 10 ppm.

WARNING

**THE AUTHORS DO NOT RECOMMEND USING A HIGH-TEMPER-
ATURE ALARM ONLY. HOWEVER, IF THAT IS DONE, MONITOR
THE AIR QUALITY CONSTANTLY.**

❑ The employer shall ensure that breathing air couplings are incompatible with outlets for
nonrespirable work site air (such as utility or plant air) or other gas systems. Never allow
any asphyxiating substance entry into breathing air lines. (The authors are aware of
several deaths due to air line respirator users using plant air. In each case the plant air
system was purged and inerted with nitrogen as part of repair. The workers died instantly!)

❑ Breathing air compressors must be marked "Breathing Air Only." No other uses, such
as pneumatic tool operation, may be powered from a "Breathing Air Only" compressor.
The use of pneumatic tools coupled to a breathing air compressor can introduce oil, used
to lubricate the tools, into the breathing air stream. Although this may not be fatal,
workers can suffer from an illness called lipid pneumonitis, or hydrocarbon or mineral
"oil pneumonia."

CAUTION

OIL-LUBRICATED BREATHING AIR COMPRESSORS ARE CURRENTLY
ALLOWED BY THE STANDARD. HOWEVER, THE AUTHORS RECOM-
MEND EXCLUSIVE USE OF OIL-FREE BREATHING AIR COMPRESSORS
FOR AIR-LINE RESPIRATORS.

If breathing air and pneumatic tool operation are required from the same air source (such
as in a remote area), the authors recommend a specially designed unit such as the Scott Air
Cart® (see Figure 6.7). This unit is specially designed so that the tool operating air and the
breathing air cannot be accidentally mixed. The use of different coupling systems makes
incorrect hose attachment impossible.

When Air-Line Respirators Should Be Used.   Air-line respirators should be worn, as a
minimum, in areas where the atmosphere is determined not to be immediately dangerous to
life or health (IDLH) but where one of the following conditions exists:

1. The requirements for using an air-purifying respirator cannot be met.
2. The work will require long-term use of ASRs.
3. Abrasive blasting or spraying is being done with or on harmful substances. Further, the
concentrations of those substances exceed the concentrations that can be safely handled
by an air-purifying respirator. Note that during abrasive blasting workers need air-line
respirators with additional protection from flying abrasive. This includes a helmet and
upper body protection and is called a Type CE respirator.

If air-line respirators are used, the supplied air source shall not be capable of being
depleted while users are depending on it. The hose length shall not exceed 90 m (300 ft)
from source to user.

WARNING

**UNLESS ACCOMPANIED BY AN EMERGENCY ESCAPE CYLINDER
WITH A MINIMUM OF 15 MINUTES OF RESPIRABLE AIR, AIR-LINE
RESPIRATORS ARE NOT ALLOWED IN IDLH CONDITIONS.**

Self-Contained Breathing Apparatus (SCBA).   Due to the relative intricacies of their operation, we will dedicate more time to SCBAs. This is not to imply that the air-line respirator
is less important. In fact, many more workers work routinely with air-line respirators than

FIGURE 6.7 Scott Air Cart. (*Courtesy Scott Health and Safety* [www.scotthealthsafety.com].)

with SCBAs. In a situation where a great deal of time (days or weeks) must be spent in a hazardous atmosphere, the air-line respirator is the only way to go for productivity, morale, endurance, and economy. However, in emergency response and in many HAZWOPER investigations, the SCBA is used almost exclusively. We will explain the three modes of SCBA operation: negative pressure or demand, positive pressure or pressure demand, and positive pressure escape or bypass.

In the *negative pressure or demand mode,* negative pressure is created inside the facepiece and breathing tubes when the wearer inhales. This negative pressure draws down a diaphragm in the regulator of the SCBA. The diaphragm depresses and opens the admission valve, allowing air to be inhaled. As long as the negative pressure remains, air flows to the facepiece. Demand mode will allow the air supply to last longer but give a lower protection factor.

⊙━━━ **SCBAs operated in demand mode only have an OSHA APF of 50!**

The problem with demand mode operation is that the wearer can inhale contaminated air through any gaps in the facepiece-to-face sealing surface or through any cracks in the hose to the regulator. For these reasons, the use of negative pressure or demand respirators is not in compliance with OSHA Regulation 29 CFR 1910.134 for atmosphere-supplying respiratory protection. Hence, these respirators have an assigned protection factor of only 50, the same as for a full-face air-purifying respirator.

An SCBA operating in the *positive pressure, pressure demand* mode maintains a positive pressure inside the facepiece at all times. The system is designed so that the inlet valve remains open until enough pressure is built up to close it. The pressure builds up because air is prevented from leaving the system until the wearer exhales. Less pressure is required to close the inlet valve than to open the spring-loaded exhalation valve.

At all times the pressure in the facepiece is greater than the ambient pressure outside the facepiece. If any leakage occurs, it is outward from the facepiece. Because of this, the positive pressure, pressure demand SCBA has an OSHA APF of greater than 1,000 and a NIOSH APF of 10,000! The drawback to this mode of use is that the air in the cylinder is depleted sooner.

There is also the *positive pressure, escape or bypass mode.* This uses the bypass position on the air supply regulator. That mode allows a much higher volume of air to flow to the facepiece. This bypass position is for emergencies requiring escape from the site. An example is when the wearer starts to detect by odor, taste, or physiological effect, which indicates hazardous materials are entering the air stream. A leaking supply hose might cause this for instance. This bypass position on the air control valve is purely for escape emergencies and should never be routinely used. Its use rapidly depletes the supply cylinder.

Components of a Typical Pressure Demand SCBA

O—π **You must know the components of an SCBA assembly: pack and harness, air cylinder (or tank), high-pressure hose, low-pressure alarm, regulator assembly, bypass or 'escape valve,' and breathing hose and facepiece (mask).**

A number of components must be fit for service and must function properly.

Pack and Harness.   A backpack frame and harness, or fanny pack, supports the cylinder and regulator weight, allowing the user to move freely. When it is properly fit, the weight of the unit will be supported on the hips, not the shoulders.

Air Cylinder (or Tank) and Mainline Valve.   OSHA requires that SCBAs carry a minimum of 30 minutes of respirable air supply. There are two common air cylinders' capacities in wide use. The 45 ft^3 capacity for a NIOSH-approved 30-minute supply and the 88 ft^3 capacity for a NIOSH-approved 60-minute supply. The tanks are filled with respirable air meeting the requirements of at least Grade D air.

The mainline valve is the valve that when opened allows air flow from the cylinder to the high-pressure hose.

There are also two cylinder sizes (for low and high pressure) in each capacity. Low pressure is about 2200 psig, and high pressure is about 4500 psig. The choice is yours. The advantage of the high-pressure models is that they are about half the size of the low-pressure models. This allows for greater maneuverability of the wearer.

Cylinders are filled through the mainline valve using a breathing-air compressor with an in-line purifying filter or a cascade (linked together) system of several large cylinders of breathing air. Should the SCBA cylinder become overfilled, a rupture disc releases the pressure. The rupture disc is located on the mainline valve, along with a cylinder pressure gauge. That gauge is required to be accurate within ±5%. Because the gauge is exposed and subject to abuse, it should be used only for judging if the cylinder is full, and not for monitoring air supply to the wearer. Also, it is behind the wearer, out of his or her view. To monitor air supply, the wearer looks at a gauge within view either on the regulator assembly, mounted on the chest harness, mounted on a flexible hose, or as a 'heads up display' in the respiratory facepiece.

WARNING

AIR CYLINDERS, OR TANKS ARE ROUTINELY RATED AS 30-MINUTE, 45-MINUTE, OR 60-MINUTE TANKS. THIS IS A GUIDELINE ONLY! THE ACTUAL BREATHING TIME DEPENDS UPON THE PHYSICAL CONDITION OF THE WEARER, THE STRENUOUSNESS OF THE WORK, THE TEMPERATURE OF THE WORK AREA, AND

OTHER FACTORS. ALL OF THESE CONDITIONS CAN GREATLY RE-DUCE THE ACTUAL AMOUNT OF TIME THE WORKER WILL BE ABLE TO BREATHE FROM A FULL CYLINDER.

Any full compressed air cylinder is considered a hazardous material. For this reason, any cylinder used with an SCBA must meet the DOT General Requirements for Shipments and Packaging (49 CFR Part 173) and Shipping Container Specifications (49 CFR Part 178).

Note

These cylinders are often erroneously referred to as "oxygen tanks." *They contain not oxygen, but compressed air.* If they were oxygen tanks, the workers could become lightheaded and could faint.

A hydrostatic test must be performed on a cylinder at regular intervals. For steel and aluminum cylinders, a hydrostatic test must be performed every five years. For composite (fiberglass, Kevlar, or other material) covering an aluminum cylinder, a hydrostatic test must be performed every three years. Composite cylinders have a DOT exemption because at this time there are no set construction requirements. The construction technology of composite cylinders reduces the weight of the cylinder and thereby the overall weight of the SCBA.

Note

OSHA requires monthly inspection and recording of dates and results for all SCBAs maintained for emergency use.

High-Pressure Hose.   The high-pressure hose connects the cylinder's mainline valve and the regulator. The hose must be connected to the valve only by hand, never with a wrench. An O-ring inside the connector ensures a good seal.

Low-Pressure Alarm.   A low-pressure warning alarm is located in-line with the high-pressure hose connected to the cylinder's valve. This alarm sounds to alert the wearer that only 20 to 25% of the full cylinder air supply is available for retreat, usually 5 to 8 minutes of respirable air.

Regulator Assembly and Bypass ('Escape') Valve.   In routine use, air travels from the cylinder through the mainline valve, and then through the high-pressure hose to the pressure-reducing regulator assembly. From the regulator assembly the air supplies the breather's facepiece. If the bypass valve is opened, the airstream bypasses the regulator assembly and travels directly through the breathing hose into the facepiece. If the mainline valve is opened, air passes through the regulator and is controlled by that mechanism. Also at the regulator is another pressure gauge that also must be accurate to ±5%. Because it is visible and well protected, this gauge is used by the wearer to monitor the air supply.

Under routine conditions, the wearer closes the bypass valve and opens the mainline valve so that air can enter the regulator. Once at the regulator, the air pressure is reduced from cylinder discharge pressure to approximately 50 to 100 psig. A pressure relief valve is located after the regulator for safety should the regulator malfunction. For the airflow rate to meet NIOSH standards, it must meet or exceed 40 L per minute. In emergency situations, when the wearer detects odors or some physiological change, the wearer opens the bypass or escape valve. The will usually allow for fresh air to be inhaled, but rapidly depletes the air supply.

Breathing Hose and Facepiece (Mask).   The breathing hose connects the regulator to the facepiece. Rubber gaskets at both ends provide tight seals. The hose is usually constructed of neoprene and is corrugated to allow stretching. When relaxed, the breathing hose is approximately 30 in. long. It can be stretched to between 5 and 6 ft for leak testing.

Above the point in the facepiece where the hose is connected is a one-way check valve. This valve allows air to flow from the breathing hose but prevents exhaled air from entering the breathing hose. Exhaled air exhausts to the atmosphere. If the check valve is not working properly, the exhaled air could be exhaled into the breathing hose and rebreathed by the wearer. Seeing that the check valves work properly is part of the checklist that must be completed before each use.

The facepiece also has a one-way check exhalation valve similar to other respirators. The exhalation valve is checked before each use during the positive and negative fit check.

If the exhalation check valve malfunctions, the wearer might breathe in contaminated air.

Inspection of SCBA Equipment.   Inspect the SCBA according to the manufacturer's recommendations as well as 29 CFR 1910.134 requirements. Check the SCBA immediately prior to use. Inspection and fit-check procedures must be followed closely to ensure safe operation of the unit.

The cylinder on a SCBA must carry the following information:

• DOT exemption (for composite cylinder)
• DOT rated pressure and capacity
• Cylinder serial number
• Manufacturer's name, symbol, and part number
• Original and latest hydrostatic test date, day, month, and year

Respirator Use Supervisor Responsibilities.   The individual in an organization who is responsible for respirator use must develop a field inspection checklist for respiratory protective equipment. That individual must institute a continuing review of the inspection procedure so as to cover all uses of respiratory protective equipment.

Safety Checks and Checklist.   Here are some of the defects to look for and safety principles to ensure in the inspection of the components of SCBAs, along with the corrective actions to be taken. In many cases it will be necessary to contact the manufacturer of the equipment or the equipment dealer for expert advice. Most repairs will require the SCBA to be sent back to the manufacturer.

WARNING

REPAIRS TO SCBAs MUST ONLY BE PERFORMED BY AUTHORIZED PERSONNEL. IF YOU ARE NOT TRAINED TO REPAIR THE SCBAs, DO NOT DO SO. YOUR "REPAIR" COULD BE DEADLY TO YOU OR A COWORKER.

❑ Check facepiece, head straps, head nets, valves, and breathing tube using the same procedures as for air-purifying respirators.
❑ Check the hood, helmet, blouse, or full suit, if applicable, and make corrections for:
  • Rips and torn seams (repair or replace)
  • Broken or worn facepiece suspension straps or nets (replace)
  • Cracks or breaks in face shield (replace)
❑ Check the air supply system and take corrective action for:
  • Breathing air quality (ensure that only grade D or better approved air sources are used)
  • Breaks or kinks in air supply hoses (replace hose)
  • Broken, deformed, or loose hose fitting(s) (replace hose fitting)
  • Proper setting of regulators and valves (use manufacturer's recommendations)
❑ Use an inspection log similar to Table 6.6 to ensure that all components of the SCBA have been fully inspected and meet the requirements for use.

TABLE 6.6 Generic SCBA Inspection Log

Inspection date _____		By _____	
Unit # _____	Cylinder # _____	Hydro test due _____	Pressure (PSI) _____
		Comments	
Facepiece	OK	Needs Repair	
High-Pressure O-ring	OK	Needs Repair	_____
Tight High-Pressure Hose	OK	Needs Repair	_____
Harness	OK	Needs Repair	_____
Regulator	OK	Needs Repair	_____
Low-Pressure Alarm	OK	Needs Repair	_____
Valves	OK	Need Repair	_____
Tested as Worn	OK	Needs Repair	_____
			_____

Inspection Schedules and Recordkeeping.   All respiratory protective equipment must be inspected:

1. Before and after each use
2. During cleaning
3. Monthly when not in use

○━━ᴨ **Self-contained breathing apparatus (SCBA) must be inspected at least monthly. SCBA cylinders must be maintained with 95% of full capacity.**

Communications.   Communications at any workplace is of great importance. Workers must be able to talk to or otherwise communicate with each other. At a HAZWOPER work site communication is of paramount importance. Because the conditions can change rapidly from a routine job to an emergency, workers and supervision must be able to be in constant contact. In confined-space work this is an OSHA requirement.

   Since HAZWOPER workers are often wearing respiratory protection, how do they communicate? For years, workers have developed and used hand signals to communicate over long distances. But that is rudimentary at best. Many respirators have voice diaphragms that aid in verbal communication, but the results are poor unless workers are standing next to each other.

To illustrate the current level of communication technology associated with respiratory protection, we have chosen the hard-wired Scott Con-Space Link™ and the wireless Scott Envoy™ Radiocom®. See Figure 6.8. The Con-Space Link™ is extremely useful in work areas where there is radio interference from the surroundings, such as in a tank or other confined space. It offers full-duplex communication (all talkers can hear each other at the same time). The Radiocom® is excellent for communications over open areas, such as hazardous waste sites. All workers and supervisors can communicate over the same channel so that information can be communicated clearly to everyone at the same time. We will see other manufacturers adopting this technology universally in the future as it is so extremely useful.

FIGURE 6.8   Scott Con-Space Link™ (top) and wireless Scott Envoy™ Radiocom® (bottom). (*Courtesy Scott Health and Safety* [www. scotthealthsafety.com].)

6.3.14 Respiratory Protective Equipment Maintenance

○─╼ᴨ **All respiratory equipment must be cleaned, sanitized, thoroughly dried, and stored properly after use.**

Respiratory protective equipment maintenance and storage must be carried out in accordance with the instructions of the equipment manufacturer. In most cases not doing so will void the equipment warranty and may put you or your coworker in jeopardy, as stated earlier.

Ongoing maintenance of respiratory protective equipment is an important part of the program. Wearing poorly maintained or malfunctioning equipment might be, in a sense, as dangerous as not wearing a respirator at all. Workers wearing malfunctioning respirators think they are protected, when in reality they are not. The consequences of this situation can be fatal.

OSHA places strong emphasis on the importance of an adequate maintenance program. All maintenance programs must follow manufacturers' instructions and must include provisions for:

- Cleaning and disinfecting of equipment
- Storage that preserves respirator quality
- Inspection for defects
- Repair that restores function to manufacturer's specifications

Cleaning and Disinfecting.   Often several people use the same respirator at different times. For that reason, it is imperative that the respirator be properly cleaned and disinfected after each use. Used and unclean respirators can transmit skin diseases from one person to another. One worker may be allergic to another's aftershave, cologne, perfume, soap, or lotion. Therefore, any cartridges or canisters must be removed and the respirator must be cleaned and disinfected after each use. In the case of air-line or SCBA respirators, the facepiece and breathing tube must be cleaned. Further, respirators might be contaminated with hazardous material. Improper cleaning may contaminate the next wearer.

Individual workers who maintain their own respirators must also be trained in the cleaning and disinfecting of respirators. Used respiratory protective equipment must be washed with mild detergent and disinfectant in warm water using a soft brush. Following cleaning and disinfecting, it must be thoroughly rinsed in clean water and then air-dried in a clean place. Care must be taken to prevent damage from rough handling. Once the respiratory equipment is completely dry, it must be stored in a clean environment, preferably out of direct sunlight. Zip-Loc® bags are ideal for this purpose.

CAUTION

THE AUTHORS RECOMMEND THAT A RESPIRATOR BE RESTRICTED TO THE SOLE USE OF ONE WORKER. WE FURTHER RECOMMEND THAT IF RESPIRATORS ARE CLEANED AND STORED TOGETHER, EACH RESPIRATOR BE INDIVIDUALLY MARKED FOR THAT WEARER.

Detergents and Disinfectants.   The authors recommend using a detergent containing a bactericide. Never use organic solvents (paint thinner, mineral spirits, kerosene, gasoline, or alcohols). They can deteriorate the rubber facepiece and may leave hazardous residue. If detergent with a bactericide is not available, a detergent may be used, *followed* by a disinfecting solution, and rinsed in potable water. Here are two disinfectant solutions that OSHA recommends to achieve the desired results:

1. Prepare a sodium hypochlorite solution (50 ppm chlorine) by adding two tablespoons of commercial chlorine bleach, such as Clorox® (5.25% sodium hypochlorite), per gallon of water. A two-minute immersion disinfects the washed respirator.

2. Prepare an aqueous solution of iodine by adding one teaspoon of tincture of iodine per gallon of water. A two-minute immersion will disinfect the washed respirator.

Using either of these two recipes will disinfect the respirator and will not deteriorate rubber or plastic parts.

Always remember to check with the manufacturer to determine the proper temperatures for the disinfecting solutions. Some manufacturers can supply you with premeasured packets of disinfectant detergent. Further, some manufacturers may require you to use their cleaning solution as a condition of warranty.

Rinsing.   The cleaned and disinfected respirators must be rinsed thoroughly in clean water (120°F maximum temperature) to remove all traces of cleaner and disinfectant. This is very important to prevent skin rashes that may be caused by residue from the cleaning or disinfecting solutions.

Drying.   The respirators can be dried with a clean towel or allowed to air-dry in a sanitary, dust-free environment. They can also be hung from a horizontal wire, nails or pegs on a wall, but care must be taken not to damage the facepieces.

CAUTION

YOU MUST ENSURE THAT THE EQUIPMENT IS COMPLETELY DRY BEFORE STORING IN A ZIP-LOC® BAG OR OTHER AIRTIGHT CONTAINER. IF THE EQUIPMENT IS STORED WET, MOLDS OR FUNGUS COULD GROW ON THE RESPIRATOR, ENDANGERING THE NEXT WEARER.

Storage of Respiratory Equipment.   Respirators must be stored so as to protect them from dust, bacteria, sunlight, heat, extreme cold, moisture, damaging chemicals (such as solvents), and physical damage. After cleaning and drying, we recommend that respirators be placed individually in sealed heavy duty Zip-Loc®-type freezer bags until needed. Store them in a single layer to prevent distortion of any rubber components. Some respirators, such as SCBAs, will have their own storage case. However, it is a good practice to put the SCBA's facepiece in its own Zip-Loc® type freezer bag for extra protection.

Store air-purifying respirators so that they are kept ready for nonroutine or emergency use. Store them flat in a cabinet in individual compartments if available. Storage cabinets must be located in noncontaminated, but readily accessible, locations.

Replacement Parts

WARNING

CONTINUED USE OF RESPIRATORS WILL REQUIRE PERIODIC REPAIR OR REPLACEMENT OF COMPONENT PARTS. A QUALIFIED INDIVIDUAL MUST MAKE SUCH REPAIRS. IF YOU ARE UNSURE OF HOW TO PROPERLY REPAIR YOUR DAMAGED OR OTHERWISE UNFIT RESPIRATOR PROPERLY OR YOU DO NOT HAVE IDENTICAL REPLACEMENT PARTS FROM THAT RESPIRATOR'S MANUFACTURER, HALT ANY REPAIR. THE OSHA REGULATIONS REQUIRE YOU TO REFUSE TO WEAR A DAMAGED RESPIRATOR UNTIL IT IS REPAIRED PROPERLY. INTERCHANGING RESPIRATOR PARTS FROM DIFFERENT MODELS FROM THE SAME MANUFACTURER OR FROM DIFFERENT MANUFACTURERS IS STRICTLY FORBIDDEN!

Replacement parts for any respirator must be those of the manufacturer of the equipment involved.

O—π **Substitution of parts from a different size, brand, model, or type of respirator will invalidate the approval of the respirator. Such action is absolutely prohibited!**

Defective air-supplying respiratory protective equipment, except the SCBA, can be repaired and worn. Regulations require SCBA equipment to be returned to the manufacturer for repair or adjustment. That is primarily because of the relative complexity of the valve and regulator assembly.

6.3.15 Written Respiratory Protection Program

The OSHA Respiratory Protection Standard (29 CFR 1910.134) lists seven elements that every respiratory protection program must contain:

1. A written plan detailing how the program will be administered
2. A complete assessment of respiratory hazards that will, or could be encountered in the workplace
3. Administrative procedures (work practices) and engineering controls (equipment) to control respiratory hazards (These procedures and controls must be employed to eliminate, limit, or reduce workers' exposures to respiratory hazards.)
4. Guidelines for the selection of respiratory protective equipment adequate for the defined hazard
5. An employee training program covering hazard recognition, the dangers associated with respiratory hazards, proper care (cleaning, disinfecting, and storage) and use of respiratory protective equipment
6. Guidelines for inspection, maintenance and repair of respiratory protective equipment
7. Medical surveillance of all affected employees
8. Recordkeeping

6.3.16 Atmosphere-Supplying Respirator Cost

Half-face air-line respirators can cost as little as $150. This is for the facepiece only and does not include the air pump or compressor, air line, valves, regulators, and filters. A complete full-face system for one worker, with 100 ft of hose, can cost as little as $900. Depending upon the application and the supplier, the full-face ASR alone can cost up to approximately $1,500.

Escape SCBAs with 5-, 10-, or 15-minute air supplies cost between $1000 and $2000. Full-size SCBAs cost between $1500 and $5000 each. Price depends upon your choice of manufacturer, high- or low-pressure system, communication devices, and air supply (30-, 45-, or 60-minute), among other factors.

We will continue to stress throughout this desk reference and manual the importance of proper care and storage of PPE. Remember, you may need it to save your life. Further, your equipment might be quite expensive.

6.4 CHEMICAL PROTECTIVE CLOTHING (CPC)

O—π **CPC is any garment, including gloves and chemical-resistant boots and coverings, that is specially designed and fabricated to reduce or eliminate the wearer's contact with hazardous materials.**

As of the writing of this book, OSHA has yet to set standards for chemical protective clothing. As you will see, there are standards for eye, head, hand, foot, respiratory, and hearing protection. Why not CPC? One reason may be that there are so many different situations, hazardous materials, suit materials, and choices of accessories. One group, the National Fire Protection Association (NFPA), has provided standards for CPC for hazardous materials emergency response operations for over a decade. The following are the most common applicable standards to use in your selection process.

- NFPA 1991: *Standard on Vapor-Protective Ensembles for Hazardous Materials Emergencies,* 2000 Edition. NFPA 1991 specifies minimum requirements for design, performance, testing, and certification of elements of vapor-protective ensembles for emergency responders to hazardous materials incidents, including chemical or biological terrorism incidents, for protection from specified chemical vapor, liquid splash, and particulate exposures. The standard also provides additional optional requirements for protection from chemical flash fire protection, liquefied gas protection, and combined chemical flash fire and liquefied gas protection.
- NFPA 1992: *Standard on Liquid Splash-Protective Clothing for Hazardous Materials Emergencies,* 2000 Edition. NFPA 1992 specifies minimum requirements for the design, performance, testing, and certification of the elements of liquid splash-protective ensembles. These include suits, gloves, and footwear. Individual protective clothing items include garments, suits, gloves, and footwear for emergency responders to hazardous materials incidents. They protect wearers from specified chemical liquid splash exposures. This standard also provides an additional optional requirement for chemical flash fire protection.
- NFPA 1993: *Support Function Protective Garments for Hazardous Chemical Operations,* 2000 Edition. Similar in content to NFPA 1991, this standard describes the minimum requirements for support personnel. This includes the minimum level of chemical protective clothing needed for personnel who do not enter the exclusion zone (see Section 7, Decon, for the description of the zones at a HAZWOPER work site). This standard specifies that workers in decon are permitted to be dressed in the next-lowest level of CPC compared to the personnel they are deconning.

Note

NFPA, ASTM, and ANSI standards are examples of what are called *consensus standards*. That means that several groups of professionals, in the NFPA case firefighting professionals, agree that the standard, if followed, will minimize risk. In most cases, compliance with consensus standards is voluntary. However, in some cases OSHA has incorporated by reference into the regulations wording from NFPA and other consensus standards. In these cases the compliance with the consensus standard is mandatory.

The Safety Wearing Apparel Group of the Industrial Safety Equipment Association (ISEA) is currently drafting a standard for CPC. It is tentatively titled ANSI/ISEA 103-200x, *"American National Standard—Classification and Performance Requirements for Chemical Protective Clothing."*

CPC is worn to prevent hazardous materials from coming in contact with the skin or eyes. It provides the protective barrier between exposed body surfaces and hazardous materials that can have a detrimental effect on the skin or eyes or that can be absorbed through those surfaces and then affect other organs. CPC is usually worn with respiratory protection. CPC ranges from simple disposable coveralls costing about $3 each to fully encapsulating suits (FES) costing up to about $7500 each.

6.4.1 Keys to Selecting CPC

A checklist for the selection of chemical protective clothing fit for the job follows:

❑ Evaluate the work environment and the work you must perform.
❑ Know the types of exposures to hazardous materials that are likely to occur.
❑ Know the effects when bodily contact is made with any of those hazardous materials.
❑ Know the physical resistance of the available CPC material.
❑ Know the chemical resistance of the available CPC material.
❑ Consider the human factors of expected physical and heat stress.
❑ Evaluate the cost associated with the equipment.

CAUTION
IF YOUR COMPANY CANNOT AFFORD THE PROPER SAFE LEVEL OF
CPC, THE PROJECT CANNOT GO FORWARD!

6.4.2 Different Classifications of CPC

CPC is classified by:

• Style (vapor or splash protection)
• Components (suit, gloves, footwear)
• Construction material(s)
• Construction method (supported or unsupported)
• Method of sealing (zippers, tape, Velcro®, among others)
• Applicability to single (disposable) or multiple use

Disposable versus multiple use is based on ease of decontamination, construction materials, and quality of construction. A disposable fully encapsulating suit (FES) is commonly considered to cost less than $50. If a suit can be decontaminated, at the discretion of the supervisor it might be considered a multiple-use suit.

It is important for HAZWOPER personnel to know what CPC products are available for their protection or that of their workers. CPC suits are classified by the hazard level from which they protect the wearer. There are four levels of protection, Level A through Level D, illustrated in Figures 6.9 through 6.12.

CAUTION
CPC LEVELS DESCRIBED HERE ARE FOR HAZWOPER WORK ONLY.
THESE ARE NOT MADE OF THE SAME MATERIALS USED IN FIRE-
FIGHTING (WITH THE EXCEPTION OF THE REFLECTOR℠ SUIT).
THESE CPC FABRICS ARE REQUIRED, HOWEVER, TO BE FIRE RESIS-
TANT. THE GENERALLY ACCEPTED DEFINITION OF FIRE RESISTANCE
IS:
1. THE FABRIC MUST NOT BURN WITHIN THREE SECONDS OF CON-
TACTING A FLAME SOURCE.
2. IT MUST NOT CONTINUE TO BURN FOR MORE THAN TEN SEC-
ONDS AFTER BEING REMOVED FROM THE FLAME SOURCE.
3. IT MUST NOT MELT AND DRIP AFTER BEING REMOVED FROM THE
FLAME SOURCE.

FIGURE 6.9   Level A fully encapsulating suit and SCBA with extra exterior gloves and boots: Kappler Responder™ (top left) and Responder™ Plus® (top right) and Reflector™ (bottom). Service: maximum breathing protection with maximum skin protection. (*Courtesy of Kappler Protective Apparel and Garments* [http://www.kappler.com/].)

○━ᴨ   **Level A CPC provides the highest level of respiratory, skin, and splash protection.**

Checklist for Level A Protection.   The following is a checklist for Level A protection:

❑ Positive pressure SCBA or airline respirator
❑ Chemical-resistant fully encapsulating suit including integral gloves and booties
❑ Chemical-resistant steel-toe and shank boots
❑ Inner and out chemical-resistant gloves

FIGURE 6.10   MSA Level B suit with air-line respirator and with exterior gloves and boots. Service: lower breathing and skin protection than Level A. (*Courtesy of Mine Safety Appliance* [www.epartner.msanet.com/].)

❑ Coveralls*
❑ Long underwear*
❑ Hard hat worn under suit*
❑ Disposable protective booties*

Choose Level A protection when any of the following conditions exist:

• At least one hazardous substance is believed to be present, but has *not* been identified.
• At least one hazardous substance has been identified. Its presence requires the highest level of protection for skin, eyes, and respiratory system. That determination can be based on either the measured (or potential for) high concentration of atmospheric vapors, gases, or particulates.

---

* Optional.

FIGURE 6.11 Level C disposable hooded suit (Kappler CPF1®, left and CPF2®, right) and full-face APR, gloves and boots. Service: APR breathing protection with low skin protection. (*Courtesy of Kappler Protective Apparel and Garments* [http://www.kappler.com/].)

FIGURE 6.12 Level D disposable coveralls (Kappler Poly-Coat®). Shown with safety glasses, boots, and gloves (normal work uniform). Service: no breathing protection, minimal skin protection. (*Courtesy of Kappler Protective Apparel and Garments* [http://www.kappler.com/].)

- The site operations and work functions involve a high potential for splash, immersion in liquid, or exposure to vapors, gases, or particulates that can be absorbed through, or are irritants to, the skin.
- Operations are being conducted in confined, poorly ventilated areas. The absence of conditions requiring Level A has not yet been determined. This may include IDLH situations, including atmospheres where oxygen concentrations less than 19.5% or greater than 23.5%.

○━π **Level B provides the highest-level respiratory protection but less skin and splash protection than Level A.**

Checklist for Level B Protection.   Level B provides the highest level of respiratory protection, but a lesser level of skin and splash protection. It consists of:

❑ Positive pressure SCBA or SAR
❑ Hooded chemical-resistant splash suit (one- or two-piece)
❑ Inner and outer chemical-resistant gloves
❑ Chemical-resistant steel-toe and shank boots
❑ Coveralls*
❑ Long underwear*
❑ Hard hat worn under suit*
❑ Disposable protective booties*

Choose Level B protection when any of the following conditions exist:

- IDLH situations including O_2 concentrations less than 19.5% or greater than 23.5%.
- The presence of incompletely identified vapors or gases is indicated by a direct-reading organic vapor detection instrument; vapors and gases are not suspected of containing high levels of chemicals harmful to skin or capable of being absorbed through the skin.
- There is a minimal splash or immersion hazard.
- You are decontaminating Level A workers (see Section 7).

The primary difference between Level A and Level B is the integrity of the seals at the wrist and ankles and the facepiece of the respirator. The seals of Level B protection may be constructed with tape. For instance, in Level B the facepiece of the respirator can be the primary barrier, whereas the primary barrier of the Level A suit must be the suit with the secondary barrier being the respirator facepiece.

Note

Level B suits can be worn in atmospheres with IDLH concentrations of specific substances that present severe inhalation hazards but do not represent a severe skin hazard.

○━π **Level C provides air-purifying respiratory protection together with some skin protection.**

Checklist for Level C Protection.   Choose Level C protection when the work site atmosphere requires air purifying respiratory protection together with some skin protection. Level C consists of:

❑ Full-face or half-face air-purifying respirator
❑ Hooded chemical-resistant clothing
❑ Inner and outer chemical-resistant gloves
❑ Chemical-resistant steel-toe and shank boots

❑ Hard hat (face shield recommended)

❑ Optional escape respirator and coveralls

Level C is used when the concentration and type of airborne substances are known and the criteria for using air-purifying respirators are met. Choose Level C protection when:

- The atmospheric hazardous materials, liquid splashes, or other direct contact will not adversely affect, or be absorbed through, any exposed skin.
- The types of atmospheric hazardous materials have been identified, concentrations have been measured, and an air-purifying respirator is available that can remove the hazardous materials.
- Level B workers are to be decontaminated.

○─π **Level D protection provides no respiratory protection and only minimal skin protection.**

Checklist for Level D Protection.   Level D provides no respiratory protection and only minimal skin protection. It consists of:

❑ Gloves

❑ Steel-toe boots

❑ Safety glasses or goggles

❑ Optional hard hat and coveralls

Level D is used only for nuisance contamination. Primarily it keeps your clothes clean. It is an all-around work uniform affording minimal protection.

Choose Level D protection when:

- The atmosphere contains no known or suspected hazards
- Work functions preclude splashes, immersion, or the potential for unexpected inhalation of, or contact with, hazardous levels of any materials.

6.4.3 Performance Requirements of Chemical-Protective Clothing

The key performance requirements of chemical protective clothing are:

- Chemical resistance
- Durability and service life
- Penetration or permeation resistance
- Flexibility and wearer comfort
- Temperature resistance
- Tolerance to decontamination

Another way of characterizing CPC is by its encapsulating properties. The Level A suit is a fully encapsulating suit (FES), a one-piece garment that completely encloses the wearer. It offers the maximum protection from airborne hazardous materials and splashed liquids.

The Levels B and C can be nonencapsulating suits, frequently called splash suits. These do not need to have a facepiece as an integral part of the suit (the facepiece of the full-face respirator is a sufficient barrier).

Nonencapsulating suits are not designed to provide maximum protection against vapors, gases, or other airborne substances, but they do provide some protection against splashes. In effect, splash suits can be made on-site by taping wrist, ankle, and neck joints to enclose the wearer totally so that no part of the body is exposed. But such self-prepared units still are not considered to be gas-tight.

This leads us to what the authors call the "duct tape dilemma." Standard duct tape has no published national standard for chemical resistance. It is, however, used almost universally for sealing CPC in HAZWOPER work.

CAUTION

WE HAVE FOUND THAT DUCT TAPE IS ALMOST UNIVERSALLY USED TO SEAL SEAMS, ZIPPERS, AND OVERLAPS ON CPC. THIS USAGE AP-PEARS TO BE SATISFACTORY DUE TO DUCT TAPE'S EXCELLENT AD-HESION AND TENSILE STRENGTH, EVEN WHEN WET. NOTE, HOW-EVER, THAT DUCT TAPE IS NOT RECOGNIZED AS CHEMICALLY RESISTANT. DUCT TAPE SHOULD NEVER BE THE PRIMARY BARRIER BETWEEN YOU AND HAZARDOUS MATERIALS.

One solution to this dilemma is Chem-Tape®. See Figure 6.13. Kappler has developed a special tape that we recommend for use instead of duct tape. It is designed and tested specifically for chemical protection. It is durable but can be torn easily for quick application. It is also bright yellow so that you can easily identify your workers who are properly protected.

Chemical-Resistant Properties.   The effectiveness of materials to protect against chemicals is based on the material's resistance to *penetration, degradation,* and *permeation.* Each of these properties must be evaluated when selecting the style of chemical protective clothing and the material from which it is made.

Penetration

○—ᴫ   **Penetration is the transport of materials through openings in a garment.**

A chemical may penetrate an item of CPC due to design or garment imperfections. Stitched seams, buttonholes, pinholes, tears, zippers, and woven fabrics can provide a path-

FIGURE 6.13 Kappler Chem-Tape®. (*Courtesy of Kappler Protective Apparel and Garments* [http://www.kappler.com/].)

way for the chemical to penetrate the garment. A well-designed and constructed garment prevents this by using self-sealing zippers, seams overlaid with tape, flap closures, and non-woven fabrics. Rips, tears, punctures, or abrasions to the garment also allow penetration.

Tests for chemical penetration rely on visual observation to ascertain whether liquid flows through imperfections (e.g., pinholes) in the garment material or through seams and closures. This test is performed by inflating the suit with air and using a soapy water solution. If no bubbles are observed, the suit passes the penetration test. If ANY bubbles are noticed, immediately take the suit out of service for disposal or proper repair.

Degradation

O—π **Degradation, or aging, is a loss of desirable physical properties with time or chemical exposure. It involves the molecular breakdown of a material due to chemical attack, heat, sunlight, or other aging effect.**

CAUTION

SOME MANUFACTURERS OF CPC STATE THAT THEIR SUITS HAVE EX-PIRATION DATES (SOMETIMES REFERRED TO AS SHELF LIFE). THAT IS, WHETHER OR NOT THEY HAVE BEEN USED, THE MATERIALS WILL LOSE IMPORTANT PROPERTIES OVER TIME. AFTER THAT TIME, THE COMPANIES WILL NO LONGER WARRANT THE EFFECTIVENESS OF THE PROTECTION OF THOSE SUITS. THIS IS OF SPECIAL CONCERN FOR COMPANIES THAT USE THEIR CPC RARELY AND MAINTAIN THEM FOR EMERGENCY USE ONLY. KEEPING TRACK OF EXPIRATION DATES DURING ROUTINE INSPECTIONS WILL ALERT ORGANIZA-TIONS TO THE NEED FOR SUIT REPLACEMENT.

Degradation is determined by observation of physical changes to the material. These changes may be observed as discoloration, gummy surface, cracks or tears in the material, or the material becoming brittle. These changes may allow permeation or penetration by a contaminant. All CPC materials age, some more rapidly than others.

The deterioration of one or more physical properties of CPC, can cause the material to crack, tear easily, or, in extreme cases, dissolve in liquid, gas, or mists. Such failures can result in toxic amounts of hazardous material being transported through the CPC to the skin. Chemical degradation data is used by industrial hygienists to eliminate those candidate CPC materials that are unsuitable for use with specific chemicals or classes of materials. Degradation-resistant materials are then evaluated for chemical permeation and chemical penetration.

Degradation test data for specific chemicals or generic classes of chemicals are available from product manufacturers, suppliers, or other sources. The published data provide the user with a general degradation-resistance rating.

Permeation

O—π **Permeation is the passage of chemicals, on a molecular level, through intact material.**

Permeation is a three-step process including absorption of the chemical into the outer surface of a material; diffusion of the chemical through the material on a molecular level; and finally desorption of the chemical as a vapor on the inside surface of the material. (See Table 6.7 for permeation data for the Kappler Responder® Plus™ suit fabrics.)

Two pieces of information are sought during permeation testing: breakthrough time and permeation rate. See below some CPC Tests Defined, below for test information.

1. *Breakthrough time* (or breakthrough detection time) is the elapsed time, expressed in minutes, between initial contact of a chemical with the outside surface of CPC and the

TABLE 6.7 Permeation Data for the Kappler Responder Suit Fabric

Chemical name	Concentration (%)	Breakthrough time normalized (min)	Permeation rate (μg/cm^2/min)	SDL (ppm)
Acetaldehyde	99	>480	ND*	0.05
Acetic acid	99	>480	ND*	1.00
Acetone	100	>480	ND*	0.13
Acetonitrile	100	>480	ND*	0.11
Acetyl chloride	95+	>240	ND*	0.01
Acrolein	95+	>180	ND*	0.03
Acrylamide	50	>480	ND*	1.0
Acrylic acid	99	>480	ND*	0.1
Acrylonitrile	99	>480	ND*	0.01
Allyl alcohol	99	>480	ND*	0.30
Allyl chloride	95+	>180	ND*	0.04
Ammonia gas	100	>480	ND*	0.088
Ammonia liquid	100	>480	ND*	0.145
Amyl acetate	99	>480	ND*	0.084
Aniline	99	>480	ND*	1.00
Antiknock compounds	63	>480	ND*	0.01
Arsine	99	>180	ND*	0.78
Benzene	100	>480	ND*	0.01
Benzonitrile	99	>480	ND*	0.01
Benzyl chloride	99	>480	ND*	0.01
Borane-pyridine complex	99	>480	ND*	0.01
Boron trichloride	100	>480	ND*	0.05
Boron trifluoride	100	>480	ND*	0.1
Bromine	100	42	>197	0.10
Bromochloromethane	95+	>180	ND*	0.01
Butadiene 1,3-	99	>480	ND*	0.066
Butanol n-	99	>480	ND*	0.072
Butyl acrylate	99	>480	ND*	0.1
Butyl ether n-	99	>480	ND*	0.025
Butyl methyl ether t-	100	>180	ND*	0.21
Butylamine	99	>480	ND*	0.01
Butylene oxide 1,2-	95+	>240	ND*	0.01
Butyraldehyde	100	>480	ND*	0.03
Calcium chloride	42% w/w in water	>240	ND*	1.00
Carbon disulfide	100	>480	ND*	1.00
Carbon tetrachloride	Unknown	>480	ND*	0.6
Chlorine dioxide	5	>480	ND*	1.00
Chlorine gas	100	>480	ND*	0.041
Chlorine liquid	100	>480	ND*	5.00
Chlorine trifluoride	100	160	>53	1.00
Chloro-1,3 Butadiene 2-	99	>480	ND*	0.01
Chloroacetic acid	99	60	63.7	0.06
Chloroacetophenone	97	>480	ND*	0.5
Chloroacrylonitrile 2-	99	>480	ND*	0.16
Chlorobenzene	99	>480	ND*	0.5
Chloroform	99	>480	ND*	0.097
Chloropicrin	98	>480	ND*	0.1
Chlorosulfonic acid	99	>480	ND*	1.00
Chromic acid	60	>480	ND*	1.0
Cresols	Mixture	>480	ND*	0.01
Crotonaldehyde	99	>480	ND*	0.1
Cyclohexane	99	>480	ND*	0.01
Cyclohexylamine	99	>480	ND*	0.03

TABLE 6.7   Permeation Data for the Kappler Responder Suit Fabric (*Continued*)

Chemical name	Concentration (%)	Breakthrough time normalized (min)	Permeation rate (μg/cm^2/min)	SDL (ppm)
Diborane	10	>480	ND*	0.02
Dichloro-2-butene 1,4-	85	>480	ND*	0.3
Dichloroethylene 1,2-	95+	>180	ND*	0.01
Dichloromethane	100	>480	ND*	0.16
Dichloropropene 1,3-	95	>480	ND*	0.2
Diethanolamine	99	>480	ND*	1.0
Diethylamine	100	>480	ND*	0.71
Diisocyanatohexane 1,6-	98	>480	ND*	1.00
Dimethyl acetamide N,N-	100	>480	ND*	26.00
Dimethyl disulfide	99	>480	ND*	0.01
Dimethyl hydrazine	98	>480	ND*	0.04
Dimethylformamide N,N-	100	>480	ND*	1.0
Dimethylsulfate	99	>480	ND*	0.4
Diphenylmethane diisocyanate 4,4-	100	>480	ND*	1.00
Epichlorohydrin	95+	>180	ND*	0.03
Ethyl acetate	100	>480	ND*	0.1
Ethyl acrylate	99	>480	ND*	0.2
Ethyl benzene	99	>480	ND*	0.01
Ethyl ether	95+	>240	ND*	0.10
Ethyl methacrylate	95+	>240	ND*	0.01
Ethyl vinyl ether	95+	>180	ND*	0.02
Ethylamine	70 w/w	>240	ND*	0.02
Ethylene	99	>480	ND*	0.02
Ethylene diamine	99	>480	ND*	0.01
Ethylene dichloride	99	>480	ND*	0.053
Ethylene glycol	95+	>240	ND*	0.12
Ethylene oxide gas	100	>480	ND*	0.21
Ethylene oxide liquid	100	>180	ND*	0.083
Ethyleneimine	99	357	.032	0.01
Ferric chloride	Saturated	>480	ND*	1.00
Ferrous chloride	Saturated	>480	ND*	1.00
Fluorine	Tech	>480	ND*	0.014
Fluosilicic acid	30	>480	ND*	1.00
Formic acid	96	>480	ND*	0.1
Freon TF	Unknown	>240	ND*	10.0
Gamma-butyrolactone	100	>480	ND*	0.02
Hexachlorobutadiene 1,3-	98	>480	ND*	0.5
Hexamethyldisilazane	100	>480	ND*	0.09
Hexane	99	>480	ND*	0.06
Hexene 1-	99	>480	ND*	0.01
Hydrazine	98	>480	ND*	0.02
Hydrazine hydrate	Unknown	>240	ND*	0.03
Hydrobromic acid	48	>480	ND*	0.19
Hydrochloric acid	37% w/w	>240	ND*	0.20
Hydrofluoric acid	49–51	>180	ND*	0.025
Hydrogen chloride	99	>480	ND*	0.056
Hydrogen cyanide	98+	>480	ND*	0.50
Hydrogen fluoride anhydrous	99	>480	ND*	0.20
Hydrogen fluoride gas	99	>480	ND*	0.01
Hydrogen peroxide	70	>480	ND*	3.00
Hydrogen sulfide	100	>180	ND*	12.00

TABLE 6.7 Permeation Data for the Kappler Responder Suit Fabric (*Continued*)

Chemical name	Concentration (%)	Breakthrough time normalized (min)	Permeation rate (μg/cm^2/min)	SDL (ppm)
Isobutane	99	>480	ND*	0.01
Isobutanol	99	>480	ND*	0.3
Isobutyl benzene	99	>480	ND*	0.01
Isophorone diisocyanate	98	>480	ND*	1.00
Isoprene	95+	>180	ND*	0.01
Isopropyl alcohol	99	>480	ND*	0.01
Isopropylamine	99	>480	ND*	0.16
JP-4	Unknown	>240	ND*	0.01
Lewisite (L) chemical agent	99	>480	N/A	N/A
Malathion	50	>480	ND*	0.04
Maleic acid	Saturated	>480	ND*	1.0
Maleic anhydride	Saturated	>480	ND*	1.00
Mercury	100	>480	ND*	0.0003
Mesityl oxide	98	>480	ND*	0.11
Methacrylic acid	99	>480	ND*	0.1
Methane	99	>480	ND*	5.0000
Methanol	100	>480	ND*	0.02
Methyl acrylate	100	>480	ND*	0.19
Methyl bromide	100	>480	ND*	0.10
Methyl chloride	100	>480	ND*	0.089
Methyl ethyl ketone	99	>480	ND*	0.19
Methyl hydrazine	98	>480	ND*	0.04
Methyl iodide	100	>480	ND*	0.01
Methyl isocyanate	Unknown	>480	ND*	1.00
Methyl mercaptan	Unknown	>480	ND*	0.8
Methyl sulfoxide	95+	>240	ND*	0.01
Methylamine (gas)	98	>480	ND*	0.01
Methylene dianiline	15	>480	ND*	0.10
Methylisobutylketone	100	>480	ND*	0.048
Mustard (HD) chemical agent	99	>480	N/A	N/A
Nerve (VX) chemical agent	99	>480	N/A	N/A
Nicotine	98	>480	ND*	0.01
Nitric acid	70	>180	ND*	0.07
Nitric acid-red fuming	90+	>180	ND*	0.089
Nitrobenzene	100	>480	ND*	1.00
Nitrogen tetroxide	Unknown	220	7	0.051
Nitrogen trifluoride	100	>480	ND*	0.10
Nitromethane	99	>480	ND*	0.31
Nitrous oxide	100	>480	ND*	0.05
Octane	99	>480	ND*	0.02
Octel antiknock compound	99	>480	ND*	0.01
Oleum	40	>480	ND*	0.06
Organo-tin paint	Unknown	>240	ND*	0.03
PCB	Unknown	>240	ND*	0.01
Pentenenitrile 2-	100	>480	ND*	0.12
Phenethyl alcohol a-	100	>480	ND*	0.02
Phenol	85	>480	ND*	0.01
Phosgene	99	>480	ND*	0.10
Phosphine	99	>480	ND*	0.03
Phosphoric acid	85	>480	ND*	1.0
Phosphorous oxychloride	99	>480	ND*	1.0
Potassium permanganate	30	>480	ND*	0.025
Propane	99	>480	ND*	0.30

TABLE 6.7   Permeation Data for the Kappler Responder Suit Fabric (*Continued*)

Chemical name	Concentration (%)	Breakthrough time normalized (min)	Permeation rate (μg/cm^2/min)	SDL (ppm)
Propionaldehyde	100	>480	ND*	2.00
Propylene oxide	95+	>480	ND*	0.03
Pseudo cumene	90	>480	ND*	0.041
Pyridine	99	>480	ND*	0.23
Sarin (GB) chemical agent	99	>480	N/A	N/A
Silane	99+	>480	ND*	0.50
Sodium cyanide	45	>480	ND*	1.0
Sodium dichromate	0.5	>480	ND*	1.00
Sodium hydrosulfide	Saturated	>480	ND*	1.00
Sodium hydroxide	50	>480	ND*	0.2
Sodium hypochlorite	5.25	>480	ND*	3.00
Soman (GD) chemical agent	99	>480	N/A	N/A
Styrene	99	>180	ND*	0.068
Sulfur dichloride	80	448	.33	0.084
Sulfur dioxide	99	>480	ND*	0.04
Sulfur hexafluoride	100	>480	ND*	0.10
Sulfur trioxide	99	90	>100	1.0
Sulfuric acid	93	>480	ND*	0.019
Tert-butyl alcohol	100	>480	ND*	0.02
Tetrabromoethane	100	>480	ND*	0.02
Tetrachloroethylene	100	>480	ND*	0.081
Tetrahydrofuran	100	>480	ND*	0.098
Tetralone	98	>480	ND*	1.00
Thionyl chloride	99	45	243	1.00
Titanium tetrachloride	100	>480	ND*	1.00
Toluene	100	>480	ND*	0.031
Toluene diisocyanate	80	>480	ND*	0.06
Tribromophenol 2,4,6-	99	<15	4.93	1.00
Trichloroethane 1,1,1-	99	>480	ND*	0.30
Trichloroethylene	95+	>240	ND*	0.01
Triethoxysilane	95	>480	ND*	0.01
Triethylamine	99	>480	ND*	0.13
Trifluoroacetyl chloride	98	>480	ND*	1.00
Trimethylphosphite	97	>480	ND*	0.11
Triphenyl phosphite	99	>480	ND*	1.00
Tungsten hexafluoride	100	>480	ND*	0.20
Turpentine	Mixture	>480	ND*	0.01
Unleaded gasoline	Unknown	>240	ND*	0.01
Vinyl acetate	95+	>180	ND*	0.01
Vinyl chloride	100	>480	ND*	0.20
Vinylidene chloride	95+	>180	ND*	0.03
Xylenes	Unknown	>180	ND*	0.036

Source: Courtesy of Kappler Protective Apparel and Garments (http://www.kappler.com/).

chemical's detection, with instrumentation, at the inside surface of the CPC film or fabric. The detection time is the time it takes for a steady flow of at least 0.1 μg/cm^2/minute to take place.

2. *Permeation rate* for a specific chemical through a specific CPC material is determined when a steady (but extremely small) flow of liquid passes through the material. The flow is through a measured area of material. Permeation rate is expressed in μg/cm^2/minute.

Both parameters are useful in selecting chemical-protective clothing. Different brands of the same generic material may have different permeation properties for the same chemicals. This means that selection of material must be based on each specific manufacturer's permeation data. Also, the chemical permeation properties of chemical mixtures must be determined by testing, not inferred from the permeation characteristics of the individual components of the mixture.

CAUTION

1. NO PROTECTIVE MATERIAL IS TOTALLY IMPERMEABLE. ALL KNOWN CPC MATERIALS HAVE PERMEATION RATES FOR HAZARDOUS MATERIALS. SO-CALLED IMPERMEABLE MATERIALS HAVE VERY LOW PERMEATION RATES FOR SPECIFIC HAZARDOUS MATERIALS.
2. NO ONE MATERIAL AFFORDS PROTECTION AGAINST ALL CHEMICALS. THERE ARE FAVORED MATERIALS FOR PROTECTION AGAINST SPECIFIC MATERIALS.
3. FOR CERTAIN HAZARDOUS MATERIALS AND CHEMICAL MIXTURES NO MATERIALS ARE AVAILABLE THAT WILL PROTECT FOR MORE THAN ONE HOUR AFTER INITIAL CONTACT.

Some CPC Tests Defined

1. Chemical permeation: ASTM F739, *"Standard Test Method for Resistance of Protective Clothing Materials to Permeation by Liquids or Gases under Conditions of Continuous Contact"*

Explanation of test method: Material is clamped between two chambers. One is filled with chemical and the other is checked for the presence of chemical. Once chemical is detected, rate of permeation through the material is measured over time.

2. Chemical penetration: ASTM F903, *"Standard Test Method for Resistance of Protective Clothing Materials to Penetration by Liquids"*

Explanation of test method: The test material is clamped over opening of a test cell that is a hollow tube. An empty cell is clamped over the first one. The test cell is filled with the liquid that will be tested for permeation rate. The liquid is then forced against the material by pressurizing the cell with approximately 5 inches of w.c. If liquid droplets are seen on the opposite side of the fabric, the material has failed the test.

3. Chemical test battery: ASTM F1001, *"Chemical Test Battery"*

This is a battery of chemicals used for permeation testing. All chemicals are reagent grade. They include: acetone, acetonitrile, carbon disulfide, dichloromethane, diethylamine, dimethylformamide, ethyl acetate, hexane, methanol, nitrobenzene, sodium hydroxide, sulfuric acid, tetrachloroethylene, tetrahydrofuran, toluene, ammonia, 1,3-butadiene, chlorine, ethylene oxide, hydrogen chloride, and methyl chloride. This test battery was chosen to provide a broad spectrum of chemical families and some of the most common industrial chemicals used today.

6.4.4 CPC Fabrication Materials

⊙━━π **CPC is manufactured from a wide variety of materials to protect against a broad spectrum of chemical and physical hazards. However, there is no single garment that will protect the wearer from all chemical and physical hazards.**

CPC is classified partly on design and partly on the material from which it is made. All CPC materials fall into two general categories: elastomers and nonelastomers.

Elastomers are polymeric (plastic-like) materials that, after being stretched, return to their original shape. Most protective materials are elastomers. Elastomers may be supported (layered onto cloth-like material) or unsupported.

Elastomers include:

Natural rubber (polyisoprene)

Polyvinyl alcohol (PVA)

Chlorinated polyethylene (Cloropel®, CPE®)

Nitrile rubber (Buna-N®, NBR, Hycar®, Paracril®, Krynac®)

Polyvinyl chloride (PVC)

Neoprene (Chloroprene®)

Butyl rubber

Viton®

Teflon®

Polyurethane

Nonelastomers are materials that have limited stretching capability and might not return fully to their original shape. Although they can be very strong, they tend to tear rather than stretch. Nonelastomers include the spun olefins Tyvek® and Tyvek®-coated garments and other materials.

Nonelastomers include:

CPF1®-CPF4®

Pro/Shield®

Tyvek®

Tyvek® QC™ (polyethylene coated)

Front Line® (Dupont Barricade® material)

Saranex™ laminated Tyvek®

Polypropylene (Responder® and ResponderPlus®)

Polyethylene

Some Common CPC Materials and Their Protective Abilities

- *Butyl.* A synthetic rubber material that offers the highest permeation resistance to gas and water vapors. Especially suited for use with esters and ketones.
- *Neoprene.* A synthetic rubber material that provides excellent tensile strength and heat resistance. Neoprene is compatible with some acids and caustics. It has moderate abrasion resistance.
- *Nitrile.* A synthetic rubber material that offers chemical and abrasion resistance—a very good general-duty glove material. Nitrile also provides protection from oils, greases, petroleum products, and some acids and caustics.

- *PVC (polyvinyl chloride)*. A synthetic thermoplastic polymer that provides excellent resistance to most acids, fats, and petroleum hydrocarbons. Good abrasion resistance.
- *PVA (polyvinyl alcohol)*. A water-soluble synthetic material that is highly impermeable to gases. Excellent chemical resistance to aromatic and chlorinated solvents. This glove cannot be used in water or water-based solutions.
- *Viton®*. A fluoroelastomer material that provides exceptional chemical resistance to chlorinated and aromatic solvents. Viton® is very flexible but has minimal resistance to cuts and abrasions.
- *SilverShield®*. A lightweight, flexible laminated material that resists permeation from a wide range of toxic and hazardous chemicals. It offers the highest level of overall chemical resistance, but it has virtually no cut resistance.
- *4H*. A lightweight, patented plastic laminate that protects against many chemicals. Good dexterity.

A variety of manufactured materials exist which are used to fabricate CPC. Each of these materials provides a certain degree of protection against a range of chemicals. However, remember that no one material affords maximum protection against all chemicals.

Properly selected CPC can minimize risk of exposure to chemical substances but may not protect against physical hazards, such as fire, radiation, or electrical shock. The need for additional personal protective equipment for multihhazard protection must also be determined.

Performance of materials requires evaluation of three interactive parameters:

1. Chemical resistance of materials
2. Physical properties of materials
3. Human factors

Physical properties of CPC are important to barrier performance. Key physical properties are resistance to flexing, tearing, abrasion, cut, and puncture. Some ergonomic evaluations, such as hand grip, involve physical properties that are governed by glove thickness. This is especially true of dexterity. Surface texture is another important property. Grip is enhanced by a rough surface. Another human factor of concern is the physiological stress to workers induced by the wearing of CPC. Worker productivity is reduced stepwise as the level of protection increases. That is, productivity declines when moving from Level D up to Level A.

Temperature also plays a role in the CPC selection process. All materials become brittle at some lowered temperature. Brittle materials crack easily. At elevated temperatures, materials may become too soft and have a tendency to rip or otherwise fail.

As you can see, the selection process is a complicated one, with many variables. To be prepared for many different types of response action, you will need to have several types of suits available.

6.4.5 Maintenance of CPC

Makers of chemical-protective clothing usually include detailed instructions for the cleaning, testing, and storage of their suits. If a high-quality fully encapsulated suit is well maintained, it can give excellent service. Good care makes sense because the average price of these suits is in the range of $500–$2,500.

For the most part, suits should be cleaned with warm soapy water, rinsed completely of all soap residue, and dried. A word here about cleaning versus decontamination. Decontamination, universally referred to as decon, is the removal of all visible hazardous material from the suit. After this process, the suit still must be cleaned. Almost every manufacturer recommends air-drying versus any application of heat. After the suit is dry, if it is an FES, it should be air-tested for the proper seal.

The manufacturer of the suit will give specific instructions on how cleaning and testing should be accomplished. You may void the suit's warranty if you do not clean and test exactly as directed.

Some manufacturers want you to use their air pumps just so there is no testing mistake. The suit is pumped up with air to about 1 to 3 in. w.c. If the suit holds that pressure over 15 minutes, it passes the airtight test. If air is leaking out of the suit, apply a soapy water solution to the suit. Where you see bubbles, there are leaks. Some manufacturers allow you to fix the leaks without voiding the warranty. Others want you to send the suit back to the factory for repair.

After cleaning and air testing, the suit must have a final inspection and then be properly stowed in accordance with the manufacturers instructions. Use this checklist to inspect the suit for:

❑ Tears
❑ Cracks
❑ Signs of swelling
❑ Signs of stiffness
❑ Signs of deterioration
❑ Signs of discoloration
❑ Nonuniform coatings
❑ Pinholes (hold up to a light source to check)

Pay special attention to seams and closures. Look closely at the armpits, knees and groin area. These areas, along with the feet, take the most abuse.

6.4.6 CPC Cost

There will continue to be a wide range of costs for CPC suits through the year 2010. Our estimates depend on type of material from which the suit is fabricated, single-use or reusable, and the manufacturer.

Level A suits will continue to cost between $50 for disposables and $2,500 for mid-high range suits, with an excellent selection in the $500–$800 range for reusable suits. (These prices do not include multiple-hazard suits, such as those rated for fire service. Multiple-hazard rated or custom-made suits can cost several thousand dollars more.) Level B splash suits will continue to cost between $50 and $300. Disposable hooded coveralls will continue to cost between $5 and $30, with most in the $7–10 range when ordered in quantity.

6.5 OTHER PPE

In the previous subsections we discussed respiratory protection and chemical protective clothing. Now we will discuss the PPE that HAZWOPER workers will use daily in the field.

Note

The authors have over 100 years of combined hazardous materials experience. We strongly recommend that the minimum PPE for any HAZWOPER site be hard hat, long-sleeve shirt, long pants, safety glasses, gloves, and safety shoes or boots.

6.5.1 Occupational Eye and Face Protection (29 CFR 1910.133)

○━━π   **More than 95% of all occupational eye injuries could have been avoided if the worker had been wearing proper eye protection.**

Prevention of eye injuries requires that all HAZWOPER workers wear protective eyewear. This includes employees, visitors, contractors, or others passing through an identified eye hazard area. To provide protection for all personnel, a sufficient quantity of goggles, face shields, or safety glasses must be available. They must meet or exceed ANSI standard Z87.1-1989, *American National Standard for Occupational and Educational Eye and Face Protection.* This is to ensure the required amount of impact protection. Individuals wearing eyeglasses must be provided with either a suitable eye protector to wear over them or prescription safety glasses.

Suitable eye protection must be worn when employees are exposed to hazards from flying particles, molten metal, acids or caustic liquids, other chemical liquids, gases, or vapors, bioaerosols, or potentially injurious light radiation.

WARNING

WORKERS ON HAZMAT DUTY MUST NEVER WEAR CONTACT LENSES. WORKERS WILL PROBABLY ENCOUNTER AIRBORNE PARTICLES, GASES, AND VAPORS THAT COULD IRRITATE OR DAMAGE THE EYES. CONTACT LENSES ACT AS A TRAP FOR THESE HAZARDOUS MATERIALS. ALSO, THEY RESTRICT THE EYES FROM THE BENEFIT OF THE EMERGENCY DECONTAMINATING PROCESS OF AN EYEWASH STATION. FURTHER, CONTACT LENSES CAN BECOME "WELDED" TO A WORKER'S CORNEA THROUGH ULTRAVIOLET LIGHT OR CHEMICAL EXPOSURE. THE LENSES THEN MUST BE SURGICALLY REMOVED—A VERY PAINFUL PROCEDURE!

Here is your safety checklist for eye and face protection:

❑ Side shields, either attached or integral to the construction of the glasses, must be used when there is a hazard from flying objects.

❑ Splash-proof goggles and face shields must be used when there is a hazard from a chemical splash.

❑ Face shields must not be worn for primary eye protection.

❑ For workers who wear prescription lenses, eye protectors shall either incorporate the prescription in the design or fit properly over the prescription lenses.

❑ For impact protection, eyewear must be permanently marked to identify the manufacturer and compliance with ANSI standard Z87.1-1989.

❑ Equipment fitted with appropriate filter lenses must be used to protect against light radiation, such as from arc welding or lasers. Generic sunglasses, even with ultraviolet protection, are not adequate light filters.

❑ Emergency eyewash facilities meeting the requirements of ANSI Z358.1, *American National Standard for Emergency Eyewash and Shower Equipment,* must be provided in all areas where the eyes of any employee or, indeed, anyone at the site, may be exposed to hazardous materials, especially corrosives. All such emergency facilities must be located where they are easily accessible in an emergency. (An eyewash station saved the sight of one of the authors!)

Prescription Safety Eyewear.   OSHA regulations require that workers who wear prescription lenses, while engaged in operations that involve eye hazards, shall wear eye protection that incorporates the prescription in its design. Otherwise, the worker shall wear eye protection that can be worn over the prescription lenses (goggles, face shields). This must not disturb the proper position of the prescription lenses or the protective lenses. Personnel requiring prescription safety glasses must be supplied with them.

Our eyes are very sensitive to injury from hazards in the environment. Anyone who has worked in industry in the past several decades is accustomed to wearing safety glasses by now. OSHA, for good reason, is very involved in eye injury prevention. Thousands of disabling work injuries per year occur due to lack of common safety glasses. There are as many excuses for not wearing them as there are types and styles of glasses. People say "They do not look cool" or "I can not see out of them" or "They are uncomfortable." Miserably poor excuses when you consider the following incident:

In 1998, Emory University agreed to pay a $66,400 fine levied by OSHA for safety violations in connection with the death of a 22-year-old laboratory assistant who died of herpes B virus after getting a few drops of rhesus monkey saliva or urine in her eye. The incident occurred while she was moving a cage. She looked inside to see if the monkey was all right. Because the virus was not thought to travel through air, the victim did not wear eye protection. The incident was felt to be too minor to be of any significance. However, symptoms developed in about 10 days. She at first seemed to respond to antiviral medication in the hospital but died a few days later. This demonstrates how what seems an insignificant incident involving the eyes can be disastrous. The excuses for not wearing eye protection are not worth the risk, are they?

That being said, the manufacturers have listened to the complaints of the workforce. They now offer comfortable, prescription and nonprescription, ANSI Z-87.1-approved safety glasses in every style and color imaginable, as Figure 6.14 illustrates.

Often there is a need to provide eye protection to visitors to the site. Since we never know whether or not the visitors have prescription safety glasses, the least expensive route is visitors' glasses. These can be worn alone or over most prescription eyewear.

Depending upon the specific requirements of your site's Health and Safety Plan (see Section 16), your glasses may have to meet the Traffic Signal Recognition (TSR) requirement of ANSI Z80.3, *American National Standard for Nonprescription Sunglasses and Fashion Eyewear Requirements.* Willson® Eclipse™ glasses are available with TSR enhancement for seeing red, yellow, and green lights.

In very unusual cases where laser light may be encountered, for instance in some surveying operations, safety glasses may have to conform to ANSI Z136.1-1986, *American National Standard for Safe Use of Lasers,* as well as ANSI Z86.1.

When there is a need for greater face and eye protection, we must use goggles and sometimes face shields. Think of your safety glasses as the last line of defense. Splash-proof goggles and a face shield are what we need when pouring liquids. See Figure 6.15.

Eye and Face Protection Cost. Basic safety glasses are the cheapest protection available. Glasses that meet the ANSI Z87.1 standard can cost as little as $2! However, numerous different types, colors, shapes, designs, and sizes are available. Even with the large selection,

FIGURE 6.14 Willson® Cruiser™ (left), Willson® Prevail® (center), and Willson® Novus™ (right) safety glasses. (*Courtesy of Dalloz Safety* [www.christiandalloz.com/].)

FIGURE 6.15 MSA model 817697 impact-resistant (left), model 817698 splash-proof goggles (center), and model 817893 industrial face shield (right). (*Courtesy of Mine Safety Appliance* [www.msasafetyworks.com].)

it is difficult to find nonprescription safety glasses that cost more than $10 per pair. The Willson® TSR™ glasses are around $10. The really expensive safety glasses are the laser-protective ones. The dedicated laser-protection glasses (for one type of laser light only) start at around $40; the broadband spectrum laser safety glasses cost around $800 per pair.

With generic glasses, the lens material is the factor that determines the price. High-quality, coated (for UV-A and UV-B protection) polycarbonate lenses will be at about the $10.00 mark.

Generic prescription safety glasses cost between $30 and $150, depending on choice of style, manufacturer, and prescription strength and lens material.

Splash-proof and impact-resistant goggles are in the $5–10 range. Generic face shields cost between $15 and $30. Specialized face shields for welding and other UV light applications range from $45 to special auto-darkening lens shields that cost approximately $500.

6.5.2   Occupational Head Protection (29 CFR 1910.135)

OSHA mandates head protection. The minimum requirements are the ANSI Z89.1-1986. They may also be required to conform to test standard Z89.1-1997, *American National Standard for Personnel Protection—Protective Headwear for Industrial Workers—Requirements*. Hard hats do not protect the wearer from all falling objects. They will protect the wearer from frequent head injury due to bumping into scaffolding, hitting the head on low-clearance entries, and glancing blows from falling objects such as hand tools and materials such as nuts and bolts.

Hard hats (Figure 6.16) are selected on the basis of the hazards from which they protect the wearer. ANSI approved hard hats fall into the following types and classes:

FIGURE 6.16 New Willson full brim (left), Willson Alpha (center), and Willson Pro-System (right) hard hats. (*Courtesy of Dalloz Safety* [www.christiandalloz.com/].)

- Type I are those helmets with a full brim.
- Type 2 are brimless helmets with a peak extending forward from the crown.
- Class G (General) helmets are intended to reduce the danger of exposure to low-voltage electrical conductors, proof tested at 2,200 volts for 1 minute, with 3 milliamps maximum leakage. These were formerly Class A helmets.
- Class E (Electrical) helmets are intended to reduce the danger of exposure to high-voltage electrical conductors, proof tested at 20,000 volts. Class E helmets are first tested for impact resistance, then tested at 20,000 volts for 3 minutes, with 9 milliamps maximum current leakage; then tested at 30,000 volts, with no burn-through permitted. (These were formerly Class B helmets.)
- Class C (Conductive) helmets, sometimes called "bump hats," offer no rated voltage protection. They are designed for light weight, comfort, and impact protection. They are used in construction, manufacturing, refineries, and where there is a possibility of bumping the head against a fixed object and impact.

The generic hard hat is hardly a fashion statement. It is worn for protection only. For that reason, many people tend not to wear them. As with the advances in safety glass design, hard hat manufacturers have added style to their inventory in an attempt to encourage compliance with the OSHA standard. One such stylish hat is the western hard hat (Figure 6.17). In areas of the country where western hats are worn daily, the HAZWOPER worker can still be in style.

Workers can also support their favorite Major League Baseball, NFL, or NCAA team or NASCAR driver with licensed hard hats as shown in Figure 6.18.

Head Protection Cost.   Hard hats generally have a small variation in unit cost. Single units cost as little as $10 to as much as $30. There is a huge dropoff in cost when they are purchased in volume. Custom-imprinted hats can be purchased for as little as $5 per hat with orders in the 2000 unit and above range.

FIGURE 6.17   Western Outlaw® hard hats. (*Courtesy of Imprint Technologies, LLC* [www.customhardhats.com/].)

FIGURE 6.18 NASCAR and team hard hats. (*Courtesy of Imprint Technologies, LLC* [www.customhardhats.com/].)

6.5.3 Occupational Foot Protection (29 CFR 1910.136)

According to OSHA, nearly 100,000 foot and toe injuries occur yearly. Three out of four of these injuries are due to lack of protective footwear.

Safety shoes or boots (Figure 6.19) must be worn on all HAZWOPER job sites. All safety footwear must comply with ANSI Z41-1991, *American National Standard for Personal Protection—Protective Footwear.* Under certain circumstances, they may also have to conform to ANSI Z41.3. *American National Standard for Men's Conductive Shoes.*

Safety shoes or boots with impact protection are required to be worn in certain work areas. They are required when carrying or handling materials such as heavy packages, objects, parts, or tools. Also included are other activities where heavy objects might fall onto the workers' feet.

Safety shoes or boots with compression protection are required for work activities involving vehicles such as fork trucks, skid loaders, and backhoes. That protection is required in other activities in which materials or equipment could potentially roll over a worker's foot, such as a 55-gallon drum. Compression protection is in the form of metatarsals. A metatarsal is a steel plate tongue that spreads the impact over a wider area of the top of the foot and lessens the impact to any one place on the foot. Safety shoes or boots with puncture protection (full steel shank) are required where sharp objects such as nails, glass, screws, rusted drums, or scrap metal could be stepped upon by workers. Notice in Figure 6.20 that the toe area of the boot appears much higher than in a normal boot. This is because of the metatarsal protection.

Safety boots and boot covers (booties—boot covers that are usually disposable) are also available in a wide variety of chemical-resistant materials. If you are wearing boots for

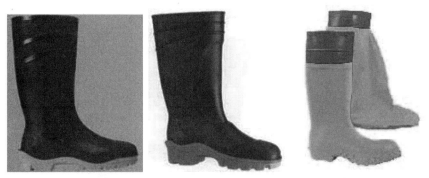

FIGURE 6.19 Bata Apollo® (left) and Mercury XCP® (center) steel-toe boots and 12-in. Latex HAZMAT boot covers (right) (*Courtesy of Kelley Supply, Inc.* [http://www.kelleysupply.com/].)

FIGURE 6.20 LaCrosse® full metatarsal guard—TracLite. (*Courtesy of Kelley Supply, Inc.* [http://www.kelleysupply.com/].)

chemical protection, make sure that they are resistant to the materials you expect to encounter. As with other PPE and CPC, boot manufacturers will also supply you with chemical-resistance data for their products.

As with other PPE, a wide selection is available. Boots are designed for every hazard imaginable. They can offer electrical protection (nonconductive fiberglass toe protection versus the more common steel toes). They also come in a wide variety of tread configurations. Special design allows easy decon of caked-on hazardous materials. They are also designed with slip-protective soles for use in wet, oily, or other slippery walking surfaces.

Safety toes and metatarsals are also available independent of the boots themselves. These are units that attach to the shoe or boot that the worker is wearing, converting them to safety shoes. There are also wide varieties of shoe and boot configurations and constructions. For work in a warehouse or factory floor, some of the shoes that resemble dress shoes or sneakers may be appropriate. However, for HAZWOPER work we need both the physical protection of approved steel-toed boots and the addition of chemical resistance. For that reason, we have illustrated this subsection with the boots that we believe are appropriate.

Remember, if you feel any itching, burning, numbness, or other than normal sensation in your feet, evacuate the area and decon your feet immediately! This may be a sign of chemical breakthrough and contamination.

6.5.4   Occupational Hand Protection (29 CFR 1910.138)

WARNING

HAND PROTECTION SHALL NOT BE REQUIRED WHERE THERE IS A DANGER OF THE HAND PROTECTION BECOMING CAUGHT IN MOVING MACHINERY OR MATERIALS. ALSO, WHILE WORKING AROUND ROTATING MACHINERY, WORKERS MUST ROLL UP THEIR SLEEVES AND REMOVE RINGS, WATCHES, AND BRACLETS. MANY PEOPLE HAVE BEEN SEVERELY INJURED WHEN THEIR CLOTHING HAS DRAWN THEM INTO MACHINERY. WHILE CLIMBING LADDERS BAREHANDED, WORKERS MUST REMOVE RINGS. THE AUTHORS KNOW SEVERAL PEOPLE WHO HAVE LOST FINGERS IN NONFATAL FALLS, BECAUSE THEIR RINGS GOT CAUGHT ON A PROTRUSION DURING THE FALL. ALSO, WHILE GLOVES ARE WORN FOR CHEMICAL PROTECTION, THE SLEEVES OF THE

WEARER'S OUTER GARMENT MUST BE WORN OUTSIDE OF THE CUFFS OF THE GLOVES OR TOPS OF THE BOOTS AND TAPED IF APPROPRIATE. THIS AVOIDS COLLECTION OF HAZARDOUS MATERIALS IN THE GLOVE OR BOOT.

When gloves are worn with long sleeves and you are working around hazardous materials, always wear the cuffs of the gloves inside the sleeves to avoid collection of material in the glove.

Suitable gloves must be worn when hazards from chemicals, cuts, lacerations, abrasions, punctures, burns (thermal or electrical), and biologicals are present. Glove selection must be based on performance characteristics of the gloves, conditions, durations of use, and hazards present. There is no one type of glove that protects against all hazardous situations.

The first consideration in the selection of gloves for protection against chemicals is to determine, if possible, the exact nature of the substances to be encountered. Read instructions and warnings on chemical container labels and MSDSs before working with any chemical. Recommended glove types are often listed in the section for personal protective equipment.

Chemicals eventually permeate all glove materials. However, they can be used safely for limited time periods if specific use and other characteristics (i.e., thickness and permeation rate and time) are known. The Site Safety Officer (refer to Section 16) or your supervisor can assist in determining the specific type of glove material that should be worn for a particular chemical.

Latex gloves are often worn when sampling. Make sure they are sufficient to protect against the hazard.

Choosing the proper glove for the job can be frustrating. Historically, the glove of choice for sampling hazardous waste has been the latex glove (Figure 6.21). However, relying on latex gloves was a fatal mistake in one case we are aware of involving methyl mercury (see Section 5). Some individuals also have latex allergies. When a latex allergic worker is exposed, the effects can range from irritation or rash outbreaks on the hand to respiratory system inflammation (anaphylactic shock) and even death.

An especially good resource for the chemical resistant properties of gloves is Best Gloves' *Guide to Chemical Resistant Best Gloves;* see www.chemrest.com/. This site allows you to select the proper glove for the task based on a number of criteria. You can search by chemical name, Chemical Abstract Service (CAS) number, chemical-by-chemical class, or for all the data on a specific glove.

FIGURE 6.21   Best® Master® natural rubber (latex) glove. (*Courtesy of Best Manufacturing* [http://www.bestglove.com/].)

If you are wearing gloves for chemical protection, make sure that they are resistant to the materials you expect to encounter. If you feel any itching, burning, numbness, or other than normal sensation in your hands or fingers, evacuate the area and decon your hands. This may be a sign of chemical breakthrough and contamination.

FIGURE 6.22   Best® Black Knight™ PVC, Best butyl rubber, Best Ultraflex® nitrile. (*Courtesy of Best Manufacturing* [http://www.bestglove.com/].)

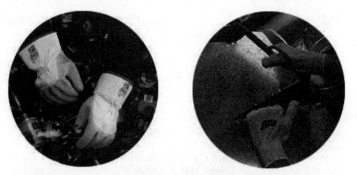

FIGURE 6.23   Best® The Original Nitty Gritty® natural rubber-coated glove (left) and NRK™ cut-resistant, Kevlar® aramid glove (right). (*Courtesy of Best Manufacturing* [http://www.bestglove.com/].)

FIGURE   6.24   Best®   Fuzzy Duck™ PVC coated glove. (*Courtesy of Best Manufacturing* [http://www.bestglove.com/].)

TABLE 6.8 Resistance to Chemicals of Common Glove Materials (E = Excellent, G = Good, F = Fair, P = Poor, NR = Not recommended)

Chemical	%	Neoprene	Nitrile	PVC	Natural rubber	Butyl rubber	Viton
Acids, mineral							
Chromic*		F	F	F	NR	P	E
Hydrochloric (HCl)	10%	G	G	G	G	—	—
Hydrochloric (HCl)	36%	F	F	P	F	E	E
Hydrofluoric	10%	G	G	G	F	G	G
Nitric*	10%	G	F	G	F	F	G
Nitric*	20%	F	F	F	P	F	G
Sulfuric*	10%	E	E	E	G	G	E
Sulfuric*	20%	E	E	E	F	G	E
Acids, organic							
Acetic	84%	F	F	E	G	G	P
Citric		G	G	G	G	E	E
Formic		G	F	G	G	E	F
Lactic	88%	G	E	E	E	E	E
Oxalic		G	G	G	G	E	E
Alcohols							
Benzyl		G	G	G	F	G	E
Ethyl		E	E	E	G	E	G
Methyl		G	F	F	G	E	P
Aldehydes							
Acetaldehyde*		G	F	G	F	E	P
Benzaldehyde		P	G	F	P	E	P
Formaldehyde*		G	G	E	G	E	P
Aliphatic solvents							
Mineral spirits		E	E	E	P	—	—
Alkalis							
Ammonium hydroxide	26%	G	E	E	G	E	G
Potassium hydroxide (KOH)	45%	G	E	E	G	E	F
Sodium hydroxide* (NaOH)	50%	G	E	G	G	E	G
Aromatic solvents							
Benzene*		P	E	P	NR	P	G
Stoddards*		G	G	F	P	P	E
Toluene*		P	F	P	NR	P	E
Xylene*		P	E	F	NR	P	E
Chlorinated solvents							
Carbon tetrachloride*		F	F	P	NR	P	E
Chlorobenzene*		P	F	P	NR	P	E
Perchloroethylene*		P	P	P	NR	P	E
Trichloroethylene*		P	P	F	NR	P	E

TABLE 6.8 Resistance to Chemicals of Common Glove Materials (E = Excellent, G = Good, F = Fair, P = Poor, NR = Not recommended) (*Continued*)

Chemical	%	Neoprene	Nitrile	PVC	Natural rubber	Butyl rubber	Viton
			Esters				
Butyl acetate*		F	F	P	P	G	P
Ethyl acetate*		F	F	P	P	G	P
			Amines				
Diethylamine		G	G	G	F	G	P
Methylamine		F	F	P	F	G	—
			Ethers				
Ethyl ether*		G	G	P	F	G	P
			Oils and fats				
Airplane hydraulic oil (Texaco BB)		F	F	P	P	P	E
Animal fats		G	G	G	P	G	E
Cutting oil (Rigid®)	10%	F	G	E	F	F	E
Linseed oil		F	G	F	P	G	E
Mineral oil		G	G	F	P	P	E
Vegetable oil		F	G	F	F	E	E
			Oxides				
Carbon dioxide		G	G	G	G	—	—
Nitrous oxide		F	F	G	F	—	—
			Ketones				
Acetone*		F	P	P	G	E	P
Methyl ethyl (MEK)*		P	NR	NR	G	E	P
Methyl isobutyl*		F	P	NR	G	G	P
			Salts, inorganic				
Copper sulfate		G	G	G	G	—	—

Source: Courtesy of Best Manufacturing (http://www.bestglove.com/).
*A known or suspected carcinogen.

Gloves, like other CPC, come in many different materials (Figures 6.22, 6.23, and 6.24), shapes, and construction. They can have no cuff, short cuffs, or cuffs that are sleeve-length. One term to be familiar with is *supported*. Gloves are supported if the chemical resistant fabric is overlaid or coated on some substrate such as cotton, leather, or other material. If the glove is unsupported, it is constructed of the chemical-resistant material alone. Supported gloves are usually more resilient to wear but not necessarily more resistant to chemical attack. They can also affect the wearer's dexterity.

Gloves can also be impregnated with resistant material. In this case, the glove itself may be cotton or man-made fiber, with the chemically resistant material absorbed into the glove material.

A glove's chemical resistance will depend on the type and thickness of the chemically resistant material of which it is constructed.

When choosing the proper glove, as with any other PPE, you must know what chemical and physical hazards you will encounter. Table 6.8 will help aid your selection.

When dealing with rusty drums and broken glass, a combination of a chemical-protective material with a puncture-protective material is required. A leather-palmed glove, or gloves constructed from the DuPont Aramid® fiber Kevlar®, may be indicated. The addition of chemical protection to cut protection may also be achieved by layering cut-protective gloves on the outside and chemical-resistant gloves on the inside.

Much of the time, HAZWOPER work may require not chemical or cut protection, but general protection of the hands from abrasions while handling containers, pallets, tools, and the like. In those cases we have found that limited-use gloves, often called disposable, are a good choice. They provide adequate protection and are inexpensive so that disposal due to contamination will not be costly.

As with all PPE, cost should be the last consideration, but it is an important one. Truly disposable nitrile or latex gloves cost around $12 per 100 pairs. A good-quality nitrile rubber reusable glove can be purchased for about $3–4 per pair. Cut-resistant Kevlar® gloves are in the $4–5 range per pair. Specialty heat- and cold-resistant gloves start at about $45 per pair.

6.5.5 Occupational Hearing Protection (29 CFR 1910.95)

O—π **Hearing loss, although the most common occupational disease, is the only one that is almost 100% preventable.**

"Work-related hearing loss is one of the most common occupational diseases in the United States," says Dr. Linda Rosenstock, director of NIOSH. NIOSH estimates that some 30 million workers are subjected to hazardous noise levels at work. NIOSH further estimates that of those workers, some 10 million have suffered some degree of permanent hearing loss. Experts agree that prolonged exposure to noise above 85 decibels (dBA) can cause hearing loss. 'Boom-box' listeners take note!

Beside the loss of our hearing, loud noises in the work area make us less aware of the activities around us. You must always be aware of your surroundings and the activities taking place there.

Not only is hearing loss common, but with the proper PPE it is also almost 100% preventable. Remember that engineering controls must be applied first. According to 29 CFR 1910.95 and 29 CFR 1926.52, the Action Level for Occupational Noise exposure is 85 dBA. Whenever employee noise exposures equal or exceed an 8-hour time-weighted average sound level (TWA) of 85 dBA measured on the A scale, the employer shall administer a continuing, effective hearing conservation program. Protection against the effects of noise exposure must be provided when the sound levels exceed those shown in Table 6.9. Hearing protection is tested in accordance with ANSI S3.19-1974, *American National Standard Method for the*

TABLE 6.9 OSHA Permissible Noise Exposures and Durations

Duration per day, hours	Sound level dBA slow response
8	90
6	92
4	95
3	97
2	100
1½	102
1	105
½	110
¼ or less	115

Source: Courtesy of OSHA.

TABLE 6.10 Common Workplace Sounds and Their dBA Rating

Sound	dBA
Weakest sound heard by the average ear	0
Whisper	30
Normal conversation	60
Ringing telephone	80
Belt sander	93
Tractor	96
Hand drill	98
Impact wrench	103
Bulldozer	105
Spray painter	105
Chain saw	110
Hammer drill	114
Pneumatic percussion drill	119
Ambulance siren	120
Jet engine at takeoff	140
12-gauge shotgun	165
Rocket launch	180

Measurement of Real-Ear Protection of Hearing Protectors and Physical Attenuation of Earmuffs.

If you are wondering how these measurements translate into your HAZWOPER working environment, consider Table 6.10, which lists common workplace sounds and their measured dBA ratings.

Now that you know the damage that common noises can cause to your hearing, what should you do about it? Hearing protection, of course! Hearing protection comes in two main types: earplugs and earmuffs (Figures 6.25 and 6.26). The noise reduction rating (NRR), or noise attenuation, of plugs or muffs (measured in dBA) depends on the manufacturer. The NRR means that, using Table 6.10, if you wear plugs with an NRR of 19 while using an impact wrench (103 dBA), your hearing will be protected to 84 dBA.

Plugs come in two subcategories, disposable and reusable. They usually have a NRR of approximately 20 dBA.

FIGURE 6.25 MSA Form Fit® disposable earplugs (left) and MSA Tri-Seal® reusable ear plugs (right). (*Courtesy of Mine Safety Appliance* [www.epartner.msanet.com/].)

FIGURE 6.26   MSA Apex 30™ (NRR = 30) earmuffs (left) and MSA Defender™ frame with Sound Blocker 26 (NRR = 26) earmuffs (right). (*Courtesy of Mine Safety Appliance* [www.epartner.msanet.com/].)

Earmuffs can be worn alone or in conjunction with earplugs. The NRR of muffs depends on the style and the manufacturer.

As you can see in Figure 6.26, muffs generally provide a higher NRR than do plugs. The combined effect of plugs and muffs used simultaneously is roughly cumulative. If you used plugs with an NRR of 20 dBA along with muffs with an NRR of 30 dBA, the cumulative effect would be a NRR of approximately 50 dBA.

A common way of ensuring that HAZWOPER workers are wearing the proper PPE is to use combination kits. As you can see in the MSA Defender™, combination kits provide head, eye and face, and hearing in one package.

A relative newcomer to the hearing protection arena is electronic earmuffs (Figure 6.27). These earmuffs, through microelectronic circuitry, attenuate instantaneous loud noises such as gunshots or pile drivers, while amplifying soft noises such as whispers. These muffs cost about five times more than standard muffs but make up for the additional cost with enhanced utility, especially for workers who already suffer from some hearing loss.

Disposable earplugs cost start at around 10 cents per pair. Earmuff costs vary widely, from around $8 to over $200 per pair.

A good rule of thumb is, if you have to raise your voice above normal conversation when speaking to someone within arm's length at the job site, you need hearing protection.

FIGURE 6.27   Peltor Tactical 6 stereo electronic earmuffs. (*Courtesy of EAR, Inc. Specialized Hearing Systems* [www.earinc.com/].)

6.5.6   Personal Fall-Arrest Systems (1926.501(6)(1) *et seq.*)

CAUTION

HERE WE WILL BE DISCUSSING PERSONAL FALL-PROTECTION SYS-
TEMS. THIS SECTION IS IN NO WAY INTENDED AS TRAINING FOR
WORKING ON, OR ERECTION AND INSPECTION OF, SCAFFOLDING!
OSHA HAS VERY SPECIFIC REGULATIONS ON ANY WORK ASSOCI-
ATED WITH SCAFFOLDING.

O—π   **The Bureau of Labor Statistics reports almost 600 occupational fatalities occurring from "Fall to Lower Level" each year. While this number is down
nearly 50% from the 1990s, falls still account for nearly one in five workplace
fatalities!**

OSHA requires that "Each employee on a walking/working surface (horizontal and vertical surface) with an unprotected side or edge which is 6 feet (1.8 m) or more above a lower
level shall be protected from falling by the use of guardrail systems, safety net systems, or
personal fall arrest systems."

Guardrails are employed on such structures as scaffolding and catwalks and at the edges
of roofs. The top rail must be 42 in. plus or minus 3 in. above the walking or working
surface (work platform). These systems must also have midrails halfway between the walking
or working surface and the top rail. Midrails may be omitted if there is a wall or parapet of
at least 21 in. below the top rail. These systems may be constructed of piping-type material
such as scaffolding or of rigid cables. Either way, they must be able to withstand a force of
200 lb, applied within 2 in. of the top edge, in any outward or downward direction, along
their entire length.

Safety nets are usually employed where there is danger of personnel, tools, or materials
falling from a work area spanning a falling hazard. The net may also serve the purpose of
protecting others below the net. These nets must be cleaned daily of tools and debris.

Guardrail systems and nets are part of engineering controls. They are so-called passive
systems. The worker does not have to wear any fall-protective device. Active systems are
the full-body harnesses with shock-absorbing lanyards, properly anchored.

O—π   **The single most important piece of fall protection PPE is the full-body harness
with shock-absorbing lanyard properly connected and anchored.**

O—π   **A personal fall-arrest system has three different components: the harness, the
connecting devices, and the anchoring (or tie-off point).**

Personal fall-arrest systems must conform with ANSI Z359.1, *American National Standard for Safety Requirements For Personal Fall Arrest Systems, Subsystems and Components.*

1. *The harness assembly* (Figure 6.28) must be capable of stopping the wearer's fall and
   must not allow the wearer to come into contact with a lower level. The harness must be
   configured in such a way that the worker cannot fall out of it. Proper fitting is essential!
   The harness is fit around the top of the worker's legs, at the waist or chest, and around
   the shoulders.

Note

So-called safety belts that only fit around the waist are no longer permitted for
fall-arresting devices. They can, however, be used for work-positioning devices.
These are the belts that electrical linemen and tree surgeons wear to hold themselves in position so that they can safely use both hands.

FIGURE 6.28 Miller® model #E552 full-body harness with shock-absorbing lanyard and beam strap anchoring device. (*Courtesy of Dalloz Safety* [www.christiandalloz.com/].)

FIGURE 6.29 Miller® self retracting lifeline and self-closing snap hook. (*Courtesy of Dalloz Safety* [www.christiandalloz. com/].)

2. *Connecting devices,* such as shock-absorbing lanyards, D-rings, snap hooks (Figure 6.29), and lanyards, must have a minimum breaking strength of at least 5000 lb. Snap hooks must be self-closing (spring-loaded). Connecting devices must be configured for easy hookup and unhooking. Tying off with a rope is not permitted. The shock-absorbing lanyard is not permitted to extend the fall more than $3\frac{1}{2}$ ft.

3. Connecting devices must be connected to an *anchoring point* capable of withstanding 5000 lb of force. This does not include scaffolding. Portable anchoring devices such as straps that fit around beams are permitted so long as they are protected with anti-chafing devices. These are pads that protect the straps from sharp edges such as on I-beams.

The D-ring connector must be placed at the back of wearers between their shoulders (Figure 6.28). The belts of the harness must be at least $1\frac{5}{8}$ in. wide. The harness is usually connected to a single use lanyard. (A shock-absorbing lanyard must be destroyed after it has arrested a fall. They are not reusable.) The lanyard connects to the harness and the anchor point (Figure 6.30) with self-closing snap rings.

These are called fall-arresting harnesses because they do not keep the worker from falling, but arrest the fall. A 6-ft worker with a 6-ft shock-absorbing lanyard will actually fall 15 ft from the anchor point. This is important to remember because if the lower level is 12 ft down, the fall arresting system will not protect the worker!

We have chosen to illustrate Miller® brand full-body harnesses (Figure 6.31) because they meet and exceed all applicable OSHA, ANSI, and CSA requirements, including ANSI Z359.1. These harnesses are comfortable and easy to wear. They do not interfere with the workers' capability to do their jobs. They come in various sizes to fit most workers. They also have models made specially for the comfort of women workers that take women's figures into consideration.

FIGURE 6.30 Miller® beam anchors. (*Courtesy of Dalloz Safety* [www.christiandalloz.com/].)

As with all PPE, they must be inspected before each use. If any defects, such as worn or missing parts, holes, cuts, or tears, are noted, the harness must be taken out of service until it can be certified safe.

WARNING

IF THE HARNESS LANYARD INCLUDES A SHOCK ABSORBER AND HAS ARRESTED A WEARER'S FALL, THE HARNESS IS NO LONGER SAFE. SHOCK ABSORBERS ARE DESTROYED AS THEY ARREST THE FALL. IF THE LANYARD IS CONNECTED TO THE HARNESS IN SUCH A WAY THAT IT CAN BE REMOVED AND REPLACED WITH A NEW ONE (AS MOST ARE), THE HARNESS ASSEMBLY MAY, AFTER THIS REPAIR, BE PUT BACK IN SERVICE.

IF THE LANYARD ASSEMBLY IS INTEGRAL TO THE HARNESS ASSEMBLY, THE ENTIRE ASSEMBLY MUST BE DESTROYED TO PREVENT FUTURE USE. IT WILL NO LONGER SAFELY PROTECT THE WEARER FROM A FALL!

According to OSHA, "Self-retracting lifelines (Figure 6.29) and lanyards which automatically limit free fall distance to two feet or less shall be capable of sustaining a minimum tensile load of 3,000 pounds applied to the device with the lifeline or lanyard in the fully extended position." These devices act like the seatbelt in your car. They allow you to work freely, but engage when they are pulled quickly as in a fall.

FIGURE 6.31 Other MILLER® full-body harnesses. (*Courtesy of Dalloz Safety* [www.christiandalloz.com/].)

Safety harnesses came originally from parachute packs. They were then adopted by auto racing and now are saving numerous lives and injuries in the workplace. Hunters use them to protect themselves from falls from tree stands.

6.6 ERGONOMICS

Ergonomics is the study of how the human body reacts to workplace conditions. In 2001, the OSHA's proposed Ergonomics Program Standard, 29 CFR 1910.900, failed to pass through Congress. This is the first time in history that a standing safety regulation has been struck. The authors are confident that an ergonomics standard will be passed in the near future.

The purpose of ergonomic equipment is to lessen or eliminate Work Related Musculo-skeletal Disorders (MSDs). These are injuries due to certain repetitive motions, awkward work placement, force, vibration, and contact stress.

Although the term *ergonomics* was not common until the 1980s, the principles go back to the late 18th century. Factory owners noted that workers who stood all day on stone floors became fatigued more easily. The factory owners then started making floors out of wooden blocks and that alleviated the fatigue somewhat. Today, we hardly ever see anyone who stands at a machine all day without ergonomic rubber mats to help with fatigue.

There are products available for aid for nearly every type of ergonomic injury. Since back injuries from improper lifting technique or attempting to lift too much weight are quite common, back support belts can help alleviate those injuries. Carpal tunnel syndrome is caused by repetitive motion of the wrist and vibrations to the hand. Various wrist braces and supports are available, as well as impact and antivibration gloves. Forearm braces help protect workers from "tennis elbow," quite common in the carpentry and construction trades. Elbow pads and supports are also available. Knee pads, for masons, concrete finishers, flooring and carpet installers and anyone else on their knees during work assist in eliminating knee pain and injury.

6.7 HAZARD ASSESSMENT FORM

In order to demonstrate that a proper hazard assessment of the workplace has been conducted, fill out the following form for each work site or each operation at a worksite. Although not strictly required by the OSHA standard, it is an excellent way to ensure and record that all known or suspected hazards were identified and that proper PPE was assigned.

Hazard Assessment Checklist and Certification Form

Date:	Location:
Assessment Conducted by:	
Specific Tasks Performed at This Location:	

Hazard Assessment and Selection of Personal Protective Equipment

1. Overhead hazards to consider include:
- ❏ Suspended loads that could fall
- ❏ Overhead beams, scaffolding, pipes, or loads that could be hit against
- ❏ Energized wires or equipment that could be hit against
- ❏ Employees working above who could drop objects on others below
- ❏ Sharp objects or corners at head level

Hazards Identified:

Head Protection

Hard Hat:	Yes	No
If yes, type:		

- **Type A** (impact and penetration resistance, plus low-voltage electrical insulation)
- **Type B** (impact and penetration resistance, plus high-voltage electrical insulation)
- **Type C** (impact and penetration resistance)

2. Eye and face hazards to consider include:
- ❏ Chemical splashes
- ❏ Dust
- ❏ Gases
- ❏ Smoke, fumes, and vapors
- ❏ Welding, chipping, and grinding operations
- ❏ Lasers/optical radiation
- ❏ Biohazards
- ❏ Projectiles

Hazards Identified:

Eye Protection

Safety glasses or goggles	Yes	No
Face shield	Yes	No

3. Hand hazards to consider include:

- ❏ Chemicals, including corrosives, solvents, and irritants
- ❏ Broken glass, nails, sharp edges, and splinters
- ❏ Temperature extremes
- ❏ Biohazards
- ❏ Exposed electrical conductors
- ❏ Sharp tools such as shears, striking tools such as chisels
- ❏ Material-handling conveyers

Hazards Identified:

Hand Protection

Gloves	Yes	No
• Chemical-resistant		
• Temperature-resistant		
• Abrasion-resistant		
• Other:		

WARNING

HAND PROTECTION SHALL NOT BE REQUIRED WHERE THERE IS A DANGER OF THE HAND PROTECTION BECOMING CAUGHT IN MOVING MACHINERY OR MATERIALS. ALSO, WHILE GLOVES ARE WORN FOR CHEMICAL PROTECTION, THE SLEEVES OF THE WEARER SHALL BE OUTSIDE OF THE CUFFS OF THE GLOVES TO AVOID COLLECTION OF MATERIAL IN THE GLOVES.

4. Foot hazards to consider include:

- ❏ Employees maneuvering heavy loads
- ❏ Punctures from nails, broken glass, sharp metal edges
- ❏ Exposed electrical conductors
- ❏ Slippery or other poor footing conditions
- ❏ Wet conditions, spilled or leaking hazardous materials
- ❏ Falling debris such as from construction or demolition

Hazards Identified:

Foot Protection

Safety shoes	Yes	No
Types: • Toe protection • Metatarsal protection • Puncture resistant (steel shank) • Electrical insulation • Other		

5. Hearing hazards to consider include:
   - ❏ Operation of heavy machinery in the area
   - ❏ Grinding, chipping or sandblasting
   - ❏ Jackhammers or hammer drills
   - ❏ Personal radios with headphones (While not necessarily a noise hazard, personal radios, CDs, and tapedecks are proven distractions and should NEVER be permitted on HAZWOPER sites.)

Hazards Identified:

Hearing Protection

Hearing Protection	Yes	No
Types: Plugs Ear Muffs Plugs and Muffs		

6. Fall protection hazards to consider include:

- ❑ Potential to fall 6 or more feet
- ❑ Distance to the lower level (if less than 15 feet from the anchor point, you must use an acceptable alternate method than the standard harness and lanyard)
- ❑ Absence of engineering controls such as work platforms, safety nets, toe boards, guardrails, and midrails
- ❑ Need for workers to climb and work at different heights

Hazards Identified:

Fall Protection

Fall protection devices	Yes	No
Types: Guardrails, midrails, toe boards, safety nets, and platforms Work positioning belts Full-body harnesses Adequate connecting devices Adequate anchoring points		

7. Other identified safety and/or health hazards:

Hazard	Recommended Protection

I certify that the above inspection was performed by me and to the best of my knowledge and ability, based on the hazards present on _____.

                                                              Date

(Signature)   (and Printed Name)

6.8 PPE SAFETY CHECKLIST

This personal protective equipment (PPE) safety checklist will help employees and supervisors follow required safety practices. The authors have attempted to include most situations. Please feel free to modify this list to make it 'site specific.'

❑ Is proper PPE provided, used, and maintained in a sanitary and reliable condition?

❑ Is PPE used to protect personnel against hazards only after engineering and administrative controls have been applied?

❑ Do supervisors inform each employee of the hazards to which he or she might be exposed, the PPE required, and how to use and maintain PPE?

❑ Do supervisors ensure that PPE is readily available, clearly identified, used for the purpose intended, fits correctly, and is properly worn?

❑ Does any of the employee's PPE interfere with the task or create additional hazards?

❑ Do employees ensure that the PPE is cleaned, repaired, stored, and disposed of as applicable?

❑ Do all employees, supervisors, administrative personnel, and visitors to the site wear hard hats and eye protection when required?

❑ Are hard hats with any defects or damage immediately removed from service?

❑ Are personnel who work around chains, belts, rotating devices, suction devices, and blowers required to cover long hair, remove rings, watches, and bracelets, and roll up their sleeves?

❑ Is protective eye and face PPE (glasses with side shields, splash proof goggles, face shields, etc.) required, provided, and worn where there is a reasonable probability of injury that can be prevented by such equipment?

❑ Does protective eye and face PPE provide adequate protection against the particular hazards for which it is designed?

❑ Does protective eye and face PPE fit snugly and not unduly interfere with the vision of the wearer?

❑ Is protective eye and face PPE durable, clean, and in good repair?

❑ Are personnel who normally wear contact lenses notified by their supervisors that contact lenses are forbidden on a HAZWOPER site?

❑ Are employees provided with goggles or glasses of one of the following types when their vision requires corrective lenses or spectacles?

  ❑ Spectacles with protective lenses that provide optical correction?
  ❑ Safety goggles that can be worn over corrective spectacles without disturbing the adjustment of the spectacles?
  ❑ Safety goggles that incorporate corrective lenses mounted behind the protective lens?

❑ Is eye protection distinctly marked to identify the manufacturer?

❑ Does eye protection meet the requirements of ANSI Z87.1-1989 and is it so marked?

❑ Are face shields used as primary eye and face protection in areas where splashing or dust, rather than impact resistance, is the problem?

❑ Is it prohibited for employees who work on or near energized electrical circuits or in flammable/explosive atmospheres to wear conductive frame eye/face protection?

❑ Is proper hand protection required for employees whose work may involve hand injury or impairment? CAUTION: Hand protection shall not be required where there is a danger of the hand protection becoming caught in moving machinery or materials.

❑ Are sleeves worn outside glove gauntlets when hazardous substances are being poured?

❑ Is protective footwear provided and worn when needed?

❑ Are requirements for protective footwear determined by the Site Safety Officer?

❑ Are conductive shoes provided to protect employees against the buildup of static electricity? CAUTION: personnel working near exposed or energized electrical circuits shall not use this type of shoe.

❑ Are nonconductive electrical-hazard safety shoes provided for employees exposed to electrical hazards? Note: other employees may wear these shoes as well.

❑ Are spark-resistant shoes provided in working areas for explosive storage, petroleum tank cleaning, and similar hazards?

❑ Are security guards and other site workers exposed to vehicular traffic equipped with reflective vests or light-reflective clothing?

❑ Are employees exposed to temperature extreme surfaces or materials provided with, and do they use, required protective clothing?

❑ Are safety belts used only for work positioning?

❑ Are body harnesses with shock-absorbing lanyards used exclusively for personal fall protection devices? (They must never be used for lifting equipment or material.)

❑ Have all employees who are required to wear personal fall protection been trained to select and wear the equipment properly?

❑ Are personal fall protection lanyards secured above the point of operation to an anchorage or structural member capable of supporting a minimum dead weight of 5400 pounds?

❑ Are lifelines that are subjected to cutting or abrasion a minimum of $7/8$-in. wire core manila rope?

❑ Are safety harness lanyards a minimum of 0.5-in. nylon or equivalent, with a maximum length to provide for a fall of no greater than 6 ft?

❑ Does the cushion part of the body belt meet the following conditions:

    ❑ Contain no exposed rivets on the inside?

    ❑ Be at least 3 in. in width?

    ❑ Be at least $5/32$ in. thick, if made of leather?

6.9 TRAINING AIDS AND ADDITIONAL RESOURCES

The authors recommend the following videos for aiding readers to see real-life examples of what they have learned in this Section:

- *Respirators: Tools for Survival*
- *Respiratory Hazards in the Workplace*
- *Air Purifying Respirators*
- *Breathing in High Hazard Zones*
- *Breathing Easy*
- *Fashions for Living*
- *Uniform for Safety*
- *From Head to Toe*

These videos are available from Safe Expectations (formerly BNA Communications, Inc.) (www.safeexpectations.com).

6.10 *SUMMARY*

An essential management function when HAZWOPER work is being performed is to see to it that workers have the proper PPE for the task. Supervision and trainers must also ensure that workers know how to select and wear PPE and that they actually do use it on the job.

All HAZWOPER personnel must understand the effective use of respirators. This is because inhalation is the chief route by which toxins enter the human body. The proper respirator can stop that entry. But "proper" is the word to be emphasized. Cartridge and canister filtering respirators are carefully designed. They are air-purifying devices fitted on respirators. They protect only against the specific contaminants for which they were designed and approved. Unless a cartridge is designed for combination protection, one kind is for particulates service and one kind is for gases and vapors. Dust masks only protect against nuisance dusts, not toxic dusts and not gases and vapors.

Cartridges, canisters, and filters of all kinds have a limited useful lifetime. Do not exceed it. When the work atmosphere is not breathable, even with all the contaminants removed, you need an atmosphere-supplying respirator. You can either carry a compressed air cylinder with you (SCBA) or have an air line supplying your breathing atmosphere.

Do not ever minimize the importance of other toxin or other contaminant entry routes. Safety glasses, goggles, and face shields protect eyes. Body suits, gloves, head covering, and boots protect against liquid splashing, tissue absorption through skin or eyes, and impacts. Body suits also protect against injection, the hazard from nails, broken glass, and all kinds of sharp objects.

Health checks before PPE use are mandatory. There may be workers available who cannot function safely with the extra physiological loads imposed by the required PPE. They must be assigned elsewhere.

Ergonomic devices help to protect against other workplace hazards that cause musculoskeletal disorders. Although these are not yet thought of as PPE, the authors are confident that they soon will be.

Everyone involved in HAZWOPER work must appreciate that PPE does not take the place of engineering controls, such as forced ventilation. Nor does PPE take the place of administrative controls, such as mandated use of the buddy system. Those controls are the first line of defense against injury from hazardous materials at work sites. When effective controls are in place, the burden of protection then rests on PPE and its proper use. "Proper use" means proper selection, proper donning and sure fit, proper decontamination (Section 7) and doffing, followed by any required maintenance and safe storage.

We cannot stress enough the value of a current PPE program. Yes, PPE is required by regulation in many circumstances. Yes, some PPE can be quite expensive. However, as a HAZWOPER manager or trainer, you must realize that your most important asset is your workers. As a HAZWOPER worker, you must realize that your most important task is to work safely and effectively.

The cost of the proper PPE will disappear on the ledger sheets compared to the cost of a single disabling work injury—one that could have been prevented by proper use of PPE.

SECTION 7
DECONTAMINATION (DECON)

OBJECTIVES

In the past several Sections you have seen what makes hazardous materials hazardous. You have learned about chemical, material, and physical hazards. In the last Section, on PPE, we presented the protective devices and clothing that shield HAZWOPER workers from attack by hazardous materials. Now you might ask, "What must I do after I have come into contact with hazardous materials?" Answering that question is a major objective of this Section.

We explain the decontamination process (decon) as a set of approved procedures for cleaning workers, their PPE, and their support equipment following contamination by hazardous materials. Many times it becomes practical to salvage materials, such as tarpaulins (tarps), following decon. An individual undergoing decon frequently must be assisted by one or more partners. Their joint goal is to ensure that as PPE is removed, contamination does not spread to any of the individuals' bodies or the environment. The decon process is designed not only for worker and equipment cleansing, but also to prevent the spread of contaminants off-site. You will see how the decon methods vary depending upon the characteristics of the contaminants.

The decon process can be as simple as washing your face and hands. But that simple task is critically important. We have found that second only to inhalation, toxins enter the body by ingestion due to contaminated hands and face. Decon can be as complex as the cleaning of large earth-moving equipment.

Decon ends when all equipment, salvageable material, and personnel have been thoroughly cleaned and personnel have changed into their street clothes. Inspectors check for the proper cleaning and storing of all material and equipment and ensure that all of the wastes generated in the decon process have been properly contained and disposed of according to regulations. As you will see in this Section, HAZWOPER personnel set up decon equipment and materials before any work at the site begins. Just about the last thing they do is to take down and remove the decon equipment and collected wastes after all site work has been completed.

We use the term *worker* in this Section in discussing the individual undergoing decon. Everyone entering the exclusion zone of a hazardous material cleanup site, whether workers, sampling technicians, engineers, managers, or visitors, must go through the entire decon process.

7.1 WHAT IS DECON?

○━╥ **Decontamination, or decon, is the approved process for removal of hazardous materials from HAZWOPER personnel, salvageable materials, and equipment.**

The process of decontamination is simple in concept. It is the last link in the personal protection chain. Decon includes methods and procedures that will keep all contaminants at the site until they are properly contained and sent off for proper disposal. Decon will keep HAZWOPER workers and equipment from translocating (tracking) the contaminants and thereby enlarging the contaminated sites. It will also help to protect the surrounding community from the hazardous materials at the site. Most importantly, it will cleanse workers' bodies so that they do not carry the contamination home. Too often in the past workers came home with asbestos, lead compounds, and other contaminants on their clothes. Then contamination spread to the family members, who then contracted illnesses. Decon of worker's equipment and PPE will make those items safe for reuse by those workers and others.

7.1.1 Division of a Hazardous Waste Site into Zones

We divide a hazardous waste site into three zones that govern decontamination. The exclusion zone, or "hot zone," is the most contaminated. The contamination reduction zone, or "warm zone," is a buffer zone where decontamination takes place. The support zone, or "cold zone," is uncontaminated and workers can safely operate there unprotected.

For safe and efficient operations, we divide a hazardous waste or emergency response site into zones. Sections 14 and 15 will describe the functions of these zones in further detail. The goals of zoning are to protect personnel and the general population and to restrict access and keep all contamination on-site. In an emergency release involving a toxic inhalation hazard vapor or gas, the exclusion, or hot zone, could extend downwind for seven miles or more! The on-site manager allows only those who are trained, prepared, and protected to enter the exclusion and contamination reduction zones. As shown in Figure 7.1, the manager maintains strict control of entry and exit through the use of access control points in the contamination reduction corridors. There may be a separate entrance and exit for personnel decon and equipment decon. There is at least one control point at the entrance, to assure workers are properly prepared for safe entry. Here, workers are logged into the work area, with time-in and required PPE carefully noted. There is at least one access control point at the exit. Here, workers are inspected to assure that they have undergone proper decon. The workers are also logged-out. These logs give accurate data as to amount of time workers were in the contaminated area. Log information is essential in calculating exposures and

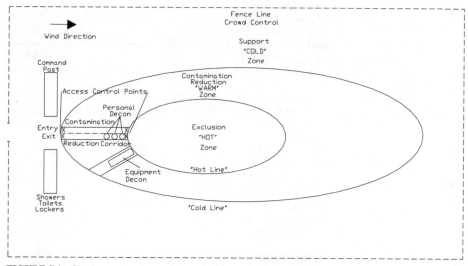

FIGURE 7.1   Site layout showing personal and equipment decon position.

assuring that proper precautions and safeguards have been taken. Remediation is the removal of contaminants from the site. This is explained further in Section 14.

The three zones are:

1. The *exclusion zone,* also referred to as the *hot zone;*

2. The *contamination reduction zone,* also referred to as the *warm zone;*

3. The *support zone,* also referred to as the *cold zone;*

The hot zone is separated from the warm zone by the *hot line.* The hot zone encompasses the entire contaminated area. Initially it may be very large. One site may have several hot zones clustered together or separated. As site work progresses and more detailed information and samples are analyzed, the hot zone(s) may shrink to a more manageable size.

The warm zone is separated from the cold zone by the *contamination reduction line,* sometimes called the *contamination control or cold line.* The warm zone is a buffer between the contaminated area and the uncontaminated surroundings. The warm zone may also change in size and layout depending on the spread of contamination, topography, and weather conditions. All access and egress to and from the hot zone is via the *contamination reduction corridor.* All access is through *access control points.* This layout limits tracking of contamination to one carefully controlled and delineated area.

In the cold zone there is no need for any PPE. This zone is the site of the Command Post as well as all support functions, including rest, eating and drinking areas, change rooms, and toilets and showers; along with decontaminated material and equipment storage.

7.1.2   How Do Workers Become Contaminated?

HAZWOPER workers leaving the hot zone must go through decontamination because they might have had vapors, mists, or particulates deposited on them, or might have been splashed by contaminants. Further, workers might have walked through contaminated liquids or solids or might have used contaminated equipment.

HAZWOPER workers responding to hazardous material emergencies or long-term clean-ups may become contaminated in a number of ways, including:

• Contacting vapors, gases, mists, or particulates in the air
• Being splashed by materials while sampling, opening containers or responding to a spill
• Walking through puddles of hazardous liquids or walking, sitting, or kneeling on contaminated soil
• Using contaminated instruments or equipment

Contaminants can be deposited on the skin, on the surface of personal protective equipment, or on tools or heavy equipment. In some cases, surface contaminants are easy to detect and remove. In other cases they may be invisible. Contaminants that have permeated into a protective material can be difficult or impossible to remove. If they are not removed by decontamination, they may continue to permeate to the inside of the material. If permeation is detected, or even suspected, the entire suit will have to be deconned inside and out. It must then be reinspected to ensure that the proper protection properties still exist.

If PPE is not properly deconned, it can cause the wearer to suffer an unexpected exposure. In cases where the PPE cannot be deconned, the PPE will have to be discarded. The authors know of a case where gloves were not properly deconned and were then worn by another worker. That worker lost the tip of one of his fingers.

Workers leaving either the hot zone or the warm zone must be thoroughly decontaminated regardless of their work purpose or duration of stay.

7.2   *TYPES OF CONTAMINATION*

7.2.1   Particulate Contaminants

Dusts and mists can cling to CPC (see Section 6), equipment, materials, and to workers themselves. These contaminants can also attach to surfaces electrostatically making them harder to remove. As a matter of fact, as we try to decon workers by dry brushing, we can actually build up the static electric charge of the suit and increase adhesion of some particulates. To counter this, we can use suits that are nonconductive or treat the CPC with an antistatic solution. Rinsing the CPC with water depletes this static charge and washes off electrostatically attached materials.

7.2.2   Gaseous and Vapor Contaminants

Gases and vapors, by their nature, usually adhere weakly to CPC. (By definition, vapors can be condensed to liquids. Humidity in the air is water vapor that can be condensed to dew. Gases cannot be condensed in the environment. Air is an example.) If a HAZWOPER worker's CPC or equipment is exposed to gaseous contaminants, any attachment is generally by absorption into another contaminant on, or permeation into, the fabric of the CPC.

7.2.3   Adhering Contaminants

Some contaminants are by nature more adhesive (sticky) than others. These properties may be temperature dependent. Some sticky contaminants, such as mud, are easily removed. Others, such as glues and epoxy resin, may not be removable at all. Removing adhesive contaminants such as tar can sometimes be achieved by freezing, in the same way you remove chewing gum from clothing. This process may make the contaminant brittle (like the chewing gum). Be aware that this procedure may ruin the desirable physical qualities of the CPC. Solvents may be used sometimes, but always check for compatibility with the CPC manufacturer. In some cases the suits cannot be deconned; then disposal is the final solution.

7.2.4   Solvents and Other Liquid Contaminants

Remove liquid contaminants from protective clothing or equipment by a combination of actions including absorption, evaporation, and cleaning with detergents. Especially with volatiles or dusty contaminants, you must ensure that unprotected workers do not inhale the contaminants during decon after the removal of respiratory protective equipment. Further, workers doing the decon must ensure that sufficient water is available to suppress any fires.

7.2.5   Acids and Bases

Thorough washing can decontaminate mild contamination caused by weak acids and bases. When more concentrated acids and bases are contaminating CPC, equipment, and materials, neutralization is required. Neutralize acids with bases. Neutralize bases with acids. Use mild citric acid and vinegar as common acids for neutralization. Common bases used for neutralization are lime, sometimes called burnt lime or quick lime (CaO), and calcium hydroxide, sometimes called slaked lime $Ca(OH)_2$.

WARNING
IN ORDER TO BE EFFECTIVE, NEUTRALIZING SOLUTIONS MAY BE STRONG ENOUGH TO HARM UNPROTECTED PERSONNEL.

TAKE SPECIAL PRECAUTIONS TO ENSURE THAT THE NEUTRAL-IZING SOLUTIONS DO NOT COME INTO CONTACT WITH THE WORKERS' SKIN. OTHERWISE, SEVERE BURNS COULD RESULT.

When using neutralizing solutions, it is best to use multiple washes of milder solutions than to use one strong (acid or base) wash cycle. Always refer to the CPC manufacturer's guidelines for proper decon solutions. Remember also that simply because an acid or a base has been neutralized (the resultant solution has a pH of approximately 7) does not mean it is nonhazardous. However, sulfuric acid neutralized with lime (calcium oxide) will result in calcium sulfate (gypsum), which is nonhazardous.

7.3 FACTORS THAT AFFECT CONTAMINATION

> **The factors that affect contamination are exposure time and the concentration of the contaminant(s).**

7.3.1 Exposure Time

The longer a contaminant is in close proximity to our bodies, our PPE, or our materials or equipment, the greater the chance of contamination. Minimize exposure time. Encourage workers to do their job and then get out of the contaminated area.

7.3.2 Concentration

The higher the concentration of a hazardous material in the work area, the greater the likelihood of worker contamination. If there are areas of high concentration of contaminants, such as pools of liquid or piles of solid material, consider performing your work from a remote area or with a remote device.

> **Even with the protection of properly selected PPE, keep as much distance as possible between you, the HAZWOPER worker, and highly contaminated areas.**

7.3.3 Decon under Hot and Cold Working Conditions

> **Temperature extremes tend to increase the possibility that hazardous materials will penetrate the CPC.**

Hot working conditions generally increase the permeation and penetration rate of contaminants such as liquids, gases, vapors, and mists. Very cold working conditions can cause some flexible CPC to become brittle and crack. Such material failures open up a direct route of entry to undergarments and also to the worker. This greatly complicates decon. In such a case, the crack, rip, or tear should be temporarily repaired with tape (see Section 6), decon of the outer CPC completed, and the worker deconned in the shower as usual. The CPC and undergarments may be salvageable after further decon and cleaning. The worker must report the incident to the Site Safety Officer in case the worker develops any reactions to the exposure. Refer to Section 6 for more details on the properties and protection provided by CPC materials.

7.3.4   CPC Pore Size and Size of Contaminant Molecules

Permeation and penetration increase as the contaminant molecule becomes smaller. Also, as the pore size of the CPC material being permeated or penetrated increases, larger contaminant molecules will be able to invade that material.

In general, gases, vapors, mists, and low-viscosity liquids have smaller molecular sizes. They tend to permeate and penetrate clothing more readily than high-viscosity (thick) liquids or solids that are of much higher molecular weight.

7.4   VARIABLES TO CONSIDER IN DESIGNING THE SITE DECON PROCESS

O━π **The entire decontamination system must be in place, understood by all personnel, and operational prior to anyone entering the exclusion zone.**

Decontamination must be a systematic process. This is true despite the degree of decon necessary at any one site. A major goal is to prevent tracking of hazardous materials to clean areas. The degree and methods of decontamination necessary are determined on a site-by-site basis. The following are the variables to take into account when designing an appropriate decon system:

- What are the materials involved?
- What is the size of the release?
- What is the nature of the release?
- What are the conditions at the site (weather, prevailing wind direction, temperature, topography)?
- How many workers will be affected?
- What facilities and utilities are available?

We stress three important points:

1. Do all decontamination in an area prepared to capture the contaminants and eliminate the chance of increasing the contaminated area. Do this by preparing the area with tarps, showers, concrete pads, basins, and pools.
2. The entire decontamination system must be in place, understood by all personnel, and operational prior to anyone entering the affected area.
3. The methods of performing the decon will depend on whether you are decontaminating personnel or equipment.

7.4.1   Emergencies and Decontamination

O━π **Decon may have to be abbreviated during medical emergencies or other site emergencies such as fire or explosion. The site workers will still have to undergo a complete decon in another location.**

If someone has a heart attack at the site, workers will usually forgo much of the decon process in order to evacuate the victim and get medical help rapidly. This might cause tracking of hazardous material outside the controlled area. Support personnel such as firefighters and EMTs might also require decon. Further, management might have to provide for decon of ambulances, fire and police vehicles, the emergency room and other equipment.

However, keep in mind that in nonemergency situations, all entrants to the hot and warm zones must exit through the entire decon process.

7.4.2 Material and Equipment Decontamination

O—π **Re-usable contaminated material and equipment must also undergo decon. But the methods, materials, and equipment used will likely be different.**

Heavy and light equipment decon requires different procedures than those for personnel. Large amounts of contaminated material may cling to the deep treads on the tires and to the bucket of a backhoe, for instance. Cleaning the material off may require a pressure washer. This could be an unsafe method for worker decon, depending on the water pressure. You can effectively decon most equipment using stronger detergents, higher-pressure water, and steam cleaning. Such methods would be hazardous for personnel decon. In addition to harming workers, such treatment could damage their CPC. Decon of heavy equipment usually takes place through a second contamination-reduction corridor.

7.5 DECONTAMINATION METHODS FOR PERSONNEL

O—π **The two methods most commonly used for personnel decontamination are mechanical removal and water washdown.**

7.5.1 Mechanical Removal

Gentle brushing, scrubbing, scraping, vacuuming, and wiping off of visible contaminants are some quick and effective mechanical ways to begin the decontamination process. Mechanical removal of contaminants is particularly useful for removing solid contaminants such as dusts and powders. Mechanical removal is excellent for removing gross contaminants. They are then collected and stored for disposal. These contaminants can be more concentrated hazardous materials. Mechanical removal will help keep the amount of contaminated wash water to a minimum. Waste minimization is paramount to a successful decon process.

7.5.2 Water Washdown

This is the most common form of decon. It may be the only type you ever use. Water, known as the universal solvent, is the most effective solution for the widest variety of contaminants. Even if the contaminant is not soluble in water, water can oftentimes rinse it off CPC surfaces. Usually, however, you must use a detergent to break the strong adhesion of high-surface tension materials to the CPC. A mild, warm detergent in water solution will take care of the vast majority of decon procedures. Consult the CPC manufacturer's literature for the proper decon procedures for that particular article of CPC. *Many manufacturers require the use of recommended, or even proprietary, solutions for cleaning their suits as a condition of warranty.*

The other reasons for the widespread use of water are that it is nonhazardous, economical, and available almost everywhere. Remember waste minimization, however. You want to use the least amount of water or water-based solutions necessary to safely remove the contamination. Also remember to segregate the decon water from the "gray water" used in a shower room. Never use a common drain. After testing and possible on-site treatment, such as filtration or neutralization, it may be permissible to dispose of decon water to a municipal sewage line along with sewage and gray water. On the other hand, decon effluent may be determined to be hazardous. In that case, commingling of hazardous with nonhazardous water waste streams will drastically increase waste volume, and therefore, disposal costs.

Personal decontamination will usually require the use of helpers to assist the contaminated worker. However, there are a number of different configurations of decon stations that the worker can use without assistance. Two such units are illustrated in Figures 7.2 and 7.3. Remember that decon must be set up and fully operational prior to any entry into the hot

FIGURE 7.2   Fendall de-fend® decon shower setup and operation. (*Courtesy of Dalloz Safety* [www.christiandalloz.com/].)

zone. Depending upon your site-specific HASP (see Section 16) and your Site Safety Officer, you will almost always have help when undergoing decon.

The de-fend® decon shower system is lightweight and compact, and can be set up in two minutes. The shower and retention pool require no assembly. Each unit requires less than five cubic feet of storage space. The shower connects to a standard garden hose fitting and is fully erect and operational within 10 seconds of the water being turned on. At a five gallon-per-minute flow rate at 40 PSI water supply pressure, the worker is ensured full decontamination. The collection pool holds 120 gallons.

Another one-man decon operation is illustrated by the Decon Cabin®. It has nine integrated nozzles in a shower curtain that are automatically erected as the cabin is inflated. The cabin can be inflated by a small blower or by using an SCBA tank. This decon station has a water consumption of approximately 12 gallons per minute.

Solubility and Insolubility in Water

🔑 If a material dissolves in water, we call it soluble in water. If it is a liquid, we sometimes say it is miscible in water. If a material will not dissolve in water, we call it insoluble in water. If it is a liquid, we sometimes say it is immiscible in water.

🔑 A contaminant's solubility in water is a key factor in determining the efficacy of the water washdown decon process.

One of the important considerations when choosing water as the decontamination liquid is the contaminant's solubility (ability to dissolve) in water. Keep in mind that *solubility* is a relative term. Some materials are soluble, some slightly soluble, and some insoluble. The

FIGURE 7.3 Decon Cabin® inflatable shower. (*Courtesy of Equipment Management Co.* [www.emc4rescue.com].)

authors use the fifteenth edition of *Lange's Handbook of Chemistry** as a handy reference for solubility data.

The following materials are *soluble* in room temperature water. That means that roughly five to several hundred grams of these materials dissolve in 100 ml of water at room temperature of about 70°F:

- All nitrates
- All acetates
- All sulfates, excepting barium and lead. Hydrogen sulfates are more soluble than the sulfates; for example sodium hydrogen sulfate (sodium bisulfate) is about twice as soluble as sodium sulfate.
- All sodium, potassium, and ammonium salts, with only a few rather uncommon exceptions
- Silver nitrate and perchlorate
- Lithium, sodium, potassium, cesium, rubidium, and ammonia hydroxide
- Sodium, potassium, and ammonium; carbonates, phosphates, chromates, and silicates

The following are *slightly soluble* in water. That means that approximately one to five grams of these materials dissolve in 100 ml of 70°F water:

- Barium, calcium, and strontium hydroxides
- The sulfates of silver, mercury, and calcium

The following are *insoluble* in water. That means they range from totally insoluble to roughly only one gram dissolving in 100 ml of 70°F water:

- All carbonates, phosphates, chromates, and silicates, are insoluble; except those of sodium, potassium, and ammonium, as well as magnesium chromate, which are soluble
- All hydroxides, except lithium, sodium, potassium, cesium, rubidium, and ammonia
- All sulfides, except lithium, sodium, potassium, ammonium, magnesium, calcium, and barium. Aluminum and chromium sulfides probably are hydrolyzed and will wash down as hydroxide precipitates.
- All silver salts, for example the sulfate, except for the nitrates and perchlorates

Always be sure to consult the material's MSDS for recommended decontamination methods.

When You Should Not Use Water (W̶)

Water should never be used as the primary decontamination solution when dealing with water-reactive materials (W̶). Where it is a concern, MSDSs warn about water reactivity.

There are some times when water should not be used for the decontamination solution. If the worker has been exposed to water-reactive materials, such as metallic lithium, sodium, or potassium, the exposure to water could generate enough heat to burn the worker or damage the worker's CPC. The violent reaction of these metals with water liberates hydrogen. This together with the reaction heat could cause fire and explosion.

HAZWOPER workers exposed to these metals or other water-reactive materials must not use water for *initial* decontamination. Dry brushing should be performed until all visible contamination is removed. Then water may be used, cautiously, for the final rinse. Take special precautions to segregate water-reactive wastes from water-containing waste streams.

---

*New York: McGraw-Hill, 1999.

WARNING

NEVER WASH DOWN A WORKER IN CPC WHO HAS VISIBLE SIGNS OF CONTAMINATION BY WATER-REACTIVE MATERIALS. THE MSDS OF EVERY WATER-REACTIVE MATERIAL WILL WARN YOU OF THIS HAZARD.

7.5.3 Absorption and Adsorption

*Ab*sorption is used in both personal and equipment decon. Clean rags and specialty absorbent towels are excellent materials for removing liquid contaminants. One thing to watch for is the compatibility of the absorbent and the contaminant. This dry decon process helps keep the cost of waste disposal down because it reduces the need for heavier dilution or washdown liquids. Some absorbents are as simple as desiccated (dried) clay (cat litter or Oil Dri®). Some are specialized to be compatible with acids, solvents, or other hazardous materials. The New Pig® Corporation stocks all types of wipes, sorbent pads, and sorbent booms for containing large areas of liquid material spills.

*Ad*sorption (see Glossary) is used in the cleanup of mercury and mercury compounds. Proprietary powders containing sulfur are used. The powders used dry, or in solution, form a coating around the minute mercury particles. They are then more easily captured by vacuuming or rinsing.

7.5.4 Neutralization

Neutralization is the process of chemically reacting a contaminant with a material of opposite pH. For instance, acid contamination can be neutralized with a weak alkali solution.

Degrading or neutralizing a contaminant can reduce it to a less harmful material or one that is more easily washable. Neutralization of an acid is commonly done by application of an alkali such as the following (listed in increasing pH):

- Sodium bicarbonate (baking soda);
- Calcium carbonate (crushed limestone);
- Calcium hydroxide (slaked lime, formed by reacting lime (CaO) with water);
- Sodium carbonate (soda ash);
- Sodium hydroxide (lye or caustic soda).

Neutralization of an alkali or base is commonly done by application of an acid such as the following (listed in decreasing pH):

- Citric acid;
- Vinegar (acetic acid);
- Dilute hydrochloric acid;

Either treatment generally produces a nonhazardous salt. However, care must be taken in choosing the neutralizing agent. It must not attack or react with the material of the CPC. Also, when using solutions other than detergent and water, make the decon team and contaminated workers aware of any potential hazards associated with the solution. Whenever acids and bases are mixed they will liberate heat. The weaker the neutralizing solution, the better the control of heat release. It is, therefore, safer to neutralize with a greater amount of dilute solution than with a small amount of a concentrated solution.

7.5.5 Isolation and Disposal

Collect for proper disposal all contamination removed from the workers, including gross material, washwater, towels, disposable PPE, and disposable sampling equipment and materials. Depending on the contaminants and their concentration in the washwater, you may be allowed to dispose of it in a sanitary drain. That water will probably require some on-site treatment such as filtering or elementary neutralization prior to least-cost disposal. However, it may still require disposal as hazardous waste. The Site Safety Officer makes that decision after testing and analysis of the wastewater stream. Never dispose of decon solutions directly down a sanitary sewer or a storm sewer without the authority of the SSO. Never mix decon solutions with "gray water" from sinks and showers. If the decon solution is deemed hazardous, the whole mixture will likely be hazardous, increasing disposal costs.

Decontamination methods are site-specific. Each site has its own contaminants, weather, topography, time restraints, and other characteristics. Each type of site requires a documented plan and training of all site workers.

7.6 DECONTAMINATION PREPARATION

Note

In this subsection we will be discussing normal decon procedures when the work has no time constraints. An emergency response situation may require an abbreviated on-site decontamination followed by further decontamination elsewhere. We will discuss this further in Section 15.

7.6.1 Factors Affecting Decon Placement and Setup

Remember that the decontamination system must be set up and fully operational before anyone is allowed to enter the hot zone.

The following is a checklist of factors that play a part in decontamination planning and site preparation:

- Amount of hazardous materials involved
- Type of hazardous materials involved
- Likely or expected exposure levels of the hazardous materials involved
- Number of on-site workers required
- Need for heavy equipment
- Prevailing wind direction
- Emergency of the situation (injuries, blocked highway, or rail line)
- Access for type(s) of equipment needed
- Availability of utilities
- Terrain
- Surrounding community
- Weather

If the site is remote, you, the HAZWOPER project manager, will probably need portable generators for power. If potable water is not available on-site, it will have to be piped or trucked in. Will some decon take place at night? If so, ensure proper lighting. Will the

TABLE 7.1   Decision Tree for Evaluating the Safety and Health Aspects of Decontamination Methods

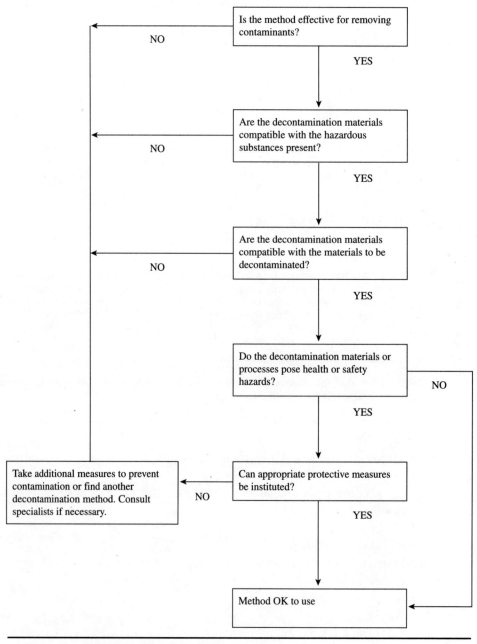

Source: Courtesy of OSHA, EPA, NIOSH, and USCG.

temperature fall below 32°F (0°C)? If so, make provisions for keeping the workers warm. Protect water or other solutions from freezing.

The importance of the following factors cannot be overemphasized when making preparations for an effective decon system:

1. If little or no site hazard information is available, then set up the initial site decon system for the worst-case scenario. As the threat is defined through investigation and identification, site decon can usually be simplified. The Site Safety Officer will modify the decon to include just those specific steps necessary to protect the workers.
2. If the terrain and weather patterns allow, place the site decon stations upwind and upgrade from the contaminated area. This will help to limit the flow of liquids, gases, and vapors from the contaminated area to the decon stations.

> **Always remember that once the workers have been decontaminated and CPC removed, they no longer have any protection against the effects of hazardous materials.**

Table 7.1 gives a sample decision tree to follow when evaluating the safety and health aspects of a decon method.

7.6.2 Identification of Hazards

As we wrote before, the amount and nature of the hazardous materials involved play a large role in setting up a decon. Material hazards (Section 3) must be identified, thoroughly investigated, and made the focus of decon planning. This information will determine the types and quantities of materials needed for the decon procedures. Other workplace hazards (Section 11) must also be investigated and minimized to the extent possible.

In an emergency response, such as to the site of a highway or rail transportation accident, downed power lines, gasoline or other fuel leaks, traffic, and crowd control all need to be addressed. Factor any such hazards into determining the location of the decon stations.

7.6.3 Personnel Requirements and Responsibilities

> **The size and number of decontamination stations needed will vary depending upon the site and the contaminants involved.**

The number of people needed for the remediation effort and the size of the contaminated area will play a major role in determining the number of dedicated decon personnel and amount of materials and supplies necessary for a satisfactory decon.

There may be a need for a dedicated workforce to operate the decon. Remediation workers, depending on the level of PPE required, will need various levels of assistance in the decontamination process. Workers in Level A suits will not be able to adequately decon themselves. Workers in Level C gear may be able to do the job themselves. See Section 6 for explanations of levels of PPE.

WARNING

REMEMBER, OSHA REQUIRES THAT A SEPARATE RESCUER BE ON STANDBY FOR EACH ENTRANT INTO THE HOT ZONE WHERE EXPOSURE LEVELS ARE MEASURED, OR EXPECTED TO BE, AT OR ABOVE THE IDLH LEVEL FOR THE CONTAMINANTS. ALSO, FOR ANY OTHER PHYSICAL OR MATERIAL HAZARD PRESENT, OR SUSPECTED, THAT RESULTS IN AN ACTUAL OR POTENTIAL IDLH SITUATION.

O——π **OSHA requires a standby rescuer for each entrant into the hot zone where the levels of contamination are measured, or suspected to be, at or above the IDLH levels.**

7.6.4 Decontamination Materials and Resources Requirements

For decontamination, the following list of project materials is considered the minimum. Conditions may dictate a much more exhaustive list of materials, supplies, and equipment. Remember, these are project materials dedicated solely to the decon process. The total list of remediation project materials, personnel, equipment, outside aid agencies, and others for remediation is much more exhaustive and will be covered in Section 14. Use the following checklist as a starting point for your decon materials and resources requirements.

❑ *Have marking tape or barricades* available to mark the boundaries of the decon area. The contamination reduction corridor must be highly visible. Certain PPE, such as Level A suits and full-face respirators, can make it difficult for workers to see clearly. They may have difficulty finding their way out of the hot zone and into the decon corridor.

Note

Some suits are available in various colors so participants can easily identify work teams performing different functions at a distance.

❑ *Make absorbent towels and pads and adsorbent materials* to capture hazardous materials readily available. Special pads, towels, and materials for dikes are made to repel water yet absorb petroleum hydrocarbons such as oils and fuels. We have found activated carbon and proprietary sorbent materials to be very useful.

❑ *A water source* and water heater, together with hoses, nozzles, valves, connectors, buckets, and sprayers. Remember, if water is not available at the site, you will have to supply it, probably via tankers.

❑ Most manufacturers of PPE recommend *soft-bristle brushes and sponges* for removing gross contamination without damaging the PPE. Dry brushes are used for gross removal, while wet brushes are used for scrubbing.

❑ Have *hard-bristle brushes* available for boot cleaning as well as decon of mobile equipment and tools. Boots and mobile equipment will be in direct contact with the contaminants on the ground and may require extra scrubbing for decon. Long-handled brushes can be especially useful for individuals cleaning their own extremities and for cleaning of hard to reach parts of heavy equipment.

❑ *Decontamination additives for cleaning water* may be mild or strong detergents. We have found that some common liquid dishwashing detergents provide very good all-around removal of solids and liquids, including oils and greases, and are still gentle enough not to harm most CPC. Specific chemical solutions might be needed to dilute, neutralize or otherwise capture the contaminant for proper disposal.

❑ *Disposal containers.* These can range from heavy duty 6 to 10-mil thick polyethylene bags to drums and even to wastewater tanks or tankers. Remember, every material used to clean the worker's suits, tools, and equipment is now considered contaminated and therefore must be treated as hazardous waste until proven otherwise. This includes all disposable suits, booties and gloves, rags, paper towels, and absorbent pads and all wastewater and other solutions. (The final determination of whether or not it is actually hazardous waste will be made after the waste is sampled.)

❑ *Washwater containment.* For personal decontamination, inflatable 'kiddie pools' are excellent. They are easily inflated and, when deflated, store in a small package. They have low walls about 12 inches high, so they allow workers to step in and out of them easily. They are also resistant to a large number of chemicals. However, you will have to determine individual chemical compatibility. They are also inexpensive. For heavy equipment, depending on the length of the cleanup, a drive-in pit, lined with a heavy-duty tarp, may suffice. Otherwise, it may be more efficient to construct a washing pit of concrete that the equipment can be driven into it easily. Whichever method is chosen, you must have sufficient storage capacity for the wastewater generated in the decon operation prior to final disposal.

❏ *Plastic tarpaulins (tarps).* Plan to cover the walking surface areas of the hot and warm zones with reinforced plastic tarps of at least 4 mil thickness. This will minimize the amount of contamination transferred to surrounding surfaces. Tarps can be purchased in custom configurations fabricated from various fiber-reinforced plastic materials. They can be an economical ground cover. Tarps can also be easily cleaned, dried, and then stored for reuse. These tarps offer a certain degree of traction but can still be slippery when wet. We recommend non-slipping soles on workers' boots.

❏ *Power.* You must provide a source of sufficient electric power. Do not assume there will be sufficient power on-site. We recommend having portable or mobile generators capable of supplying 200% of your estimated power usage. OSHA requires that you have the outdoor field power sources, such as extension cords, protected by ground fault circuit interrupters (GFCIs).

❏ *Air compressor.* If you are using airline respirators, remember that you must have a *breathing air* compressor. If you are only using pneumatic tools, an ordinary compressor will suffice. If both types of compressors are on-site, the authors strongly advise that the coupling devices be incompatible so that a breathing line cannot accidentally be connected to the pneumatic tool compressor. Air line respirators are frequently used by decon personnel so that they are not hampered by respirator time restraints.

❏ *Lighting.* Prepare for lighting to accommodate operation around the clock. It is always best to have some lighting available for emergencies, just in case a planned daylight operation does extend into the night. Telescoping quartz-halogen lighting systems are ideal.

❏ *Personal hygiene facilities.* At the very minimum, plan for having a port-a-potty on-site. For longer-term operations (longer than one day), these facilities must include a shower and isolated changing facility with lockers for clean clothes. OSHA requires separate facilities for men and women. In 29 CFR 1910.141, OSHA requires toilets (1 per 15 workers), potable water, hot water, soap, and towels to be available for personnel on hazardous waste cleanup sites.

❏ *Options for larger and longer-term operations.* There are a number of trailers on the market today that have everything you need in one unit: changing facilities, hot and cold running water, heat and air conditioning, showers, and lockers. They are ideal for remote locations and for temperature extremes. A number of these trailers have a contaminated side and a clean side. This is an especially useful arrangement during inclement weather. Trailer facilities can also serve as a rest area for workers as well as an emergency treatment area while awaiting medical help. They can be rented or purchased.

Set up decon materials and equipment, mark the on-site zones, and have all personnel understand the site-specific decon methods before allowing workers to contact any hazardous materials.

Note

Proper preplanning is essential to setting up a safe and operable decon. It is always better to be overprepared than to come up short. The authors have observed many delays in operation due to assumptions made by decon crews. Power and water supply are the two most common utilities we have found to be lacking. Management had assumed that power and water could be shared with a nearby facility. They were denied that access. Work was delayed for several days until those utilities were provided.

7.7 *OPERATIONS IN DECONTAMINATION PROCESS STATIONS*

Personal decontamination is conducted at a series of stations (Table 7.2 and Figure 7.4) in the contamination reduction corridor (CRC) previously shown in Figure 7.1. They begin with gross contamination removal and end with a clean worker at the support zone. All workers who exit the hot zone must pass through decon.

The only exception to this rule occurs when a worker suffers a life-threatening illness or injury in the hot zone for which immediate medical attention is required. In this case, the decon process will be abbreviated. The Emergency Medical Technicians (EMTs) must be provided PPE to avoid their contamination. The emergency treatment room must be ready to deal with a contaminated person. Both the ambulance and emergency treatment area will have to be inspected for contamination. If contamination is found, they must be decontaminated after the incident.

Equipment decontamination may take place in a single station. The equipment may be scrubbed down or pressure-washed two or three times until it is found to be clean by an appropriate inspection. It may then be removed to the equipment staging area until it is needed again.

All personnel conducting the decon may be in the next-lower level of PPE (refer to Section 6) than the people being cleaned. For instance, Level B workers can decon Level A workers, and Level C workers can decon Level B workers. The decon personnel collect all salvageable materials and equipment and ensure cleanliness and readiness for storage. Then they store the materials and equipment for the next work period. All decon personnel, after they have helped the last worker through decon, must themselves also go through the complete decon process.

Decon personnel may be dressed in the next-lower level of protection than the workers that they are deconning. For instance, to decon Level A workers, decon workers can be in Level B protection.

TABLE 7.2 Decon Stations and Their Corresponding Activities (Courtesy of OSHA, EPA, NIOSH, and USCG.)

Station Number	Corresponding Activities
Station 1: segregated equipment drop	1. Deposit equipment used on site (tools, sampling devices and containers, monitoring instruments, radios, clipboards, etc.) on plastic drop cloths or in different containers with plastic liners. During hot-weather operations, a cool-down station may be set up within this area.
Station 2: boot cover and glove wash	2. Scrub outer boot covers and gloves with decon solution or detergent/water.
Station 3: boot cover and glove rinse	3. Rinse off decon solution from station 2 using copious amounts of water.
Station 4: tape removal	4. Remove tape around boots and gloves and deposit in container with plastic liner.
Station 5: boot cover removal	5. Remove boot covers and deposit in container with plastic liner.
Station 6: outer glove removal	6. Remove outer gloves and deposit in container with plastic liner.
Station 7: suit and boot wash	7. Wash encapsulating suit and boots using scrub brush and decon solution or detergent/water. Repeat as many times as necessary until the CPC appears clean.
Station 8: suit and boot rinse	8. Rinse off decon solution using water. Repeat as many times as necessary until all visible solution is removed.
Station 9: cylinder change	9. If an air cylinder change is desired, this is the last step in the decontamination procedure. Air cylinder is exchanged, new outer gloves and boot covers donned, and joints taped. Worker returns to duty.
Station 10: safety boot removal	10. Remove safety boots and deposit in container with plastic liner.
Station 11: fully encapsulating suit and hard hat removal	11. Fully encapsulated suit is removed with assistance of a helper and laid out on a drop cloth or hung up. Hard hat is removed. Hot-weather rest station may be set up within this area for personnel returning to site.
Station 12: SCBA backpack removal and body wash down	12. While still wearing facepiece, remove backpack and place on table. Disconnect hose from regulator valve. Wash down complete body with facepiece on.
Station 13: inner glove wash	13. Wash with decon solution that will not harm the skin. Repeat as often as necessary.
Station 14: inner glove rinse	14. Rinse with water. Repeat as many times as necessary.
Station 15: facepiece removal	15. Remove facepiece. Deposit in container with plastic liner. Avoid touching face with fingers.
Station 16: inner glove removal	16. Remove inner gloves and deposit in container with liner.
Station 17: inner clothing removal	17. Remove clothing and place in lined container. Do not wear inner clothing off-site, since there is a possibility that small amounts of contaminants might have been transferred in removing the fully encapsulating suit.
Station 18: shower or field wash	18. Shower if highly toxic, skin-corrosive or skin-absorbable materials are known or suspected to be present. Wash hands and face if shower is not available.
Station 19: redress	19. Put on clean clothes.

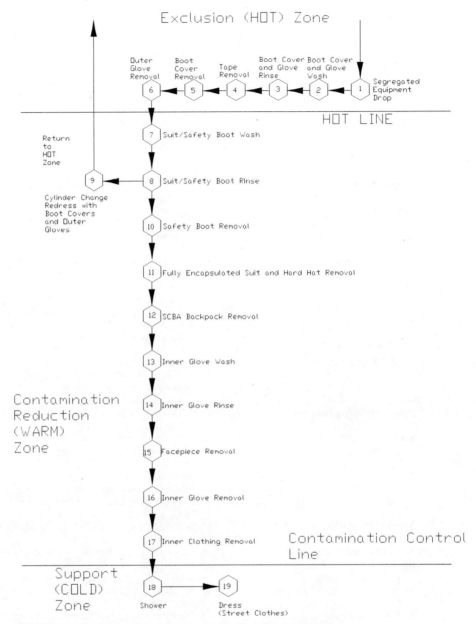

FIGURE 7.4   Stations in personal decontamination from Level A protection.

7.7.1 Personal Decontamination for Level A Protection

The following are the step-by-step decon procedures. Refer back to Figure 7.4 for a graphic representation of the personal decon process for Level A protection.

Maximum Measures for Level A Decontamination.   In the cold zone, workers put on their clean street clothes. Then they can rest, eat, and drink nonalcoholic beverages while waiting to return to the work zone. At this point, if workers are working in Level A or Level B CPC and unusually hot or cold working conditions are present, medical monitoring may take place. Such testing is advisable to ensure that workers are physically capable of returning to work. Rehydration is very important at this point. Whether in winter or summer, workers in Level A or B CPC will lose tremendous amounts of water and are in real danger of dehydration. Potable water must be available. Encourage workers to drink enough to sustain them during their work period. See Section 12, "Sampling and Monitoring" for a discussion of 'real-time' medical monitoring equipment.

TABLE 7.3 Suggested Stocking of Equipment and Materials for Decon Stations (Courtesy of OSHA, EPA, NIOSH, and USCG.)

Station 1:
    a. Various-size containers
    b. Plastic liners
    c. Plastic drop cloths

Station 2:
    a. Containers (kiddie pools)
    b. Decon solution or detergent in water
    c. Two or three long-handled, soft-bristled scrub brushes

Station 3:
    a. Containers (kiddie pools) and low-pressure pump spray unit
    b. Decon solution
    c. Two or three long-handled, soft-bristled scrub brushes

Station 4:
    a. Containers (kiddie pools)
    b. Plastic liners

Station 5:
    a. Containers (kiddie pools)
    b. Plastic liners
    c. Bench or stools

Station 6:
    a. Containers (kiddie pools)
    b. Plastic liners

Station 7:
    a. Containers (kiddie pools)
    b. Decon solution or detergent water
    c. Two or three long-handled, soft-bristled scrub brushes

Station 8:
    a. Containers (kiddie pools) and low-pressure pump spray unit
    b. Water
    c. Two or three long-handled, soft-bristled scrub brushes

Station 9:
    a. Changed air tanks or facemasks and proper respirator cartridges, depending on conditions
    b. Tape
    c. Boot covers
    d. Gloves

Station 10:
    a. Containers (kiddie pools)
    b. Plastic liners
    c. Bench or stools
    d. Boot jack

Station 11:
    a. Rack
    b. Drop cloths
    c. Bench or stools

Station 12:
    a. Table

Station 13:
    a. Basin or bucket
    b. Water
    c. Small table

Station 14:
    a. Water
    b. Basin or bucket
    c. Small table

Station 15:
    a. Containers (kiddie pools)
    b. Plastic liners

Station 16:
    a. Containers (kiddie pools)
    b. Plastic liners

Station 17:
    a. Containers (kiddie pools)
    b. Plastic liners

Station 18:
    a. Water
    b. Soap
    c. Small table
    d. Basin or bucket
    e. Field showers
    f. Towels

Station 19:
    a. Dressing trailer is needed, especially in mixed-gender operations and inclement weather
    b. Tables
    c. Chairs
    d. Lockers
    e. Disposable cloths

Note

This is the minimum decon process at a site where there is a possibility of tracking hazardous material off-site. If the Site Safety Officer deems it necessary, steps may be added to this process. Conversely, steps may be removed from this process in less grossly contaminated sites. Some slightly contaminated sites may have as little as one-step decon, where workers wear a disposable oversuit and no respiratory protection. In this case, they simply remove the oversuit and bag it for proper disposal.

Equipment Needed to Perform Maximum Decontamination Measures for Levels A, B, and C. Stations correspond to Table 7.2 and Figure 7.4. Table 7.3 lists suggested stocking of the stations.

Personal Decontamination for Level B Protection. Minimum required decon for Level B protection is identical to that of Level A except the splash suit can be substituted for the FES.

Note

If workers are using SCBAs, they will have to follow the same path described in Section 7.4.1 for Level A workers with regard to exchanging empty air cylinders for full cylinders.

Personal Decontamination for Level C Protection. The minimum required decon for Level C is identical to Levels A and B, with the exception that the worker is using an air-purifying respirator (APR) instead of an air-supplying respirator (ASR). As described above, workers needing new respirator cartridges will have to follow the same path as described for both Level A and Level B—when they exchange air cylinders.

7.7.2 Heavy Equipment Decon

Heavy equipment such as bulldozers, trucks, backhoes, skid loaders, drilling, and GeoProbe® equipment can be difficult to decontaminate. The design and construction of the decontamination trough or pad must take into consideration protection against overspray by a pressure washer to an adjacent zone. The pad must have the strength to accommodate the heaviest equipment on the site. The pad requires a collection sump to collect all of the contaminated cleaning and rinsing liquids. A sump pump serves to transfer the liquid periodically to a storage tank or tanker for transport to disposal.

Experience has taught us that the most effective means of cleaning heavy equipment is by washing it with a pressure washer on a sloped-concrete or plastic-covered pad. A detergent-water solution must be followed by a thorough water rinse. Particular attention should be given to tires, tracks, scoop, augers, drills, stabilizer pads, and other parts that have directly contacted contaminated surfaces.

Fortunately, heavy equipment will not have to be deconned every day. Usually, it can wait until the equipment use is no longer necessary in the hot zone. Heavy equipment will have to be deconned if it is being relocated across a cold zone to another hot zone.

After completing decon, use wipe tests or direct-reading instruments to determine whether the equipment is indeed clean. Make available storage tanks for appropriate treatment systems and for temporary storage of contaminated wash and rinse solutions.

Use this checklist of supplies and resources needed for heavy equipment decon:

❑ *A concrete pad or plastic covered, diked area* with a dedicated sump at the lowest point, will have to be constructed so that machinery and equipment can be driven or loaded

onto the area for cleaning. The pad or diked area must be capable of withstanding the weight of the largest piece of machinery to be decontaminated.

❑ *An enclosure* can be fabricated from poly curtains. A specially constructed booth may be necessary to contain overspray from pressurized sprays.

❑ *A sump pump* for draining the sump of contaminated wash and rinse solutions and storage tanks. The sump pump must be capable of pumping all anticipated contaminants. It is advisable to use a pump capable of pumping at least one-inch solids. (A strainer for the pump is advisable. Larger solids can be shoveled out of the sump periodically and placed into drums for proper disposal.)

❑ *Hoses and couplings* matched to the sump pump outlet and the inlet of the contaminated solution holding tank. Make sure there are sufficient lengths of hose and that their internal diameter will also handle one-inch solids.

❑ *Long-handled, hard-bristled brushes* for general exterior cleaning. Wooden-handled brushes should be avoided as they are very hard to decontaminate. Nylon handled and bristled brushes work well.

❑ *Washing and neutralizing solutions* chosen based on the type of contamination. These solutions must be of sufficient strength to work effectively. With heavy equipment, stronger solutions may be used than with PPE.

❑ *Sufficient water supply and rinsing solutions* based on the type of contaminant and of sufficient volume to supply the chosen pressure washer.

❑ *Pressure washer(s)* capable of generating at least 10,000 psig. Select a pressure washer that has a long wand (at least four feet in length) to help direct the stream into areas that cannot be reached by hand. One good practice for removing heavy, caked-on mud is to saturate the equipment with water from a hose first to loosen the dried mud. Follow this with the pressure washer for best results.

❑ *Shovels and scrapers* for dislodging contaminated material from tires, treads, tracks, and the underside of equipment. These shovels are also used for cleaning large solids off the pad and out of the sump after the equipment has been cleaned.

❑ *Containers* to hold solid contaminants, including contaminated soil removed from the equipment. Open-mouth drums work well. Use small buckets in the cleaning of the inside of equipment's operating compartment that has been contaminated by the operator.

❑ *Whisk brooms,* foxtails, and dustpans for cleaning gross dry contamination from the operator areas of equipment.

❑ *Prepare large containers* such as enclosed dumpsters and tankers for temporary storage. Use tankers also for disposal of contaminated wash and rinse solutions. Also use dumpsters for damaged or heavily contaminated parts that cannot be salvaged, contaminated material, and equipment to be discarded.

❑ *Ensure that there is proper PPE for use by the decon crew.* This will usually include, depending on the nature of the contaminants, full Level A PPE, rain suits, or splash suits. Workers performing decon of heavy equipment must go through personal decon after cleaning the equipment.

7.7.3 Minimize Decon and Disposal Effort

Take the following steps to minimize worker exposure to hazardous materials at work sites and ease final disposal of wastes.

1. *Reduce direct contact with contaminants* and their containers by using remote handling equipment such as remote-sampling, drum-puncturing, and drum-handling equipment. (See Section 13 for information on drum-handling equipment.)

2. *Protect monitoring instruments* from contamination by covering them with plastic or using plastic bags. Access to sample ports can be cut in the plastic.

3. *Wear disposable CPC* such as outer gloves and boot protection. If practical, have workers wear disposable suits.

4. *Contain contamination* where possible. For example, use plastic tarps on soil piles and overpacks on 55-gallon drums.

5. *Train workers in hygienic work practices.* This includes strict limitation in touching, stepping in, kneeling on, or sitting on obvious areas of contamination such as puddles and waste piles.

6. Whenever possible, *deposit suspected material in containers suitable for shipping.* That way, if the material is found to be hazardous, it does not need to be handled again and repackaged for transportation. Ensure that all containers are labeled ''Hazardous Waste'' until the contents are proven otherwise. Deposit contaminated liquids in bung-type 55-gallon drums, or tankers; solids in 55-gallon open-mouth drums, intermediate bulk containers (IBCs) capable of holding much more than drums, or enclosed roll-off dumpsters.

7.8   TRAINING AIDS AND ADDITIONAL RESOURCES

The authors recommend the following video for aiding readers to see real-life examples of what they have learned in this Section:

• *DECON Revised,* available from Safe Expectations (formerly BNA Communications, Inc.) (www.safeexpectations.com).

7.9   SUMMARY

We cannot overemphasize the importance of workers using prescribed decontamination (decon) procedures and methods. Every year we receive horror story reports about workers accidentally carrying hazardous materials off-site due to ineffective decon. Their families then unknowingly face the hazards.

Management must ensure the preparation of a written and approved decon plan describing the procedures, equipment, and methods prior to worker training. No worker may enter the hot zone without that training. No worker may enter the hot zone until the decon system is approved for operation by the Site Safety Officer or Decon Officer. If necessary, dedicated decon personnel must also be in place.

The site decon system described in this Section focuses on operations at a typical Superfund or brownfield remediation site such as described in Section 14. A permanently established decon system, such as would be applicable to a TSDF or where hazardous materials are being processed or are occurring as wastes (Section 13), would be much more specialized. Both the workers and the decon personnel in that case would be much more familiar with the required minimum methods and procedures. They would also know much more about specific successful cleaning materials.

A decon system at an emergency response site (Section15) is something different altogether. There is usually little opportunity to set up a system with the detail and care we have described. The responders will establish the decon zones as best they can. Of course, the responders would wear the same required PPE. Upon leaving the hot zone they will be cleaned by other personnel as rapidly as possible, on the spot. Many times the washdown gets mingled with the spill material that almost always exists. The emergency responders must bring in their vehicles everything needed for their own decon.

SECTION 8

LABELS, PLACARDS, AND OTHER IDENTIFICATION

OBJECTIVES

Signage is a worldwide way of giving people critical and timely information. Labels and placards and other identification give life- and time-saving information to HAZWOPER personnel on a daily basis. *Signage* is the broad term that includes a number of ways of presenting identification, cautions, and warnings. These kinds of information can be displayed for the reading at distances from inches or a few feet (markings and labels) to a hundred yards or more (some placards and signs). Much study has been devoted to finding the best kinds of signage to give critical safety information almost instantaneously and make it easier to understand. Our objective in this Section is to present to all personnel involved with hazardous materials and their wastes the variety of signage they will see and what each kind means.

Signs have traditionally been alphanumeric. You will notice graphics being used more frequently in recently prepared signs. They have the advantage of being independent of the viewer's language. Even nonreaders understand them readily. Understanding graphics and pictorial signage is becoming particularly important in hazardous materials and hazardous waste operations. Throughout this Section, you will begin to appreciate the wealth of information that you can glean at a glance from the labels, placards, and other identification presented here.

This Section will also explain the use of the *Emergency Response Guidebook,* included in the Companion CD, which gives you the most important safety information about the contents of any container that is properly placarded. Another objective, then, is to know what kinds of signage must be used to meet OSHA and DOT requirements. You need to know where signage must be placed on containers, how it must be maintained, when it must be removed, and the proper signage to use depending upon the type of container.

8.1   THE CRITICAL NEED FOR SIGNAGE

O—π   Properly placed signs and placards give the HAZWOPER worker and especially the emergency responder at the time of the incident the ability to identify the contents of a container immediately.

Hazardous materials are often stored and transported in large quantities. Sometimes a large number of containers with incompatible contents are stored or shipped together. It is crucial for HAZWOPER workers as well as emergency response personnel to be able to assess the situation as quickly as possible. An accidental release of these materials presents a potential hazard to workers, the public, and the environment. Such an incident is managed more safely when the hazardous material is specifically identified. Further, that identification must rapidly lead the viewer to an understanding of the hazards of a release and the recommended countermeasures.

8.2 THE FIVE WAYS OF IDENTIFYING HAZARDOUS MATERIALS CONTAINERS

⊙━🔑 **The five ways of identifying the contents of a hazardous material container are markings, labels, DOT placards, signs, and manifests.**

Markings are brief notations of the contents of a container, applied directly to the container. If the marking is applied to another material (usually plastic or paper) before it is applied to the container, it is considered a label. Many times drums, boxes, and other small containers are marked. Marking is usually the least informative method of signage.

Labels usually provide more extensive descriptions of the contents of a container than markings. OSHA and the DOT require labels on containers of hazardous materials. OSHA, the DOT, and the EPA all require labels on hazardous waste.

CAUTION

OSHA DOES NOT REQUIRE LABELING OF CONTAINERS OF HAZARDOUS MATERIALS IF THE CONTENTS WILL BE USED BY THE WORKER IN THAT WORK SHIFT. THE AUTHORS BELIEVE THIS TO BE A LOOPHOLE IN THE REGULATIONS. THE AUTHORS STRONGLY RECOMMEND LABELING ALL CONTAINERS, EVEN IF THE CONTENTS ARE GOING TO BE USED WITHIN A FEW HOURS. WE KNOW OF AN INCIDENT WHERE A MAINTENANCE MECHANIC USED AN OLD SODA CAN TO CONTAIN SOME XYLENE AS A DEGREASER. A COWORKER ACCIDENTALLY DRANK SOME OF THE XYLENE, THINKING IT WAS HIS CAN OF SODA. THE COWORKER SAID THAT UPON INGESTION THE BURNING SENSATION CAUSED HIM TO FEAR DEATH! HE DID RECOVER, HOWEVER, FOLLOWING HOSPITAL TREATMENT. WE REITERATE THAT NO FOOD OR DRINK MAY BE PRESENT IN THE WORK AREA.

To be a proper DOT shipping label, the label must, at a minimum, contain the proper shipping name, the hazard class or division, the UN/NA identification number, and the packing group, as appropriate. Desirably, they contain additional information such as emergency contacts, safe handling requirements, PPE requirements, and disposal information. They are printed or written on labeling stock, and then affixed on individual containers of material. Container examples are: cans, bottles, drums, boxes, crates, Intermediate Bulk Containers (IBCs), tanks, and bins. Labels placed on hazardous material containers that are transported must contain information identical to that on the material's DOT placard. Labels on containers that are transported must remain on the container, and remain legible, until the container is cleaned completely of all residue of the material.

DOT/UN/NA hazardous materials placards are primarily used to identify materials in transit by rail or truck. As readers will see, trucks and railcars do not require labeling the same as small packages. These placards cover the entire cargo carried by the vehicle. The remainder of the shipping information is contained in the manifest (see below). Some hazardous materials may be transported by air. However, air carriers are not required to placard the planes themselves.

The widely recognized graphics of these placards are also used, in the form of labels, to identify the contents of almost any type of container, including drums, tanks, and cylinders containing liquids and gases.

Signs are descriptions of the contents of a storage or operations area, along with precautions for entry. This includes PPE requirements, incompatibilities that can result from various classes of materials being stored in the area, and more. Signs are usually made of a weather-resistant, resilient material such as metal or plastic. They are affixed to storage tanks, doors, walls, or fences in and around storage and operations areas.

Manifests are documents having detailed descriptions of the contents of a carrier in transit. They are also called "shipping papers" or "way bills." For tractor-trailers, all shipping papers must be carried in a compartment in the driver's side door. On rail shipments, these papers must be carried in the engine compartment with the engineer and freight conductor.

A special type of manifest, called a Uniform Hazardous Waste Manifest, is required for the shipment of hazardous waste. This document must accompany all shipments of hazardous waste. This special manifest contains vital information for the First Responder at the scene of an accident or spill. (See Section 13.)

Unfortunately, the contents of storage containers, tank trucks, or railcars may be incorrectly identified or characterized. Records or shipping papers may be inaccessible. Even without such information, experienced HAZWOPER personnel at times must assess potential hazards and their seriousness.

CAUTION

DO NOT RELY SOLELY ON A MANIFEST OR PLACARD WHEN RE-SPONDING TO AN EMERGENCY. ALWAYS PROCEED WITH CAUTION! MARKINGS CAN BE MISSING, INCOMPLETE, OR INCORRECT. SAMPLING AND MONITORING (SECTION 12) ARE YOUR ONLY ROUTE TO BECOMING ABSOLUTELY SURE OF THE MATERIAL WITH WHICH YOU ARE DEALING.

OSHA regulations require that placards and labels on containers of hazardous materials be maintained and legible until such time as the container is clean and free of the potential hazard.

Note

OSHA regulations require employers having items such as packages, transport vehicles, freight containers, motor vehicles, or rail freight cars containing hazardous material to display the required signage—that is, marking, placarding, or labeling in accordance with the U.S. Department of Transportation's (DOT) Hazardous Materials Regulations. Such employers are required to maintain the required signage on the items until they are sufficiently cleaned of residue and purged of vapors to remove any potential hazards.

8.3 THREE MAJOR SYSTEMS OF SIGNAGE

The three major systems of signage are the NFPA 704M® System, the HMIS®, and DOT placards.

There are two main systems (NFPA and HMIS®) for labeling the containers of the hazardous material. There is one major placarding (DOT/UN/NA) system, affixed to the exterior of transport vehicles and railcars, that identifies hazardous material in transit. Each of these systems can assist responders by presenting additional information when responders must deal with a hazardous material incident quickly and safely.

1. The National Fire Protection Association (NFPA) 704M® system is used on storage tanks and smaller containers at fixed facilities.

2. The Hazardous Materials Identification System (HMIS®) is used on containers of products that come from manufacturers that are members of the Paint and Coatings Council.

3. The DOT/UN/NA placarding system is used mainly on trucks, railcars, intermodal containers, and tanks transported in commerce in North America, including Canada and Mexico. UN/NA stands for United Nations/North America. It is also used as a labeling system for smaller containers of hazardous materials while in shipment and at storage and use facilities.

8.3.1   NFPA 704M® SYSTEM

⊙━ᴨ **The NFPA 704M® system gives a HAZWOPER worker a quick visual indication of the hazards of the contents of a container involved in a fire.**

The purpose of the NFPA 704M (see Figure 8.1) labeling system is to provide a way of quickly identifying the various fire- and health-related hazards associated with a particular material. Firefighters need a quick visual information source indicating the contents of storage tanks and containers. Based on that almost instantaneous assessment, the firefighters know to fight the fire or pull back to a better control position. The NFPA 704M®, also known as the "fire diamond," is most commonly found on bulk storage containers (tanks and bins), but it is also widely used on drums, on all sizes of containers, and as signage for areas. So widespread is the NFPA 704M® fire diamond that you will find the fire diamond information on most MSDSs. The NFPA fire diamond system has been in effect for over 25 years and is the most widely recognized signage by HAZWOPER and firefighting personnel.

⊙━ᴨ **The NFPA 704M® System uses the following four colors to indicate the type of hazard: blue (health), red (flammability), yellow (reactivity), and white (special hazards).**

⊙━ᴨ **The NFPA 704M® System uses the numbers 0 to 4 to indicate the severity of the hazard. Zero indicates little or no fire-related hazard, and 4 indicates the greatest level of fire-related hazard.**

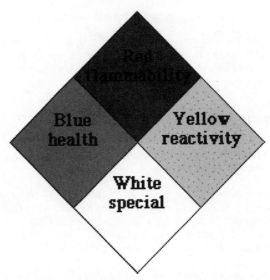

FIGURE 8.1   The NFPA 704M safety diamond.

NFPA 704M® is a standardized system that uses numbers and primary colors on a label to define the basic hazards of a specific material. Health (blue), flammability (red), and reactivity (yellow) are identified and rated on a scale of 0 to 4 depending on the degree of hazard presented. Zero represents no hazard and 4 represents extreme hazard. The ratings for individual substances can be found in the NFPA *Guide to Hazardous Materials.** Another reference, *Sax's Dangerous Properties of Industrial Materials,*** contains the NFPA ratings for individual substances. Such information is useful not only in emergencies but also during long-term remedial activities when extensive evaluation must be completed.

Note

Hazard ratings are written in the shaded diamond corresponding to the hazard.

Flammability Hazards (Red)

- *Hazard rating 4:* Materials that will rapidly or completely vaporize at atmospheric pressure and normal ambient temperature, or materials that are readily dispersed in air and that will burn readily. Rating 4 includes liquids with a flash point below 73°F and a boiling point below 100°F.

- *Hazard rating 3:* Liquids and solids that can be ignited under almost all ambient temperature conditions. This includes liquids with a flashpoint below 73°F and a boiling point above 100°F, or liquids with a flash point above 73°F but not exceeding 100°F and a boiling point below 100°F.

- *Hazard rating 2:* Materials that must be moderately heated or exposed to relatively high ambient temperatures before ignition can occur. These are liquids with flash point above 100°F but not exceeding 200°F.

- *Hazard rating 1:* Materials that must be preheated before ignition can occur; liquids that have a flash point above 200°F.

- *Hazard rating 0:* Materials that will not burn.

Health Hazards (Blue)

- *Hazard rating 4:* Materials that on very short exposure could cause death or major residual injury.

- *Hazard rating 3:* Materials that on short exposure could cause serious temporary or residual injury.

- *Hazard rating 2:* Materials that on intense or continued, but not chronic, exposure could cause incapacitation or possible residual injury.

- *Hazard rating 1:* Materials that on exposure could cause irritation but only minor residual injury.

- *Hazard rating 0:* Materials that on exposure under fire conditions will offer no hazard beyond that of ordinary combustible material.

Reactivity Hazards (Yellow)

- *Hazard rating 4:* Materials that in themselves are readily capable of detonation, explosive decomposition, or reaction at normal temperatures and pressures.

---

*Fire Protection Guide to Hazardous Materials, 12th ed., Quincy, Mass.: National Fire Protection Association, 1997.

****R. J. Lewis, *Sax's Dangerous Properties of Industrial Material,* 10th ed., New York: Wiley-Interscience, 2000.

- *Hazard rating 3:* Materials that in themselves are capable of detonation or explosive decomposition or reaction but require a strong initiating source, that must be heated under confinement before initiation, or that react explosively with water.
- *Hazard rating 2:* Materials that readily undergo violent substance change at elevated temperatures and pressures; that react violently with water, or that may form explosive mixtures with water.
- *Hazard rating 1:* Materials that in themselves are normally stable but that can become unstable at elevated temperatures and pressures.
- *Hazard rating 0:* Materials that in themselves are normally stable, even under fire exposure conditions, and are not reactive with water.

Special Hazards (White)

ACID	Acid
ALK	Alkali
COR	Corrosive
OXY	Oxidizer
W	Reacts with water
☣	Biohazard
☢	Radioactive

8.3.2 Hazardous Materials Identification System® (HMIS®)

◖—π **The HMIS® is very similar to the NFPA 704M® and is used on containers of paints and allied products.**

◖—π **The major difference between the HMIS® and the NFPA 704M® system is the white category. For HMIS®, white stands for PPE required.**

Note

Do not confuse the Hazardous Materials Identification System® (HMIS®) with the Department of Defense Hazardous Materials Identification System (HMIS). Yes, both names are the same. However, the former is a registered trademark of the National Paint and Coatings Association and latter is a U.S. government computer database.

The purpose of the HMIS® system (see Figure 8.2) is to aid employers in day-to-day compliance with OSHA's Hazard Communication Standard (HAZCOMM). (See the Glossary, Appendix A.) The program uses a numerical hazard rating system (similar to the NFPA 704M®), labels with colored bars (same colors as the NFPA 704M®), and training materials to inform workers of substance hazards in the workplace. Personal protective equipment information (in place of "special hazards" on the NFPA 704M® label) is supplied to give employees information needed to protect themselves from hazardous materials they might encounter on the job.

HMIS LABEL

HEALTH HAZARD

4 - Deadly
3 - Extreme Danger
2 - Hazardous
1 - Slightly Hazardous
0 - Normal Materials
* Chronic Hazard

REACTIVITY HAZARD

4 - May Detonate
3 - Shock & Heat
    May Detonate
2 - Violent Chemical
    Change
1 - Unstable if Heated
0 - Stable

HEALTH

FIRE

REACTIVITY

PPE

FIRE HAZARD

4 - Very Flammable
3 - Readily Ignitable
2 - Ignited with Heat
1 - Combustible
0 - Will not Burn

PERSONAL
PROTECTIVE
EQUIPMENT
RECOMMENDATIONS

FIGURE 8.2 Hazardous Materials Identification System® (HMIS®). Health-Blue, Fire-Red, Reactivity-Yellow, and PPE-White.

8.3.3 DOT Placarding System (Material Identification System for Transport and Stationary Containers and Storage Areas) and UN / NA Numbering System

○━π **The DOT placarding system is required for any hazardous material offered for transport.**

A unified placarding system is one in which all the DOT placarding information is shown in all formats. See Figure 8.3.

○━π **All DOT-regulated hazardous materials also have a four-digit UN/NA number. This might indicate a specific material, that material at a certain concentration, or a class of materials.** ·

○━π **With the exception of oxygen and radioactive materials, the text that indicates the hazard is not required.**

Each hazardous material that is transported within North America (Mexico, the United States, and Canada) is assigned to a hazard class, a division, and a packing group along with a four-digit United Nations/North America UN/NA number from the Hazardous Materials Table (49 CFR 172.101). The Hazardous Materials Table is available at www.rspa.dot.gov and on the Companion CD.

FIGURE 8.3 Example of a unified placarding system.

CAUTION

DO NOT CONFUSE THE FOUR-DIGIT UN/NA NUMBER WITH THE EPA'S FOUR-CHARACTER, ALPHANUMERIC HAZARDOUS WASTE CODE. UN/NA NUMBERS ARE ALWAYS DIGITS, WHILE THE EPA HAZARDOUS WASTE CODES ALL BEGIN WITH A LETTER (D, F, K, P, OR U), FOLLOWED BY THREE DIGITS.

Materials are classed as to the specific transportation and storage hazard they present. The nine hazard classes are: Explosives, Gases, Flammable Liquids, Flammable Solids, Oxidizers, Poisonous Materials, Radioactive Materials, Corrosive Materials, and Miscellaneous Dangerous Goods.

○━π **The DOT placarding system requires shippers to label their material containers with one of the nine hazard classes; or as "Other Regulated Materials."**

○━π **The nine DOT hazard classes are: Explosives; Gases; Flammable Liquids; Flammable Solids, Spontaneously Combustible Materials, and Materials that are Dangerous When Wet; Oxidizers and Organic Peroxides; Poisonous Materials; Radioactive Materials; Corrosive Materials; and Miscellaneous Dangerous Goods.**

Note

Some materials, such as unapproved (by DOT) explosives, are so sensitive that their transport is forbidden. Other materials are prohibited shipments by certain modes of transportation, such as by Passenger Aircraft. *The fine for such shipments is $27,000 per unit shipped.* Remember the ValuJet plane crash in the Florida Everglades, with the loss of over 100 lives? Outdated oxygen generators that were required by law to have been disposed of as hazardous waste were illegally shipped as cargo. The sodium chlorate in the generators in the plane caught fire and caused the crash.

Some products may present *multiple hazards.* For instance, calcium hypochlorite (dry), if inhaled or ingested by humans, can cause serious internal burns. Is it classified as a corrosive? NO. When it reacts with other materials, it releases poison chlorine gas. Is it classified as a poisonous material? NO. Although calcium hypochlorite can be corrosive and can react violently to produce toxic chlorine gas, it has a more immediate transportation and storage threat. It is a strong oxidizer. When it comes into contact with numerous other materials, such as fuels, it can create explosive reactions. It is therefore classed as an oxidizer. In the case of multiple hazards, consult the Hazardous Materials Table to determine which hazard takes precedence.

The *class* of a material is shown as the number (1–9) at the bottom of the diamond-shaped placard. The class may also be displayed in text across the center of the diamond, with the number of the class at the bottom. The class may also be displayed as the colored placard with icon (for instance, white flames on a red placard) with the class number at the bottom.

Within each hazard classification are *divisions.* The division of the class gives further information about the hazard of the material. You must understand the individual class divisions. The hazard class division's lowest number is the worst *physical hazard* and the highest is the least physical hazard. This does not mean they are less hazardous to humans upon exposure! For example, look at the gases class divisions below under Class 2: Gases. (There, Division 2.1 members have a less toxic effect on humans than do Division 2.4.)

Also, within the classes there are up to three Packing Groups (I, II, and III). Packing Group I represents the greatest risk (the most regulated), Packing Group II represents a moderate risk (moderately regulated), and Packing Group III represents the least risk and is the least regulated.

All DOT-regulated hazardous materials containers, including tankers, cargo trailers, bulk trailers, intermodal containers, and rail cars, must be placarded. If shipped to or from Mexico or Canada, they must be placarded with the four-digit UN/NA number of their contents. That number may be displayed alone (black numerals on an orange background) or on the placard, describing the class of the material.

The UN/NA number will either be displayed as an oblong placard by itself or across the middle of the graphic placard. Figure 8.4 shows the proper display of the UN/NA number panel for Paint, Flammable (black numbers on an orange background). The number will direct responders to either the exact substance or the class of substances to which they are responding. As can be seen in Figure 8.3, any of the three placards shown is acceptable for a shipment of LPG. All the necessary information for the correct marking, labeling, placarding, and manifesting of a shipment of a hazardous material is found in the D.O.T.'s Hazardous Materials Tables. The information the tables include for each material is:

- Symbol
- Description, proper shipping name
- Hazard class or division
- Identification numbers
- PG (packing group)
- Label codes
- Special provision
- Packaging (173.***)(*** are the different subsections of 49 CFR 173)
  1. Exceptions
  2. Non-bulk
  3. Bulk
- Quantity limitations
  1. Passenger aircraft/rail
  2. Cargo aircraft only
- Vessel stowage
  1. Location
  2. Other

The four-digit UN/NA number also corresponds to information in the *2000 Emergency Response Guidebook* (ERG), which will direct responders to the correct sources for assistance as well as emergency response procedures. This will be discussed in Section 8.4. The ERG is included in the Companion CD.

Note

Because of their widespread acceptance, easy identification, and familiarity, these placards, as the NFPA 704M® system, are used extensively as signage at facilities. They are seen on fixed facility tanks, as signs on fences, and at doors and on walls.

FIGURE 8.4 Four-digit UN/NA number panel for flammable paint.

The following is a list of the DOT/United Nations/North America (UN/NA) hazard classes, divisions, and, where applicable, compatibility groups and their corresponding descriptions and placards. Circled numbers in Figures 8.5 through 8.13 correspond to the ERG guide number pages (see Section 8.4). Beginning in this Section the classes will be discussed individually, in detail.

- *Class 1: Explosives*
  - Division 1.1: Explosives with a mass explosion hazard
  - Division 1.2: Explosives with a projection hazard
  - Division 1.3: Explosives with predominantly a fire hazard
  - Division 1.4: Explosives with no significant blast hazard
  - Division 1.5: Very insensitive explosives
  - Division 1.6: Extremely insensitive explosive articles
- *Class 2: Gases*
  - Division 2.1: Flammable gases
  - Division 2.2: Nonflammable gases
  - Division 2.3: Poison gas
  - Division 2.4: Corrosive gases
- *Class 3: Flammable liquids*
  - Division 3.1: Flash point below −18°C (0°F)
  - Division 3.2: Flash point −18°C and above, but less than 23°C (73°F)
  - Division 3.3: Flash point 23°C and up to 61°C (141°F)
- *Class 4: Flammable Solids; Spontaneously Combustible materials; and materials that are dangerous when wet*
  - Division 4.1: Flammable solids
  - Division 4.2: Spontaneously combustible materials
  - Division 4.3: Materials that are dangerous when wet
- *Class 5: Oxidizers and organic peroxides*
  - Division 5.1: Oxidizers
  - Division 5.2: Organic peroxides
- *Class 6: Poisons and etiologic materials*
  - Division 6.1: Poisonous materials
  - Division 6.2: Etiologic (infectious) materials
- *Class 7: Radioactive materials*
  - Any material or combination of materials that spontaneously gives off ionizing radiation. Covered materials have a specific activity greater than 0.002 microCuries per gram ($\mu C/g$).
- *Class 8: Corrosives*
  - A material, liquid or solid, that causes visible destruction or irreversible alteration to human skin, or a liquid that has a severe corrosion rate on steel or aluminum.
- *Class 9: Miscellaneous*
  - A material that presents a hazard during transport but that is not included in any other hazard class (such as a hazardous substance or a hazardous waste).
- *ORM (A–E): Other regulated material*
  - See Section 8.5.10 for more detail.

Class 1: Explosives (Orange Placard with Black Graphics and Text, See Figure 8.5)

- *Division 1.1:* Explosives with a mass explosion hazard
- *Division 1.2:* Explosives with a projection hazard

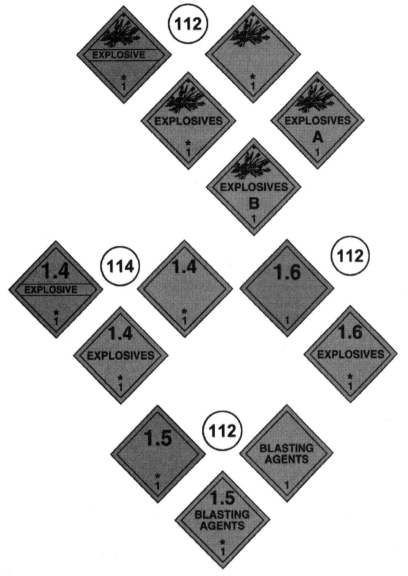

FIGURE 8.5 The Class 1 explosive placards.

- *Division 1.3:* Explosives with predominantly a fire hazard
- *Division 1.4:* Explosives with no significant blast hazard
- *Division 1.5:* Very insensitive explosives; blasting agents
- *Division 1.6:* Extremely insensitive detonating articles

   The DOT defines explosives as any substance, compound, mixture, or device capable of producing an explosive-pyrotechnic effect, with substantial instantaneous release of heat and gas. Examples are nitroglycerin, fireworks, blasting caps, Christmas cracker snaps, igniters, fuses, flares, and ammunition.

- *Hazard division 1.1, 1.2, and 1.3 placards:* For appropriate compatibility groups. The *
  shall be replaced by the appropriate division number.
- *Hazard division 1.4, 1.5, and 1.6 placards:* For compatibility group B, C, D, E, F, G, S,
  and N. The * shall be replaced by the appropriate hazard class number and division number.

Class 2. Gases (See Figure 8.6) (See Divisions for Placard Descriptions)   Gases compressed, liquefied, or dissolved under pressure. This includes permanent gases, which cannot be liquefied at ambient temperatures; liquefied gases, which become liquid under pressure

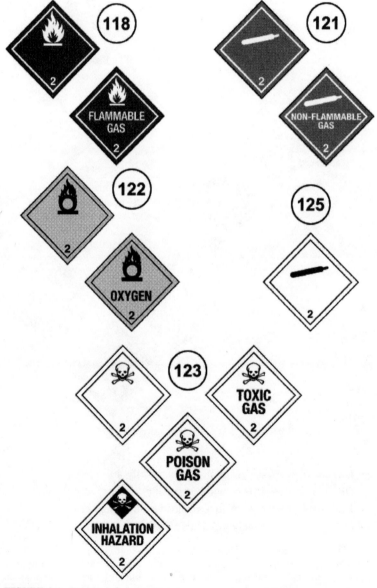

FIGURE 8.6   The Class 2 gases placards.

at ambient temperatures; and dissolved gases, which are dissolved under pressure in a solvent.

- *Division 2.1: Flammable gases* (red placard with white graphics and text).
  - Examples of flammable compressed gases are hydrogen, ethane, methane, propane, butane, cigarette lighters, and gas cylinders for camping stoves. Placard 1001 lb or more gross weight of flammable gas.
- *Division 2.2: Nonflammable, nontoxic compressed gases* (green placard with white graphics and text.) Exceptions are oxidizer gases, yellow background with black graphics, except oxygen, which is a yellow placard with black text that must say "OXYGEN"!)
  - Examples of nonflammable, nontoxic compressed gases are oxygen, carbon dioxide, nitrogen, neon, fire extinguishers containing such gases, and aerosols. Placard 1,001 lb or more aggregate gross weight of nonflammable gas.
- *Division 2.3: Gases toxic by inhalation* (white placard with black graphics and text).
  - Examples of toxic compressed gases are chlorine and the fluorine-containing gases used in printed circuit board manufacture.
- *Division 2.4: Corrosive gases* (*Canada*) (white placard with black graphics and text).
  - Examples of corrosive compressed gases are hydrogen chloride, hydrogen sulfide, and sulfur dioxide.

Class 3: Flammable Liquids and Combustible Liquids (U.S.) (Red Placard with White Graphics and Text, See Figure 8.7)

- *Flammable and combustible liquids:* Liquids, mixtures of liquids, or liquids containing solids in solution or suspension that give off a flammable vapor. Any liquid with a closed-cup flash point below 140°F is prohibited.
  - *Examples:* Acetone, benzene, cleaning compounds; gasoline; lighter fluid; paint thinners and removers, petroleum; solvents. Placard 1001 lb or more flammable liquid. Placard a combustible liquid when transported in a packaging exceeding 10 gallon rated capacity, in a cargo tank or a tank car. Note that a FLAMMABLE placard may be substituted for the COMBUSTIBLE placard on a cargo tank and portable tank in highway transportation.

Class 4: Flammable Solids; Spontaneously Combustible Materials; Dangerous-When-Wet Materials, (See Figure 8.8)

- *Division 4.1:* Flammable solids (red and white vertical stripes with black graphic and text.) Placard 1001 lb or more gross weight of flammable solid.

FIGURE 8.7 The Class 3 flammable and combustible liquids placards.

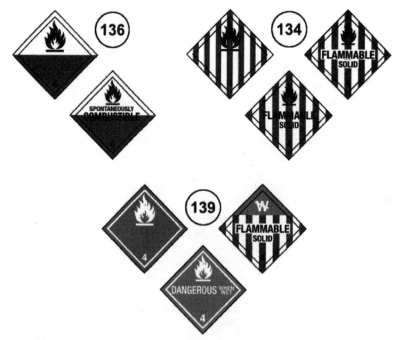

FIGURE 8.8   The Class 4 flammable solids, spontaneously combustible materials, and dangerous-when-wet materials placards.

- *Division 4.2:* Spontaneously combustible materials (white top, red bottom, with black graphics and text.)
- *Division 4.3:* Dangerous-when-wet materials (blue background with white graphics and text or top one-third white on blue and bottom two-thirds red and white vertical stripes.)
- *Flammable solids:* Solid materials that are liable to cause fire by friction, absorption of water, spontaneous substance changes, or retained heat from manufacturing or processing, or that can be readily ignited and burn vigorously.
  - Examples: Matches (any type including safety), calcium carbide, cellulose nitrate products, metallic magnesium, nitrocellulose-based film, phosphorus, potassium; sodium, sodium hydride, zinc powder; zirconium hydride.

Class 5: Oxidizers and Organic Peroxides (*Yellow Placard with Black Graphics and Text, See Figure 8.9*)

- *Division 5.1:* Oxidizers (placard 1001 lb or more gross weight of oxidizing material).
- *Division 5.2:* Organic peroxides (placard 1001 lb or more gross weight of organic peroxide).
- *Oxidizing substances and organic peroxides:* Though not necessarily combustible themselves, these substances may cause or contribute to combustion of other substances. They may also be liable to explosive decomposition, react dangerously with other substances, and be injurious to health.

FIGURE 8.9   The Class 5 oxidizers and organic peroxides placards.

• Examples: bromates, chlorates, components of fiberglass repair kits, nitrates, perchlorates; permanganates, peroxides.

Class 6. Toxic Materials and Infectious Substances (*White Placard with Black Graphics and Text, See Figure 8.10*)

• *Division 6.1:* Toxic materials (placard 1001 lbs or more gross weight of poison B.)
• *Division 6.2:* Etiologic (infectious) substances.

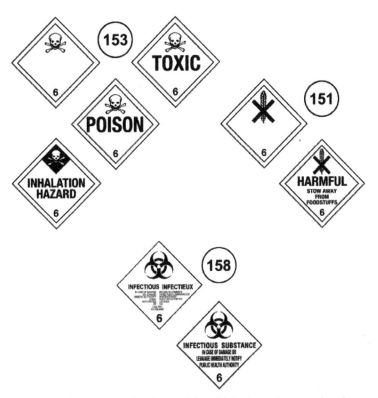

FIGURE 8.10   The Class 6 toxic materials and infectious substances placards.

FIGURE 8.11 Class 7 radioactive materials placards.

- *Toxic (poisonous) and infectious substances and other medical substances:* Substances liable to cause death or injury if swallowed or inhaled, or by skin contact. Substances containing microorganisms or their toxins that are known or suspected to cause disease.
  - Examples: Arsenic, beryllium, cyanide, fluorine, hydrogen selenide, mercury, mercury salts, mustard gas, nitrobenzenen, nitrogen dioxide, pathogenic material, rat poison, serum, vaccines.

Class 7: Radioactive Materials (*White Placard with Black Text, or Bottom Half White, Top Half Yellow—MUST SAY RADIOACTIVE! See Figure 8.11*)

- *Radioactive material:* Any material with a specific activity greater than 74 kilobecquerels (kBq) per kg (0.002 μC/g) (See Section 3.6 and Appendix A.)
  - Examples: Fissile material (Uranium 235,etc.), radioactive waste material, uranium or thorium ores. (Placard any quantity of packages bearing the yellow RADIOACTIVE III Label.)

Class 8: Corrosive Materials (*Top Half White, Bottom Half Black with Black Graphics and Reverse Text, See Figure 8.12*)

- *Corrosives:* Substances that can cause severe damage by substance action to living tissue, other freight, or the means of transport.
  - Examples: Aluminum chloride, caustic soda, corrosive cleaning fluid, corrosive rust remover/preventative; corrosive paint remover; electric storage batteries, hydrochloric acid, nitric acid, sulfuric acid.

FIGURE 8.12 Class 8 corrosive materials placards.

Class 9: Miscellaneous Dangerous Goods (White Placard with Black Graphics and Text Except "!" and "DANGEROUS," Which Are Red on White, See Figure 8.13)

Note

The DANGEROUS placard is intended for situations when you are shipping two or more hazardous materials listed on the Hazardous Materials Tables. As an example, let us look at flammable liquids and corrosives. Supposing neither one of these hazardous materials you are offering for shipment equals or exceeds 2200 lb (1000 kg). You could then ship 1500 pounds of flammable liquid and 1500 pounds of corrosive liquid with a DANGEROUS placard and be in compliance with DOT's regulations.

- *Miscellaneous dangerous materials:* Substances that present dangers not covered elsewhere. Category includes environmentally hazardous substances, elevated temperature material, some hazardous wastes, and marine pollutants.
  - Examples: Asbestos, dry ice (solid carbon dioxide), and strongly magnetized material.

FIGURE 8.13 Class 9 miscellaneous dangerous goods placards.

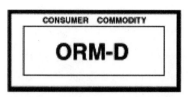

FIGURE 8.14   ORM placard (ORM-D).

Other Regulated Materials (*ORMs. See Figure 8.14*)   Other regulated materials (ORMs) are materials that are required to be packaged and transported in accordance with the DOT Hazardous Materials Tables but are not included in the nine hazard classes. They are known as ORM-A through ORM-E and are described below. The exception is if the package that contains the hazardous material is in a small consumer-type size. Then the rules that apply to shipping hazardous materials do not apply.

- *ORM A:* A material that has an anesthetic, irritating, noxious, toxic, or other similar property and that can cause extreme annoyance or discomfort to passengers and crew in the event of leakage during transportation (49 CFR 173.500(b)(1))
- *ORM B:* A material (including a solid when wet with water) capable of causing significant damage to a transport vehicle from leakage during transportation (49 CFR 173.500(b)(2))
- *ORM C:* A material that has other inherent characteristics not described as an ORM A or ORM B but that make it unsuitable for shipment unless properly identified and prepared for transportation (49 CFR 173.500(b)(4))
- *ORM D:* A material, such as a consumer commodity, that presents a limited hazard during transportation due to its form, quantity, and packaging (49 CFR 173.500(b)(4))
- *ORM E:* A material that is not included in any other hazard class but is subject to the requirements of 49 CFR 173.500 and includes hazardous waste.

8.3.4   EPA Hazardous Waste Labels

The EPA requires a hazardous waste label to be placed on any container of material that the generator (see Section 2.2) has determined is a hazardous waste. This will be discussed in fuller detail in Section 13. However, because it is a labeling requirement, we will introduce the subject here.

The EPA requires that a hazardous waste label be placed on any container immediately before ANY hazardous waste is placed inside. That is, if a worker puts any amount on the material in a drum, even a cup full, that drum must have an EPA hazardous waste label affixed to the side of the drum between the chimes. At this time, California, Ohio, New Jersey, and South Carolina also require specific state hazardous waste labels.

The Federal EPA hazardous waste label must contain the following information:

- Generator's name
- EPA ID number (not required for small quantity generator)
- Address
- EPA/DOT proper shipping name
- Hazard class
- UN/NA number
- Constituents (as percent)

- EPA waste code
- Reportable quantity (RQ)
- Accumulation start date (when the first material was put in the drum. Large quantity generators have 90 days from the accumulation start date to properly dispose of the waste in an EPA Approved landfill. This rule and the exceptions will be discussed in greater detail in Section 13.)
- Emergency Contact Information (our example uses the Coast Guard's National Response Center, NRC)

Figure 8.15 is an example of a properly filled out EPA hazardous waste label.

The EPA requires that a hazardous waste label be placed on any container immediately before ANY hazardous waste is placed inside.

Note

If the label is for the purposes of accumulation only, then the label can be printed in black and white. For shipping purposes, the label must be in color, yellow body with red border. Text can be in red or black. The information shown in Figure 8.15 is for illustration purposes only. The format of the label can take a number of configurations as long as the required information is present.

HAZARDOUS WASTE

FEDERAL LAW PROHIBITS IMPROPER DISPOSAL
If found, contact the nearest police or public safety authority, and the Washington State Dept. of Ecology or the Environmental Protection Agency

Generator Name	**ABC Chemical Co.**	EPA ID # **WA1234567890**

Address **123 Any Street**   City **Anytown**   State **WA**   ZIP **00000-0000**

EPA/DOT Shipping Name	**Sulfuric Acid, 52%**	Hazard Class	**8: Corrosive material**

UN/UNA No. **1830**
(   )

Constituents **(Sulfuric Acid 52% Water 48%)**

REPORTABLE QUANTITIES
"RQ" IN POUNDS (Identified in 40 CFR Subchapter J, Part 302, Table 302.4)
☐1 ☐10 ☐100
☒1000 ☐5000
"RQ" _____ LBS.

EPA Waste Code/s and/or Characteristic/s
(D003)

Manifest Document #   **1234567**

| Start Date of Accumulation | **1-Jan-02** | ☒LQG (90 Day) Date **3-1-02** |
| | | ☐MQG (180 Day) Date _____ |

In the event of a spill or release of this hazardous waste, contact the US Coast Guard National Response Center at 1-800-424-8802 for information and assistance.

FIGURE 8.15 A properly filled out EPA hazardous waste label (for a generator in Washington State) for sulfuric acid.

8.4   EMERGENCY RESPONSE GUIDEBOOK (ERG)

    O━━π   **The ERG gives HAZWOPER responders the immediate information needed to protect people and the environment from accidental releases of hazardous materials.**

One of the essential references for HAZWOPER workers is the *Emergency Response Guidebook* (see Figure 8.16), formerly known as the *North American Emergency Response Guidebook*. This book is easily understood once the reader understands how it is organized and how the data are presented. All HAZWOPER personnel need to spend enough time reviewing the ERG's contents that they can rapidly search it to find the information they need.

Each material or group of closely related materials is assigned a specific four-digit number called an I.D. number. The I.D. number is the same as the UN/NA number. Each material or group of materials is also assigned a three-digit guide number. As you will see, the guide numbers guide the HAZWOPER responder on how to deal with the hazard. The Guidebook is fully self-explanatory, but do not wait for an emergency to learn how to read it! The authors have found this to be a handy exercise. We recommend that you purchase a paper copy of the ERG by contacting the DOT at www.hazmat.dot.gov/gydebook.htm. In fact, as EPA First Responders, we carry the book, binoculars, and a reflective safety vest in all our vehicles. Then, if we witness an accident, we can pull over to a safe spot, don the vest in our vehicle to make us highly visible at the roadside, and try to spot any placards on material carriers. The easiest way to become comfortable with ERG use is to keep the ERG paper copy in your vehicle. Watch for placarding on different modes of transport. Practice consulting the ERG to determine the nature of the cargo.

This subsection will give an overview of the contents and how to read the ERG. Refer to the Companion CD for a complete, searchable ERG.

Note

Some readers will remember this as the *North American Emergency Response Guidebook* (NAERG).

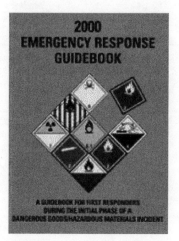

FIGURE 8.16   The *2000 Emergency Response Guidebook* (ERG).

8.4.1 Table of Contents of the 2000 ERG

RESOURCES

- ERG 2000 User's Guide
- Emergency Response Telephone Numbers
- Protective Clothing
- Criminal/Terrorist Use of Substance/Biological Agents
- Glossary
- Background Information on the TII&PAD (see Section 8.4.5)
- Publication Data

IDENTIFICATION

- How to Use the Guidebook During an Incident Involving Dangerous Goods
- Hazard Classification System
- Shipping Documents (Papers)
- Rail Car Identification Chart
- Road Trailer Identification Chart
- Hazard Identification Codes Displayed on some Intermodal Containers
- Table of Placards and Initial Response Guides to Use On-scene—Introduction
- Identification Number Index
- Name of Material Index
- List of Dangerous Water-Reactive Materials

ACTIONS

- Safety Precautions
- Protective Actions
- Protective Action Decision Factors to Consider
- Who to Call for Assistance
- Fire and Spill Control
- Guide Number Index
- Table of Initial Isolation and Protective Action Distances (TII&PAD)—Introduction
- How to Use the Table

  For quick identification and actions to be taken, based upon type and scale of emergency, the ERG is divided into the following four major sections, each cross-referenced with the other three sections. This allows a HAZWOPER worker or responder to collect essential information about the material using the placards on the container or vessel.

1. *ID Number* (*UN/NA number*): This is the index of hazardous materials as they appear in the yellow section of the Guidebook (see Figure 8.17), ordered by UN/NA (ID number). The guide number for each material is accessible from here. This section is in numerical order.

   O━π The ERG yellow pages identify hazardous material names by UN/NA numbers in numerical order.

ID No.	Guide No.	Name of Material
1823	154	Caustic soda, solid
1823	154	Sodium hydroxide, dry
1823	154	Sodium hydroxide, bead
1823	154	Sodium hydroxide, flake
1823	154	Sodium hydroxide, granular
1823	154	Sodium hydroxide, solid
1824	154	Caustic soda, solution
1824	154	Sodium hydroxide, solution
1825	157	Sodium monoxide
1826	157	Nitrating acid, spent
1826	157	Nitrating acid mixture, spent
1827	137	Stannic chloride, anhydrous
1827	137	Tin tetrachloride
1828	137	Sulfur chlorides
1828	137	Sulphur chlorides
1829	137	Sulfur trioxide
1829	137	Sulfur trioxide, inhibited
1829	137	Sulfur trioxide, stabilized
1829	137	Sulfur trioxide, uninhibited
1829	137	Sulphur trioxide
1829	137	Sulphur trioxide, inhibited
1829	137	Sulphur trioxide, stabilized
1829	137	Sulphur trioxide, uninhibited
1830	137	Sulfuric acid
1830	137	Sulfuric acid, with more than 51% acid
1830	137	Sulphuric acid
1830	137	Sulphuric acid, with more than 51% acid
1831	137	Oleum
1831	137	Oleum, with less than 30% free Sulfur trioxide
1831	137	Oleum, with less than 30% free Sulphur trioxide
1831	137	Oleum, with not less than 30% free Sulfur trioxide
1831	137	Oleum, with not less than 30% free Sulphur trioxide
1831	137	Sulfuric acid, fuming
1831	137	Sulfuric acid, fuming, with less than 30% free Sulfur trioxide
1831	137	Sulfuric acid, fuming, with not less than 30% free Sulfur trioxide
1831	137	Sulphuric acid, fuming
1831	137	Sulphuric acid, fuming, with less than 30% free Sulphur trioxide
1831	137	Sulphuric acid, fuming, with not less than 30% free Sulphur trioxide
1832	137	Sulfuric acid, spent
1832	137	Sulphuric acid, spent
1833	154	Sulfurous acid
1833	154	Sulphurous acid
1834	137	Sulfuryl chloride
1834	137	Sulphuryl chloride
1835	153	Tetramethylammonium hydroxide
1836	137	Thionyl chloride
1837	157	Thiophosphoryl chloride
1838	137	Titanium tetrachloride
1839	153	Trichloroacetic acid
1840	154	Zinc chloride, solution
1841	171	Acetaldehyde ammonia
1843	141	Ammonium dinitro-o-cresolate
1845	120	Carbon dioxide, solid
1845	120	Dry ice
1846	151	Carbon tetrachloride
1847	153	Potassium sulfide, hydrated, with not less than 30% water of crystallization
1847	153	Potassium sulfide, hydrated, with not less than 30% water of hydration
1847	153	Potassium sulphide, hydrated, with not less than 30% water of crystallization
1847	153	Potassium sulphide, hydrated, with not less than 30% water of hydration
1848	132	Propionic acid
1849	153	Sodium sulfide, hydrated, with not less than 30% water
1849	153	Sodium sulphide, hydrated, with not less than 30% water
1851	151	Medicine, liquid, poisonous, n.o.s.
1851	151	Medicine, liquid, toxic, n.o.s.
1854	135	Barium alloys, pyrophoric
1855	135	Calcium, metal and alloys, pyrophoric
1855	135	Calcium, pyrophoric
1855	135	Calcium alloys, pyrophoric
1856	133	Rags, oily
1858	126	Hexafluoropropylene
1858	126	Refrigerant gas R-1216
1859	125	Silicon tetrafluoride
1859	125	Silicon tetrafluoride, compressed
1860	116P	Vinyl fluoride, inhibited
1862	129	Ethyl crotonate
1863	128	Fuel, aviation, turbine engine
1864	128	Gas drips, hydrocarbon
1865	131	n-Propyl nitrate
1866	127	Resin solution
1867	133	Cigarettes, self-lighting
1868	134	Decaborane
1869	138	Magnesium
1869	138	Magnesium, in pellets, turnings or ribbons
1869	138	Magnesium alloys, with more than 50% Magnesium, in pellets, turnings or ribbons
1869	138	Magnesium scrap
1870	138	Potassium borohydride
1871	170	Titanium hydride
1872	141	Lead dioxide
1872	141	Lead peroxide
1873	143	Perchloric acid, with more than 50% but not more than 72% acid
1884	157	Barium oxide
1885	153	Benzidine
1886	156	Benzylidene chloride
1887	160	Bromochloromethane
1888	151	Chloroform
1889	157	Cyanogen bromide
1891	131	Ethyl bromide
1892	151	ED
1892	151	Ethyldichloroarsine
1894	151	Phenylmercuric hydroxide
1895	151	Phenylmercuric nitrate
1897	160	Perchloroethylene
1897	160	Tetrachloroethylene
1898	156	Acetyl iodide
1902	153	Di-(2-ethylhexyl)phosphoric acid
1902	153	Diisooctyl acid phosphate
1903	153	Disinfectant, liquid, corrosive, n.o.s.

Page 46   Page 47

FIGURE 8.17   The ERG yellow pages—materials indexed by UN/NA number. (*Courtesy of ERG2000.*)

2. *Name of material:* This is the index of hazardous materials as they appear on the blue section of the Guidebook (see Figure 8.18), indexed alphabetically by name of material. The guide number and UN/NA (I number) are also here for each material.

○━ㅈ   **The ERG blue pages identify hazardous materials by name alphabetically.**

3. *Guide number:* The orange section of the Guidebook (see Figure 8.19) is considered the most important section of the Guidebook because this is where all safety actions and recommendations are provided. The orange section tells us what to do in hazardous materials emergencies. It is made up of individual guides each providing safety recommendations and emergency response information to protect workers and the public. Each guide is designed to cover a group of substances that possess similar chemical and toxicological characteristics. Each is laid out identically to the others so that information is immediately accessible and there is no confusion. As we know, confusion just adds to the emergency. The guide pages are laid out in this order:

Potential Hazards

Health
Fire or Explosion
Public Safety

Protective Clothing
Evacuation (large spill or fire)
Emergency Response

Fire (Small, large, or involving Tanks or Car/Trailer Loads)
Spill or Leak
First Aid

Name of Material	Guide No.	ID No.
Sulphur dioxide, liquefied	125	1079
Sulphur hexafluoride	126	1080
Sulphuric acid	137	1830
Sulphuric acid, fuming	137	1831
Sulphuric acid, fuming, with less than 30% free Sulphur trioxide	137	1831
Sulphuric acid, fuming, with not less than 30% free Sulphur trioxide	137	1831
Sulphuric acid, spent	137	1832
Sulphuric acid, with more than 51% acid	137	1830
Sulphuric acid, with not more than 51% acid	157	2796
Sulphuric acid and Hydrofluoric acid mixtures	157	1786
Sulphurous acid	154	1833
Sulphur tetrafluoride	125	2418
Sulphur trioxide	137	1829
Sulphur trioxide, inhibited	137	1829
Sulphur trioxide, stabilized	137	1829
Sulphur trioxide, uninhibited	137	1829
Sulphur trioxide and Chlorosulphonic acid mixture	137	1754
Sulphuryl chloride	137	1834
Sulphuryl fluoride	123	2191
Tabun	153	2810
Tars, liquid	130	1999
TDE (1,1-Dichloro-2,2-bis (p-chlorophenyl)ethane)	151	2761
Tear gas candles	159	1700
Tear gas devices	159	1693
Tear gas grenades	159	1700
Tear gas substance, liquid, n.o.s.	159	1693
Tear gas substance, solid, n.o.s.	159	1693
Tellurium compound, n.o.s.	151	3284
Tellurium hexafluoride	125	2195
Terpene hydrocarbons, n.o.s.	128	2319
Terpinolene	128	2541
Tetrabromoethane	159	2504
1,1,2,2-Tetrachloroethane	151	1702
Tetrachloroethane	151	1702
Tetrachloroethylene	160	1897
Tetraethyl dithiopyrophosphate	153	1704
Tetraethyl dithiopyrophosphate mixture, dry or liquid	153	1704
Tetraethyl dithiopyrophosphate and gases, in solution	123	1703
Tetraethyl dithiopyrophosphate and gases, mixtures	123	1703
Tetraethyl dithiopyrophosphate and gases, mixtures, or in solution (LC50 more than 200 ppm but not more than 5000 ppm)	123	1703
Tetraethyl dithiopyrophosphate and gases, mixtures, or in solution (LC50 not more than 200 ppm)	123	1703
Tetraethylenepentamine	153	2320
Tetraethyl lead, liquid	131	1649
Tetraethyl pyrophosphate, liquid	152	2783
Tetraethyl pyrophosphate, liquid	152	3018
Tetraethyl pyrophosphate, solid	152	2783
Tetraethyl pyrophosphate and compressed gas mixtures	123	1705
Tetraethyl pyrophosphate and compressed gas mixtures (LC50 more than 200 ppm but not more than 5000 ppm)	123	1705
Tetraethyl pyrophosphate and compressed gas mixtures (LC50 not more than 200 ppm)	123	1705
Tetraethyl pyrophosphate mixture, dry	152	2783
Tetraethyl silicate	132	1292
1,1,1,2-Tetrafluoroethane	126	3159
Tetrafluoroethane and Ethylene oxide mixture, with not more than 5.6% Ethylene oxide	126	3299
Tetrafluoroethylene, inhibited	116P	1081
Tetrafluoromethane	126	1982
Tetrafluoromethane, compressed	126	1982
1,2,3,6-Tetrahydro-benzaldehyde	132	2498
Tetrahydrofuran	127	2056
Tetrahydrofurfurylamine	129	2943
Tetrahydrophthalic anhydrides	156	2698
1,2,3,6-Tetrahydropyridine	129	2410
1,2,5,6-Tetrahydropyridine	129	2410
Tetrahydrothiophene	129	2412
Tetralin hydroperoxide	145	2136
Tetramethylammonium hydroxide	153	1835
1,1,3,3-Tetramethylbutyl hydroperoxide	145	2160
1,1,3,3-Tetramethylbutyl peroxy-2-ethylhexanoate	148	2161
Tetramethylmethylenediamine	132	9069
Tetramethylsilane	130	2749
Tetranitromethane	143	1510
Tetrapropyl orthotitanate	128	2413
Textile treating compound or mixture, liquid (corrosive)	154	1760
Thallium chlorate	141	2573
Thallium compound, n.o.s.	151	1707
Thallium nitrate	141	2727
Thallium sulfate, solid	151	1707
Thallium sulphate, solid	151	1707
4-Thiapentanal	152	2785
Thia-4-pentanal	152	2785
Thickened GD	153	2810
Thioacetic acid	129	2436
Thiocarbamate pesticide, liquid, flammable, poisonous	131	2772
Thiocarbamate pesticide, liquid, flammable, toxic	131	2772
Thiocarbamate pesticide, liquid, poisonous	151	3006
Thiocarbamate pesticide, liquid, poisonous, flammable	131	3005
Thiocarbamate pesticide, liquid, toxic	151	3006
Thiocarbamate pesticide, liquid, toxic, flammable	131	3005
Thiocarbamate pesticide, solid, poisonous	151	2771
Thiocarbamate pesticide, solid, toxic	151	2771
Thioglycol	153	2966
Thioglycolic acid	153	1940
Thiolactic acid	153	2936
Thionyl chloride	137	1836
Thiophene	130	2414
Thiophosgene	157	2474
Thiophosphoryl chloride	157	1837
Thiourea dioxide	135	3341
Thiram	151	2771
Thorium metal, pyrophoric	162	2975
Thorium nitrate, solid	162	2976

FIGURE 8.18   The ERG blue pages—materials indexed alphabetically by name of material. (*Courtesy of ERG2000.*)

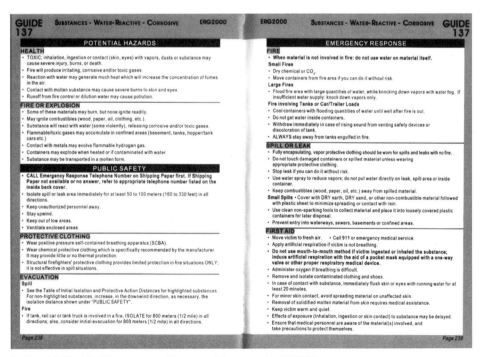

FIGURE 8.19   The ERG orange guide pages. (*Courtesy of ERG2000.*)

O━π   **The ERG orange pages are the hazardous materials guide number section. These guides tell the HAZWOPER responder what to do at an incident.**

4. *Table of Initial Isolation and Protective Action Distances (TII&PAD).* The green section of the Guidebook (see Figure 8.20) consists of a table that lists, in ID number (UN/NA) order, only those substances that are poisonous by inhalation (TIH, toxic-by-inhalation hazards). This table provides two different types of recommended safe distances: initial isolation distances and protective action distances. These TIH substances are clearly listed for easy identification in both numeric (yellow section) and alphabetic (blue section) lists of the Guidebook.

O━π   **The ERG green pages show the downwind distances from which people must be restricted following a hazardous material release.**

This section will also give different distances for spills from railcars (due to the large quantity of material involved) and different distances for daytime and nighttime incidents.

This last part, TII&PAD for day versus night, bears explaining. For most spills on a large scale, the environment and weather conditions play a large part in emergency response and remediation efforts. On a clear dry day, spills of highly volatile materials will tend to dry up and blow away, as the saying goes. On a rainy day, those same materials will tend to be contained in a smaller area. Further, ground and groundwater at the site will be more affected. Rain tends to cleanse the atmosphere and drive pollutants to the ground. Windy conditions tend to dilute and disperse local atmospheric contamination. Those two conditions will modify the following statements.

At night, a weather phenomenon called *inversion* characteristically occurs. It is called "inversion" because the air temperature increases with height above the earth's surface, up to some much higher altitude. Then the temperature starts decreasing again to the cold temperatures of the upper atmosphere and space. That is, inversion is the opposite of the more usual, sunny daytime condition, called lapse, in which the air temperature decreases with height all the way up to space.

An inversion layer is a stable layer of the atmosphere. It will tend to hold any escaping gas or vapors close to the ground, causing any contamination to spread further along the ground and the lowest part of the atmosphere. Obviously, this leads to a greater potential for impact on the surrounding community.

Inversion conditions are typically sought by military groups for poison gas attacks. These conditions hold the gas close to the ground. You can readily understand why inversion conditions increase the hazard of spills of hazardous materials that tend to vaporize.

You can easily detect whether inversion or lapse conditions exist if you can see any kind of stack emitting visible exhaust, such as a white or gray vapor trail. If the exhaust plume leaving the stack flows horizontally, or even downward (when there is no, or slight, wind), you can expect inversion conditions locally. If the plume rises upward, you are witnessing lapse conditions. Practice watching plumes from stacks so that you can learn to make an off-the-cuff estimate of local atmospheric conditions.

O━π   **The ERG green pages also list the water-reactive materials that produce toxic gases.**

New in the 2000 ERG is the Table of Water-Reactive Materials Which Produce Toxic Gases (Figure 8.21). This section immediately follows the TII&PAD section and is also in the green pages. It gives the HAZWOPER responder the toxic-by-inhalation (TIH) gases produced when certain materials come into contact with water. This list is extremely valuable to responders, as many of them are firefighters whose main weapon is water. The material names are given in I.D. number order. The gases created in the water reaction are listed by chemical formula, with a reference at the bottom of each page giving the chemical name for the formula. Toxic gases included are Br_2 (bromine), Cl_2 (chlorine), HBr (hydrogen bromide), HCl (hydrogen chloride), HCN (hydrogen cyanide), HF (hydrogen fluoride), HI (hydrogen iodide), H_2S (hydrogen sulfide), NH_3 (ammonia), PH_3 (phosphine), SO_2 (sulfur dioxide), and SO_3 (sulfur trioxide).

DANGEROUS

TABLE OF INITIAL ISOLATION AND PROTECTIVE ACTION DISTANCES

ID No.	NAME OF MATERIAL	SMALL SPILLS (From a small package or small leak from a large package)				LARGE SPILLS (From a large package or from many small packages)			
		First ISOLATE in all Directions		Then PROTECT persons Downwind during-		First ISOLATE in all Directions		Then PROTECT persons Downwind during-	
		Meters	(Feet)	DAY Kilometers (Miles)	NIGHT Kilometers (Miles)	Meters	(Feet)	DAY Kilometers (Miles)	NIGHT Kilometers (Miles)
1818	Silicon tetrachloride (when spilled in water)	30 m	(100 ft)	0.2 km (0.1 mi)	0.3 km (0.2 mi)	125 m	(400 ft)	1.3 km (0.8 mi)	3.4 km (2.1 mi)
1828	Sulfur chlorides (when spilled on land)	30 m	(100 ft)	0.2 km (0.1 mi)	0.3 km (0.2 mi)	60 m	(200 ft)	0.5 km (0.3 mi)	1.0 km (0.6 mi)
1828	Sulfur chlorides (when spilled in water)	30 m	(100 ft)	0.2 km (0.1 mi)	0.2 km (0.1 mi)	60 m	(200 ft)	0.6 km (0.4 mi)	2.3 km (1.4 mi)
1828	Sulphur chlorides (when spilled on land)	30 m	(100 ft)	0.2 km (0.1 mi)	0.3 km (0.2 mi)	60 m	(200 ft)	0.5 km (0.3 mi)	1.0 km (0.6 mi)
1828	Sulphur chlorides (when spilled in water)	30 m	(100 ft)	0.2 km (0.1 mi)	0.2 km (0.1 mi)	60 m	(200 ft)	0.6 km (0.4 mi)	2.3 km (1.4 mi)
1829 1829 1829 1829 1829 1829 1829 1829	Sulfur trioxide Sulfur trioxide, inhibited Sulfur trioxide, stabilized Sulfur trioxide, uninhibited Sulphur trioxide Sulphur trioxide, inhibited Sulphur trioxide, stabilized Sulphur trioxide, uninhibited	60 m	(200 ft)	0.3 km (0.2 mi)	1.1 km (0.7 mi)	305 m	(1000 ft)	2.1 km (1.3 mi)	5.6 km (3.5 mi)
1831 1831 1831 1831 1831 1831 1831	Oleum Oleum, with not less than 30% free Sulfur trioxide Oleum, with not less than 30% free Sulphur trioxide Sulfuric acid, fuming Sulfuric acid, fuming, with not less than 30% free Sulfur trioxide Sulphuric acid, fuming Sulphuric acid, fuming, with not less than 30% free Sulphur trioxide	60 m	(200 ft)	0.3 km (0.2 mi)	1.1 km (0.7 mi)	305 m	(1000 ft)	2.1 km (1.3 mi)	5.6 km (3.5 mi)
1834	Sulfuryl chloride (when spilled on land)	30 m	(100 ft)	0.2 km (0.1 mi)	0.2 km (0.1 mi)	30 m	(100 ft)	0.3 km (0.2 mi)	0.6 km (0.4 mi)
1834	Sulfuryl chloride (when spilled in water)	30 m	(100 ft)	0.2 km (0.1 mi)	0.2 km (0.1 mi)	125 m	(400 ft)	1.1 km (0.7 mi)	2.4 km (1.5 mi)
1834	Sulphuryl chloride (when spilled on land)	30 m	(100 ft)	0.2 km (0.1 mi)	0.2 km (0.1 mi)	30 m	(100 ft)	0.3 km (0.2 mi)	0.6 km (0.4 mi)
1834	Sulphuryl chloride (when spilled in water)	30 m	(100 ft)	0.2 km (0.1 mi)	0.2 km (0.1 mi)	125 m	(400 ft)	1.1 km (0.7 mi)	2.4 km (1.5 mi)
1836	Thionyl chloride (when spilled on land)	30 m	(100 ft)	0.2 km (0.1 mi)	0.5 km (0.3 mi)	60 m	(200 ft)	0.5 km (0.3 mi)	1.1 km (0.7 mi)
1836	Thionyl chloride (when spilled in water)	30 m	(100 ft)	0.2 km (0.1 mi)	1.0 km (0.6 mi)	335 m	(1100 ft)	3.2 km (2.0 mi)	7.1 km (4.4 mi)
1838	Titanium tetrachloride (when spilled on land)	30 m	(100 ft)	0.2 km (0.1 mi)	0.2 km (0.1 mi)	30 m	(100 ft)	0.3 km (0.2 mi)	0.8 km (0.5 mi)
1838	Titanium tetrachloride (when spilled in water)	30 m	(100 ft)	0.2 km (0.1 mi)	0.3 km (0.2 mi)	125 m	(400 ft)	1.1 km (0.7 mi)	2.9 km (1.8 mi)
1859 1859	Silicon tetrafluoride Silicon tetrafluoride, compressed	30 m	(100 ft)	0.2 km (0.1 mi)	0.5 km (0.3 mi)	60 m	(200 ft)	0.5 km (0.3 mi)	1.6 km (1.0 mi)
1892	ED (when used as a weapon)	30 m	(100 ft)	0.3 km (0.2 mi)	0.8 km (0.5 mi)	125 m	(400 ft)	1.3 km (0.8 mi)	2.6 km (1.6 mi)
1892	Ethyldichloroarsine	30 m	(100 ft)	0.2 km (0.1 mi)	0.3 km (0.2 mi)	60 m	(200 ft)	0.5 km (0.3 mi)	1.0 km (0.6 mi)
1898	Acetyl iodide (when spilled in water)	30 m	(100 ft)	0.2 km (0.1 mi)	0.2 km (0.1 mi)	60 m	(200 ft)	0.6 km (0.4 mi)	1.6 km (1.0 mi)
1911 1911	Diborane Diborane, compressed	30 m	(100 ft)	0.2 km (0.1 mi)	0.3 km (0.2 mi)	95 m	(300 ft)	1.0 km (0.6 mi)	2.7 km (1.7 mi)
1923 1923 1923	Calcium dithionite (when spilled in water) Calcium hydrosulfite (when spilled in water) Calcium hydrosulphite (when spilled in water)	30 m	(100 ft)	0.2 km (0.1 mi)	0.2 km (0.1 mi)	30 m	(100 ft)	0.3 km (0.2 mi)	1.1 km (0.7 mi)

"+" means distance can be larger in certain atmospheric conditions

Page 128

Page 129

FIGURE 8.20 The green pages are the Table of Initial Isolation and Protective Action Distances. (*Courtesy of ERG2000.*)

TABLE OF WATER-REACTIVE MATERIALS WHICH PRODUCE TOXIC GASES

Materials Which Produce Large Amounts of Toxic-by-Inhalation (TIH) Gas(es)
When Spilled in Water

ID No.	Guide No.	Name of Material	TIH Gas(es) Produced
1834	137	Sulfuryl chloride	HCl SO_3
1834	137	Sulphuryl chloride	HCl SO_3
1836	137	Thionyl chloride	HCl SO_2
1838	137	Titanium tetrachloride	HCl
1898	156	Acetyl iodide	HI
1923	135	Calcium dithionite	H_2S SO_2
1923	135	Calcium hydrosulfite	H_2S SO_2
1923	135	Calcium hydrosulphite	H_2S SO_2
1939	137	Phosphorus oxybromide	HBr
1939	137	Phosphorus oxybromide, solid	HBr
2004	135	Magnesium diamide	NH_3
2011	139	Magnesium phosphide	PH_3
2012	139	Potassium phosphide	PH_3
2013	139	Strontium phosphide	PH_3
2442	156	Trichloroacetyl chloride	HCl
2495	144	Iodine pentafluoride	HF
2576	137	Phosphorus oxybromide, molten	HBr
2691	137	Phosphorus pentabromide	HBr
2692	157	Boron tribromide	HBr
2806	138	Lithium nitride	NH_3
2977	166	Radioactive material, Uranium hexafluoride, fissile	HF
2977	166	Uranium hexafluoride, fissile containing more than 1% Uranium-235	HF
2978	166	Radioactive material, Uranium hexafluoride, non-fissile or fissile excepted	HF

Chemical Symbols for TIH Gases:

Br_2	Bromine	HF	Hydrogen fluoride	PH_3	Phosphine
Cl_2	Chlorine	HI	Hydrogen iodide	SO_2	Sulfur dioxide
HBr	Hydrogen bromide	H_2S	Hydrogen sulfide	SO_2	Sulphur dioxide
HCl	Hydrogen chloride	H_2S	Hydrogen sulphide	SO_3	Sulfur trioxide
HCN	Hydrogen cyanide	NH_3	Ammonia	SO_3	Sulphur trioxide

Use this list only when material is spilled in water.

TABLE OF WATER-REACTIVE MATERIALS WHICH PRODUCE TOXIC GASES

Materials Which Produce Large Amounts of Toxic-by-Inhalation (TIH) Gas(es)
When Spilled in Water

ID No.	Guide No.	Name of Material	TIH Gas(es) Produced
2978	166	Uranium hexafluoride, fissile excepted	HF
2978	166	Uranium hexafluoride, low specific activity	HF
2978	166	Uranium hexafluoride, non-fissile	HF
2985	155	Chlorosilanes, flammable, corrosive, n.o.s.	HCl
2985	155	Chlorosilanes, n.o.s.	HCl
2986	155	Chlorosilanes, corrosive, flammable, n.o.s.	HCl
2986	155	Chlorosilanes, n.o.s.	HCl
2987	156	Chlorosilanes, corrosive, n.o.s.	HCl
2987	156	Chlorosilanes, n.o.s.	HCl
2988	139	Chlorosilanes, n.o.s.	HCl
2988	139	Chlorosilanes, water-reactive, flammable, corrosive, n.o.s.	HCl
3048	157	Aluminum phosphide pesticide	PH_3
3049	138	Metal alkyl halides, n.o.s.	HCl
3049	138	Metal alkyl halides, water-reactive, n.o.s.	HCl
3049	138	Metal aryl halides, n.o.s.	HCl
3049	138	Metal aryl halides, water-reactive, n.o.s.	HCl
3052	135	Aluminum alkyl halides	HCl
9191	143	Chlorine dioxide, hydrate, frozen	Cl_2

Chemical Symbols for TIH Gases:

Br_2	Bromine	HF	Hydrogen fluoride	PH_3	Phosphine
Cl_2	Chlorine	HI	Hydrogen iodide	SO_2	Sulfur dioxide
HBr	Hydrogen bromide	H_2S	Hydrogen sulfide	SO_2	Sulphur dioxide
HCl	Hydrogen chloride	H_2S	Hydrogen sulphide	SO_3	Sulfur trioxide
HCN	Hydrogen cyanide	NH_3	Ammonia	SO_3	Sulphur trioxide

Use this list only when material is spilled in water.

FIGURE 8.21 Table of water-reactive materials which produce toxic gases. (*Courtesy of ERG-2000.*)

CAUTION

A MISTAKE HAS BEEN FOUND IN THE ENGLISH VERSION OF THE PRINTED VERSION OF THE GUIDEBOOK. PLEASE CORRECT THE ENTRY ON PAGE 360, UNDER THE ID NO. 1162 (FIRST ENTRY ON THE TABLE). THE *GUIDE NO.* IN THE GUIDEBOOK IS 151. **THE CORRECT** *GUIDE NO.* PAGE SHOULD BE **155** AND NOT 151. THIS ERROR IS **ONLY PRESENT** IN THE ENGLISH VERSION OF THE PRINTED GUIDEBOOK. THE COMPANION CD VERSION ***DOES NOT HAVE*** THIS ERROR. PLEASE CORRECT IN YOUR HARD COPY OF THE GUIDEBOOK.

8.5 PLACEMENT OF PLACARDS AND LABELS ON STORAGE CONTAINERS

The container (package) labels and markings required by DOT must remain and be maintained by the end user. That is, until the container is sufficiently cleaned of residue and purged of vapors to remove any potential hazards.

29 CFR 1910.1201, "*Retention of DOT markings, placards and labels,*" requires:

(a) Any employer who receives a package of hazardous material which is required to be marked, labeled or placarded in accordance with the U.S. Department of Transportation's Hazardous Materials Regulations (49 CFR Parts 171 through 180) shall retain those markings, labels and placards on the package until the packaging is sufficiently cleaned of residue and purged of vapors to remove any potential hazards.

(b) Any employer who receives a freight container, rail freight car, motor vehicle, or transport vehicle that is required to be marked or placarded in accordance with the Hazardous Materials Regulations shall retain those markings and placards on the freight container, rail freight car, motor vehicle or transport vehicle until the hazardous materials which require the marking or placarding are sufficiently remove to prevent any potential hazards.

(c) Markings, placards and labels shall be maintained in a manner that ensures that they are readily visible.

The term *employer* as used in the OSHA regulations includes ANY HAZWOPER organization or entity. Conforming with this particular regulation can seem like a major task, especially when, on a long-term cleanup operation or at a TSDF, there may be tens of thousands of containers, including packages, bottles, bags, boxes, cans, drums, cylinders, tanks, containing hundreds or thousands of materials.

Through experience, we find that HAZWOPER workers who perform weekly inspections of the storage area using a checklist can quickly and accurately discharge this duty.

Hazardous materials container identification is discussed in Sections 8.2–8.5. Once a material becomes a hazardous waste due to being no longer usable, being consolidated with other wastes, or no longer being needed, it must be labeled as a hazardous waste. See Sections 13 and 14 for further information on hazardous waste accumulation, storage, and transportation requirements.

8.5.1 Drum and Small Container Marking and Labeling

Legal drum labeling can vary widely. For instance, it is legal to label a drum of material with the name of the contents and the DOT hazard class. However, most companies are adding more information to their labeling.

As we see, the OSHA regulations require the retention of DOT shipping labels on the container until the container has been sufficiently cleaned so that it contains no residue. This regulation, when followed, is extremely helpful to the HAZWOPER worker.

CAUTION

JUST BECAUSE A CONTAINER IS LABELED AS MATERIAL X, FOR EXAMPLE, THE HAZWOPER WORKER CANNOT ALWAYS ASSUME THE CONTAINER CONTENTS REALLY ARE MATERIAL X. THE ONE POSSIBLE EXCEPTION TO THIS RULE IS IF THE CONTAINER IS OBVIOUSLY IN UNOPENED CONDITION. AS A RULE, ALL CONTAINERS FOUND AT WASTE SITES WILL HAVE TO BE SAMPLED TO CONFIRM THE PROPERTIES OF THEIR CONTENTS. THIS IS CALLED WASTE CHARACTERIZATION.

Depending on the supplier of the material, the container label can be anywhere between marginally useful and extremely helpful. For instance, a "properly" labeled drum might have the information in Figure 8.22.

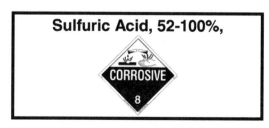

FIGURE 8.22   A poorly but legally labeled container.

To the experienced HAZWOPER worker, this may be enough information to make any decisions about compatibility, storage, consolidation, PPE, spill cleanup, emergency actions, and waste disposal. However, to the inexperienced worker, this might mean very little.

Manufacturers are increasingly adding more detail to their labels. The HMIS® label (Section 8.4) is a good example. Another good example is our MSDS (Section 9) material. A commendable, and particularly informative, label would look something like Figure 8.23.

Note the wealth of information from the label of this material that would otherwise have to be retrieved from an MSDS.

Sulfuric Acid 52-100%

ABC Chemical Company
123 Fourth Street
Anytown, USA 00000-0000

For Technical Information: (123) 456-7891
For Transportation Emergencies Call:
CHEMTREC 1-800-424-9300
For MSDS: (123) 456-7892

CORROSIVE
8

NFPA Ratings: Health: **3** Flammability: **0** Reactivity: **2** Other: **Water reactive (W)**
Label Hazard Warning:
POISON! DANGER! CORROSIVE. LIQUID AND MIST CAUSE SEVERE BURNS TO ALL BODY TISSUE. MAY BE FATAL IF SWALLOWED OR CONTACTED WITH SKIN. HARMFUL IF INHALED. AFFECTS TEETH. WATER REACTIVE. CANCER HAZARD. STRONG INORGANIC ACID MISTS CONTAINING SULFURIC ACID CAN CAUSE CANCER. Risk of cancer depends on duration and level of exposure.
Label Precautions:
Do not get in eyes, on skin, or on clothing. Do not breathe mist. Keep container closed. Use only with adequate ventilation. Wash any exposed skin thoroughly with copious amounts of water, for 15 minutes. Do not add water to sulfuric acid. A violent reaction occurs that causes eruption of acid. To dilute sulfuric acid, cautiously add it to water until the temperature rise ceases.
Label First Aid:
In all cases call a physician immediately. In case of contact, immediately flush eyes or skin with plenty of water for at least 15 minutes while removing contaminated clothing and shoes. Wash clothing before re-use. Excess acid on skin can be neutralized with a 2% bicarbonate of soda solution. If swallowed, DO NOT INDUCE VOMITING. Give large quantities of water. Never give anything by mouth to an unconscious person. If inhaled, remove to fresh air. If not breathing, give artificial respiration. If breathing is difficult, give oxygen.

FIGURE 8.23   A well-labeled container.

8.5.2 Tank Labeling for Fixed Facilities

Tanks and large containers at fixed facilities are usually labeled with the NFPA 704M® system. They will probably also be labeled with the name of the material and DOT/UN/NA placards.

Tanks at fixed facilities (storage tanks and bins) are usually labeled with the NFPA 704M® system (see Section 8.3.1). Understand that they will probably also be labeled with the name of the material and DOT/UN/NA placards. Figures 8.24 to 8.26 are examples of labeling locations, descriptions, and diagrams. They are from the ERG2000 and the United States Fire Administration (USFA) *Hazardous Materials Guide for First Responders.* The ERG2000 is available in a complete, searchable version on the Companion CD. Some of the content descriptions have been updated for this manual. A searchable version of the USFA Guide is available on their website at www.usfa.fema.gov/hazmat/reference.htm.

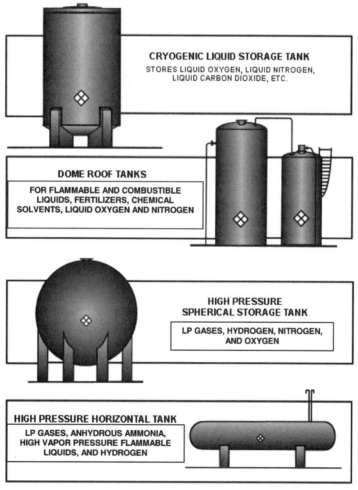

CRYOGENIC LIQUID STORAGE TANK
STORES LIQUID OXYGEN, LIQUID NITROGEN, LIQUID CARBON DIOXIDE, ETC.

DOME ROOF TANKS
FOR FLAMMABLE AND COMBUSTIBLE LIQUIDS, FERTILIZERS, CHEMICAL SOLVENTS, LIQUID OXYGEN AND NITROGEN

HIGH PRESSURE SPHERICAL STORAGE TANK
LP GASES, HYDROGEN, NITROGEN, AND OXYGEN

HIGH PRESSURE HORIZONTAL TANK
LP GASES, ANHYDROUS AMMONIA, HIGH VAPOR PRESSURE FLAMMABLE LIQUIDS, AND HYDROGEN

FIGURE 8.24   Examples of types of tanks, their contents and labeling. (*Courtesy of the USFA* Hazardous Materials Guide for First Responders.)

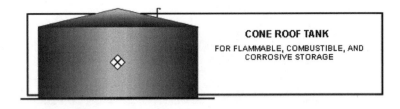

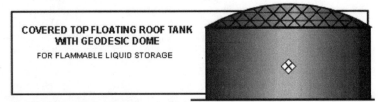

FIGURE 8.25   Examples of types of tanks, their contents and labeling. (*Courtesy of the USFA* Hazardous Materials Guide for First Responders.)

PORTABLE TANKS
FOR TRANSPORTING BULK SOLIDS, LIQUIDS,
AND GASES. ALSO KNOWN AS
INTERMEDIATE BULK CONTAINERS (IBC'S)

3 TYPES: 1. METAL
          2. PLASTIC WITHIN METAL FRAME
          3. REINFORCED CARDBOARD

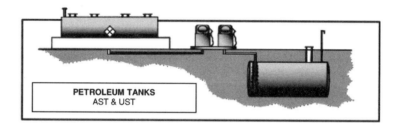

PETROLEUM TANKS
AST & UST

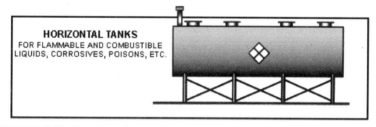

HORIZONTAL TANKS
FOR FLAMMABLE AND COMBUSTIBLE
LIQUIDS, CORROSIVES, POISONS, ETC.

FIGURE 8.26 Examples of types of tanks, their contents and labeling. (*Courtesy of the USFA* Hazardous Materials Guide for First Responders.)

Note

Notice that the facility in which the tank in Figure 8.27 is placed uses both the DOT/UN/NA placarding system and the NFPA 704M® labeling system. It also includes additional information for PPE requirements first aid, and spill remediation. Also note that the labels are maintained in clean, legible condition. The contents label and NFPA label are appropriately large enough to be read at distances of about 50 feet.

FIGURE 8.27 One of the authors inspecting a properly labeled stationary tank. (*Courtesy of U.S. Filter, Wilmington, Delaware, Facility.*)

8.6 *PLACEMENT OF PLACARDS AND LABELS ON CONTAINERS IN TRANSIT*

Containers used for transporting hazardous materials, such as tankers, bulk containers, dry cargo trailers, or railcars, whether or not in transit, must be placarded in specific locations on the container. Those placards must remain in place until the container has been cleaned of all material hazards.

This section will show the DOT Classes of various tankers (liquids, liquefied gases, high-pressure gases, and molten solids such as molten sulfur), tube trailers (high-pressure gases), dry bulk material trailers (fine solids such as grains and portland cement), and box trailers (smaller containers, often mixed loads). Figures 8.28 to 8.37 illustrate the proper placement of DOT placards on container vehicles in transport.

Note

Transfer of molten materials allows easier unloading and use of the liquid at the customer's facility. Usually the customer wants the material molten for easier use in the process. Molten sulfur, for example, is shipped to customers who store it as liquid in steam-heated tanks. Then they pump the sulfur as a liquid through a jet with air oxidation to produce sulfur dioxide for further chemical process use.

Hazard-class placards are required on any transport vehicle or freight container that is:

• Holding quantities more than 2200 lb (one long ton or metric ton) of one class or division of a hazardous material. This is a general rule that holds true for most materials. There are other materials that are restricted to lower quantities.

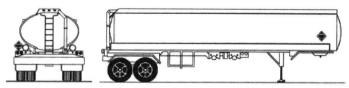

DOT 406/MC-306 ATMOSPHERIC PRESSURE TANK TRUCK
9,000 GALLONS CAPACITY
GENERAL PURPOSE CARGO

OPS Pressure Less than 3 PSI	Gasoline
Typical Maximum Capacity 9000 Gallons	Fuel Oil
New Tanks Aluminum	Alcohol
Older Tanks Steel	Other Flammable/Combustible Liquids
Oval Shape/Multiple Compartments	Liquids
Recessed Manholes/Rollover Protection	Liquid Fuel Products
Bottom Valves	(In Noncoded Tankers)
Will Likely Have Vapor Recovery.	

FIGURE 8.28 DOT 406/MC 306 nonpressurized liquid tank. (*Courtesy of ERG2000.*)

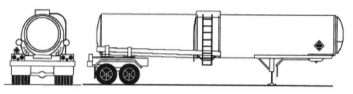

DOT 407/MC-307 LOW-PRESSURE TANK TRUCK
6000-7000 GALLONS CAPACITY
TRANSPORTS CHEMICALS, FLAMMABLE AND COMBUSTIBLE LIQUIDS

OPS @ 25-40 PSI	Flammable Liquids
Typical Maximum Capacity 6000 Gallons	Combustible Liquids
May Be Rubber Lined/Steel	Acids
Single or Double Top Manhole	Caustics
Single Outlet Discharge for Each	Poisons
Compartment at Bottom (Midship or Rear)	
Typically Double Shell	
Stiffening Rings	
Rollover Protection	
May Be Multiple Compartments	
Horseshoe Or Round Shaped	
Unit Pictured Is Insulated and Covered	
With Smooth Metal Skin	
Tank Has Several Stiffening Rings	

FIGURE 8.29 DOT 407/MC 307 low-pressure chemical tank. (*Courtesy of ERG2000.*)

- Required to carry special placards for materials that are poisonous if inhaled.
- Required to carry special placards to identify transport vehicles or freight containers that may pose a hazard because they contain fumigants that are poisonous when inhaled.

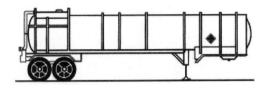

OPS Pressure Less than 75 PSI Typical Maximum Capacity 6,000 Gallons May Be Rubber Lined/Steel Stiffening Rings and Rollover Protection Splash Guard Provides Rollover Protection Top Loading at Rear or Center Loading Area Typically Coated with Corrosive Resistant Material Small Diameter for Length (Tube Shaped) Typical Single Compartment	Corrosive Liquids Typically Acids

FIGURE 8.30   DOT C-312 corrosive liquid tanker (*Courtesy of ERG2000.*)

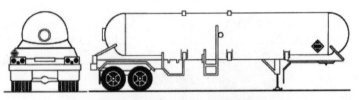

MC-331 HIGH-PRESSURE TANK TRUCK
11,500 GALLONS CAPACITY
TRANSPORTS LP GAS AND ANHYDROUS AMMONIA

OPS Pressure up To 300 PSI Typical Maximum Capacity 11,500 Gallons Single Steel Compartment/Non/Insulated Bolted Manhole At Front or Rear Internal and Rear Outlet Valves Typically Painted White or Other Reflective Color May Be Marked Flammable Gas and Compressed Gas Round/Dome Shaped Ends	Pressurized Gases & Liquids Anhydrous Ammonia Propane Butane Other Gases That Have Been Liquefied under Pressure

FIGURE 8.31   DOT MC-331 high-pressure tanker. (*Courtesy of ERG2000.*)

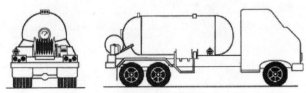

BOBTAIL TANK
LOCAL DELIVERY OF LP GAS AND ANHYDROUS AMMONIA

FIGURE 8.32   Bobtail tanker. (*Courtesy of ERG2000.*)

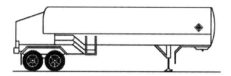

MC-338 CRYOGENIC LIQUID TANK TRUCK
WELL-INSULATED, DOUBLE-WALLED "THERMOS BOTTLE" DESIGN
TRANSPORTS LIQUID NITROGEN, OXYGEN, AND CARBON DIOXIDE AMONG
OTHERS

OPS at Less than 22 PSI Well insulated Thermos Bottle like Steel Tank May Have Vapor Discharging from Relief Valves Loading/Unloading Valves Enclosed at Rear May Be Marked "Refrigerated Liquid" Round Tank with Same Type of Cabinet at Rear	Liquid Oxygen Liquid Nitrogen Liquid Carbon Dioxide Liquid Hydrogen Liquid Helium Other Gases That Have Been Liquefied by Lowering Their Temperature

FIGURE 8.33 DOT MC-338 cryogenic liquid tanker. (*Courtesy of ERG2000.*)

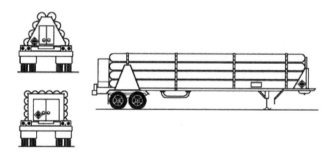

COMPRESSED GAS TRAILER
3000-5000 PSI
TRANSPORTS COMPRESSED GAS

OPS at 3000-5000 PSI (Gas Only) Individual Steel Cylinders Stacked and Banded Together Typically Will Have Overpressure Device for Each Cylinder Bolted Manhole at Front or Rear Valving at Rear (Protected) Manufacturer Name May Be on Cylinders, i.e., AIRCO, Air Products, Air Liquide, etc. Flat Truck with Multiple Cylinder Stacked in Modular or Nested Shape	Helium Hydrogen Methane Oxygen Other Gases

FIGURE 8.34 DOT compressed gas tube trailer. (*Courtesy of ERG2000.*)

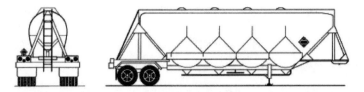

OPS AT Less THAN 22 PSI	Calcium Carbide
Typically Not under Pressure	Oxidizers
over the Road	Corrosive Solids
Top Side Manholes	Cement
Bottom Valves/Air Assisted	Plastic Pellets
Loading/Unloading	Fertilizers
Shapes Vary, but Will Have Hoppers	

FIGURE 8.35   DOT dry bulk cargo tanker. (*Courtesy of ERG2000.*)

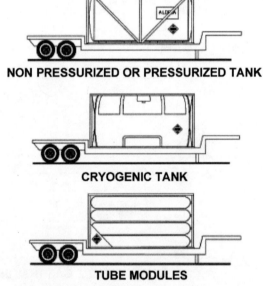

NON PRESSURIZED OR PRESSURIZED TANK

CRYOGENIC TANK

TUBE MODULES

FIGURE 8.36   Intermodal containers (all three types may also be found on railcars or aboard ships). (*Courtesy of ERG2000.*)

Placarding for loads of quantities less than these, and mixed loads, are often simply placarded as "DANGEROUS." For that reason, DANGEROUS loads can often be the most difficult to respond to in case of an emergency.

All of these containers must carry the correct DOT/UN/NA placards that correspond to their contents. Liquid containers are, for the most part, for single materials. However, they can also be compartmented to carry different materials without mixing.

You will also note the mandatory locations of the placards. All of these trailers must be placarded on the right and left sides towards the front of the trailer and driver's side rear. It is not mandatory that the front of the trailer be placarded, as long as the trailer remains connected to the tractor at all times, in which case the tractor will be placarded in the front.

ROAD TRAILER IDENTIFICATION CHART*

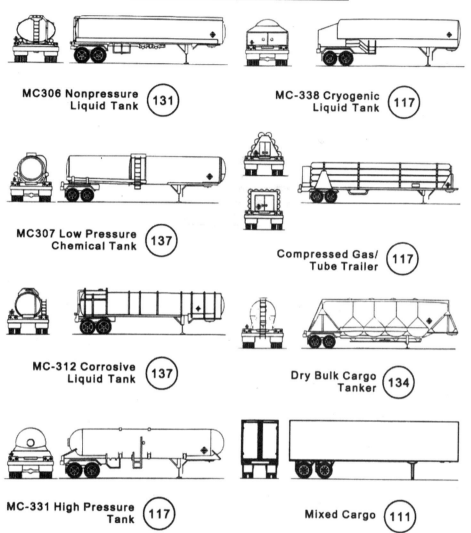

CAUTION: This chart depicts only the most general shapes of road trailers. Emergency response personnel must be aware that there are many variations of road trailers, not illustrated above, that are used for shipping chemical products. The suggested guides are for the most hazardous products that may be transported in these trailer types.

*   **The recommended guides should be considered as last resort if product cannot be identified by any other means.**

FIGURE 8.37   Several common types of over-the-road trailers and their corresponding ERG guide numbers (circled) (*Courtesy of ERG2000.*)

If the trailer will be disconnected from the tractor while it contains hazardous material, the front of the trailer must also be placarded.

8.7   LABELING FOR RAILCARS

A brief description of the labeling of tank railcars is included in this manual to demonstrate the differences between over-the-road markings and railcar markings. Railcars are placarded the same as over-the-road vehicles.

They also must have additional permanent markings painted directly on the car. These markings must include the car's unique rolling stock number, capacity, and tare (unloaded weight) among others. These markings can become quite extensive if the railcar is designed to carry refrigerated (cryogenic) liquid, high-pressure gasses, or extremely hazardous materials such as hydrogen cyanide. In the latter cases, the permanent markings will include hydrostatic test information such as test weight and date. They will also include the railcar's American Association of Railroads (AAR) Certification Number.

Understand that responding to a railcar incident is reserved for well-trained, experienced First Responders. That response is not usually within the job description of an individual only having HAZWOPER 40-hour training. HAZWOPER personnel need many more hours of specialized training before responding to a rail incident. We include this section for information and reference purposes only. Railcars can carry approximately 15 times more capacity than a tractor-trailer. That means in case of an incident involving spills, they have a much greater potential for hazardous material exposure to the responders and the community. Figures 8.38 to 8.48 show the different types of railcars, what cargo they are designed to carry, and their proper markings and placarding.

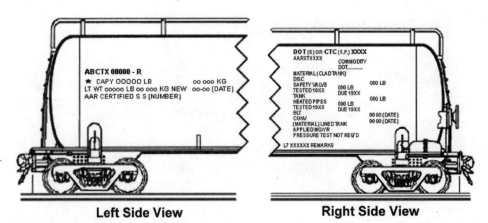

Left Side View   **Right Side View**

FIGURE 8.38   Typical placement of required information on a railcar: left side view and right side view. (*Courtesy of the USFA* Hazardous Materials Guide for First Responders.)

FIGURE 8.39 Further examples of typical placement of required information on a railcar: back end view (note the placement of the DOT/UN/NA placard just to the left of the bottom rung of the ladder.) (*Courtesy of the USFA* Hazardous Materials Guide for First Responders.)

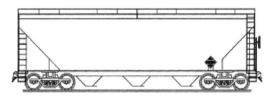

COVERED HOPPER
CARRIES CALCIUM CARBIDE, CEMENT, GRAIN

FIGURE 8.40 Covered hopper. (*Courtesy of the USFA* Hazardous Materials Guide for First Responders.)

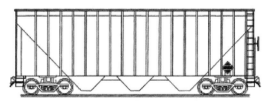

OPEN-TOP HOPPER
CARRIES COAL, ROCK, SAND

FIGURE 8.41 Open-tp hopper. (*Courtesy of the USFA* Hazardous Materials Guide for First Responders.)

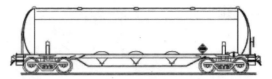

PNEUMATIC HOPPER
CARRIES PLASTIC PELLETS, FLOUR, OTHER FINE-POWDERED MATERIALS

FIGURE 8.42   Pneumatic hopper. (*Courtesy of the USFA* Hazardous Materials Guide for First Responders.)

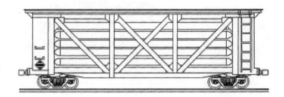

TUBE CAR
CARRIES HELIUM, HYDROGEN, METHANE, OXYGEN, OTHER GASES

FIGURE 8.43   Tube car. (*Courtesy of the USFA* Hazardous Materials Guide for First Responders.)

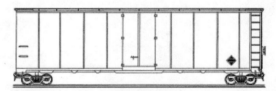

BOX CAR
CARRIES ALL TYPES OF MATERIAL AND FINISHED GOODS

FIGURE 8.44   Box car. (*Courtesy of the USFA* Hazardous Materials Guide for First Responders.)

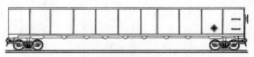

GONDOLA
CARRIES SAND, ROLLED STEEL AND OTHER PRODUCTS AND MATERIALS THAT
DO NOT REQUIRE PROTECTION FROM THE WEATHER

FIGURE 8.45   Gondola. (*Courtesy of the USFA* Hazardous Materials Guide for First Responders.)

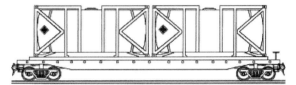

FLAT/BED CAR WITH INTERMODAL TANKS
CARRIES VARIOUS PRODUCTS IN CONTAINERS, I.E., ONE-TON CHLORINE CYLINDERS, INTERMODAL CONTAINERS (SHOWN), LARGE VEHICLES, OTHER COMMODITIES THAT DO NOT REQUIRE PROTECTION FROM THE WEATHER

FIGURE 8.46 Flatbed car with intermodal tanks. (*Courtesy of the USFA* Hazardous Materials Guide for First Responders.)

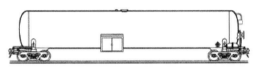

CRYOGENIC CAR
CARRIES LIQUID OXYGEN, LIQUID NITROGEN, LIQUID CARBON DIOXIDE, LIQUID HYDROGEN, OTHER GASES THAT HAVE BEEN LIQUEFIED BY LOWERING THEIR TEMPERATURE

FIGURE 8.47 Cryogenic car. (*Courtesy of the USFA* Hazardous Materials Guide for First Responders.)

RAIL CAR IDENTIFICATION CHART*

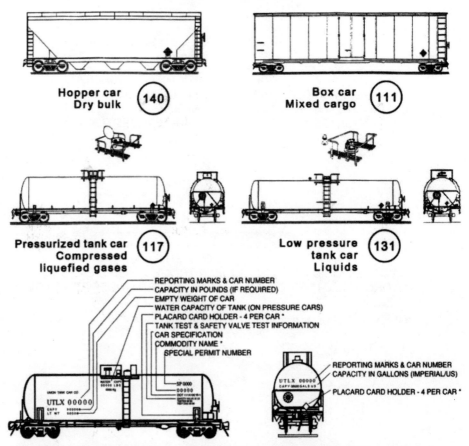

CAUTION: Emergency response personnel must be aware that rail tank cars vary widely in construction, fittings and purpose. Tank cars could transport products that may be solids, liquids or gases. The products may be under pressure. It is essential that products be identified by consulting shipping documents or train consist or contacting dispatch centers before emergency response is initiated.

The information stenciled on the sides or ends of tank cars, as illustrated above, may be used to identify the product utilizing:

a.   the commodity name shown; or

b.   the other information shown, especially reporting marks and car number which, when supplied to a dispatch center, will facilitate the identification of the product.

*   **The recommended guides should be considered as last resort if product cannot be identified by any other means.**

FIGURE 8.48   Examples of typical placement of required placarding on railcars. (*Courtesy of ERG2000.*)

8.8   GENERAL WORKPLACE SIGNS

O—ᴨ   **General workplace signs can be either alphanumeric or pictograms.**

O—ᴨ   **General workplace signs are divided into five classes: Danger, Caution, Notices, Fire and Emergency, and Safety advisories.**

ANSI has standardized some signs that you will see in work areas. Here are some examples of worded signs. There are also graphic counterparts for almost all of these signs, but, they are not as well standardized.

DANGER signs always use red ovals and black text on a white background. They indicate an immediate hazard to life or health of the worker. See Figure 8.49.

CAUTION signs use black text on a yellow background. They indicate a potential threat to personnel or property. Some caution signs you might see are shown in Figure 8.50.

NOTICE signs state company policy (see Figure 8.51). They may be precautionary or simply stating policy. They usually have a blue NOTICE bar over the message that is black text on a white background.

Signs for *Fire* exits, doors, extinguishers, and other fire related messages are always red and white (see Figure 8.52). They provide vital information in the event of a fire.

SAFETY advisory signs (see Figure 8.53) usually have a green SAFETY bar over black text on white.

FIGURE 8.49   ANSI standardized Danger signs. *Caution* signs use black on a yellow background. They indicate a potential threat to personnel or property.

 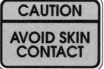

FIGURE 8.50 ANSI standardized Caution signs.

FIGURE 8.51 ANSI standardized Notice signs.

FIGURE 8.52 ANSI standardized fire-related signs.

FIGURE 8.53 ANSI standardized safety signs.

FIGURE 8.54 International caution signs.

Pictograms (see Figure 8.54) (words expressed in pictures) are not well standardized, but they do manage to relay their messages. They are part of the move toward international signage that has made its way into U.S. highway signs. Pictogram signs include "H" for hospital, knife and fork for restaurant, and a gas pump for gas station.

In case you were not able to understand these pictograms, they stand for:

#1 Danger—Corrosive Material

#2 Danger—Explosive Material

#3 Danger—Toxic Material

#4 Danger—Unguarded Opening in Floor

#5 Emergency Eyewash Station

FIGURE 8.55 A spectacular boiling liquid expanding vapor explosion (BLEVE) at a chemical plant. (*Courtesy of the EPA.*)

#6 Danger—Unguarded Machinery

#7 No Smoking, Matches, or Open Lights

Figure 8.55 illustrates what can occur if signage in the workplace is ignored. It is a picture of a spectacular BLEVE at a chemical plant. Obey those signs!

8.9   TRAINING AIDS AND ADDITIONAL RESOURCES

The authors recommend the following video for aiding readers to see real-life examples of what they have learned in this Section:

• Marking, Labeling, and Placarding-HazMat for Transportation

This video is available from Coastal Safety and Environmental (www.coastal.com).

8.10   SUMMARY

Properly placed labels, placards, and other identification give HAZWOPER personnel and emergency responders the ability to immediately identify the contents of a container. The five ways of identifying the contents of a hazardous material container are markings, labels, DOT placards, signs, and manifests. OSHA regulations require that placards and labels on containers of hazardous materials be maintained and legible until such time as the container is clean and free of the potential hazard.

The three major systems of signage are the NFPA 704M® system, the HMIS®, and DOT placards. The NFPA 704M® system gives a HAZWOPER worker a quick visual indication of the hazards of the contents of a container involved in a fire. The system uses the following four colors to indicate the type of hazard: Blue (health), Red (flammability), Yellow (reactivity), and White (special hazards). The NFPA 704M® system uses the numbers 0 to 4 to indicate the severity of the hazard, with 0 being little or no hazard and 4 being the greatest level of hazard.

The HMIS® is very similar to the NFPA 704M® and is used on containers of paints and allied products. The major difference between the HMIS® and the NFPA 704M® system is the White category. For HMIS®, white stands for PPE required.

The DOT placarding system is required for any hazardous material offered for transport. The DOT placarding system requires shippers to label their material containers with one of the nine hazard classes or as "Other Regulated Materials." The nine hazard classes are Explosives, Gases, Flammable Liquids, Flammable Solids, Oxidizers, Poisonous Materials, Corrosive Materials, and Miscellaneous Dangerous Goods. All DOT-regulated hazardous materials also have a four-digit UN/NA number. This might be a specific material, that material at a certain concentration, or a class of materials. With the exception of oxygen and radioactive materials, the text that indicates the hazard is NOT REQUIRED. All DOT-regulated hazardous materials containers, including tankers, cargo trailers, bulk trailers, intermodal containers, and railcars, must be placarded. If shipped to or from Mexico or Canada, they must be placarded with the four-digit UN/NA number of its contents. That number may be displayed alone (black numerals on an orange background) or on the placard, describing the class of the material.

The EPA has further labeling requirements for hazardous waste. It requires that an EPA hazardous waste label MUST be affixed to any container immediately before ANY hazardous waste is deposited inside. Hazardous waste labels are REQUIRED to have extensive information about their contents, ownership (generator), dates of accumulation, reportable quantities (RQs), EPA ID number (hazardous waste code), proper shipping name, hazard class, and emergency contact information.

The *Emergency Response Guidebook* (ERG) gives HAZWOPER responders the immediate information needed to protect people and the environment from accidental releases of hazardous materials. The ERG is reproduced in the Companion CD. The ERG yellow pages identify hazardous material names by UN/NA numbers in numerical order, in order of entry. The ERG blue pages identify hazardous materials by name alphabetically. The ERG orange pages are the hazardous materials guide number section. These guides provide vital information for HAZWOPER responders at the scene of an incident. The ERG green pages show the downwind distances from a release area to which people must be restricted due to the spread of hazardous vapors or gases. The ERG green pages also list the water-reactive materials that produce toxic gases when they come into contact with water; and the gases they produce.

The container (package) labels and markings required by DOT must remain and be maintained by the end user. That required maintenance extends until the container is sufficiently cleaned of residue and purged of vapors to remove any potential hazards.

Containers labeled according to regulations can have a wide variability in the information presented on their labels. For instance, it is legal to merely label a container of material with the name of the contents and the DOT hazard class. However, many companies, acting with Responsible Care®, are adding more information to their labeling. Tanks and large containers at fixed facilities are usually labeled with the NFPA 704M® system. They will probably also be labeled with the name of the material and DOT/UN/NA placards. Containers transporting hazardous materials, such as tankers, bulk containers, dry cargo trailers, or railcars, whether traveling or not, must be placarded in specific locations on the container. Those placards must remain in place until the container has been cleaned of all material hazards.

SECTION 9
MATERIAL SAFETY DATA SHEETS (MSDSs) AND INTERNATIONAL CHEMICAL SAFETY CARDS (ICSCs)

OBJECTIVES

In this Section our main objective is to show readers what they can learn from the primary source of user information about materials in the workplace: the Material Safety Data Sheet (MSDS). U.S. chemical manufacturers and importers are required to prepare MSDSs for the products they sell. These sellers are required to provide MSDS copies with the first shipment of a product and any time the constituents of the product change. They must also provide a copy to anyone who requests one.

Proprietary ingredients need not be included. However, any hazards those ingredients add to the final product must be explained. Proprietary ingredients MUST be released to medical professionals treating an exposed worker. These MSDS requirements are all specified to comply with 29 CFR 1910.1200, the Hazard Communication Standard (HAZCOMM), and the Community Right-to-Know Act (RTK).

We describe the individual sections of the MSDS, both OSHA-suggested and optional. OSHA does not specify a mandatory form that the MSDS must follow. The most widely accepted form is contained in ASTM publication E1628-2000, *Standard Practice for Preparing Material Safety Data Sheets to Include Transportation and Disposal Data for the General Services Administration*. The authors feel that this is the most comprehensive, understandable, and helpful form.

Readers involved in international trade will see another form of material safety information. To familiarize those readers with another format that is gaining in usage, we discuss the International Chemical Safety Card (ICSC). This form of reference is used for chemical materials in European Union (EU) countries and other parts of the world.

9.1 WHY WE HAVE THE MSDS SYSTEM

The MSDS is the primary source of information concerning the safe use, storage, disposal, and emergency procedures following a release, chiefly for any chemical material manufactured, distributed, or imported by U.S. companies.

According to 29 CFR 1910.1200(a)(1), the purpose of OSHA's Hazard Communication Standard (HCS, or HAZCOMM) is to "ensure that the hazards of all chemicals produced or imported are evaluated, and that information concerning their hazards is transmitted to employers and employees. This transmittal of information is to be accomplished by means of comprehensive hazard communication programs, which are to include container labeling and other forms of warning, Material Safety Data Sheets and employee training."

⊙━π **OSHA's Hazard Communication Standard (HAZCOMM) made the MSDS mandatory.**

All industries must comply with HAZCOMM regulations. Unfortunately, this is all the hazardous materials (HAZMAT) training some people will ever get. HAZCOMM's implementation of the MSDS is particularly important to HAZWOPER workers because of the wide variety of materials and material mixtures HAZWOPER personnel can come into contact with on a daily basis.

HAZCOMM is based on two simple concepts:

1. That employees have both a need and a right to know the hazards and identities of the chemicals they are exposed to when working

2. They also need to know what protective measures must be taken to prevent adverse effects due to exposure

The core of HAZCOMM is the Material Safety Data Sheet, or MSDS. Although the regulation specifically cites chemicals, materials of many kinds not universally considered chemicals are now shipped with an MSDS. Knowledge acquired under HAZCOMM is bound to help employers provide safer work practices and workplaces for employees. When employers have more extensive information about the chemicals being used, they can take steps to reduce worker exposures, substitute less hazardous materials as product ingredients and processing aids, and establish proper hazardous waste disposal practices.

9.2   WHO PREPARES THE MSDS?

⊙━π **U.S. chemical manufacturers and importers are required to prepare MSDSs for the products they sell.**

Chemical manufacturers and importers must evaluate the hazards of the chemicals they produce or import. Using that information, they must then prepare labels for containers and MSDSs.

Chemical manufacturers and importers of hazardous chemicals are required to provide the appropriate labels or markings and MSDSs to the employers to whom they supply the chemicals. Suppliers must send a properly completed MSDS at the time of the first shipment of a chemical. They must send a new MSDS with the next shipment after an MSDS update that contains new and significant information about the hazards of products being shipped. Suppliers must also send an MSDS, free of charge, any time anyone requests it.

A great deal of care must be taken in the preparation of the MSDS so that a reader can rely on the information. According to OSHA, the user has no independent duty to analyze the chemical or evaluate the hazards of its use.

CAUTION

AS A HAZWOPER WORKER YOU MUST ALWAYS BE ON YOUR GUARD WHEN DEALING WITH WASTES. THE SAME IS TRUE OF CONTAMINATED PRODUCTS. THE MSDS THAT YOU READ FOR ANY PRODUCT WILL HAVE BEEN PREPARED FOR THAT SPECIFIC MATERIAL OR COMBINATION OF MATERIALS AS MANUFACTURED. AN MSDS WILL NOT HAVE BEEN PREPARED FOR CONTAMINATED WASTES AND COMBINATIONS OF WASTES. THEREFORE, USE SOUND JUDGMENT AND SEEK HELP FROM SPECIALISTS TO DETERMINE THE HAZARDS OF HANDLING MIXTURES OF MATERIALS AND WASTES.

9.3   *WHERE CAN I GET AN MSDS?*

○━π   **OSHA requires the manufacturer or importer of products to supply MSDSs for all of their chemical products. They must supply them, free of charge, to anyone who requests them.**

OSHA requires information to be prepared and transmitted with all initial hazardous materials shipments. The MSDS covers both physical hazards (such as flammability) and health hazards (such as irritation, lung damage, and cancer), among others. Many materials and chemicals used in the workplace have some hazard potential and are therefore covered by the rule. There are some consumer-type exceptions, such as drug preparations, alcoholic beverages, vitamins and cosmetics.

9.3.1   Manufacturers

All manufacturers and importers (material suppliers) are required to supply you with MSDSs of their products upon request. Many suppliers are using a "fax-back" system. Others will send MSDSs in the mail. However, most companies will give you access to a technical representative who will answer your specific questions regarding any of their products.

9.3.2   The Internet

○━π   **The Internet and "fax-back" services are the quickest way to receive an MSDS.**

In this Internet-dependent world, the quickest, easiest way to get the information you need is from the Internet MSDS databases. Millions of MSDSs are immediately available to you. The following URLs are particularly helpful:

- *MSDS Search 2000,* http://www.msdssearch.com/
- *Cornell University,* http://msds.pdc.cornell.edu/msdssrch.asp
- *Vermont Safety Information Resources, Inc.* (SIRI), http://hazard.com/msds/
- *Interactive Learning Paradigms, Inc.,* http://www.ilpi.com/msds/index.html

You will also most likely find MSDSs at the supplier's website. Thousands of companies are publishing their MSDSs on the Internet. Look for links to MSDSs or to product information.

9.3.3   CHEMTREC

In case of an emergency, many organizations will rely on the assistance of CHEMTREC (the Chemical Transportation Emergency Center). There is no charge for this service. It is affiliated with the American Chemistry Council (ACC), formerly the Chemical Manufacturers Association (CMA). Emergency responders do not need to be registered with CHEMTREC to access their information. It is provided as a public service and is available 24/7/365 toll-free at 800-424-9300.

CHEMTREC has an extensive database of 24-hour contacts, including chemical manufacturers, shippers, carriers, emergency response contractors, and other organizations that have information and resources. Their database of nearly 3 million MSDSs provides detailed information about specific chemical or hazardous materials products. Due to the increased use of high technology such as GPS (see Section 12), and their communications with the ACC's member companies, CHEMTREC can give up-to-date information about chemical shipments anywhere in the United States.

Note

The time to obtain and read an MSDS is neither during nor after an exposure. All hazardous materials in your workplace must be included in a file of MSDSs. You must be allowed to read, and you must understand, all the MSDSs for every hazardous material you, the people you train, or the people you supervise might be exposed to or use.

9.4   INFORMATION CONTAINED IN AN MSDS: AN EXAMPLE

⊙━🔑   **Although OSHA permits flexibility in the amount and type of information that must be present in an MSDS, certain minimum information is required. The authors recommend following the guidelines in the relevant ASTM standard.**

A Material Safety Data Sheet is designed to provide both workers and emergency personnel with the proper procedures for handling or working with that material. Although OSHA does not yet standardize the contents of MSDSs, the generally accepted format and contents are included in ASTM E1628-2000, *Standard Practice for Preparing Material Safety Data Sheets to Include Transportation and Disposal Data for the General Services Administration.* This 16-section MSDS format also has been developed and is documented in International Labour Organization (ILO), International Organization for Standards (ISO), American National Standards Institute (ANSI), and European Union (EU) guidelines and standards.

A properly prepared MSDS, following ASTM guidelines, includes information such as shown in the following 16 subsections. Where a given characteristic has little relevance to a material's hazardous properties, you might find the statement "no information found." Such is the case with evaporation rate for the sulfuric acid example given in Section 9.6.

9.4.1   Title

This section will contain:

- The company name
- The material name
- Further manufacturer's identification, such as stock code or part number
- MSDS number and effective date. If the preparation or review date is not recent (within the last five years), consult the manufacturer for a later release. Revised OSHA PELs along with new research data may have been released in the interim. Remember that PELs have been dropping.

9.4.2   Product Identification

This section of the MSDS contains the following information:

- *Synonyms.* Frequently there is more than one name commonly given to widely used materials. They may be names with which you are more or less familiar. This helps in identifying materials in older packages and materials used by various organizations each having their own terminology. For instance, the synonyms for sulfuric acid are battery acid, hydrogen sulfate, oil of vitriol, and sulphuric acid.
- *CAS number.* This is a unique numerical identifier for a specific chemical. The CAS number is a unique identifier for practically any known chemical. The Chemical Abstract Service

(CAS), a service of the American Chemical Society, has all the information needed to answer any question you might have about any chemical material. Their database contains information on over 29 million chemical materials.

- *Molecular weight,* or *formula weight.* This is the sum of the atomic weights of the elements that make up a molecule. Atomic weights and molecular weights have no units. They are all relative to the atomic weight of carbon, taken as about 12. Knowing the molecular weight (MW) is useful in making estimations of properties. Hydrogen (MW = 2) is a light gas, water (MW = 18) is a liquid, table salt (MW about 58) is a solid. MW is used in making many kinds of calculations.

- *Chemical formula.* If the product is a single chemical, this gives the kinds and numbers of atoms in the chemical's molecule. If the product is a mixture, the chemical formulas of each ingredient are given. The chemical formula for water is H_2O. It tells those trained in the interpretation of chemical formulas a lot about what to expect in the properties of that chemical and its reactions with other materials.

- *Product codes.* These indicate a manufacturer's specific designation, or designations, for the MSDS product. That is especially important if there is a wide range of products with similar, but not identical, compositions. This MSDS section, or the one at Section 9.4.16 below, will tell you who to contact, and how, if you have any questions. This section will also tell you if the information is current. If the preparation date is not recent (less than five years old), the exposure information in Section 9.4.3 below may have to be updated.

9.4.3 Composition / Information on Ingredients

This section will contain the following additional ingredient information:

- A statement telling which, if any, ingredients are proprietary.

Note

If an ingredient is proprietary, neither its CAS number nor its formula will be listed. However, complete identity must be made available to health professionals in the case of an emergency.

- Ingredients and sequence, from largest percentage to smallest percentage. Note that this is the kind of information you are given in consumer products, such as breakfast cereal. The MSDS, however, gives information on the hazards of the material as a waste, such as SARA land ban material and CERCLA hazardous waste listing.

- National registry numbers of the ingredient, including CAS and NIOSH Registry of Toxic Effects of Chemical Substances (RTECS) numbers.

- A statement as to whether or not the ingredients are hazardous.

9.4.4 Hazards Identification

In this section you will find more detailed hazards information, including exposure pathways, health effects of exposure, protective equipment required during normal use and in emergencies, and recommended storage conditions.

- Hazards of accidental human exposure.
- One or more methods of hazards rating, such as:
  - NFPA 704M® ratings
  - Company hazard ratings (i.e., J. T. Baker SAF-T-DATA®)

- PPE requirements for safe use
- Recommended storage data and conditions
- Health effects of exposure, including by inhalation, ingestion, skin contact, eye contact, and chronic exposures, and effects on preexisting conditions, among others. These effects may include carcinogenicity rated by OSHA, the International Agency for Research on Cancer (IARC), or the National Toxicology Program (NTP).

9.4.5 First Aid Measures

This section includes specific instructions for first aid when dealing with an accidental exposure to the MSDS material, including aid such as removing victims to fresh air or helping them to an eyewash station or a safety shower. If applicable, other aid is recommended that might prevent permanent injury to the victim. For a better understanding of any toxic hazards you see listed in an MSDS, refer to Section 5. Do not become a victim yourself! Only attempt to rescue someone if you know how to protect yourself in so doing.

WARNING

FIRST AID MEASURES ARE ONLY FOR IMMEDIATE USE WHILE AWAITING THE ARRIVAL OF RESCUE PERSONNEL. ALWAYS CONTACT THE EMS AS SOON AS AN ACCIDENTAL EXPOSURE OCCURS. *NEVER* TRY TO ADMINISTER MEDICAL AID TO A VICTIM UNLESS YOU HAVE BEEN TRAINED TO PROVIDE THAT AID.

9.4.6 Firefighting Measures

WARNING

NEVER ATTEMPT TO FIGHT A FIRE UNLESS YOU ARE EXPRESSLY TRAINED TO DO SO. THIS BOOK DOES NOT TRAIN YOU AS A FIREFIGHTER! ONLY ATTEMPT TO CONTROL A FIRE IF YOU CAN DO SO WITHOUT PUTTING YOURSELF OR OTHERS IN DANGER. EVERY ORGANIZATION MUST TRAIN ALL EMPLOYEES IN THOSE SAFE ACTIONS THEY CAN TAKE TO SUMMON HELP, HOW TO CONTROL A FIRE, AND WHEN AND HOW TO EVACUATE.

This section of the MSDS includes the following:

Fire and Explosion Data.   Here you will find information on what will likely occur when this material is involved in a fire. In the case of sulfuric acid, finely divided materials may be ignited. Contact with most metals will give off explosive hydrogen gas.

Fire-extinguishing Media.   This section describes the type of fire extinguisher or other material necessary to put out or control a fire involving the MSDS material. This is very important information to have in emergencies. Most people usually think of using water to extinguish fires. Water is generally a good extinguisher. Containers of most acids must be cooled with water so as not to explode, or "cook off," thereby increasing the hazard. *Cooking off* is a term HAZWOPER responders and firefighters use to describe catastrophic rupture of the bottom of a drum of material caused by overheating. The drum can then actually launch into the air, up to several hundred feet, due to the rocket-like exhaust of the drum's contents from its bottom. The movie *Backdraft* showed some spectacular views of drums doing just that.

If the fire involves open containers of acid or other water-reactive material, using water could cause the acid to react violently and splash the responders, also adding to the hazard.

If part of your job description is to respond to small fires, know what types of extinguishers are available, how they operate, how to use them, on what materials they can be safely used, and when to use them. The authors recommend the PASS method:

Pull the pin from the handle, thereby allowing the extinguisher to be activated.

Aim the nozzle at the base of the fire. Always attempt to position yourself upwind of the fire.

Squeeze the handle lever, activating the extinguisher.

Sweep the nozzle of the extinguisher from side to side, equally distributing the contents on the base of the fire.

Sometimes it is best simply to let a fire burn out while shutting off any additional material flow to the fire and controlling its spread.

Special Information.   This area will tell you about the unusual media and equipment necessary to fight a fire safely involving this specific material. Here you will find guidance for actions that might not necessarily occur to you. These are special precautions that will help you deal safely with a fire involving the MSDS material. Some materials, such as sulfuric acid, pose hazards even when their containers appear empty.

9.4.7 Accidental Release Measures

This section will tell you how to control spills of this material safely. You will want to fully understand this section so that you do not make matters worse. For instance, you may normally clean up spilled materials such as oils with sawdust. In the case of our example substance, sulfuric acid, this section will explain that adding sawdust could cause a fire!

9.4.8 Handling and Storage

You will note, as you work around various hazardous materials, that there are usually special precautions for handling and storage. You will also note certain universal rules such as "Do not store acids with oxidizers, fuels, or combustible materials" and "Do not store fuels with oxidizers."

In this section you will receive explicit instructions as to the proper safe handling and storage of the MSDS material.

9.4.9 Exposure Controls and Personal Protection

Here is the section that answers the most frequently asked question: How do I protect myself from being overexposed to this material? Here you will find:

- *Exposure limits:* OSHA PELs, NIOSH RELs and ACGIH TLVs® among others. See Section 5 for more detailed descriptions.
- *Recommended mechanical controls:* Ventilation systems, for example.
- *Respiratory, skin, and eye protection:* See Section 6 for a detailed discussion of PPE.

9.4.10 Physical and Chemical Properties

All materials have properties that determine the type and degree of the hazard they represent. You must understand each of these properties. Further, you must be aware of hazardous conditions in case of accidental release of the material or accidental contact by the material with other materials. This section will include:

Appearance and Odor. Sulfuric acid appears as a clear, oily liquid. It may be odorless to some people, and to others it might have a somewhat salt-like or stinging odor.

Solubility. Solubility is the tendency of a solid, liquid, gas or vapor (the solute) to dissolve in a solvent (on an MSDS the solvent is usually water). The MSDS may give solubility in another material, such as methanol. The general rule is "Like dissolves like." That is, low-boiling point, high-evaporation rate liquids tend to dissolve like materials of higher boiling point and lower evaporation rate. As an example, petroleum solvents, such as gasoline or kerosene, tend to dissolve petroleum tars and asphalt.

The solubility of a material is important when dealing with materials that have accidentally come in contact with each other, such as in a spill on water. Solubility may be said to be complete when all of a material will dissolve in the solvent, as is the case with sulfuric acid in water. A material is insoluble when it can be physically mixed with a solvent but remains undissolved and will eventually separate from the solvent.

If a material spilled on water is insoluble and unreactive with water, it will either sink or float. It will sink if its specific gravity is greater than one; it will float if its specific gravity is less than one. Solubility is usually given in parts per million (ppm), milligrams per liter (mg/L), or grams per liter (g/L) when the solute will dissolve only to a certain extent in the solvent. For sulfuric acid, no solubility data are given. This is because sulfuric acid is infinitely soluble in water; there is no solubility limit.

Density or Specific Gravity (SpG). The density of a substance is its weight per unit volume, commonly expressed in grams per cubic centimeter (g/cc) or pounds per cubic foot (lb/ft³). The density of water is 1 g/cc or 62.4 lb/ft³.

Specific gravity is the ratio of the density of a substance (at a given temperature) to the density of water, assigned a SpG of 1.0, at a stated temperature. Usually that is the temperature of water's maximum density (4°C). SpG is expressed without units since it is a ratio of two densities (the density units cancel out). As stated above, if the SpG of a material is greater than 1 (the SpG of water) and the material is only slightly soluble in water, it will sink in water. The material will float on water if its SpG is less than 1.

In the case of a spill, this knowledge of a liquid's density, combined with its water solubility, will tell you whether the liquid will float on water, like fuel oil, or sink in water, like mercury (SpG = 8.0). This will help you decide how to clean up a water-borne spill.

pH. This is a measure of the materials' acidity (low pH) or alkalinity (high pH) on a scale of 1 to 14, with 7 being neutral. Both low- and high-pH materials tend to be corrosive and can react violently with other materials. See Section 3 for a more detailed discussion of pH.

Percent Volatiles by Volume. Volatiles, as you will remember from Section 3, are materials that tend to evaporate or vaporize easily and are often flammable. In general, materials with higher percentages of volatiles have higher vapor pressures. Materials with higher vapor pressures evaporate more readily following a spill.

Boiling Point and Melting Point. The boiling point is the temperature at which liquid changes to vapor—that is, the temperature where the pressure of the liquid equals atmospheric pressure. The opposite change in phases is the condensation point. Handbooks usually list temperatures as degrees Celsius (°C) or Fahrenheit (°F).

The lower the boiling point, the easier it is for a chemical to vaporize. The warming temperature at which a solid changes phase to a liquid is the melting point. This temperature is also the freezing point, since a liquid being cooled at that temperature will change phase to a solid. The proper terminology depends on the direction of the phase change. (The three phases of materials are solid, liquid, and vapor or gas.) The vapor state of flammable liquids is the most hazardous state. Vapors travel almost invisibly. They can create flammability and explosion hazards at considerable distances from a liquid spill.

Boiling point is generally dependent on density or specific gravity. As specific gravity increases, boiling point generally increases.

Vapor Density.   Vapor density is a method of rating relative weights of gases and vapors much as specific gravity does for liquids and solids. It is a specific measurement having no units of measurement.

The density of a gas or vapor can be compared to the density of air in the ambient atmosphere. Air has been assigned a vapor density value of 1.0. If the vapor density of a vapor or gas is greater than 1.0, the vapor or gas will tend to settle to the lowest point in the surrounding atmosphere. This could be the ground or down a manhole, for instance. If vapor density is close to 1.0, it will tend to mix with the air at the site. If the vapor density is lower than 1.0, the vapor will tend to disperse in the atmosphere. Dense vapors collecting at a height of approximately 1 ft from the surface in a workspace may not be immediately noticed. This can create an unexpected hazard should some task require lying prone on the floor. This is because routine personal monitoring many times takes place only at the erect workers' breathing zone.

WARNING

DENSE VAPORS CREATE THREE MAJOR HAZARDS:
1. IF THE VAPOR DISPLACES ENOUGH AIR TO REDUCE THE AT-MOSPHERIC CONCENTRATION OF OXYGEN BELOW 19.5%, AS-PHYXIA MAY RESULT.
2. IF THE VAPOR IS TOXIC, RESPIRATOR PROTECTION MAY BE REQUIRED EVEN IF THE ATMOSPHERE IS NOT OXYGEN-DEFICIENT.
3. IF A SUBSTANCE IS EXPLOSIVE AND DENSE, THE EXPLOSIVE HAZARD MAY BE CLOSE TO THE GROUND. GASOLINE VAPORS ARE A GOOD EXAMPLE.

Vapor Pressure.   The pressure exerted by the vapor from a liquid against the inside of a closed container, or to the atmosphere, is called vapor pressure. It is temperature-dependent but not pressure-dependent. That is, as temperature increases, so does the vapor pressure, regardless of the pressure in a container or in the atmosphere. With rising temperature in a container of liquid, more liquid evaporates, causing a greater pressure to build up in the container. Obviously, if this process is not relieved, at some point the container will fail!

There is also a relationship between boiling point and vapor pressure. The lower the boiling point of the liquid, the greater the vapor pressure it will exert at a given temperature. Values for vapor pressure are usually given in millimeters of mercury (mm Hg) at a specific ambient temperature (usually near 70°F).

The standard sea level pressure is considered to be 760 mm Hg. Therefore, when a liquid develops a vapor pressure of 760 mm Hg, it is at its boiling point at sea level. The vapor pressure of acetone, a common solvent, is 70 mm Hg at 32°F and nearly 3000 mm Hg at 212°F.

Evaporation Rate.   Evaporation rate is the rate at which one material will evaporate, or vaporize, as compared to a reference material at a given temperature, usually 70°F. A common reference material is butyl acetate (BuAc). The reference material is assigned an evaporation rate of 1.0.

Using the reference material BuAc (1.0), a solvent such as hexane has a very fast evaporation rate of 8.4. Water has a very slow evaporation rate of 0.3. These rates are temperature-dependent: the higher the temperature, the faster the rate. Think of an open pot full of water. It could take weeks to evaporate, depending on the temperature and relative humidity. However, put that same pot on the stove with heat, but not enough to boil, and the water may evaporate in less than an hour.

There is a direct relationship between evaporation rate and vapor pressure. Materials with a high vapor pressure will tend to have a high evaporation rate; materials with a low vapor pressure will tend to have a low evaporation weight.

There is also a relationship between evaporation rate and boiling point. The lower the boiling point the higher the evaporation rate, and the higher the boiling point, the lower the evaporation rate.

Understanding evaporation rate will help you determine how quickly a material may become a vapor and become an inhalation hazard or cause other hazards such as fire.

Here you will very likely also find the following (if applicable):

Flash Point.   The minimum temperature at which a substance produces enough flammable vapors to ignite is its flash point. If the vapor does ignite, combustion can continue as long as the temperature remains at or above the flash point. The relative flammability of a substance is based on its flash point.

As far as immediate hazards are concerned, fires and explosions are two of the easiest to prevent and the hardest to stop once they have started (see Section 3). Materials with a low flash point (less than 100°F) are considered highly flammable. Under ambient conditions, a simple electrical spark, discarded cigarette, or static electricity could potentially cause a fire or explosion.

Decomposition Temperature.   This is the temperature at which the molecules of materials begin to break apart, forming two or more other molecules. This may also help to define the stability of a material. Solutions of hydrogen peroxide (H_2O_2) tend toward instability. Regardless of the concentration, hydrogen peroxide slowly decomposes to form water (H_2O) and oxygen (O_2). The lower the decomposition temperature of the material, the less stable it will tend to be.

HAZWOPER workers must understand the chances of decomposition of materials they handle. The decomposition products may be more hazardous than the original material. Further, any heat and gases produced can cause other hazards.

Viscosity.   Viscosity is the property of a liquid whereby it resists flow. Water is relatively free flowing, that is, it is not considered viscous. Molasses and heavy oils, on the other hand, are viscous liquids.

Viscosity is temperature-dependent: the higher the temperature, the lower the viscosity. Honey, when heated, will lose viscosity and become almost as free-flowing as water. On the other hand, well-formulated industrial lubricants will retain their desirable viscosity properties over a broad range of temperatures. This is what makes most motor oils such good lubricants. They can flow and provide protection at relatively low temperatures, yet still cling to the motor parts, flow satisfactorily, and protect the motor from damaging friction at relatively high temperatures.

Viscosity is commonly measured in centipoise units (cP). The viscosity of water at room temperature is about 1.0 cP. A 50% sugar in water solution has a viscosity at room temperature of about 15 cP.

In an emergency, knowing the viscosity of a spilled material relative to that of water will help you estimate how far a spill might travel over time.

Corrosion Rate.   This is the speed with which a material will corrode, or eat through, a metal container or attack structural materials. Often expressed in inches per year (IPY) for high corrosion rates and mils per year (MPY) for slow corrosion rates. A common corrosive, sulfuric acid must be stored in glass, stainless steel, or plastic containers. Sulfuric acid would quickly corrode a steel drum.

Some of the other factors that affect corrosion are temperature and the presence of oxygen and humidity. Higher values of these factors tend to speed corrosion. Where there are dissimilar metals in contact with each other and where metals are in contact with salt water, electrochemical reactions can occur that speed corrosion.

9.4.11 Reactivity Data

Stability. This subsection explains the ability of a material to remain unchanged over time. A material is stable if it remains in the same form under expected and reasonable conditions of storage or use. The following are the kinds of information to expect to find if the material tends to be reactive. See Section 3 and Section 4 for explanations of problems that can occur when incompatible materials contact each other.

Hazardous Decomposition Products. As a material decomposes, several things can occur. It can release or absorb heat. It can revert to the materials from which it was formed, which may or may not be hazardous. Or it can form other materials, which may or may not be hazardous.

In this section of the MSDS you will find what hazardous products the manufacturer or importer knows will be formed due to various decomposition processes. In the case of sulfuric acid, heating to decomposition gives off toxic oxides of sulfur: sulfur dioxide (SO_2) and sulfur trioxide (SO_3), collectively represented as (SO_x).

Hazardous Polymerization (Will/Will Not) Occur. Polymerization is the process of forming a polymer (e.g., polyethylene or polyvinyl chloride) by combining large numbers of monomers into long chains. Some forms of uncontrolled polymerization can be extremely hazardous. Some polymerization reactions can release considerable heat, can generate enough pressure to burst a container, or can be explosive. Some chemicals can polymerize upon contact with water, air, or other common chemicals. Inhibitors can be added to materials to reduce or eliminate the possibility of uncontrolled polymerization.

Mixtures of Part A and Part B epoxy paint, accidentally collected as wastes in the same drum, will undergo hazardous polymerization with the resultant release of a great deal of heat.

Incompatibilities. Incompatible materials react with each other, usually liberating heat. This may be a violent reaction or cause the release of hazardous reaction materials. Sulfuric acid is incompatible with a great number of other materials. Oxidizers are incompatible with all fuels. (See Section 4.5.)

Conditions to Avoid. These can include exposure of the material to any of the incompatible materials listed, heat or cold, moisture or air, and atmospheric pollutants.

9.4.12 Toxicological Information

For an understanding of the science of toxicology and the physiological hazards associated with hazardous materials, including nomenclature, please see Section 5.

9.4.13 Ecological Information

This section of the MSDS will give information on a material's *environmental fate,* meaning what will likely happen in the environment if an accidental release of the material occurs. This section may also give information on the material's toxicity to aquatic life. It is important to understand the information in this section in order to appreciate the need to dispose of any waste material properly.

9.4.14 Disposal

In the disposal section of the MSDS you may find the most diverse information. You will almost always find the text "Dispose of excess or waste material in accordance with federal, state and local regulations." This is what we call catch-all phrasing. What we would like to see here is the exact regulatory citations to consult for the proper safe disposal of the excess or waste material.

Pay careful attention to this portion of the MSDS. Disposal can be tightly regulated. This is explained in greater detail in Section 12.

9.4.15 Transport Information

This section will include domestic and international transportation information, including proper shipping name, hazard class, UN/NA number, and packing group. For a more detailed description of packaging and labeling of hazardous materials, please see Section 8.

9.4.16 Other Information

In this section of the MSDS you may find information that does not readily fit into one of the other categories. That does not mean that the information is of less use. In our example MSDS, this section is highly informative. It might contain:

- NFPA ratings
- Label hazard warning
- Label precautions
- Label first aid
- Product use
- MSDS revision information
- Preparer and contact number

The fictitious manufacturer in the example has chosen this section as a capsule of the most important information contained in the MSDS.

9.5 OSHA MSDS FORM

The following is a generic form from OSHA. This form will allow competent persons to draft MSDSs for materials at their site. This form will bring the company into compliance with the Hazard Communication Standard (29 CFR 1910.1200). However, the authors recommend that MSDSs contain additional data. See Section 9.6 for an example of what the authors feel is a complete MSDS.

Material Safety Data Sheet
U.S. Department of Labor

May be used to comply with
OSHA's Hazard Communication Standard,
29 CFR 1910.1200. Standard must be
consulted for specific requirements.

Occupational Safety and Health Administration
(Non-Mandatory Form)
Form Approved
OMB No. 1218-0072

IDENTITY (*As Used on Label and List*)	Note: Blank spaces are not permitted. If any item is not applicable, or no information is available, the space must be marked to indicate that.

Section I

Manufacturer's Name	Emergency Telephone Number
Address (*Number, Street, City, State, and ZIP Code*)	Telephone Number for Information
	Date Prepared
	Signature of Preparer (*optional*)

Section II—Hazard Ingredients/Identity Information

Hazardous Components (Specific Chemical Identity; Common Name(s))	OSHA PEL	ACGIH TLV	Other Limits Recommended	%(*optional*)

Section III—Physical/Chemical Characteristics

Boiling Point		Specific Gravity ($H_2O = 1$)	
Vapor Pressure (mm Hg)		Melting Point	
Vapor Density (AIR = 1)		Evaporation Rate (Butyl Acetate = 1)	
Solubility in Water			
Appearance and Odor			

Section IV—Fire and Explosion Hazard Data

Flash Point (Method Used)	Flammable Limits	LEL	UEL
Extinguishing Media			
Special Firefighting Procedures			
Unusual Fire and Explosion Hazards			

Section V—Reactivity Data

Stability	Unstable		Conditions to Avoid	
	Stable			
Incompatibility (*Materials to Avoid*)				
Hazardous Decomposition or Byproducts				
Hazardous Polymerization	May Occur		Conditions to Avoid	
	Will Not Occur			

Section VI—Health Hazard Data

Route(s) of Entry:	Inhalation?	Skin?	Ingestion?

Health Hazards (*Acute and Chronic*)

Carcinogenicity:	NTP?	IARC Monographs?	OSHA Regulated?

Signs and Symptoms of Exposure

Medical Conditions

Generally Aggravated by Exposure

Emergency and First Aid Procedures

Section VII—Precautions for Safe Handling and Use

Steps to Be Taken in Case Material is Released or Spilled

Waste Disposal Method

Precautions to Be Taken in Handling and Storing

Other Precautions

Section VIII—Control Measures

Respiratory Protection (*Specify Type*)		
Ventilation	Local Exhaust	Special
	Mechanical (*General*)	Other
Protective Gloves	Eye Protection	
Other Protective Clothing or Equipment		
Work/Hygienic Practices		

9.6   AN EXAMPLE OF AN EXHAUSTIVE MSDS

Since sulfuric acid is the most widely used chemical, as measured by tons produced annually in the United States (not including some petroleum products such as fuel oil and gasoline), we have prepared its MSDS as an example.

TABLE 9.1   A Properly Written MSDS

ABC Chemical Co., Inc.
123 Chemical Drive
Anytown, U.S.A. 12345
24 Hour Emergency Phone
(123) 456-7890

National Response Center (NRC)
(1-800) 424-8802
CHEMTREC (1-800) 424-9300
National Response Centre (Canada)
CANUTEC (613) 996-6666

Non-emergency Technical Assistance (123) 456-7899

SULFURIC ACID, 50–100 %

MSDS Number: S12345—Effective Date: 03/13/02

1. Product Identification

Synonyms: Oil of vitriol; battery acid, sulphuric acid aqueous.
CAS No.: 7664-93-9
Molecular Weight: 98.08
Chemical Formula: H_2SO_4 and H_2O
Product Codes: ABC XXXX, ABC YYYY, ABC ZZZZ

2. Composition/Information on Ingredients

Ingredient	CAS No	Percent	Hazardous
Sulfuric Acid	7664-93-9	52–100%	Yes
Water	7732-18-5	0–48%	No

TABLE 9.1   A Properly Written MSDS (*Continued*)

3. Hazards Identification

EMERGENCY OVERVIEW

POISON! DANGER! CORROSIVE. LIQUID AND MIST CAUSE SEVERE BURNS TO ALL BODY TISSUE. MAY BE FATAL IF SWALLOWED OR CONTACTED WITH SKIN. HARM-FUL IF INHALED. AFFECTS TEETH. WATER REACTIVE. CANCER HAZARD. STRONG INORGANIC ACID MISTS CONTAINING SULFURIC ACID CAN CAUSE CANCER. Risk of cancer depends on duration and level of exposure.

POTENTIAL HEALTH EFFECTS

Inhalation: Inhalation produces damaging effects on the mucous membranes and upper respiratory tract. Symptoms may include irritation of the nose and throat, and labored breathing. May cause lung edema and a medical emergency.

Ingestion: Corrosive. Swallowing can cause severe burns of the mouth, throat, and stomach, leading to death. Can cause sore throat, vomiting, and diarrhea. Circulatory collapse with clammy skin, weak and rapid pulse, shallow respiration, and scanty urine may follow ingestion or skin contact. Circulatory shock is often the immediate cause of death.

Skin Contact: Corrosive. Symptoms of redness, pain, and severe burn can occur. Circulatory collapse with clammy skin, weak and rapid pulse, shallow respiration, and scanty urine may follow skin contact or ingestion. Circulatory shock is often the immediate cause of death.

Eye Contact: Corrosive. Contact can cause blurred vision, redness, pain and severe tissue burns. Can cause blindness.

Chronic Exposure: Long-term exposure to mist or vapors may cause damage to teeth. Chronic exposure to mists containing sulfuric acid is a cancer hazard.

Aggravation of Preexisting Conditions: Persons with preexisting skin disorders or eye problems or impaired respiratory function may be more quickly susceptible to the effects of the substance.

4. First Aid Measures

Inhalation: Remove to fresh air. If not breathing, give artificial respiration. If breathing is difficult, give oxygen. Call a physician immediately.

Ingestion: DO NOT INDUCE VOMITING. Give large quantities of water. Never give anything by mouth to an unconscious person. Call a physician immediately.

Skin Contact: In case of contact, immediately flush skin with plenty of water for at least 15 minutes while removing contaminated clothing and shoes. Call a physician immediately. Excess acid on skin can be neutralized with a 2% solution of bicarbonate of soda. Wash clothing before reuse.

Eye Contact: Immediately flush eyes with gentle but large stream of water for at least 15 minutes, lifting lower and upper eyelids occasionally. Call a physician immediately.

5. Firefighting Measures

Fire: Concentrated material is a strong dehydrating agent. Reacts with organic materials and may cause ignition of finely divided materials on contact.

Explosion: Contact with most metals causes formation of flammable and explosive hydrogen gas.

Fire-Extinguishing Media:
Dry chemical, foam or carbon dioxide. DO NOT use water on sulfuric acid! However, water spray may be used to keep fire-exposed, sealed containers cool.

TABLE 9.1 A Properly Written MSDS (*Continued*)

Special Information: In the event of a fire, wear full protective clothing and NIOSH-approved self-contained breathing apparatus with full facepiece operated in the pressure demand or other positive pressure mode. Structural firefighter's protective clothing is ineffective for fires involving this material. Stay away from sealed containers.

6. Accidental Release Measures

Ventilate area of leak or spill. Wear appropriate personal protective equipment as specified in Section 8. Isolate hazard area. Keep unnecessary and unprotected personnel from entering. Contain and recover liquid when possible. Neutralize with alkaline material (soda ash, lime), then absorb with an inert material (e.g., vermiculite, dry sand, and earth), and place in a chemical waste container. Do not use combustible materials, such as sawdust. Do not flush to sewer! U.S. Regulations (CERCLA) require reporting spills and releases to soil, water, and air in excess of reportable quantities. The toll free number for the U.S. Coast Guard National Response Center is (800) 424-8802.

7. Handling and Storage

Store in a cool, dry, ventilated storage area with acid-resistant floors and good drainage. Protect from physical damage. Keep out of direct sunlight and away from heat, water, and incompatible materials. Do not wash out container and use it for other purposes. When diluting, always add the acid to water; never add water to the acid. When opening metal containers, use nonsparking tools because of the possibility of hydrogen gas being present. Containers of this material may be hazardous when empty since they retain product residues (vapors, liquid); observe all warnings and precautions listed for the product.

8. Exposure Controls/Personal Protection

AIRBORNE EXPOSURE LIMITS: For Sulfuric Acid:

- OSHA Permissible Exposure Limit (PEL): 1 mg/m^3 (8 hour TWA).
- ACGIH Threshold Limit Value (TLV®): 1 mg/m^3 (8 hour TWA), 3 mg/m^3 (30 minute STEL)

VENTILATION SYSTEM: A system of local and/or general exhaust is recommended to keep employee exposures below the airborne exposure limits. Local exhaust ventilation is generally preferred because it can control the emissions of the contaminant at its source, preventing dispersion of it into the general work area. Please refer to the ACGIH document *Industrial Ventilation: A Manual of Recommended Practices,* most recent edition, for details.

PERSONAL RESPIRATORS (NIOSH-APPROVED): If the unprotected exposure limit is exceeded, a full-facepiece respirator with an acid gas cartridge and dust and mist filter may be worn up to 50 times the exposure limit, or the maximum use concentration specified by the appropriate regulatory agency or respirator supplier, whichever is lowest. For emergencies or instances where the exposure levels are not known, use a full-facepiece positive-pressure, air-supplied respirator. WARNING: Air-purifying respirators do not protect workers in oxygen-deficient atmospheres.

SKIN PROTECTION: Wear impervious protective clothing, including boots, gloves, lab coat, apron, or coveralls, as appropriate, to prevent skin contact.

EYE PROTECTION: Use chemical safety goggles and/or a full face shield where splashing is possible. Maintain eyewash fountain and quick-drench facilities in work area.

9. Physical and Chemical Properties

APPEARANCE: Clear oily liquid.

ODOR: Odorless to some people. Others detect a stinging or salt-like odor.

SOLUBILITY: Completely miscible with water, exothermic reaction (generates heat).

SPECIFIC GRAVITY: 1.84 (98%), 1.40 (50%), 1.07 (10%)

TABLE 9.1   A Properly Written MSDS (*Continued*)

pH: 1 N solution (ca. 5% w/w) = 0.3; 0.1 N solution (ca. 0.5% w/w) = 1.2; 0.01 N solution (ca. 0.05% w/w) = 2.1 (Note: w/w means weight of acid per weight of solution.)

% VOLATILES BY VOLUME @ 21C (70F): No information found.

BOILING POINT: approximately 290°C (554°F) (decomposes at 340°C)

MELTING POINT: approximately 10°C (100%), −32°C (93%), −38°C (78%), −64°C (65%)

VAPOR DENSITY (AIR = 1): 3.4

VAPOR PRESSURE (MM HG): 1 @ 145.8°C (295°F)

EVAPORATION RATE (BuAc = 1): No information found.

10. Stability and Reactivity

STABILITY: Stable under ordinary conditions of use and storage. Concentrated solutions react violently with water, spattering and liberating heat.

HAZARDOUS DECOMPOSITION PRODUCTS: Toxic fumes of oxides of sulfur when heated to decomposition. Will react with water or steam to produce toxic and corrosive fumes. Reacts with carbonates to generate carbon dioxide gas, and with cyanides and sulfides to form poisonous hydrogen cyanide and hydrogen sulfide respectively.

HAZARDOUS POLYMERIZATION: Will not occur.

INCOMPATIBILITIES: Water, potassium chlorate, potassium perchlorate, potassium permanganate, sodium, lithium, bases, organic material, halogens, metal acetylides, oxides and hydrides, metals (yields hydrogen gas), strong oxidizing and reducing agents, and many other reactive substances.

CONDITIONS TO AVOID: Heat, moisture, and incompatibles.

11. Toxicological Information

TOXICOLOGICAL DATA: Oral rat LD_{50}: 2140 mg/kg; Inhalation rat LC_{50}: 510 mg/m^3/2H; standard Draize, eye rabbit, 250 μg (severe); investigated as a tumorigen, mutagen, reproductive effector.

CARCINOGENICITY: Cancer Status: The International Agency for Research on Cancer (IARC) has classified "strong inorganic acid mists containing sulfuric acid" as a known human carcinogen (IARC category 1). This classification applies only to mists containing sulfuric acid and not to sulfuric acid or sulfuric acid solutions.

	Cancer Lists		
	NTP Carcinogen		
Ingredient	Known	Anticipated	IARC Category
Sulfuric Acid (7664-93-9)	No	No	None
Water (7732-18-5)	No	No	None

TABLE 9.1   A Properly Written MSDS (*Continued*)

12. Environmental and Ecological Information

ENVIRONMENTAL FATE: When released into the soil, this material may leach into groundwater. When released into the air, this material may be removed from the atmosphere to a moderate extent by wet deposition. When released into the air, this material may be removed from the atmosphere to a moderate extent by dry deposition.

ENVIRONMENTAL TOXICITY: LC_{50} Flounder 100 to 330 mg/L/48 hr aerated water/Conditions of bioassay not specified; LC_{50} Shrimp 80 to 90 mg/L/48 hr aerated water/Conditions of bioassay not specified; LC_{50} Prawn 42.5 ppm/48 hr salt water/Conditions of bioassay not specified. **This material may be toxic to aquatic life.**

13. Disposal Considerations

Whatever cannot be saved for recovery or recycling should be handled as hazardous waste and sent to an RCRA-approved incinerator or disposed of in an RCRA-approved waste facility. Processing, use, or contamination of this product may change the waste management options. State and local disposal regulations may differ from federal disposal regulations. Dispose of container and unused contents in accordance with federal, state, and local requirements.

14. Transport Information

DOMESTIC (LAND, DOT)

PROPER SHIPPING NAME: SULFURIC ACID (WITH MORE THAN 50% ACID)
HAZARD CLASS: 8
UN/NA: UN1830, Packing Group: II
INFORMATION REPORTED FOR PRODUCT/SIZE: 800 LB

INTERNATIONAL (WATER, International Maritime Organization—I.M.O.)

PROPER SHIPPING NAME: SULPHURIC ACID (WITH MORE THAN 51% ACID)
HAZARD CLASS: 8
UN/NA: UN1830, Packing Group: II
INFORMATION REPORTED FOR PRODUCT/SIZE: 800 LB

15. Regulatory Information

Chemical Inventory Status - Part 1

Ingredient	TSCA	EC	Japan	Australia
Sulfuric Acid (7664-93-9)	Yes	Yes	Yes	Yes
Water (7732-18-5)	Yes	Yes	Yes	Yes

Chemical Inventory Status - Part 2

Ingredient	Korea	Canada DSL	Canada NDSL	Phil.
Sulfuric Acid (7664-93-9)	Yes	Yes	No	Yes
Water (7732-18-5)	Yes	Yes	No	Yes

TABLE 9.1 A Properly Written MSDS (*Continued*)

Federal, State & International Regulations - Part 1

Ingredient	SARA 302		SARA 313	
	RQ	TPQ	LIST	Chemical Catg.
Sulfuric Acid (7664-93-9)	1000	1000	Yes	No
Water (7732-18-5)	No	No	No	No

Federal, State & International Regulations - Part 2

Ingredient	CERCLA	RCRA 261.33	TSCA 8(d)
Sulfuric Acid (7664-93-9)	1000	No	No
Water (7732-18-5)	No	No	No

Chemical Weapons Convention: No TSCA 12(b): No CDTA: Yes
SARA 311/312: Acute: Yes Chronic: Yes Fire: No Pressure: No
Reactivity: Yes (Pure / Liquid)

Australian Hazchem Code: 2P

Poison Schedule: No information found.

WHMIS: (Canadian *Workplace Hazardous Materials Information System*—similar to the OSHA *Hazard Communication Standard.*)
This MSDS has been prepared according to the hazard criteria of the Controlled Products Regulations (CPR) and the MSDS contains all of the information required by the CPR. (The CPR is roughly the equivalent of the U.S. DOT's Hazardous Materials regulations.)

16. Other Information

NFPA RATINGS: Health: **3** Flammability: **0** Reactivity: **2** Other: **Water reactive**

LABEL HAZARD WARNING: POISON! DANGER! CORROSIVE. LIQUID AND MIST CAUSE SEVERE BURNS TO ALL BODY TISSUE. MAY BE FATAL IF SWALLOWED OR CONTACTED WITH SKIN. HARMFUL IF INHALED. AFFECTS TEETH. WATER REACTIVE. CANCER HAZARD. STRONG INORGANIC ACID MISTS CONTAINING SULFURIC ACID CAN CAUSE CANCER. Risk of cancer depends on duration and level of exposure.

LABEL PRECAUTIONS: Do not get in eyes, on skin, or on clothing. Do not breathe mist. Keep container closed. Use only with adequate ventilation. Wash exposed skin thoroughly after handling. Do not contact material with water.

LABEL FIRST AID: In all cases call a physician immediately. In case of contact, immediately flush eyes or skin with plenty of water for at least 15 minutes while removing contaminated clothing and shoes. Wash clothing before reuse. Excess acid on skin can be neutralized with a 2% bicarbonate of soda solution. If swallowed, DO NOT INDUCE VOMITING. Give large quantities of water. Never give anything by mouth to an unconscious person. If inhaled, remove to fresh air. If not breathing, give artificial respiration. If breathing is difficult, give oxygen.

PRODUCT USE:
Laboratory Reagent.

TABLE 9.1   A Properly Written MSDS (*Continued*)

REVISION INFORMATION:

MSDS Section(s) changed since last revision of document include: X.

DISCLAIMER:

ABC Chemical Co, Inc. provides the information contained herein in good faith but makes no representation as to its comprehensiveness or accuracy. This document is intended only as a guide to the appropriate precautionary handling of the material by a properly trained person using this product. Individuals receiving the information must exercise their independent judgment in determining its appropriateness for a particular purpose. ABC Chemical Co., Inc. makes no representations or warranties, either express or implied, including without limitation any warranties of merchantability, fitness for a particular purpose with respect to the information set forth herein or the product to which the information refers. Accordingly, ABC Chemical Co, Inc. will not be responsible for damages resulting from use of or reliance upon this information.

Prepared by: ABC Chemical Co., Inc.

9.7   INTERNATIONAL CHEMICAL SAFETY CARDS

The World Health Organization (WHO), the International Labour Organization (ILO), the United Nations Environment Programme (UNEP), and the Commission of the European Communities (CEC) are working together on the International Programme on Chemical Safety (IPCS). This program is producing the International Chemical Safety Card (ICSC), the worldwide equivalent of the MSDS. ICSCs are copyright IPCS and CEC.

ICSCs are available online at http://www.cdc.gov/niosh/ipcs/ipcs0000.html#S, and on the Companion CD to this book.

9.7.1   NIOSH Comparison of MSDSs and ICSCs

In NIOSH's own words:

> MSDSs and the ICSCs are not the same. The MSDS, in many instances, may be technically very complex and too extensive for shop floor use, and secondly it is a management document. The ICSCs, on the other hand, set out peer-reviewed information about substances in a more concise and simple manner. While not a legal document, the ICSC is an authoritative document emanating from WHO/ILO/UNEP.
>
> This is not to say that the ICSC should be a substitute for an MSDS. Nothing can replace management's responsibility to communicate with workers on the exact chemicals, the nature of those chemicals used on the shop floor and the risk posed in any given work place. Indeed, the ICSC and the MSDS can even be thought of as complementary. If the two methods for hazard communication can be combined, then the amount of knowledge available to the safety representative or shop floor workers will be more than doubled. The ICSC could serve as a model for disseminating chemical safety information to workers.

To further disassociate the compilers from the law each ICSC comes with the following disclaimer:

IMPORTANT LEGAL NOTICE

Neither the CEC or the IPCS nor any person acting on behalf of the CEC or the IPCS is responsible for the use which might be made of this information. These cards contain the collective views of the IPCS Peer Review Committee and may not reflect in all cases all the detailed requirements included in national legislation on the subject. The user should verify compliance of the cards with the relevant legislation in the country of use.

9.7.2 The Authors' Comparison of MSDSs and ICSCs

All that being said, how exactly do the two information sources differ? This can best be explained by a comparison of the two on some important issues. The authors feel that the MSDS is more comprehensive in its content as well as its direction to the reader. Where the ICSC is considered by some to be more user-friendly, we feel that with the proper training (HAZCOMM per 29 CFR 1910.1200), the North American worker is better served by an MSDS following the ASTM standard.

We do not feel this way simply because we are North Americans. Our view is based purely on a comparison. As Table 9.3 shows, the sections of both of the sources are relatively identical in content. Each might have more or less description than its counterpart.

Our critique of the two systems is that the MSDS gives specific directions for contacting the manufacturer, ChemTREC, and other response organizations and personnel, while the ICSC gives more generic advice. Also, a complete MSDS offers regulatory advice and proper disposal information. This is not to say that the MSDS does not have its faults. One section, "Disposal Information," is notorious for being poorly written. A large number of MSDSs simply state: "Dispose of (material) in accordance with federal, state, and local regulations."

This is really no help at all. We, the readers, know we have to abide by disposal regulations. What we need is guidance at least on exactly those federal laws with which we have to comply.

9.7.3 A Typical International Chemical Safety Card

The ICSC is included here as an example of what other countries are doing to answer the "What if?" questions. An example of an ICSC is shown as Table 9.2. A side-by-side comparison of the information on an MSDS and an ICSC is shown as Table 9.3.

TABLE 9.2 A Typical International Chemical Safety Card for Sulfuric Acid

---

**SULFURIC
ACID**

ICSC: 0362
October 2000

Sulfuric acid 100%
Oil of vitriol

CAS No: 7664-93-9
RTECS No: WS5600000
UN No: 1830
EC No: 016-020-00-8

H_2SO_4
Molecular mass: 98.1

TABLE 9.2 A Typical International Chemical Safety Card for Sulfuric Acid (*Continued*)

Types of Hazard/ Exposure	Acute Hazards/Symptoms	Prevention	Fire Fighting
FIRE	Not combustible. Many reactions may cause fire or explosion. Gives off irritating or toxic fumes (or gases) in a fire.	NO contact with flammable substances. NO contact with combustibles.	NO water. In case of fire in the surroundings: powder, AFFF, foam, carbon dioxide.
EXPLOSION	Risk of fire and explosion on contact with base(s), combustible substances, oxidants, reducing agents or water.		In case of fire: keep drums, etc., cool by spraying with water but NO direct contact with water.
EXPOSURE		**PREVENT GENERATION OF MISTS! AVOID ALL CONTACT!**	**IN ALL CASES CONSULT A DOCTOR!**
Inhalation	Corrosive. Burning sensation. Sore throat. Cough. Labored breathing. Shortness of breath. Symptoms may be delayed (see Notes).	Ventilation, local exhaust, or breathing protection.	Fresh air, rest. Half-upright position. Artificial respiration if indicated. Refer for medical attention.
Skin	Corrosive. Redness. Pain. Blisters. Serious skin burns.	Protective gloves. Protective clothing.	Remove contaminated clothes. Rinse skin with plenty of water or shower. Refer for medical attention.
Eyes	Corrosive. Redness. Pain. Severe deep burns.	Face shield, or eye protection in combination with breathing protection.	First rinse with plenty of water for several minutes (remove contact lenses if easily possible), then take to a doctor.
Ingestion	Corrosive. Abdominal pain. Burning sensation. Shock or collapse.	Do not eat, drink, or smoke during work.	Rinse mouth. Do NOT induce vomiting. Refer for medical attention.

Spillage Disposal	Packaging & Labelling	
Consult an expert! Evacuate danger area! Do NOT absorb in sawdust or other combustible absorbents. (Extra personal protection: complete protective clothing including self-contained breathing apparatus). Do NOT let this chemical enter the environment.	C Symbol R: 35 S: (1/2-)26-30-45 Note: B UN Hazard Class: 8 UN Pack Group: II	Unbreakable packaging; put breakable packaging into closed unbreakable container. Do not transport with food and feedstuffs.

Emergency Response	Storage
Transport Emergency Card: TEC (R)-10B NFPA Code: H 3; F 0; R 2; W	Separated from combustible and reducing substances, strong oxidants, strong bases, food and feedstuffs, incompatible materials. See Chemical Dangers. May be stored in stainless steel containers. Store in an area having corrosion resistant concrete floor.

TABLE 9.2 A Typical International Chemical Safety Card for Sulfuric Acid (*Continued*)

Important Data	
Physical State; Appearance	**Routes of exposure**
COLOURLESS, OILY, HYGROSCOPIC LIQUID, WITH NO ODOUR.	The substance can be absorbed into the body by inhalation of its aerosol and by ingestion.

Emergency Response	Storage

Important Data	
Chemical dangers	**Inhalation risk**
The substance is a strong oxidant and reacts violently with combustible and reducing materials. The substance is a strong acid, it reacts violently with bases and is corrosive to most common metals forming a flammable/explosive gas (hydrogen—see ICSC 0001). Reacts violently with water and organic materials with evolution of heat (see Notes). Upon heating, irritating or toxic fumes (or gases) (sulfur oxides) are formed.	Evaporation at 20°C is negligible; a harmful concentration of airborne particles can, however, be reached quickly on spraying.
	Effects of short-term exposure
	Corrosive. The substance is very corrosive to the eyes the skin and the respiratory tract. Corrosive on ingestion. Inhalation of an aerosol of this substance may cause lung edema (see Notes).
Occupational exposure limits	**Effects of long-term or repeated exposure**
TLV: 1 mg/m^3 (as TWA); 3 mg/m^3 (as STEL) A2 sulfuric acid contained in strong inorganic acid mists (ACGIH 2000). MAK: 1 mg/m3; inhalable fraction of aerosol (1999).	Lungs may be affected by repeated or prolonged exposure to an aerosol of this substance. Risk of tooth erosion upon repeated or prolonged exposure to an aerosol of this substance. Strong inorganic acid mists containing this substance are carcinogenic to humans.

Physical Properties	Environmental Data
Boiling point (decomposes): 340°C Melting point: 10°C Relative density (water = 1): 1.8 Solubility in water: miscible Vapour pressure, kPa at 146°C: 0.13 Relative vapour density (air = 1): 3.4	The substance is harmful to aquatic organisms.

NOTES

The symptoms of lung edema often do not become manifest until a few hours have passed and they are aggravated by physical effort. Rest and medical observation are therefore essential.

NEVER pour water into this substance; when dissolving or diluting always add the acid slowly to the water.

Other UN numbers: UN1831 Sulfuric acid, fuming, hazard class 8, subsidiary hazard 6.1, pack group I; UN1832 Sulfuric acid, spent, Hazard class 8, Pack group II.

IPCS

International

Programme on

Chemical Safety

TABLE 9.2   A Typical International Chemical Safety Card for Sulfuric Acid (*Continued*)

Prepared in the context of cooperation between the International Programme on Chemical Safety
and the European Commission

© **IPCS 2000**

LEGAL NOTICE
Neither the EC nor the IPCS nor any person acting on behalf of the EC or the IPCS is responsible for
the use which might be made of this information.

9.7.4   A Side-by-Side Comparison of the MSDS and the ICSC

TABLE 9.3   A Side-by-Side Comparison of the MSDS and the ICSC

International Council of Chemical Associations (ICCA) Headings of Material Safety Data Sheets	International Programme on Chemical Safety (IPCS) Headings of International Chemical Safety Cards
1. Chemical product identification and company identification	1. Chemical identification
2. Composition/Information on ingredients	2. Composition/formula
3. Hazards identification	3. Hazard identification from fire and explosion and from exposure by inhalation, skin, eyes and ingestion, and prevention measures (with personal protective equipment)
4. First aid measures	First aid measures
5. Firefighting measures	Firefighting measures
6. Accidental release measures	4. Spillage, disposal
7. Handling and storage	5. Storage
	6. Packaging, labeling, and transport
8. Exposure controls/personal measures	See 3. above
	7. Important data:
See 15. below	Occupational exposure limits
9. Physical and chemical properties	See 8. below
10. Stability and reactivity	Physical and chemical dangers
11. Toxicological information	Routes of exposure
	Effects of short- and long-term exposure
See 9. above	8. Physical properties
12. Ecological information	9. Environmental data
13. Disposal considerations	See 4. above
14. Transport information	See 6. above
15. Regulatory information	See 7. above
	10. Notes
16. Other information	11. Additional information

9.8 TRAINING AIDS AND ADDITIONAL RESOURCES

The authors recommend the following videos for aiding readers to see real-life examples of what they have learned in this Section:

* *Understanding MSDS*

This video is available from Safe Expectations (formerly BNA Communications, Inc.) (www.safeexpectations.com)

* *Material Safety Data Sheets—Read it Before You Need It!*

This video is available from Coastal Safety and Environmental (www.coastal.com).

9.9 SUMMARY

We have endeavored to show readers what they can learn from the Material Safety Data Sheets (MSDS) for a given material. The MSDSs are the primary source of information concerning the safe use, storage, disposal, and emergency procedures following a release of materials in the workplace. MSDSs are chiefly required for any chemical material manufactured, distributed, or imported by U.S. companies. MSDS requirements are all specified to comply with 29 CFR 1910.1200, the Hazard Communication Standard (HAZCOMM), and the Community Right-to-Know Act (RTK).

OSHA does not specify a mandatory format that the MSDS must follow. Although OSHA permits flexibility in the amount and type of information that must be present in an MSDS, certain minimum information is required. The most widely accepted form is contained in ASTM publication E1628-2000. The authors feel that this is the most comprehensive, understandable, and helpful form and recommend its use.

Proprietary ingredients need not be included. However, any hazards those ingredients add to the final product must be explained. Proprietary ingredients must be released to medical professionals treating an exposed worker.

Manufacturers, distributors, and importers (the sellers) are required to provide MSDS copies with the first shipment of a product and any time the constituents of the product change. They must also provide a copy to anyone who requests an MSDS; free of charge. Sellers must evaluate the hazards of the chemicals they produce or import. According to OSHA, the user has no independent duty to analyze the chemical or evaluate the hazards of its use.

To familiarize readers with another format for material safety data, one that is gaining in usage in other parts of the world, we presented the elements of the International Chemical Safety Card (ICSC). This form of reference is used for chemical materials in European Union (EU) countries and elsewhere.

The World Health Organization (WHO), the International Labour Organization (ILO), the United Nations Environment Programme (UNEP), and the Commission of the European Communities (CEC) are working together on the International Programme on Chemical Safety (IPCS). This program is producing the International Chemical Safety Card (ICSC), comparable to the MSDS.

The authors feel that the MSDS is more comprehensive in its content as well as the direction it provides to U.S. HAZWOPER personnel. With the proper training, these workers are better served using an MSDS following the ASTM standard.

SECTION 10
CONFINED SPACES

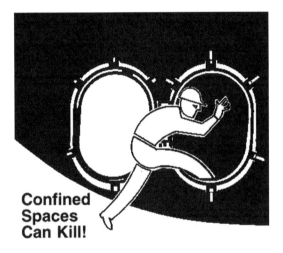

**Confined
Spaces
Can Kill!**

OBJECTIVES

The major objective of this Section is to explain to HAZWOPER personnel the hazards, atmospheric testing, entry, rescue, and working requirements for confined spaces and permit-required confined spaces, routinely called permit spaces. Permit spaces are a class of confined spaces requiring a specially prepared permit and its authorization before entry. The OSHA regulations pertaining to confined spaces are found in 29 CFR 1910.146.

Every year OSHA reports cases of injury and death resulting from confined spaces incidents. This is clear evidence that the hazards of confined spaces entry and work practices are not taken seriously enough by workers and supervisors.

As a HAZWOPER worker you will at one time or another probably be required to enter a permit space. For you as an Emergency Responder, detailed entry and victim rescue (extraction) from permit spaces also will be part of your additional training and periodic drills.

We find that too frequently in HAZWOPER worker training, confined spaces training is omitted. True, it is not part of the strictly required training under OSHA regulations. However, based upon long experience in HAZWOPER activities, both in training and in the field, the authors feel compelled to include this section as an integral part of a manual and desk reference. We have witnessed too many common misconceptions about confined spaces work. Another objective of Section 10 is to support the OSHA mission to reduce confined spaces incidents. The high number of casualties, injuries, and close calls yearly reported by OSHA, caused by insufficient training in confined space entry and rescue, is daunting. A sobering thought is the fact that well-meaning rescuers frequently become victims themselves.

In this Section you will learn how to define confined spaces and permit spaces. You must also understand the additional planning and preparation required for entry into a permit space. This Section will also describe the training necessary for the five main players on a permit space entry team: the confined spaces Supervisor, the confined spaces Inspector, the confined spaces Entrant, the confined spaces Attendant, and the confined spaces rescue personnel.

10.1 THE PERMIT-REQUIRED CONFINED SPACES (C/S) RULE (29 CFR 1910.146)

The final rule for General Industry C/S entry became effective in 1999. It amended 29 CFR 1910.146 of 1993 to provide for enhanced employee participation in the employer's permit space program.

OSHA specified that employers must provide certain employees with additional observation opportunities. Authorized permit space Entrants or their authorized representatives, usually union representatives, are required to be given an opportunity to observe any testing of the space that is conducted prior to, or subsequent to, entry. OSHA believed that these revisions were necessary to ensure that permit space Entrants, whose work often requires entry into potentially life-threatening atmospheres, have that additional information. This is information that could protect the workers and their coworkers from C/S hazards.

OSHA has the goal of ensuring that the preentry testing has been done properly; that the respirators and other personal protective equipment are appropriate; and that the entrants understand the nature of the hazards present in the space.

10.2 THE FIRST QUESTION TO ASK

The first question you must always ask yourself is, "Must I, or the people I supervise, enter the space in order to do the job safely?" If there is any other reasonable way to accomplish the assigned task without entering the C/S, DO IT THAT OTHER WAY!

☞ **Never enter a C/S if your task can be accomplished from the outside.**

Remote sampling is an example of another way. Ask yourself, "Can I lower a sampling container or probe into the space to collect the needed sample?" If at all possible, always work from the comparative safety of the outside of the C/S. There are a number of reasons for this. Entry into a C/S takes extra time, manpower, and equipment. Even a comparatively simple, short-duration job takes on greater proportions once C/S entry is involved. This is not to imply that the extra time and effort is not well spent. Unprepared entry may spell the difference between life and death!

10.3 A TRAGIC INCIDENT

Consider the following real-life example reported by NIOSH in a review of C/S fatalities. A two-person crew, a father and son, were assigned the job of detecting the sludge level in a sewage digester tank. Without further instruction, they opened the top of the tank and lowered an unprotected light bulb on an extension cord into the tank to see the level. The light bulb broke. The methane gas and air atmosphere in the tank exploded. The father and son were killed instantly.

As you read through this section, you will learn a number of required steps that were not taken by these two men. These steps, if taken in the proper sequence, would have, in all likelihood, saved their lives.

10.4 HOW OSHA DEFINES CONFINED SPACES

OSHA defines a confined space as any space that the Entrants can get into but that has limited, possibly small, openings for entry and exit. The space is not designed, nor intended for, continuous worker occupancy. Due to the space size, the configuration, and the service it is intended to provide, the Entrants may be in immediate proximity of a hazard without the ability to rescue themselves. Emergency rescue must be prepared and ready to act. These spaces oftentimes have unfavorable natural ventilation. This allows hazardous atmospheres to accumulate and concentrate. Always provide a fail-safe supply of breathing air for Entrants.

Consider what you know about a space you are thinking of entering or one for which you are supervising Entrants. Use the following checklist in your decision-making process.

10.4.1 Confined Space Determination Checklist

1. Is the space in question large enough that any of the assigned Entrants and Rescue workers can enter bodily with ease?

❑ YES
❑ NO

2. Is the space configured so that a worker can enter to perform the specific work task inside?

❑ YES
❑ NO

3. Does the space in question have limited or restricted means for entry or exit?

❑ YES
❑ NO

4. Is the space designed for continuous worker occupancy?

❑ YES
❑ NO

If your answers were YES, YES, YES, AND NO, then the space is a C/S by OSHA definition.

Examples of C/Ss include manholes, storage tanks, vats, holds of ships, process vessels, mixers, pits, silos, boilers, ventilation and exhaust ducts, sewers, tunnels, underground utility vaults, and pipelines (see Figure 10.1). Even a 55-gallon drum can be considered a C/S. This is because any part of your body that breaks the plane of the opening is exposed to the contents. These are all examples of C/Ss that a HAZWOPER worker could encounter. Figure 10.2 shows a properly labeled permit space.

Section 10.6 describes what to do when you determine that a space proposed for entry is a permit space.

⚷ **You must know the difference between a C/S and a permit-required C/S.**

10.5 HOW OSHA DEFINES PERMIT-REQUIRED CONFINED SPACES

Permit-required C/Ss (permit spaces) are C/Ss as defined above. In addition, they contain, or have a potential to contain, one or more of the following hazards:

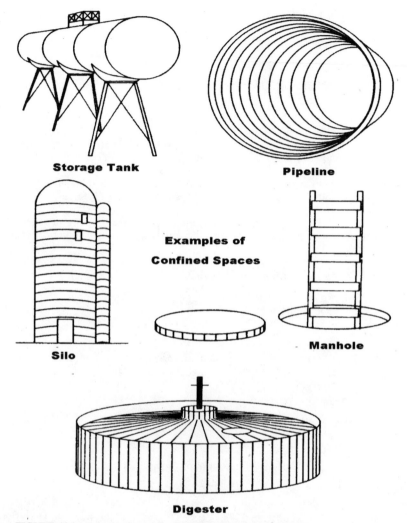

FIGURE 10.1   Examples of confined spaces. (*Courtesy of NIOSH.*)

1. A hazardous atmosphere
2. A material that has the potential for engulfing an Entrant
3. An internal configuration such that an Entrant could be trapped or asphyxiated by inwardly converging walls or by a floor that slopes downward and tapers to a smaller cross-section
3. Broken, deteriorated, or missing ladders
4. Spaces containing any other recognized serious safety or health hazard

WARNING

IN THE OPINION OF THE AUTHORS, THE MAJORITY OF C/Ss SHOULD BE DESIGNATED AS PERMIT SPACES. OFTEN THEY ARE NOT DESIGNATED AS SUCH. TREAT ALL C/Ss AS PERMIT SPACES UNLESS SPECIFICALLY INSTRUCTED OTHERWISE BY THE C/S SU-

FIGURE 10.2 One of the authors at a permit-required confined space opening. (*Courtesy of U.S. Filter.*)

PERVISOR. EVEN THEN, ALWAYS BE AWARE OF THE POTENTIAL HAZARDS THAT COULD TURN THAT SPACE INTO A PERMIT-REQUIRED SPACE!

10.5.1 Is This Space a Permit-Required Confined Space?

Consider carefully your answers to ALL of the following questions. Then check YES or NO.

1. Does the space in question contain a hazardous atmosphere or hazardous substances?

   ❏ YES
   ❏ NO

2. Does the space contain a material that has the potential for *engulfing* an Entrant? That is, does the space contain or have the possibility of filling with finely divided solids such as sawdust or filling with liquid?

   ❏ YES
   ❏ NO

3. Does the space have an internal configuration that could cause an Entrant to be trapped or asphyxiated by inwardly converging walls? (That is, could the Entrant become trapped such that chest expansion upon breathing would become difficult or impossible?)

   ❏ YES
   ❏ NO

4. Does the space have an internal configuration of a floor that slopes downward and tapers to a smaller cross-section?

   ❏ YES
   ❏ NO

5. Does the space contain any other known safety or health hazard?
   ❑ YES
   ❑ NO

If you answered YES to one or more of these questions, your C/S is a permit-required C/S by OSHA definition. Figure 10.3 shows OSHA's decision tree for permit spaces. Study it carefully before supervising or performing such an entry.

O━π **You must understand the hazards of C/Ss: hazardous atmospheres or substances, potential for engulfment or entrapment, or other detectable safety or health hazards.**

10.6   PERMIT SPACE HAZARDS

Examples of the hazards of permit spaces follow:

- *Hazardous atmospheres* are those that are flammable, toxic, irritant, or asphyxiating atmospheres. All of these conditions require specific PPE (Section 6).
- *Hazardous liquids on solids* left in the space can cause either chronic or acute exposures. Again, you need appraisal of the proper PPE for the job.
- *Engulfment* can occur when the potential exists for liquids, sludge, or finely divided solid materials to come pouring into the space. This can also occur when bridging of solids occurs. Solids are not always firmly packed into a space such as a powder bin. There may be air gaps bridging within the powder. When workers step into such a bin, they immediately fall down into the powder and are engulfed and then asphyxiated. Engulfment followed by drowning is also a serious hazard in low-density liquids. Do not count on being able to swim or float in aerated water, such as in a sewage treatment lagoon.
- *Entrapment or entanglement* hazards include converging walls or surfaces, crevices, pits, or unguarded machinery with belts or bands that could trap arms, legs, hands, feet, or the whole body.
- *General safety hazards* include:
   1. Unguarded rotating or moving machinery that could lead to crushing, entrapment and entanglement.
   2. Unguarded holes, tripping hazards, or otherwise unstable footing.
   3. Inadequate lighting.
   4. Unprotected live electrical transmission equipment such as exposed busses posing shock or electrocution hazards.
   5. The potential for drowning (especially in low-density liquids, such as some solvents or aerated sewage treatment liquids) or sudden flooding (such as in a storm sewer).
   6. A particularly difficult entryway such as a very small hole or vertical entry without a ladder. If so, question whether or not rescue can be performed efficiently and safely for all.

Figure 10.4 shows common atmospheric and physical C/S hazards.

10.7   HOW TO SET UP A CONFINED SPACES ENTRY PROGRAM PLAN

O━π **As a C/S entry team member (Supervisor, Inspector, Entrant, Attendant, or Rescue Team), you must understand all aspects of the C/S entry program.**

DANGER
CONFINED SPACE
ENTRY BY
PERMIT ONLY

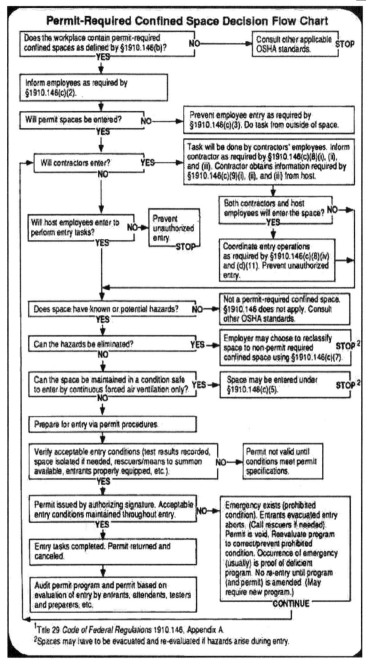

Permit-Required Confined Space Decision Flow Chart

Does the workplace contain permit-required confined spaces as defined by §1910.146(b)? **NO** → Consult other applicable OSHA standards. **STOP**

YES ↓

Inform employees as required by §1910.146(c)(2).

Will permit spaces be entered? **NO** → Prevent employee entry as required by §1910.146(c)(3). Do task from outside of space.

YES ↓

Will contractors enter? **YES** → Task will be done by contractors' employees. Inform contractor as required by §1910.146(c)(8)(i), (ii), and (iii). Contractor obtains information required by §1910.146(c)(9)(i), (ii), and (iii) from host.

NO ↓

Both contractors and host employees will enter the space? **NO** →

YES ↓

Coordinate entry operations as required by §1910.146(c)(8)(iv) and (d)(11). Prevent unauthorized entry.

Will host employees enter to perform entry tasks? **NO** → Prevent unauthorized entry **STOP**

YES ↓

Does space have known or potential hazards? **NO** → Not a permit-required confined space. §1910.146 does not apply. Consult other OSHA standards.

YES ↓

Can the hazards be eliminated? **YES** → Employer may choose to reclassify space to non-permit required confined space using §1910.146(c)(7). **STOP** [2]

NO ↓

Can the space be maintained in a condition safe to enter by continuous forced air ventilation only? **YES** → Space may be entered under §1910.146(c)(5). **STOP** [2]

NO ↓

Prepare for entry via permit procedures.

Verify acceptable entry conditions (test results recorded, space isolated if needed, rescuers/means to summon available, entrants properly equipped, etc.). **NO** → Permit not valid until conditions meet permit specifications.

YES ↓

Permit issued by authorizing signature. Acceptable entry conditions maintained throughout entry. **NO** → Emergency exists (prohibited condition). Entrants evacuated entry aborts. (Call rescuers if needed). Permit is void. Reevaluate program to correct/prevent prohibited condition. Occurrence of emergency (usually) is proof of deficient program. No re-entry until program (and permit) is amended. (May require new program.)

YES ↓

Entry tasks completed. Permit returned and canceled.

↓

Audit permit program and permit based on evaluation of entry by entrants, attendants, testers and preparers, etc.

CONTINUE

[1] Title 29 *Code of Federal Regulations* 1910.146, Appendix A.
[2] Spaces may have to be evacuated and re-evaluated if hazards arise during entry.

FIGURE 10.3   Permit-required confined space decision tree. (*Courtesy of NIOSH.*)

Loose, granular material stored in bins, silos and hoppers, such as
grain, sawdust, coal, or similar material, can engulf and then suffocate
a worker. The loose material can 'crust' or 'bridge' over in the storage
bin and break loose under the weight of the worker.

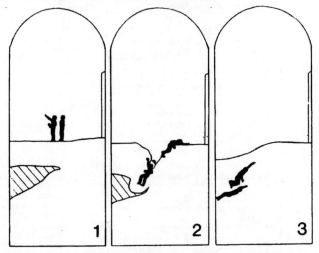

The Hazard of Engulfment is Unstable Material

FIGURE 10.4   Examples of atmospheric and physical confined space
hazards. (*Courtesy of NIOSH.*)

○━━ᴨ **As a C/S Entry Team member, you must know when a permit is required and
how it is issued.**

○━━ᴨ **As a C/S Entry Team member, you must understand the C/S entry plan and
the responsibilities of the other Entry Team members.**

To meet OSHA requirements, a C/S entry program must include a written plan and
certified training. As a part of training, this plan must be read (or read to), and understood

by, all involved personnel. These are the supervisors, workers, and contractors who will need to attend to Entrants and enter or perform rescue from permit spaces. Make the program a so-called living document. That is, periodically (at least yearly) provide updating and revising. Figure 10.5 shows a C/S tripod system used for vertical entry, exit, or rescue. Use the following checklist as you prepare the plan. The following parts of this Section provide detailed explanations of checklist items.

10.7.1 Confined Spaces Entry Program Plan Checklist

❑ Explain the duties and responsibilities of all members of the C/Ss team.

❑ Identify all C/Ss at the site.

❑ Determine whether or not they are permit spaces.

❑ Describe special procedures for IDLH atmospheres.

❑ Assess the hazards of each space.

❑ Describe all the hazard controls available to the members of the team.

❑ Describe the entry permit system. Include the permit form itself and step-by-step instructions for the proper filling out of the permit form, along with the names of the personnel authorized to complete and sign the permit forms and where copies of the forms are maintained.

❑ List the step-by-step procedures for opening and entering a C/S.

❑ List the step-by-step procedures for atmospheric testing, including the testing process, types of testing meters used by the Inspector, pre- and postcalibration of those meters, their proper use and maintenance, and record keeping.

❑ Describe the procedures for isolation of the C/S work area from all recognized hazards, including the lock-out/tag-out procedures.

❑ Explain the required warning signs along with a description of their placement at all permit-required C/Ss.

FIGURE 10.5   MSA® Lynx® portable entry and rescue tripod system. (*Courtesy of Mine Safety Appliances, Co.* [www.epartner.msanet.com].)

❑ Describe the initial training that each member of the team must complete before being allowed to perform the functions of Inspector, Entrant, Attendant, C/S Supervisor, or C/S Rescue Team member. Include the frequency of retraining, process of certification and recertification, training materials or service used, the process used to decertify a worker for cause, and record keeping.

❑ Explain the duties and responsibilities of the Rescue Team. Where available, many companies will choose an outside public or private rescue service.

10.7.2 OSHA-Required Special Procedures for IDLH Atmospheres

Note
See Section 5.24.4 for an explanation of IDLH.

To prepare for the special hazards of IDLH atmospheres, use the following checklist. Appendix F on the Companion CD is the NIOSH Chemical Listing for Immediately Dangerous to Life and Health (IDLH) Concentrations.

❑ Entrant(s) must be equipped with pressure demand or other positive pressure SCBAs, or a pressure demand or other positive pressure supplied-air respirator with auxiliary SCBA.

❑ In addition to the Attendant one employee or, when needed, more than one employee is located outside the IDLH atmosphere for rescue purposes.

❑ Visual, voice, or signal line communication is maintained between the employee(s) in the IDLH atmosphere and the Attendant located outside the IDLH atmosphere.

❑ The Rescue Team located outside the IDLH atmosphere are trained and equipped to provide effective emergency rescue.

❑ The employer or designee is notified before the employee(s) located outside the IDLH atmosphere enter the IDLH atmosphere to provide emergency rescue.

❑ The employer or designee, authorized to do so by the employer, once notified, provides necessary assistance appropriate to the situation.

❑ Rescuers located outside the IDLH atmospheres are equipped with:
   1. Pressure demand or other positive pressure SCBAs, or a pressure demand or other positive pressure supplied-air respirator with auxiliary SCBA; and either
   2. Appropriate means of rescue, or retrieval equipment, for removing any entrant from a hazardous atmosphere where retrieval equipment would contribute to the rescue of the employee(s) and would not increase the overall risk resulting from entry; or
   3. Equivalent means for rescue where retrieval equipment is not required.

10.7.3 Permit Space Team Members

As a permit space team member, you must know your duties and responsibilities and those of all the other members of the permit space entry team.

There must always be at least one outside Attendant even if there is only one Entrant into the C/S.

Each confined space team member must understand the duties and responsibilities of all team members. Team members are the Inspector, the Entrant(s), the Attendant, the C/S Supervisor, and Rescue personnel.

DANGER
CONFINED SPACE
ENTRY BY
PERMIT ONLY

The Duties and Responsibilities of the Confined Spaces Inspector. It is the duty of the Inspector, upon completing a permit space inspection, to authorize the proposed work plan and safety measures. Otherwise, the inspector requires work or safety changes and another inspection before authorization for worker entry.

The Inspector is the first person on-the-scene. The Inspector is sometimes classified as a gas-free engineer or gas-free engineering technician. "Gas-free," in this context, means the designated individual either certifies that the space has acceptable breathing air or specifies that contamination is present and then specifies the required PPE. Historically, this has been the most important aspect of the Inspector's job. See Figure 10.6. The Inspector, who is the first person to be exposed to any hazards contained in the space, must exercise extra caution. The Inspector must have completed current C/S training requirements and be qualified by documented training and experience to be the Inspector.

Use the following checklist to determine whether you, or the employee you designate, is a qualified Inspector. The Inspector must:

❑ Be able to recognize actual or potential physical, chemical, and atmospheric hazards and their sources—that is, hazards that are, or are likely to be, present in the permit space. These are hazards that require procedures to minimize or eliminate them altogether (such as all safety and rescue equipment and lock-out/tag-out devices).

❑ Have thorough knowledge of the selection, calibration, operation, and data logging of the appropriate atmospheric testing equipment. (Atmospheric testing equipment and sampling strategies are covered in depth in Section 12.)

FIGURE 10.6 The Inspector monitoring air inside a confined space from the entrance. (*Courtesy of EPA.*)

❑ Have a complete knowledge of sampling strategies and techniques, including testing stratified atmospheres and allowing sufficient response time for monitoring meters to collect and analyze the samples.

❑ Know how to physically inspect the entire permit space, including the atmospheres at all levels.

❑ Be knowledgeable about all available engineering controls (such as ventilation equipment) and know when to require those controls as a condition of entry.

❑ Be knowledgeable about the step-by-step procedures for atmospheric testing.

❑ Know the meaning and proper application of PELs, TLVs®, LELs, and UELs.

❑ Have the capabilities of a competent person, that is, a person who has the knowledge to specify the equipment and procedures necessary to perform a job safely as well as the authority to act on that knowledge.

WARNING

THE INSPECTOR MUST BE FULLY AWARE OF THE SCOPE OF THE WORK OPERATION THAT WILL TAKE PLACE INSIDE THE PERMIT SPACE. THE INSPECTOR MUST ALSO BE AWARE OF HAZARDS THAT ARE PRESENT OUTSIDE THE PERMIT SPACE. FOR INSTANCE, WELDING OR PAINTING OPERATIONS OUTSIDE THE SPACE MAY AFFECT THE ATMOSPHERE INSIDE THE SPACE.

TEAM MEMBERS MUST UNDERSTAND THAT IF THE SCOPE OF WORK CHANGES AND UNEXPECTEDLY A WELDER, PAINTER, OR OTHER TRADESPERSON IS REQUIRED TO WORK IN THE SPACE, THAT NULLIFIES THE PERMIT AND ANOTHER INSPECTION WILL BE REQUIRED. IN THAT CASE, THE INSPECTOR, AFTER REINSPECTION, MAY REQUIRE ADDITIONAL SAFETY EQUIPMENT AND MANPOWER. ADDED WELDING WORK WILL REQUIRE POINT-SOURCE VENTILATION AND A FIRE WATCH WITH AN EXTINGUISHER AT THE WORK SITE IN THE SPACE. AN ADDED PAINTING OPERATION WILL REQUIRE GENERAL AREA VENTILATION.

The Duties and Responsibilities of the Confined Spaces Entrant.   Use the following checklist to determine if all Entrants are qualified for duty. Have the Entrants:

❑ Completed current C/S training?

❑ A full understanding of the configuration of the space?

❑ The ability to recognize potential C/S hazards?

❑ Read and understood the entry permit requirements?

❑ The ability to stay alert to the hazards that could be encountered in a confined space?

❑ Understanding of how to use the equipment as required by the permit?

❑ Understanding of the importance of immediately exiting the C/S when:

    1. Ordered to do so by the Attendant?
    2. Automatic alarms sound?
    3. They perceive an unexpected dangerous condition?
    4. They notice physiological stresses or unexpected changes in personal health or that of coworkers (e.g., dizziness, blurred vision, shortness of breath)?

The Duties and Responsibilities of the Confined Spaces Attendant.   The Attendant must have completed current C/S training. The Attendant must remain stationed at the entrance to the permit space continually, as long as there are Entrants in the space. Use the following checklist to determine if a candidate Attendant is qualified. Is the Attendant:

☐ Knowledgeable about the configuration of the space?

☐ Knowledgeable about, and able to recognize, potential C/S hazards?

☐ Prepared to maintain a sign-in/sign-out log with a count of all persons in the C/S and ensure that all entrants sign in/sign out?

☐ Able to monitor activities surrounding the space to ensure the safety of entry personnel?

☐ Knowledgeable about the requirement to report unsafe conditions or unauthorized entry immediately to the Entrant and the Supervisor, in that order?

☐ Able to maintain effective and continuous communication with Entrants during C/S entry?

☐ Capable of ordering personnel to evacuate the C/S upon:

  1. Observing a condition that is not allowed on the entry permit?
  2. Noticing the Entrants acting strangely, possibly as a result of exposure to hazardous substances?
  3. Noticing a situation outside the C/S that could endanger entry personnel?
  4. Noticing within the C/S a hazard that has not been previously recognized or taken into consideration?
  5. Having to leave the work station?
  6. Having to focus attention on the rescue of personnel in some other C/S that the Attendant is monitoring?
  7. Capable of immediately summoning the Rescue Team if Entrant rescue becomes necessary?
  8. Capable of keeping unauthorized persons out of the C/S without leaving the post?

WARNING

AT NO TIME MAY THE ATTENDANT ENTER THE SPACE OR LEAVE HIS OR HER POST, FOR ANY REASON, INCLUDING RESCUE OF THE ENTRANT—THAT IS, WITHOUT THE ATTENDANT EITHER HAVING THE ENTRANTS EVACUATE THE SPACE OR BEING RELIEVED OF ATTENDANT DUTY BY ANOTHER TRAINED ATTENDANT.

The Duties and Responsibilities of the Confined Spaces Rescue Team

Note

The Attendant and Entrant can be seen in Figure 10.7 wearing belt-attached escape cylinders. The Entrant is seen wearing an air-line respirator with air supplied by a cart at the entrance. The Inspector is directly behind the Entrant, with his gas meter on a strap around his neck, while an additional Entrant is prepared for entry beside him.

The permit space Rescue Team must have completed current C/S training, plus additional specialized permit space rescue techniques training. They must be physically fit and have the strength to retrieve an unresponsive worker. They must have a complete knowledge of all the special extraction gear necessary for C/S rescue.

OSHA requires employers whose employees have been designated to provide permit space rescue and emergency services to take the following measures:

1. Provide affected employees with the personal protective equipment (PPE) needed to conduct permit space rescues safely and train affected employees so they are proficient in the use of that PPE, at no cost to those employees.

2. Train affected employees to perform assigned rescue duties. The employer must ensure that such employees successfully complete the training required to establish proficiency as an authorized Entrant.

FIGURE 10.7 A vertical confined space entry or rescue using nonentry retrieval tripod, Scott AIR PAK air-line respirators supplied by portable Scott Air Cart, and Scott SKA-PAK escape cylinders. (*Courtesy Scott Health and Safety* [www.scotthealthsafety.com].)

3. Train affected employees in basic first aid and cardiopulmonary resuscitation (CPR). The employer must ensure that at least one member of the Rescue Team or service is available holding a current certification in first aid and CPR.

4. Ensure that affected employees practice making permit space rescues at least once every 12 months. They do this by means of simulated rescue operations in which they remove dummies, manikins, or actual persons from the actual permit spaces or representative permit spaces. Representative permit spaces must, with respect to opening size, configuration, and accessibility, simulate the types of permit spaces from which rescue is to be performed.

Note

Often the expertise necessary to form an adequate Rescue Team is not available to the employer. In such cases, the use of either a private Rescue Team or a qualified public Rescue Team such as a local fire department may be required. If this is the case, additional coordination and training must be conducted with the outside agency to ensure that they will be able to respond to any emergency that might evolve in the specified time frame. Rescue personnel must be able to respond to the site in a timely manner. OSHA states that the host company must supply the Rescue Team. That is, a contractor is not obligated to provide a Rescue Team, even though conducting C/S operations.

There has been much discussion about the term *timely*. Many have argued that if a rescue cannot be carried out within four to six minutes, the Entrant could die or suffer irreversible harm. OSHA believes that in order for a person to suffer that degree of harm, the atmosphere would have to be an IDLH atmosphere. OSHA further states that entering an IDLH atmosphere is already covered in the respiratory standard. A fully outfitted Rescue Team must support that entry.

Evaluation and selection of a C/S rescue service (private or public) is a serious undertaking that deserves much thought and input from all concerned parties. Evaluate the prospective rescue service's ability to:

- Carry out rescue-related tasks knowledgeably and expediently
- Make the proper selection and use of all necessary C/S rescue equipment
- Rescue Entrants expediently and effectively from the particular permit space or types of permit spaces identified at your site

Select a rescue team or service from those evaluated that:

- Has the capability to reach the victim(s) within a time frame that is appropriate for the permit space hazard(s)
- Is equipped for, and proficient in, performing the needed rescue services
- Is informed of, and able to deal safely with, the hazards they may confront when called on to perform rescue at the site

As the host site supervisor, you must provide the Rescue Team or service selected with preparatory access to all permit spaces from which rescue may be necessary. This allows the rescue service to develop appropriate rescue plans and practice rescue operations.

Nonentry Rescue. If your organization should decide to use nonentry rescue, the following guidelines must be followed:

- Retrieval systems must be used whenever an authorized Entrant enters a permit space. If retrieval equipment would increase the overall risk of entry or would not contribute to the rescue of the entrant, a Rescue Team is required. Retrieval systems must meet the following requirements:
- Each authorized Entrant shall use a chest or full-body harness, which must have a retrieval line attached at the center of the Entrant's back near shoulder level, above the Entrant's head, or at another point that presents a profile small enough for the successful removal of the Entrant.
- 'Wristlets' (wrist attachments to rescue lines) may be used in lieu of the chest or full-body harness if the use of a chest or full-body harness creates a greater hazard and the use of wristlets is the safest and most effective alternative.
- The other end of the retrieval line must be attached to a mechanical device or fixed point outside the permit space. It must be positioned in such a manner that rescue can begin as soon as the rescuer becomes aware that rescue is necessary.
- A mechanical device must be available to retrieve personnel from all vertical entry permit spaces more than 5 ft (1.52 m) deep.

Possibly some of our recommendations will appear controversial. The authors believe that a standby Rescue Team is an ideal situation, but also understand that it may be impractical for an uncomplicated permit space entry. However, timely first aid service, with training in rescue, is a requirement for all work sites.

The Confined Spaces Supervisor. The C/S Supervisor must have completed current confined space training providing a thorough knowledge of all aspects of the C/S program. Furthermore, the Supervisor must be able to show, through documented training and experience, an awareness and understanding of all aspects of the job descriptions of the Inspector, Entrant, Attendant, and Rescue personnel. The C/S Supervisor has the ultimate responsibility for the safety and health of the members of the C/S team. One of the worst mistakes a

Supervisor can make is to assign too many inspections to one Inspector on one shift. The Supervisor must be knowledgeable as to the amount of time it takes to safely inspect the assigned spaces thoroughly and not assign more than that specific Inspector can handle. See the "Carnac Syndrome," described in Section 10.8.2.

Coordination between a Host Company and Its Contractors.   Coordination between two or more employers who have employees entering a particular permit space is required by 1910.146(c)(8)(iv), (c)(9)(ii), and (d)(11). The host employer who arranges for a permit space entry by contractor employees has a duty to instruct the contractor on the hazards that make the space a permit space. The contractor who will have employees enter the permit space is responsible for obtaining that information prior to entry. All employers who will have employees in the permit space must have procedures to coordinate entry operations. For example, that responsibility extends to determining operational control over the space, employee training, and provisions for rescue and emergency services. Any one of the employers having employees enter the permit space could have operational control over the permit space. According to OSHA, all parties (host employer and contractors) retain responsibility for the protection of their own employees even though all the employers have agreed to a specific permit space-controlling employer.

CAUTION

THERE SHOULD BE ABSOLUTELY NO DOUBT BY ANY PERMIT SPACE ENTRANT, ATTENDANT, OR ENTRY SUPERVISOR REGARDING WHO THE CONTROLLING EMPLOYER IS AND WHOSE POLICY AND PERMIT SPACE PRACTICES ARE TO BE FOLLOWED.

One of the most troublesome areas in the design, implementation, and successful (read "safe") operation of a C/S program is coordination between host companies and contractors. This is universally true throughout corporate safety programs.

A host company has a measure of control over its own employees, and an ability to train them, that it does not have over contractors. Contractors often are only on-site for a matter of hours, days, or weeks. The following general attributes of contractor personnel can spell disaster.

- They can be under extreme time constraints for completion of their task.
- The host company for whom they are working does not, for the most part, train them.
- They are relatively unfamiliar with the physical layout of the facility.
- They can be relatively unfamiliar with the emergency procedures of a host facility. Warning alerts such as bells, horns, flashing lights, and whistles can prove somewhat puzzling. Contractors can also be unfamiliar with the emergency escape routes and muster areas. (Muster areas are the safe places where evacuated workers gather for a head count during emergencies. Supervision then determines whether or not they have to begin immediate search and rescue.)

The problems can be further exacerbated by confusion over who has control of the worksite. OSHA requires that the host employer who arranges for a permit space entry by a contractor's employees have a duty to instruct the contractor on the hazards and other factors that make the space a permit space. The contractor who will have employees enter the permit space is responsible for obtaining that information prior to entry.

As contracting-out services gain popularity daily with corporate America, companies must plan carefully for these outside personnel in their safety program. Companies deal with contractors in various ways. Some companies may show their contractors a short video, or worse yet, simply have each contractor read and sign a short outline of the host company's safety program. The authors have worked a few times at such facilities and have been appalled at these short-sighted and dangerous practices.

The authors strongly urge a more comprehensive approach. Some companies require a week-long safety indoctrination into the host company's safety policy and procedures. Others require that contract workers show either a 10-hour or 30-hour OSHA Outreach Training course completion card for Construction or General Industry safety. The authors feel that either of these requirements fulfills more closely the minimum level of training for contractors prior to the specialized permit space entry training.

10.7.4 Identification and Evaluation of Permit and Non-Permit Confined Spaces

All confined spaces at a site must be identified. Further, once a space is designated a permit space, it must be labeled accordingly and designated on a map of the site. This includes giving each permit space a unique identification. The authors agree that, if feasible, the configuration of the space should also be mapped. This can often be accomplished by utilizing as-built drawings, or sketches drawn by prior Entrants.

A satisfactory sign indicating that a C/S is a permit-required C/S is shown in Figure 10.8. The signage must be consistent throughout the facility. This includes size, exact wording, and coloring ("DANGER" in white on red background, other lettering in black on white background). Signs must be permanently mounted at all entrances to the space. They must be placed so that entry without viewing the sign would be impossible. Ideally this is directly above or beside the entrance. Danger signs must never be mounted on a removable hatch or opening door. Danger signs must also be maintained in legible condition (not painted over, for instance) and mounted in such a place as to not hinder entry or exit or be obscured or defaced by equipment such as ventilation tubing.

Note

Manhole covers in roadways are a notable exception. There is no signage required because they are universally considered permit-required C/Ss.

Each space must also be individually evaluated for actual or potential hazards. The determination must then be made as to whether the space is a permit space. The determination that a space does not require a permit for entry must not be taken lightly. It must be supported by documentation provided by a Certified Gas Free Engineer, Certified Industrial Hygienist (CIH), Certified Safety Professional (CSP), or other occupational safety and health professional. This is for the safety of all workers and the legal protection of the facility.

Increasingly, more companies are considering all their C/Ss permit spaces. They acknowledge that eliminating errors and providing a higher level of employee safety outweigh the extra cost of this decision.

Hazard Assessment for Permit Spaces.   All possible hazards must be identified before proceeding with the opening, atmospheric testing, and subsequent entry into the permit space. The following are the likely hazards and conditions to be considered when assigning or preparing for work in a permit space:

FIGURE 10.8 OSHA-approved permit space danger sign.

Atmospheric hazards:

1. Oxygen content
   - Oxygen deficiency (less than 19.5%)
   - Oxygen enrichment (greater than 23.5%)
2. Explosive gases/vapors
3. Toxic gases/vapors/mists
4. Fumes, dusts, mists, fogs, and smoke

Biological agents:

1. Raw sewage
2. Biohazard materials
3. Vermin (such as rats, fleas, lice, ticks)
4. Poisonous animals, snakes, and spiders
5. Dead and decaying animals

Physical and environmental safety hazards, such as:

1. Entry and exit (access and egress) difficulties
2. Configuration problems (some spaces are wide open, some are like a maze)
3. Ventilation system requirements
4. Unguarded machinery
5. Unblanked (blinded) piping
6. Unbraced and unblocked distribution systems
7. Presence of residual chemicals or hazardous materials
8. Exposed electrical transmission equipment
9. Poor visibility
10. Physical obstacles
11. Uneven, rough, unsupported, or slippery walking or working surfaces
12. Temperature extremes
13. High humidity
14. Loud noise
15. Vibration
16. Radiation

Type of work to be performed, such as:

1. 'Hot work' (including welding, burning, chipping, grinding, sanding, or any other work that is likely to produce a spark or generate temperatures greater than 400°F)
2. Chemical, steam, or high-pressure cleaning
3. Painting

Human factors, including:

1. Claustrophobia
2. The mental and physical condition of workers

To summarize, there are two broad categories of hazard that an entrant is likely to encounter in a permit space. First, beware of *hazardous atmospheres,* including those in which flammable, toxic, irritant, and asphyxiating conditions are present. Also beware of chronic and acute exposure to hazardous materials (see Section 5). Second, be on guard for *general safety hazards,* including difficult entry and exit, the potential for engulfment in finely divided solid materials, entrapment, drowning (especially in low-density liquids or sudden flooding in a storm sewer), crushing, entrapment and entanglement in unguarded machinery, and electrocution.

What You Can Learn from Previous Use of a Space.   The fact that any space has been previously designated a permit space means that some hazard exists or is likely to exist. Much of the hazard assessment can be performed without even going into the space. Learn as much as you can about previous uses of a space to guide your permit preparation. For example, consider the following:

- What was stored last in the space? For example, wet activated carbon, used in water purification, when stored in an enclosed space will adsorb oxygen, rendering the space oxygen deficient.

- What operations were last performed in the space? If a tank was just steam cleaned and is still hot, will there be an oxygen deficiency? Also, if a steel space was recently sand-blasted and then closed, the production of rust (iron oxide) is an oxidation reaction. Rusting can, in as little as a few days, reduce the oxygen content to as low as 2–3%! Further, if painting occurred in the space and the paint was not allowed to fully dry or cure before the space was closed, there may be an oxygen deficient, explosive or toxic atmosphere in the space.

When was the space used last? If a space has been closed for a long period and even if it only contained a small amount of water and organic sludge, it might now contain an explosive atmosphere of methane or explosive or toxic levels of hydrogen sulfide.

What is the nature and design of internal machinery and equipment in the space? Because the space was not designed for human occupation, normal safeguards may not be present. Rotating machinery may not be guarded, posing an entrapment or entanglement hazard. The same is true for lack of protection from electrical busses.

☛ **Safeguards that are second nature outside of a C/S will often not be in place inside the space.**

As you can see, the hazards of many spaces can be assessed without even opening the space. Entry will be discussed in more detail in a later subsection.

Improper Work Practices That Have Resulted in Horror Stories.   The following are routinely observed improper work practices that have led to injuries and deaths in C/Ss:

- Entering a C/S without testing the atmosphere

- 'Inerting' a C/S such as to force a liquid out of a tank and not ventilating it prior to entry

- 'Pressing up' a C/S (in order to test for leaks) using nitrogen gas and not ventilating it before entry

- Complete lack of, or insufficient, ventilation

- Using oxyacetylene or oxy-Mapp gas torches, hoses, valves, and regulators in a C/S without periodically checking for leakages

- Using oxygen to ventilate a C/S

- Not investigating the effect of stirring up sludges in a C/S

- Using improper respiratory protection

- Not checking nearby processes for possible release of toxic or flammable material

- Welding in a tank without checking neighboring compartments
- Not blanking, blinding, or disconnecting lines entering a tank
- Not performing a thorough lock-out/tag-out procedure (see Section 11)
- Leaving a C/S that has been tested safe for entry and reentering it later without retesting
- Bypassing proper rescue procedures to perform a "heroic" rescue

Hazard Controls.   All known hazards of the C/S to be entered must be eliminated to the extent feasible. Identification of hazards is the responsibility of both the Inspector and the Entrant. The two main hazard controls, ventilation and hazardous energy control, are described below.

○━π   **Never enter an IDLH atmosphere to perform routine tasks.**

Ventilation with Fresh, Breathing Air.   Because of poor natural ventilation, higher concentrations of toxic or flammable gases or vapors are likely to exist in a C/S than at an open site. Natural ventilation will almost always be absent, particularly if there is only one entrance into a space. In addition, C/Ss frequently contain other hazardous or suffocating gases. For example, hydrogen sulfide, carbon monoxide, and methane are often found in sewers. Organic material in enclosed spaces tends to consume the oxygen in the surrounding air as it decomposes. This produces an oxygen-deficient atmosphere.

A common contributor to oxygen reduction is the high iron content in structural steel. A common steel tank can be sandblasted perfectly clean and then closed up tight. If there is sufficient moisture in the air in the tank, the tank surfaces will oxidize to form iron oxide (rust). This process has been found to lower the oxygen content in a tank to less than 3% in only two weeks! Every pound of iron oxidizing to rust will remove nearly 200 ft^3 of oxygen from the air. This means that a 1000 ft^3 steel tank C/S, initially full of fresh air, could be nearly devoid of oxygen if just 2 lb of iron rusted.

Purging and inerting (see Glossary) are other common practices that will require ventilation before entry. Process vessels and tanks are often purged of gases, vapors, or mists. The atmosphere is then replaced by nitrogen, an inert gas. Unfortunately, it is also an asphyxiating gas. This process is used frequently to prepare the vessel for refilling, exterior maintenance, or repair such as welding. An inerted vessel must be adequately ventilated with fresh air in order to ensure safe entry.

○━π   **Forced exhaust ventilation of a C/S is the safest way to remove a hazardous atmosphere.**

There are three types of ventilation: supply, exhaust, and natural. Supply ventilation is the blowing of outside air *into* a C/S forcing out bad air. Exhaust ventilation is sucking the bad air *out of* the space, allowing clean, breathable air in. This includes local exhaust ventilation. Local exhaust ventilation removes the contaminants (welding fumes, or paint vapors for instance) from their source so they do not impact the rest of the space. Natural ventilation is the free movement of air within a space due to wind, pressure differential, or chimney or stack effect. When good natural ventilation conditions exist, simply opening up a space and allowing it to air out may be sufficient. In situations where poor or no natural ventilation occurs, use exhaust or supply mechanical ventilation.

WARNING

ALTHOUGH BOTH SUPPLY AND EXHAUST VENTILATION ARE ALLOWED BY OSHA, THE AUTHORS' EXPERIENCE LEADS US TO RECOMMEND EXHAUST VENTILATION ONLY. SUPPLY VENTILATION CAN CAUSE A FIRE, ONCE STARTED, TO SPREAD MORE RAPIDLY THROUGHOUT THE SPACE, MAKING EXITING AND RESCUE FROM THE SPACE THAT MUCH MORE DIFFICULT.

🔑 **OSHA requires four air changes per hour in the workplace. When you are determining ventilation requirements, the volume of air requiring movement multiplied by four air changes per hour must be included in your calculations.**

The design of a C/S ventilation system can be a difficult task. Use this checklist to ensure that your design includes the standards and variables you must consider when designing a C/S ventilation system.

Confined Spaces Ventilation Checklist

❑ Step 1: Determine, by atmospheric and visual inspection, what, if any, contaminants are inside or outside of the space. Will any contaminants be expected to be generated inside or outside the space, such as paint vapors and mists, welding, or other operations? If not, go to Step 2. If so, go to Step 3.

❑ Step 2: If the above does not require ventilation, and natural ventilation has been determined to be adequate, STOP HERE—you do not require additional ventilation.

❑ Step 3: You have determined that ventilation is required. Follow the rest of this checklist.

❑ Step 4: If the Inspector determines that ventilation is required to keep the proper level of breathing oxygen, and acceptable levels of LEL/LFL (see Section 3.2.4) and toxics in the space, the ventilation MUST be powered by an uninterruptible power source (UPS).

❑ Step 5: Determine the volume of the space.

❑ Step 6: Determine the amount of ventilation needed to supply the OSHA-required four air changes per hour. (To ensure proper ventilation, multiply the volume of the space times four. That is the amount of air your blower will have to move every hour. The authors always use 80% of the rated capacity of a blower as a safeguard.

❑ Step 7: Use large enough-diameter flexible ventilation tubing, at least 6 in. inside diameter, and install in such a way as to prevent kinks and tight bends. Those measures reduce air flow friction and help ensure adequate flow. Also, where possible, install the flexible tubing so that it does not become a tripping hazard or become crushed or broken. Tying the tubing up off the floor of the space is best.

❑ Step 8: Determine the configuration of the space. Every C/S is not a simple box or tank. There are also spaces containing baffles, piping and electrical systems, and labyrinthine paths. These are particularly hard to ventilate properly. Large or segmented C/Ss often require several ventilation points.

❑ Step 9: Determine the number of entryways into the space. If there is only one opening, exhaust ventilation should be used, exhausting bad air from the farthest point in the space.

❑ Step 10: Determine if there are multiple openings. Some of those openings may have to be blanked off to assure that bad air is being exhausted, not mostly fresh air from one of those openings. If there are two openings at opposite ends of the space and both openings are in a fresh air environment, exhausting from one opening will sometimes give ideal ventilation.

CAUTION

FOR BETTER ESCAPE OPTIONS, EXTRA OPENINGS SHOULD ONLY BE BLANKED OFF WITH AN EASILY REMOVABLE BARRIER ('SOFT COVER') SUCH AS POLYETHYLENE AND DUCT TAPE. THIS WILL NOT ALLOW THE AIR IN BUT CAN BE EASILY REMOVED FOR ESCAPE PURPOSES. IF ANY OF THESE EXTRA OPENINGS ARE ON WALKING SURFACES, THEY WILL HAVE TO BE GUARDED WITH RAILS TO PROTECT OTHERS FROM FALLING THROUGH THE 'SOFT COVER'.

- ☐ Step 11: Determine the type of atmosphere that is being exhausted from the space.
- ☐ Step 12: Does the atmosphere of the space contain flammable gas or vapor above the UEL/UFL?
- ☐ Step 13: If the answer to the above is YES, then the ventilation will eventually dilute the atmosphere until it is in the explosive range. In this case, explosion-proof ventilation will be required. Special care must also be taken to ensure that the exhausted, flammable gases or vapors do not cause a hazard outside the space.
- ☐ Step 14: Does the atmosphere in the space contain any toxic gases or vapors?
- ☐ Step 15: If so, ensure that the ventilation stream is exhausted where it will not harm other workers outside the space.
- ☐ Step 16: Determine what work will be conducted inside the space.
- ☐ Step 17: Will that work contribute to the contamination of the atmosphere in the space? Examples are welding and painting.
- ☐ Step 18: If so, take extra precautions such as point-source ventilation.
- ☐ Step 19: Will work activities being conducted outside the space, such as spray painting, cause a possible hazardous atmosphere inside the space? You must be careful not to draw a hazardous atmosphere into the space.
- ☐ Step 20: If so, then you will need to extend make-up air ventilation tubing to an area free of that contamination.
- ☐ Step 21: You must also be aware of "short circuiting," that is, drawing make-up air into the space from the air exhausted from the space. This concentrates the contaminants in the space. For that reason, exhaust from the space must be downwind a minimum of 50 feet from the make-up air intake source.
- ☐ Step 22: Are there internal combustion engines (such as vehicles, backhoes, compressors, and generators) in the work area? None of these may be permitted to exhaust within 50 feet from any opening into the C/S, or make-up air intake source.

An Example of a Confined Space Tragedy. Two workers were attempting to dewater an underground utility vault using a gasoline-powered pump. As the work progressed, they needed more hose to reach farther into the space. Instead of going back to their shop to obtain more lengths of hose, they moved the pump right next to the manhole. Both workers succumbed to carbon monoxide poisoning and were later found dead at the site.

The most important things to remember about C/S ventilation are that adequate fresh air must be supplied to all workers, allowing no dead spots, and that if ventilation is required to maintain acceptable atmospheric conditions within the space, it must have an uninterruptible backup power source (UPS), in case of power failure.

Control of Hazardous Energy. C/Ss are often prone to hazardous energy release without warning. Because the space is not intended for human habitation, it may not be designed with the normal safeguards that an occupied space would contain. Electric transfer busses are often left exposed, rotating machinery is often unguarded, and pipes often empty their contents into the space without warning. For this reason, every organization needs a hazardous energy control system. The methods and procedures used are commonly referred to as lock-out/tag-out. Please see Section 11 for a detailed explanation of a lock-out/tag-out program.

C/Ss Preentry Checklist. Use the C/Ss preentry checklist (Table 10.1) to aid supervision and the entry team members in assuring that safety precautions are met. Copy the checklist. Add questions to the checklist to customize it to the particular space being considered. For any NO answer, take corrective action to convert that answer to YES before you allow any entry.

TABLE 10.1 Confined Spaces Preentry Checklist

❏ Are appropriate atmospheric tests performed to check for oxygen deficiency, flammability, and toxic substances concentrations in the C/S before entry?

❏ Is either sufficient natural, or mechanical, ventilation provided prior to C/S entry?

❏ Are C/Ss thoroughly emptied of any corrosive or hazardous substances such as acids or caustics before entry?

❏ Are all lines to a C/S containing inert, toxic, flammable, or corrosive materials valved off and blanked or disconnected and separated before entry?

❏ Are all impellers, agitators, or other moving parts and equipment inside C/Ss locked out and blocked and chocked if they present a hazard?

❏ Is adequate illumination provided for the work to be performed in the C/S?

❏ Will the atmosphere inside the C/S be frequently tested or continuously monitored during conduct of work?

❏ Is there an assigned Attendant outside of the C/S whose sole responsibility is to watch the work in progress, sound an alarm if necessary, and render assistance?

❏ Is the Attendant appropriately trained and equipped to handle an emergency?

❏ Are rescue personnel prohibited from entering the C/S without lifelines and respiratory equipment, no matter what emergency occurs?

❏ Is approved respiratory equipment provided and required to be worn if the atmosphere inside the C/S cannot be made acceptable?

❏ Is all portable electrical equipment used inside C/Ss either grounded and insulated or equipped with ground fault protection?

❏ Before gas welding or burning is started in a C/S, are hoses 'drop tested' (checked for leaks), and compressed gas cylinders (except fire extinguishers) forbidden inside of the C/S?

❏ If employees will be using oxygen-consuming equipment, such as gas- or oil-fired space heaters, torches, and furnaces, in a C/S, is sufficient fresh air provided? That is, sufficient to ensure combustion without reducing the oxygen concentration of the atmosphere below 19.5% by volume?

❏ Whenever combustion-type equipment is used in a C/S, are provisions made to ensure the exhaust gases are vented outside of the enclosure (such as extended exhaust hoses)?

❏ Is each C/S checked for decaying vegetation or animal matter that may produce methane?

❏ Is the C/S checked for possible industrial waste having toxic properties?

❏ If the C/S is below the ground and near areas where motor vehicles will be operating, is it possible for vehicle exhaust or carbon monoxide to enter the space?

Procedures for Atmospheric Testing of Confined Spaces.   Atmospheric testing is required for two distinct purposes: evaluation of the atmospheric hazards of the permit space and verification that acceptable conditions exist for entry into that space. The reference for these required procedures is 29 CFR 1910.146, Appendix B.

1. *Evaluation testing:* The atmosphere of a permit space must be analyzed using equipment that can identify and evaluate any hazardous atmospheres that might occur. Then permit entry procedures can be developed for that space. Analyzing equipment must be calibrated both before and after testing, and the results logged. Evaluation and interpretation of these data and development of the entry procedure must be performed by, or reviewed by, a technically qualified professional. Examples are a Certified Industrial Hygienist (CIH), Registered Safety Engineer, and Certified Safety Professional.

○━╥   **You must know the atmosphere testing order: first oxygen, then LEL for flammables, then toxics.**

2. *Verification testing.* Following operation of the proposed ventilation method, test the atmosphere of a permit space for residues of all contaminants identified by evaluation testing. Permit-specified equipment must be used to determine that residual concentrations at the time of testing and entry are within the range of acceptable entry conditions.

Testing order will be: first oxygen, second flammables, and finally toxics. (29 CFR 1910.146 (cX5)(llXC) and (d)(5)(iii)). Results of testing (actual concentrations in percent, ppm, or ppb) must be recorded on the permit.

Note

Refer to the sample permit forms in Table 10.2. This testing order is specified due to the fact that the meters used to test the atmosphere require an acceptable amount of oxygen in order to perform properly, the reason being that they have been calibrated in a normal oxygen-content atmosphere containing 20.9% oxygen.

WARNING

IF THE TEST ORDER IS NOT FOLLOWED, THE RESULTS OF THE LEL OR TOXICS READINGS MAY BE MEANINGLESS.

3. *Duration of testing.* Take all measurements for at least the minimum response time of the test instrument specified by the manufacturer, which is the time it takes for the instrument to give acceptably accurate readings.

4. *Testing stratified atmospheres.* Remember that gases heavier than air, such as hydrogen sulfide, seek the lowest level, while gases lighter than air, such as hydrogen, will be found at the highest levels. Gases with approximately the same weight as air, such as carbon monoxide, will be found, if at all, in the middle of this stratification.

Note

An example is a cave in Europe known as the "Cave of the Dead Dog." People could visit the cave with no problem, but dogs that wandered in were found dead. Carbon dioxide and other toxins more dense than air tended to form a lower layer in the cave atmosphere, too low to be a danger to erect people.

When monitoring for entries involving a DESCENT into atmospheres that may be stratified, the atmosphere must be tested at a distance of approximately 4 ft (1.22 m) in the direction of travel and to each side. If a sampling probe is used, the Inspector's rate of progress shall be slowed to accommodate the sampling speed and detector response.

5. Periodically retest to verify that the atmosphere remains within acceptable entry conditions.

10.8 CONFINED SPACE ENTRY

○━━π **You must know the definition of C/S entry, whether or not the space is a permit space. Entry is defined as when any part of your body breaks the plane of the opening.**

Safe procedures for entry into a permit-required C/S are quite specific. They are found in 29 CFR 1910.146.

TABLE 10.2 Two Examples of C/S Entry Permits Meeting OSHA Requirements

C/S ENTRY PERMIT VERSION ONE

Date and Time Issued: _____ Date and Time Expires: _____
Job site/Space I.D.: _____ Job Supervisor: _____
Equipment to be worked on: _____ Work to be performed: _____
Attendant personnel: _____
  1. Atmospheric Checks: Time   _____
               Oxygen   _____%
               Explosive _____% LEL
               Toxic   _____PPM

  2. Inspector's signature: _____

  3. Source isolation (No Entry):   N/A Yes No
     Pumps or lines blinded, disconnected, or blocked   ()   ()   ()

  4. Ventilation Modification: N/A Yes No
    Mechanical Exhaust   ()   ()   ()
    Mechanical Supply   ()   ()   ()
    Natural Ventilation only   ()   ()   ()

  5. Atmospheric check after isolation and Ventilation:
    Oxygen _____%   > 19.5%
    Explosive_____% LEL   < 10%
    Toxic _____PPM   < 10   PPM hydrogen sulfide
Time _____
Inspector's signature: _____

  6. Communication procedures: _____
_____

  7. Rescue procedures: _____
_____
_____
_____

  8. Entry, Attendant, and Rescue Personnel: Yes No
    Successfully completed required training?   ()   ()
    Is it current?   ()   ()

  9. Equipment: N/A Yes No
    Direct reading gas monitor
     Calibrated and tested?   ()   ()
    Safety harnesses and lifelines
     for Entry and Attendant personnel   ()   ()   ()
    Hoisting equipment   ()   ()   ()
    Powered communications   ()   ()   ()
    SCBAs for Entry and Attendant personnel ()   ()   ()

    Protective Clothing   ()   ()   ()

All electric equipment listed
Class I, Division I, Group D
and Non-sparking tools   ()   ()   ()

TABLE 10.2   Two Examples of C/S Entry Permits Meeting OSHA Requirements (*Continued*)

10. Periodic atmospheric tests:

Oxygen	____%	Time ____	Oxygen	____%	Time ____
Oxygen	____%	Time ____	Oxygen	____%	Time ____
Explosive	____% LEL	Time ____	Explosive	____% LEL	Time ____
Explosive	____% LEL	Time ____	Explosive	____% LEL	Time ____
Toxic	____%	Time ____	Toxic	____%	Time ____
Toxic	____%	Time ____	Toxic	____%	Time ____

We have reviewed the work authorized by this Permit and the information
contained herein. Written instructions and safety procedures have been
received and are understood. Entry cannot be approved if any squares are marked
in the ''No'' column. This Permit is not valid unless all appropriate items are
completed.

Permit Prepared By: (Inspector) _____

Approved By: (C/S Supervisor) _____

Reviewed By (C/S Team):

_____   _____

(printed name)   (signature)

Keep a copy of this Permit at job site. Return job site copy to management files
following job completion.

C/S ENTRY PERMIT VERSION 2

PERMIT VALID FOR 8 HOURS ONLY. ALL COPIES OF PERMIT WILL REMAIN AT JOB SITE UNTIL
JOB IS COMPLETED

DATE: - - SITE LOCATION and DESCRIPTION _____

PURPOSE OF ENTRY _____

SUPERVISOR(S) in charge of Entry Team Type of Team Phone #

_____

_____

COMMUNICATION PROCEDURES _____

RESCUE PROCEDURES (PHONE NUMBERS AT BOTTOM) _____

_____

REQUIREMENTS COMPLETED	DATE	TIME
Lock Out/Tag-Out/De-Energize Try-out	____	____
Line(s) Broken-Capped-Blanked	____	____
Purge-Flush and Vent	____	____
Ventilation	____	____
Secure Area (Post and Flag)	____	____
Breathing Apparatus	____	____
Resuscitator - Inhalator	____	____
Attendant on Duty	____	____
Full Body Harness w/ ''D'' ring	____	____
Emergency Escape Retrieval Equip	____	____
Lifelines	____	____
Fire Extinguishers	____	____
Lighting (Explosion Proof)	____	____
Protective Clothing	____	____
Respirator(s) (Air Purifying)	____	____
Burning and Welding Permit	____	____

Note: Items that do not apply enter N/A in the blank.

TABLE 10.2 Two Examples of C/S Entry Permits Meeting OSHA Requirements (*Continued*)

**RECORD CONTINUOUS MONITORING RESULTS EVERY 2 HOURS

CONTINUOUS MONITORING**	Permissible								
TEST(S) TO BE TAKEN	Entry Level								
PERCENT OF OXYGEN	19.5% to 23.5%	___	___	___	___	___	___	___	___
LOWER FLAMMABLE LIMIT	Under 10%	___	___	___	___	___	___	___	___
CARBON MONOXIDE	+35 PPM	___	___	___	___	___	___	___	___
Aromatic Hydrocarbon	+ 1 PPM * 5PPM	___	___	___	___	___	___	___	___
Hydrogen Cyanide	(Skin) * 4PPM	___	___	___	___	___	___	___	___
Hydrogen Sulfide	+10 PPM *15PPM	___	___	___	___	___	___	___	___
Sulfur Dioxide	+ 2 PPM * 5PPM	___	___	___	___	___	___	___	___
Ammonia	*35PPM	___	___	___	___	___	___	___	___

(Add any other detected or expected)

*Short-term exposure limit: Employee can work in the area up to 15 minutes.

+ 8 hr Time Weighted Avg.: Employee can work in area 8 hrs (longer with appropriate respiratory protection).

REMARKS: _____

Inspector ID	Instrument(s) Used	Model &/or Type	Serial &/or Unit #
_____	_____	_____	_____
_____	_____	_____	_____

AN ATTENDANT IS REQUIRED FOR ALL C/S WORK

Attendant	ID	Confined Space Entrant(s)	ID	Confined Space Entrant(s)	ID
_____	_____	_____	_____	_____	_____
_____	_____	_____	_____	_____	_____

SUPERVISOR AUTHORIZING - ALL CONDITIONS SATISFIED _____

TITLE/PHONE _____

AMBULANCE XXXX FIRE XXXX Safety XXXX Inspector XXXX/XXXX

10.8.1 Opening a Permit-Required C/S; Inspection and Initial Testing

A C/S that has been closed for some period of time is one of the most potentially dangerous to test and enter. As we have discussed earlier, in the absence of airflow, oxygen can be consumed and flammable and toxic gases can be generated. This deterioration of the atmosphere can take place in a few days.

Ideally, upon opening a C/S, immediately ventilate it for a minimum of eight hours. Depending upon the size and configuration of the space, more or less time may be satisfactory.

Most facilities require that a senior Inspector or the C/S Supervisor make the initial inspection. Depending upon the contents stored in the space, the initial inspection may require the use of an atmosphere-supplying respirator. If IDLH conditions are known or suspected, the ASR and a stand-by Rescue Team are an OSHA requirement.

Once the initial inspection is completed and all safety requirements are met, the routine entry procedures may then proceed.

10.8.2   The Permit System

The permit system is the heart of a successful (read "safe") C/S program. Copies of permits must be maintained for at least one year! This is because of the latent reactions due to exposure to some toxics. It is preferable to maintain a log book also, so that dates, times, space designations, atmospheric readings; and Inspectors', Attendants', and Entrants' names can be easily located for a specific entry.

CAUTION

BECAUSE INSPECTORS OFTEN ARE REQUIRED TO INSPECT THE SAME SPACES ON A DAILY BASIS OFTEN FOR MONTHS AT A TIME, THE SUPERVISOR MUST BE AWARE OF THE *"CARNAC SYNDROME."* THIS IS NAMED AFTER JOHNNY CARSON'S PORTRAYAL OF *"CARNAC THE MAGNIFICENT,"* WHO COULD READ THE CONTENTS OF A SEALED ENVELOPE.

SOMETIMES, DUE TO TIME CONSTRAINTS, A HISTORY OF SATIS-FACTORY INSPECTIONS OF THE SAME SPACE, OR SHEER LAZINESS, THE INSPECTOR MERELY UPDATES THE PERMIT WITHOUT INSPECT-ING THE SPACE AT ALL. THIS PRACTICE IS EXTREMELY DANGEROUS AND IS STRICTLY FORBIDDEN. THE AUTHORS BELIEVE THAT IF THIS PRACTICE IS DISCOVERED, THAT, AT THE VERY LEAST, THE DISCI-PLINARY ACTION SHOULD INCLUDE REMOVAL OF THE OFFENDING INSPECTOR FROM THE CONFINED SPACES PROGRAM. EACH TIME A SPACE IS TO BE CERTIFIED, IT MUST BE INSPECTED AS IF IT WERE THE FIRST TIME.

The atmosphere in the space must be tested prior to anyone, including the Inspector, entering the space. For an unprotected worker to enter a C/S, the atmosphere must contain between 19.5% and 23.5% oxygen, less than 10% of the LEL of any flammable contami-nants, and no toxics. Testing results by calibrated instrumentation must be written on the permit.

O—π   **The Inspector must fill out the permit form (see Table 10.2).**

O—π   **Allow no unanswered lines on the permit form. If a particular question is not applicable, enter "N/A" to show that the question has been considered.**

O—π   **OSHA requires that the Entrant and the Attendant, or their designees, be al-lowed to witness the preentry inspection.**

Note

Since the lives of the Entrant, and possibly the Attendant, depend on the In-spector's correct performance of the inspection, the authors strongly suggest that all persons signing the entry permit be present at the inspection. This may be physically impossible for a Supervisor who oversees numerous entries per day. In that case, the Supervisor should perform spot checks routinely.

The Inspector must test the atmosphere in all areas of the C/S and also inspect the space for physical hazards such as exposed electrical busses, unguarded rotating machinery, broken or weak ladders, and unguarded open holes or pits, among other hazards. The results of the inspection must be logged on the permit form. No entry is permitted until all required safety precautions have been met. Some of the required equipment might be lighting, fall-protection harnesses, PPE, including respiratory protection; radios, or other evacuation signal devices, real-time atmospheric monitoring instruments, and rescue equipment (retrieval gear).

A C/S entry permit has at least three copies. The original (hard copy) is posted at the designated entrance to the space. The C/S Supervisor maintains one copy, and one copy is maintained at the safety office.

The Inspector, Entrant, and Attendant must sign the permit, acknowledging that they have read (or have had read to them), understand, and will abide by the contents and precautions listed on the permit. The permit must also be signed by the C/S Supervisor, attesting awareness of the entry and that all the parties involved are properly trained to undertake the work. OSHA does not require the Supervisor to be present during testing, entry, or performance of the assigned work.

The minimum number of personnel on a permit space entry is two: one Entrant and one Attendant.

Remember that conditions can change in the space in a way not planned for in the permit. If this occurs, the entrants must immediately evacuate the space and the permit becomes Void.

Additional engineering controls may have to be instituted. Any reentry will have to include the retesting of the space. Table 10.2 provides two examples of suggested permit forms. Supervisors should make additions or changes to the chosen form to make it apply more specifically to the spaces at the worksite.

10.8.3 An Example of An Exemplary Confined Spaces Program

The U.S. Navy has one of the most successful C/S testing and entry programs in the world. This is because it is so thorough. Mr. Angelo Marchiano, Environmental Program Manager for the Navy's Northern Division-Naval Facilities Engineering Command, was one of the pioneers of the program. As head of the Gas Free Engineering program at the venerable Philadelphia Naval Shipyard, Mr. Marchiano made many improvements to the C/S permitting process that were adopted by the Navy worldwide.

His first improvement was the design of a comprehensive entry permit form (such as the ones in Table 10.2) that included a check-off list of all forseeable conditions that could exist in a space and safeguards that had to be taken.

Another important innovation was the expansion of the use of colored tags at all entry sites. Mr. Marchiano feels that the tags give workers a quick visual indicator of the overall safety of the space. This was extremely important because there is only one copy of the permit and that is posted at the main entrance. Colored tags are posted at all entrances.

The Navy uses the following color code for tags at C/S entrances:

Red tag: Space unsafe for entry and unsafe for hot work

Yellow tag: Space safe for entry but unsafe for hot work

Blue tag: Space safe for entry but hot work by 'work site permit' only

Green tag: Space safe for entry and safe for hot work

White tag: Space inerted. Unsafe for entry

One of the authors worked under this system. For a period of approximately two decades, over a million permit-required C/Ss were entered. The required work was performed without a single incident due to the permitting system. With such a stellar record for safety, we strongly encourage all C/S program administrators to implement these proven practices of an inclusive permit form and easily identifiable placarding or tags into their own programs.

10.9   *TRAINING*

Training, retraining, and periodic drills are the only ways to ensure that you and your workers will maintain the excellent safety record of your safe C/S Program.

OSHA requires training of all personnel who might become involved in C/S work. This training must be conducted regularly and documented. Records must be maintained for a minimum of three years. (The authors suggest that all training records be maintained at least during the period of employment.)

The employer must provide training to ALL workers who work in and around permit and non-permit-required C/Ss. Workers must understand the hazards of such work and possess the skills necessary for the safe performance of their duties concerned with C/Ss.

10.9.1   When to Train

Workers must be trained before:

- The employee is first assigned C/S duties
- When there is a change in those assigned duties.

  Workers must be retrained:

- Whenever there is a change in permit space operations that presents a hazard for which an employee has not been trained
- Whenever the employer has reason to believe either that there are deviations from the permit space entry procedures required in this Section or that there are inadequacies in the employee's knowledge
- Following 'close-call' situations.

The training must ensure employee proficiency in the duties required by Section 10 and must introduce new or revised procedures, as warranted, for full compliance with Section 10 requirements.

10.9.2   General Training for All Employees

All employees who will enter, attend, or work near C/Ss must be trained in the C/S program.

Supervisors, Inspectors, Entrants, Attendants, and Rescue personnel must be adequately trained in their job-specific duties prior to any C/S entry. In most programs, all C/S workers will be trained in all facets of the C/S program for general knowledge and interchangeability of job functions. The possible exceptions to this rule are when a dedicated team of Inspectors serves that function only and when the facility uses a dedicated Rescue Team, either in-house or contracted. In that case, the other team members need only receive their job-specific training.

General training must include the following:

- Explanation of the general hazards associated with C/Ss
- Discussion of specific C/S hazards associated with the facility, location, or operation
- Reason for, proper use, and limitations of PPE and other safety equipment required for entry into C/Ss
- Explanation of permits and other procedural requirements for conducting a C/S entry
- A clear understanding of what conditions would prohibit entry
- How to respond to emergencies

- Duties and responsibilities as a member of the C/S entry team
- Description of how to recognize symptoms of overexposure to probable air contaminants in themselves and coworkers and method(s) for alerting Attendants

Conduct refresher training yearly to maintain employee competence in entry procedures and precautions.

Conduct remedial training whenever the program supervisor has reason to believe that any entry team member requires remedial training. Close-calls situations are a good example.

Also, conduct debriefing of ALL team members after any C/S incident, regardless of injuries or severity. Such "lessons learned" sessions are often the best way to drive home important points.

10.9.3   Goals of Confined Space Team Members' Training

The following five subsections provide a wrap-up of the training required for the five key team components of the permit-required confined space team. We recommend the permit-required option. Logically, the training must cover, as a minimum, all of the required knowledge for team members. As members gain actual C/S experience in different spaces and receive yearly refresher training, they develop a much deeper understanding of their duties and responsibilities. They will know, almost instinctively, how to protect other team members and themselves from injury and illness. Members must be able to prove that competence by passing examinations prepared by trainers.

Training for Inspectors.   The Inspector must be trained to:

- Recognize actual or potential physical, chemical, or atmospheric hazards and their sources that are present, or likely to be present, in the permit spaces of all assigned sites
- Be thoroughly knowledgeable in the selection, calibration, operation, and data logging of the appropriate atmospheric testing equipment
- Have a complete knowledge of sampling strategies and techniques
- Understand the procedures for atmospheric testing
- Personally and physically inspect each entire permit space, including the atmospheres at all levels
- Understand any required engineering controls, such as ventilation equipment
- Understand the meaning and proper application of PELs, TLVs®, LELs, and UELs
- Be thoroughly knowledgeable in the availability, selection, and use of all required PPE and mechanical controls at the site
- Be a competent person, that is, a person who has the knowledge to specify the equipment and procedures necessary as well as the authority to act on that knowledge

Training for Entrants.   The Entrant must be trained to:

- Have a full understanding of the configuration of the space
- Have the ability to recognize potential C/S hazards
- Read and understand the entry permit requirements
- Have the ability to stay alert to the hazards that could be encountered in a confined space
- Understand how to use the equipment as required by the permit
- Understand the importance of immediately exiting the C/S when:
  - Ordered to do so by the Attendant
  - Automatic alarms sound

- He or she perceives an unexpected dangerous condition
- He or she notices physiological stresses or unexpected changes in personal health or that of coworkers (e.g., dizziness, blurred vision, shortness of breath)

Training for Attendants.   An Attendant must be trained to:

- Understand the configuration of the space
- Understand and be able to recognize potential C/S hazards
- Maintain a sign-in/sign-out log to account for all persons in the C/S and ensure that all Entrants sign in and sign out upon entering and exiting the space
- Monitor surrounding activities to ensure the safety of entry personnel
- Immediately summon the Rescue Team if Entrant rescue becomes necessary
- Keep unauthorized persons out of the C/S by ordering them away from the entrance
- Notify supervision and all entrants of any unauthorized entry that cannot be prevented
- Maintain effective and continuous communication with entry personnel during C/S entry, work, and exit
- Order personnel to evacuate the C/S upon:
  - Observing a condition that is not allowed on the entry permit
  - Noticing the Entrants acting strangely, possibly as a result of exposure to hazardous substances
  - Noticing a situation outside the C/S that could endanger entry personnel
  - Noticing, within the C/S, a hazard that has not been previously recognized or taken into consideration
  - Having to leave the Attendant work station
  - Having to focus attention on the rescue of personnel in some other C/S that the Attendant must monitor

Training for Emergency Rescue Team.   Rescue Team training must include:

- A rescue plan and procedures developed for each type of C/S that the team must service
- Proper use of all emergency rescue equipment
- Competency in first aid and CPR techniques
- How to locate workers in all assigned unique C/S configurations, in order to minimize rescue time under difficult rescue conditions, such as smoky atmospheres

Note

The entire training for a C/S rescue team is exhaustive. We have given some of the highlights here. Members of the Rescue Team require the highest level of training. For this reason, many facilities employ specially trained private or public rescue teams.

10.9.4 Training for C/S Supervisors

The C/S Supervisor must:

- Have completed current general C/S training
- Have a thorough knowledge of all aspects of the C/S program
- Be able to show, through documented training and experience, awareness of and understanding of all aspects of the requirements of the job descriptions of the Inspector, Entrant, Attendant, and rescue personnel.

The C/S Supervisor has the ultimate responsibility for the safety and health of the members of the C/S team.

Note

Our expertise in the field of permit spaces comes, in large part, from one author's participation in the U.S. Navy Gas Free Engineering Program as an Inspector. This program has one of the best safety records in the world. We have found a wide range in the degree of C/S training at various facilities. The Navy requires 80 hours of initial training, including hands-on exercises under a Certified Gas Free Engineer. In addition, for the Inspector Trainee to certify permit spaces as a Gas Free Engineering Technician, the Trainee must undergo an additional 160 hours of training under the supervision of a Certified Gas Free Engineering Technician. This adds up to 240 hours of training, with an 8-hour annual refresher!

Training Documentation and Verification. Yearly assessment of the effectiveness of employee training must be conducted by the C/S Supervisor. Training sessions must be repeated as often as necessary to maintain an acceptable level of personnel competence.

10.10 *TRAINING AIDS AND ADDITIONAL RESOURCES*

The authors recommend the following videos for aiding readers to see real-life examples of what they have learned in this Section.

- *Confined Spaces, Deadly Places*
- *Survival by Permit*

These videos are available from Safe Expectations (formerly BNA Communications, Inc.) (www.safeexpectations.com).

- *Confined Space Entry—Permit Required!*
- *Atmospheric Testing*
- *Confined Space Ventilation*
- *Confined Space Hotwork—Checklist for Safety*
- *Confined Space Non-Entry Rescue*
- *Confined Space Rescue*
- *Confined Space Case Histories*

These videos are available from Coastal Safety and Environmental (www.coastal.com).

10.11 *SUMMARY*

In this Section you have learned what defines the difference between a C/S and a permit-required (permit) space.

Never enter a permit space without the assurance that it has been properly inspected. Always use all required safety equipment and follow the permit to the letter.

Conditions may change in the space. As the Entrant, you must know when to evacuate!

We have shown a broad range of hazards that may be encountered. However, as we have seen in Section 6, the most common hazards are atmospheric contamination or lack of oxygen. Depending on the prior use of the space, current operations, or lack of natural ventilation, these hazards may be toxic, explosive, or irritant; indeed, any atmospheric hazard may exist. We also discussed the various physical hazards that may exist. Remember that the space was not designed for human occupancy, so normal machine and electrical guarding may not exist.

We have discussed the proper order for atmospheric testing of C/Ss and the reason for that order. Test for oxygen first because it is needed for breathing and because your instrument was calibrated in a normal oxygen atmosphere. Oxygen content must be between 19.5% and 23.5%. Otherwise, the meter's other readings will be incorrect. Test for flammables (LEL) second. This is because fire is the second-largest atmospheric threat. Check for toxins third. This will require some knowledge of the spaces' prior contents. However, at a minimum, test for carbon monoxide and hydrogen sulfide first, then other toxins that might be suspected.

Remember to test for stratified atmospheres. Be sure to allow ample time for your meter to collect and analyze the sample. Whether seconds or minutes, every analyzing instrument takes some time to indicate or record accurately.

You must understand the duties of all members of the C/Ss Team and who the members are: Inspector, Entrant, Attendant, Supervisor, and Rescue Team. You may be specially trained to fulfill any one of these team roles or, indeed, all of them. The minimum number of individuals that must be present at a confined space site during entry work is two: one Entrant and one Attendant.

Remember that more than half the injuries and fatalities that occur in C/Ss occur to well-meaning rescuers. The Attendants must never leave their post without evacuating the Entrants first. They must resist the urge to enter as helpers and instead summon the Rescue Team. Otherwise, an Attendant could become another victim!

Adherence to the permit system is essential. It is not only an OSHA requirement, but it provides for a unified, consistent method for all team members to follow. Know how the permit system works.

Finally, training, retraining, and careful work experience in confined spaces are the only ways to build and maintain a safe permit-required C/S program.

SECTION 11
OTHER WORKPLACE HAZARDS

OBJECTIVES

Our primary objective in Section 11 is to alert HAZWOPER personnel to hazards in any workplace that are not necessarily caused by hazardous materials. Our secondary objective is to complete the explanations of the prominent kinds of other workplace hazards, their effects, and how HAZWOPER workers can protect themselves day by day.

Our goal in this reference work is to lead readers in a stepwise fashion through the regulations, hazards recognition, and methods of protection of self and coworkers to the kinds of work HAZWOPER personnel do on a daily basis.

As examples, in Section 3 we discussed material hazards such as fire, explosion, toxicity, reactivity, and more. In Section 4 we showed the tremendous hazards that could result if incompatible materials came into contact with each other. In Section 5 we introduced the science of toxicology, including how toxins enter the body and some examples of likely health effects.

In Section 6 we presented PPE from our vantage point as HAZWOPER workers, supervisors, engineers, and trainers over many years. For example, readers learned the different types of PPE to protect their breathing, ranging from dust masks to SCBAs. By now we trust you can make a reasonable selection of the items of PPE that will help protect you from material hazards. (In Section 11 you will be introduced to other equipment that can protect you from falls, for example.)

Section 7 described how to decontaminate yourself, or others, in the event of exposure to material hazards. In Section 8 you learned the value and proper usage of signs, labels, and placards. You saw examples, their uses, and required placement. In Section 9 Material Safety Data Sheets (MSDSs) were discussed in detail. That Section gave you a thorough understanding of what information must be made available to you before you are asked to work around hazardous materials.

In Section 10 we began our discussion of hazards other than material hazards with the subject Confined Spaces. Sections 10 and 11 both deal with other workplace hazards. Due to the severity of the consequences and the great number of misconceptions about confined spaces, not to mention the large number of accidents that occur each year, we decided confined spaces deserved a separate section.

Again, in this Section our objective is to explain the nonmaterial hazards that are associated with HAZWOPER work. Many of these hazards are not unique to work at brownfields, TSDFs, or where hazardous materials are used in processing or are encountered in emergency response; but they need explanation here none the less.

To work safely, you must be able to recognize all the potential hazards involved in your work assignment. Always know where you are in relationship to other workers, machinery, and operations at the site. A final objective of this Section is to enable you, and the people you train or who work for you, to take a proactive approach to safety. You will see that simply by being observant, you can make your life, and the lives of all your site coworkers, safer.

Recently, some workers digging a ditch were concentrating on the excavating and not the overhead power lines. The arm of the backhoe connected with a 13,200-volt power line, the bucket was blown off the backhoe, and the operator was killed instantly. Not taking in the big picture can be disastrous.

The following Sections of the book (Sections 12 through 16) deal with the kinds of daily work and planning that our readers will have to either do, train others to do, or supervise.

○━━π **OSHA has determined that three out of four accidents that happen in the workplace each year are preventable.**

11.1 ACCIDENTS, INJURIES, AND THEIR CAUSES

○━━π **An accident is an undesirable, unplanned event that can result in physical harm, environmental harm, damage to property, or interruption of the work schedule.**

In Section 6 we discussed the excellent attributes of personal protective equipment. PPE will help protect you from many workplace hazards. Nothing will take the place of your constant vigilance. PPE is the final solution for protection from many hazardous materials. It does not protect you from many other workplace hazards or thoughtless acts.

11.1.1 Major Incidents Not Involving Hazardous Materials

○━━π **Overexertion, falls, and being struck by an object are major causes of worker injuries and illnesses.**

Liberty Mutual Group, a workers' compensation insurance company, prepared the Liberty Mutual Workplace Safety Index for 1998. It lists the 10 leading causes of injuries and illness that account for 86% of the $38.7 billion in wage and medical payments employers paid in 1998.

Liberty Mutual used its own claims data and findings from the Bureau of Labor Statistics and the National Academy of Social Insurance. According to the *Safety Index*, overexertion, falls, and being struck by an object are among the leading causes of workplace accidents. When the indirect costs of workers' compensation claims are added to the $38.7 billion in direct costs identified by the report, the total economic burden of workplace injuries and illness is far greater, with estimates ranging up to $155 billion per year.

Gary Gregg, a Liberty Mutual Executive, cited three benefits from improved workplace safety:

1. Reduction in employee pain and suffering
2. Avoidance of the direct cost of workplace injuries, such as wage replacement payments and medical care expenses
3. Saving of the indirect cost of avoided accidents, lower employee morale, lost productivity, and the cost of hiring or training overtime or temporary replacement workers

Table 11.1 lists the prime causes of injuries from other workplace hazards. These are injuries that resulted in employees missing five or more days of work in a year.

The Bureau of Labor Statistics Survey of Occupational Injuries and Illnesses from 1999 reported 5.7 million injuries and illnesses in private industry. That equates to a rate of 6.3 cases per 100 full-time workers—quite a sobering statistic.

TABLE 11.1 Leading Causes and Costs of Workplace Injuries and Illness

Accident causes	Percent of workers' compensation direct cost paid in 1998	Estimated workers' compensation direct cost nationwide
Overexertion—injuries caused by excessive lifting, pushing, pulling, holding, carrying, or throwing of an object	25.57%	$9.8 billion
Falls on same level	11.46%	$4.4 billion
Bodily reaction—injuries resulting from bending, climbing, loss of balance, and slipping without falling	9.35%	$3.6 billion
Falls to lower level, such as falling from a ladder or over a railing	9.33%	$3.6 billion
Being struck by an object, such as a tool falling on a worker from above	8.94%	$3.4 billion
Repetitive motion	6.10%	$2.3 billion
Highway accidents	5.46%	$2.1 billion
Being struck against an object, such as a carpenter walking into a door frame	4.92%	$1.9 billion
Becoming caught in or compressed by equipment	4.176%	$1.6 billion
Contact with temperature extremes that results in such injuries as heat exhaustion, frost bite or burns	.92%	$.3 billion
All accident causes	100.00%	$38.7 billion

Source: Courtesy Liberty Mutual Workplace Safety Index.

11.1.2 Unique Risks of HAZWOPER Personnel

○━π **HAZWOPER workers may face unusual or unique risks that other industry workers will not face.**

HAZWOPER workers in some cases face a higher risk of accident and injury than the typical industrial employee. They must deal with hazardous situations when an unplanned incident occurs at an unfamiliar site. Then site information is incomplete. Further, the incident can be beyond their complete control. HAZWOPER workers often work in emergency situations where time is critical in the protection of life, environment, or property. In short, response workers are subject to many different outside forces that can increase their chance of being involved in an accident.

Note further from Table 11.1 that overexertion was cited as the single cause of the largest number of incidents. HAZWOPER workers can be required to don masks and fully encapsulated chemical protective clothing. Depending a good bit upon temperature and humidity, these workers can be much more liable to overexertion stress than the overall working population.

The best protection is hazard recognition and having been trained in the safe response to hazards. Rather than go into great detail in this section on accident prevention, we refer the reader to the Health and Safety Plan (HASP) as part of Section 16. That HASP is exhaustive. Use discretion. You may download a template, or model plan, from the Companion CD.

Parts may be deleted if they do not apply to your particular operation. However, we believe that that model can guide you to satisfaction of OSHA's requirements.

An accident may be the result of an unsafe act. Always climb a ladder using the three-point climbing technique: you have two hands and two feet; keep at least three of them firmly placed on the ladder at all times. Trouble might result from not wearing your hard hat and safety glasses. Or it might be due to an unsafe condition such as a deteriorated ladder, unguarded rotating machinery, or a toxic atmosphere. These situations can cause multiple hazards. One individual's unsafe act can result in an unsafe condition for someone else.

11.1.3   What Are Proactive and Reactive Safety?

O—π   Work safety is the condition of being personally protected from, and not causing others, hurt, injury, or loss.

Dictionary definitions generally read something like, "Safety is the condition of being safe from undergoing or causing hurt, injury, or loss." Others must take care so that actions to protect or reduce accidents for one worker do not set up conditions for subsequent accidents for other workers.

When PPE was first becoming popular, the wearers and their employers thought of the gear as some sort of magic protection. Some employers saw little need to eliminate potential hazards because their workers were "protected." Experience has shown us the fallacy in this sort of thinking.

Comprehensive planning for workers' safety requires us to be both *proactive* and *reactive*.

Proactive safety requires that workers and supervisors alike analyze a task and then plan to do it in such a way as to eliminate any known or potential hazard. Proactive safety involves all of the safety training and operating instructions, and assigned work uniform and PPE, an individual receives based upon the kinds of work to be performed. Further, it includes ensuring the proper operation of all engineering controls that can affect the work. To summarize, being proactive means planning ahead of time as to how we will prevent, or react to, hazards. It means taking preventive action before a threat arises, actively seeking out unsafe conditions and acts and correcting them before an accident can occur.

O—π   Proactive safety means planning and preparing ahead of time for protection against threats to safe working conditions.

Reactive safety follows proactive safety. At that point, we have already greatly reduced work hazards by following all of the safety training principles and planned work operating instructions. Then, as a reaction to specific incidents, we make use of the more incident-specific kinds of PPE, such as CPC for full body protection. We also use rescue personnel, other shielding, robotic devices, and any other support personnel and protective equipment that will help keep us out of harm's way.

O—π   Reactive safety means taking those specific measures, including use of incident-specific personal protective equipment, rescue personnel, devices, and support personnel that will further reduce hazards and injuries.

11.2   ACCIDENT CAUSES—UNSAFE CONDITIONS AND UNSAFE ACTS

O—π   To reduce workplace hazards greatly, we must eliminate unsafe conditions and reduce unsafe acts.

There are two main approaches to reducing or preventing accidents: elimination of unsafe conditions and reduction of unsafe acts.

11.2.1   Eliminate Unsafe Conditions

We must work diligently to locate conditions that can contribute to an accident. Then we must either remove those conditions or shield worker exposure to them. The following is a checklist of unsafe conditions we have frequently found in facility inspections. They are usually caused by lack of: proper signage, engineering controls, housekeeping, or personal hygiene.

❏ Missing or poorly maintained warning and caution signs

❏ Missing or incorrect labeling and placarding of materials containers, from bottles and drums to stationary tanks

❏ Unenclosed live electrical circuits

❏ Unrestricted access to hazardous material storage areas

❏ Unrestricted access to high-voltage switching panels

❏ Unguarded rotating machinery

❏ Insufficient forced air ventilation to maintain a clean atmosphere

❏ Missing guardrails, midrails, and toeboards around openings in floors and on scaffolding

❏ Insufficient worker personal hygiene

❏ Uninsulated very hot and very cold pipelines and vessels that workers can contact

❏ Inadequate housekeeping leading to slipping or tripping hazards, and trash and unused materials accumulation presenting fire-spreading hazards

❏ Insufficient protection of pressurized cylinders from damage

HAZWOPER supervision and workers must be especially vigilant in finding, and then eliminating unsafe conditions during emergency response operations. Hazards under such conditions may not be readily apparent.

11.2.2   Reduce Unsafe Acts

Each worker must make a conscious effort to work safely despite possible adverse conditions of the work environment. A high degree of safety awareness must be maintained so that safety factors involved in a task become an integral part of that task. The following is a checklist to help alert workers, supervisors, and trainers to unsafe acts that invariably lead to accidents.

❏ Overexertion. Working beyond your endurance ability due to high heat and humidity or the continual burden of heavy loads (approximately 50 lb)

❏ Falls due to not being observant of a tripping or slipping hazard

❏ Bodily reactions to prevent falls, resulting from too fast movement over uneven surfaces, resulting in strains, sprains, and other injuries

❏ Falls from higher levels due to not practicing ladder safety and not using guardrails and handrails or fall-protection harnessing

❏ Removing guards from operating rotating machinery or working around rotating machinery without guards

❏ Working in an area where required ventilation has failed

❑ Failing to follow lock-out/tag-out procedures

❑ Not being observant of other work going on at your site; being unaware of the chance of workers dropping tools from higher levels; being unobservant of the chance of being struck by materials carried by other workers or by powered equipment or vehicles.

○━π *Engineering controls include machine guarding, ventilation, "dead-man" switches, fall protection, and lock-out/tag-out procedures.*

Workers must know the importance and function of the engineering controls in the workplace, including machine guarding, ventilation, "dead-man" switches, fall protection, and lock-out/tag-out procedures. Lock-out/tag-out will be discussed in Section 11.12. When workers observe any engineering controls that are not functioning as designed or are being bypassed, they must report their findings to the supervisor immediately.

○━π **Bypassing engineering controls, such as machine guarding, on energized machines for convenience, or even for repair, is illegal.**

WARNING

WORKERS ARE FREQUENTLY TEMPTED TO IGNORE OPERATING PROCEDURES OR REMOVE ENGINEERING CONTROLS IN ORDER TO MAKE SOME FORM OF CORRECTION TO EQUIPMENT OR PROVIDE MINOR CLEANING. THEY MUST RESIST THAT TEMPTATION!

Workers must be attuned to the routine sights and sounds of the workplace. They must not ignore changes in these conditions. Changes in operating sounds of machinery nearly always precede machine damage and then possibly catastrophic failure. Workers must be trained to immediately report anything out of the ordinary. The best practice is to keep logs of all kinds of work so that observations may be preserved. Review and comparisons over time can reveal impending hazards.

Another area that needs mention is horseplay. Everyone likes to hear a good joke. Sometimes we need a sense of humor just to get through the day. Workers, trainers, and supervisors alike must understand, however, that there is no place for practical jokes or horseplay in HAZWOPER work. We can work safely around hazardous materials, but that requires strict adherence to all the safety rules and regulations.

○━π **Workers in general, and HAZWOPER workers in particular, must be keenly aware of their surroundings. As much as practical, they must keep all of their senses unhampered in order to avoid harm.**

11.3 CHARACTERIZING THREATS NOT DUE TO HAZARDOUS MATERIALS

Several factors distinguish the hazardous waste site environment from other occupational situations involving hazardous substances. One important factor is the uncontrolled condition of many sites. Uneven load-bearing capabilities of site surfaces have led to accidents involving rollover of backhoes and cranes. Many injuries and deaths yearly are caused by collapse of unsound excavation walls on to workers.

The combination of all these conditions results in a working environment that is characterized by numerous and varied hazards that may:

• Pose an immediate danger to life or health

• Not be immediately obvious or identifiable

- Vary according to the location on the site and the task being performed
- Change as site work progresses

General categories of hazards that may be present at a site are described below. In approaching a site, it is prudent to assume that all these hazards are present, until site characterization has shown otherwise. A site health and safety program (see Section 16) must provide comprehensive instructions. It must tell how to deal with general potential hazards as well as to provide specific protection against individual known hazards. It should be periodically updated with new information as site changes evolve.

A convenient system for characterizing hazards is the *hazardous energy exchange concept.* Injuries are produced when a harmful amount of energy is transferred from outside sources to the human body. Your site should have a written hazardous energy control program. The type of energy transferred classifies these hazardous energy sources:

- *Kinetic/mechanical:* These are "striking" or "struck by" injuries. Examples are slips, falls to a lower level, being struck by a vehicle, and disintegration of rotating machinery.
- *High pressure.* These are injuries due to pressure vessel, pipeline, or component rupture; or shock wave from an explosion.
- *Thermal:* Thermal burns are caused by very cold and very hot surfaces.
- *Electrical:* Shocks and burns are received from body contact with faulty wiring or downed power lines. Other causes are contact of workers' equipment, such as a backhoe, aluminum ladder, or scaffolding contacting an uninsulated power line. Overload discharges at power panels can be particularly hazardous.
- *Acoustical:* Explosions, loud machinery, and noisy operating equipment can permanently damage your hearing. Further, they can make you less aware of your surroundings.
- *Radiation:* Ionizing and nonionizing radiation (discussed in depth in Section 3).

Other site conditions that do not fit well into these common energy transfer categories but are extremely important are:

- *Biological:* Poisonous and harmful plants, insects, animals, blood-borne pathogens, and disease-producing organisms (some of these were discussed in Section 5).
- *Confined spaces:* Spaces that are big enough for personnel to enter but are not made for continuous human occupancy. They have limited means of entry and exit and may have unfavorable natural ventilation. Many of these spaces contain known or suspected material or physical hazards. Confined spaces are discussed in detail in Section 10.
- *Hypothermia and hyperthermia:* The body's loss of its ability to regulate its internal temperature. The inability of the body to heat itself is hypothermia, and to cool itself is hyperthermia.

11.4   *KINETIC/MECHANICAL HAZARDS*

Kinetic energy (which causes kinetic hazards) is the energy possessed by a moving body. A compressed spring is an example of a device having stored or *potential energy.* When you release it and it springs away, it expends kinetic energy. Kinetic energy can do work, as in a spring-wound device such as a mechanical clock. A pile driver, when it is raised, has gained potential energy. When it drops, its potential energy becomes kinetic energy, and when it strikes the pile, driving it into the ground, it is doing useful work. When we fall from a ladder, the impact that absorbs our kinetic energy causes us bodily injury.

Mechanical hazards are present when the potential exists for personnel to be struck by, run over, caught in, or pulled into operating or malfunctioning equipment.

11.4.1   Slips, Trips, and Falls

⊶π   **Following overexertion, slips, trips, and falls are the leading cause of disabling work injuries (DWIs) in American industry.**

⊶π   **A sloppy work area without proper housekeeping is an accident waiting to happen.**

Many times HAZWOPER work sites are not like factory floors. You may be working on a steep incline, around unguarded holes, pits, trenches, or wells. Poor footing, puddles, deteriorated drums, exposed nails on pallets, and broken glass are other common hazards.

29 CFR 1910.21 to 1910.38 are the OSHA regulations on walking surfaces. Slips, trips, and falls are OSHA's number one injury group each year. (Note that slips, trips, and falls are a combination of three of the four top categories in Table 11.1.) Thousands of these individual accidents occur, and millions of work hours are lost each year. OSHA states that about 75% of them are preventable.

As with all safety considerations, being aware of your surroundings is paramount. HAZWOPER workers must be constantly aware of what they are stepping upon, what they are holding onto, and what operations are taking place all around them.

The walking surface of the job site may be littered with pools of unknown liquids, drums in various stages of deterioration, broken pallets, exposed nails, and, most assuredly, uneven footing. Such sites would be difficult enough to navigate under normal conditions. Now add the required PPE and you have new challenges. The full-face respirator you might be wearing will limit your peripheral vision. You will need to look around much more to become aware of surrounding hazards.

When you are climbing, always remember to use the three-point rule. Always keep both hands and one of your feet, or both feet and one of your hands on the ladder at all times. This means don't try to carry something in one hand when climbing a ladder. When working at high levels, you must be tied off securely (see Section 6.5.6). Use the 4:1 (see Figure 11.1) ratio for proper ladder placement.

The same 4:1 ratio is used for rolling scaffolding. The height of the work platform cannot exceed four times the minimum base dimensions. Remember, NEVER move a rolling scaffold with a worker on the platform. It takes very little force to topple rolling scaffolding that is topheavy with a worker.

⊶π   **One of the most overlooked, yet important, warning signs you will see is "Watch Your Step."**

Use the following checklist for frequent review or audit at any facility. It lists the major contributors to slips, trips, and falls.

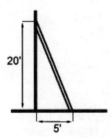

FIGURE 11.1   Proper placement of a ladder.

❑ Debris remaining from incomplete housekeeping can cause tripping or fire hazards or hide other hazards.

❑ Wet or otherwise slippery work surfaces, not cordoned off until their condition is corrected, are major slipping hazards.

❑ Aisles and passageways not kept clear could block access or escape routes in case of emergency and cause falls.

❑ Aisles and walkways must be obvious and marked for easy visibility. Masks, face shields, and other PPE usually limit worker visibility to some extent. Some facilities mark "Aisle" with caution tape so that workers do not accidentally wander off the walkways.

❑ Holes in the walking surface, without guards installed or otherwise made safe, must be cordoned off until repaired.

❑ Standard guardrails must be provided wherever aisle or walkway surfaces are elevated more than 30 in. above any adjacent floor or the ground. When working at low elevations, we are not always conscious of being elevated and often step off a platform and fall unexpectedly.

❑ Toe boards at least 4 in. high must be installed around the edges of scaffolding or permanent floor openings where persons may pass below the opening. Toe boards are reminders that workers are approaching a hole. Toe boards also protect those working below from being struck by material accidentally kicked off the higher level.

❑ Changes in height of the walking surface of one-half inch are enough to cause tripping.

Avoid these other kinetic/mechanical hazards:

❑ Aisles or walkways must not pass near operating machinery such that collision, entanglement, or pinching could occur.

❑ Do not allow repairing of unblocked and unchocked conveyers, such as material elevators. Potential (stored) energy could be released, providing a kinetic hazard that could crush workers.

❑ Forklift, skid loader, and backhoe operators and operators of other vehicular machinery must be well trained to avoid close proximity to other workers. Likewise, workers must be well trained to maintain a safe distance from such machinery.

11.5   *HIGH-PRESSURE HAZARDS*

○─π **Pneumatic hazards generally involve the uncontrolled release of compressed gases.**

High-pressure hazards can be divided into two broad categories: *pneumatic* and *hydraulic*. Pneumatic hazards generally involve the uncontrolled release of compressed gases at 150 psig (pounds per square inch, gauge) or higher. Pneumatic air is used for spraypainting, sandblasting, pneumatic tools, and pumps, among other equipment. Pneumatic systems are broadly used in hazardous atmospheres where electrical power would add to the hazard of the operation.

○─π **Hydraulic hazards generally involve the uncontrolled release of liquids such as water or oil, under tremendous pressure.**

Hydraulic hazards involve the uncontrolled release of liquids or oils under tremendous pressure, sometimes at tens of thousands of psig. The hydraulic jack at your gas station is a good example. Through the use of hydraulic power, by pumping the handle of the jack, the mechanic can raise part of your several thousand-pound car into the air.

Another common use of hydraulics is the pressure washer. Using only water, but at pressures up to 50,000 psig, these devices are widely used for washing. The spray quickly loses power as distance from the nozzle increases. However, within about 3 ft of that nozzle, this pressurized water can cut a two-by-four in half! In fact, water-jet cutting using pressures exceeding 100,000 psig is widely used. Water jets can cut metals such as aluminum up to one 1 ft thick and steel up to 4 in. thick.

Great care must be taken when working with and around compressed gases. That includes air, steam lines, and other containers or equipment under pressure. OSHA tells us how to work with compressed gases and compressed air systems at 29 CFR 1910.166 to 1910.169.

○━π   **High-pressure gas cylinders that are improperly stored or handled can become deadly projectiles.**

Cylinders containing compressed gases such as oxygen and nitrogen, often referred to as "bottles," can hold pressures up to 4500 psi. There have been incidents in which the valve on the top accidentally broke off, leaving a nozzle-like hole. The cylinders took off like a rocket. One flew over 100 ft, even crashing through a concrete block wall.

At a HAZWOPER work site, make sure cylinders, whether in use or waste, are stored properly. Make sure they cannot be easily knocked over. See that the protective cylinder end cap is screwed in place. Compressed gas cylinders must always be stored in an upright position and be chained or otherwise restrained to a wall or other structure, something that is capable of holding the cylinder upright should it become involved in a minor collision. When moving cylinders, never roll them. When at all possible, use the cylinder cart designed for that purpose

Don't allow improper use of unregulated compressed air lines. Workers have been known to use them for blowing dust off a work uniform. This can result in lacerations, material being driven into the skin, and, believe it or not, eyes being blown out of their sockets. Use a blow-down diffuser at the end of the air line. Then the pressure is regulated to around 30 psig and is diffused so as not to pose a hazard. When using live air blow-down to clean up the work area, always be sure to wear the proper eye and respiratory protection.

When work involving opening a portion of a hydraulic or pneumatic system is being done, make sure that the pressure is released. Then purge the system if necessary and see that any connected cylinders, pipes, or vessels are blanked off or vented to atmosphere.

11.6   THERMAL AND ELECTRICAL HAZARDS

Two thermal hazards that might be present on a HAZWOPER site are uninsulated steam lines and stored cryogenic liquids. Leaking steam lines are doubly dangerous. You know you can get a painful burn from a steam leak in the form of a jet. You may not know that many times you cannot see the jet at the point of leakage. It is only visible some distance away, where condensation forms a fog-like stream. You will probably hear the hiss of leaking steam. When you do, don't approach any further. Find some way to shut off the steam supply farther away.

WARNING

A HIGH-PRESSURE STEAM LEAK (OVER 200 PSIG) WILL MAKE A VERY LOUD SOUND, BUT THE LEAKING STEAM MAY BE INVISIBLE. THIS IS BECAUSE THE STEAM IS ESCAPING SO QUICKLY IT HAS NOT YET STARTED TO CONDENSE. NEVER ATTEMPT TO LOCATE A HIGH-PRESSURE STEAM LEAK WITHOUT THE PROPER EQUIPMENT. SEEK A WAY TO SHUT OFF THE STEAM SUPPLY AT A DISTANCE.

Don't get the idea that cryogenic storage tanks or highway tankers, which might be labeled as containing refrigerated liquid, contain something that might be stored in a refrigerator. Actually those liquids, usually nitrogen or oxygen, are at a temperature about 300°F colder than freezer temperatures! If any of that liquid were to get on your skin, you would suffer severe frostbite. At first such contact feels like a burn. That is why it has been called a "cold burn." Those liquids are stored in specially insulated, double-walled stainless steel tanks to form a sort of large Thermos® bottle. Rubber, plastics, and even plain (carbon) steel can shatter almost like glass when immersed in those liquids and then subjected to stress.

Quite severe burns can also come from contact or near contact (arcing) with energized electrical equipment. Electric shock burns are a secondary hazard to electrocution. All cause severe burns immediately upon unprotected contact. Well-insulated PPE (usually containing Nomex® fabric) will help protect workers who energize high-voltage equipment. HAZ-WOPER workers need to realize that unless they are trained for high voltage work, they must stay well away from electrical panels where energizing and deenergizing of high-voltage circuits is done.

⊶ You need to know which burns can be treated at the work site and which cannot.

11.6.1 Degree Classifications of Burns

- *First degree burns* are the equivalent of "sunburns" caused by unprotected skin exposure to the sun and to arc welding. The symptom is reddened, darkened, or very warm skin. Note that first degree burns, although having done the least physical damage to the skin layer, are widely considered the most painful to the victim.
  - First aid: Wash in cool clean water. When dry, cover the affected area with a clean cloth, such as a tee shirt.
- *Second degree burns* show up as blistered skin. Much like a very bad sunburn, these are serious injuries requiring medical attention.
  - First aid: Wash in cool, clean water. Apply dry, sterile gauze to the affected area until proper medical treatment can begin. Do not use grease, butter, vasoline, or any skin softening agent (see Warning next page).
- *Third degree burns* are identified as gray or black charred skin. Blistered areas often surround it. The injured person may slip into shock. Third degree burns always require immediate professional medical treatment. Although these burns cause the greatest amount of harm to the skin and underlying tissue, they are often the least painful. This is due to the fact that the nerve endings have been damaged so much so that the injured person can no longer feel the pain. These burns often require months to heal and the application of skin grafts. The recovery to third degree burns is extremely painful.
  - First aid: do not apply water; it may force dirt and debris into the wound. Cleaning third degree burns is a task for professionals. Cover the affected area with clean, dry gauze or cloth. Try to keep the victim conscious and as comfortable as possible until medical help arrives.

11.6.2 Types of Burns

Types of burns are determined by the source of the hazardous energy causing the burn.

- *Flame burns* are caused by clothing catching fire or exposure to open flames.
- *Scalds* are caused by hot liquids or steam.

- *Contact burns* are caused by body contact with very hot objects, such as uninsulated steam lines.
- *Chemical burns* are caused by contact with a corrosive chemical, such as battery acid.
- *Electrical burns* are caused by contact with noninsulated live wires or unprotected electrical outlets. Please note that as electrical voltage and amperage increase, contact may not be necessary in order to sustain a shock or burn. At high power levels, an arc can jump to ground several feet away!
- *Ultraviolet burns* are caused by overexposure to the sun. They can also be caused by UV lamps (used in some polymer curing processes) and welding arcs.

As you will note, all burns, with the exception of an unforeseen accident (a steam pipe exploding for instance), can be prevented by wearing the proper PPE or by using engineering controls to isolate the hazard (insulating a pipe for instance).

WARNING
NEVER APPLY BUTTER, GREASE, OR OTHER CREAMS TO A BURN AREA. ASK A HEALTH PROFESSIONAL ABOUT TREATMENT. MEDICINES SUCH AS SILVER SULFADIAZINE, SOLD UNDER THE TRADENAMES SSD CREAM®, SILVADENE®, AND THERMAZENE® ARE USED BY DOCTORS FOR THE TREATMENT OF BURNS. IT IS USED TO PREVENT AND TREAT INFECTIONS OF SECOND- AND THIRD-DEGREE BURNS. IT KILLS A WIDE VARIETY OF BACTERIA. SINCE IT IS A SULFA DRUG, THOSE ALLERGIC TO SULFA DRUGS MAY NOT BE CANDIDATES FOR ITS USE. SILVER SULFADIAZINE SHOULD ONLY BE APPLIED UNDER DOCTORS' ORDERS.

11.7   *ELECTRICAL SHOCK HAZARDS*

Electric shocks from relatively low-voltage and low-amperage systems can hurt or kill you. OSHA's regulations on electrical safety-related work practices are found at 29 CFR 1910.331-335.

HAZWOPER work sites are often out-of-doors or in wet working environments. All electrical equipment must be connected to the electrical source via a ground fault circuit interrupter (GFCI or GFI).

Always remember to keep electrical equipment in good condition. Never use electrical equipment with worn or frayed wiring. To prevent someone else being hurt by an unsafe electrical cord, cut off the plug if the cord has visible conductors (you can see the wires). Always use a ground fault circuit interrupter (GFCI or GFI) when working outdoors or in a damp or wet environment. If the equipment is accidentally grounded, a properly working GFCI will prevent shock by nearly instantaneously shutting off the power.

Use this checklist as part of preparation for working on or near electrically energized equipment.

❑ Treat deenergized electrical equipment and conductors as energized until lock-out/tag-out, test, and grounding procedures are implemented. Where appropriate, discharge the stored charge in a capacitor.

❑ Work only on electrical equipment and conductors that are deenergized unless your supervisor can demonstrate that deenergizing introduces additional or increased hazards or is not feasible due to equipment design or operational limitations. Then only work in that environment with extreme caution.

❑ Lock-out/tag-out and ground equipment (where appropriate) before work commences.

❑ Wear protective clothing and equipment and use insulated tools in areas where there are potential electrical hazards.

❑ Deenergize, or visibly guard and temporarily insulate, uninsulated overhead power lines whenever contact with them is possible.

❑ Check and double-check the safety regulations when a ladder or parts of any vehicle or mechanical equipment structure will be elevated near energized overhead power lines. Call your local electric utility for assistance. People standing nearby on the ground may be particularly vulnerable to possible injury.

11.8   ACOUSTICAL (NOISE) HAZARDS

Hearing loss is the only 100% preventable occupational disease!

With the wide range of hearing protection we have available to us now, there is no excuse for routine hearing loss. OSHA requires that employees be placed in a hearing conservation program if they are exposed to average noise or sound levels of 85 dBA or greater during an 8 hour workday. See Table 11.2 for the noise levels of some common sounds. A commonly used measurement of noise levels is the decibel A scale (dBA). Meters costing less than $100 easily measure noise levels.

Sound pressure levels (noises to our ears) are measured on a logarithmic scale in decibels on the A scale (dBA). That means that an increase of 10 dBA is perceived as a doubling of sound intensity. From Table 11.2, an ambulance siren would sound twice as loud as a chainsaw. The human ear has a remarkable range of loudness sensitivity. A sharply painful sound

TABLE 11.2   Noise Levels Measured on the dBA Scale

Sound	dBA
Weakest sound heard by the average ear	0
Whisper	30
Normal Conversation	60
Ringing telephone	80
Belt sander	93
Tractor	96
Hand drill	98
Impact wrench	103
Bulldozer	105
Spraypainter	105
Chain saw	110
Hammer drill	114
Pneumatic percussion drill	119
Ambulance siren	120
Jet engine at takeoff	140
Twelve-gauge shotgun	165
Rocket launch	180

Source: NIOSH.

of about 140 dBA is about 10 million times as intense as the least audible sound. The average person can distinguish sound pressure level differences of about 3 dBA.

When you are working in environments where the noise level is higher than 85 dBA on an 8 hour time-weighted average, OSHA requires hearing protection. Read more about hearing and its protection in Section 6 of this manual.

Sources of loud noises commonly encountered by HAZWOPER personnel at hazardous waste sites include:

- Generators for electric power
- Air compressors for pneumatic tools
- Breathing air compressors for air-line respirators
- Heavy mobile equipment, including skid loaders, backhoes, and forklifts
- Mechanized sampling equipment such as the GeoProbe (to be described in Section 12)

○─⚷ **Hearing loss is not the only hazard from loud noises. Inattention and inability to hear surrounding work or audible warnings are other hazards.**

When working around equipment or machinery that generates loud noises, even with hearing protection, some workers tend to "tune out." You will often see workers in these areas almost in a daze. This is extremely unsafe! When working with proper hearing protection, you must be more alert, as you have greatly reduced your ability to hear warnings. Now you will have to look around more often to be aware of your surroundings. You may also not be able to hear audible alarms, bells, or whistles. It is always recommended that there be both an audible and visual signal for every emergency. Such signals as prearranged hand signals, flags, and flashing strobe lights are good examples.

A problem that contributes to the inattention hazard is the fact that most heavy equipment operators assume that workers see or hear them coming and will get out of the way. This is a serious error. The authors know of an example of a terrible accident involving too-close operation of a vehicle to a pedestrian, combined with pedestrian inattention. A forklift operator was trying to drive a forklift very slowly past a crowd of workers just coming off shift work. One of the workers stumbled into the path of the forklift, and his leg was caught between the dual front tires. Even though the operator immediately applied his brakes, the momentum of the forklift ripped off the worker's leg.

11.9 RADIOACTIVE HAZARDS

○─⚷ **The EPA recommends that all work on a site should cease if radiation readings are 2 mRem above background levels, until an in-depth assessment can be made by a health physicist.**

The subject of radiation was covered in depth in Section 3. It is both a physical hazard, via radiation, and a chemical one. Less than one gram of inhaled plutonium dust is lethal! The dangers of radiation are many, including cancers, burns, and death in severe cases of overexposure. However, you can rest assured that if you are called to work at a radioactive waste site, you will receive plenty of extra training. In the early part of the twentieth century, there were very few controls on the use of radioactive material and its waste disposal. It does show up at hazardous waste sites from time to time. One source was tritium used in phosphorescent paint. If you see this symbol:

make sure you back off until the proper assessment of the site has taken place.

WARNING

NEVER OPEN ANY CONTAINER THAT BEARS THE RADIOACTIVE WARNING SYMBOL. FIRST HAVE A PROPERLY EQUIPPED MONITORING TECHNICIAN DETERMINE THAT DANGEROUS NUCLEAR RADIATION IS ABSENT. WHERE RADIATION IS SUSPECTED, *THE EPA RECOMMENDS THAT ALL WORK CEASE UNTIL A HEALTH PHYSICIST CONDUCTS A SURVEY IF RADIATION LEVELS REACH 2 mRem/HOUR.*

11.10 BIOLOGICAL HAZARDS

HAZWOPER workers can protect themselves from many biological hazards by wearing the proper basic work uniform, including hard hat, safety glasses, gloves, long-sleeve shirts, long pants, and safety shoes or boots.

Hazardous waste sites are usually, but unfortunately not always, in out-of-the-way places. No one wants to live near one. This is often referred to as the NIMBY (Not in my back yard) syndrome. These sites are often overgrown, so an early HAZWOPER task can be to clear the area. You need good visibility over the entire site as part of the analysis of potential material hazards. During this process, and indeed during the entire project time at the site, the possibility exists of contact with poisonous plants. Poison sumac, poison ivy, and poison oak can all have disastrous effects on workers who are allergic to the toxins discharged by those plants. These plants must be collected safely but then must *not* be burned, as the smoke will also carry the toxins.

Unexpected encounters with mosquitoes, bees, wasps and hornets, fleas, ticks, spiders, rats, and snakes are a real possibility. Sometimes dead animal life will be found at the site. During the initial investigation the site assessor needs to determine what the animals died from. Most times the death was from some natural cause. In other instances, it might have been caused by contact with material from the site. In rare cases, these animals, especially birds, may have been carriers of an unusual biohazard such as West Nile Virus.

Each encounter brings with it a varying degree of hazard, from nuisance to lethal. Where these hazards are a likelihood, weed- and pest-elimination specialists must first investigate the site and eliminate the hazards.

With AIDS, tuberculosis, and other blood-borne pathogens included with all types of hospital chemotherapeutic and even household waste, there is the distinct possibility that a HAZWOPER worker could encounter these and other pathogens. If you see this symbol:

make sure you are properly protected according to the instructions in Section 6.

For minimum personal protection at HAZWOPER sites, the authors recommend wearing hard hat and safety glasses, long-sleeve shirt and long pants, gloves, and safety shoes or preferably boots.

11.11 HYPOTHERMIA AND HYPERTHERMIA

Hypothermia and hyperthermia occur when your body loses its ability to regulate its internal temperature. These medical conditions traditionally occur due to extremes in environmental temperature and humidity, wearing of PPE, and physical health. The last includes impaired physical fitness, poor diet, alcohol consumption, and the side effects of certain prescription and nonprescription drugs. Age can be a factor when it leads to diminished physical endurance.

11.11.1 Hypothermia

⊶ **Hypothermia is the condition when your body is unable to keep itself warm.**

When your body can no longer keep itself warm in a cold atmosphere, your internal temperature will begin to drop below 98.6°F. Hypothermia occurs when your internal temperature is less than 95°F. When working in a cold environment, make sure you or your personnel have adequate clothing to maintain body warmth. Layering clothes is one of the best methods of protection. Your comfort and safety can be ensured by adding or removing clothing layers as needed.

- *Workplace cause:* Working in a cold environment, and especially a wet or high relative humidity atmosphere, most commonly causes hypothermia.
- *Symptoms:* A progressive decrease in mental ability occurs, including confusion and lethargy, as well as shivering, low blood pressure, and coma. In some cases the victim may lose the shivering reflex.
  - *First aid:* Take immediate actions to make the individual warm and dry. If the individual is unconscious or unable to communicate (unresponsive), seek medical help immediately.

11.11.2 Hyperthermia

⊶ **Hyperthermia is the body's inability to cool itself.**

On August 1, 2001, we were all saddened to hear of the death of Korey Stringer, the promising young, Minnesota Vikings All-Pro tackle. Mr. Stringer was in the best physical condition of his career. He began to get nauseous during practice the day before. The temperature was only 90°F, but the relative humidity was around 80%, giving a heat index of 110. He was vomiting when they finally took him to the hospital. His internal temperature was 108°F! He died without reviving.

Your body cools itself by sweating. It is actually cooling by evaporation. The more we sweat, the more water we must consume. Sweat also contains electrolytes (sodium, potassium, and calcium), giving it its salty taste. To maintain health, they must also be replaced.

Hyperthermia is more common than hypothermia and is also more life threatening. In the case of hyperthermia, the body has lost its ability to cool itself and the core (internal body) temperature can rise to above 106°F. Hyperthermia is a real danger to the HAZWOPER worker due to the confinement of the additional PPE that the worker frequently wears. The physical conditions for HAZWOPER workers inside their PPE can be much like that of firefighters in their PPE. Firefighters wear relatively heavy PPE (called "turn-out gear") for protection from flame and high temperature. They are also often burdened with SCBAs, heavy hoses, and strenuous climbing. Death by heart attack caused by hyperthermia and overexertion is a leading cause of their loss of life.

Hyperthermia can be avoided by following this checklist:

❑ Become acclimatized (accustomed to the surrounding conditions) gradually.
❑ Take frequent breaks in a cool or shaded area.
❑ Drink plenty of fresh water.

❑ Use electrolyte-replenishing drinks such as Gatorade® in moderation. Drink one glass of Gatorade® or equivalent for every four to eight glasses of fresh water.

❑ As a fitness practice, reduce or eliminate all alcoholic beverages. Of course, do not consume alcoholic beverages before reporting to the site or at the site.

❑ Reduce or eliminate tea, coffee, and all caffeinated soft drinks. The caffeine they contain is a mild diuretic. It drives liquid from your system.

❑ Consider shifting work hours to nighttime, if practical, to avoid the heat of the day.

O—π The progressive stages of hyperthermia are heat cramps, heat exhaustion (heat stress), and heat stroke.

Hyperthermia, if left unchecked, progresses through three distinct phases:

1. Heat cramps and heat rash:
- Workplace cause: strenuous physical activity for an extended period at ambient temperatures above 75°F and high humidity or while wearing protective clothing.
- Symptoms: muscle cramps and profuse sweating, usually with normal body temperature. Affected individuals can develop an irritating, itchy rash.
- First Aid: Rest, drink nonalcoholic and caffeine-free beverages, and preferably cool, NOT cold, water. Do not drink cold water; it will increase the intensity of any stomach cramps that might be present.

2. Heat exhaustion (sometimes referred to as heat stress):
- Workplace causes: Strenuous physical activity at above 80°F ambient temperature and high humidity (above 70%) or while wearing protective clothing, and dehydration from sweating.
- Symptoms: Fatigue, lightheadedness, confusion, nausea, vomiting, headache, tachycardia, hyperventilation, low blood pressure, normal or slightly elevated temperature, and profuse sweating (or no sweating at all).
- First aid: rest in a cool area and elevate the legs above the head, drink nonalcoholic and caffeine-free beverages, preferably cool, not cold, water. Cool, wet cloths can be applied to the areas of the body where the greatest blood flow is closest to the skin (neck, armpits, and groin). If the victim is unconscious or unresponsive, get emergency medical help immediately.

O—π *Heat stroke is always a medical emergency!*

3. Heat stroke (always a medical emergency):
- Workplace causes: Heat stroke results from excessive strenuous physical activity in a very hot environment. Dehydration occurs from profuse sweating.
- Symptoms: A sufferer will have hot, dry skin, loss of consciousness or confusion, hallucinations, seizures, other neurologic symptoms.
- First aid: Get emergency medical treatment immediately! Remove the victim's excess clothing. Provide rest in a cool area and elevate the feet above the head. Provide cool water or nonalcoholic or caffeine-free beverages. Apply cool, wet cloths to the areas of the body where greatest blood flow is closest to the skin (neck, armpits, and groin). Slow fanning might help.
- Do not use ice: This will cause shivering and will constrict the blood vessels and slow blood flow. It could also cause the victim to go into shock. Try to keep a victim calm and responsive. He or she will often feel sleepy and want to nap. Gently keep the victim awake by talking to him or her, assuring him or her that help is on the way. The emergency medical technicians can help a conscious, responsive victim much more effectively.

WARNINGS

1. NEVER GIVE ANYTHING TO AN UNCONSCIOUS PERSON BY MOUTH.

2. IF A VICTIM IS UNRESPONSIVE (CANNOT TALK OR REASON) OR IS UNCONSCIOUS, *IT IS ALWAYS A MEDICAL EMERGENCY. GET MEDICAL HELP IMMEDIATELY.*

11.12 *CUTS, PUNCTURE WOUNDS, AND ABRASIONS*

Minor cuts and abrasions are common occurrences requiring first aid at the work site. On a HAZWOPER site, without proper treatment, they can become serious health problems very quickly. A cut or puncture wound gives hazardous materials an immediate entry route into the blood system. Although proper PPE can reduce the incidence of cuts and puncture wounds, they still happen.

- First aid: Many small cuts and abrasions can be treated at the site. Wash the affected area with copious amounts of clean water and an antiseptic scrub solution, such as Betadine®, or equivalent. Cover the small cut with an adhesive bandage (BandAid® or equivalent). If the cut has been contaminated with a hazardous material, the worker should seek medical treatment.
- In the case of a deep cut or puncture wound, apply pressure to the area and elevate to reduce blood loss. Deep cuts and puncture wounds from work at a hazardous waste site must be treated as a medical emergency.

Note

Have another worker collect as much information as possible about the material the injured worker was working with at the time of the incident. This will help the EMTs and doctors treat the wound properly.

11.13 *LOCK-OUT/TAG-OUT: PREVENTING HAZARDOUS ENERGY RELEASE*

⚷ **Lock-out/tag-out is the single most important component of an effective hazardous energy control program.**

Since many accidents occur when workers are repairing machinery, systems, or components, we must use a hazardous energy control program, otherwise known as lock-out/tag-out. A properly implemented, obeyed, and enforced lock-out/tag-out program will prevent many injuries. This program will help safeguard employees from the unexpected startup of machines or equipment or release of hazardous energy while they are performing servicing or maintenance.

11.13.1 Lock-out and Tag-out Defined

⚷ **Lock-out is the physical act of placing a padlock at a deenergized energy source so that it is impossible to reenergize without first removing the lock.**

Lock-out prohibits the operation of an energy source or control point. An energy source may be a disconnect switch, a power panel, or a power cord, among others. A control point may be a control panel, a start or an on button or switch, or a pneumatic, hydraulic, or process liquid or gas pressurized valve. The OSHA standards apply to any source of mechanical, hydraulic, pneumatic, chemical, thermal, electrical, or other energy source.

To demonstrate the importance of lock-out, consider the following. OSHA estimates that compliance with the lock-out standard prevents about 122 fatalities, 28,400 lost workday injuries, and 31,900 non-lost workday injuries each year.

> **Tag-out is the placement of a DANGER tag at a deenergized energy source. It is a warning to everyone that energizing the source can injure workers and can damage equipment.**

11.13.2 Lock, Tag, and Try

> **Although OSHA recognizes both lock-out and tag-out as independent means of hazardous energy control, the authors strongly recommend the use of both lock-out and tag-out in conjunction with each other.**

First the assigned lock-out/tag-out worker completes the task of deenergizing the system. That means applying locks and tags on switches in the off position, closed valves, and other hazardous energy control devices. Then the worker must attempt to energize the equipment. This is called "Lock, Tag, and Try."

In some instances, either by design or by accident, equipment or a system is "back-fed" with energy. This means that the equipment or system is energized from two or more sources. For instance, in your home you may have a light switch at the bottom of the stairs and one at the top. They both might energize the same light fixture. Therefore, to deenergize the fixture you would have to deenergize both sources.

In cases of multiple energizing sources, there will have to be multiple lock-out locks and tag-out tags applied, one for each energy source. The lock or locks used must have only one key, which must remain in the control of the authorized employee until the operation is complete.

The exception here is shift work when the job will continue into the next work shift. In that case, the authorized employees will exchange locks with their relief at the lock-out source, when, and only when, the relieving worker has been thoroughly instructed as to what is locked-out, why the equipment is locked out, and the procedure for safely unlocking and reenergizing the equipment.

Whether using locks or tags or both, the device must have the following information attached to it:

- The name of the worker applying the lock-out device
- The location and identification of the equipment, machine, or system the device is locking out from operation
- The date and time the device was applied
- The phone number, department name, or other contact information for locating the authorized worker who applied the device

11.13.3 Energy-Isolating Devices

> **Energy-isolating devices include, but are not limited to, manually operated electrical circuit breakers, disconnect switches, line valves, pipe blanks and blocks, and chocks.**

An energy-isolating device is any mechanical device that physically prevents the transmission or release of energy. These include, but are not limited to, manually operated electrical circuit breakers, disconnect switches, line valves, pipe blanks, and blocks.

There are two types of energy-isolating devices: those capable of being locked and those that are not. The standard differentiates between the existence of these two conditions and the use of tag-out when either condition exists.

Note

The federal government is already mandating that lock-out capability be designed into newly manufactured machinery. The reason they still accept tag-out alone as a method of hazardous energy control is the large number of industrial machines still in use that were built before the standard and were not constructed with lock-out capability.

Portable energy-isolating devices that can be used to lock out include:

- "Clamshells," plastic devices that cover a valve wheel on a gate or globe valve so that it cannot be operated
- Devices for locking out ball valves
- Devices for locking out the male end of a power cord, circuit breakers, and almost any other kind of electrical equipment
- Devices to lock out nearly any other blanking, blocking, or chocking situation that arises

When the energy-isolating device cannot be installed, the employer must use tag-out. When using tag-out, the employer must comply with all tag-out-related provisions of the standard. In addition to the normal training required for all employees, here is a checklist specific to tags. The employer must train the workers in the following limitations of, and requirements for, tags:

- Tags are essentially warning devices affixed to energy-isolating devices; they do not provide the physical restraint of a lock-out device.
- When a tag is attached to an isolating means, it is not to be removed except by the person who applied it, and it is never to be bypassed, ignored, or otherwise defeated.
- Tags must be legible and be understandable by all employees.
- Tags and their means of attachment must be made of materials that will withstand the environmental conditions encountered in the workplace.
- Tags may give some workers a false sense of security; they are only one part of an overall energy control program.
- Tags must be securely attached to the energy-isolating devices so that they cannot be detached accidentally by nearby work.

CAUTION

IF THE ENERGY-ISOLATING DEVICE IS LOCKABLE, THE EMPLOYER MUST USE LOCKS UNLESS IT CAN BE DEMONSTRATED THAT THE USE OF TAGS WOULD PROVIDE PROTECTION AT LEAST AS EFFECTIVE AS LOCKS AND WOULD ENSURE FULL EMPLOYEE PROTECTION.

11.13.4 Blocking and Chocking; Blanking or Blinding

Blocks and chocks must be used when potential energy cannot be relieved even when the energy source has been locked out and tagged out. A good example is when working on elevating systems. Even though they are locked and tagged out, there is still a potential for falling. In this case, blocks will be used to keep the equipment from falling onto the workers. The blocks must be of sufficient strength to hold the entire weight of the falling piece.

If piping systems cannot be disconnected and valves do not hold (positively stop flow), the pipes must be blanked or blinded. Even if valves seem to hold, you should also use pipe blanks. Blanks, also called blinds, are disks that can be inserted into flanged piping systems. They effectively cap the pipe at that point, thereby stopping the flow of the material being transferred through the pipe. If feasible, pipes should be disconnected completely.

An interesting and relatively new method of blanking is by the use of cryogenics. This process, called freeze plugging, is used for water pipelines that require service or replacement. Extremely cold liquid nitrogen ($-320°F$) is applied to the outside of the pipe, and an ice plug is formed on the inside of the pipe, stopping flow. Service can then be carried out on a piping system that is still under pressure.

CAUTION

THIS METHOD SHOULD BE USED WITH GREAT CARE ON ANY PLAS-TIC, IRON, OR CARBON STEEL PIPING. THOSE MATERIALS ARE ALL PRONE TO SHATTERING AT TEMPERATURES COLDER THAN ABOUT $0°F$ TO $-100°F$. ASTM RECOMMENDED STAINLESS STEELS ARE SUIT-ABLE FOR $-320°F$ SERVICE.

11.13.5 Accidents Due to Improper Lock-out / Tag-out Procedures

OSHA lists the five main causes (the "fatal five") of lock-out/tag-out injuries as:

1. Failure to stop the equipment before attempting repair or clearing (removing debris) from the equipment
2. Failure to disconnect the equipment from its power source
3. Failure to dissipate (bleed off, discharge, neutralize, or vent) stored energy
4. Accidental restarting of equipment while work is in progress
5. Failure to alert fellow workers and ensure a safe work environment before restarting equipment

11.13.6 Specialized Devices for Lock-out and Signage for Tag-out

Remember that a lock-out lock only has one key, and that key must be under the control of the authorized worker at all times.

Figure 11.2 shows typical padlocks for lock-out purposes. There are many kinds of lock-out devices, other than padlocks, available for just about any energy-isolating purpose you can imagine. A specialized lock-out device, when placed in service to lock a valve closed, will prevent opening of the valve until the lock-out device is removed.

Tag-out is the placement of a DANGER tag at a deenergized energy source. Its purpose is to warn other workers that energizing the source can injure another worker and can also damage equipment. Figure 11.3 shows a personalized danger tag. An effective practice is to provide authorized workers with these personalized, laminated DANGER tags. This is especially helpful at large facilities in personalizing the potential hazard caused by defeating lock-out/tag-out. The kits are inexpensive, self-laminating, resistant to moisture, and durable. All you need is a picture of the employee.

FIGURE 11.2 American lock-out locks. (*Courtesy American Lock Company.*)

FIGURE 11.3 A generic danger tag. (*Source: NIOSH.*)

CAUTION

THE OSHA STANDARDS REFER TO LOCK-OUT AND TAG-OUT AS TWO SEPARATE PROCEDURES FOR CONTROLLING THE HAZARDOUS RE-LEASE OF ENERGY. HOWEVER, THE AUTHORS OF THIS BOOK STRONGLY RECOMMEND THAT THE TWO PROCESSES BE USED IN CONJUNCTION WITH EACH OTHER. WHILE LOCK-OUT IS THE STRONGER OF THE TWO ACTIONS, OSHA HAS ALLOWED FOR TAG-OUT TO BE USED ALONE, WHEN EQUIPMENT WAS MANUFACTURED WITH NO PROVISIONS FOR LOCKOUT. OSHA DOES REQUIRE THAT ALL NEW EQUIPMENT BE EQUIPPED WITH LOCK-OUT CAPABILITIES. WITH SO MANY DIFFERENT DEVICES AVAILABLE ON THE MARKET FOR LOCK-OUT OF CONTROL PANELS, VALVES, ELECTRICAL PAN-ELS, SWITCHES, AND OTHER ENERGIZING DEVICES, THERE IS REALLY NO EXCUSE FOR NOT USING BOTH LOCK-OUT AND TAG-OUT.

11.13.7 What the OSHA Standards Require of a Hazardous Energy Control Program

The standards require that covered facilities having energy-operated equipment and those having piping and control valving shall have a written hazardous energy control program in place. The following components are required:

1. Documented description of the practices and procedures necessary to shut down and lock out/tag out applicable equipment.
2. A written policy in force requiring that employees receive training in their role in the lock-out/tag-out program. That policy shall mandate that periodic inspections are conducted to maintain and enhance the energy control program.

Before service or maintenance is performed on machines or equipment, the machines or equipment must be turned off and disconnected from the energy source. All stored energy (such as capacitors in electrical circuits and hydraulic pressure in hydraulic lines) must be relieved, if possible, and the energy-isolating device must be locked-out or at least tagged-out. Energy relief can include static discharge of capacitors, venting a tank, bleeding a pressurized line, and allowing a spring to travel to its extended position.

11.13.8 Exceptions to the Rule

Lock-out/tag-out need not be practiced under two conditions:

1. While servicing or maintaining cord-and-plug connected electrical equipment, provided that the equipment is unplugged from the energy source. Further, the plug must remain under the exclusive control of the employee performing the work.
2. During "hot tap" operations that involve transmission and distribution systems for gas, steam, water, or petroleum products when they are performed on pressurized pipelines. It must be proven that continuity of service is essential, shutdown of the system is impractical, and employees are provided with alternative protection that is equally effective.

Note

There are lock-out devices for corded plugs. The authors strongly advise their use in all situations.

11.13.9 Lock-out/Tag-out during Servicing and Maintenance Operations

Servicing and maintenance activities, such as lubricating, cleaning, or part replacement, might take place during production. In that case, the employees performing those activities may be subjected to hazards that are not encountered as part of the production operation itself. Workers engaged in such activities are covered by lock-out/tag-out when any of the following conditions occur:

- The worker must either remove or bypass machine guards or other safety devices, resulting in exposure to hazards at the point of operation.
- The worker is required to place any part of his or her body into a danger zone associated with a machine operating cycle.

In the above situations, the equipment must be deenergized and locks or tags (we recommend both) must be applied to the energy-isolation devices.

- Use lock-out/tag-out when other servicing tasks occur. Examples are setting up equipment and making significant equipment adjustments. Workers performing such tasks are required to lock-out or tag-out if unexpected energizing or startup of the equipment can injure them.

11.13.10 Lock-out/Tag-out Written Operating Instructions and Procedures

The facility must have documented procedures. The written procedures must identify the information that the authorized employee must know to control hazardous energy. The authorized employee is the one who locks out machines or equipment in order to perform the servicing or maintenance. If this information is the same for various machines or equipment, then a single energy control procedure may be sufficient. There might be other conditions, such as multiple energy sources ("back-feeding"), different connecting means, or a particular sequence that must be followed during equipment shutdown. Then the employer must develop separate energy control procedures to protect operators.

The energy control procedures must outline the scope, purpose, authorization, rules, and techniques that will be used to control hazardous energy sources. This means that the instructions must clearly state how compliance will be enforced.

Here is a checklist for supervision to determine if the minimum requirements for written lock-out/tag-out procedures are being met. Ask yourself whether they contain:

❑ A statement on how the procedures shall be used

❑ The procedural steps needed to shut down, isolate, block, and secure machines or equipment

❑ The steps designating the safe placement, removal, and transfer of lock-out/tag-out devices and who has the responsibility for them

❑ The specific requirements for testing machines or equipment to determine and verify the effectiveness of locks, tags, and other energy control measures

❑ Instructions as to how the employer, or an authorized employee, must notify affected employees before lock-out or tag-out devices are applied and after they are removed from the machine or equipment

❑ Definition of an authorized employee as one who is in control of lock-out/tag-out devices

❑ Definition of an affected employee, one whose job requires operation or use of equipment on which servicing or maintenance is being performed under lock-out, or who works in an area in which such servicing or maintenance is being performed

Step 1: Application of Controls and Lock-out/Tag-out Devices.   The established procedure of applying energy controls includes the specific elements and actions that must be implemented in sequence:

1. *Prepare* for shutdown, notify other workers who depend upon or work near machine operation.
2. *Shut down* the machine, system, or equipment by turning off power, if applicable.
3. *Isolate* the source of energy by block, blind, bleed, purge, or other means.
4. *Apply the lock-out and/or tag-out device.*
5. *Render safe* all stored or residual energy, such as by safely discharging capacitors.
6. *Verify* the isolation and deenergization of the machine or equipment; check for no electrical current flow or storage; or no fluid, gas or material flow, or placement of chocks or blocks to prevent motion.

Step 2: Removal of Locks and Tags.   Following completion of servicing or maintenance and before lock-out or tag-out devices are removed and energy is restored to the machine or equipment, the authorized employee(s) must ensure that the following are done.

1. *Inspect* the work area to ensure that nonessential items have been removed. See that machine or equipment components are intact and capable of operating properly.
2. *Check* the area around the machine or equipment to ensure that all employees have been safely positioned or removed.
3. *Make sure* that ONLY the authorized employees who attached the locks or tags remove them. In the very few instances when this is not possible, the device may be removed under the direction of the employer, provided that the employee doing the removing strictly adheres to the documented procedures that follow the standard.
4. *Notify* affected workers *after* removing locks or tags and before starting equipment or machines.

11.13.11 Additional Safety Requirements

Special procedures must be included in the operating instructions when:

1. Machines need to be tested or repositioned during servicing
2. Outside (contractor) personnel are at the work site
3. A group (rather than one specific person) performs servicing or maintenance
4. Shifts or personnel changes occur during servicing or maintenance

11.13.12 Testing or Positioning of Machines

OSHA allows the temporary removal of locks or tags and the reenergizing of the machine, system or equipment ONLY when necessary under special conditions, such as when power is needed for the testing or positioning of machines, equipment, or components. The reenergizing must be conducted in accordance with the following steps:

1. Clear the machines or equipment of tools and materials.
2. Remove workers from the machines or equipment area.
3. Remove the lock-out or tag-out devices as specified.
4. Energize and proceed with testing or positioning.
5. Deenergize all systems, isolate the machine or equipment from the energy source, and reapply lock-out or tag-out devices as specified.

11.13.13 Outside Personnel (Contractors)

The on-site employer and any contractors must inform each other of their respective lock-out or tag-out procedures. Each employer must ensure that his or her personnel understand and comply with all restrictions and prohibitions of the other employer's energy control program.

11.13.14 Group Lock-out / Tag-out

A crew, craft, department, or other group sometimes performs servicing or maintenance. It must use a procedure that affords affected employees a level of protection equal to that provided by use of a personal lock-out or tag-out device.

Occasionally more than one authorized worker has to lock-out an individual source. The device used in this case is an 8- or 10-space hasp. This device is applied to the power source. Then 1-to-10 separate locks can be applied to the hasp. This keeps any other authorized worker from reenergizing the system until the last workers have completed their work and removed their locks.

Sometimes several authorized workers must isolate and lock out a large number of sources, requiring multiple locks. This is called group lock-out. In this case, all workers put their personal locks on a group lockbox with all their other lock-out keys inside. Then all systems can only be energized after the last lock is taken off the box by the last worker and all locks removed from the system.

11.13.15 Shift Operations

In shift operations, departing authorized workers shall remove their locks only after the relieving authorized workers have a full understanding of the work

to be accomplished. They must understand where and why the energy sources are locked out. Further, the departing authorized workers shall remove their locks only after the relieving authorized workers have placed their locks on the devices, or have observed same.

During shift operations either maintain continuous control of the energy-isolating devices or require that the oncoming shift verify deenergization and the lock-out/tag-out procedure. In the latter case, the relieved workers remove their locks and/or tags only after their relief has placed their locks and/or tags in place.

11.13.16 Lock-out/Tag-out Checklist

To ensure that proper lock-out/tag-out procedures are being adhered to in your facility, use the following checklist. Add and delete items as appropriate for your facility. First be sure you have identified and labeled all sources of hazardous energy. Before beginning work where there could be a hazardous energy release, are authorized employees in your facility trained to do the following tasks?

❑ Notify all affected employees relying on the energy source.

❑ Deenergize all sources of hazardous energy.

❑ Disconnect or shut down engines or motors.

❑ De-energize electrical circuits.

❑ Disconnect, block or blind fluid (gas or liquid) flow in hydraulic, pneumatic, or process piping systems.

❑ Block machine parts against motion.

❑ Block or dissipate stored energy.

❑ Discharge capacitors.

❑ Release or block springs that are under compression or tension.

❑ Vent or drain gases and fluids from pressure vessels, pipes, and tanks so the pressure at the point where work is being done is at atmospheric. Never vent toxic, flammable, or explosive vapors or gases directly into the atmosphere. Never allow liquids to spill or drain without being captured.

❑ Lock out and tag out all forms of hazardous energy including electrical breaker panels and control valves. (See Section 11.13.1.)

❑ Make sure that only one key exists for each of the authorized employees' assigned locks and that only that employee holds that key.

❑ Verify by test or observation that all energy sources are deenergized.

❑ On pressurized lines, do not trust the gauge. Always open pressurized lines slowly to bleed off the pressure. Gauges are notorious for becoming stuck, especially if they are always reading a relatively constant pressure.

❑ Upon completion of work on a system, inspect the work before removing an assigned lock and activating the equipment.

❑ Make sure that only authorized employees remove their assigned locks.

❑ Make sure that all affected workers are clear of danger points before reenergizing the system. If the work was on a pressurized system, bring the pressure up slowly, always watching for leaks in the system.

CAUTION

WHEN AN AUTHORIZED EMPLOYEE HAS LOST A LOCK-OUT KEY, THE LOCK MAY BE REMOVED BY CUTTING IT LOOSE. IN CERTAIN SITUATIONS THE LOCK MAY BE OPENED BY OTHER THAN THE AUTHORIZED EMPLOYEE. EXTREME CARE MUST BE EXERCISED. THE ASSIGNED SITE SAFETY OFFICER MUST CERTIFY IN WRITING HAVING PERSONALLY INSPECTED THE SYSTEM. FURTHER, THE OFFICER MUST CERTIFY THAT THE SYSTEM IS PROPERLY REPAIRED AND READY FOR OPERATION. THEN THE LOCK MAY BE REMOVED BY CUTTING.

11.14 TRAINING AIDS AND ADDITIONAL RESOURCES

The authors recommend the following videos for aiding readers to see real-life examples of what they have learned in this Section.

- *Fall Protection in the Workplace*
- *Lock-out Tag-out: When Everyone Knows*
- *What's Wrong with This Picture?*

These videos are available from Safe Expectations (formerly BNA Communications, Inc.) (www.safeexpectations.com).

- *Slips, Trips, and Falls, Real, Real Life*
- *Heat Stress—Don't Lose Your Cool*
- *Beat the Heat—Preventing and Treating Heat Disorders*
- *Lockout/Tagout, an Open and Shut Case*
- *Lockout/Tagout, Real, Real-Life*
- *Fire Safety, Real, Real-Life*
- *Fire! in the Workplace*

These videos are available from Coastal Communications (www.coastal.com).

11.15 SUMMARY

Section 11 has presented the kinds of hazards that HAZWOPER personnel might face in any site or facility; hazards not necessarily caused by hazardous materials. We have pointed out that the costs and painfulness of workplace accidents and injuries, not necessarily related to hazardous materials, are staggering. Our definition of an accident is an undesirable, unplanned event that can result in physical harm, environmental harm, damage to property, or interruption of the work schedule.

We trust readers will agree that these hazards are not unique to work at Superfund sites, brownfields, TSDFs, where hazardous materials are used in processing, or in emergency response. Unfortunately these other hazards are common to all types of facilities where people work.

Sometimes forgotten in the rush to complete tasks is the fact that noise is dangerous. When surrounding noise is above the 85 dBA level, see that hearing protection is worn. Remember, hearing loss is the only 100% preventable occupational disease!

Early in the Section we summarized findings of a leading insurer that overexertion, falls, and being struck by an object are among the leading causes of workplace accidents.

In Section 6 we presented PPE from the point of view of protection of HAZWOPER workers from the effects of hazardous materials. In this Section you were introduced to other equipment that can protect you from other workplace hazards. It is a dilemma that the more PPE a worker needs to wear for protection from hazards, the more likely the wearer is to suffer from overexhaustion. And overexhaustion has been cited as an important cause of workplace illnesses.

Obviously judgment, training, and experience are the keys to success. They involve knowing the correct PPE to wear and knowing just how long the wearer can be expected to work at a task in the surrounding conditions. We have explained the stresses of working in high heat and humidity together with practices that can avoid having those stresses lead to illnesses.

Deaths and many injuries occur to workers yearly because they do not recognize all the potential hazards involved in their work assignment. As a life-saving rule, always know where you can safely be in relationship to other workers, machinery, and operations at the site. Electrocutions occur when workers' equipment strikes uninsulated power lines. Do not work with electrical equipment in the field that is not connected to electric power via a ground fault circuit interrupter. Watch out for very cold and very hot surfaces. You can be burned by both!

Workers slip, trip, and fall due to uneven surfaces, debris, and unsafe climbing practices. Warning signage, maintenance, regular housekeeping, and training significantly reduce the incidence of "slip, trip, and fall" accidents. An accident that appears to be occurring more frequently than in the past is the overturning of vehicles at sites with uneven and varying stability soil conditions. You must know the conditions of your walking and driving surfaces.

We stressed that everyone in the reader's organization, you, the people you train, or the people who work for you, must take a proactive approach to safety. Unsafe conditions and unsafe acts that must be eliminated from the workplace were checklisted in this Section.

Engineering controls are equipment designs or modifications put into effect to enhance workplace safety. Machine guards, forced air ventilation, and lock-out devices are prime examples. All workers must be aware of the importance of hazardous energy control. Energy-isolating devices include, but are not limited to, manually operated electrical circuit breakers, disconnect switches, line valves, pipe blanks and blocks, and chocks.

Lock-out and tag-out are vital control practices for all organizations. When properly practiced by well-trained workers, lock-out and tag-out prevent electric shock and electrocution. They protect against engulfment of workers when working at an opened pipeline. They also protect against entrapment or injury by equipment mistakenly being started during maintenance work, as well as a host of other potential incidents.

Remember that OSHA has determined that three out of four accidents that happen in the workplace each year are preventable.

SECTION 12
SAMPLING AND MONITORING

OBJECTIVES

Absolutely the first step in preparation for dealing with potentially hazardous materials threats at an unfamiliar site is to learn the identities, concentrations, and distributions of any such materials. Satisfying these needs for critical site information is the goal of sampling and monitoring. Broadly speaking, this activity includes the analysis and evaluation of all those findings. The objective of Section 12 is to explain how all that is done.

The first seven sections of this manual began with the laws and regulations affecting work at hazardous waste operations. Then we covered the kinds of hazards you can expect to encounter in such work. That was followed by how such encounters can affect your health if you do not wear the recommended protective equipment. We trust readers agree that cleaning up carefully after site work (decon) is as important as careful work on the site.

The previous four sections of this book told about further ways to protect personnel you train or supervise, as well as yourself. These include various kinds of signage that can alert you to the presence of hazardous substances on site or in transport. Then we introduced the MSDSs, which have the backing of law. MSDSs tell you what you need to know about the specific characteristics of materials, all for your own protection. We explained where to find detailed information about practically any material, even if you know only its name, CAS number, or UN/NA number.

Sections 10 and 11 explained things that HAZWOPER personnel need to know, whether project manager or field worker, about common workplace hazards. You need to be alert to the nonmaterial hazards present at almost all sites. This is true whether you work at a TSDF, hazardous materials processing company, Superfund or brownfield, or become involved in an emergency response. We started with the kind of site that we believe is the least understood and least appreciated but that can present considerable danger: confined spaces. This was followed by a summary of other workplace hazards that we all face at one time or another.

With that background understood and appreciated, our objective now is to present information about the way you can approach work with hazardous materials with the confidence that you have the full armor of technology on your side. That is, when you have adequate sampling and monitoring to determine the nature and extent of the hazards you face.

A troubling thought frequently comes to the mind of personnel assigned to an unfamiliar site: "How will I know whether or not a hazardous waste threat exists when I dispatch my workers or I arrive myself at the site?" Answering that question to the best of their ability, and sometimes with limited equipment, is properly the task of those charged with sampling and monitoring.

In Section 12 you will learn that sampling means taking an instrument reading at a site or collecting a physical sample of a gas, liquid, semisolid (sludge), or solid. In the case of any of these media, you may need to carry a physical sample off-site for laboratory analysis. However, each of these media can be analyzed (to some extent) in the field.

Monitoring usually implies an ongoing process. At some hazardous waste sites monitoring has been in progress for decades! During monitoring, sampling and analysis may take place continuously. Usually, short-term monitoring lasts hours or days. Where weeks, months, or years are involved, samples are collected and analyzed in batches, according to a timed schedule, and charted. The question frequently to be answered by monitoring is, "Is the contamination at the site being reduced?" If not, more aggressive remediation action is required.

As described in Sections 5 and 6, the hazards that HAZWOPER personnel are most likely to encounter will be toxic gases and vapors, the so-called inhalation hazards. Remember that they will first attack your breathing passages. Our objective is to concentrate on the most commonly used methods of analysis to determine the presence of breathing, fire, and explosion hazards using air sampling and monitoring instruments. We also look toward the future of HAZWOPER field analysis by presenting the GC and GC/MS methods that we predict will be widely used in the next several decades.

Following that, we deal with sampling of liquids and solids. For the most part, solids and liquids are sampled in the field for later laboratory analysis. There are field test kits that analyze for various contaminants in solids and liquids. Their accuracy is usually sufficient only for screening purposes. Investigations that must be accurate and detailed require laboratory testing in accordance with EPA analytical test methods. Analyses most often take place in an EPA certified laboratory.

As you have already seen, HAZWOPER personnel sample and monitor for critical reasons. We want to know the level of PPE necessary for our protection. We must know the characteristics of the materials we will encounter so we will know how to handle them properly. We need to know the extent of any contamination: Is it contained or is it spreading further into the environment? If it has been released to the air, water or soil, how far has it spread? The objective is to answer these questions as well as is practical before beginning to formulate a comprehensive cleanup plan. That is true whether the site is a Superfund site, a brownfield, a TSDF, a hazardous materials processing plant, or an emergency response site.

> **Always understand what your instrumentation will and will not measure. There is no single instrument that will detect every constituent in the atmosphere, water, or soil.**

12.1 SAMPLING AND MONITORING PROTOCOLS

> **Any samples you take or monitoring that you perform must follow strict protocols. This is to ensure that each time a sampling or monitoring event takes place, it is performed in the same manner. This is the only way to ensure accurate results.**

A sampling and monitoring protocol is an established system of rules and chain of procedures for collecting and analyzing specific samples. HAZWOPER management must ensure that it has the proper protocol in its quality assurance/quality control (QA/QC) manual. Scientists have known for many years that the accuracy and precision of any test results are rated on their repeatability. The chosen protocol must be followed to the letter to ensure repeatability and overall validity. Failure to adhere to the assigned protocol leads to skewed and questionable results. EPA's guidelines can be found in their publication QA/R-5, *EPA Requirements for Quality Assurance Project Plans for Environmental Data Operations.*

There will be different protocols depending on the media to be sampled: air, water, other liquids, or solids. There will also be different protocols depending on the analytes (chemical materials or elements) you believe might be in the samples. Your state's Environmental or Natural Resources division may also have guidelines to follow that are more stringent than those of OSHA or EPA.

Be familiar with two major sources of guidance. The first is the OSHA *Sampling and Analytical Methods,* found in the Companion CD. The second is the EPA's Office of Solid Waste Management (OSW) SW-846, *Test Methods for Evaluating Solid Waste, Physical/*

Chemical Methods. The complete text can be found in searchable format at www.epa.gov/epaoswer/hazwaste/test/sw846.htm.

12.2 DATA LOGGING, CHAIN OF CUSTODY, AND FORENSIC EVIDENCE

⊙━π **A chain-of-custody form must accompany every environmental sample you take for analysis.**

⊙━π **Every time you monitor the environment of a work site, the measurements must be data logged electronically in the monitoring equipment and written in a logbook. Even if your monitoring equipment has data-logging capabilities, you must make a habit of keeping a logbook including the other pertinent information about the monitoring event.**

As the individual charged with sampling and monitoring, when you take each sample, either by a direct reading instrument or as a physical sample, you must record identifying information. Here is a generic checklist of data-logging information to be recorded.

❑ The unique identifying number for the sample taken or read. Label a physical sample.

❑ The time, date, location, and method of sampling together with your signature.

❑ The approximate amount of a physical sample collected.

Note

Most analytical lab protocols will require a minimum amount (grams for solids, liters for liquids and air) of any physical samples. Their analytical procedures require at least that minimum for accuracy and precision in their analyses.

❑ Temperature, relative humidity, and weather conditions. (These may all be important to the analyst and for the accuracy of field instruments.)

❑ Type, manufacturer, model, and serial number of any instrument used.

❑ Calibration information for any instrument used. (Data analysts use this information to help determine measurement accuracy. Also, use any of the instrument's technical literature in order to perform any necessary data correction.)

❑ If you have taken a physical sample, you will need to fill out a chain-of-custody form for the sample. This form is usually provided by the analytical lab. When properly used, this form must record, without any unaccounted for times, all people who have had custody (personal charge) of the sample. All EPA-approved labs require chain-of-custody forms. Figure 12.25 shows a chain-of-custody form. (See page 12.45)

12.2.1 The Value of Forensic Evidence

⊙━π **Forensic evidence is evidence that is legally defensible in a court of law.**

According to EPA QA/G-5, *EPA Guidance For Quality Assurance Project Plans,* sample custody procedures are necessary to prove that the sample analytical data correspond to the sample collected, if data are intended to be legally defensible in court as evidence. In a number of situations, a complete, detailed, unbroken chain-of-custody will allow the documentation and data to substitute for the physical evidence of the samples (which are often hazardous waste) in a civil courtroom. Some statutes or criminal violations may still necessitate that the physical evidence of sample containers be presented along with the custody and data documentation.

Note

Chain of custody is an important feature of evidence that can be considered forensic evidence. True forensic evidence is that which will stand up in court. Remember some instances in the widely publicized O. J. Simpson trial. At times the prosecution attempted to present physical evidence on the record that they felt was critically important. The defense countered that there were lapses in the chain of custody. Where that contention was upheld, the evidence was flawed forensically, and of much less value.

Although many of the devices incorporate data-logging by design, we must continue to use sampling and monitoring logbooks as backup. All trainers, supervisors, and workers must be conscientious at all times about keeping accurate records of all of these events.

12.3   AIR SAMPLING AND MONITORING

12.3.1   Analyzing for Vapors and Gases in the Atmosphere

⊶ **Inhalation is the most common route of entry of hazardous material into the body. We cannot see or smell many vapors and gases. Those factors make air sampling and monitoring two of the most important tasks HAZWOPER personnel must perform.**

⊶ **In any atmospheric conditions worldwide, gases such as oxygen and nitrogen in the air, as well as hydrogen and helium, always remain gases.**

⊶ **All vapors condense to liquid at the proper lower atmospheric temperature. Vapors that easily condense at lower temperatures to liquids and then vaporize easily as the temperature rises are considered volatile.**

In any air-sampling or monitoring project the first task is to determine what kind of hazardous materials vapors or gases are present. We are interested in what hazards are present in the atmosphere of the work site. Vapors are given off from liquids. The warmer the ambient temperature, the greater the vaporization from liquids. An example is humidity forming in the atmosphere from rainwater. Also in the atmosphere, if it is cold enough, vapors condense back to liquids (dew forms on the grass from condensation of the water in humid air.)

⊶ **Vapors, such as humidity (moisture) in the air, condense to liquid at the proper lower temperature (like dew that forms when the temperature drops in the evening). Condensed (liquid) materials that are volatile (such as water or gasoline) vaporize easily with rising temperatures.**

⊶ **We learn what elements and chemicals are present in a material sample by qualitative analysis. For example, we learn whether or not an air sample contains carbon monoxide. This is also called detection. Quantitative analysis then tells us how much, if any, carbon monoxide is present in the air sample.**

We cannot get the valid sampling and monitoring data we seek unless we use instruments or devices that are sensitive to, and calibrated for, the actual vapors or gases present. Our first task is referred to as detection. Its goal is to tell us enough about the kinds of materials, or the specific materials, that are present in the atmosphere. That is the *qualitative* type of analysis. This information is essential so that we may select a suitable measuring device. The device will require calibration so that the concentration data we measure will be accurate. When we determine accurate concentration data (ppm, ppb, or %, for example) we are doing *quantitative* analysis. When we know the identity of materials in the atmosphere, and their concentration, we can compare our findings with the published regulatory data to tell us what kind of threat, if any, we face.

If this is not done properly, we might get readings for a toxic gas or vapor that are 10 times too high or 10 times too low! Since all or part of this analysis must be done in the field, each instrument or device must have characteristics of portability, ruggedness, and ease of operation for a given project.

There are at least four acceptable categories of methods used in monitoring air quality at hazardous waste sites or in emergency response situations. Additional methods and improvements in detection, sensitivity, portability, and economy are being introduced yearly. However, we have chosen to explain the methods currently in wide use, as well as those we predict will find growing use in the first decade of the 21st century. Some of the methods have been in use for decades. The methods are:

1. Chemical detector tubes

2. Portable meters and detectors

3. Portable gas chromatographs (GC) or gas chromatograph-mass spectrometer combinations (GC-MS)

4. Other laboratory instruments, modified for field use

We will discuss the advantages and limitations of each of these methods. As you will see, to be properly armed to seek out atmospheric contaminants, you will need to have access to more than one method of detection and measurement. Sample canisters and bags that can transport materials for laboratory analysis without introducing contamination or allowing loss of components are growing in importance.

For more detailed information than that presented here, consult the companion CD section "*OSHA Sampling and Analytical Methods*." Note that OSHA is uniquely interested in worker protection from air contamination. Remember that breathing is the most common route of contaminant entry into the human body. (See Sections 5 and 6.)

12.3.2 Performance Characteristics Required of Portable Air-Sampling and Monitoring Instruments

> **The performance characteristics required of portable sampling and monitoring instruments are portability, rugged construction, ease of operation, accuracy, precision, and inherently safe construction.**

To be useful and produce reliable results in detecting and measuring hazardous atmospheres, air-sampling and monitoring devices must be:

• Portable and rugged

• Easy to operate (especially while sometimes cumbersome PPE is being worn)

• Able to collect samples and provide data with the desired accuracy and precision

• Inherently safe

Portability means having a unit weight of less than about 30 lb, being no more bulky to carry than a small suitcase, being battery powered by commonly available batteries, and being readily rechargeable.

Rugged construction is sometimes overlooked. Shocks from dropping or bumping in the field are routine occurrences. Operating temperature ranges from about −20°F to 120°F must be expected. High humidity (over 80%), rain, and mild acid or alkali conditions all can be expected at one time or another.

Easy operation is important. For instruments, a large, easily operated keypad for input, and easily viewable digital readout are practically essential. There will be enough challenges at the site to contend with without fumbling with troublesome operation. Furthermore, you want operator training to involve the least complication. With that training, improper operation and incorrect data entry must be highly unlikely.

Accuracy of the measured data means that the data you read on the instrument are acceptably close to the true values; that is, the data is not skewed toward higher or lower values than the true values.

Precision means that when you take multiple readings of the same sample at the same point, they will only vary by a tolerable amount. The "scatter" in that data will be low. When that is the case, it is easier to detect small, true differences between samples.

Explosion-proof or intrinsically safe ratings mean that an instrument or device is designed and constructed so that, as long as it is undamaged, it cannot generate a hazardous discharge or spark, and that it is so certified. Underwriters Laboratory, Inc. (UL) and Factory Mutual Research Corp. (FM) issue the approvals and certification.

⊙━┓ **Know the difference between explosion-proof and intrinsically safe. Explosion-proof instruments are encased in a housing that, when undamaged, will not allow an explosive atmosphere to enter the instrument. Intrinsically safe instruments generate no electrical sparks or arcs during normal operation.**

WARNING
IMPROPER USE OF EITHER AN EXPLOSION-PROOF OR AN INTRINSICALLY SAFE DEVICE CAN RESULT IN AN EXPLOSION OR FIRE HAZARD. FOR INSTANCE, DROPPING AN EXPLOSION-PROOF INSTRUMENT AND BREAKING ITS PROTECTIVE CASING MIGHT CAUSE AN EXPLOSION OR FIRE IN AN EXPLOSIVE ATMOSPHERE. ALSO, CARELESSLY BANGING AN INTRINSICALLY SAFE INSTRUMENT AGAINST A SPARKING METAL SURFACE IN AN EXPLOSIVE ATMOSPHERE MIGHT CAUSE AN EXPLOSION OR FIRE.

After an explanation of the National Electrical Code classes, divisions, and groups, further information will be given on instrument design to avoid ignition incidents.

12.3.3 Types of Hazardous Atmospheres

Depending upon the HAZWOPER worker's background, the term *hazardous atmosphere* conjures up situations ranging from toxic air contaminants to flammable atmospheres.

In Section 3 we described the varieties of toxic materials. In Section 5 we explained the human health hazards they present.

In Sections 12.3.5 and 12.3.6 we will give more information about the wide variety of flammable and explosive atmospheres that HAZWOPER personnel may encounter. Refer to Section 3.2.4 for further explanations of flammability and LFL, UFL, LEL, and UEL.

12.3.4 NEC Criteria for Hazardous Atmospheres

For National Electrical Code (NEC) purposes, an atmosphere is hazardous if it meets the following criteria:

• It is a mixture of any flammable material in air having a concentration that is within the material's flammable range, that is, between the material's lower flammable limit (LFL) and its upper flammable limit (UFL). These are nearly the same as the lower explosive limit (LEL) and upper explosive limit (UEL).

• There is the potential for an ignition source to be present.

- A fire at one point can easily propagate beyond where it started. An example is the ignition of heavier-than-air flammable vapors. Flame can rapidly spread great distances through low-lying, invisible vapor clouds.

NEC Classes, Divisions, and Groups.   To describe hazardous atmospheres adequately the NEC categorizes them according to their classes, divisions, and groups.*

Classes

- Class I: Flammable vapors and gases such as gasoline and hydrogen
- Class II: Combustible dusts such as coal and grain
- Class III: Ignitable fibers

   Divisions.   *Division* describes the location of generation and release of the flammable material.

- *Division 1* is a location where the generation and release are continuous, intermittent, or periodic into an open, unconfined area under normal conditions.
- *Division 2* is a location where the generation and release are only from ruptures, leaks, or other failures from closed systems or containers.

Groups

- *Group A, B, C, D:* Chemicals having similar flammability characteristics
- *Group E, F, G:* Dusts having similar combustible properties

Selected Class I Chemicals by Groups

   Group A atmospheres:
      Acetylene
   Group B atmospheres:
      1,3-butadiene
      Ethylene oxide
      Formaldehyde (gas)
      Hydrogen
      Manufactured gas (containing greater than 30% H_2 by volume)
      Propylene oxide
      Propyl nitrate
      Allyl glycidyl ether
      N-butyl glycidyl ether
   Group C atmospheres:

Acetaldehyde	Ethylene glycol	Nitropropane
Carbon monoxide	Epichlorohydrin	Tetrahydrofuran
Crotonaldehyde	Ethylene	Triethylamine
Dicyclopentadiene	Ethyl mercaptan	Ethylene glycol
Hydrogen selenide	Hydrogen cyanide	Diethyl ether
Hydrazine	Di-isobutyl amine	Hydrogen sulfide
Methylacetylene	Morpholine	Tetraethyl lead

---

*Source: National Fire Protection Association, ANSI/NFPA 497M, *Manual for Classification of Gases, Vapors, and Dusts for Electrical Equipment in Hazardous (Classified) Locations,* 1986.

Group D atmospheres:

Acetone	Methane	Methyl ethyl ketone
Methanol	Acrylonitrile	Benzene
Ammonia	Naphtha	Styrene
Propane	Butane	Vinyl Chloride
Chlorobenzene	Acetonitrile	

Selected Class II Chemicals by Groups

Group E, conductive dusts: Atmospheres containing metal dusts, including aluminum, magnesium, and their commercial alloys, and other metals of similarly hazardous characteristics.

Group F, semivolatile dusts: Atmospheres containing carbon black, coal, or coke dust with more than 8% volatile material.

Group G, nonconductive dusts: Atmospheres containing flour, starch, grain, carbonaceous materials, chemical thermoplastic, thermosetting, and molding compounds. One of the authors experienced a grain dust explosion that demolished a silo in the city of Philadelphia.

WARNING

ALL OF THE ABOVE EXAMPLES OF DUSTS ARE MATERIALS THAT PROBABLY YOU KNOW WILL BURN. HOWEVER, SPECIAL CARE MUST BE GIVEN TO ANY WORK WITH DUSTS. THE AUTHORS KNOW OF AN EXPLOSION INVOLVING PORTLAND CEMENT DUST. IT OCCURRED DURING THE BOTTOM UNLOADING OF PORTLAND (DUSTY) CEMENT FROM A TRAILER. THE UNGROUNDED TRAILER AND PART OF THE ADJACENT BUILDING WERE TOTALLY DEMOLISHED. FORTUNATELY, NO ONE WAS PRESENT AT THE MOMENT OF EXPLOSION. THE BEST PROTECTION AGAINST SUCH AN INCIDENT IS SECURE EQUIPMENT GROUNDING OF THE VEHICLE BEING UNLOADED. THAT WILL PREVENT STATIC ELECTRIC CHARGE BUILDUP AND THEN SPARK DISCHARGE IGNITION.

Remember the fire triangle from Figure 3.1. The proper combination of fuel and oxygen (in air), together with heat (from a spark), causes fire or explosion. Although we commonly consider portland cement to be a nonfuel material, it can contain small percentages of organic materials that are fuels, that is, that can burn. A spark discharge in that air-fuel mixture could have caused the explosion. There is another possibility. An onboard blower on a bulk material trailer is routinely used to provide the aeration, vibration and directional airflow needed to unload dry bulk materials. In those cases fuel or other flammable vapors from surrounding contaminated air, may be drawn in with the unloading air, and a spark could cause that mixture to explode. For HAZWOPER personnel, the lesson learned is that when transferring dusty dry material or flammable liquids, you should always use the protection of secure equipment grounding.

Bonding and Grounding

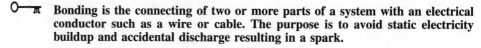 **Bonding is the connecting of two or more parts of a system with an electrical conductor such as a wire or cable. The purpose is to avoid static electricity buildup and accidental discharge resulting in a spark.**

The two types of grounding are earth grounding and equipment grounding. Earth grounding gives electrical overloads, such as lightning strikes, a safe path

to the earth. Equipment grounding protects both people (electrical shocks) and equipment (electrical shorts) from hazardous electrical discharge.

Bonding is the process of securely connecting two or more parts of a system with an electrical conductor such as a wire or braided cable. Bonding must be used, for instance, when pouring flammable liquids from one container to another. As liquids and dusts fall through air, they create static electricity. If that electricity is discharged as a spark, an explosion can occur. Then what about filling my gas tank at the gas station, you might wonder. The nozzle of the filling hose contacts the metal of your gas tank-filling spout. The hose itself has a wire wrapped around it that completes the circuit back to the pump. Disastrous fires have occurred when plastic gasoline cans were filled while in the plastic bed liner of pickup trucks. Static discharge sparks ignited the fuel vapors.

The NEC defines a ground as "a conducting connection, whether intentional or accidental between an electrical circuit or equipment and the earth, or to some conducting body that serves in place of the earth." *Earth grounding* is an intentional connection from a circuit conductor (usually the green colored or plain copper ground wire in a three-wire conductor) to a ground conductor (usually a copper rod) driven into the earth. *Equipment grounding* ensures that electrical equipment within a structure is properly grounded. These two grounding systems are required by code to be kept separate except for a connection between the two systems to prevent differences in potential (overload) from a possible flashover from a lightning strike. The purpose of a ground is to protect people (from electrical shock) and equipment. It is also to provide a safe path for the dissipation of fault currents (electrical shorts), static discharges, and interference.

12.3.5 Characteristics Designed into Sampling Instruments to Avoid Ignition Incidents

There are six prime requirements for instruments that can avoid ignition incidents, some or all of which are engineered into the design of each sampling device or instrument:

1. Intrinsically safe
2. Explosion-proof
3. Purged system
4. Non-incendiary
5. Dust-ignition-proof
6. Certified

An instrument that is certified as explosion-proof, intrinsically safe, or purged for a given class, division, and group is certified as not contributing to ignition. Of course, it must be used, maintained, and serviced according to the manufacturer's instructions. Explosion-proof is the only rating approved for use in explosive hazardous locations. Class I, Divisions I, II, Group D, and Class II, Divisions I, II, Group G, as defined in the *National Electrical Code, Article 500.*

Any instrument is not, however, certified for use in atmospheres other than those indicated. All certified instruments must be marked to show class, division, and group. Any manufacturer wishing to have an electrical sampling or monitoring instruments certified must submit a prototype to an approved laboratory for testing. If the unit passes, that model is certified as submitted.

However, the manufacturer agrees to allow the testing laboratory to randomly check the instrument manufacturing plant at any time, as well as any marketed units, for strict compliance with the certification. Furthermore, any change in the unit requires the manufacturer to notify the test laboratory. The laboratory can continue the certification or withdraw it until the modified unit can be retested.

Neither OSHA nor NFPA performs certification testing. Underwriters Laboratory Inc. (UL) or Factory Mutual Research Corp. (FM) perform certification testing. Currently, these are the only two testing labs whose certification is recognized by OSHA.

12.3.6 Oxygen, Flammability, and Multigas Monitors

Note

These instruments and devices are basic to the most widely used methods for determining the degree of hazard that HAZWOPER personnel encounter on the job. They are at times called indicators, meters, monitors, or detectors. In most cases they provide solutions to three of our main concerns about our working atmosphere. Our first concern is to ensure a breathable atmosphere. Our second concern is to avoid a flammable atmosphere. Finally, we want to know about the presence of common toxic materials in the atmosphere.

The fact that in most cases these instruments can provide that atmospheric information and be hand-held is of great importance. Sampling and monitoring personnel are frequently weighted down with bulky, heavy PPE and other required working tools. Where less common toxic vapors and gases are suspected to be present, we look toward the capabilities of specialized detector tubes and instruments described in later subsections.

Oxygen Indicators.   Oxygen indicators are used to evaluate atmospheres for the following conditions:

- *Oxygen content for respiratory purposes.* Pure dry air contains about 20.9% oxygen. OSHA requires that the oxygen content of breathing air be between 19.5% and 23.5%, by volume (not by weight). If the oxygen content falls below 19.5%, the atmosphere is *oxygen-deficient* and entrants need supplied-air respirators.
- *Increased risk of combustion.* Oxygen concentrations above 23.5% are considered *oxygen-enriched*. Oxygen-enriched atmospheres increase the speed of flame-spread once a fire starts. (Remember the deaths of the three astronauts from a flash fire in the capsule at the Cape. That tragedy occurred during oxygen-enriched [approximately 60%] atmospheric tests in the flight crew's compartment.)
- *Presence of contaminants.* A decrease in oxygen content can be due to the consumption (by combustion or an oxidation reaction such as rusting) of oxygen or the displacement of air by a chemical vapor or gas. If it is due to consumption, then the concern is the lack of oxygen. If it is due to displacement, then a material is present that could be flammable or toxic in addition to lowering the oxygen content.
- *Use of other instruments.* Most instruments require sufficient oxygen for operation. For example, most combustible gas indicators do not give reliable results at oxygen concentrations below 5%. Many of these instruments' LEL readings will begin to degrade as soon as the atmospheric oxygen content begins to drop below the concentration at which the instrument was calibrated (20.9%). Also, the inherent safety approvals for instruments are for normal atmospheres and not for those that are oxygen-enriched.

⚷ **Always check oxygen (O_2) content in the atmosphere before checking for flammability and other atmospheric contaminants.**

Oxygen meters are available in pocket-size units weighing approximately 4 oz. We have chosen the ToxiRAE PLUS® monitor as an example (see Figure 12.1). It provides an accuracy of 0.1% in a range from 0–30% O_2. It has high (greater than 23.5%) and low (less than 19.5%) oxygen level audible alarms of around 96 dBA (about as loud as a power drill.)

You can check an oxygen meter's operation in the field by first exposing to an uncontaminated atmosphere and expecting to read 20.9% oxygen. To check the meter's operation

FIGURE 12.1 ToxiRAE PLUS® oxygen gas monitor. (*Courtesy of RAE Systems, Inc.* [www. draesystems.com].)

further, you can take a breath and exhale into the probe. The air you exhale will have an oxygen content of approximately 10–13%. This will actuate the meter's low alarm (set at 19.5%). If this simple field check does not give you these results, the meter must be recalibrated.

Combustible Gas Indicators.   Combustible gas indicators (CGIs) measure the concentration of a flammable vapor or gas in air. They usually indicate test results as a percentage of the lower explosive limit (LEL) of the calibration gas. Frequently methane (the chief component of natural gas) is the calibration gas. The management of some facilities prefers the use of lower flammability limits (LFL). Most times the LEL and LFL are rather close together. The LEL of methane is 5%, and its LFL is 6%. Thus, if the methane concentration were 5%, the CGI would read 100% of the LEL.

Most safety managers will not allow workers in atmospheres that are anywhere near the LEL (or LFL). OSHA's safety rule is intended to prevent workers from entering an atmosphere that is richer in flammable gas than 10% of the LEL (or LFL). This would mean, for methane contamination, entrance would not be allowed if the LEL reading were 10% or greater on the indicator (that is, 0.5% methane or greater in the atmosphere).

CGIs are used to determine whether organic chemical vapors are present and normally read in percent of the LEL. There are several types of CGIs, which analyze for LEL differently. The most common types use a metal oxide sensor to measure the catalytic reaction between the chemical and air and transform that information into usable data (% of LEL).

The response of the instrument is temperature-dependent. If the temperature at which the instrument is zeroed differs from the sample temperature, the accuracy of the reading is affected. Higher temperatures raise the temperature of the filament and produce a higher than actual reading. Lower temperatures reduce the reading. It is best to calibrate and zero the instrument at the sample temperature.

The instruments are accurate only in normal oxygen atmospheres. Oxygen-deficient atmospheres will produce lowered readings. Also, the safety guards that prevent the combustion source from igniting a flammable atmosphere are not designed to operate in an oxygen-enriched atmosphere. LEL—reading instruments can give accurate readings up to 30% oxygen.

Sulfur compounds and silicone compounds may foul the filament. Acid gases, such as hydrogen chloride and hydrogen fluoride, can corrode the filament.

Using the current microelectronics available, there are meters for O_2, LEL, carbon monoxide (CO), hydrogen sulfide (H_2S), and a host of other gases. The gases can be measured by pocket-size, single-gas meters weighing around 4 oz.

Combustible gas indicators also come in pocket-size units. The ToxiRAE PLUS® Combustible LEL Monitor (Figure 12.2) has a response time t_{90} of 15 seconds, a range of 0 to 100% of LEL, and an accuracy of 1%. t_{90} means the instrument reads 90% of what it will ultimately read (if you wait a longer time) in the number of seconds stated.

You can check an LEL meter's operation in the field by first exposing to an uncontaminated atmosphere and expecting to read 0.0% LEL. To check the meter's operation further, you can expose the probe to the organic vapor from a permanent marker, such as a Sharpie®. This will cause the meter to register an LEL reading. If this simple field check does not give you these results, the meter must be recalibrated.

FIGURE 12.2 ToxiRAE PLUS® combustible LEL monitor. (*Courtesy of RAE Systems, Inc.* [www.raesystems.com].)

WARNING

DO NOT USE A BUTANE LIGHTER TO PERFORM THE ABOVE FIELD CHECK. BUTANE LIGHTERS ARE NOT PERMITTED AT HAZARDOUS WASTE SITES, NOR ARE ANY OTHER TYPES OF LIGHTERS OR MATCHES.

Multigas Meters

Multigas meters are the most common type of work-site air monitoring instrument. They will measure oxygen content as well as LEL. They come in a wide variety of gas- and vapor-detecting combinations. One of the most common configurations of meters has sensors for O_2, LEL, CO, and H_2S. However, the latter two sensors can be replaced with two others chosen from a variety of available sensors.

Multigas meters test for a wide variety of hazardous vapors and gases. These meters have become very popular tools. There is a growing market for affordable, durable, easily operated meters for use in confined space entry, at hazardous material sites, and during emergency response actions. These meters may be customized by the addition or substitution of a variety of modules.

For instance, if a response worker needs a hand-held meter for testing the atmosphere of a sewer, an excellent choice of meter is one that, fitted with the proper modules, measures O_2, LEL, H_2S, and CO.

Acceptable oxygen content must be determined first for two reasons:

1. To find whether or not breathing oxygen content is present (between 19.5% and 23.5% oxygen)
2. Because the meter will not function properly if the O_2 content is outside this range

Since an LEL reading is likely to occur in sewer gas due to the presence of methane caused by the anaerobic (oxygen-deficient) digestion of organic material, the LEL reading

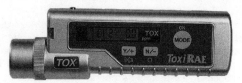

FIGURE 12.3 ToxiRAE PLUS® Toxic Gas Monitor. (*Courtesy of RAE Systems, Inc.* (www.raesystems.com.)

TABLE 12.1   ToxiRAE PLUS® Toxic Gas Monitor. Model PGM-35 Gas, Range, Resolution, and Response Times

Sensor	Symbol	Range	Resolution	Response time (t_{90})
Carbon monoxide	CO	0–500 ppm	1 ppm	20 sec
Hydrogen sulfide	H_2S	0–100 ppm	1 ppm	30 sec
Sulfur dioxide	SO_2	0–20 ppm	0.1 ppm	15 sec
Nitric oxide	NO	0–250 ppm	1 ppm	20 sec
Nitrogen dioxide	NO_2	0–20 ppm	0.1 ppm	25 sec
Chlorine	Cl_2	0–10 ppm	0.1 ppm	60 sec
Hydrogen cyanide	HCN	0–100 ppm	1 ppm	60 sec
Ammonia	NH_3	0–50 ppm	1 ppm	150 sec
Phosphine	PH_3	0–5 ppm	0.1 ppm	60 sec

Source: Courtesy of RAE Systems, Inc. (www.raesystems.com).

is necessary. Remember that OSHA requires that the LEL reading be below 10% for normal work to take place.

H_2S is also a product of the anaerobic digestion of organic material. Since H_2S is both toxic and flammable, an H_2S reading is essential. Remember that the IDLH level for H_2S is only 100 ppm!

CO is also both toxic and flammable. A product of incomplete combustion, it is present in motor vehicle exhausts and often in sewers. The IDLH level is only 1200 ppm.

The ToxiRAE PLUS® toxic gas monitor can continuously monitor for carbon monoxide, hydrogen sulfide, sulfur dioxide, nitric oxide, nitrogen dioxide, chlorine, hydrogen cyanide, ammonia, and phosphine.

With interchangeable sensors, it can monitor for CO, H_2S, sulfur dioxide (SO_2), nitric oxide (NO), nitrogen dioxide (NO_2), and chlorine (Cl_2). With noninterchangeable sensor units, it can monitor for hydrogen cyanide (HCN), ammonia (NH_3), and phosphine (PH_3). Notice the knurled, removable cap at the left end of the ToxiRAE PLUS® in Figure 12.3. It is removable for replacement of a defective sensor or to exchange for optional sensors. This can be performed by the customer, and the unit need not be sent back to the manufacturer. See Table 12.1 for Range, Resolution, and Response Times for this instrument.

12.3.7   Measuring Hazardous Atmospheres via Detector Tubes

Colorimetric gas detector tubes are a relatively inexpensive method of detecting and getting an approximate measure of concentration of a wide range of hazardous materials in the air. You get the wide range by using a variety of tubes. Most tubes detect only a single gas or vapor. They are single-use, disposable devices.

Advantages and Disadvantages of Detector Tube Use.   Hazardous atmosphere sampling and monitoring can involve detection and measurement of a wide variety of common and not so common hazardous gases and vapors. They can be members of many different chemical families, which presents major chemical analysis complications. The requirements in these cases extend far beyond the capabilities provided by the oxygen indicators, the combustible gas indicators, and even the multigas meters discussed in the previous subsection.

Detector tubes, also called gas detector tubes, Dräger® tubes, or colorimetric tubes, are one-time use, disposable, very easy to use devices. In many cases they can provide the level of detection and measurement required. They are an inexpensive alternative to more costly instruments. However, they do not provide the accuracy and precision of the portable photoionization detector (PID), gas chromatograph (GC), or gas chromatograph/mass spectrometer (GC/MS), covered in later subsections.

Dräger®, and Other Gas-Detector Tubes.   See Figure 12.4. Gas-detector tubes are often generically called Dräger® tubes or colorimetric tubes. The German company Dräger® invented the system more than 60 years ago. Although many other companies, such as MSA and Sensidyne, offer a wide range of gas-detection tubes, the name Dräger® remains attached, much as the Kleenex® name is used when referring to other facial tissue.

○──ᴨ   **Gas-detector tubes are extremely useful when screening to detect hazardous gases and vapors in an atmosphere before making concentration measurements. By using enough different tubes, you can learn which hazardous vapors or gases are present. Then you can select the best meter to get much higher accuracy in concentration measurement.**

Accuracy of Gas-Detector Tubes.   Gas-detector tubes have the advantage of ease of use following a brief training period. As their name implies, they are for detecting gases and vapors. They can provide rough concentration readings, when the proper tube is used, in a matter of minutes. They have the disadvantage of low accuracy and precision in measuring concentrations. In the past, NIOSH tested and certified detector tubes that were submitted to them. For the tubes they tested they certified the accuracy to be ±35% at concentrations of one-half the PEL and ±25% at one to five times the PEL. NIOSH has discontinued testing and certification. Special studies have reported concentration measurement errors of 50% and higher for some tubes.

These results vary widely from manufacturer to manufacturer. Dräger® offers a wide selection of tubes for sampling of about 500 materials and certifies the accuracy of around 150 of the tubes.

These tubes can detect either chemical families or individual chemicals. Most suppliers of gas-detection tubes offer systems for simultaneous sampling through multiple tubes. This acts as a decision tree, allowing the sampling technician to screen for a range of chemicals. As the sampling progresses, the range narrows until an individual contaminant is being sampled for its concentration.

FIGURE 12.4 A selection of Dräger® tubes. (*Courtesy of Drager Safety* [www. draeger.com].)

The Measurement Method Using Gas Detector Tubes.   Measurement requires at least one detector tube, a hand-operated or battery powered pump, and connecting fittings. A gas-detector tube is a thin glass straw containing specific reagents. It is hermetically sealed to maintain the purity of the contents. To collect a sample, the technician snaps the straw at both ends, inserts the tube into the pump, and draws the required volume of (contaminated) air through the tube. The contents are various chemical reagents that change color when exposed to specific gases or vapors. That color is then visually checked against a standard, and the comparison indicates the approximate concentration of the gas or vapor being measured.

Some tubes determine concentration by how far a color change travels up the tube, the printed concentration being read off the tube at the color interface. Others determine concentration by the intensity of a color change. In either case, the concentration is indicated directly on the tube or by comparison to an accompanying chart.

The chemical reactions involved in the use of the tubes are affected by heat and moisture. It is important to keep these tubes at a moderate temperature, up to about 75°F, or refrigerated during storage.

The following is a checklist of considerations that technicians must keep in mind when interpreting the changes they see in the gas detector tubes.

❑ Some tubes do not have a prefilter to remove humidity and may be affected by high humidity.

❑ Some tubes will respond to interfering gases or vapors. Read the manufacturer's instructions carefully.

❑ The chemical reactants used in the tubes deteriorate over time. Thus, the tubes are assigned a shelf life, varying from one to three years. Check that the shelf-life date of a tube has not expired.

❑ Interpretation of results can be a problem. Some color changes are diffuse, not clear-cut; others may have an uneven end point.

❑ The total volume to be drawn through the tube varies with the tubes. The volume to be drawn is given as the number of pump strokes or seconds needed, that is, the number of times the piston or bellows is manipulated manually, or the number of seconds or minutes an electric pump is run. (Figure 12.5 is an example of a manual pump. Manual pumps are the ones used most commonly.)

❑ The air being sampled does not instantaneously go through the tube. It may take one to two minutes for each volume (stroke) to be completely drawn. Therefore, sampling times can vary from 1 to 30 minutes per tube. This can make the use of detector tubes time-consuming.

We include detector tubes here because of their widespread use in HAZWOPER work. They have been popular in sampling and monitoring for decades. They are inexpensive to use compared to other instrumentation. They are lightweight, most of the weight being in the pump. They provide the accuracy sufficient for screening purposes. Also, the technology is constantly being improved.

FIGURE 12.5 RAE® Colormetric tube hand pump. (*Courtesy of RAE Systems, Inc.* [www.raesystems.com].)

With hundreds of tubes available to test for specific, and sometimes very unusual, gases and vapors, these tubes fill a special detection niche. We anticipate that gas-detector tubes will continue to be a common method of sampling for toxic and contaminated atmospheres for some time to come.

12.3.8   Ionization Detectors

O—π   **Many gases and vapors, particularly the vapors of volatile organic compounds (VOCs), have their ionization potential (IP) listed in the NIOSH Pocket Guide in electron volts (eV). (See the Companion CD.)**

O—π   **If you know that a single contaminant is in the atmosphere, say benzene, you can find its IP (9.24 eV) in the Guide. Then, using a PID containing a UV lamp that emits rays with a power just above the IP of benzene, you can accurately measure its concentration in the atmosphere. You do that by using the PID manufacturer's correction factor (CF) for benzene to convert from the calibration gas reading to the correct benzene reading in ppm.**

O—π   **If there is a mixture of three kinds of vapors in the atmosphere, all with IPs lower than the PID's installed lamp, the concentration you read in ppm will be for a total of all three in terms of the calibration gas.**

FID, PID, and CDID Types Compared.   At least three types of ionization are methods used in hand-held detector instruments, based upon flame, photo, and corona discharge ionization, respectively. Flame ionization detectors (FIDs) were the only type in wide use several decades ago. There must be tens of thousands still in use. They are rugged and reliable. However, they are somewhat bulky to carry. They usually require a hydrogen gas source to generate a flame with air. Sample gas molecules that are fragmented in passing through the flame (ionized), form the ions that are measured in the detector.

Photo ionization detectors (PIDs) have been in use for a few decades. See Figures 12.6 and 12.7. They are now the most widely used of the three types. They are less bulky and lighter weight than the FIDs. PIDs are reliable and now accurate in the ppb range.

The newest method, corona discharge ionization detectors (CDIDs), as the name implies, use a high-energy corona to ionize sample molecules. The CDID instruments are intended to detect VOC concentrations in the ppb range with IPs up to 11.5 eV. They also detect concentrations from 0.01 to 10,000 ppm.

Note that all three types depend upon ionization of sample molecules. They will all use the same reference tables, widely available, for the characteristic ionization potentials (IPs)

FIGURE 12.6   ppbRAE® PID. (*Courtesy of RAE Systems, Inc.* [www.raesystems.com].)

FIGURE 12.7   CDRAE® corona discharge ionization detector. (*Courtesy of RAE Systems, Inc.* [www.raesystems.com].)

of chemicals. That is how they identify individual chemicals. In the next subsection we will focus on PIDs, currently the most widely used of these three types of detectors.

12.3.9 Photoionization Detectors: Method of Operation

○—π **The photoionization detector (PID) is one of the most widely used air sampling instruments in HAZWOPER work.**

○—π **The PID is more expensive to buy than colorimetric tubes. However the PID has a long-lifetime lamp as compared to the single-use tubes.**

○—π **PID detector use requires more training than use of detector tubes.**

○—π **PIDs can detect and measure concentration of a range of hazardous gases, achieving much higher accuracy than detector tubes, with a single instrument.**

These devices are sometimes called organic vapor analyzers (OVAs). They are most commonly used for the field detection of organic chemical vapor contaminants in the atmosphere.

○—π **Organic chemicals are those containing carbon, and they almost always contain some hydrogen in their molecular structure, with some exceptions. Calcium and magnesium carbonates, such as found in limestone, are inorganic chemicals containing carbon.**

Organic chemicals are those containing carbon, and they almost always contain some hydrogen in their molecular structure. They may also contain other elements such as chlorine, nitrogen, and sulfur. They are largely the product of the petroleum and petrochemical industries. However, over the past 100 years they have spread to become either raw, in-process, maintenance, product, or byproduct materials of almost every industry. Think of all those wonder pharmaceuticals that you inhale or swallow. Practically all of them are organic chemicals, or, more accurately, synthetic organic chemicals.

Later you will see that PIDs also detect a much smaller number of inorganic chemicals; ammonia being one of them.

A PID consists of:

- A data read-out method (meter, screen, and sometimes a port for data download to a computer or printer)
- A rechargeable battery as a power supply
- A probe and an air pump for capturing the vapor sample
- An ionization source consisting of a UV lamp
- Electronics for signal processing (newer models contain microprocessor chips for on-board data storage and other functions)
- An alarm that can be set to warn of hazardous vapor concentrations

PIDs detect and measure concentrations of gas and vapor molecules in air. We will concentrate on vapors, which are produced chiefly by the vaporization of volatile liquids. In HAZWOPER work we frequently encounter VOCs. Typically they are all kinds of liquid fuels, solvents, components of paints and other coatings. They also come from many kinds of chemical wastes.

PIDs depend upon a high-energy ultraviolet (UV) light source to ionize sample molecules. "Photo" is Greek for "light." Rays from the energetic UV lamp ionize vapor molecules, producing a positively charged fragment (+) and an electron (−). Lamp energy is rated in electron volts (eV). Lamps are produced that generate radiation in several energy levels, one characteristic level for each lamp.

See Figure 12.8. Those ionized molecular fragments flow to oppositely charged electrodes in the detection system. Depending upon the amount of ionization that occurs, there is more or less current flow through the detector. That causes the concentration of vapor to be detected and output to the display. Photoionization detectors are calibrated with a calibration gas specified by the manufacturer.

In this hand-held VOC detection device, the sizes of units for field use are being reduced while data collection and data presentation features are being increased. This size reduction is happening in much the same way as in the computer field. In fact, marrying the detector to the computer promises greatly expanded data analysis and presentation advances.

Note that the sample to be ionized is drawn into the instrument and passes over the lamp. The ionized fragments pass between the electrodes in the detector. Upon exiting the detector, the vapor or gas largely reforms into the original molecules.

○──π **The ionization potential is that energy in electron volts (eV) needed to free an electron from a molecule. This produces a positively charged fragment and the negatively charged electron.**

Factors Affecting PID Operation.   Dust in the atmosphere can collect on the lamp and block the transmission of UV light, causing a reduction in instrument reading. Humidity can cause two problems. When a cold instrument is taken into a warm, moist atmosphere, the moisture can condense on the lamp. Like dust, this will reduce the available light. Moisture in the air also reduces the ionization of molecules and causes a reduction in concentration reading.

Since an electric field is generated in the sample chamber of the instrument, radio-frequency (RF) interference from power lines, transformers, generators, radio wave and microwave transmission may produce a false response. Careful record- and log-keeping, including the weather at time of day and any possible interference at the site when you take the sample, is essential.

As the lamp ages, the intensity of the light decreases and the response declines. These changes will be detected during calibration, and adjustments can be made. However, the lamp will eventually burn out and require replacement.

○──π **Always understand what your instrumentation will and will not measure.**

What Does a PID Measure?   The largest group of chemicals measured by a PID is the organics, chemicals whose molecules contain carbon atoms. They also nearly always contain hydrogen (H) atoms. Frequently organic molecules contain oxygen (O), nitrogen (N), or sulfur (S) atoms. All ionization devices measure only the vapor state, not liquids or solids.

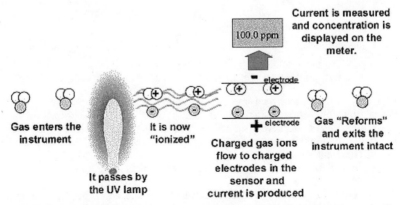

FIGURE 12.8   UltraRAE® PID internal operation. (*Courtesy Christopher Wrenn, Application Note #203, RAE Systems, Inc., Sunnyvale, CA* [www.raesystems.com].)

Organic vapors that can be detected and measured include the following chemical families:

- *Aromatics:* chemicals containing a benzene ring, including benzene, ethyl benzene, toluene and xylene. These are chemicals found in gasoline and many petroleum products.
- *Ketones and aldehydes:* chemicals with a C=O double bond, including acetone, methyl ethyl ketone (MEK), and acetaldehyde. These are materials used as solvents and chemical intermediates.
- *Amines:* chemicals containing nitrogen, such as methylamine, found in decaying meat and fish; used in production of other chemicals.
- *Chlorinated hydrocarbons:* trichloroethylene (TCE), perchloroethylene ("perc"), chemicals found in solvents and cleaning wastes. This family also contains the chlorofluorocarbon refrigerants such as some freons.
- *Sulfur compounds:* mercaptans. These are the odorant chemicals added to natural gas. Without them you couldn't smell a natural gas leak! The most commonly found sulfur compound is hydrogen sulfide (H_2S). It has the "rotten egg" odor.
- *Unsaturated hydrocarbons:* e.g., butadiene and isobutylene, synthetic rubber ingredients.
- *Alcohols:* chemicals with a hydroxyl group (OH) attached to a carbon atom in a compound. Examples are methanol found in antifreeze and the ethanol in beverage alcohol.
- *Saturated hydrocarbons:* e.g., butane (in lighters), isooctane in gasoline, and cetane in diesel fuel.
- *Inorganic vapors,* which characteristically do not contain carbon, that are detected include:
  - *Ammonia* (NH_3), used as fertilizer and in window cleaners.
  - *Gases used in semiconductor chip manufacturing,* including arsine and phosphine.
  - *Nitric oxide* (NO), a component of smog. Also found in all combustion exhausts, along with other nitrogen oxides, referred to as NO_X.
  - *Bromine and iodine* (Br_2 *and* I_2), used in chemical manufacturing, and found in small amounts in all seawater.

What PIDS Do Not Measure

- *Nuclear radiation.*
- *The gases found in air,* N_2, O_2, and CO_2.
- *Common toxics,* carbon monoxide (CO), present in auto exhausts; sulfur dioxide (SO_2), the sharp odor from a burning match.
- *Natural gas,* even though it contains the saturated hydrocarbons methane, ethane, and propane. (Their IPs are too high for the available UV lamps, and the lamps are not made more powerful because they would then ionize the gases in air and cause serious interference.)
- *Acid gases,* hydrochloric (HCl), used in concrete cleaners, hydrofluoric (HF), used in glass etching, and nitric (HNO_3), widely used in the chemical industry.
- *Ozone* (O_3), a gas found in the Earth's ozone layer, also formed by reactions of internal combustion engine exhausts with oxygen and sunlight.
- *Nonvolatiles* (*they give off very little vapor*), PCBs, formerly used as dielectric (insulating) fluids in power transformers and fluorescent lamp transformers; lube greases; and waxes.

Photoionization Considerations and Limitations.

The ability to detect the molecules of a sample chemical, using a device that relies on an ionization detector, depends on the device's ability to ionize the molecules. Photoionization power is rated in electron-volts (eV). As can be seen in Table 12.2, too great a lamp power will cause problems. There is a lamp power limit imposed by the components of air. That is to say, the lamp cannot be too energetic.

TABLE 12.2 Ionization Potentials (IP) of Selected
Chemicals

Gases and vapors	Ionization potential (eV)
Hydrogen cyanide	13.9
Carbon dioxide	13.8
Methane	13.0
Hydrogen chloride	12.5
Water	12.6
Oxygen	12.1
(Highest power common lamp)	(11.7)
Chlorine	11.5
Propane	11.1
Hydrogen sulfide	10.5
Hexane	10.2
Ammonia	10.1
Vinyl chloride	10.0
Acetone	9.7
Benzene	9.2
Phenol	8.5
Ethylamine	8.0

Source: Courtesy of NIOSH.

Notice from Table 12.2 that if the lamp had a power just over 12 eV it would cause ionization and detection interference from the common atmospheric gases oxygen and carbon dioxide. Further, the common atmospheric water vapor (humidity) would ionize and interfere with the readings for the gas or vapor contaminants.

The energy values of lamps include 8.3, 8.4, 9.5, 10.2, 10.6, 10.9, 11.4, 11.7, and 11.8 eV. Not all lamps are available from a single manufacturer. Some PIDs have more than one lamp installed at a time. Lamps burn out just like light bulbs and must be replaced. Lamps must be cleaned on a regular basis so they can deliver their full energy output. That makes for accurate analyses.

One use of the different lamps is for selective determination of chemicals. For example, if a spill of propane and vinyl chloride were to be monitored with a PID, the first check would be to see if they could be detected. The ionization potential (IP) of propane is 11.1 eV, and the IP of vinyl chloride is 10.0 eV. To detect both, a lamp with energy greater than 11.1 eV is needed (like one rated at 11.7 or 11.8 eV). If the highly hazardous vinyl chloride were the only organic vapor of concern, then a lamp with an energy greater than 10.0 but less than 11.1 eV (such as 10.2 or .10.6 eV) would have been used. In that case, the propane molecules would neither be ionized nor detected. Thus, propane would not interfere with the vinyl chloride readings.

Benzene has an IP of 9.2 eV . It can be ionized by the next-higher power UV lamp, a standard 10.6 eV lamp. Methylene chloride has an IP of 11.32 eV. It can only be ionized by the highest power common lamp, an 11.7 eV lamp. Carbon monoxide has an IP of 14.01 eV and cannot be ionized by the commonly available PID lamps. IPs can be found in the NIOSH Pocket Guide, PID manufacturer's literature, and some chemical reference texts. The NIOSH Pocket Guide is included in the Companion CD.

Ionization Potential of Selected Gases and Vapors.   As shown in Table 12.2, it takes a specified minimum ionization potential by the lamp to cause the gas or vapor molecule to break into positively and negatively charged fragments, or ions. Traditionally, the unit has a separate sensor unit because the four common lamp power ratings available: 8.3, 9.5, 10.2 (standard), and 11.7 eV all require separate electronic circuits. To change the energy of ionization, the whole sensor must be switched, not just the lamp. Lamps are replaceable upon failure.

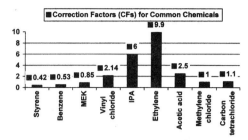

Note: Styrene, Benzene, MEK, Vinyl Chloride, IPA, and Ethylene CFs are for a 10.6 eV Lamp; Acetic acid, Methylene chloride, and Carbon tetrachloride CFs are for an 11.7 eV Lamp

FIGURE 12.9 PID correction, or relative response, factors. (*Courtesy of RAE™ Systems, Inc., Sunnyvale, CA* [www.raesystems.com].)

What Is a Correction Factor?   Correction factors (CFs), also known as relative response factors, are an important correction to make when using PIDs (see Figure 12.9). They are a measure of PID sensitivity to a particular gas. CFs permit calibration on one gas while directly reading the concentration of another, eliminating the need for multiple calibration gases.

CF Measures Sensitivity.   The lower the CF, the more sensitive the PID is to a gas or vapor. In Figure 12.9, benzene has a CF of 0.53. The PID is roughly 18 times as sensitive to benzene as it is to ethylene, which has a CF of 9.9. In general, it is acceptable to use PIDs to measure gases with CF's up to 10. PID manufacturers publish CF lists, and some integrate this information into a microprocessor installed in the PID.

CF Adjusts Sensitivity when Measuring Pure Compounds.   CFs are scaling factors used to adjust the sensitivity of the PID to measure a particular vapor directly compared to the calibration gas. (Note that isobutylene in this case is loosely referred to as a gas. It is actually a liquid that boils to a vapor at about 20°F.) For example, a PID is nearly twice as sensitive to benzene (CF = 0.53) as it is to its calibration gas of isobutylene (CF = 1.00). Therefore, if we are measuring 1 ppm of benzene, after calibrating on isobutylene, we have the following options:

- We will see approximately 2 ppm on the display of the PID. If we multiply this reading by 0.53, we will get the true reading of benzene.

- We can apply the CF of 0.53 during calibration so that the display will automatically read in ppm benzene.

- Microprocessor-equipped PIDs can automatically store and apply many correction factors automatically.

By using a CF of 0.53, we can reset the internal scale of the PID to read 1 ppm of benzene even though we calibrated on isobutylene.

PID manufacturers determine CF by determining a PID's response to a known concentration of target gas. Correction factors tend to be instrument- and/or manufacturer-specific, so it is best to use the CFs from the manufacturer of the PID. Therefore, it may be best to choose a PID manufacturer with the largest listing of CFs.

How to Determine If a PID Can Measure a Particular Vapor

1. Is the IP of the vapor less than the eV output of the lamp?
  - Yes, according to the NIOSH table. *Go to step 2.*
  - No. Then a PID with that lamp cannot measure that vapor.

- Don't know. Check the PID manufacturer's literature for the lamp rating. Compare that with the NIOSH table IP value for the vapor. If the vapor IP value is less than that of the lamp rating, you can measure the vapor.

2. Is the CF less than 10?
   - Yes. A PID is an appropriate way of measuring that vapor.
   - No. A PID is not an accurate means of measuring that vapor. The PID could still be a way of detecting gross contamination, such as in leak detection.
   - Don't know. Most PID manufacturers can help.

PID Lamps, 9.8 and 10.6 eV versus 11.7 eV. At first glance, it may appear that to measure the broadest range of gases with a PID, an 11.7eV lamp should be used instead of a 10.6eV lamp. However, the following must be considered:

- Nine point eight and 10.6 eV lamps are more specific and more accurate for vapors with lower IPs.
- Nine point eight and 10.6 eV lamps are less expensive than 11.7 eV lamps.
- Eleven point seven eV lamps generally have a shorter life than 9.8 or 10.6 eV lamps. 11.7 eV bulbs should only be used when compounds with IPs over 10.6 eV are expected (e.g., methylene chloride, chloroform, and formaldehyde).

PID Calibration and Correction Factors (or Relative Response) of Vapors. Photoionization detectors are calibrated to a single chemical's vapor. The instrument's response to chemicals other than the calibration vapor can vary. Table 12.3 shows the correction factors of several chemicals for a specific PID.

At high concentrations, the instrument response can decrease. While the response may be linear (i.e., 1-to-1 response) from 1 to 600 ppm for an instrument, a concentration of 900 ppm may only give a meter response of 700.

Calibrating the PID and Adjusting to a Known Chemical. When an unknown chemical release is approached, the PID is set to its calibration gas. Once the chemical is identified by means of placard, manifest, waybill, or other means (see Section 8), the PID sensitivity can be adjusted for that chemical vapor. For example, if we calibrate on isobutylene and happen to measure a benzene leak of 1 ppm, the PID will display 2 ppm because it is twice as sensitive to benzene as it is to isobutylene. Once we have identified the leak as benzene, then the PID scale can be adjusted using a benzene CF. Then the PID will accurately read 1 ppm when exposed to 1 ppm of benzene.

O—π **Remember that no CF is applied until a chemical vapor is definitely identified.**

TABLE 12.3 Correction Factors for Selected Chemicals Using the HNu Model P1101 with 10.2 eV Probe Calibrated to Benzene

Chemical	CF or relative response
M-Xylene	1.12
Benzene	1.00
Phenol	0.78
Vinyl chloride	0.63
Acetone	0.50
Hexane	0.22
Phosphine	0.20
Ammonia	0.03

Source: Courtesy of HNu® Systems, Inc., Newton, MA (www.hnu.com).

Setting the PID Alarm for a Single Gas or Vapor

- Identify the chemical.
- Set the PID correction factor for that vapor using the PID manufacturer's listing.
- Find the exposure limit(s) for the vapor (OSHA/NIOSH/ACGIH and Section 5.23.1).
- Set the PID alarm according to the exposure limit.

PID Alarms for a Gas/Vapor Mixture with Constant Make-up. Often HAZWOPER incidents do not involve a single chemical. Rather, they involve a mixture of toxic chemicals. The "witches' brew" of toxic vapors that forms requires greater care in determining alarm setpoints. If the mixture is identifiable, then the individual chemicals and their concentrations should be easily determined. Alarm setpoints are usually based on the concentration of the most toxic compound. Many times this determination is as simple as reading the MSDS sheet for a compound like paint, which often contains significant amounts of solvents like xylene and toluene, both with TWA values of 100 ppm. (Refer to Section 5.24 for TWA and PEL explanations.)

Do the following:

- Find the average makeup of the mixture.
- Determine the most toxic vapor.
- Base setpoints on the most toxic vapor.

For example, after determining that oxygen is in the breathing range and that we are detecting far lower vapor concentrations than the LEL, we can focus on toxicity hazards in the presence of gasoline vapors. While the typical TWA for gasoline is 300 ppm and the STEL is 500 ppm, we can set alarms based upon the relative concentration of chemicals in gasoline.

Gasoline is a mixture of hydrocarbons and some oxygen-containing ethers and possibly ethyl alcohol. It contains hydrocarbons such as isooctane, which has a 100 octane rating. There are smaller amounts of benzene, ethyl benzene, toluene, and xylene, totaling roughly 15%. The gasoline oxygenate methyl tert butyl ether (MTBE), which is being removed from gasoline, also makes up approximately 15%. Most components of gasoline are readily ionizable by a PID. The PID will be measuring a total of all ionized chemicals as ppm equivalent of the PID's calibration standard. The PID manufacturer will recommend a standard for gasoline measurement. First we will focus on benzene, which is by far the most toxic.

Regular, unleaded gasolines contain less than about 5% benzene. Benzene's PEL is only 1 ppm. Benzene is a listed human carcinogen by IARC (see Section 3.1.3). Therefore, in a worst-case scenario where gasoline has 5% benzene content, 100 ppm of gasoline vapor means that you are exposed to as much as 5 ppm of benzene! Then 20 ppm of gasoline would be an appropriate level to start using respiratory protection. This is summarized as follows:

- Gasoline contains as much as about 5% benzene.
- Benzene is carcinogenic (PEL = 1 ppm).
- One hundred ppm of gasoline could contain as much as 5 ppm benzene.
- Set the PID high alarm at 20 ppm gasoline to stay below 1.0 ppm benzene.
- Set the PID low alarm at 10 ppm gasoline to warn of approaching the PEL and therefore approaching the need for respiratory protection.

PID Usage in Initial Hazard and PPE Needs Assessment. When approaching a potential HAZWOPER incident, the responder must make a PPE decision. Some potential incidents may not be an incident at all and may not require any PPE. Some incidents may initially appear to have no contamination yet require significant levels of PPE.

○━🔑 **No single sampling or monitoring device will provide all the answers about hazards and their concentrations in the atmosphere.**

However, the PID is an excellent aid in this decision-making process. For many incidents, the PID lets the responder identify the presence or absence of potentially toxic gases or vapors. Sometimes it does this by detecting a family of chemicals, rather than an individual chemical.

The following information gleaned from an article by Christopher Wrenn, Product Application Manager of RAE® Systems, Inc., is included here with the permission of RAE® Systems. Visit the RAE Systems® site (www.raesystems.com) for more systems details.

A HAZWOPER certified contractor was called by railroad personnel to respond to a leaking chemical tank car. It was a high-humidity (95% RH) summer day. According to the manifest, the tank car was loaded with benzene. Due to the carcinogenic nature of benzene (PEL of 1 ppm), the contractor chose to dress out in Level A PPE (see Section 6.)

However, because it was a hot summer day, this potentially exposed the responders to heat stress injuries. In the assessment of the leaking tank car, it was found that the puddle under the car was coming from moisture condensation and dripping from the tank underside (see Figure 12.10.)

The resulting pool did not contain benzene. The tank car had been loaded at 65°F. The high ambient temperature combined with relative humidity above 95% produced a puddle of water, not benzene.

Using a PID would have helped the contractor determine whether there was an ionizable vapor present. Because the manifest identified the tank car contents as benzene and benzene is readily ionizable, the contractor could have ruled out the presence of benzene vapors using a PID. This would have reduced the cost of the response and prevented the potential of heat-stress injuries from dressing out workers in full Level A PPE.

The PID as a HAZWOPER Emergency Response Tool.   The previous subsections give a broad overview of the capabilities of the traditional PID. As the PID becomes one of the instruments of choice for HAZWOPER workers, we feel inclusion of a more detailed description of its capabilities and usage is in order.

Photoionization detectors (PIDs) can accurately detect and measure low levels (0–2000 ppm) of hazardous VOCs and other toxic vapors and gases. This makes PIDs valuable tools for making HAZWOPER decisions, including:

* Initial assessment of PPE requirements
* Leak detection
* Site perimeter establishment and maintenance
* Spill delineation
* Decontamination progress
* Remediation Accomplishment

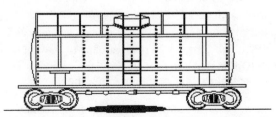

FIGURE 12.10   Railcar leaking benzene? (*Courtesy of RAE® Systems, Inc., Sunnyvale, CA* [www.raesystems.com].)

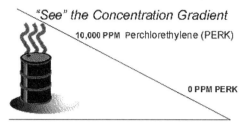

FIGURE 12.11 Vapor concentrations usually decrease considerably as distances increase from the source. (*Courtesy of RAE Systems®, Inc., Sunnyvale, CA* [www.raesystems.com].)

Using a PID to Ensure Decontamination. Hazardous materials often splash, spill, or condense on responders. For ionizable compounds like fuels and other VOCs, a PID is a quick and effective means of determining if a responder requires decontamination. Then after decon the PID can determine if the cleansing has been complete. This may make it easier for a HAZWOPER team to make a decision whether to reuse a valuable encapsulation suit.

Often at a fuel-spill incident, gasoline splashes on a firefighter's flame-retardant turnout gear (clothing). Absorbed gasoline will compromise the flame-retardant properties of turnout gear. The PID will quickly respond to such contamination and identify this dangerous condition. If it is detected, the turnout gear can be properly laundered and retreated before being used in firefighting.

Leak Detection with a PID. Often the source of a leak is not readily apparent. The source must be located before leakage can be stopped. Any time a gas or vapor is released into air it disperses away from the source of the leak (see Figure 12.11). As the gas or vapor disperses, it is diluted by ambient air until at some point the gas or vapor cannot be detected. This process establishes a concentration gradient. The concentration of the gas or vapor is greatest at the source of the leak. The concentration is effectively zero somewhere along the path of dispersion.

The PID allows us a convenient method to measure and track concentration gradients back to the source of a leak.

As shown in Figure 12.12 for a liquid spill, say of diesel fuel, in an area where there is significant standing water, a PID can spot a diesel puddle. Then limited absorbent can be applied directly to the diesel fuel itself.

Perimeter Monitoring with a PID. HAZWOPER personnel arriving at the site of a release assess the incident. They establish a site perimeter based upon the surrounding environment, toxicity of any released gas or vapor, temperature, wind direction, and other factors.

FIGURE 12.12 Using a PID to find oil amid an oil-water puddle. (*Courtesy of RAE Systems, Inc., Sunnyvale, CA* [www.raesystems.com].)

However, people without a high degree of experience may control perimeters. As conditions change, perimeters often are not adjusted because perimeter workers do not have the experience to recognize that the conditions have changed. The experienced HAZWOPER personnel typically are focused on the problem of dealing with the release. Therefore, perimeter workers and bystanders are often unprotected from changing conditions that may require movement of a perimeter farther away from the release site. A PID in the hands of those controlling a perimeter line gives them the easily understandable information to adjust the perimeter in response to changing conditions.

Data Logging for Forensic Purposes.   Data logging PIDs provide HAZWOPER supervisors with documentation of exposure levels and provide evidence to justify evacuations should they be required. However, many HAZWOPER teams only data log incidents at times when the PID records atmospheric contamination. This misses an important use of data logging. Many times a negative result on a data log is more beneficial than a positive result. Saving a "nondetect" can help to establish quickly that a release was promptly and properly contained. This can save time and money if a release or spill ever results in legal action.

Photoionization Detectors Compared to Gas Chromatographs

- A single-lamp PID, without modification, has no separation capability. When only a single known PID measurable vapor is present, that PID can report its concentration.

- More than one vapor can be present within the PID's detectable range of IPs. That PID reports a total concentration of the mixture. It does not measure the individual vapors.

- The gas chromatograph, with its separation column and detector, has both separation and detection capability. It can detect dozens of vapors and accurately measure their concentrations in a mixture.

Gas chromatography will be explained in the following subsections, together with the substantial special benefits it provides. PIDs provide excellent accuracy and sensitivity in the detection and measurement of families or classes of gases and vapors, especially the vapors from organic chemicals. However, without the separation column of the gas chromatograph, and with only one UV lamp, the PID cannot analyze (report, identify, and determine concentrations) for more than one gas or vapor at the same time.

Gas chromatographs detect and measure tens to hundreds of individual vapors in one analysis. (The GC can be fitted to handle liquid samples by vaporizing them before analysis.) PIDs measure in the 0.1–2000 ppm range, with 0.1 ppm resolution. Note, however, that the corona discharge ionization detector is designed to detect concentrations at the ppb levels, and in the range of 0.1 to 10,000 ppm. Gas chromatographs can reach down to the ppt levels and measure all concentrations.

The advantages of the PID are that it is small, lightweight, and far less expensive than a portable GC. The portable GCs are roughly 20 to 30 times larger and heavier than PIDs. PIDs are continuous monitors. Where a single known vapor is present, they provide instantaneous detection and ppm concentration readings to workers. GCs, on the other hand, take batch samples that must then pass through the separation columns before passing through the GC's detector and providing readout. A GC sample can be quite small, say several milliliters.

Personal PIDs.   Personal PIDs are pocket-size instruments weighing as little as 6 oz. They can be clipped onto a belt or pocket. OSHA acknowledges that they are useful devices to improve worker protection while reducing costs of unnecessary PPE and to aid the selection of appropriate PPE in hazardous waste operations.

Figure 12.13 shows one of the ToxiRAE PLUS® PID Monitor PGM-30® series models. This little device can detect VOCs in the 0–99.9 ppm range with an accuracy of 0.1 ppm.

FIGURE 12.13 ToxiRAE PLUS® PID monitor model PGM-30. (*Courtesy of RAE Systems, Inc., Sunnyvale, CA* [www.raesystems.com].)

In the 100–2000 ppm range its accuracy is 1 ppm. These personal PIDs monitor continuously with a 10-second response time.

The ToxiRAE® PID can be used:

• In the field when proper factors of safety are built into the alarm settings. The unit will require a trained person to determine the alarm settings and interpret the output

• As a personal monitor to indicate when PPE should be upgraded

• As a means to alert the industrial hygienist of a potential problem

12.3.10 The Portable Gas Chromatograph (GC)

○━π **The portable GCs can collect gas and vapor samples and analyze them for multiple analytes in the field.**

○━π **The portable GCs can measure gases and vapors down to the ppb range and can measure concentrations from zero to 100%.**

○━π **GCs can also analyze readily vaporized liquids by preheating before entry into the GC column.**

Background.   Gas chromatography analyzers have been developed over the past 50 years or so. For most of that time they have been specialized laboratory instruments. Now portable instruments designed for field use are being developed, especially for both screening purposes and final determinations of contaminant concentrations.

The portable instruments and the laboratory instruments each have their own role to play in the identification and quantification of volatile organic compounds (VOCs), particularly the hazardous members of that family. The portable GCs collect samples and analyze them in the field. The laboratory analyses require that reliable samples be collected in the field and transported to the laboratory unchanged by loss or contamination.

We will explain the use of the highly specialized Summa®-type of canister for collection and transport to the laboratory of truly representative field samples. See page 12.29. "Use of SUMMA®-Type Canisters for Preferred Collection of Air Samples." The laboratory GC analyses have the advantage of highly controlled conditions and the ability to use large-volume, heavyweight, equipment and instrumentation. All of those factors contribute to the ability of the laboratory GC to provide the most sensitive detection of chemicals in a sample and the highest accuracy in determining their concentrations.

We pay special attention to the portable GCs because they are probably the most complicated (but valuable) devices that HAZWOPER personnel will be using. HAZWOPER personnel will also be collecting samples for laboratory analyses to be conducted by other technicians.

The mass spectrometer (MS) is another specialized laboratory instrument that has been brought into field use as the detection section of a GC. This has created the portable GC/MS, an extremely useful (and expensive) sampling and monitoring tool. Worldwide effort is being applied to make GC/MS analyzers of much lighter weight. The U.S. Army is calling for the development of hand-held devices. The GC/MS is described in Section 12.3.11.

"Gas" is in the gas chromatograph's name because it can only separate gases (actually gases and vapors) for analysis in its internal analyzing column. (It vaporizes liquid samples before entry into its column.) Also, entirely different devices are used in laboratories that perform liquid chromatography. ("Chroma" is in the name because in earlier, somewhat related devices, colored liquid mixtures could be separated to reveal individually colored chemicals.) "Graphy" refers to the ability to write, or show, the separated materials.

How Does Gas Chromatography Work?   In gas chromatography analysis, the operator injects a sample of a gas, vapor, or liquid mixture into the inlet end of the GC's column. Samples are relatively small. Injection may be by means of a hypodermic needle. The column is a long (up to 50 m, or so), coiled tube. It can be several millimeters in inside diameter and have specially treated granular media inside.

Optionally, the column can be fine-bore (inside diameter less than 1 mm) capillary tubing with a special coating on the inside wall. The length can be 50 m or more. In operation, a pure inert 'carrier gas,' such as nitrogen or helium, flows through the column continuously. It carries the sample mixture through the column.

Ideally, the molecules of each sample component will be delayed differently in their travel through the column. The delay depends upon each component's differing absorption and desorption by the granular media or the coating on the wall of the tubing. More volatile components, such as those in gasoline, will tend to be absorbed and then desorbed quickly and will travel through the column and exit rapidly. Less volatile molecules in the sample mixture, such as lube oil molecules, will be absorbed and more slowly desorbed.

Thus, individual components will exit the column at different times. Those exit times are characteristic for each chemical species, for a specific column, and for its operation. The time it takes for an individual chemical to travel through a specific column is called its retention time. Methane (natural gas) will have a shorter retention time than say, octane, (a gasoline constituent).

A specially designed detector identifies each batch of separated chemical molecules exiting the column. A variety of ionization, far ultraviolet absorbance, thermal conductivity, electron capture detectors, and others are in use. EPA laboratories and others favor the mass spectrometer as the detector for certain analyses. That combined instrument is referred to as the GC/MS. The output signals from any of these detectors are usually recorded as peaks on a graph. The area under a peak reveals that chemical's concentration in the sample. Other types of analysis outputs are available.

All GC analyzers take snapshot analyses. They analyze a sample collected at one point in time, as opposed to continuous second-by-second, on-line monitoring as done, for example, by a PID (as described above under Photoionization Detectors Compared to Gas Chromatographs.) GCs can be designed to run an analysis in a matter of seconds or minutes and then be ready to run the next analysis. In that case, they are analyzing in a batch-continuous mode.

Laboratory Gas Chromatography.   The portable GC instruments, to be described later, do not replace laboratory instrumentation. The portable instruments do provide highly valuable field screening information that would be impossible for the lab instruments. Laboratory GC instruments, however, are highly developed to perform difficult analyses on complex samples under highly controlled conditions. Results of work with the laboratory instruments provides guidance for the outfitting of components for portable GCs to increase their detection capability and accuracy.

Refer to EPA's Compendium Method TO-14A, *Determination of Volatile Organic Compounds (VOCs) in Ambient Air Using Specially Prepared Canisters with Subsequent Analysis by Gas Chromatography,* for the details of laboratory methods for the determination of toxic organic compounds in air samples. TO-14A was prepared by the Center for Environmental Research Information, Office of R&D, U.S. EPA, Cincinnati, OH 45268.

Use of SUMMA®-Type Canisters for Preferred Collection of Air Samples

O—π   **SUMMA®-type canisters are specially prepared sample collection canisters that prevent loss of sample integrity. They prevent contamination from impu-**

rities in the container, sample loss due to leakage or adsorption on the walls of the canister, and sample decomposition due to light exposure.

Note that TO-14A refers to collection of air samples "using specially prepared canisters" (see Figure 12.14.) There has been quite a bit of study on the best methods for collecting air samples in the field that will be truly representative. Past methods of field sample collection have been subject to contamination by the collection containers, problems in capturing samples, and problems in protecting samples until their analysis in the lab.

The 6-L, stainless steel, canister-based method for collecting field samples of VOCs has proven to be a widely used approach. It is based on research and evaluation performed since the early 1980s that has verified the lack of contamination by the canisters and the sample stability of VOCs in canisters. EPA initially summarized the canister-based method in the original method TO-14 document. The canister-based method is now considered a better alternative than the solid sorbent-based methods.

As you will see from the NIOSH Pocket Guide on the companion CD, there are still over 20 types of adsorbent and absorbent methods in wide use for collecting samples for laboratory analysis. However, the SUMMA® canister method is considered to have parts per trillion by volume (pptv) detection limits for samples of typically 300–500 mL of whole air.

EPA's method TO-14A was originally based on collection of whole air samples in SUMMA® passivated stainless steel canisters, but it has later been generalized to include other specially prepared canisters. The EPA provides procedures for sampling into canisters to final pressures both above and below atmospheric pressure. This sampling is referred to as pressurized and subatmospheric pressure sampling.

This canister-collection method is applicable to specific VOCs that have been tested and determined to be stable when stored in pressurized and subatmospheric pressure canisters. However, numerous VOCs have shown successful storage stability in canisters. Measurements at the ppbv level are considered accurate.

This method applies under most conditions encountered in sampling of ambient air into canisters. Under conditions of normal usage for sampling ambient air, most VOCs can be recovered from canisters near their original concentrations after storage times of up to 30 days.

HAZWOPER personnel could be assigned to collect both subatmospheric pressure and pressurized samples. Both modes typically use an initially evacuated canister and pump-ventilated sample line during sample collection. Pressurized sampling requires an additional pump to provide positive pressure to the sample canister. A sample of ambient air is drawn through a sampling train. It has components that regulate the rate and duration of sampling into a preevacuated, specially prepared, passivated canister.

Passivation is part of the special preparation of the inside of the SUMMA®-type canister. It helps to prevent sample materials from being adsorbed on the interior wall of the canister.

FIGURE 12.14 SUMMA®-type canisters.
(*Courtesy of Meriter* [www.canister.com].)

Otherwise part of the sample would be held inside the sample container and lost from the analysis.

After the air sample is collected, the canister valve is closed. An identification tag is attached to the canister, and a chain-of-custody (COC) form is completed to accompany the canister. Upon receipt at the laboratory, the canister tag data are recorded, the COC is completed, and the canister is attached to the analytical system.

As required during analysis, water vapor is reduced in the gas stream by a Nafion® dryer. The VOCs are then concentrated by collection in a cryogenically cooled trap. (The VOCs are condensed to liquids and the air gases are exhausted.) Liquid nitrogen can be used as the cryogen. The cryogen is then removed and the temperature of the trap is raised. The VOCs are vaporized, injected into the GC column for separation, and then detected by one or more detectors for identification and determination of their concentrations.

Individual states are adopting EPA's air-sampling method, using the SUMMA®-type canister of sample collection, into their own VOC and petroleum hydrocarbon air sampling methods.

Plastic Air-Sampling Bags for Screening Studies

O—π **A widely used and inexpensive alternative to canister collection of gas and liquid samples in the field is the use of specially designed plastic bags. They provide somewhat less integrity than the canisters, depending on the sample being collected.**

A much more common, and far less expensive, method of collecting air samples uses specialized plastic bags. This method is not expected to collect and deliver to the lab samples preserved as carefully as with the SUMMA® method.

The most common air (and liquid and solid) sampling bags for screening studies are made from DuPont Tedlar® polyvinyl fluoride (PVF) film (Figure 12.15.) This is a special type of plastic that offers low permeability so that there is minimal sample loss. These bags are usually clear but are also offered in black for use with photoactive analytes. The bags come in varying sizes from 0.5–100 L. Sample size will depend on the requirements of the lab. In general, larger samples yield more accurate results.

These bags are most commonly used for collecting air samples to be analyzed in a lab by a GC. They are fitted with a stainless steel sampling port that is then connected to the sampling pump for filling. These ports are self-sealing to avoid sample loss.

Development of the Portable GC for Field Use.   A few decades ago, GC instruments were exclusively laboratory research devices. Then models were developed that could be used on the factory floor. They analyze in-process materials. Over the past decade or two, portable GCs have been developed. Portable GCs can analyze gases, vapors, or vaporized liquid samples in the field. They can detect the components in field samples and record their concentrations down to the ppb levels. Further development will stretch that capability to the ppt levels.

It has been a feat of science and engineering to convert the laboratory GC analyzer, or even the factory floor type, to a portable model for field use. It can have multiple separation

FIGURE 12.15 DuPont Tedlar® air-sampling bags. (*Courtesy of Ben Meadows Company* [www.benmeadows.com].)

columns and multiple specialized detectors that provide highly accurate measurements of a specific kind of vapors. However, it is still a challenge to produce a full-featured, portable GC for less than about $30,000. Most units cost above $50,000. Another continuing challenge is to make the portable GC light enough (say 20 lb), and compact enough to be carried continuously by a HAZWOPER field worker.

That said however, the portable GC is a remarkable analyzer. It can be used in the field to provide a wide range of sampling and monitoring analyses, together with detection accuracy, that cannot be provided by any other field-operating device.

The Portable Gas Chromatograph.   We have chosen, as an excellent example of this class of instruments, the Hnu™ Model 311-D Portable Gas Chromatograph, as shown in Fig. 12.16. This microprocessor-based instrument is designed to perform analyses in the field. The microprocessor controls both the GC and the communication between the user and the GC. Operators communicate with the 311-D using the unit's internal firmware via the on-board keypad. Optionally, you can use the external PeakWorks® software, which operates with a Pentium® or later model PC and a Windows™ 95, 98, Me, XP, or later operating system.

The Model 311-D is a completely self-contained GC for field use. There it operates using line power, a portable generator, or a battery pack via an inverter.

Major GC Components.   The major components of the portable GC are:

- A carrier gas supply and sample pump
- Interchangeable columns, providing the separation capability
- A temperature-controlled oven to contain the column(s) for accuracy and precise measurements
- Multiple or interchangeable detectors for measurement at ppb levels of specific families of vapors
- A microprocessor, keyboard input, LCD and printer-plotter output
- Optional download to a computer for further data analysis

The automatic sampling mode of operation allows unattended monitoring of ambient air by automatically injecting air (or other gas stream) samples into the analytical columns.

Carrier Gas Supply and Sample Pump.   The on-board carrier gas cylinder and hydrogen supply (for FID or FPD detectors, as described later) are the only gas supplies required for operating the Model 311-D. Flow rates for the on-board carrier gas cylinder typically vary from 10–80 mL/minute, depending on the application. This cylinder provides carrier gas for 8 hours of operation at a flow rate of 60 mL/minute (or 24 hours at 15 mL/minute, etc.).

FIGURE 12.16 The HNu® Model 311-D portable gas chromatograph. (*Courtesy of Process Analyzers, Walpole, MAb* [www.hnu.com].)

The carrier gas cylinder needs to be filled with instrument-quality nitrogen (99.999%) or helium.

The pump pulls sample into the GC at a flow rate of approximately 250–300 mL/minute (without tubing). Any length of tubing attached to the "sample in" port decreases the flow rate of the pump. The pump is capable of drawing sample through approximately 100 ft of ¼ in. or ⅛ in. tubing from a remote location.

The Analyzing Columns, Separating the Components of a Sample Mixture.

Two columns in the GC, the precolumn and analytical columns, perform the separation of the chemical components in a sample. The precolumn separates undesirable components (interferences) from a sample so that only the compounds of interest pass through it into the analytical column. The precolumn is typically a short piece of the same type of material as the main (longer) analytical column.

The analytical column performs the major separation of the components in a sample so that an independent measurement can be made of one or more specific compounds.

The columns can be any type of packed (¼, ⅛, or 1/16 in.) or capillary (0.18–0.32 mm I.D. or 0.53 mm I.D.) columns suitable for the application. Packed columns are coiled with a diameter of about 6 in. Capillary columns must be coiled with a radius no greater than 6½ in. The capillary column can be about 30 m in length.

The column that does the separation of components in the GC needs to be kept at a reproducible constant temperature, or at temperatures that rise in a carefully programmed manner. Once you have settled on the kind of vapors you want to capture (say perchloroethylene and related solvent vapors) and installed the recommended column, the oven temperature governs how long an analysis will take.

The higher the temperature, the faster will be each analysis. Your goal, as the operating technician, is to see clearly separated peaks on the plotter output for all the components in the sample. The time for any single peak to be output, measured from the time the sample was injected into the column, is called the elution time, or sometimes the residence time. Too low a temperature (too near ambient) might be too difficult to control. It would also take an unnecessarily long analysis time. Too high a temperature will rush the sample too fast through the column and will cause two or more peaks to be blurred together. That would confuse the results.

Function of the Temperature-Controlled Oven.

The oven contains the heating elements, fan, and columns and is large enough (about 300 in.^{3}) to accommodate one precolumn and up to two analytical columns (packed or capillary).

The oven can be operated at a chosen temperature sufficiently higher than ambient, or temperature-programmed to 200°C. Temperature programming is a scheme used to speed some analyses. In some cases, components of a sample, say high-molecular weight, low-volatility vapors, might take an exceptionally long time to exit the column compared to the other low-molecular weight components. The plotted output would then show no peaks for a longer time than necessary.

To speed up this analysis, the temperature may be programmed to rise to a higher value after the low-molecular weight components have been eluted (exited the column). That will speed up the elution of the higher molecular weight components. The time that any component of a sample takes to travel through the column is its own characteristic retention time.

Detectors for the Gas Chromatograph.

The Model 311-D is available with six detector options: photoionization (PID), far ultraviolet (FUV), flame ionization (FID), flame photometric (FPD), thermal conductivity (TCD), and electron capture (ECD). The 311's dual-detector capability allows any two of the six detectors to be run separately or in series. All six detectors are interchangeable by the user. The detectors quantitatively measure the concentration of the components in the sample.

Photoionization Detector.   This is the same type of detector used in the hand-held PID, as discussed in Section 12.3.9. Photoionization occurs when a molecule absorbs a photon (light energy) of sufficient energy, creating a positive ion and an electron. The sample drawn into the ion chamber is exposed to photons generated by the ultraviolet lamp. Molecules in the sample with ionization potentials less than or equal to the energy level of the lamp are ionized.

A positively biased, accelerator electrode repels these ions, causing them to travel to the collecting electrode. There an analog signal proportional to the concentration of the sample is generated. The signal is amplified to provide an analog output (peaks on a graph versus elution time) for graphic recording or electronic integration.

Far Ultraviolet Absorbance Detector (FUV).   The far ultraviolet detector (FUV) provides a nearly universal response (except for the noble gases: helium [He], neon [Ne], argon [Ar], krypton [Kr], xenon [Xe], and radon [Rn]) to organic and inorganic compounds at low parts per million (ppm) levels.

Flame Photometric Detector (FPD).   With the flame photometric detector (FPD), the sample is burned in a hydrogen-rich flame, which excites sulfur or phosphorus to a low-lying electronic level. This is followed by a resultant relaxation to the ground state with a corresponding emission of a blue (S) or green (P) photon. This type of emission is termed chemiluminescence. The emission is at wavelengths of 394 nm for sulfur and 525 nm for phosphorus.

Flame Ionization Detector (FID).   With the flame ionization detector (FID), ionization of organic compounds occurs when the carbon-carbon bonds are broken via a thermal process in the flame, resulting in the formation of carbon ions. These ions are collected in the flame by application of a positive potential to the FID jet. The ions are pushed to the collection electrode, where the current is measured. The response (electric current) is proportional to the concentration and is measured with an electrometer/amplifier. The FID is a mass-sensitive detector, the output of which is directly proportional to the ratio of the compound's carbon mass to the total compound mass.

Thermal Conductivity Detector (TCD).   The thermal conductivity detector (TCD) measures differences between the thermal transfer characteristics of the gas and a reference gas, generally helium. However, hydrogen or nitrogen can be used depending on the application. The sample and reference filaments are two legs of a Wheatstone bridge.

A constant current is applied resulting in a rise in filament temperature. As the sample passes through the detector, the resistance changes as the reference gas is replaced by the sample that has a lower thermal conductivity. This difference in resistance is proportional to the concentration. The response is universal since the detector responds to any compound that conducts heat. The minimum detection limit is in the 100–200 ppm range. The maximum concentration is 100% so that the TCD is not overpowered by high concentrations of detected compounds.

Electron Capture Detector (ECD).   The electron capture detector (ECD) provides excellent sensitivity for halogenated (containing fluorine, chlorine, bromine, or iodine) compounds. Detection limits are in the low parts per billion (ppb) and parts per trillion (ppt) range for many compounds. The ECD consists of a sealed nickel 63 (Ni-63) radioactive source, a sample chamber, and a polarizing voltage source. Carrier gas flowing through the sample chamber is ionized by the Ni-63 radioactive source. This creates positively charged ions that are collected and measured, providing a background current. The background current is kept constant by a polarizing voltage that pulses at variable frequencies.

As a sample enters the chamber, its electrons bind to any free electrons, creating negatively charged ions, resulting in a decrease in the background current. The electronics compensate for this decrease in current by increasing the voltage pulse frequency. The increase in the pulse frequency is measured and is used to generate a voltage output proportional to the concentration of the sample. The signal is amplified to provide an analog output for graphic recording or integration.

Microprocessor and Data Input and Output.   The internal firmware allows the Model 311-D to be operated at its full analytical capacity with full printer/plotter capabilities, without the need for a separate computer. The firmware allows the programming and storing of all operational parameters into the GC. These parameters are stored in battery-backed RAM chips in the GC until it is programmed with new values.

A membrane-type keypad containing 20 keys with audible response and a liquid crystal display (LCD), located on the top right side of the GC, operates the firmware. The 10 numeric and 10 function keys allow the user to program operating parameters as well as choose functions that control the detector(s) and the printer/plotter. The LCD shows current GC conditions as well as prompts and user entries.

The built-in plotter provides a printout of a chromatogram (one detector at a time) and complete data reports, including peak height and areas, concentrations, alarm levels, time-weighted average (TWA), and short-term exposure limits.

The Peakworks™ software program offers the same analytical features of the on-board firmware, with the added flexibility of automatic data collection, integration, and storage within a personal computer for further data reduction. All data and methods can be stored on disk and downloaded to spreadsheet or database programs for report generation or to a printer.

Lab Preparation and Field Operation.   Outfitting of the GC for a particular sampling mission is usually performed in a laboratory. A technician selects and installs the desired column and detector for the specific analyses to be conducted in the field. The technician then inputs the programming that completes the GC setup for operation.

To conduct analyses in the field then, after setup and supplying power, the operator doing sampling need only input some keystrokes to start operation. The operator then injects the air samples or sets the instrument for automatic sampling. The results are captured by the microprocessor memory or printed out on the plotter.

12.3.11   The Portable Gas Chromatograph/Mass Spectrometer

O——π   **The portable GC/MS is the single most accurate field sampling and monitoring instrument available.**

We have chosen the HAPSITE® (Inficon, East Syracuse, NY) as an example of the use of mass spectrometry detection combined with the separation powers of the gas chromatograph. Figure 12.17 shows the instrument with its probe.

M. Jankowski and K. Meekin of Los Alamos National Laboratory have reported on extensive HAPSITE use in sampling and monitoring (www.esh.lanl.gov). The instrument provides confirmatory test results immediately on-site, making it ideal for hazardous waste site investigations and emergency response. They found that they could carry HAPSITE to the test site and it endured harsh testing environments. Its critical components are resiliently mounted to withstand the rough handling and shocks typical of fieldwork.

The HAPSITE® weighs about 35 lb. Figure 12.18 illustrates hazardous waste site sampling and analysis. The instrument has an isothermal GC with a methyl silicone column and a quadrupole mass spectrometer (MS). The column is coiled and 30 m in length, with 0.32 mm inside diameter and 1.0 mm outside diameter. The MS was programmed to scan from masses 45–300 atomic mass units (amu). The vacuum system for the MS uses a nonevaporable getter (NEG) pump. This is a sorption pump with no moving parts. The NEG pump offers an effective solution to pumping hydrogen by the use of an activated gettering alloy (16% aluminum and 84% zirconium) deposited onto a preformed carrier plate. The alloy requires activating under vacuum each time pumping is required. Electrical heating of the

FIGURE 12.17   HAPSITE® portable GC/MS. (*Courtesy of Inficon, East Syracuse, NY* [www.inficon.com].)

carrier plate controls the activation process. To maintain pumping performance, the pumping material requires periodic regeneration by heating to approximately 400°C.

The instrument contains everything needed to sample, monitor, and analyze VOC samples in the field. Its direct-sampling system makes testing much easier because it eliminates the problems associated with conventional collecting, shipping, storing, and analyzing of samples. Technicians collect verifiable test results on-site, eliminating costly retesting. The compact construction of the instrument is illustrated in Figure 12.19.

Jankowski and Meekin reported that the HAPSITE® instrument is easy to use in the field. A technician selects a method, and HAPSITE® sets and monitors all the operating conditions, signaling when to start sampling. They found that the instrument finished the sampling process quickly and displayed reliable results on the instrument's front panel.

HAPSITE®'s Windows-based software allows customized data analysis and methods development. To review data saved on the instrument's hard drive, the operator downloads to a diskette or connects to a PC. The software gives users the ability to modify or create new sampling, operating, and analysis methods.

Using the GC/MS in the MS-Only Mode.   Jankowski and Meekin reported further on an optional methodology using the GC/MS. They used the HAPSITE device in the mass spectroscopy-only mode to identify and quantify mixtures of volatile organic compounds in the

FIGURE 12.18   HAPSITE® portable GC/MS. (*Courtesy of Inficon East Syracuse, NY* [www.inficon.com].)

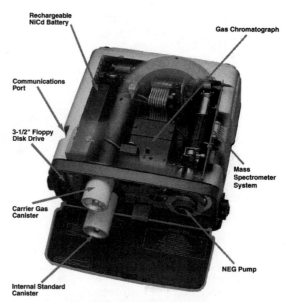

FIGURE 12.19 View of HAPSITE's® internal components. (*Courtesy of Inficon, East Syracuse, NY* [www.inficon.com].)

field. They bypassed the gas chromatograph and sent a suspected source of VOCs directly into the quadrupole MS. They found that reduced the time between sampling and the reporting of results to as little as 10 seconds. When a mixture of VOCs was sampled in this mode, after 10 seconds the instrument displayed the compound of highest concentration for which it was calibrated. Subsequent to this initial report, a full Microsoft Excel® report was retrievable to determine whether the other compounds calibrated for were present.

Two separate scenarios were created with the instrument calibrated and programmed to search for three compounds whose TLVs ranged from 1 ppm to 10 ppm: 1,2-dichloroethane, chloroform, and carbon tetrachloride. The first scenario involved sampling a mixture of nine different VOCs with concentrations ranging from 1.09–27.1 ppm. Of the three calibrated compounds, only CCl_4 was present in the mixture (2.2 ppm). CCl_4 was shown in the initial display at 1.6–2.0 ppm. A final laboratory report gave 1.7 ppm.

The second scenario consisted of sampling a mixture of 5 ppm CCl_4, 7.5 ppm chloroform, and 10 ppm 1,2-dichloroethane. The instrument initially reported 1,2-dichloroethane at 10–12 ppm. The final report listed 6.5 ppm CCl_4; 7.1 ppm chloroform; and 12.3 ppm 1,2-dichloroethane.

For each scenario, the instrument correctly identified the VOCs for which it was calibrated, produced no false positives, and quantified all calibrated compounds within 83% of their true values. The investigators concluded that this instrument's speed, specificity, and accuracy in identifying and quantifying VOCs in the MS mode makes it an extremely valuable field tool.

Using a PID for Detection and a GC/MS for Positive Identification and Concentration Measurement. The Los Alamos investigators conducted monitoring for volatile organic compounds (VOCs) during hazardous waste drum opening and borehole modification. A photoionization detector (PID) was first used to indicate the presence of VOCs. The GC/MS was then used to identify the unknown VOCs. For those compounds that were calibrated, concentrations were also determined.

For the opening of a drum to verify its contents, the paperwork identified the contents only as "alcohols." Without it being known which alcohols were present or what their PID response factors were, appropriate personal protective equipment (PPE) could not be assigned.

The drum opening was performed under local exhaust ventilation. A reading of 450 ppm was obtained from a PID with a 10.6 eV lamp. GC/MS results identified benzene, toluene, and methylene chloride as the compounds with the most significant concentration at 1, 19, and 109 ppm, respectively. PID hold limits and PPE were then assigned so the drum could be removed from local exhaust ventilation to verify contents.

Monitoring for VOCs was required during modification of a waste disposal pit borehole due to previous PID readings. In addition, it was possible that the borehole was exhaling. A 10.6 eV PID produced readings between 60 and 90 ppm, and GC/MS identified tetrachloroethylene and 1,1,1-trichloroethane as the VOCs at concentrations of 1 and 9 ppm. The work was allowed to continue as originally planned.

The Los Alamos investigators concluded that the use of this instrument makes it possible to save time and money by not having to wait for laboratory analysis or use PPE when it is not necessary.

CG/MS Identification of Organic Sulfur Compounds. Inficon® (www.infocon.com), in an Application Note, *Determining Sulfur Caustics at a Paper Processing Facility,* explained GC/MS use to determine worker exposure to sulfur compound vapors. For the mixture of gases present at the site, the investigators felt the sample collection in containers would result in reactions and sample loss. An industrial hygienist was able to carry a HAPSITE® portable GC/MS to a caustic tank, analyze the gases in the condensate vapors, and accurately determine the compounds present. Methyl and ethyl mercaptan were identified as the major sulfur compounds in the samples using the National Institute of Standards and Technology (NIST) library search program. The NIST Standard Reference Data—Mass Spectral Library may be viewed at (www.nist.gov/srd/nist1a.htm). The chromatogram for the vapors is shown in Figure 12.20.

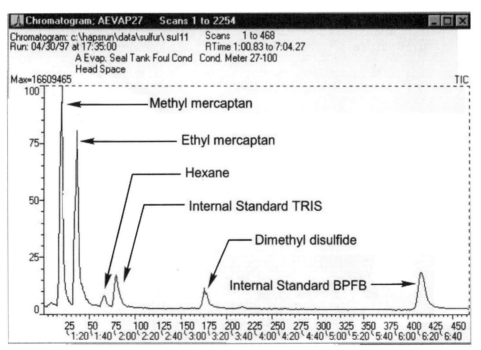

FIGURE 12.20 Chromatogram of sulfur compound vapors from a condensate tank. (*Courtesy of Inficon, East Syracuse, NY* [www.inficon.com].)

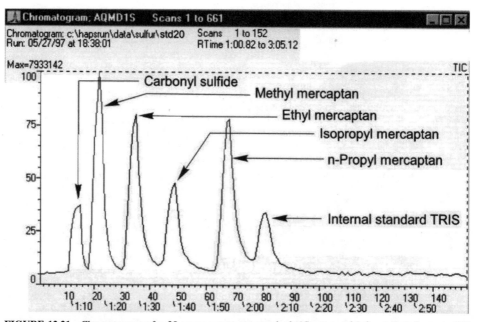

FIGURE 12.21   Chromatogram of a 20 ppmv mercaptan standard. (*Courtesy of Inficon, East Syracuse, NY* [www.inficon.com].)

Figure 12.21 is a chromatogram of a 20 ppmv mercaptan standard used to confirm the NIST identification and calculate concentrations in the measured sample. The concentrations in the sampled tank were 16 ppmv of methyl mercaptan and 17 ppmv of ethyl mercaptan.

Use of the NIST Mass Spectral Database for Confirmaton of Analyses.   As an example of the use of the MS capabilities of HAPSITE®, the headspace above a container of unknown chemical was sampled. Within five minutes of taking the sample, the NIST mass spectral database search routine had confirmed the compound as 1, 1, 1-trichloroethane. Figure 12.22 shows (upper) the NIST spectra for 1,1,1-trichloroethane, and (lower) the spectra for the unknown compound. Notice the matches at atomic mass units 61 and 97, confirming that the unknown is 1, 1, 1-trichloroethane.

Figure 12.22 shows how the GC/MS adds MS spectral identification to GC's retention time identification for the positive confirmation of analytes. It also shows how the instrument is able to identify unknowns by searching the mass spectrum of an unknown against a reference library of mass spectra such as the NIST Mass Spectral Database.

EPA's Environmental Technology Verification (ETV) Program has confirmed that the HAPSITE® instrument can produce field results directly comparable to reference laboratory results.

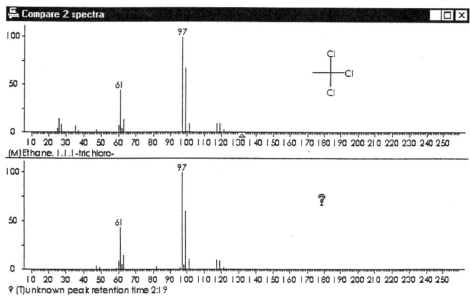

National Institute of Standards and Technology (NIST) mass spectral search results for GC/MS peak at 2 minutes 19 seconds, confirmed as 1,1,1-Trichloroethane.

FIGURE 12.22   (Upper) the NIST spectra for 1,1,1-trichloroethane; (lower) the spectra for the unknown.

12.3.12   Recommended for Further In-Depth Reading

Chou, J., *Hazardous Gas Monitors: A Practical Guide to Selection, Operation and Applications,* New York: McGraw-Hill and SciTech Publishing, Inc., 2000.

White, L. T., *Hazardous Gas Monitoring: A Guide for Semiconductor and Other Hazardous Occupancies,* Norwich, NY: William Andrew / Noyes, 2000.

12.3.13   Personal Air Monitoring

Personal air monitoring can take several forms. Particulate contaminants are collected on filters that are later analyzed by a laboratory. Chemical vapor, gas, and mist exposure can be monitored by a third party, by a personal PID, or by inexpensive detector badges.

A wide variety of HAZWOPER personal monitoring instruments and methods are available, far more than we could adequately cover within this book. Examples of site and personal monitoring instruments such as the PID were described in Section 12.3.9.

Stricter personal monitoring is required by OSHA, for instance, when working on a lead- or asbestos-abatement project. The abatement may only take place when the area to be abated is segregated from the rest of the facility by a containment. The containment is constructed of polyethylene sheeting that is at least 6 mils thick. The entire containment area is kept under negative pressure using specially designed HEPA (now referred to as 'P-100') filtered exhaust units called air filtration devices or AFDs. The workers along with all other entrants into the area must wear proper PPE, usually Tyvek® coveralls and APRs. This OSHA monitoring, as it is called, may not be performed with an instrument, such as a fibrous aerosol monitor (FAM).

For both asbestos and lead remediation, workers wear a battery-powered pump on their belts. The pump pulls air through a hose connected to a special filter contained in a cassette. This assembly is called the sampling train. Workers wear this train for the entire shift, with the inspector noting when they are working in the contaminated area and when they are on a break. Careful documentation of a worker's work function (demolition and removal, among others) is essential. At the end of the shift, the cassette is removed from the sampling train and sent to a lab for microscopic analysis. Air samples must be taken in the workers' breathing zones. The work area, and random areas outside the work area, are also sampled. This sampling is required to ensure that at no time was the air contaminated to the point that the worker's PPE was insufficient. It is also a good indicator that the workers were using the proper work methods and engineering controls so that no one outside the containment was contaminated.

These samples must be analyzed by laboratories with the following accreditation: National Voluntary Laboratory Accreditation Program (NVLAP) for lead and NIOSH Proficiency Analytical Testing Program (PAT) for asbestos. The lab reports counts of fibers per cubic centimeter (f/cc) for asbestos and micrograms per cubic meter ($\mu g/m^3$) for lead.

Another type of personal monitoring, used to detect worker's exposure to gases, vapors, and mists, uses colorimetric badges. These show color changes or numerical output of total exposure, much as the colorimetric tubes show concentration. The badges are simply clipped onto the worker's uniform, similar to an ID badge. They have sampling times ranging from 8 to 48 hours. They are divided into two broad types:

1. Warning badges alert wearers that the PEL for the chosen chemical has been exceeded. Workers must evacuate the area.

2. Monitoring badges give a clear readout of the exposure over the wearing time as a time-weighted average (TWA).

Colorimetric badges are available to detect approximately 100 different chemicals. They can be an inexpensive way to determine a worker's exposure. They require no lab analysis fees and delay. The wearer can read them at any time during the shift. However, as with the colorimetric tubes, they do not have the accuracy of correctly calibrated instruments. Further, wearers must know the contaminant of concern, as they are used for single chemical detection only.

12.4 LIQUIDS SAMPLING

O—π **There are a number of different sampling techniques for liquid samples. Know which of the available techniques are satisfactory for the desired level of data confidence your sampling and monitoring program requires.**

O—π **Liquid samples, as with all other samples taken for laboratory analysis, must be accompanied by a completed chain-of-custody form.**

As a HAZWOPER worker or supervisor, you will often be working around and with liquids that are of unknown origin and composition. Is that just rainwater in that puddle, or is it a toxic liquid leaking from those rusted and bulging drums? The only way we can know for sure is to sample and analyze. There are several simple-to-read colorimetric test strips available that cover the most common hazards. On-site test strips are available.

• PCB-contaminated oils and solvents can be detected by Dexsil® Clor-N-Oil 50®.

• Acids and bases can be detected by Colorphast® pH test strips.

• Organic peroxides can be detected by EM Quant® peroxide strips.

• Most common chlorinated pesticides can be detected by Agri-Screen® and the Cholinesterase® test, among others.

These test strips will instantly tell you if you are dealing with one of these potential health or chemical reaction hazards.

We should reiterate from Section 3 that no one should touch, step in, or kneel on any unknown material at a hazardous waste site or during a response action. How, then, do we tell what the material is? We sample it. (Here we will focus on liquids sampling. Section 12.5 explains solids sampling.) There are several approaches to this procedure. If we have access to some of the air-sampling devices discussed in the previous parts of this section, we can take a reading and see if we get any results from liquid vaporization. As you have already seen, this can be a laborious process if we don't have any indication of the material's identity.

What if the instrument you are using, a PID for instance, does not provide any reading? Does that mean the material is harmless? Absolutely not! The material might be one that the instrument doesn't detect. The material might be so cold that it is not giving off enough vapor for measurement, or the instrument may not be calibrated for that material. How do we solve this dilemma? We carefully take a sample of the liquid and send it to the lab for analysis.

Many workers think that liquid sampling is a 'no-brainer'. Far from it. Sample integrity is essential in the HAZWOPER world. We do not simply dip a jar into a pool of liquid, screw on the cap, and send it to the lab.

12.4.1 Liquids-Sampling Procedures

Samples must be taken in clean jars or bottles. For photoreactive analytes such as VOCs, the jars or bottles must be amber colored. Some labs still provide glass sample containers, although most are now using Nalgene® or other appropriate plastics because they will not break. The containers are usually specially prepared and provided by the laboratory that will perform the analyses. For some samples, the sampling containers will be empty. For other samples, such as for metals analysis, the containers will arrive at the site containing a small amount of acid or some other reagent. These chemicals are used to fix an analyte in the sample. The samples must then be refrigerated for preservation. Never pour this solution out! It is there for the specific purpose of protecting the sample. All environmental samples have a hold time; make sure you don't exceed it. If you exceed these hold times (2 to 180 days, depending on the sample type and analysis method), you will need to resample.

> **Hold time is the amount of time you are allowed to store a sample of a material before it must be analyzed. There may be storage condition restrictions such as temperature. If either the time or the condition restrictions are exceeded, the sample is no longer valid.**

Hold times for wastes to be analyzed by TCLP (see Section 12.5.2) must not be exceeded. They are as follows:

- Volatiles—14 days
- Semivolatiles—14 days
- Mercury—28 days
- Other metals except mercury—180 days

WARNING

BE CAREFUL WHEN USING PRETREATED SAMPLE CONTAINERS. THE CHEMICALS THEY CONTAIN, USUALLY STRONG ACIDS, ARE EXTREMELY CORROSIVE. MAKE SURE TO WEAR THE PROPER PPE AND AVOID SPLASHING OR OVERFILLING THE CONTAINER.

FIGURE   12.23 CON-BAR® Type 302 stainless steel sample bomb. (*Courtesy of CONBAR Environmental Products* [www.conbar.com].)

If the liquid you wish to sample is in a small enough container, you can sometimes pour it directly into the sample container. Be careful not to contaminate the sample with material from the lip of the container.

If you are sampling from a pool, a body of water, or a test or monitoring well, you may use a pump or bailer. A bailer is a container that you lower into a well to collect a sample after the well has been developed (see below). Never scoop up a liquid sample with the sample container. This practice is forbidden. Your project's sampling protocol may allow you to reuse bailers after proper cleaning. However, they are often single-use items that must be discarded after one use.

If you are sampling monitoring wells, depending on your sampling strategy, you may have to purge the well first. This is the process of well development. You develop a well by pumping out a predetermined number of well volumes, typically three to five, so that the sample is not stagnant from the well, but fresh from the water-bearing layer desired.

The water sample is then taken by lowering a bailer or sample bomb down into the water. Bailers are open-topped containers that range in size from one-half to 3 in. diameter and 6 in. to 4 ft in length. They are offered in unweighted, weighted, and double-weighted models. Double weighted models are recommended for more viscous liquids.

Sample bombs (see Figure 12.23) are cylindrical units typically constructed of brass, bronze, stainless steel, high-density polyethylene, polypropylene, or Teflon.® They typically range in capacity from 4 to 16 fluid ounces. They are designed to collect liquid samples from storage tanks, lakes, ponds, manholes, wells, and other hard-to-reach places. They have a plunger assembly that will activate automatically when the sampler contacts a hard surface such as the bottom of a tank. They can also be manually activated by the operator at various depths with the use of a pull line.

Remember that most all states will not allow you simply to dump your purge water back down the well or on the ground after sampling. The purge water will have to be tested and it may require treatment prior to disposal.

12.4.2 Sampling Liquids from Drums

⊙━🔑 **The two most common devices used to sample liquids from drums are the drum thief and the coliwasa.**

In HAZWOPER work there are distinct phases of finding waste drums and determining what they contain. That process will be described in greater detail in Section 13. The phases are locating, excavating, identifying, sampling, and characterization.

One of the most common chores encountered by the HAZWOPER worker is the need to sample drums and large containers, as shown in Figure 12.24. Note that the workers are

FIGURE 12.24 Using a drum thief to sample a liquid unknown. (*Courtesy of the USEPA.*)

taking a liquid sample from a solids (open-mouth) drum. A HAZWOPER worker can never assume the contents of a container by the container type! Trainers, supervisors, and workers must be familiar with two sample-drawing implements. One is the drum thief (shown in Figure 12.24), and the other is the coliwasa (composite liquid waste sampler).

A drum thief is a hollow glass, HDPE, or stainless steel tube, like a big straw. One end is placed into the drum or other container that you wish to sample. You slowly lower the thief into the drum, allowing the liquid level outside the thief and inside the thief to equilibrate. After a sufficient volume of sample material is collected, you hold your thumb over the end outside the drum and draw it out of the drum or other container. Your sample will be held in place by the vacuum that you have created.

To fill your sample container, place the lower end of the drum thief into the bottle and take your thumb off the other end. It sounds simple, but it requires practice. Having some rags around to wipe up spill is essential. Drum thieves are usually single-use items because they are notoriously hard to decontaminate. Consult your sampling protocol about reuse. The drum thief is usually discarded as contaminated by simply being left in the sampled container.

A drum thief works well with low-viscosity liquids, like water and solvents, but not too well with thicker liquids such as paint and thick oils. In those cases, your best bet is a coliwasa, which is similar to a drum thief but requires a different mode of operation.

The coliwasa typically has a larger bore than the drum thief and is therefore better suited for more viscous liquids. The design is different in that the tube has a plug at the bottom. To operate, the plug is opened by a handle that is attached by a shaft that runs down the center of the tube. The coliwasa is then lowered slowly into the container. Some patience is necessary with viscous liquids such as waste paint and thick oils. To remove a composite sample (one from all stratified layers of the container), you must allow the levels to equilibrate. Once a sample is contained in the tube, the rod is lifted and the plug then holds the sample in the coliwasa. Again, having an adequate supply of rags is essential. Coliwasas are also very difficult to decontaminate but are more expensive than the drum thief. There are also versions that use a plunger to draw a sample up into the tube. These are especially useful for extremely viscous liquids. Using one to draw a stratified sample does, however, take some practice.

Again, consult your site sampling protocol about decontamination and re-use.

Whether sampling is done using a bailer, sample bomb, drum thief, coliwasa, or any other implement, special care must be taken in decontaminating any that are reusable. The smallest residue of sample material, detergent, or water left in any of these implements can contaminate the next sample taken and invalidate your sampling data. All of these implements are made in disposable configurations. For best results, use them for one sample and then dispose of them.

12.4.3 Guidelines for Sampling Liquids

Here is a checklist of guidelines to follow when taking liquid samples. Drum handling and sampling will be covered in more detail in Section 13.

❏ Always use the proper container for the sample that is being taken. They will usually be labeled, but if you are uncertain about a selection, ask your supervisor.

❏ Never allow the sample to come in contact with your skin. There are two reasons for this:

1. The contaminant may be toxic, corrosive or otherwise injurious to you.
2. Contact with your skin may contaminate the sample, especially if you have a foreign substance such as soap residue on your hands.

❏ Never allow the material to come into contact with your PPE. This can cause cross-contamination with other samples and cause you to become contaminated with the material.

❏ Be careful not to contaminate the sample with dirty sampling equipment.

❏ Immediately clean any spilled material from the side of the sampling container. This material may be injurious to you or your fellow workers, may be corrosive and compromise its outer container, or may contaminate other samples.

❏ Immediately seal the sample container. Some labs, including EPA-approved labs, require a chain-of-custody seal to be applied to the top and side of the container. If it is broken by anyone other than the analytical lab, the sample is rejected.

Note

The authors recommend use of a chain of custody protocol on all sampling whether or not it is required by the analytical lab. Chain of custody will be required by all EPA- and American Industrial Hygiene Association (AIHA)-accredited labs.

❏ Immediately label your sample.

❏ Immediately log your sample. Figure 12.27 is a groundwater sampling log sheet suggested by EPA. For your particular organization you will probably want to make modifications. You will probably want to add an instruction sheet to the log sheet. When the sampling technician returns with the sheet filled in properly, will there still be any nagging questions about what happened during the sampling?

❏ Make sure that you immediately, properly, and completely fill out the chain-of-custody records for each sample. (The longer you wait to do this, the harder it is to remember the facts surrounding your sampling.) Figure 12.25 is an example of an EPA-suggested chain-of-custody record.

❏ Refer to more detailed procedures in Section 13 of this book for the latest approved methods for sampling drums.

❏ Make sure to note the date and time the sample was taken. There are specific hold times for most water samples, for instance.

❏ Ensure that samples that are held are stored properly. Water samples must often be refrigerated or kept on ice until delivery to the lab. This inhibits bacterial growth that can confuse lab results. Water samples are often contained in brown bottles to reduce reactions caused by sunlight.

Note

It can be very costly to have to return to a site and resample due to your first sample being rejected by the lab for expired hold times. The same can be said for rejected samples that were not stored properly, have incomplete or incorrect

| EMSL ANALYTICAL | CHAIN OF CUSTODY | | LEAD |

EMSL Rep:	Paul Nyfield	DATE:	
Your Company Name:	EMSL Analytical	EMSL-Bill to:	*SAMPLE*
Street:	107 Haddon Ave.	Street:	
Box #:		Box #:	
City/State:	Westmont, NJ Zip: 08108	City/State:	Zip:
Phone Results to:	**Paul Nyfield**	Fax Results to:	**Paul Nyfield**
Telephone #:	609-858-4766	Fax #:	609-858-4766
Project Name/Number:		Purchase Order #:	

MATRIX	METHOD	INSTRUMENT	mdls	TAT
Lead Chips*	SW846-7420 or AOAC 5.009 (974.02)	Flame Atomic Absorption	0.01% ++	
Lead Wastewater	SW846-7420	Flame Atomic Absorption	0.4 mg/l water 50 mg/kg (ppm) soil	
Lead Soil +	or SW846-6010	ICP	0.1 mg/l water 10 mg/kg (ppm) soil	
Lead in Air***	NIOSH 7082	Flame Atomic Absorption	5 ug/filter	
	or NIOSH 7300	ICP	3.0 ug/filter	
Lead in Wipe▲ ☐-ASTM	SW846-7420	Flame Atomic Absorption	10 ug/wipe	
☐-non ASTM	or SW846-6010	ICP	3.0 ug/wipe	
TCLP Lead **	SW846-1311/7420	Flame Atomic Absorption	0.4 mg/l (ppm)	
	or SW846-6010	ICP	0.1 mg/l (ppm)	
Lead in Air ****	NIOSH 7105	Graphite Furnace Atomic Absorption	0.03 ug/filter	
Lead Wastewater	SW846-7421	Graphite Furnace Atomic Absorption	0.003 mg/l (ppm) water	
Lead Soil +			0.3 mg/kg (ppm) soil	
Lead in Drinking Water (check state Certification Requirements)	EPA 239.2	Graphite Furnace Atomic Absorption	0.003 mg/l (ppm)	
Total Dust	NIOSH 0500-0600	Gravimetric Reduction	0.0001g	

TAT (Turnaround) - Same day, 24 hr - 1 Day, 2 Days, 3 Days, 4 Days, 5 Days, 6-10 Days
*, **, ***, ****, +, ++   Please Refer to Price Quote
▲ if no box is checked, non-ASTM is assumed

SAMPLE #	LOCATION	Air volume, L Area, in²	LAB #

Relinquished By; (Person) _____ Date: _____

Received at EMSL By: _____ Date: _____

Received at EMSL By: _____ Date: _____

Note: Please duplicate this form and use additional sheets if necessary.

FIGURE 12.25 A chain-of-custody form for lead samples. (*Courtesy of EMSL® Analytical, Inc.* [www.emsl.com/index.html].)

information on the chain-of-custody papers, or are mislabeled. The authors have had experience with companies that have incurred over $100,000 in additional sampling and analytical charges due to these common oversights or mistakes.

12.4.4 Monitoring Wells—Boring Logs, Sampling Logs, Field Analysis, and the Traditional Approach to Sampling

Oπ **Monitoring wells are the principal means of collecting data concerning the characteristics of subsurface water.**

As we discussed above, one of the tasks HAZWOPER personnel may routinely be asked to accomplish is monitoring well sampling. These wells are customarily installed in and around factories and other sites where hazardous materials and wastes have been found. They also surround all hazardous waste landfills and many sanitary landfills. Their purpose is to monitor the quality of the groundwater below the site. They may be used to track the spread of a plume, which is a clearly delineated underground area of contamination. They are also quite often used to monitor the progress and the efficacy of a remediation effort.

A typical monitoring well boring log is shown in Figure 12.26. Although you may not be responsible for the analysis of the well boring, it is important to know how to read a boring log because it gives us important data about the makeup of the subsurface strata at that particular spot on the site.

A well boring log tells us a number of things. It tells us at exactly what depth different strata are located. Depending on the makeup of that strata, we can then determine whether contaminants of concern will travel easily through the ground under the site. For instance, if the well drillers struck limestone at 1 foot below grade (FBG), which continued for 10 more feet, we know that almost no contaminants can travel through that limestone, as limestone is not a porous rock. If, however, we see sandy loam for the first 20 FBG and then a limestone layer, we can assume that contaminants penetrating through the surface could travel underground quite well, at least until they encountered the limestone.

The next thing boring logs tell us is the depth to groundwater, assuming the well is dug to sufficient depth to encounter groundwater. If groundwater is struck, and we are dealing with contaminants that are lighter than and immiscible with water, we can determine where they will float. In Figure 12.26, for example, groundwater is at 2.1 meters below grade. Dark gray limestone was struck at 3.9 meters after gravelly and sandy clay was penetrated.

The screened interval of the well is the length of the well lining that has slots cut in the side to allow water into the well. Groundwater has seasonal high and seasonal low levels. Well screening is placed to include from above the seasonal high to below seasonal low level so that the monitoring well will always have water in it for sampling.

Well grouting is a near-neutral pH bentonite seal extending down 0.4 meters below grade. The bentonite seal keeps nutrients and other pollutants from the surface from entering the well bore hole. The well diameter is 8 in. The annulus (around the 2-in. diameter well pipe) is filled with gravel pack to allow for free flow of water. The well pipe is screened to allow water entry. The screened interval in Figure 12.26 is from 1.4 to 4.4 meters below grade.

Figures 12.27 and 12.28 show EPA-recommended log sheets for monitoring well sampling and field analysis. Typically, the field analysis of monitoring well samples will include:

- Date and time sample was taken
- Pump rate (if the sample was taken from a pump in or above the well)
- A visual description of the sample (clear, cloudy, stratified)
- pH
- Temperature
- Electrical conductivity (a measure of dissolved solids, measured in millivolts [mVs])
- Total dissolved solids (measured in ppmw)
- Dissolved oxygen (measured in ppm or milligrams per liter and % saturation)
- Turbidity (the measurement of suspended solids, measured in nephelometric turbidity units [NTUs])

To eliminate the necessity of carrying bulky instruments into the field, many of the parameters listed above can be measured with pocket-sized combination instruments. For instance, you might choose an ORION Portable Multi-Parameter® meter to measure pH, conductivity (in mV), °C, salinity, and dissolved oxygen.

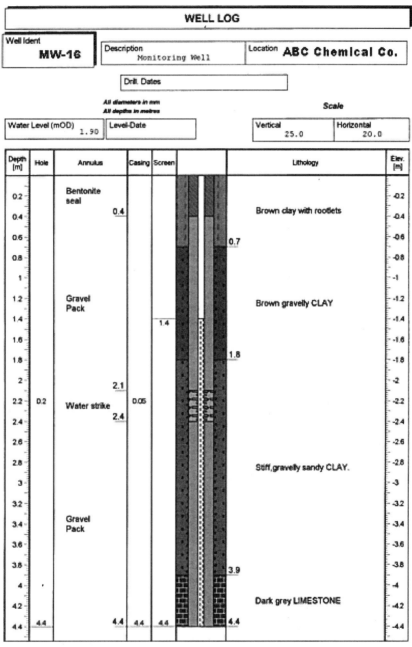

FIGURE 12.26   Typical monitoring well (MW) boring log.

Ground Water Sampling Log

Project _____ Site _____ Well No. _____ Date_____
Well Depth _____ Screen Length _____ Well Diameter _____ Casing Type _____
Sampling Device _____ Tubing type _____ Water Level _____
Measuring Point_____ Other Infor_____

_____

Sampling Personnel_____

Time	pH	Temp	Cond.	Dis.O_2	Turb.	[ ]Conc			Notes

FIGURE 12.27   A generic groundwater sampling log sheet. (*Courtesy of USEPA.*)

Monitoring wells along with other types of wells and their uses as part of remediation methods will be discussed further in Section 14.

HAZWOPER workers assigned to well sampling and monitoring must maintain accurate logs (written records) of those activities.

When wastes and other materials are dumped, buried, or have leaked onto the ground, they do not, for the most part remain there. As rainwater percolates through the soil, these materials, or some of their components, are usually carried along with the water flow. Eventually these materials migrate to the groundwater. As we covered earlier, when discussing chemical properties, some of these materials will mix fairly well with water. Others will

Ground Water Sampling Log (with automatic data logging for most water quality parameters)

Project _____ Site _____ Well No. _____ Date _____

Well Depth _____ Screen Length _____ Well Diameter _____ Casing Type _____

Sampling Device _____ Tubing type _____ Water Level _____

Measuring Point _____ Other Infor _____

Sampling Personnel _____

Time	Pump Rate	Turbidity	Alkalinity	[ ] Conc	Notes

FIGURE 12.28 Another generic groundwater sampling log sheet. (*Courtesy of USEPA.*)

float on top of the water and some will sink because they are relatively immiscible with (insoluble in) water.

Groundwater flows somewhat like a wide underground river—not as fast, in a known direction nonetheless. The direction and rate of flow vary from location to location. These variables are affected by season (which determines groundwater depth), topography of the site, and lithology (the geology of the strata under the ground) of the site, among other factors. Water ultimately flows downhill or toward a lower pressure.

Monitoring well sampling regulations vary from state to state, but in general the process has, in the past, been a laborious one. In a typical scenario, a monitoring well must first be purged or 'developed'. To purge the well, the samplers first have to pump out a certain amount of stagnant liquid. This is usually three to four well volumes. That is to say, that they had to figure the volume of the well above the liquid level, multiply it by three or four, and then pump out that much liquid before sampling the well. This process is still required in most states.

The rationale is that the liquid that is stagnant in, or has migrated to, the well (because it is an area of lower pressure) is not a representative sample of the groundwater as a whole. The contaminants of concern might have evaporated or otherwise degraded while stagnant in the well. Therefore, we must purge.

When we say that the process is laborious, we mean that it can take a two-person crew an entire day to sample two wells. There is also a great deal of equipment that needs to be set up. The crew will need a power supply, pump or compressor, hoses, and a drum or drums to receive the waste liquid. In most states, when you are monitoring for contaminants, the waste liquid generated during purging may not be poured on the ground or back down the well. It must be treated as potentially hazardous until it can be analyzed and proven otherwise. As you can see, sampling of monitoring wells can be much more labor-intensive than you might expect.

12.4.5 Sampling at Discrete Depths—An Alternative Approach to Traditional Well Sampling

O—π **Discrete-depth sampling of wells has been gaining acceptance and approval in the environmental regulatory community due to the accuracy, repeatability, and simplicity of the procedure.**

Under EPA's Tiered Approach to Corrective Action Objectives (TACO) the contaminant concentrations of discrete samples are compared to the applicable groundwater remediation objective(s). Discrete samples are samples taken from specific depth intervals below grade. The site will have achieved compliance if the analytical results from each sample do not exceed the applicable remediation objective(s).

Sampling groundwater from discrete (or specific) levels (intervals) within the screened portion of well casing has been accomplished by several methods. At sites where discrete-level sampling has been successful, a detectable stratification of water chemistry existed within screened portions of the wells. This fact implies that little vertical mixing occurred within the screened intervals. Further, the water at discrete well depths had equilibrated to the groundwater chemistry at those same depths.

Discrete-level sampling strategies are appropriate whether or not a vertical concentration gradient exists in the well column. Further, lateral groundwater flow is generally sufficient to reestablish lateral and vertical chemical equilibrium within minutes to hours after being disturbed, after any previous activities that would tend to mix the water in the well column, such as purging.

Discrete-level sampling is useful to delineate areas of concentrated contaminants and plume boundaries in vertical profile. Horizontal concentration data combined with vertical profile concentration data and subsurface flow data provide a powerful and accurate analytical tool that we can use to determine the actual subsurface conditions at an environmental investigation or remediation site.

Stratification can only be detected if the water column remains relatively undisturbed during sampling. Where discrete-level sampling techniques are used, they preclude purging.

A unique sampling device for discrete-level sampling is the Kabis® sampler from Sibak Industries Ltd., Inc.® These units sample discretely at any desired depth. That is, they will only sample from the depth at which you want a sample taken. No well purging is required. They sample directly into a standard sample container so that there is no possibility for cross-contamination when handling the samples. The samplers are also easily decontaminated after use, with simple materials and methods. Furthermore, they will sample directly through free product as viscous as heavy-end petroleum products, such as Number 4 fuel oil, that might be contaminating the well water.

12.4.6 How the Kabis® Sampler Works

The sampler is attached to a measuring tape or any metered lowering device and is manually lowered into the well. The size, orientation, and construction of the inlet and exhaust ports of the sampler are such that it does not fill while it is being lowered down through the water column in the well. When the sampler is held stationary at the desired sampling depth, it begins to fill under hydrostatic pressure. As the sampler is removed from the well, its design does not allow water from other levels to cross-contaminate the sample. Figure 12.29 shows two models of Kabis® samplers, both in and out of their carrying case.

In the foreground, the Model I (right), is capable of taking one 40-mL volatile organic analysis (VOA) sample. The Model III (left) is capable of taking three 40-mL VOA samples. The measuring stick represents 1-foot scale. The Model I is for 2-in. wells, and the Model III is for 4-inch and larger wells.

The Kabis® Cone Cap™ is made to fit standard 40-mL sample vials. Its construction ensures that there will be no air bubble in the sealed sample. This zero-headspace sample container prevents any loss of volatile matter during sample storage and transfer prior to analysis.

The Kabis® Sampler is currently being used by the U.S. EPA, the DOE, several state departments of environmental protection and natural resources, and numerous large corporations and environmental engineering firms.

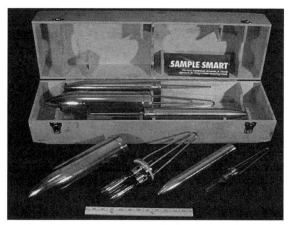

FIGURE 12.29 Kabis® discrete level samplers. (*Courtesy of SIBAK Industries Ltd., Inc.* [www.sibak.com].)

12.5   *SOLIDS SAMPLING*

12.5.1   Soil Grab Samples

On some occasions, such as determining whether equipment has been completely deconned or there are other surfaces to be sampled, you might have to take a wipe sample. The requirements for this protocol vary from contaminant to contaminant, as does the analysis for the analytes. Usually a wipe sample is taken over a known area (square inches to square feet). The sample wipe is placed into the lab's container, labeled, and shipped with the chain-of-custody sheet to the lab for analysis.

HAZWOPER workers usually will be concerned with contaminated soil sampling. Many sampling tasks will involve unlabeled drums, contaminated containers, or piles of suspect materials. Solids samples are often referred to as "grab samples." This misnomer reflects years of incorrect sampling techniques. The term came from workers grabbing a handful of soil or other solids and putting it in a sample container. Never do this! Always remember that our duty when sampling such material is to provide a sample that will allow an accurate determination of the site properties. This includes never allowing cross-contamination from sampling implements and PPE such as gloves. As we emphasized previously, never allow suspect material to come into direct contact with your body. If possible, keep material being sampled off of your PPE.

Following project protocol (discussed further in Section 14), you will usually take grab samples with a clean scoop, an auger, a coring device, or a small spade. You will deposit the samples in containers suitable for delivery to analysis. They range from glass and plastic jars to Zip-Loc®-type bags provided by the analytical lab. All of the rules regarding packaging, labeling, chain of custody and safety emphasized previously also apply here.

When you are sampling certain areas of site soil (areas of abnormal staining, distressed vegetation, or evidence of dead animal life), you will need to take discrete samples. These are samples taken from a specific spot in the soil.

Composite Samples.   Your site sampling protocol will direct you as to where and what type of samples to take. When it comes to soil remediation, which we will cover in more depth in Section 14, there is often room for negotiation with the regulators. Many times, depending upon the specific remediation project and its intended outcome, you may be allowed to apply site-specific standards, that is, standards that differ from published federal and state regulators' standards. This will, of course, depend on which of the 50 states has project oversight, the intended use of the site after remediation, and the contaminants of concern.

For soil sampling, you may be allowed to take composite samples. To take a composite sample, you sample representative areas of the remediation site. Then you physically mix these soil samples in the same container. The number of samples you collect depends on the site-specific standards under which the remediation is performed. Sites might involve leaking underground storage tanks (LUSTs), Superfund site remediation, RCRA closure, and corrective action.

CAUTION

COMPOSITING OF SOIL SAMPLES TO BE ANALYZED FOR VOCS IS NOT PERMITTED. THAT PROCESS TENDS TO FREE VOCS FROM SAMPLES AND GIVE RESULTS THAT ARE NOT REPRESENTATIVE OF ACTUAL SITE CONDITIONS.

12.5.2   Sampling Wastes to Determine Disposal Method: The Toxicity Characteristic Leaching Procedure

The toxicity characteristic leaching procedure (TCLP) is the EPA test method used to determine if a waste is to be characterized as toxic or nontoxic waste. If waste is found to be nontoxic (passes the TCLP test), it may be disposed of in a sanitary landfill. Otherwise it will have to be either treated (see Section 13) or disposed of in an EPA-regulated hazardous waste landfill.

The TCLP procedure consists of the following steps:

1. *Filtration of liquid wastes:* For liquid wastes (i.e., those containing less than 0.5% dry solid material), the waste, after filtration through a $0.6–0.8$-μm glass fiber filter, is defined as the TCLP extract. For wastes containing greater than or equal to 0.5% solids, the liquid, if any, is separated from the solid phase and stored for later analysis.

2. *Particle size reduction:* Prior to extraction, the solid material must pass through a 9.5-mm (0.375-in.) standard sieve. It must have a surface area per gram of material equal to or greater than 3.1 cm^2 or be smaller than 1 cm in its narrowest dimension. If the surface area is smaller or the particle size larger than described above, the solid portion of the waste is prepared for extraction by crushing, cutting, or grinding the waste to the surface area or particle size described above. (Special precautions must be taken if the solids are prepared for organic volatiles extraction.)

3. *Extraction of solid material:* The solid material from Step 2 is extracted for 18 hours with an amount of extraction fluid equal to 20 times the weight of the solid phase. The extraction fluid employed is a function of the alkalinity of the solid phase of the waste. A special extractor vessel is used when testing for volatile analytes.

4. *Final separation of the extraction from the remaining solid:* Following extraction, the liquid extract is separated from the solid phase by filtration through a $0.6–0.8$-μm glass fiber filter. If compatible, the initial liquid phase of the waste is added to the liquid extract and these are analyzed together. If incompatible, the liquids are analyzed separately and the results are mathematically combined to yield a volume-weighted average concentration.

5. *Testing (analysis) of TCLP extract:* Inorganic and organic species are identified and quantified using appropriate analytical methods.

6. *Results of analysis:* The sample is said to fail the TCLP if the concentration of the contaminant for which the sample was analyzed exceeds the EPA's allowable concentration (see Table 12.4.) The waste must then be properly disposed of in a hazardous waste landfill or treated with some available technique to the point that it is no longer hazardous.

Note

As previously stated, state environmental agencies may have more strict guidelines than the EPA. There are also site-specific guidelines that may allow higher levels of contamination. That depends upon the intended use of the property. All of this information must be taken into account when deciding how to interpret the results of the TCLP.

TABLE 12.4   Maximum Concentration of Contaminants in Leachate test for Toxicity Characteristic[a]

Contaminant	EPA hazardous waste number	Minimum concentration in leachate for toxicity characteristic (mg/L)
Arsenic	D004	5.000
Barium	D005	100.000
Benzene	D018	0.500
Cadmium	D006	1.000
Carbon tetrachloride	D019	0.500
Chlordane	D020	0.030
Chlorobenzene	D021	100.000
Chloroform	D022	6.000
Chromium	D007	5.000
O-Cresol	D023	200.000
M-Cresol	D024	200.000
P-Cresol	D025	200.000
Cresol	D026	200.000
2,4-D	D016	10.000
1,4-Dichlorobenzene	D027	7.500
1,2-Dichloroethane	D028	0.500
1,1-Dichloroethylene	D029	0.700
2,4-Dinitrotoluene	D030	0.130
Endrin	D012	0.020
Heptachlor and its hydroxide	D031	0.008
Hexachlorobenzene	D032	0.130
Hexachlorobutadiene	D033	0.500
Hexachloroethane	D034	3.000
Lead	D008	5.000
Lindane	D013	0.400
Mercury	D009	0.200
Methoxychlor	D014	10.000
Methyl ethyl ketone	D035	200.000
Nitrobenzene	D036	2.000
Pentachlorophenol	D037	100.000
Pyridine	D038	5.000
Selenium	D010	1.000
Silver	D011	5.000
Tetrachloroethylene	D039	0.700
Toxaphene	D015	0.500
Trichloroethylene	D040	0.500
2,4,5-Trichlorophenol	D041	400.000
2,4,6-Trichlorophenol	D042	2.000
2,4,5-TP (Silvex)	D017	1.000
Vinyl chloride	D043	0.200

[a]Wastes having a leachate containing concentrations above these levels are considered characteristic hazardous wastes because of toxicity.

12.5.3   Geoprobe® Sampling of Subsurface Soil

○──ᴨ   **Percussion probing machines are a highly portable, quick method of taking a physical subsurface soil sample.**

A Geoprobe® is a hydraulically powered percussion/probing machine designed specifically to provide cores of subsurface soil. Figure 12.30 shows a Geoprobe® operator obtaining

FIGURE 12.30 Subsurface sampling using a Geoprobe® at a housing redevelopment site.

core samples at one of the housing redevelopment sites where one of the authors was working. Core sample analysis determines whether subsurface soils contain any hazardous or other material foreign to the native soil of the area. The first Geoprobe® brand machine was built for the Environmental Protection Agency in 1988.

Soil-probing techniques can be thought of as a subcategory of what are commonly referred to as direct push techniques. Direct push refers to tools and sensors that are pushed into the ground, without the use of drilling, to remove soil or to make a path for a tool. A Geoprobe® relies on a relatively small amount of static (vehicle) weight combined with percussion as the energy for advancement of a 'tool string.' Figure 12.31 shows technicians performing a preliminary examination of a core sample as they load it into a sample bag for laboratory analysis.

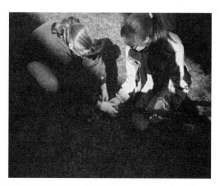

FIGURE 12.31 Examination of a several foot-long Geoprobe® core sample.

Using a Geoprobe®, you can:

- Drive tools to obtain continuous soil cores or discrete soil samples.
- Drive samplers to obtain groundwater samples or vapor samples.
- Insert permanent sampling implants and air sparging points.
- Drive a conductivity sensor probe to map subsurface lithology.
- Install small-diameter monitoring wells.
- Penetrate surface pavements 12 in. (305 mm) or more in thickness and place probes beneath them.

This equipment is typically used for site investigations to depths of 30 to 60 ft (9 to 18 m), depending upon soil conditions. However, with the development of improved equipment, this range will increase. A probing tool is also available to make a continuous log of soil conductivity and probe penetration rates. Some further Geoprobe® features include the following:

- No cuttings (surface debris, or spoil) are produced during the sampling process.
- Probing is fast. Typical penetration rates are from 5 to 25 ft (2 to 8 m) per minute.
- Mobilization is quick and economical.
- The sampling process is fast. Typically 20 to 40 sample locations can be completed per day.
- Probing machines are easy to operate and relatively simple to maintain.
- Probing tools create small-diameter holes that minimize surface and subsurface disturbance.
- Geoprobe® machines fold compactly and store in cargo vans or truck toppers where the unit and tools can be locked and secured.
- Geoprobe® machines have lower capital costs and are more economical to operate than rotary drilling machines. The level of effort and labor required for Geoprobe® operation is much less than for conventional drilling.

12.5.4 X-ray Fluorescence Spectrometry (XRF) for Site Screening for Heavy Metals

X-ray fluorescence spectrometry is a portable, noninvasive method of determining the presence or absence of RCRA-regulated metals, along with other metals, on-site.

X-ray fluorescence spectrometry (XRF) is a nondestructive method for the detection of elements in solids and liquids. The sample can be analyzed *in situ* without a physical sample being removed. An x-ray beam irradiates the sample. The beam emanates from a radioactive isotope source. Irradiation of the sample causes the emission of fluorescent luminescence, caused by the absorption of radiation at one wavelength, followed by nearly immediate reradiation, usually at a different wavelength.

The XRF operates something like a camera. When a camera shutter is opened, it allows visible light waves to strike the film. To detect and measure concentrations of a heavy metal, the XRF opens a shutter and the radioactive source emits invisible X-rays. The X-rays pass through less dense material in the sample but are stopped by some more dense elements. These elements then fluoresce radiation back to the XRF detector. The circuitry of the XRF measures the fluoresced radiation and determines the measurable elements and their concentrations in the sample.

The spectrum of emitted X-rays is detected using an energy dispersive (ED-XRF) detector. The elements in the sample are identified by the frequency of the emitted X-rays (qualitative

TABLE 12.5 Elements Detected and Required Sources

Element set	Required source
Cr, Mn, Fe, Co, Ni, Cu, Zn As, Se, Pb, Hg, Rb, or, Mo	109Cd, 10 mCi (370 MBq) standard activity
Cd, Ba, Ag, Sn, Sb, plus others	241Am, 14 mCi (520 MBq) standard activity
K, Ti, Sc, V	55Fe, 18 mCi (670 MBq) standard activity

Source: Courtesy of NITON® Corporation (www.niton.com / index.html.)

analysis), while the concentrations of the elements are determined by the intensity of those X-rays (quantitative analysis). XRF is a bulk analysis technique with the depth of sample analyzed varying from approximate 50 μm to 1 cm, depending on energy of the emitted X-ray and the sample composition. For a table of elements commonly detected, see Table 12.5.

Often there will be a need to determine the presence of heavy metals (the eight RCRA-regulated metals) on a Superfund or brownfields site before any remediation work can begin. The PID/FID and GC/MS, described in Sections 12.3.9 through 12.3.11, are designed for analysis of chemicals that are vapors or can easily be volatilized. The RCRA-regulated metals, with the exception of mercury, are extremely difficult to volatilize.

A valuable tool for RCRA-regulated metals detection is the XRF. Although originally a laboratory instrument like the GC/MS, the portable XRF has been in the field for some time now. HUD and a number of states have written their lead abatement guidelines around XRF technology. The XRF can easily detect the amount of lead in lead-based painted structures with acceptable accuracy. But XRFs are not solely for the lead abatement field anymore.

We have chosen to illustrate this point with NITON® Corporation's XL-700™ Series multielement analyzers. See Figure 12.32. They are the latest in high performance and portability for XRF analyzers for on-site metal contamination testing. Many federal, state, and private regulators, institutions, organizations, and companies use NITON® analyzers.

FIGURE 12.32 The NITON XL-700® Series hand-held XRF. (*Courtesy of NITON Corporation* [www.niton.com / index.html].)

NITON's XL-700® Series multielement analyzers can provide simultaneous analysis of up to 25 elements, including all eight of the RCRA metals.

NITON XL-700® Series analyzers are rugged, versatile, easy to use, and economical to operate. Superior performance NITON XL-700® Series Multielement analyzers report concentration levels for elements and 95% precision values in ppm. Results are continuously updated, so you can end tests when you reach the levels of precision you need for target analytes.

In the readout of the Niton XL-700® XRF seen in Figure 12.33, for instance, you see a nominal 65-second readout of barium (Ba) at 1020 ppm with an accuracy of ±81 ppm. You also see that antimony (Sb), lead (Pb), cesium (Cs), and tellurium (Te) are all below detectable limits.

Employing three interchangeable radioactive sources, the XL-700® Series analyzers provide immediate results for up to 25 different elements and can be factory-calibrated for a wide variety of sample types.

Note

To be certified to use an XRF instrument, the user must take an instructional course, usually offered by the manufacturer and usually lasting only one day. The reasons for training are to explain operation, including safe use of the XRF containing a radioactive source. Misuse of the instrument could cause injury. Also note that if your work with an XRF takes you across state lines, state or federal regulations require you to notify the state you are entering that you are transporting a radioactive source. These notification requirements differ depending upon the state and the isotope. We recommend that you consult the instrument manufacturer for guidance.

WARNING

THE XRF'S RADIOACTIVE SOURCE EMITS POWERFUL X-RAYS. EXTREME CARE MUST BE TAKEN WHEN USING THE UNIT AROUND OTHER WORKERS TO PREVENT INJURY. THESE RAYS CAN PENETRATE LOW-DENSITY SUBSTRATES, SUCH AS DRY-

FIGURE 12.33 The NITON XL-700® Series hand-held XRF readout. (*Courtesy of NITON Corporation* [www.niton.com/index.html].)

**WALL. ALWAYS USE CAUTION WHEN ACTIVATING THE INSTRU-
MENT—MAKE SURE THAT THERE ARE NO OTHER WORKERS ON
THE OPPOSITE SIDE OF THE STRUCTURE YOU ARE ANALYZING.**

12.6 GROUND-PENETRATING RADAR; NONINVASIVE SUBSURFACE INVESTIGATIONS

○━π **Ground-penetrating radar is an excellent method for "seeing" underground.
HAZWOPER personnel use it principally for locating buried drums, tanks, and
other containers.**

Ground-penetrating radar (GPR) is a noninvasive (you do not have to dig holes or pits)
electromagnetic instrument for subsurface exploration, characterization and monitoring. It
consists of a portable radar transceiver with an antenna. It is often used to locate buried
utilities, pipes, drums, and USTs, among other items. It may be operated from the surface
by hand or vehicle (the most common way for HAZWOPER sites), in boreholes or wells,
from aircraft, and even from satellites. To date, it has the highest resolution of any method
for "seeing" underground. It may have resolution down to centimeters under certain conditions.

Resolution (how well you can distinguish objects) is controlled by wavelength of the
radar wave in the ground. As frequency increases (shorter wavelength), so does resolution.

For our HAZWOPER purposes, it is used to detect subsurface, man-made anomalies,
which are buried items of different size, shape, and density. They differ in their response to
the GPR from the surrounding soil, sand, or rock. GPR can also report the depth of a buried
item. *Detection* is dependent on the size, shape, makeup, and orientation of the objects
relative to the receiving antenna of the GPR. Because our HAZWOPER sites are often on
urban fill, this can be troublesome. However, a host of computer programs are available that
allow for modeling the underground strata. Typically, GPR is used to examine directly under
the surface to as deep as 100 ft, depending upon the type and densities of the subsurface
strata.

A typical GPR unit is about the size of a lawnmower. In taking readings, it is towed
across the surface of the ground. Sometimes the readout is easy to interpret, such as when
a steel drum is buried horizontally under 5 ft of sand. At other times, such as when a drum
is buried with other fill such as construction debris, the interpretation is much more difficult.

Basically, GPR gives clues about where to dig to investigate underground material that
might be hazardous. Please see Section 14 for a more detailed discussion of GPR.

12.7 OTHER SITE MONITORING INSTRUMENTATION AND METHODS

As we emphasized in Sections 10 and 11, there are other workplace hazards that can adversely affect HAZWOPER workers. Here we will touch briefly on some additional instrumentation that can warn of potential health problems and provide basic site data. Noise,
temperature, relative humidity, and wind conditions are important variables to consider at
the site.

Noise is measured with a decibel meter giving results in dBA. These results will tell us
whether we need hearing protection. As described in Sections 6 and 11, generally speaking,
noise levels at the 85 dBA level for a worker require hearing protection. The use of the
Geoprobe®, described in Section 12.5.3, is an example of work that might well require
hearing protection for the operator.

We measure temperature and relative humidity conveniently with a sling psychrometer or other devices. This instrument gives both a reading of temperature and relative humidity. Both values can be important to the individual analyzing samples in the laboratory.

⊙━━πℓ High relative humidity (greater than 90%) will adversely affect the accuracy of portable devices such as the PID and colorimetric tubes.

Relative humidity is measured by the difference between the wet bulb temperature and the dry bulb temperature. These are two independent thermometers mounted on the psychrometer. The wet bulb is kept wet by a water-moistened cloth covering the bulb. Slinging the psychrometer in the air causes the wet bulb to become cooler by water evaporation from the cloth covering. But of course, at high relative humidity there is very little evaporative cooling and thus little difference between wet bulb and dry bulb temperatures.

This is an accurate instrument that has no electrical requirements. There are other, battery-powered wet bulb thermometers on the market, but they do not offer any great improvement in accuracy.

The higher the relative humidity at a given temperature, the more oppressive the climate feels to us. Due to evaporative cooling from our skin, the less humid it is, the cooler we feel at a given temperature. HAZWOPER workers wearing PPE are much more sensitive to these conditions. Table 12.6 shows relative humidity readings for warm working conditions at the mid-80s°F level.

Another important consideration at the job site, especially in response actions, is the wind speed and direction. These conditions are measured by a weather vane, which indicates the wind direction, and a velometer, which tells us wind speed. Wind speed is essential when we need to determine the wind chill index, which indicates how much colder it feels to us than the dry bulb temperature indicates.

Rapid changes in wind and other weather conditions at a site can drive the decision-making process about continuing or ceasing work. Weather changes can affect access and emergency exit routes. Thus, many sites now use portable weather stations.

Mini Mitter, Co., Inc., offers a real-time physiological monitoring called VitalSense℠. This system is excellent for heat stress applications where workers are wearing Level A and Level B CPC. Their website is www.minimitter.com/vitalsen.htm. This device provides monitoring from a PC of up to 10 workers for heart rate, temperature, and activity using unobtrusive probes attached to the worker. The system also has user-configurable alarms that are both audible and visual. The data are also logged on the PC for purposes of adjusting work/rest cycles and adjusting levels of CPC (if safe), among others.

TABLE 12.6   Relative humidity chart for 30°C (86°F) ambient temperature

Difference between dry bulb and wet bulb temperatures	Relative humidity
None	100%
0.5°	96%
1.0°	93%
1.5°	89%
9.0°	44%
9.5°	42%
14.5°	19%
15.0°	17%
18.0°	5%

Source: Courtesy of NASA.

12.7.1 Remote Sampling

⚷ Remote sampling should be used whenever available, for the safety of the worker.

We cannot emphasize enough the need for HAZWOPER workers to keep their distance when dealing with unknowns (material and containers from which they cannot glean any identifying information).

When we speak of remote sampling, we are speaking of a wide range of possible techniques that isolate the HAZWOPER worker from the contamination. When sampling the atmosphere for contaminants, we can attach to our sampling instrument or media (colormetric tubes, O_2/LEL meter, PID, GC, etc.) one of several types of long-range (20–30 ft) probes for horizontal or altitude sampling. We can also drop a long (up to 100 feet) hose down a well, shaft, or other space to take our sample. This eliminates the need for a long climb into what might be a hazardous atmosphere.

The EPA's Office of Research and Development has investigated remote subsurface sampling. One method uses thermal extraction for the sampling of VOCs and semi-VOCs. At depths of up to 70 feet, a probe heats the surrounding soil to a temperature of 400°C (752°F). At this temperature all VOCs are volatilized and pumped to the surface. There they are freeze-trapped (the hot vapors are condensed) and then fed into a GC/MS for laboratory-quality results. This method is purported to also extract traditionally hard-to-volatize compounds such as PCBs and polycyclic aromatic hydrocarbons (PAHs).

Remote sampling of liquids and other materials in containers and drums takes us into a different arena. That rusty drum may be full of dry sawdust or highly reactive picric acid; how can you tell? In Section 14 we will be discussing some of these methods with respect to their part in overall site characterization.

For HAZWOPER sites where you really don't know what you are getting into but where serious hazards are suspected, the state-of-the-art technology is robotic manipulation and sampling. Figure 12.34 illustrates an all-terrain robot that is radio-controlled, with a manipulating grappler, two closed-circuit TV cameras, two high-intensity lights, and the capability of carrying multiple sampling probes and devices. Remote-controlled robots (autonomous vehicles) have been used to inspect the interior of pipelines, sample the ocean floor, find land mines and collect and detonate them, and fight fires remotely, and have also been used by police bomb squads.

Robotic vehicles are now in use by the DOE, able to travel to a sampling site, open a container, and use a variety of sampling and monitoring methods to characterize the material. These are especially useful when dealing with sites capable of producing IDLH conditions, unknown radiation exposure, unexploded ordinance, or shock-sensitive material. In the near

FIGURE 12.34 An example of a remote sampling radio-controlled all-terrain robot. (*Courtesy of NASA.*)

future, as development continues (and the prices drop), we will see these robots used more often in HAZWOPER work.

12.7.2 Emergency Response Kits

Emergency response kits allow the HAZWOPER worker to determine rapidly the presence of most hazardous materials, or at least material families, in the air, liquids, and some solids; without the need for several different instruments.

Sometimes you will not have the luxury of sampling the media and sending it to the lab for analysis. There are a number of manufacturers of emergency response kits. A rather complete kit we have seen is the HazCat® Chemical Identification Kit™ from ChemSafe, CC. It is used by the HazMat team at Los Alamos National Laboratory. These kits are designed for use by the non-professional. As you can see in Figure 12.35, this extensive test kit is portable. It is packaged in a toolbox that measures 19 × 11 × 11 in. According to the supplier, the kit:

- Contains everything needed for field testing of gases, liquids, and solids, using the optional Sensidyne® or Dräger® gas-detection tubes
- Identifies or categorizes over 1000 hazardous and nonhazardous substances, including flammables, corrosives, caustics, poisons, metals, paints, plastics, pesticides, oxidizers, explosives, water-reactives, and asbestos
- Differentiates between incompatible subsets of oxidizers
- Identifies secondary hazards (it is not enough to know the general hazard class of an unknown; for example, certain flammable liquids or oxidizers are also skin contact poisons)
- Identifies a wide variety of poisons, including arsenic, lead, strychnine, aniline, mercury, phenol, and PCBs
- Identifies explosive oxidizers such as benzoyl peroxide, ammonium nitrate, and ammonium perchlorate
- Identifies multiple-component unknowns
- Identifies nonhazardous materials as such, expediting cleanup operations

FIGURE 12.35   HazCat® chemical identification kit. (*Courtesy of ChemSafe®, CC* [www.chemsafe.co.za].)

Chemicals and materials that can be positively identified by the HazCat® chemical identification kit are: acetaldehyde, acetic acid, acetone, aluminum, ammonia, ammonium nitrate, ammonium perchlorate, aniline, arsenic, arsenic acid, asbestos, benzoyl peroxide, bleach, boric acid, boron, cadmium, calcium carbide, calcium hydroxide, calcium metal, calcium sulfate, carbon disulfide, carbonated water, cement, chlorobenzene, chrome, chromic acid, chromium trioxide, cobalt, detergent, diesel fuel, dimethyl sulfate, ether, ethylene glycol, fiberglass, flour, formaldehyde, gasoline, hydrazine, hydrofluoric acid, hydrogen peroxide, hydrogen iodide, iodine, iron, kerosene, latex paint, lead, lime, lithium hydroxide, lithium metal, magnesium, mercury, methyl ethyl ketone, methyl ethyl ketone peroxide, nitric acid, oxalic acid, paint stripper, PCBs, perchloric acid, phenol, phosphoric acid, phosphorous pentoxide, plastics/plastic resins, picric acid, potassium cyanide, potassium hydroxide, potassium metal, propyl chloride, pumice, salicylic acid, shellac, silica, silver, sodium cyanide, sodium hydroxide, sodium metal, strychnine, sugar, sulfur, sulfuric acid, tetrahydrofuran, thionyl chloride, thiophene, turpentine, urea, urethane plastic, urine, wax, wood and wood products, and zinc.

Functional groups that can be positively identified by the HazCat® chemical identification kit are acetates, acrylates, alcohols, aldehydes, amines, BTEX hydrocarbons, carbamates, carbonates, cellosolves, chlorinated hydrocarbons, chlorocyanurates, glycols, isocyanates, ketones, mercaptans, nitriles, organometals and organophosphates, polyols, and thionated pesticide.

When you see the range of chemicals, materials, and chemical groups that can be field analyzed, you begin to appreciate the power of such a field monitoring kit.

12.8 GLOBAL POSITIONING SYSTEM, GEOGRAPHIC INFORMATION SYSTEMS, AND THEIR USE IN SAMPLING AND MONITORING

The Global Positioning System (GPS) can be used by personnel operating portable GPS transceivers to get the exact coordinates of any point on earth. The GPS, together with a geographic information system (GIS) computer database, allows personnel to record sampling points that can be displayed on downloaded, customized topographical maps. This is a major improvement over hand-drawn maps.

The Global Positioning System (GPS) was developed in the 1970s by the U.S. Department of Defense, principally the Army and Air Force. Originally, for reasons of national security and antiterrorist activity, the Air Force used what was called selective availability. That meant that the military could scramble the satellite transmissions so that nonmilitary GPS units (receivers) would become inaccurate. The military accuracy was within about 10 m^2 of a true point, while the signals to civilian models were adjusted so that accuracy was only within about 100 m^2.

Selective availability was turned off in 2000, so that civilian GPS models are now as accurate as the old military models. However, the military still reserves the right to scramble signals in cases of national emergency.

There are 24 satellites in the system, orbiting above the earth at 20,200 km (12,500 miles). They simultaneously broadcast the time and their exact position to portable units that receive and analyze that satellite data. At any point on the earth a user can receive the data from at least four satellites.

The GPS unit can either be hand-held for fieldwork or vehicle mounted. That can be used for tracking hazardous material shipments. Receiving directions or other aid while traveling, such as through the use of OnStar®, has become quite popular. The GPS provides the user with rather precise latitude and longitude as well as altitude and speed of travel. The GPS creates way points when the users stop and check their location. The newer receivers can store hundreds of these points, thereby giving users a directional map of where they have been. Since timing is precise, the GPS can also calculate speed.

A geographic information system (GIS) is a computer system capable of assembling, storing, manipulating, and displaying geographically referenced information, that is, identified according to locations on the earth. GIS is a sort of revved-up system of creating topographical maps. The GIS can process information that the user desires and that is available and output it for printing on a map. Many states already have their GIS systems up and running, Massachusetts being one of the first.

Now consider the potential of combining GPS and GIS technology. With the reduced size and enhanced data-logging capabilities of monitoring instruments, you have state-of-the-art data collection. Where previously we had to hand-draw maps of sampling locations, we can now match data logged sampling events with GPS way points to give a precise map of the sampling locations. Combine that with PC downloadable information, with your state's GIS, and you have precision mapping of all your sample locations.

12.9   TRAINING AIDS

Sampling and monitoring equipment, training manuals, and computer resources on their proper selection, calibration, use, and maintenance are available from:

Delphian Corporation (www.delphian.com)

Quest Technologies, Inc. (www.quest-technologies.com)

Mine Safety Appliance (MSA) (www.msanet.com)

RAE Systems (www.raesystems.com)

Hnu Process Analyzers (http://www.hnu.com)

NITON Corporation (www.niton.com)

ChemSafe, CC (www.chemsafe.co.za)

Meriter (http://www.canister.com)

Ben Meadows Company (www.benmeadows.com)

Geoprobe Systems (geoprobesystems.com)

The authors recommend the following video for aiding readers to see real-life examples of what they have learned in this Section:

• *Atmospheric Testing,* available from Coastal Communications (www.coastal.com).

12.10   SUMMARY

Sampling and monitoring have been explained as critically important to the safe and successful initiation of HAZWOPER activity where there are potentially hazardous material threats. As project team members, we want entry to a site to be on the basis of having acceptable knowledge of such threats. Then, as site work progresses, HAZWOPER management must ensure that it has the proper protocol for continued sampling and monitoring in their quality assurance/quality control (QA/QC) manual.

That protocol had better be an established system of rules, practices, and chain-of-custody procedures. It must cover collecting, analyzing, and presenting the results of vapor (or gas), liquid, and solids sampling that can qualify as forensic evidence. Forensic evidence is especially valuable because it will be accepted in a court proceeding along with legal testimony.

From such efforts team members want to know, with satisfactory assurance, the identities, concentrations, and distributions of any hazardous wastes. When cleanup efforts are completed, HAZWOPER project managers, property owners, and regulators must all be confident that contamination has been eliminated or at least reduced to satisfy regulatory requirements. The success of such a project rests heavily on effective sampling and monitoring.

Sampling has been explained as the art and science of taking instrument readings, or collecting physical samples, of a gas, liquid, semisolid (sludge), or solid. In the case of any of these media, HAZWOPER personnel might need to take a physical sample off-site for laboratory analysis. This must be done carefully, using a chain-of-custody system to ensure sample integrity. However, each of these media can be analyzed (to some extent) in the field.

Monitoring has been defined as an ongoing process. At some sites monitoring has been in progress for decades! During monitoring, sampling and analysis may take place continuously. This would usually be for a short term of hours or days of monitoring. Where weeks, months, or years are involved, samples are collected and analyzed in batches, according to a timed schedule. The question frequently to be answered by monitoring is: "Is the contamination at the site being reduced?" If not, more aggressive remediation action is required.

Our explanations of sampling and monitoring have focused largely on atmospheric testing. This is because the greatest threat to HAZWOPER personnel comes from atmospheric contaminants. Detection and concentration measurement methods ranging from the simplest (chemical detector tubes) to the most sophisticated (GC/MS) have been described. HAZWOPER personnel must know the usage of oxygen meters that will tell if an atmosphere contains enough breathing oxygen. Next they need to know the usage of flammability, or LEL, meters to know whether the atmosphere is flammable or explosive. Finally, they must know the usage of detector tubes, badge-type sensors, multigas meters, and PIDs for the screening detection of atmospheric contaminants.

An understanding of the correct methods for collection of field samples is critical. We have described the simplest method of collection in plastic bags. It is important also to understand the use of SUMMA® canisters, or their equivalent, for collecting, and transporting to the lab, the most representative field samples.

We have presented the portable GC and portable GC/MS as instruments on the forefront of the state of the art. HAZWOPER personnel might come in contact with them infrequently. However, it is important to have some introductory understanding of GC and GC/MS components and technology. They are going to become much more widely used in the first decades of the 21st century. They will increasingly fill the role of being replacements for laboratory analyses. We want readers to have a basic understanding of their operation and usefulness.

Liquids sampling and monitoring are especially important where groundwater testing is required. We have described the importance of knowing that when you sample from a monitoring well, you are getting true and unadulterated samples. We explained the function of the Kabis® sampler in collecting discrete-depth samples. When samples are taken to a lab for analysis, workers must ensure careful preservation in storage and transit and accurate chain-of-custody entries.

In addition to grab-sampling for solids sampling, we have explained the use of the Geoprobe® as an efficient method of obtaining cores of subsurface soil. With modifications, the Geoprobe® can obtain subsurface water and vapor samples. This small, truck-mounted device replaces much larger and heavier drilling equipment. It can be transported to sites that naturally have space, weight, and surface disruption limitations.

We introduced the GPS and GIS systems of positioning and mapping to alert HAZWOPER personnel to the fascinating changes underway for more accurately documenting site sampling and monitoring activities.

SECTION 13

WORK AT HAZARDOUS WASTE GENERATORS AND AT TREATMENT, STORAGE, AND DISPOSAL FACILITIES (*TSDFs*)

OBJECTIVES

Comparing HAZWOPER Working Environments

Section 13 explains the varied site operations and the duties and responsibilities of HAZWOPER personnel at hazardous waste generators and TSDFs. These are two of the four broad types of work that a 40-hour HAZWOPER certified worker might be engaged in on a day-to-day basis. We explain the kinds of work that require certification, and then the kinds of work not specifically requiring certification. Where there are gray areas of uncertainty in the regulations, we usually recommend certification or a significant portion of the 40-hour training. These are work areas where there is the potential for contact with hazardous materials, but not routine, daily handling of hazardous materials.

In both of the work environments described in this Section, workers generally report to one facility day after day and hopefully year after year. Many times that is not true of the site remediation work described in Section 14. It is definitely not true of emergency response work described in Section 15.

Varieties of Hazardous Waste Generators

Our first objective in Section 13 is to describe what is quite possibly the most common HAZWOPER work environment. It occurs in a facility that is dependent upon raw materials or processing aids or produces a product or byproduct some part of which becomes hazardous waste for various reasons. The facility is called a generator. A generator could be the local dry cleaner having waste solvents. It could be a painting contractor having waste paints or an auto body repair shop having waste resins. These would most likely be small-quantity generators. Although the authors consider it wise to have key operators in such facilities 40-hour HAZWOPER certified, that is not required by federal regulations.

In much larger facilities (large quantity generators), say a production plant for many tons per year of pesticides or chlorinated or heavy metal compounds, there might be a need for a number of people to be 40-hour HAZWOPER certified. The authors have worked at several large quantity generators. Some personnel at those facilities will be managing hazardous wastes routinely. Workers will, however, be dealing largely with the same or similar hazardous materials and wastes on a daily basis. Rightly, they, more than workers in the other types of work described in Sections 14 and 15, will be expected to have more specific knowledge about the particular hazards they face. They will be expected to know more about the selection and proper fitting of their own particular PPE, for example.

Varieties of TSDFs

Our second objective of this Section is to describe the type of work that takes place in the environment of a treatment, storage, and disposal facility (TSDF). The authors have worked in three large TSDFs, two in the eastern and one in the western United States. Our work involved inspection and characterization of a wide variety of wastes, determination of treatment options and transportation needs, and design and installation of environmental controls.

Challenges of TSDF Work

The day-to-day analytical and compatibility determinations of incoming hazardous wastes to a TSDF and the control of facility emissions can be quite challenging. However, selection of treatment steps for wastes detoxification in a safe and economical manner is even more challenging. Since we were dealing with wastes, almost always considered low-value materials, economic considerations were right up there just below safety in everyone's mind. In two of the facilities we dealt routinely with cyanide wastes. This is the same class of materials used in the Nazi death chambers! Safety was definitely consideration number one for us. Our facilities managed treatment, storage, and disposal.

However, there are TSD facilities that focus only on treatment while providing essential storage. Others offer only storage, no treatment or disposal. Then there are those that provide little or no treatment. They act as the final disposal site for the wastes that they collect. They are EPA-permitted hazardous waste landfills, the "grave" in the often referred to "cradle-to-grave" regulation of hazardous wastes. Each type of site and work provides its own challenges.

For the TSDF category, training specific to any new types or mixtures of hazardous wastes, prior to the arrival of such wastes at the facility, is in order. Further, TSDF personnel training usually must be more extensive than that at a generator's site. All TSDF work involves hazardous wastes.

The legislation and regulations governing generation, transportation, treatment, storage, and disposal of hazardous wastes were introduced in Sections 1 and 2. The authors, based on their experiences, encourage a much wider participation in 40-hour HAZWOPER certification than required by regulations. Employers' goals from this wider participation are to receive the benefits of reduced employee lost time through injuries and illnesses, reduced environmental liability, and insurance cost savings.

13.1   WHO GENERATES HAZARDOUS WASTES?

We all do. We trust that individuals and homeowners will dispose of hazardous wastes in the manner recommended on the packaging of most materials we buy. So the question becomes, "Who is subject to hazardous waste disposal regulations?" The answer is, more places than you would ever suspect. There are about 100,000 small and large quantity hazardous waste generators in the United States. It seems that in most workplaces there is the possibility of contact with hazardous materials, especially when you consider all the materials discussed in Sections 2 through 5 of this book. And the hazardous materials ultimately generate some hazardous wastes. The questions of whether they are a type of regulated hazardous wastes, and in the quantity that is regulated, were partly answered in Section 2 of this book.

Remember that OSHA defers to EPA in the definitions of "regulated hazardous wastes." OSHA defers to DOT when transportation of hazardous wastes becomes involved. Any time a hazardous waste moves from one owner or controller to another, that is "transportation." There are about 500 companies involved in hazardous waste transportation in the United States.

O—π **Transportation, in the regulations, occurs any time a hazardous waste moves from one owner or controller to another. If you, as a transporter, receive hazardous waste and its manifest from an owner (generator), you become a controller (have control of the waste) doing hazardous waste transportation.**

13.1.1 Exclusions from Regulated Hazardous Wastes

O—π **When we refer to hazardous waste we generally mean EPA-regulated hazardous waste. Some major exclusions are household wastes and domestic sewage, agricultural wastes used as fertilizers and irrigation return flows, and cement kiln dust and mining wastes remaining at the site.**

Some wastes are excluded from the EPA definition of regulated hazardous waste. Remember that OSHA accepts that EPA definition. Also remember that some wastes are excluded from the definition of solid waste and therefore from the hazardous waste regulations. Excluded from the definition of solid waste are domestic sewage, irrigation return flows, and *in situ* mining wastes. Also, household wastes, agricultural wastes used as fertilizers, and cement kiln dust are examples of wastes that are excluded from the definition of hazardous waste.

That does not mean those wastes are not covered under some other federal, state, or local regulation as far as disposal is concerned. Further, it does not mean that these excluded wastes (or unregulated wastes) are not harmful to humans. Sewage sludge is not a regulated hazardous waste. However, it is very hazardous for unprotected humans to contact. Sewage treatment plant workers are not required to have 40-hour HAZWOPER certification. However, they are covered by state and local regulations for safety and health training and are licensed by the state. For example, they are inoculated for protection against tetanus, and hepatitis.

13.2 CLASSIFICATIONS OF HAZARDOUS WASTE GENERATORS

O—π **According to EPA, any facility that produces or generates hazardous waste is a hazardous waste generator.**

Any facility that, in the course of its operations, produces or generates hazardous wastes is called a generator according to the EPA definition accepted by OSHA. A generator could be the local dry cleaner having waste solvents. It could be a painting contractor having waste paints or an auto body repair shop having waste resins. These would most likely be small quantity generators. Although the authors consider it wise to have key operators in such facilities 40-hour HAZWOPER certified, this might not be required in many cases.

In much larger facilities (large quantity generators), say a production plant for many tons per year of pesticides or chlorinated or heavy metal compounds, there might be a requirement for a number of people to be HAZWOPER certified. Personnel at those facilities will be managing hazardous wastes routinely.

O—π **The three classes of hazardous waste generators are large quantity generators (LQGs), small quantity generators (SQGs), and conditionally exempt small quantity generators (CESQGs).**

13.2.1   Employee Training Needs and Hazardous Waste Generators Defined

The detailed references for these definitions are in Appendix D on the Companion CD, HAZWOPER Standard, 29 CFR 1910.120. They are also in EPA's *Standards Applicable to Generators of Hazardous Waste,* 40 CFR 262. For over a decade, one of the authors worked with 40-hour HAZWOPER certification in a government facility that was a large quantity generator. During that period, he also received the required 8-hour yearly refresher training and many hours of additional specialized training.

Employee Training at Hazardous Waste Generators.   The HAZWOPER standard does not specifically state that 40-hour HAZWOPER certification is required of employees at a generator facility. However, the authors recommend such certification for all employees who come in contact with, or might come in contact with, hazardous materials. That includes both workers who contact hazardous materials or their containers and their direct supervisors. To complete certification, the training must be followed by three days of supervised on-site work.

Where hazardous materials wastes exist but there is low probability of employee contact with the hazardous materials, we recommend an abbreviated version of the 40-hour training, a 24-hour training period covering the same topics, but in less detail. That must be followed by one day of supervised work involving the hazardous waste.

There are some overlapping regulations between OSHA, EPA, and DOT. However, OSHA focuses on work environments and operating procedures that might affect employee safety and health. EPA focuses on site administration, operations, and procedures that might affect the population at large and the environment. DOT focuses specifically on preventing all kinds of problems that might occur during transportation of all kinds of materials; especially hazardous materials.

Basic Requirements of Large Quantity Generators (LQGs) and Small Quantity Generators (SQGs).   Both large and small quantity generators must meet basic requirements. Then there are additional requirements that differ for large and small quantity generators. We will also define conditionally exempt small quantity generators (CESQGs).

Both LQGs and SQGs must be registered with EPA and have an ID number. HAZWOPER workers must store materials in secure containers. Incompatible wastes must be stored such that reactions do not occur between them. Wastes must be labeled, marked, and placarded properly for storage and transportation. That means the containers shall be clearly labeled "Hazardous Waste." HAZWOPER workers must inspect wastes regularly and inspections must be recorded to determine that containers and storage area integrity are being maintained.

O—🔑   **Stored hazardous waste is referred to as accumulation. The amount of hazardous waste accumulation is regulated. The time that any waste is permitted to accumulate from the first waste added to the time the waste must be disposed of is also strictly regulated.**

The hazardous waste label must show an accumulation start date, then the number of days allowed storage time, and finally the date by which the waste must be accepted at a disposal facility. These dates must be monitored carefully. The start date is the first date that *any waste* was deposited in the container, not the date when it was filled. The date for shipment to, and acceptance by, the disposal facility must take into account any delays that might occur in transportation or delays in acceptance by the disposal facility. Any shipment must be to a permitted disposal facility. Generators must use the manifest system and keep manifests and test reports, as proof of operations according to requirements, for at least three years.

Large Quantity Generators (LQGs)

O—🔑   **LQGs generate 2200 lb or more of hazardous waste, or 2.2 lb or more of acutely hazardous waste per calendar month. Their total accumulation amount is not limited.**

These are facilities that generate 2200 lb (1000 kg, a metric ton) or more of hazardous waste per month, or 2.2 lb (1 kg) or more of acute hazardous waste per month. Examples of acute hazardous wastes are some pesticides, toxins, and arsenic and cyanide compounds.

Personnel in contact with hazardous wastes at LQGs are required to have safety and health training under EPA's hazardous materials generator regulations. OSHA also requires their training. OSHA does not specifically cite 40-hour HAZWOPER training as fulfilling the LQG personnel training requirement. However, in the opinion of the authors, 40-hour HAZWOPER training does fulfill those requirements. It has been our experience that the most reputable LQGs have relied on such training as the core of their training for workers in contact with hazardous wastes.

> **LQGs generally may not accumulate a hazardous waste for longer than 90 days from the time that the first waste is added to the container until the container is accepted at a TSDF.**

LQGs may not store, or accumulate, individual hazardous waste streams for greater than 90 days. However, if they must transport the waste over 200 miles to a TSDF or landfill, LQGs are allowed a 270-day storage time limit. Beyond these limits the generator may request a 30-day extension of time from the Regional EPA Administrator if there are unforeseen shipping delays. At LQGs ignitable or reactive wastes must be stored at least 50 ft within the facility's boundary line.

An LQG must file an exception report for late or missing manifests. An LQG must meet the personnel training requirements, including documentation of training. Emergency or incident response must be planned and employees trained in its operation. That includes making arrangements with local fire and police departments, hospitals, emergency response contractors, and equipment suppliers. Those organizations must have documentation explaining emergency arrangements, hazards of materials handled, and layout of the generator's facility. Sufficient response equipment must be available, regularly tested and maintained, including:

- Telephones or portable two-way radios
- An internal communication or alarm system
- Fire and spill control equipment, including fire extinguishers, hoses, sprinklers, neutralizing agents, spill adsorbents, overpack drums, and standby 55-gallon drums

Small Quantity Generators (SQGs)

> **SQGs generate more than 220 lb, but less than 2,200 lb of hazardous waste, or up to 2.2 lb of acutely hazardous waste per calendar month.**

SQGs are facilities that generate more than 220 lb (100 kg) of hazardous waste per month but less than 2,200 lb (1,000 kg). They may not generate more than 2.2 lb (1 kg) of acute hazardous waste per month. They must fulfill the basic requirements (see previous subsection). They may not exceed the 13,200-lb (6,000-kg) total waste accumulation and 180-day storage time limits. However, as with LQGs, if they must transport over 200 miles to a disposal facility, they are allowed a 270-day storage time limit. Beyond these limits the generator may request a 30-day extension of time to solve unexpected difficulties from the Regional EPA Administrator.

> **SQGs may not accumulate more than 13,200 lb of hazardous waste at any one time. The accumulation time limit is 180 days.**

For emergency planning the SQG must have at least one employee or a designee with authority as Emergency Coordinator (EC) on 24-hour call and prepared to follow emergency procedures as specified in 40 CFR 262.34(d)(5), including taking necessary steps to address spills and fires and notifying the National Response Center (24-hour number: 800-424-8802). We recommend that the EC be 40-hour HAZWOPER certified.

Conditionally Exempt Small Quantity Generators (CESQGs)

⊶𝜋 **CESQGs generate less than 220 lb of hazardous waste, or less than 2.2 lb of acutely hazardous waste per calendar month. They may not accumulate more than 2,200 lb at any one time.**

In addition to the basic requirements and according to 40 CFR 261.5, CESQGs may only generate less than 220 lb of hazardous waste per month and less than 2.2 lb of acute hazardous waste per month. They must perform hazardous waste determinations, and they may not accumulate greater than 2,200 lb at any time. They must ensure delivery of their hazardous waste to a permitted disposal facility and keep records documenting that disposal. We recommend that at least one key worker be 24-hour HAZWOPER trained.

Might a Facility Be a Generator and a TSDF?   The answer is Yes. The generator in this case would most probably be an LQG. This would usually be a facility that, in the normal course of operation, produces (or generates) large amounts of hazardous wastes.

Rather than relying on other facilities to treat and dispose of its wastes, such an organization might opt to treat and dispose of its own wastes. Treating might not be too complicated. It might be as straightforward as using ozone, or ultraviolet assisted ozonation, to destroy the toxic nature of the wastes, followed by neutralization. That might render the final waste suitable for disposal with the sanitary wastes.

Another example of generator disposal is hospital disposal of hazardous biological wastes, or biohazards. Many hospitals have found it practical to incinerate or sterilize their hazardous wastes. They install and operate on-site, economical, incinerator or steam sterilization packaged systems. These in-house treatment and disposal systems save the practicing hospital potentially large storage, transportation, and external organization disposal costs.

In addition to those considerations, the generator organization practicing treatment and disposal simplifies all of its cradle-to-grave concerns. There is no possibility that, during shipping of hazardous materials from the generator to disposal, an accidental loss of the material will occur that must be tracked down. Further, there is no possibility that a chosen disposal facility will foul up on the actual disposal. That causes problems that will come back to haunt the generator at some time in the future. **Remember, joint and several liability, which, freely stated, means that anyone having ownership or control at any time of an improperly managed hazardous waste can be held 100% responsible for its improper disposal (and subsequent clean up) at any time in the future.**

Another generator might have high confidence in dealing with a reputable and competent TSD facility, one that has the proper (and possibly quite expensive) processing equipment to treat and dispose of the generator's wastes. Such might be the case with wastes such as the PCBs. Incineration seemed a best treatment and disposal for many PCBs wastes. However, their complete destruction called for quite high-temperature incineration, using temperatures that were beyond the capability of many common incinerators. The special residence times and temperatures required for PCBs' complete destruction resulted in the need for expensive incineration systems. Generators avoided those capital costs by outsourcing treatment and disposal.

13.3   KEY PERSONNEL DUTIES AND RESPONSIBILITIES AT A GENERATOR FACILITY

13.3.1   Owners', Operators', and Managers' Duties and Responsibilities

The individual or individuals in charge of a generating facility, whether carrying the title "Owner," "operator," or "manager," must accept the duty to see that the facility operates according to regulations. That might appear crystal clear, but every year we hear about people bearing one of those titles or another related supervisory title being fined or going to jail for having violated OSHA, EPA, or DOT regulations involving hazardous waste.

The individual punished sometimes feels it was someone else's responsibility for the violation. Sometimes that is an absentee owner or operator, who feels removed from the facility operations. The individual managing day-to-day operations may not fully sense the duty and responsibility of meeting all regulations, regardless of any instructions, stated or implied, to the contrary from an owner. However, ignorance of the regulations is no excuse for noncompliance.

> **Generator facility owners are responsible for establishing standard operating procedures (SOPs) that, when followed, satisfy regulations. Further, they must ensure that employees follow the SOPs as a condition of employment.**

Ownership, management, and all supervisory levels are responsible for establishing standard operating procedures and seeing to it that all employees follow them as a condition of employment. These must be procedures that satisfy the regulations for the particular classification of facility registry. (These classifications are explained in Section 13.3.)

There is another way of stating major management responsibility. As Dr. Edwards Deming, the foremost quality expert emphasized, management is responsible for establishing the operating system for any organization. Once the system is properly established, it will only be necessary for the employees to do their best at working within the system as willing workers. It is nearly impossible for the employees to produce more desirable work results than the structure of the system allows. The system, its standard operating procedures, and employer enforcement will make or break the success of the generator's hazardous waste operations in the eyes of OSHA, EPA, and DOT. Any auditors or inspectors from those organizations will want to see documentation of effective procedures and demonstration that they are being followed.

Management also has the duty to train all employees who contact, or might contact, hazardous wastes, for their own protection and that of their coworkers and for environmental protection. Training must be in waste hazards, health effects of unprotected contact with wastes, proper personal protection, emergency actions, and much more. As previously stated, it is the authors' opinion that the fundamental training that management must supply to meet those requirements is the 40-hour HAZWOPER training. Obviously, supervisors of such employees require the same basic training. When management also has the required Health and Safety System as part of its operating system, as described in Section 16, the long-term health of employees is preserved.

Management in generator organizations has the responsibility to provide for proper storage of wastes to prevent incompatibility problems and provide for effective firefighting. See Section 4. Further, management has the responsibility to maintain proper labeling, placarding, and other identification on stored wastes and wastes to be shipped. See Section 8. They may assign these tasks as duties of HAZWOPER employees.

13.3.2 Supervision's and Workers' Duties to Minimize Hazardous Wastes

> **All generator employees have the duty to minimize hazardous waste by careful operation according to SOPs and by suggesting recycle and reuse opportunities to supervision as well as less harmful substitutes.**

It is the duty of all employees, and especially HAZWOPER employees of an organization, continually to work toward hazardous waste minimization. This includes recycling, reuse and reducing hazardous wastes by substituting less hazardous ingredients. It is common sense that, from an economic viewpoint, this should be a goal of any organization generating hazardous waste. For safety and environmental reasons it is also a goal of OSHA and EPA. Landfill space near major generators is rapidly filling up. On hazardous waste manifests, supervision at the generator organizations must signify that they are taking all reasonable steps to minimize hazardous wastes.

Recycling and reuse are the most desirable ways of reducing wastes. In the past, recycling was always practiced where wastes contained obvious economic value. For decades in photo processing and jewelry manufacturing the silver in wastewaters has been recovered. Gold has long been recovered from plating and jewelry manufacturing wastewaters. Now the chief driving force for recovery of all kinds of materials is not necessarily their economic value, but rather the reduction in the costs and potential problems of waste disposal. Substituting less hazardous water-based paints for oil-based paints is an example of reducing the hazardous waste stream.

In almost any facility that works with toxic materials, where powders and dusts are created, supervision should not allow dry sweeping and the airborne spread of particulates. Wherever hazardous wastes are generated, floor drains should not be connected directly to total facility waste treatment or municipal waste lines. Drainage of hazardous waste must be collected and treated, or safely stored, while it is most concentrated. Then organizations achieve least overall waste disposal cost, least chance of worker contact, and least chance of release to the environment.

13.3.3 Employees' Duties and Responsibilities

The HAZWOPER employees' duties at generator facilities of any size are similar. They must understand the characteristics and proper on-site transportation, packaging, labeling, and storage of any wastes they handle.

On-site transportation frequently means the movement of 55-gallon (208-L) steel or plastic drums. See Section 14.15.7 for details on the safe handling of these containers.

Packaging for Shipment. Packaging of wastes for shipment from the generator facility requires determining whether it is being shipped to a recycler, a treatment facility, or a landfill for final disposition. The results of the TCLP determine whether a waste must be disposed of in a hazardous waste landfill. Section 12.5.2 described the TCLP test. A common shipment to a recycler is liquid with fuel or lubricating value. If there is fuel value, and the liquid is acceptably low in content of water and halogens (usually chlorine-containing solvents), it is acceptable to recyclers for their treatment and sale as substitute boiler fuels. The determination as to whether spent lubricants can be recycled is more complex but is always worth investigating. Shipment of liquids to recyclers or TSDFs can be by over-the-road tankers.

If packaging is for shipment to a treatment facility, the waste may be liquid or solid, so long as it is properly packaged and labeled. It is desirable that the container be filled, but it could be a partially filled drum of liquid or solid. Shipment of containers to landfills for disposal requires that HAZWOPER employees ensure that the contents are solid waste and that the containers are filled. In landfilling situations, the term *solid waste* means that the packaging, drum or otherwise, does not contain free-standing liquid.

The Paint Filter Test; Landfill Rejection of Liquids. Employees receiving a semisolid material must determine if it is free-standing using the paint filter test. A paint filter is the kind of fine-mesh filter used in paint stores to filter any surface film or other solid contaminants from paint. At the generator's site, employees receiving a semisolid material must place samples of the material in a paint filter and watch to see if any liquid drips through the filter. If any does, the waste is considered as having free-standing liquid and cannot be sent to a landfill in that condition. To package it for landfill disposal, it first must be solidified such that it can pass the paint filter test. That will require some additive to, or treatment of, the waste so that it thickens, gels, or solidifies.

> **Landfills generally will not accept liquid wastes in any container, or unfilled containers. To be accepted, liquids must first be solidified and containers must be full to within 2 in. from the top of a drum. Half empty containers must be filled with inert material before shipment.**

Landfills will not receive liquids in degradable containers. Corrosion or rupture of the container in the landfill would lead to liquid travel through the landfill mass. This can produce undesirable leachate and the potential for incompatibility reactions with other wastes and can lead to voids and undesirable settling of the landfill.

HAZWOPER employees must also ensure that if only part of a drum is filled with waste and the decision is made to ship the drum to a landfill, the remaining space in the drum must be filled with a compatible or inert material. Oil Dri® (desiccated clay absorbent) has been used for this purpose. Most landfills will not accept containers with voids.

13.3.4 Employees Need More Familiarity with MSDS Information

In the experience of the authors, many workers still do not appreciate the benefits of knowing more about the materials with which they work. Workers in all industry, laboratories, and medical services must know more about those raw materials, products, equipment, catalysts, processing aids, and more with which they come in contact.

> Hazardous wastes are either listed or characteristic. If you can identify the waste or the process the waste came from, you can consult the EPA lists to determine if the waste is listed. You can check the hazardous characteristic using a pH meter, a flammability test, and a toxic vapor meter.

OSHA has mandated that all information relevant to worker contact with such materials must be explained in the MSDS collection for the facility. Such information must be presented to employees as part of HAZCOMM, the Hazard Communication Standard, 29 CFR 1910.1200. The authors have routinely detected that workers do not understand the implications of much of the data given in an MSDS. Full interpretation of MSDSs has been given in Section 9 of this book.

13.4 HOW HAZARDOUS WASTES ARE GENERATED

When we consider hazardous materials processes and the hazardous wastes they will ultimately generate, we tend to think of hazardous feed materials or hazardous materials that are produced. But there are many manufacturing processes, industries, and services that use hazardous materials as processing aids, or even as part of the operating equipment. There are those facilities that generate hazardous byproducts and hazardous wastes from nonhazardous materials.

13.4.1 Raw Materials and Process Materials as Sources of Hazardous Wastes

The generators we think of first are those that use lead, mercury, arsenic, chlorine, and other toxic elements in their raw materials. However, processing materials or aids can also introduce hazardous materials. In the production of PCB-containing electrical components, a manufacturer found that process water intake to the plant, coming from a large river was a source of PCBs. The raw process water had a higher level of PCB contamination than was allowed by the facilities discharge permit for wastewater.

EH&S personnel in the food and beverage industries should be aware of the contents of raw materials. We cannot always be sure that apparently pure raw materials are free of hazardous materials. Newly dug drinking water wells in India have produced water containing naturally high and toxic amounts of arsenic. Their drinking water wastes would obviously be hazardous wastes. This has prompted reduction in the allowable amount of arsenic in U.S. drinking water. See Section 5.5 for further examples.

Improper processing of infected animal carcasses and then their use as raw materials for livestock feed supplements led to the "mad cow" disease scare in England and Europe.

13.4.2   Hazardous Wastes from Process Equipment

One example of a process that in the United States is being displaced by other technology is the mercury cell production of chlorine and caustic soda from salt. The mercury is in the process equipment. It is used as a fluid cathode in the electrolytic cells. In the past, small amounts of mercury compound wastes were discharged from such plants. They found their way into water bodies and then into fish. Some bioaccumulation occurred, together with production of organic mercury compounds. A particularly bad actor was found to be methyl mercury. People eating contaminated fish might suffer an illness.

Metal fines generated in the milling of food and other materials to obtain powders can be a source of iron-, chromium-, and nickel-contaminated materials.

13.4.3   Hazardous Wastes Generated during Nonhazardous Product Manufacture

Suprisingly, organics and titanium tetrachloride, during high-temperature oxidation to produce titanium dioxide as part of bright white pigment production, have led to dioxin-contaminated wastes.

13.4.4   Hazardous Wastes Generated during Waste Treatment

Chemists and chemical engineering employees must be alert to the possibility that materials can be made hazardous by some common treatments. Nontoxic biphenyl has been used in carpet manufacturing. Nontoxic wastes from that manufacturing became toxic during routine chlorination of the wastewater. The biphenyl was converted to polychlorinated biphenyls (PCBs). Within any facility, it is a responsibility of EH&S staff, and the research capabilities they can draw upon, to investigate continually the raw materials and their impurities, processes, byproducts, products, and wastes to determine whether new hazards are emerging.

Dioxins, some of the most toxic chlorinated organic chemicals, hazardous at the ppt levels, are being found in a wide range of combustion emissions. Power plant and incinerator stacks, as well as other high-temperature reactions with chlorine and organics, seem to be able to yield troubling dioxin concentrations. Since the use of PVC, CPVC, and other chlorinated organic materials in packaging and construction materials is so widespread, it is not unusual to find dioxins in incinerator emissions. Further research is needed. However, it is believed that incinerator temperatures in excess of 2,200°F, and residence time of at least three seconds, can destroy any dioxin formation. Apparently, if destruction can take place at less than 1,000°F, such as by catalytic action, dioxin formation is suppressed.

13.5   KINDS OF HAZARDOUS WASTE GENERATORS

There are hundreds of kinds of hazardous waste generators. A couple of examples here might suffice. For a more complete appreciation of the diversity in hazardous waste generators, refer to Appendix I on the Companion CD. Appendix I gives descriptions of listed hazardous wastes from processes not specific to any particular manufacturing process (F waste codes) and hazardous wastes from specific manufacturing processes (K waste codes). You will note breaks in the numerical series in the lists. These were waste codes that have been delisted.

13.5.1   Pesticide, Fungicide, and Rodenticide Wastes

Workers in pesticide, fungicide, and rodenticide production must know that waste from raw materials, products, and byproducts all can be quite hazardous. Especially well-trained

HAZWOPER workers are needed here. It is hard to find a powerful agent in this class that efficiently removes pests but does not affect humans.

For example, the pesticides and their wastes that are organic phosphorus-containing compounds can attack the human central nervous system. They are somewhat distant cousins to the family of phosphorus-containing nerve gases first developed by the Germans for use in World War II. A goal of waste treatment for the pesticides and waste nerve gases is to break down the molecule such that inorganic phosphate is formed, possibly suitable as fertilizer raw material.

Laboratory R&D personnel in coordination with toxicology researchers and EH&S investigators in organizations doing this work must study thoroughly the properties and use of any new compounds synthesized. Arsenic compounds, containing their own unique hazards, were used for pest control before DDT was introduced. Some readers may remember how DDT was hailed as the miracle insecticide that saved millions of lives by preventing typhus and malaria outbreaks in the mid-20th century. Then P. Mueller won the Nobel Prize in Medicine in 1948 for his discovery of the modern synthesis of DDT and its powerful effectiveness as an insecticide against mosquitoes. Early on it was found to have low water solubility and to be readily absorbed by insects, causing their rapid death. However, in the following decades DDT was found to be toxic to humans and to bioaccumulate in the fatty tissue of fish, animals, and humans. As a further disappointment, some insect resistance was also found to be developing.

13.5.2   Hazardous Metal Fabrication and Mining Wastes

Metal-fabrication industries can be the source of antimony, beryllium, chromium, nickel, and other metal hazardous wastes. Mining wastes have long been the source of many illnesses and diseases and the cause of bad-tasting, high-iron and sulfur content groundwater contamination. The use of cyanides in gold recovery and jewelry manufacturing has been widespread. The authors, as HAZWOPER employees, have worked in two TSDF facilities that specialized in treating metal cyanide toxic wastes.

Compounds of nearly all of the toxic metals have been used in an endless list of consumer products. Lead compounds are the first that come to mind. The wastes from those industries must be carefully managed. It is a prime duty of scientists and engineers in these industries to develop innovative ways to reduce such hazardous wastes and improve recycling. Another kind of duty is that of finding substitute nonhazardous materials that are practical for use in production processes.

13.6   *WORK AT A TREATMENT STORAGE AND DISPOSAL FACILITY (TSDF)*

Note

For references to any regulations required for TSDF operation, refer to CFR Guide To Hazardous and Solid Waste Regulations, under Section 1.4.1 of this book.

13.6.1   What is a TSDF?

A full-service TSDF is a facility that can accept specified wastes from generators. It then treats the wastes to destroy the hazardous properties, stores received and in-process materials, and then disposes of its wastes as recyclable materials, liquids to municipal treatment, and solids to landfills.

O—π   **The function of a TSDF is to reduce the volume and toxicity of hazardous wastes deposited in hazardous waste landfills.**

There are about 1600 TSDFs in 54 states, territories, and Indian nations in the United States. EPA's regulations for TSDFs are covered in 40 CFR Part 264/265, Subparts A through E. Although there are some exceptions to the regulations covering TSDFs, as we will discuss in later parts of this Section, TSDFs are defined as facilities that treat, store, and dispose of hazardous wastes. Generators typically send their hazardous waste to TSDFs. When the TSDFs are truly disposal facilities, they are the end of the road for RCRA's cradle-to-grave tracking requirements for hazardous waste.

The HAZWOPER standard requires only 24 hours of relevant training for TSDF employees. The authors, however, recommend that full 40-hour HAZWOPER training be the basic training for TSDF employees who contact, or might contact, hazardous waste. We have seen too many incidents where full 40-hour HAZWOPER training, and more, could have saved injuries.

The RCRA Act (see Section 1.4.1) requires treatment storage and disposal facilities to obtain a permit for operation. The TSDF must submit an application describing the facility's operation. There are two parts to the RCRA permit application. Part A defines the processes to be used for treatment, storage, and disposal of hazardous wastes, the design capacity of such processes; and the specific hazardous wastes to be handled at the facility. Part B requires detailed site-specific information such as geologic, hydrologic, and engineering data. The following terms are used in describing TSDFs.

"Facility" Defined.   A *facility* includes all the land and any of its structures and appurtenances on or in the land used for treating, storing, or disposing of hazardous waste. A single facility may consist of several types or combinations of operational units. If a facility is engaged in either treatment, storage, or disposal or any combination of the three, it must be in compliance with 40 CFR Part 264/265, Subparts A through E.

"Treatment" Defined.   *Treatment* is defined as any method, technique, or process designed to change the physical, chemical, or biological character or composition of any hazardous waste. This can include energy or resources recovery from the waste. The treatment may be as simple as neutralization, or it may be such that it will render wastes nonhazardous, or less hazardous, safer to transport, store, or dispose of, or better suited for recovery.

Note

The authors' HAZWOPER experience in three TSDFs has taught them that analytical determinations of incoming hazardous wastes can be quite challenging. However, following that, selection of treatment steps for wastes detoxification in a safe and economical manner is even more challenging. Since we were dealing with wastes, almost always considered low-value materials, economic considerations were right up there just below safety in everyone's mind. We dealt with cyanide wastes. Safety was definitely consideration number one for us.

"Storage" Defined.   *Storage* means holding hazardous waste for a temporary period. Its purpose is to keep the hazardous waste safe from any release until it is treated, disposed of, or stored elsewhere.

"Disposal" Defined.   A *disposal* facility is any site where hazardous waste is intentionally placed and at which the waste will remain, even after site closure.

TSDF is EPA's term grouping all such facilities together for the purpose of discussing the regulations. As a HAZWOPER-certified individual, you may work at a hazardous waste storage facility, at a treatment and disposal facility, or at any other part or combination of the three types of facilities. Each type of facility must abide by the regulations and obtain and maintain an operating permit.

13.6.2 A TSDF as a Generator of Hazardous Waste

If a TSDF initiates a waste shipment, a new manifest must be prepared to comply with Part 262 standards (Sections 264/265.71(c)). This could be the case if the TSDF, in the course of its business, treated specific wastes that did not include chlorinated solvents. But if the TSDF discarded such solvents, which were listed in 40 CFR Part 261 and were wastes from degreasing TSDF equipment, it would be considered a generator and it would have to comply with the regulatory requirements applicable to generators in Part 262.

Compliance with the Part 262 regulations can include an accumulation area under Section 262.34 that is exempt from permitting. This exempt accumulation area would be available only to hazardous wastes that are generated by the TSDF on-site.

13.6.3 A TSDF as a Transfer Facility for Hazardous Waste

TSDFs may also serve as transfer facilities according to Section 260.10. In that case they are permitted to hold waste that is appropriately packaged in accordance with DOT regulations for up to 10 days, provided the TSDF is not the final destination, or designated facility, for that waste.

13.6.4 A TSDF as a Treatment, Recycling, Storage, and Disposal Facility

We can expect a growing emphasis on recycling by TSDFs, along with the rest of industry, in the 21st century. HAZWOPER employees can make significant contributions to this effort by continually considering other possible uses for the wastes they manage. Approximately 85% of lead used in the United States each year goes into the manufacture of lead-acid storage batteries. TSDFs specializing in receiving waste auto batteries are doing serious recycling.

RSR Company reports that it recycles more than a third of the U.S. domestic batteries generated as waste each year. In its recovery process the liquid electrolyte (dilute sulfuric acid) is separated from the solid battery case. The recycled plastic is provided for reuse to plastic compounders and fabricators for use in new battery cases and other consumer products. Purification and then neutralization with caustic soda converts the acid to sodium sulfate for use in the pulp and paper industry. The sulfate is also used as a stabilizer in common laundry detergents.

Smelting processes and refining steps produce pure lead and lead alloys from the lead components separated from the polypropylene plastic battery cases. The refined lead is interchangeable with primary (mined) lead. The recycled lead is supplied to manufacturers and used in all segments of the lead industry. That includes battery manufacturing, lead chemicals, solder for computer circuits, shielding for X-ray machines, soundproofing in architecture, and oxides for the glass in computer monitors. Due to this lead recycling, use of primary lead has decreased by over 50% over the past several decades. Recycled lead now represents 85% of domestic production.

13.7 PERMITTED TSDFs; OWNERS' OR OPERATORS', MANAGERS', AND EMPLOYEES' DUTIES AND RESPONSIBILITIES

Note
In the past there has been an "interim status facility" designation for TSDFs.

This will eventually be phased out. They will either conform to the regulations as permitted facilities or be closed down.

⊙—π   **A TSDF must be permitted for operation by the EPA.**

Permitted facilities are facilities that have applied for, and been granted, EPA permits to operate as a TSDF. Operation without a permit is illegal. Every facility owner or operator has the duty to apply for an EPA ID number. Once registered, they have the responsibility to ensure that hazardous wastes under their control are properly identified and handled. They must ensure that facilities are secure and operating properly and that personnel are trained in hazardous waste management. It is the owner's or operator's duty to ensure that any managers whom they entrust with those critical duties are actually performing them according to regulations. This means that the managers must have the delegated authority to carry out their responsibilities. The following are typical work tasks for HAZWOPER employees:

• Conduct of waste analyses (characterization)
• Installation of security measures
• Monitoring and control of waste storage and treatment
• Conduct of training by qualified employees and faithful attendance at training by trainees
• Proper management of ignitable, reactive, toxic and incompatible wastes
• Prevention of unlawful emissions from the facility

Owners are responsible for the safe and environmentally responsible operation of their facility. EPA regulates TSDFs so as to ensure that owners or operators establish the necessary procedures and plans to run a facility properly and handle any emergencies or accidents. Owners and all levels of management personnel need 40-hour HAZWOPER training if they intend to inspect facilities where they come in contact with, or might come in contact with, hazardous wastes.

13.7.1 Design Requirements for TSDFs

Facility-specific performance standards and design and operating requirements are incorporated into the permit for TSDFs. For example, tanks storing hazardous waste must be designed to specifications found in 40 CFR Part 264. The permit language in the regulations is general. It serves as a guideline for permit writers in setting the specific design and operating requirements through use of best engineering judgment.

Facility owners and operators must take steps to prepare for and prevent accidents. These are requirements intended to minimize the possibility and effects of a release, fire, or explosion. An emergency action plan must be developed and documented (see Section 16 for details).

13.7.2 Manifest System: Record-Keeping and Reporting

⊙—π   **All hazardous waste being transported must be accompanied by a Uniform Hazardous Waste Manifest. The manifest must include a waste characterization sheet.**

Some 92,000 businesses in the United States conduct hazardous waste manifest-related activities. In 2001, the EPA proposed to improve the Uniform Hazardous Waste Manifest system by automating procedures and standardizing the manifest form. EPA claims waste handlers and states could save considerable manifest-related costs without compromising hazardous waste management. The proposed changes would allow waste handlers the option of using an electronic manifest. Standardization of forms and procedures would be included.

The Manifest System Audit Trail has been described in detail in Section 2.3.7.

The new standard format could be used in all states. These changes would eliminate the different manifest forms that are currently required by many authorized states. A waste handler with multistate operations could register and use its own manifest forms everywhere it does business. EPA would oversee the process by which states, waste handlers, or other printers would be registered to print the standard form according to EPA's specifications.

The Uniform Hazardous Waste Manifest must be returned from the TSD facility owner or operator to the generator. (A copy of the manifest form is shown in Figure 2.1.) That completes the manifest loop (40 CFR Part 262). The TSDF is responsible for ensuring that the waste received is the same as the waste described on the manifest. Discrepancies must be reported to the EPA in 15 days. Other required record-keeping includes operating records, biennial reports, unmanifested waste reports, and reports on releases, groundwater monitoring, and closure.

The paperwork required by Part 264/265, Subpart E is designed to track hazardous waste from cradle to grave. The manifest system tracks each shipment of hazardous waste, while the operating records and biennial reports summarize facility activity over time.

TSDFs that accept waste from off-site are the final signatories to the manifest. When a manifested waste shipment is received, all copies of the manifest are signed and dated by the facility owner/operator. The transporter and the TSDF keep signed copies of the manifest, and a copy is sent to the generator within 30 days to verify acceptance of the waste (Sections 264/265.71). If the owner/operator of a TSDF must send the waste on to another TSDF for further treatment or disposal, a new manifest with a new designated facility must be initiated.

Upon receipt of a manifested waste, the owner or operator of a TSDF must determine whether the manifest accurately describes the waste it accompanies. Any discrepancies in weight (for bulk shipments, over 10%), piece count (for batch or containerized waste shipments, one container per truckload), or waste type are considered significant and should be noted on all copies of the manifest at the time of signature. The owner/operator must try to reconcile the discrepancy with the transporter or generator promptly. Any discrepancies not resolved within 15 days of waste receipt must be reported to the Regional Administrator with an explanatory letter and a copy of the manifest (Sections 264/265.72).

If a TSDF accepts waste from off-site without a manifest, an unmanifested waste report must be prepared in accordance with Sections 264/265.76. The report must be submitted to the Regional Administrator within 15 days of receiving the waste (Sections 264/265.76).

13.8   KEY PERSONNEL: DUTIES AND RESPONSIBILITIES AT A TSDF

Personnel with a sound understanding of the methods and instrumentation for analyzing complex mixtures of inorganic and organic chemicals to determine their composition are a prime requirement for the first stage of work at a TSDF. This is followed by the need for chemists and chemical engineers skilled in knowing the safe and efficient ways of detoxifying and rendering less hazardous, nonhazardous, or recyclable the wastes accepted for treatment. To keep the equipment and instrumentation and control systems operating (pumps, agitators, blowers, dryers, process and plant exhaust scrubbers, and effluent) and well maintained, mechanical, instrumentation, and environmental specialists are needed. All of these key personnel are bound to have contact with hazardous materials. OSHA mandates that they receive special training.

13.8.1   Analytical Chemists and Characterization of Wastes Arriving at the TSDF for Treatment

The first duty of TSDF employees when hazardous waste shipments arrive, is to ensure, through counting and weighing containers, that the amount of waste on the manifest is the amount of waste received.

○──π   **The second duty of employees at a TSDF is to ensure, through characterization, that the type of hazardous waste manifested is the type of waste received.**

When a transporter arrives at the TSDF with a load of hazardous waste, and the proper manifest to transfer control of the waste to the TSDF, the first priority of the TSDF is to determine whether to accept control and ownership. In the ideal situation, the TSDF analytical chemists have previously accepted many loads from this transporter having the same waste characterization.

The analytical chemists, while not required to have full 40-hour HAZWOPER certification, must at least have special laboratory safety training in the handling and analysis of the variety of hazardous wastes that the TSDF will accept. In the event of a plant incident, they may be required to evacuate the premises according to a prepared emergency action plan (see Section 16).

Chiefly the analytical chemists and the treatment plant engineers and operators, but also any other employees contacting incoming wastes to the TSDF, must know what they can learn from the waste characterization sheets that accompany the wastes. However, the chemists will have prescribed tests to run to determine that the waste is actually very similar to what has been received and treated successfully on previous occasions. If it is a new kind of waste, but judged to be treatable, acceptance testing will be more extensive.

Most TSDFs will not even schedule a transporter to arrive with a load until it has been determined that the load falls within the TSDF's guidelines for acceptable waste loads. The TSDF might say, for example, that it will accept metal plating wastewaters containing cyanides, chromium, nickel, zinc, and other metals. However, the TSDF might have a policy of not accepting wastes containing radioactive materials, heavy sludge loads, or significant organic chemical content (solvents, polymers, resins, insecticides, or pharmaceuticals).

Not until the analytical chemists and management approve of the acceptance will the transporter be allowed to unload to a specific receiving tank at the TSDF. There may be several receiving tanks, depending upon fundamental characteristics of the incoming loads (acids, alkalis, sludges, and others). Obviously, the more routine the shipments of any specific waste, the faster this acceptance procedure will go. In one of the TSDFs where an author worked, there was the practice of TSDF personnel going to the generator's facility and identifying, and then labeling and sealing, waste containers. This allowed quick acceptance and unloading at the TSDF. In the case of a first-time delivery of even what ultimately turns out to be an acceptable type of waste, hours or a day of wait time might be required of the transporter. For the transporter, the worst-case scenario is the ultimate refusal by the TSDF to accept the waste due to some characteristic that the TSDF cannot handle.

Here are typical acceptance steps that a TSDF might require to have completed before any waste unloading from a bulk liquid wastewater transporter will be allowed. A small sample of the water-based fluid waste is brought to the laboratory. The waste is mixed with reactants to be used in the treatment process. The mixture is observed for reactions such as gas or odor formation, heat generation, color changes, and precipitation. The results of these tests are then passed to the chemists or engineers charged with the duty of determining the treatment steps.

○──π   **Following characterization of an incoming waste load, TSDF personnel must determine the facilities capability of treating the waste before it is unloaded. They must accept the waste only after they have prescribed the available treatment steps that can lead to regulated disposal of all residuals.**

13.8.2   Chemical Engineers and Treatment Specialists

Chemical engineers, chemists, and treatment specialists are charged with the duty of determining the treatment reactions and the proper sequences to detoxify or convert the hazardous wastes to less hazardous, nonhazardous, or recyclable materials. Together with mechanical engineers and instrumentation specialists, all three skills and any others they may require must join in the effort to design treatment equipment before startup. In the continuing effort of observing operating results, they must act to modify any equipment as dictated by operating experience.

For modern plants, instrumentation specialists must design and troubleshoot the operation of process control computers and programmable logic controllers. They will be sensing and then controlling reactor or storage vessel temperature, pressure, pH, liquid levels, flow rates, and other critical parameters. The difference between TSDF treatment plant operation and that of a chemical manufacturing plant is the much greater diversity in the raw materials that the TSDF receives.

As an example, treatment steps might be planned to minimize reactions that produce gaseous or vapor emissions. A further goal might be to maximize reactions that separate dissolved or suspended solids from a water base. The treatment process for metal plating wastes must be capable of destroying cyanides. Oxidation by hypochlorites to yield carbon dioxide and nitrogen could be used. Acids and alkalis would need to be neutralized. Dissolved metals could be precipitated as the elemental metals themselves, or as insoluble salts. Powerful oxidants such as hydrogen peroxide and ozone could be used to destroy complexes that tend to keep metals in solution.

Following these reactions, the mixtures could be separated to produce three products: (1) powdered or granular metal-bearing wastes suitable for recycling and refining, (2) other detoxified solids suitable for landfilling, and (3) water wastes acceptable to the municipal sewage system.

In the opinion of the authors, all of the personnel with the skills mentioned above must be 40-hour HAZWOPER certified, as a minimum. Should any of those personnel only be required for consulting or equipment repair for several days, an abbreviated training focusing on the area of their work will be adequate. They must, however, be restricted to that planned area and work task.

13.8.3   Treatment Process Supervisors and Operators

The treatment process supervisors and operators are the personnel charged with the duty of operating the process according to the standard operating procedures (SOPs). The engineering and operating supervisory personnel prepare these procedures. They benefit greatly from review input from the process operators themselves. The procedures tell, step by step, what the operators must do to operate safely and efficiently all the treatment, storage, and disposal processes. It is the duty of the supervisors to see that the operators follow the procedures as a condition of employment.

The kinds of operation to be covered are preparation for startup, routine startup, routine or normal operation, conditions causing abnormal operation and how to take corrective measures, routine shutdown, and emergency shutdown.

With all of the potential for unexpected reactions and accidental creation of hazardous environments in a TSDF facility, there should be no question about these supervisors and process operators being 40-hour HAZWOPER certified as basic job training.

13.8.4   An Author's Memorable Incidents while Working at a TSDF

You can get some appreciation of the need for 40-hour HAZWOPER training, understanding of chemicals, their reactions, and how to manage them by the engineers, supervisors, and process operators at a TSDF from the following narratives.

Waste Incompatibility Followed by Rapid Evolution of Nitrogen Oxides.   (One of the authors, reporting as "we," for management, was newly employed as Technical Director of a Part B TSDF.) The facility received metal-bearing, waste liquids (acid, alkaline and cyanide) via tank truck. The facility's processing included treatment for metals removal and reclamation and subsequent sewer disposal of treated effluent.

Bench scale chemical analyses and compatibility testing were performed on all wastes received prior to acceptance. Facility procedures required checks to determine that the wastes were acceptable for facility processing. Further, laboratory personnel had to determine if there were any potential adverse reactions or incompatibilities between newly arriving wastes and stored wastes. Over the years there had been only a few incidents where this testing failed to identify a problem. One such incompatibility incident was memorable to the author. The mechanism of this particular reaction was never completely understood.

A tank truck load of nitric acid, plating-rack, rinse water, containing dilute nitric acid (less than1%) and primarily copper compounds was received and offloaded without incident to the selected temporary holding tank. Prior to this load being transferred to a storage tank pending treatment, a sample of this waste and one of the previous waste stored in the storage tank were tested for chemical compatibility in the laboratory. The previous waste in the storage tank was a tin-lead plating rinse water. No adverse reactions were found. Treatment plant operators were authorized to transfer the load from the holding tank to the storage tank.

The storage tank was a 5,000-gallon, welded, chemical-resistant polypropylene, vertical-cylindrical tank. It was contained in a 110% capacity concrete dike. Double-diaphragm pumps were used for transferring waste liquids, which often contained sludges as well. The storage tank had been emptied of its previous contents, the tin-lead plating rinse water, leaving only a heel of 2 to 3 in. in the bottom of the tank (perhaps 100 gallons). The tank had a 4-in. vent line to a chemical scrubber and an 18-in. access port on the top.

Approximately five to ten minutes after initiating the transfer (at approximately 100 gpm), the plant operator noticed a wisp of orange-brown vapor escaping from the open access port at the top of the tank. As he went to investigate, the vapor deepened in color to a reddish brown and became noticeably thicker. The operator immediately set off the warning alarms and shut off the transfer pump. Until this time, the ambient plant ventilation system was adequately removing the vapor. But as the operator approached the access port, he immediately noticed the acrid odor of nitric acid and nitrogen oxides. As he closed the access port, he noticed that the vapor was hot, and, peering into the tank, he saw the liquid bubbling. The exterior tank wall at and below the liquid level was too hot to touch. In addition, the vapor continued to grow denser and darker brown.

It appeared that an uncontrolled exothermic reaction had been initiated, resulting in the rapid decomposition of nitric acid to nitrogen oxides vapors. We needed to bring this reaction quickly under control. The tank was showing signs of bulging due to heat stress, and we were concerned that it might fail. The polypropylene was rated to perhaps 250°F. Tank failure would release the mixture into the containment dike and require evacuation of the plant. Neutralization with sodium hydroxide was ruled out due to the additional heat of reaction it would contribute. The tank wall temperature was estimated to be 180 to 200°F.

We decided to use cooling and immediately sent plant personnel to a local convenience store to buy all of the block ice available. Personnel equipped with respirators carefully placed 10-lb blocks of ice in the tank through the access port. After we had introduced 10 to 12 blocks (100 to 120 lb) of ice into the approximately 600 to 800 gallons of liquid, the fume generation reduced markedly and stopped after 10 minutes. The mixture was allowed to cool to ambient temperature overnight and was treated without incident in the morning. The remaining waste from the holding tank was transferred to an emptied and triple rinsed storage tank without incident.

The cause of the uncontrolled reaction is not fully known. A subsequent search of the chemical literature turned up some references to the ability of tin ions to catalyze the degradation of nitric acid to nitric oxides under certain conditions. However, the concentrations referred to were significantly higher than that in the rinse water. We hypothesized that there may have been a layer at the tank bottom of residual tin-lead sludge from the tank's previous contents. That layer might have been dissolved by the added nitric acid rinse water and formed a stratified, concentrated layer at the bottom of the tank. That mixture then could have initiated the reaction. The exothermic reaction then would have increased temperature and reaction rate until it became uncontrolled. Such a sludge-liquid reaction would account for why the lab had failed to observe any reaction when combining the two liquid streams during compatibility testing.

One lesson learned was that we had to think creatively in terms of available chemical reactants. None of our standard chemical reagents was useful because all would have increased the heat in the tank, almost certainly leading to tank failure. We had not previously considered ice as a treatment chemical. After this incident, the plant staff made sure to know where to obtain large quantities of block ice if necessary! However, to my knowledge, this reaction never occurred again. From that point on, nitric acid-containing wastes, regardless of concentration, were designated as incompatible with tin-lead wastes.

Uncontrolled Dispersion of Dense White Smoke.   The second memorable incident was the generation of chemical smoke. The TSDF's designers were German. They had designed the facility to use 18% hydrochloric acid (more commonly used in Europe) instead of 90% sulfuric acid (more commonly used in the United States). This had been an annoyance factor in routine processing, leading to corrosion of unprotected steel fixtures and others like them. But the hydrochloric acid had been generally serviceable. Another contributing factor to the incident that occurred was that the pH control for the treatment tanks was initially manually operated, a common practice in Europe at the time, where typically more labor and fewer automatic controls were used, as opposed to the United States, where pH adjustment has long been performed using automatic controllers.

However, the day the plant operator attempted to neutralize a treatment tank filled with ammonia-copper, etchant rinse water, the undesirability of HCl was visible to all. Due to the extremely high concentrations of the ammonia etchant used, the spent rinse waters are consequently also of high concentration. They contain up to 5 grams per liter of copper, and thus make a valuable waste stream. This also means the concentration of ammonia is high, and likewise the pH. It may have been that the operator was trying to bring the pH down too quickly, and because of the dilute HCl the reagent pumps were higher rate than would be typical for the more concentrated sulfuric acid. In any event, although the pH was still elevated, the first indication the operator had of a problem was when someone reported that a dense white cloud was emitting from the scrubber stack. Outside the facility it was slowly settling to the ground like fog. The plant operator immediately shut off the HCl pump and checked the inspection port of the scrubber. The interior was opaque white!

Within 15 minutes, the outside parking lot was enveloped in a dense white cloud that was spreading to the properties of several nearby businesses. Visibility in the plant was approaching zero. All nonoperating personnel had been evacuated upwind. Facility personnel wearing SCBAs began adding water to the treatment tank to dilute the contents and bring the reaction under control. Then the local police arrived. Despite protests from plant operators, the police demanded at gunpoint that all personnel evacuate the plant. Plant operators were not allowed to return until the fire department HAZMAT team arrived and took control. By this time the local television station had arrived and was taping the incident for the evening news.

After repeated assurances that the ammonium chloride fume was nontoxic, the crowds of nearby residents and onlookers eventually left. The rise in wind velocity with evening gradually dispersed the chemical fog. Postincident review identified the use of hydrochloric rather than sulfuric acid as a potential liability, due to the prevalence of ammonia compounds in the plating wastes treated by the facility.

This incident, as well as the nitrogen oxides vapor incident described previously, also demonstrated the need for a dramatic increase in plant exhaust scrubber capacity. That resulted in a major plant upgrade. Two of the authors designed and engineered the change from a single-stage 700-cfm scrubber to a two-stage (neutralization-oxidation) 20,000-cfm Catenary Grid® scrubber installation. That facility exhaust control system has operated for over 10 years without subsequent incident.

13.9   EPA's RCRA AND OSHA's TSDF REGULATIONS

There are several areas of EPA's RCRA TSDF regulations that overlap with OSHA's HAZWOPER regulations. HAZWOPER regulations require a training program, contingency plan, and provisions for preparedness and worker health and safety preservation. HAZ-WOPER requirements differ from the provisions for TSDFs because OSHA regulations are designed to protect the worker. EPA focuses on the environment. Some of the requirements are similar, however. Complying with the regulations under OSHA may at least partially satisfy the RCRA requirements for training, contingency plan, preparedness, and prevention.

Congress directed EPA to establish a system for issuing TSDF permits (RCRA Section 3005). Not only must a TSDF comply with the standards of Part 264/265, but an owner or operator needs to obtain a permit under Part 270 to engage in hazardous waste management. A permit is an authorization, license, or equivalent control document issued by EPA or an authorized state to implement the TSDF requirements. TSDF permits are facility-specific and are issued after a documentation and review process.

13.9.1 Hazardous Waste Generators and Transporters; TSDF Permits Not Required

Generators and transporters are not required to obtain permits for several reasons: they generally handle smaller amounts of waste for shorter periods of time, and their operations are more changeable, which is not conducive to the long and detailed process of permitting. A TSDF, on the other hand, is a permanent facility in operation for the purpose of making a profit from waste management.

This is not to say that generators cannot be permitted TSDFs. Many corporations generate so much hazardous waste that it only makes good fiscal sense to have a captive TSDF. These facilities do not usually take in waste from other generators but simply deal with all the hazardous wastes generated at the site.

13.9.2 Permits by Rule

Certain facilities that have permits under other environmental laws may qualify for a special form of a RCRA permit, known as a permit by rule. Essentially, a facility with a permit under another environmental law that meets the conditions in Section 270.60 is exempt from the substantive requirements of Part 264/265. Sections 264.1(c), (d), and (e) state that the Part 264 standards apply to permit-by-rule facilities (ocean disposal, underground injection, publicly owned treatment works [POTWs]) only to the extent that they are included in a RCRA permit by rule granted under Part 270. Part 265 is different in that only ocean disposal and POTWs are exempt from regulation (Sections 265.1(c)(1) and (3)). Hazardous waste injection facilities are subject to interim status regulation under Part 265, Subpart R. Any treatment or storage at a TSDF prior to placement in facilities exempt under permit by rule is subject to Part 264/265 requirements. In addition, sludge generated at a POTW is a solid waste and may be characteristically hazardous, making the owner/operator a hazardous waste generator.

13.9.3 Conditionally Exempt Small Quantity Generator Waste

Facilities that only treat (including recycling), store, or dispose of waste generated by conditionally exempt small quantity generators (CESQGs) regulated under Section 261.5 are excluded from Part 264/265 (Sections 264.1(g)(1)/265.1(c)(5)). According to Section 261.5(g)(3), such facilities must be either permitted, licensed, or registered by the state to: handle nonhazardous nonmunicipal or municipal solid waste, qualify as a recycling facility, or be subject to the universal waste handler or destination facility regulations under Part 273.

13.9.4 Recyclable Materials

According to Sections 264.1(g)(2) and 265.1(c)(6), owners or operators managing the following recyclable materials are subject to the facility standards of Part 264/265 only to the extent that the Part 266 or Part 279 recycling regulations refer back to them:

- Materials used in a manner constituting disposal
- Hazardous waste burned in boilers and industrial furnaces
- Precious metals that are recycled
- Spent lead-acid batteries that are reclaimed
- Used oil that is recycled

Owners or operators managing the following recyclable materials are not subject to Part 264/265 (Sections 264.1(g)(2)/265.1(c)(6)):

- Industrial ethyl alcohol that is reclaimed
- Scrap metal
- Fuels produced from the refining of oil-bearing hazardous wastes
- Oil reclaimed from oil-bearing hazardous wastes

Although these facilities may generate a hazardous waste stream as part of their recycling operations they are not subject to generator regulations that limit the amount of hazardous materials they may store, nor are they subject to storage time limits. They are, however, subject to storage care, labeling, packaging and manifesting requirements.

13.10   EXCEPTIONS TO THE HAZARDOUS WASTE REGULATIONS

Exceptions to the hazardous waste regulations include totally enclosed treatment or neutralization of wastes as part of an individual process; wastewater treatment units where effluents meet Clean Water Act (CWA) standards; treatment of emergency response wastes, and a few universal wastes.

13.10.1   Totally Enclosed Treatment

Totally enclosed treatment facilities are excluded from Part 264/265 (Sections 264.1(g)(5)/265.1(c)(9)). A totally enclosed treatment facility means a facility that is directly connected to an industrial production process and is constructed and operated in a manner that prevents the release of any hazardous waste or any constituent thereof into the environment during treatment (Section 260.10). The exemption for totally enclosed treatment facilities applies only to the enclosed unit. Effluent from the facility is regulated when the waste entering the totally enclosed treatment facility is derived from listed waste or when the effluent is characteristically hazardous.

13.10.2   Elementary Neutralization

According to Sections 264.1(g)(6)/265.1(c)(10), elementary neutralization units (ENUs) are excluded from Part 264/265. In order to qualify as an ENU, the unit must be a container, tank, tank system, transport vehicle, or vessel. It must only neutralize wastes that are hazardous solely for the characteristic of corrosivity (Section 260.10). Neutralization in a surface impoundment or any other land-based unit is subject to regulation.

13.10.3   Wastewater Treatment Units

Wastewater treatment units are also excluded under Part 264/265 (Sections 264.1(g)(6)/265.1(c)(10)). To meet the exclusion, these units must meet all three parts of the definition of a wastewater treatment unit in Section 260.10. A wastewater treatment unit must:

1. Have a discharge subject to Clean Water Act (CWA) pretreatment standards or permitting requirements (National Pollution Discharge Elimination System, NPDES) under CWA Section 402 or 307(b)

2. Properly manage hazardous wastewater or hazardous wastewater treatment sludge and

3. Meet the definition of a tank or tank system

Note

Since only the unit is exempt from regulation, any hazardous sludge generated in the wastewater treatment unit is subject to regulation when it is removed from the tank.

13.10.4 Emergency Response

Sections 264.1(g)(8) and 265.1(c)(11) exclude emergency response actions to contain or treat immediately a spill of hazardous waste or a material that becomes a hazardous waste when spilled. EPA does not define "immediately." However, any hazardous waste generated must be managed in accordance with Part 262, and any treatment or storage after the emergency situation has passed is subject to full Subtitle C regulations. If the activity does not fall within the scope of this exclusion, an emergency permit of 90 days or less may be required (Section 270.61).

13.10.5 Transfer Facilities

Manifested wastes that are in transit and stored in containers by a transporter for less than 10 days at a transfer facility in accordance with regulations are not subject to generator or TSDF standards. A transfer facility is any transportation-related facility, including loading docks, truck parking areas, and storage areas, as well as any area where shipments of hazardous waste are held during the normal course of transportation.

13.10.6 Universal Wastes

There are four different materials defined as universal wastes:

1. Lead-acid and nickel-cadmium waste batteries
2. Agricultural pesticides
3. Waste thermostats
4. Lamps containing mercury, lead, and metal halides

Again, as with the other types of TSDFs, there are small and large quantity facilities that accumulate universal waste lamps but do not treat, recycle, or dispose of them. They are handlers of the lamps. Universal waste handlers can be exempt from the large and small generator holding or storage regulations limiting the amounts of materials that may be stored as well as the time period allowed for storage. Regulations do require that spent lamps be managed in a way that prevents releases of mercury or other hazardous constituents to the environment during accumulation, storage, and transport. Handlers may accumulate universal waste lamps for one year. If the lamps are stored for longer than one year, the handler must be able to demonstrate that such accumulation is solely for the purpose of accumulating such quantities of universal waste as are necessary to facilitate proper recovery, treatment, or disposal.

13.10.7 Waste Analysis Planning and Performance at TSDFs

The waste analysis plan (WAP) must, at a minimum, contain the following basic elements:

- The parameters to be analyzed
- Testing and analytical methods used
- Sampling methods used to obtain representative samples
- Frequency of waste reevaluation
- For off-site TSDFs, the waste analyses that generators have agreed to supply
- Procedures to ensure that the waste received at the off-site TSDF matches the identity of the waste designated on the accompanying manifest.

   TSDFs need to verify the composition (i.e., hazardous constituents and characteristics) of incoming waste in order to treat, store, or dispose of the waste properly. The WAP outlines the verification procedures, including specific sampling methods, necessary to ensure proper treatment, storage, or disposal (Sections 264/265.13). The WAP must be written and kept on-site.
   Before owners or operators treat, store, or dispose of any hazardous waste, they must obtain a detailed chemical and physical analysis of a representative sample of the waste. They get this information by sampling and laboratory analysis or through acceptable knowledge. Acceptable knowledge is defined broadly to include process knowledge. This is the same as getting data from existing published or documented waste analyses or studies. Further, it includes waste analysis data obtained from the generator or through the facility's own records of analyses.

13.10.8 Frequency of Analysis

The waste analysis must be performed initially upon receipt of a new waste and repeated periodically to ensure that the information on a given hazardous waste label and on the manifest is accurate and up to date. At a minimum, the waste analysis must be repeated:

1. When the TSDF is notified, or has reason to believe, that the process or operation generating the hazardous wastes has changed
2. When inspection indicates that the hazardous waste received does not match the information on the accompanying manifest

   Off-site combustion facilities must characterize all wastes prior to burning to verify that permit conditions will be met.

13.11 *SECURITY PROVISIONS AT TSDFs*

Security provisions are intended to prevent accidental entry and minimize the possibility of unauthorized entry of people or livestock onto the active portion of the facility (Sections 264/265.14). Unless the owner/operator of a facility demonstrates to the Regional Administrator that livestock or unauthorized persons who enter the facility will not be harmed or cause any portion of the regulations to be violated, the facility must install the following security measures (Sections 264/265.14(a)(1) and (2)):

1. A 24-hour surveillance system that continuously monitors and controls entry onto the active portion of the facility (e.g., television monitoring, guards)

or

2. An artificial or natural barrier that completely surrounds the active portion of the facility (e.g., fence), and a means to control entry to the active portion at all times via gates or entrances (if the active portion is located at a larger facility that has barriers and a means to control entry that meet the above standards, the active portion does not need its own system)

and

3. A sign reading "Danger—Unauthorized Personnel Keep Out" at each entrance to the active portion. The sign must be written in English and any other language that is predominant in the area surrounding the facility. It must be legible from a distance of 25 ft. Alternative language conveying the same message may be used.

13.12 INSPECTION ACTIVITY AND DOCUMENTATION REQUIREMENTS

The owner or operator must visually inspect the facility for malfunction, deterioration, operator errors, and discharges (Sections 264/265.15). The inspection provisions are to be carried out according to a written inspection schedule that is developed and followed by the owner/operator and kept at the facility. The schedule must identify the types of problems to be looked for and set the frequency of inspection, which may vary. Areas subject to spills, such as loading and unloading areas, must be inspected daily when in use. Unit-specific inspections or requirements also must be included in the schedule (e.g., Section 265.226 for surface impoundments).

The owner or operator must record inspections in a log or summary and must remedy any problems identified during inspections. The records must include the date and time of inspection, the name of the inspector, notation of observations, and the date and nature of any necessary repairs or other remedial actions and must be kept at the facility for three years.

Performing weekly hazardous waste inspections is one of the simplest ways you, as an operator, can protect your facility from a leak or spill, as well as to meet new container regulations. If done correctly, your effort will prevent potential releases to the environment before they occur, ensure that wastes are identified properly, and see that wastes are shipped off-site before your accumulation time is up.

Here is a model checklist. EPA refers frequently to "accumulation" rather than "storage." Storage seems to imply long-term or indefinite time periods for retention of materials. Accumulation is meant to stress the building up of materials over time and that there is a limit to the amount of accumulation and the time for accumulation that can take place.

HAZARDOUS WASTE ACCUMULATION CHECKLIST

WEEKLY INSPECTION CHECKLIST

Inspection for the week of _____ to _____

Accumulation

_____ Are all drums and containers marked with a hazardous waste label?

_____ Are all drums and containers marked with a DOT placard label, if appropriate?

_____ Are all drums marked with the accumulation start date?

_____ Are there any drums that are near or have exceeded the 90/180-day timeframe?

_____ Are all drums marked with the proper waste code(s)?

_____ Are all containers closed?

_____ Are all drum labels visible and readable?

_____ Do all aisles have one pallet only, with drum or other container labels facing the aisle?

_____ Are all drums and containers in good condition?

_____ Are there 30 inches of aisle space between rows of containers?

_____ Are any drums leaking?

Sumps

_____ Are sumps clean and free of contamination, spills, leaks, and standing water?

Safety Equipment

_____ Are fire extinguishers charged?

_____ Are spill kits stocked?

_____ Is the first aid cabinet stocked?

_____ Is the emergency shower and eyewash station functioning properly?

_____ Are the emergency communication devices operating properly?

_____ Is emergency response information posted near all communication devices?

Secondary Containment

_____ Is the secondary containment free of standing water, cracks, or other failures?

Comments

Describe the actions that you took to correct any deficiencies noted above, and the date the actions were taken

_____

_____

Printed Name _____ Signature _____

Date _____ Time _____

13.13 PERSONNEL TRAINING AND RECORDS RETENTION

Personnel at TSDFs must successfully complete a program of classroom instruction or on-the-job training in compliance with Sections 264/265.16. OSHA initial training for TSDF employees is 24 hours and refresher training is 8 hours annually. Employees who have received the initial training, and passed the subject examination, are to be given a written certificate attesting that they have successfully completed the necessary training.

OSHA requires the following minimum training:

- Safety and health program (see Section 16.4)
- Hazard communication program (see Section 9)
- Medical surveillance program (see Section 16.6)
- Decontamination program (see Section 7)

- New technology program; explanation of new processes, methods, or equipment
- Material handling program (see Section 3)
- Emergency response program (see Section 15)

The emergency response program includes:

- Emergency response plan
- Elements of an emergency response plan
- Pre-emergency planning and coordination with outside parties
- Personnel roles, Lines of authority, Training, and Communication
- Emergency recognition and prevention
- Safe distances and places of refuge
- Site security and control
- Evacuation routes and procedures
- Decontamination procedures
- Emergency medical treatment and first aid
- Emergency alerting and response procedures
- PPE and emergency equipment
- Critique of response and follow-up

At a minimum, the training must focus on effective response to emergencies. The authors recommend 40-hour HAZWOPER certification, followed by the yearly 8-hour refresher.

The training program must be completed six months from the date the facility is subject to Part 264/265 or Part 266 regulation, or six months after the date a worker is newly employed. New employees are required to work under supervision until their training is complete. Facility personnel must take part in an annual review of their initial training.

Training-related documents and records must be kept at the facility. These must include a job title for each person and the name of the employee filling that position. Also, a written job description is needed for each position and records documenting that the employee holding that position has completed the training or job experience satisfactorily. Finally, the files must contain the training records on current personnel and past employees for three years.

> ⚷ **OSHA requires initial training and yearly refresher training for employees of TSDF facilities, together with documentation of that training. Employees must be trained and competent to accomplish the job description for the position they fill.**

13.14 PREPAREDNESS AND EMERGENCY PLANNING AND PROCEDURES

The preparedness and prevention standards are intended to minimize and prevent emergency situations at TSDFs. Facilities must be operated and maintained in a manner that minimizes the possibility of a fire, explosion, or any unplanned release of hazardous waste to air, soil, or surface water.

The regulations require maintenance of equipment, alarms, minimum aisle space, and provisions for contacting local authorities. Specifically, Sections 264/265.32 mandate that a facility must have an internal communication or alarm system, a phone or radio capable of summoning emergency assistance, firefighting equipment, and adequate water supply. Sections 264/265.33 and 264/265.34 require that this equipment be maintained and tested regularly and that all personnel have access to an alarm system or emergency communication device.

Facilities must also have provisions for contacting local authorities that might be involved in emergency responses at the facility. The local authorities must be familiar with the facility and properties of the hazardous waste(s) handled at the facility (Sections 264/265.37). Local authorities include police, fire department, hospitals, and emergency response teams. Where more than one local authority is involved, a lead authority must be designated. Where state or local authorities decline to enter into such arrangements, the owner and operator must document the refusal in the operating record (Sections 264/265.37(b)).

A copy of the emergency planning documents must be maintained at the facility and provided to all local authorities that might have to respond to emergencies (Sections 264/265.53).

The plans must be reviewed and amended when the applicable regulations or facility permits are revised, when the plan fails in an emergency, or when there are changes to the facility. Plans must also be amended when there is a change in the list of emergency coordinators or the list of emergency equipment (Sections 264/265.54).

13.14.1 Emergency Coordinator (EC)

The TSDF must have an Emergency Coordinator available either on-site or by phone contact at all times to take charge of documented emergency procedures should an incident occur.

The Emergency Coordinator (Sections 264/265.55) is responsible for assessing emergency situations and making decisions to respond. At least one employee must be either on the facility premises or on call 24 hours a day to fill this role. This person must have the authority to commit the resources needed to carry out the contingency plan.

Although it is not required by OSHA, the authors recommend 40-hour HAZWOPER certification as minimum training for Emergency Coordinators. Read the following paragraphs to appreciate the authors' recommendation.

13.14.2 Emergency Procedures

In the event of an imminent or actual emergency situation, the Emergency Coordinator must immediately activate internal facility alarms or communication systems and notify appropriate state and local authorities. In cases where there is a release, fire, or explosion, the emergency coordinator must immediately:

- Identify the character, exact source, amount, and extent of any released materials
- Assess possible hazards to human health or the environment
- Determine whether or not the emergency threatens human health or the environment outside of the facility, and whether evacuation of local areas may be advisable. If so, the coordinator must notify appropriate authorities and either the designated government official for the area or the National Response Center

During an emergency, measures must be taken to ensure that additional fires, explosions, and releases do not occur, recur, or spread. If the facility stops operation, the Coordinator, operating in a safe manner, must monitor for leaks, pressure buildup, gas generation, or ruptures in containers, valves, pipes, or other equipment (Sections 264/265.56(a)–(f)).

13.14.3 Postemergency Procedures

After an emergency, any residue from the release, fire, or other event must be treated, stored, or disposed of according to all applicable RCRA regulations. The facility may end up assuming generator status for management of these residues. The Coordinator must ensure that all emergency equipment is cleaned and fit for use before operation is resumed. The owner/operator must document, in the facility operating record, incidents that required the implementation of the contingency plan. Within 15 days of the incident, the owner/operator must submit a written report describing the incident to the EPA Regional Administrator (Sections 264/265.56(g)–(j)).

O⎯π **TSDF owners must ensure that detailed records of daily operations are maintained and kept available for review by regulators. This must include records of any incident. Every two years, a written report to the EPA is required.**

13.15 REQUIRED OPERATING RECORDS FOR THE TSDF

Until closure, the owner or operator are required to keep a written operating record on-site describing all waste received, methods and dates of treatment, storage, and disposal, and the wastes' location within the facility as detailed in Appendix I of Part 264/265 (Sections 264/265.73). All information should be cross-referenced with the manifest number. The operating record also must include waste analysis results, details of emergencies requiring contingency plan implementation, inspection results (for three years), groundwater monitoring data, land treatment and incineration monitoring data, and closure and postclosure cost estimates.

13.15.1 Biennial Report

Biennial reports must be filed with the Regional Administrator by March 1 of each even-numbered year, covering the facility's activities for the previous year (Sections 264/265.75). For example, the biennial report covering years 2002 and 2003 activities would be due March 1, 2004. The facility report (Form 8700-13A/B) is sent by the facility to the region or the authorized state.

13.15.2 Reports of Releases, Fires and Explosions, and Groundwater Contamination

Other reports that must be made to the Regional Administrator include, but are not limited to, reports of releases, fires and explosions, groundwater contamination and monitoring data, and facility closure (Sections 264/265.77). Releases may also trigger Comprehensive Environmental Response, Compensation, and Liability Act (CERCLA) and Emergency Planning and Community Right-to-Know Act (EPCRA) reporting, such as releases of greater than the Reportable Quantity (RQ) or the Threshold Quantity (TQ) of a listed hazardous material.

13.15.3 Record Availability

Sections 264/265.74 specify that all records and plans must be available for inspection. Required record retention periods are automatically extended during enforcement actions or as requested by the Administrator. When a TSDF landfill certifies closure, a copy of records of waste disposal locations and quantities must be submitted to the Regional Administrator and the local land authority.

13.16 DISPOSITION OF WASTES; THE TOXICITY CHARACTERISTIC LEACHING PROCEDURE (TCLP)

All HAZWOPER personnel must be dedicated to the proper disposition of wastes that they encounter. Wastes can contain a hazardous material and still not have to be sent to an EPA-regulated hazardous waste landfill. The question is, Can the hazardous material be expected to migrate through the landfill and possibly be carried off-site into groundwater?

Note

See Section 12.5.2 for TCLP details.

The toxicity characteristic leaching procedure (TCLP) is designed to simulate the leaching a waste will undergo when it is disposed of in a sanitary landfill. When a waste material passes the TCLP test and is not a characteristic or listed waste, it may be disposed of in a sanitary or municipal solid waste landfill.

13.17 REGULATORY DEFINITION OF A TOXICITY CHARACTERISTIC

Please refer to Section 12. Under the toxicity characteristic, a solid waste exhibits the characteristic of toxicity if the TCLP extract from a subsample of the waste contains any of the contaminants listed in Table 12.4 at a concentration greater than or equal to the respective value given in that table. If a waste contains <0.5% filterable solids, the liquid waste itself, after filtering, is considered to be the extract for the purposes of analysis.

Under the Land Disposal Restrictions program, a restricted waste identified in 40 CFR 268.41 may be land disposed only if a TCLP extract of the waste, or a TCLP extract of the treatment residue of the waste, does not exceed the values shown in the "Constituent Concentration in the Waste Extract" (CCWE) table in 40 CFR 268.41 (Table 12.4 in Section 12) for any hazardous constituent listed in Table CCWE for that waste. Note that for landfill disposal, the liquid would still have to be gelled or solidified to restrict mobility.

13.18 DISPOSAL METHODS

Final disposal methods focus on waste volume reduction to lessen transportation and land filling costs, high temperature and chemical destruction, and stabilization processes. Simple filling of waste containers is the most widely used method. However, for some special cases, more complex methods have been proposed. The authors recommend 40-hour HAZWOPER training for workers performing these tasks.

13.18.1 Supercompaction

The supercompaction process can receive drums of sorted nonflammable, nonexplosive solid waste. The drums of waste are punctured. Then they are compacted by a powerful hydraulic press that controls the shape of the resultant supercompacted "puck" through the use of a mold. Because drums entering the supercompactor will have been previously characterized and vented, puncturing and compaction within the supercompactor are not likely to present an explosion hazard. The supercompactor and its associated air pollution control system must be designed to accommodate such overpressures should they occur. Under the extreme pressure of supercompaction, any gas or vapor emitted must be vented and processed through the facility air pollution control system. The volume reduction for each drum is dependent on the drum contents and packing fraction but is expected to be an average of 65–85%. The pucks would be placed into a puck drum located in the postcompaction area. The puck drums would then be transferred to the macroencapsulation process. The puck drum would be the final waste form's outermost container.

13.18.2 Macroencapsulation

Waste is fed into the macroencapsulation process in two forms: containers of pucks and noncompactible debris waste. The grout or cement used in the macroencapsulation process is poured into the puck drum, thus stabilizing the noncompactible waste or pucks in the final waste form container. Grouted drums are lidded and allowed to cure at a drum cure area.

The macroencapsulation system could be used to encapsulate pucks or large pieces of metal debris not suitable for compaction. After curing for a few days, the final waste form containers would be examined and certified for final disposal at a landfill.

13.18.3 Incineration

Incineration is generally the favored method of thermal treatment where temperatures no greater than 2400°F are required. Where varied wastes are to be incinerated, feed rates and composition are tuned for optimum plant performance and maximum safety. Wastes destined for incineration are crushed and shredded by size-reducing equipment located at the head of the incineration process. All corrosive and reactive wastes are pretreated to neutralize and stabilize them prior to thermal treatment. The size-reduced waste is conveyed to a waste hopper, where it is held until it is fed at a controlled rate into the incinerator feed system. A typical high-efficiency incinerator might be a dual-chamber, auger hearth, gas-fired system. The incinerator feed stock is continuously supplemented by propane or other fuel to maintain combustion temperatures and flame residence times. They are kept in ranges adequate to ensure full thermal destruction of all waste materials and their intermediate combustion products. The primary combustion chamber might operate at 1500 to 1600°F and the secondary chamber at 2200 to 2400°F.

The incinerator air pollution control system should include a combination of dry filtration and wet scrubbing systems. The incinerator primary combustion chamber and secondary combustion chamber are integral to the control of airborne emissions from the incinerator. They volatilize and burn waste organic matter. The downstream pollution control systems must further control pollutants present in the offgas prior to release to the atmosphere. Downstream equipment might consist of all or a part of the following: saturation quencher for cooling, venturi scrubber, two absorbers in series, condensing wet electrostatic precipitator, offgas reheater, redundant first-stage HEPA filtration, carbon adsorbers, redundant second- and third-stage HEPA filtration, associated pumps and blowers, and an exhaust flue.

13.18.4 Brine Evaporation

The main component of a facility brine-reduction system is an evaporator used to dry the scrubber brine blowdown generated from the incinerator. The evaporator can also process inorganic liquid wastes from other areas of the TSDF. Process brines and other liquids are accumulated and stored in three brine mix tanks. During operation, one of these tanks is used to collect the brine blowdown, one is sampled and stabilized, and one tank is feeding the evaporator.

13.18.5 Vitrification

Vitrification is an energy-intensive process that is used only for hazardous radioactive, metal, and inorganic waste that must be securely contained in a mass that has little chance of any future interaction with the environment. Feed waste is placed in a hopper and held until fed at a controlled rate into the vitrification unit. A high-temperature melter, such as a direct current arc melter, might be used. Glass-forming materials are continuously fed with the waste to enhance the glass quality of the final waste form. The vitrification offgas treatment system in a complete and large system design might include a gas cooler, cyclone separator, two parallel trains of high-temperature filters, heat exchangers, several HEPA filters in series, and several parallel main blowers to maintain the melter at a constant negative pressure. Efficiency of the cyclone for 10-micron-diameter particles could only be 80–85%. The high-temperature filter can be designed to collect more than 99% of all particles greater than 0.5 microns in diameter. The HEPA filters are 99.97% efficient for 0.3-micron particles.

13.18.6 Chemical Processing

Chemical processing is any process that removes or changes an unwanted characteristic of the waste using a chemical reaction. It may include several different types of reactions, ranging from neutralization of acids and bases and selective oxidation and reduction reactions to amalgamation of mercury, or many other reactions. Section 14.19.7 describes the use of Fenton's reagent as a potential oxidation method for destruction of organic chemical wastes that are very resistant to other chemical destruction methods.

13.18.7 Microbiological Reactions

As the name implies, this is the use of living organisms to induce reactions that detoxify toxic characteristics of the waste. The mechanisms and characteristics of these reactions are explained in Section 14.19.1. Microbiological reactors can have the living cells coated on rotating plates in a flowing wastewater stream. When the microbes are properly matched to the organic hazardous wastes in the water and the residence time in the reactor is sufficient, the microbes will use the wastes as a food source, providing waste destruction.

13.19 POSTCLOSURE MONITORING

TSDFs and other units that store and use hazardous materials have three options under RCRA/CERCLA for closure: clean closure, modified closure, and landfill closure. Modified closure and landfill closure options also can be used to accommodate RCRA/CERCLA integration needs. (RCRA does not address petroleum hydrocarbons and CERCLA does. CERCLA does not address radioactive wastes and RCRA does.) EPA has moved towards a joint RCRA/CERCLA approach to address all contaminants to the best possible protection of people and the environment.

Under the Pollution Liability Insurance and Risk Retention Act, to start a business as a TSDF, you must be able to show that you have the financial backing, or the equivalent in insurance, to open, operate, and close your facility properly. Once a facility is closed, the owner must be able to prove, through postclosure monitoring, that the facility no longer poses a health or environment risk.

13.19.1 Clean Closure

Clean closure is accomplished when cleanup levels as prescribed by regulatory agencies have been achieved. It is accomplished by verifying that the potentially dangerous constituents treated, stored, and/or disposed of at the TSD unit being closed are not present above cleanup levels for those potential contaminants.

Cleanup levels are based on equations and exposure assumptions for the various contaminants. For noncarcinogens, the principal variable relating human health to cleanup levels will often be the oral reference dose. For carcinogens, the cancer slope factor (see Section 5) is often the basis for determining human health effects and is a measurement of risk per unit dose. The oral reference dose and cancer slope factor are chemical-specific and are obtained from the Integrated Risk Information System (IRIS) database (www.epa.gov/iris). Cleanup levels are usually based on values that are current at the time of approval of closure documentation.

Protection of human health and the environment are accomplished by removing or treating all hazardous waste constituents at a TSDF, that is, by treating to concentration levels that are not a threat to human health or the environment.

13.19.2 Modified Closure

If hazardous waste constituents present at the TSDF are above background levels but below industry-permitted levels, then a modified closure option could be used. Requirements for a modified closure are specified in the facilities permit. These requirements usually include at least the following:

- Provision of institutional controls for a minimum period of time, usually five years
- The requirement to conduct periodic assessments of the TSDF to determine the effectiveness of the closure controls
- Development of a postclosure permit application, including final status postclosure groundwater monitoring
- Selection of a cleanup option with consideration of the potential future site use for that TSDF

13.19.3 Landfill Closure

A landfill closure occurs when dangerous waste constituents are left at the TSDF unit in concentrations that are above permitted industrial levels.

When waste or contamination is left in place, the submittal of postclosure documentation is required. This documentation would contain a RCRA-compliant landfill cover design and a postclosure soil and groundwater sampling and monitoring plan. The postclosure sampling and monitoring plan would describe how the covered TSD unit would be monitored and maintained to ensure protection of human health and the environment. Regulations require monitoring and maintenance for at least 30 years, unless a shorter time period is approved by the EPA. That shorter time period must be shown to be sufficient to protect human health and the environment. Requirements for a landfill closure are contained in the TSDF permit.

13.20 TRAINING AIDS AND ADDITIONAL RESOURCES

The authors recommend the following videos for aiding readers to see real-life examples of what they have learned in this Section.

- *Drum Beat*
- *Packaging Hazardous Waste*
- *Uniform Hazardous Waste Manifest*
- *Lab Packing*
- *Disposal Facility*

These videos are available from Safe Expectations (formerly BNA Communications, Inc.) (www.safeexpectations.com).

- *Drum Handling*
- *Lab Packs*
- *Hazardous Waste Transportation*
- *Hazardous Waste Manifests*
- *HazMat Shipping Papers*
- *HazMat Transportation: What You Don't Know Can Hurt You*
- *Marking, Labeling & Placarding: HazMat for Transportation*

These videos are available from Coastal Communications (www.coastal.com).

13.21 SUMMARY

We have described the quite varied conditions and operations at hazardous waste generators and at TSD facilities and the duties and responsibilities of their HAZWOPER personnel. In both of the two work environments in this Section, workers generally report to one site day after day.

HAZWOPER Work at a Hazardous Waste Generator

First we described what is probably a quite common work environment. It occurs in a generator facility. There are a total of nearly 100,000 generators, large and small quantity, in the United States. They are facilities dependent upon raw materials or processing aids that are hazardous. Alternatively, there may be hazardous wastes generated from nonhazardous materials during the normal course of operations. Some personnel at those facilities will be managing hazardous wastes routinely. Those hazardous waste workers will, however, be dealing largely with the same or similar hazardous materials and wastes on a daily basis. Rightly, they, more than the emergency response workers described in Section 15, will be expected to have specific knowledge about the particular hazards they face. They will be expected to know more about the selection and proper fitting of their own particular PPE, for example.

The question of whether any given facility is a defined generator of hazardous waste and HAZWOPER certification is required of specific personnel was partially answered in Sections 1 and 2 and further amplified in this Section.

HAZWOPER Work at a TSDF

The name of the second kind of facility regularly doing HAZWOPER work, "Treatment, Storage, and Disposal Facility" (TSDF), tells it all. That site receives hazardous wastes as its raw material or process feed material. The facility then can be organized and licensed to do all or parts of TSD. After treatment, and generally short-term storage, the TSDF disposes of its wastes in a regulated manner.

Recommended Training

The work functions and recommended training for key employees at a generator and at a TSDF have been explained. For the TSDF category, training specific to any new types or mixtures of hazardous wastes, prior to the arrival of such wastes at the facility, is in order. Further, personnel training must be more extensive than that at a generator. All TDSF work involves hazardous wastes. Some TSDFs act more as hazardous waste-collection sites while doing only limited kinds of treatment. Such facilities require procedures for disposing of their hazardous wastes. Other TSDFs act as the final disposal site for the wastes that they collect. They are the "grave" in the often referred to "cradle-to-grave" monitoring and regulation of hazardous wastes.

Duties and Responsibilities of Management and Employees

Also explained in this Section were the special duties and responsibilities of owners and operators representing management and those of employees who must follow standard operating procedures, and the safety procedures learned in their training. Landfills, either sanitary (municipal solid waste—garbage and trash) or regulated hazardous waste, will not accept liquids in any container. How to deal with these containers to make them acceptable was discussed.

Uniform Hazardous Waste Manifest System

We emphasized the importance of following the Uniform Hazardous Waste Manifest system. The manifest paperwork must follow, with signatures of all entities having control of those wastes from generator to final disposition. Then the generator must receive and file a copy that records all that history.

Recycling and Waste Destruction

The importance of recycling was explained, as opposed to destruction or landfilling of hazardous wastes. The rules limiting waste accumulation were explained, together with a model inspection checklist to be used when monitoring waste accumulation.

Common and specialized waste disposal methods were explained that apply to waste destruction and treatment for land filling.

SECTION 14

SUPERFUND SITES AND BROWNFIELDS: SITE INVESTIGATION, CONTROL, AND REMEDIATION

OBJECTIVES

This section focuses on two HAZWOPER work categories: the cleanup of Superfund sites and of those sites that are referred to frequently as "Brownfields."

Hazardous materials must be removed from those sites so that they can be developed from an unusable condition, or usable with serious limitations, to usable for immediate real estate development. Brownfields may contain wide varieties of wastes. Cleanup may take anywhere from days to decades to accomplish.

Our objective in Section 14 is to explain that for an individual, subject to HAZWOPER regulations, whether laborer, technician, engineer, or manager, there are three broad activities or phases you can become involved in and for which you must be prepared: site investigation, site control, and site remediation.

Before setting foot on a hazardous waste site, in addition to being HAZWOPER trained and certified, you must have a minimum of information about site characteristics. That means what has happened over the known history of the site and what is going on when you become involved. A health and safety system must be in operation for all sites. A major feature of that system is the site-specific health and safety plan (HASP), which is fully described in Section 16. You need to have a model plan on hand, such as the one in Appendix K, when you begin site investigation. Use that model plan to begin determining how you will modify it to produce a site-specific HASP.

When the site has been identified and under investigation for periods ranging from weeks to years, you will have the advantage of reviewing or learning the results of a large body of collected information. Site investigation requires at least a Phase I environmental site assessment (ESA). Depending upon specific site conditions, a Phase II ESA may be required.

With those ESAs, you can prepare yourself, or your team if you are the manager, and then enter the site when ready. You will find this investigation activity exceptionally challenging when you are introduced to it during emergency response, as described in Section 15.

Site control concerns the *controlling of access* to the site. That means having adequate site security. As we all know, nothing attracts kids like a large pile of dirt. If that pile is contaminated, so will the children become who play in it. Following investigation, your HAZWOPER responsibility is to engage in control. Under HAZWOPER, among other duties, this means the control of the spread or dissemination of hazardous material by not having it removed from the site in an unauthorized manner.

Control is the longest-duration activity in most cases. Control includes initially fencing off the site and segregating it into zones according to their approved activities until remediation is completed and the fencing may be removed.

Remediation, the most labor-intensive activity, includes the collection, accumulation, treatment, packaging, manifesting, and shipping of the hazardous waste off-site for appropriate treatment or to an EPA-approved hazardous waste landfill. Remediation may also include the use of technologies that completely treat all of the waste without ever moving it from the site.

Remediation is completed when the responsible authorities are satisfied that no further hazardous materials remediation work needs to be done. Furthermore, the authorities must agree to attest that the site is available for specified real estate use. Remediation can take decades!

14.1   FORTY-HOUR HAZWOPER CERTIFICATION AND KEY PERSONNEL AT SUPERFUND AND BROWNFIELD SITES

> **All personnel performing site-remediation duties with hazardous wastes at Superfund sites require 40-hour HAZWOPER training.**

Almost everyone who steps foot on a Superfund site or contaminated brownfield site must be at least 40-hour HAZWOPER trained. Wallet cards are usually supplied by the training company to certify that a worker is trained and up to date with refresher courses. 40-hour HAZWOPER certification was initially designed specifically for work at Superfund sites. The training is so helpful and protective that, as you have read in Section 13, it is becoming almost a U.S. standard for all employees who work with hazardous materials and wastes. Key personnel at Superfund and Brownfield sites include:

- Project manager (PM)
- Site safety officer (SSO)
- Site supervisors
- Decon officer
- Decon personnel
- Remediation personnel
- Equipment operators
- Sampling and monitoring technicians
- Engineering, geology, hydrology, and toxicology consultants as needed
- Site security

Why do we say that 40-hour HAZWOPER certification is "almost" a standard requirement for personnel on a hazardous waste site? Both EPA and OSHA agree that there may be site tasks requiring less than the 40-hour training. An example might be a monitoring well crew. Let us say that the crew has to sample monitoring wells near the periphery of a Superfund site. They are not required to enter the work zones *per se,* but they will be on the site. OSHA considers them adequately trained with at least the 24-hour course.

The authors recommend, however, that the HAZWOPER 40-hour training be the minimum training for anyone working on or near any project that uses hazardous materials or creates hazardous waste or where there is the potential for a release of hazardous materials.

14.2   WHAT IS A SUPERFUND SITE?

> **A Superfund site is a parcel of real property that has undergone a specific series of research and sampling efforts required by the EPA for NPL listing.**

Listed sites contain one or more specific public health hazards that must be remediated.

A Superfund site is real property where either hazardous materials are being released into the environment or there is the substantial threat that they will be. These are *uncontrolled* sites as distinct from operating hazardous waste facilities. Uncontrolled sites are covered by CERCLA (Superfund). Operating hazardous waste management facilities (controlled sites) are covered by RCRA. A Superfund site has a distinct life cycle, from listing through remediation to deletion from the National Priorities List (NPL). That cycle is the topic of Section 14.3. *An example of an uncontrolled site would be an illegal dumpsite where hazardous materials were deposited.*

14.2.1 Exposure Pathways

Our concern about an illegal dumpsite is that an exposure pathway might have been completed. An exposure pathway has five elements:

1. A source of contamination (such as a ruptured drum of hazardous waste)
2. Contact with an environmental medium (soil around the drum and storm water draining into a creek)
3. A point of exposure (such as a creek)
4. A defined route of exposure (from the drum to the creek)
5. A receptor population (fish in creek, humans eating fish; or humans using the water as a drinking water source, or pollution of sensitive environmental areas)

A completed exposure pathway (CEP) is an exposure pathway that links a contaminant source to a receptor population. According to researchers Barry L. Johnson and Christopher T. DeRosa of the Agency for Toxic Substances and Disease Registry (ATSDR),

> Uncontrolled hazardous waste sites are a major environmental and public health concern in the United States and elsewhere. The remediation of, and public health responses to, these sites is mandated by the federal Superfund statute. Approximately 40,000 uncontrolled waste sites have been reported to U.S. federal agencies. About 1,300 of these sites constitute the current National Priorities List (NPL) of sites for remediation. Findings from a national database on NPL sites show approximately 40% present completed exposure pathways, though this figure rose to 80% in 1996. Data from 1992 through 1996 indicate 46% of sites are a hazard to public health. Thirty substances are found at 6% or more of sites with completed pathways. Eighteen of the substances are known human carcinogens or reasonably anticipated to be carcinogenic. Many of the 30 substances also possess systemic toxicities. The high percentage of sites with completed exposure pathways and the toxicity potential of substances in these pathways show that uncontrolled hazardous waste sites are a major environmental threat to human health.*

14.2.2 How Many Superfund Sites Are There?

The EPA's Office of Technology Assessment (OTA) estimates that there could be nearly 440,000 hazardous waste sites in the United States that could require remediation. EPA's inventory of uncontrolled hazardous waste sites, called the Comprehensive Environmental Response, Compensation, and Liability Information System (CERCLIS), lists currently uncontrolled waste sites. EPA has stated that approximately 2,000 Superfund sites have been satisfactorily remediated. A large facility might be the location of more than one Superfund site. One of the authors worked at such a facility that had five such sites. There are also

---

*B. L. Johnson and C. T. DeRosa, ''The Toxicologic Hazard of Superfund Sites,'' *Reviews on Environmental Health,* vol. 12, no 4, pp. 235–251.

state "superfund" sites that do not qualify for the NPL but that the state has deemed as requiring cleanup.

14.3   THE LIFE CYCLE OF A SUPERFUND SITE

O—⚲   **Superfund sites must undergo a specific nine-step process starting with the preliminary assessment and, following successful cleanup, ending with site deletion from the NPL.**

O—⚲   **Sites that, when discovered, pose an immediate threat to people or the environment are waived from the nine-step process and are cleaned up under Superfund's Emergency Response program.**

The Superfund cleanup process to be conducted by 40-hour HAZWOPER certified workers begins with site discovery or notification to EPA of possible releases of hazardous substances. Sites are discovered by various parties, including citizens, state agencies, and EPA regional offices. Once discovered, sites are entered into the CERCLIS. EPA then evaluates the potential for a release of hazardous substances from the site through these nine steps in the Superfund cleanup process:

1. Preliminary Assessment/Site Inspection (PA/SI)
2. Hazard Ranking System (HRS) scoring
3. NPL site listing process
4. Remedial Investigation/Feasibility Study (RI/FS)
5. Record Of Decision (ROD)
6. Remedial Design/Remedial Action (RD/RA)
7. Construction Completion (CC)
8. Operation and Maintenance (O&M)
9. NPL site deletions, removal of sites from the NPL

EPA uses these steps to determine and implement the appropriate response to threats posed by releases of hazardous substances. Releases that require immediate or short-term response actions are addressed under the Emergency Response program of Superfund.

14.3.1   Preliminary Assessment and Site Inspection

O—⚲   **After a site has been brought to the attention of the EPA, the first step toward NPL listing is to perform a preliminary assessment. This is similar to a Phase I Environmental Site Assessment.**

The Preliminary Assessment (PA) and Site Inspection (SI) are used by EPA to evaluate the potential for a release of hazardous substances from a site.

Preliminary Assessment (PA).   This is a limited-scope investigation performed on every CERCLIS site. It is the Superfund version of a Phase I environmental site assessment (ESA).

Note

The authors strongly recommend that a Phase I ESA be accomplished and its results carefully evaluated before purchase of any real estate. For an explanation of the Phase I ESA and examples of problems arising from its omission, see www.nstengineers.com/ESAs.htm.

PA investigations collect readily available information about a site and its surrounding area. There will be a site visit, photographs, records search, and interviews. No physical sampling of the site will take place at this time. The PA is designed to distinguish, based on limited data, between sites that pose little or no threat to human health and the environment and sites that might pose a threat and require further investigation. The PA also identifies sites requiring assessment for possible emergency response actions. If the PA results in a recommendation for further investigation, a site inspection is performed. The EPA publication *Guidance for Performing Preliminary Assessments under CERCLA*, EPA 9345.0-01A, and the electronic scoring program, PA-Score, provide more information on conducting PAs. PA-Score can be downloaded for free at www.epa.gov/superfund/resources/pascore/#down.

> **Based on the Preliminary Assessment (PA), either a site is scheduled for a site inspection or no further action is taken.**

Site Inspection (SI)

> **A Superfund Site Inspection (SI) is similar to a Phase II Environmental Site Assessment (ESA). At this point physical sampling takes place.**

The SI identifies sites that enter the NPL site listing process and provides the data needed for Hazard Ranking System (HRS) scoring and documentation. The SI is the Superfund equivalent of a Phase II environmental site assessment. HAZWOPER investigators typically collect environmental and waste samples to determine what hazardous substances are present at a site. They determine whether these substances are being released to the environment and assess whether they have reached nearby targets. The SI can be conducted in one stage or two.

The first stage, or *focused* SI, tests hypotheses developed during the PA and can yield information sufficient to prepare an HRS scoring package. If further information is necessary to document an HRS score, an *expanded* SI is conducted. The EPA publication *Guidance for Performing Site Inspections under CERCLA; Interim Final*, EPA 9345.1-05, provides more information on conducting SIs.

> **If physical sampling shows contamination above regulatory levels, the site information is used to perform hazard ranking. If no contamination is found or contamination is below regulatory levels, no further action is required to be taken.**

14.3.2 The Hazard Ranking System (HRS)

> **The HRS score determines whether or not a site will be added to the National Priorities List (NPL) sites.**

Information collected during the PA and SI is used to calculate an HRS score. Sites with an HRS score of 28.50 or greater are eligible for listing on the NPL and require the preparation of an HRS scoring package.

The Hazard Ranking System (HRS) is the principal mechanism EPA uses to place uncontrolled waste sites on the NPL. It is a numerically based screening system that uses information from initial limited investigations (the preliminary assessment and the site inspection) to assess the relative potential of sites to pose a threat to human health or the environment.

HRS scores do not determine the priority in funding EPA remedial response actions. That is because the information collected to develop HRS scores is not sufficient to determine either the extent of contamination or the appropriate response for a particular site. The sites with the highest scores do not necessarily come to the EPA's attention first. EPA relies on more detailed studies in the Remedial Investigation/Feasibility Study (RI/FS) that typically follows listing.

The HRS uses a structured analysis approach to scoring sites. This approach assigns numerical values to factors that relate to risk, based on conditions at the site. The factors are grouped into three categories:

1. Likelihood that a site has released or has the potential to release hazardous substances into the environment
2. Characteristics of the waste, such as toxicity; and waste quantity
3. People or sensitive environments (receptors) affected by an actual or potential release

Four pathways can be scored under the HRS:

1. Groundwater migration (drinking water)
2. Surface water migration (drinking water, human food chain, sensitive environments)
3. Soil exposure (resident population, nearby population, sensitive environments)
4. Air migration (population, sensitive environments)

After scores are calculated for one or more exposure pathways, they are combined to determine the overall site score.

The electronic scoring system, Preliminary Ranking Evaluation score (PREscore), can be used to do the scoring calculations. PREscore Version 4.1 is available for free download from www.epa.gov/superfund/resources/prescore/prescr41.htm.

If all pathway scores are low, the site score is low. However, the site score can be relatively high even if only one pathway score is high. This is an important requirement for HRS scoring because some extremely dangerous sites pose threats through only one pathway.

Note that the EPA reserves the right to determine the final HRS score.

For more information, please consult EPA 9345.1-07, *The Hazard Ranking System Guidance Manual, Interim Final.*

14.3.3   NPL Site Listing Process

○━πι   **The ranking of a site on the NPL is determined by the degree of threat the site poses to the community and the environment.**

Our estimate is that there were about 2000 NPL sites in the spring of 2002. Sites are listed on the NPL upon completion of Hazard Ranking System (HRS) screening and public solicitation of comments about the proposed site and after all comments have been addressed.

The NPL primarily serves as an information and management tool. It is a part of the Superfund cleanup process. The NPL is updated periodically.

Section 105(a)(8)(B) of CERCLA, as amended, requires that the statutory criteria provided by the HRS be used to prepare a list of national priorities among the known releases or threatened releases of hazardous substances, pollutants, or contaminants throughout the United States. This list, which is Appendix B of the National Contingency Plan (NCP), is the NPL.

The identification of a site for the NPL is intended primarily to guide EPA in:

• Determining which sites warrant further investigation to assess the nature and extent of the human health and environmental risks associated with a site
• Identifying what CERCLA-financed remedial actions may be appropriate
• Notifying the public of sites EPA believes warrant further investigation
• Serving notice to potentially responsible parties (PRPs) that EPA may initiate CERCLA-financed remedial action

Inclusion of a site on the NPL does not in itself reflect a judgment of the activities of its owner or operator. It does not require those persons to undertake any action, and does not assign liability to any person. The NPL serves primarily informational purposes, identifying for the states and the public those sites or other releases that appear to warrant remedial actions.

14.3.4 Remedial Investigation/Feasibility Study (RI/FS)

○──π **The RI/FS is performed to gather even more detailed information about the site as well as determine which type of remediation technology or combination of technologies will work best. It also determines approximate costs of remediation.**

After a site is listed on the NPL, a Remedial Investigation/Feasibility Study (RI/FS) is performed at the site. The RI/FS is initially funded by the EPA. Later, the EPA seeks PRPs to pay for that work, as well as continuing cleanup work.

During the RI data are collected to:

- Characterize site conditions
- Determine the nature of the waste
- Assess risk to human health and the environment
- Conduct treatability testing to evaluate the potential performance and cost of the treatment technologies that are being considered

The FS is used for the development, screening, and detailed evaluation of alternative remedial actions.

The RI and FS are conducted concurrently. Data collected in the RI influence the development of remedial alternatives in the FS, which in turn affect the data needs and scope of treatability studies and additional field investigations. This phased approach encourages the continual scoping of the site characterization effort, which minimizes the collection of unnecessary data and maximizes data quality.

The RI/FS process includes these five phases:

1. Scoping
2. Site characterization
3. Development and screening of alternatives
4. Treatability investigations
5. Detailed analysis

For more information, please consult EPA 9355.3-01, *Guidance for Conducting Remedial Investigations and Feasibility Studies under CERCLA, Interim Final.*

14.3.5 Record of Decision (ROD)

○──π **After the RI/FS has been completed, the EPA issues a Record of Decision (ROD).**

The record of decision (ROD) is a public document that explains which cleanup alternatives will be used to clean up a Superfund site. The ROD for sites listed on the NPL is created from information generated during the RI/FS.

This ROD is given a public comment period, during which the public, along with environmental organizations, can pose questions and comment either for or against the ROD.

The ROD will include a summary of all investigative information gathered from the site so far. It will include details about the prior uses, along with the intended reuse of the site. It will also include agreed upon site-specific levels of contamination that will be considered acceptable based on the intended reuse. As you will see, some remedial action technology choices may satisfy the regulators but still leave the site with strictly defined development uses and deed restrictions.

14.3.6   Remedial Design and Remedial Action (RD/RA)

○━━πτ   **Once the ROD is finalized, the site is ready for RD/RA (cleanup). It is conducted according to a Remedial Action Work Plan (RAWP).**

Remedial design (RD) is the phase in Superfund site cleanup where the technical specifications for cleanup remedies and technologies are designed. Remedial action (RA) follows the remedial design phase and involves the actual construction or implementation phase of Superfund site cleanup. The RD/RA is based on the specifications described in the ROD. All work on the site is based on the Remedial Action Work Plan (RAWP).

14.3.7   Construction Completions

EPA has developed a construction completions list (CCL) to simplify its system of categorizing sites and better communicate the successful completion of cleanup activities. Sites qualify when:

- Any necessary physical construction is complete, whether or not final cleanup levels or other requirements have been achieved
- EPA has determined that the response action should be limited to measures that do not involve construction
- The site qualifies for deletion from the NPL

Inclusion of a site on the CCL has no legal significance to the site owner or PRPs.

14.3.8   Operations and Maintenance (O&M)

○━━πτ   **Operations and maintenance (O&M) is the longest phase at a Superfund site. O&M at many of these sites frequently operates over decades.**

Operation and maintenance (O&M) activities ensure that the RAWP for the selected remedy for a site is followed. O&M measures are initiated by the federal or state regulators after the remedy has achieved the Remedial Action objectives and remediation goals outlined in the ROD. The site is then deemed to be operational and functional (O&F) based on state and federal agreement. For Superfund-lead sites, remedies are considered O&F either one year after construction is complete or when the remedy is functioning properly and performing as designed, whichever is earlier. Remedies requiring O&M measures include landfill caps, gas-collection and leachate systems, groundwater extraction treatment, groundwater monitoring, and surface water treatment. Overall supervision for such work will usually be 40-hour HAZWOPER certified personnel who are professional geologists or professional engineers.

Once the O&M period begins, the state or potentially responsible party (PRP) is responsible for maintaining the effectiveness of the remedy. O&M monitoring includes four components:

1. Inspection
2. Sampling and analysis

3. Routine maintenance

4. Reporting

O&M activities are usually required for sites where cleanup proceeded through landfill capping activities, groundwater activities, or monitored natural attenuation (MNA). This is the process by which contamination is reduced in concentration over time, through evaporation, absorption, adsorption, degradation, or dilution.

O&M is not usually considered a phase of a cleanup when the remedy is excavation and off-site treatment of all contaminated media. In that case, construction completion, including sampling and monitoring records proving that all contamination has been reduced to below regulatory-PRP agreed-upon levels, is the end of the project and the site is ready for deletion from the NPL and closure.

14.3.9 How Sites Are Deleted from the NPL

EPA may delete a final NPL site if it determines that no further response is required to protect human health or the environment. Under Section 300.425(e) of the NCP, a site may be deleted where no further response is appropriate if EPA determines that one of the following criteria has been met:

- EPA, in conjunction with the state, has determined that responsible or other parties have implemented all appropriate response action required.

- EPA, in consultation with the state, has determined that all appropriate Superfund-financed responses under CERCLA have been implemented and that no further remedial work by responsible parties is appropriate.

- A remedial investigation has shown that the release poses no significant threat to public health or the environment and therefore further remedial measures are not appropriate.

Since 1986, EPA has followed these procedures for deleting a site from the NPL:

- The Regional Administrator approves a close-out report that establishes that all appropriate response actions have been taken or that no action is required. The Regional Office obtains state concurrence.

- EPA publishes a notice of "intent to delete" in the *Federal Register* and in a major newspaper near the community involved. A public comment period is provided.

EPA responds to the comments and, if the site continues to warrant deletion, publishes a deletion notice in the *Federal Register*.

Sites that have been deleted from the NPL remain eligible for further Superfund-financed remedial action in the unlikely event that conditions in the future warrant such action. Partial deletions can also be conducted at NPL sites.

Beginning in 1999, EPA adopted a streamlined approach to the process of deleting sites from the NPL, called a Notice of Direct Final Action to Delete. With this approach, the Notice Of Intent to Delete (NOID) and the Notice Of Deletion (NOD) are published in the *Federal Register* on the same date. This combination of steps reduces the amount of time it takes to finalize a site deletion. Using this process, EPA will publish both a NOID and a notice of direct final action to delete in the *Federal Register*. The notice of direct final action to delete includes the statement that the direct final action will become effective unless EPA receives significant adverse or critical comments during the 30-day public comment period. If no significant adverse or critical comments are received, the deletion will become effective without any further EPA action and the site will be deleted from the NPL. The site is then clean.

Should significant adverse or critical comments be received within the comment period, EPA will publish a notice of withdrawal of the direct final rule, prepare a response to the comments received, and continue with the rule-making process on the basis of the proposal to delete and the comments received. Should the comments be considered substantive, additional work might have to be conducted at the site.

14.4 FORTY-HOUR HAZWOPER CERTIFICATION AND WORK AT BROWNFIELDS

O━π **Since data regarding a brownfield site are often vague and incomplete, the authors recommend 40-hour HAZWOPER training for all workers contacting, or who might contact, hazardous waste.**

O━π **Brownfields are parcels of real property unutilized or underutilized due to the actual or perceived presence of environmental contaminants.**

Brownfields are parcels of real property unutilized or underutilized due to the actual or perceived presence of environmental contaminants. They are present mostly in urban areas and are usually attributed to the relocation of facilities. The facility may have, in the past, improperly used, stored, or disposed of hazardous materials at or on the site. The site may contain one or more leaking underground storage tanks (LUST) sites.

Due to the enormous amount of unutilized real property in the U.S. because of this stigma, along with the dwindling amount of virgin land, so-called brownfield initiatives are being promoted by governments all over the country, especially in major urban areas. During 2001, Governor Ruth Ann Minner of Delaware, representing all 50 Governors, stressed the importance of these issues to a congressional committee. The authors have worked at several dozen brownfield sites. As fast as some sites are identified and remediated, new ones seem to be identified.

You may well ask, what does that have to do with HAZWOPER work? Remember that these properties are unutilized or underutilized based on actual or perceived contamination. The authors strongly advise that any site workers, including those who are initially sampling and monitoring the site, be 40-hour trained. Why? Because during their investigation they may become personally exposed to hazardous materials. Further, through their investigation they may cause the environment to be exposed to those same materials. For example, a crew takes direct push core samples of the site prior to a GPR survey. (See Section 12 for definitions of these technologies.) Several of their sample probes pierce deteriorated subsurface drums. The contents of these drums are then released into the environment to contaminate site soil and possible groundwater.

Properly trained HAZWOPER workers would not, we like to believe, make that mistake.

14.5 BROWNFIELDS

"On the brownfields of yesterday, we will build the green industries of tomorrow."

—George W. Bush, President of the United States, from the *Remarks by the President to the U.S. Conference of Mayors National Summit*, April 5, 2001

The interest in redeveloping brownfields into productive useful property is increasing as urban sprawl continues. No one wants to live or work next to a derelict factory or deserted commercial property. The solution, for many families and businesses, is to move farther out of town. Thus more irreplaceable farmland, forests and other groundcover and habitat is lost, the tax base for the urban area declines, and along with it, the infrastructure and municipal services degrade. How do we stop this destruction of greenfields and urban sprawl downward spiral? By redeveloping the brownfields into useful properties.

As incentives such as real estate tax abatement, low-cost redevelopment loans, and sometimes relaxation of regulatory standards become available, these sites now exhibit a new potential for real estate and industry developers.

The steps of the Superfund program are codified in the regulations. The steps of brownfield remediation are not so rigidly defined. There is more room to maneuver. We categorize these steps, broadly, as Environmental Site Assessment (ESA), Phases I, II, and III. In this section, we will discuss not the redevelopment of the site but the preparation of the site to make it ready for redevelopment. To be sure, the proposed reuse of the site will drive the cleanup of the site. For example, if the proposed reuse of the site is as a playground for children, the remediation will surely be held to higher environmental standards and more stringent regulations than if the site were to be developed as a parking lot and mall. In some instances, such as the latter, depending on the extent of the contamination, the redevelopment could be part of the remediation.

A case in point: a developer wants to build a large shopping center over ground that is highly contaminated with lead. To remove all of the soil to regulatory acceptable levels would cost more than the property is worth. With the aid of a properly executed ESA, the developer negotiates with the state regulators. The engineers design the new facility so that the lead-contaminated soil cannot come into contact with the surface or groundwater. The new shopping center construction then caps the soil. An eyesore has been removed from the community, jobs and infrastructure have been created, tax base has increased, and the area has been revitalized.

14.6 SITE INVESTIGATION

O—π **Prior to any work on brownfields or any other potential remediation site, a site investigation must be conducted.**

Our advice to all individuals or corporations worldwide is that negotiations for real estate transactions should not proceed without, at a minimum, a properly executed Phase I ESA. We believe Phase I ESAs must be modeled after, and incorporate, ASTM Standards E-1527-00, *Conduct of Environmental Site Assessments,* and E-1528-00, *Transaction Screen Analysis.* This basic practice assists a property purchaser in satisfying one of the requirements to qualify for the innocent landowner defense to RCRA liability.

The authors believe that this practice and its authorized procedures constitute *"all appropriate inquiry into the previous ownership and uses of the property consistent with good commercial or customary practice,"* as defined by CERCLA in 42 USC 9601(35)(B).

Personnel conducting Phase I ESAs must be environmental professionals who have successfully completed approved training and demonstrated their ability to perform this task. The authors have found that many personnel conducting Phase I ESAs are not HAZWOPER-trained. The authors prefer to have HAZWOPER-trained personnel do Phase I ESAs in light of the potential hazards of walking over unfamiliar and potentially hazardous property. Further, they are far better prepared to detect problems and make recommendations than untrained personnel.

We believe proper training in the use of the above-cited ASTM methods allows personnel to provide a comprehensive coverage of Phase I objectives. Further, it provides a logical sequence for reporting of all findings in a standardized format. The goal of the format is to facilitate review, evaluation, and approval of a Phase I report. The details of this practice (investigation methods, minimum reporting requirements, explanations of terminology, and approved report format) must be incorporated in a set of standard procedures for organizations performing Phase I studies.

14.6.1 Definition of ESA Phase I

A complete environmental site assessment can involve three phases, depending upon the size, type, construction and location of the property; the past and present intended use(s) of the property; and availability and access to complete records.

A Phase I ESA determines, for a parcel of real estate, the *recognized environmental conditions,* that is, the presence, or likely presence, of any hazardous substances or petroleum products on the parcel. In effect, Phase I asks, "Are such substances present under conditions that indicate an existing release, a past release, or a material threat of a release of the substances into structures on the property. Further, are there releases into the ground, groundwater, or surface water of the property?" Phase I investigation does this by accomplishing due diligence in:

1. Making a visual inspection of the property, including walking over the entire site, ideally with the owner, manager, or user present to answer questions
2. Assembling a comprehensive photographic log
3. Conducting interviews with the owner, manager, or user of all adjacent properties
4. Thoroughly reviewing all practically reviewable records pertaining to the property and surrounding properties within ASTM radii (this may incorporate a report from such services as EDR-Sanborn)
5. Preparing a comprehensive written report

14.6.2 Limitations to Phase I Scope

The scope of this practice includes research and reporting requirements that aid the client in qualifying for the innocent landowner defense. Therefore, sufficient documentation of all sources, records, and resources utilized in conducting the inquiry required by this practice must be provided in the written report.

CAUTION

THERE ARE SOME COMPANIES WHO SAY THEY PROVIDE A PHASE I ESA AT LOW COST. WHAT WE HAVE OBSERVED THEY ARE ACTUALLY DOING IS RELYING ON THE REPORT FROM A COMPANY SUCH AS EDR-SANBORN. ALTHOUGH SUCH MATERIAL IS BASIC TO MOST PHASE I REPORTS, AND EDR-SANBORN IS THE NATION'S LEADER IN PROVIDING SUCH INFORMATION, IT IS NOT A SUBSTITUTE FOR THE SITE VISIT, INTERVIEWS, PHOTOGRAPHS, AND WORKING THROUGH THE TRANSACTION SCREEN ANALYSIS WITH THE PRESENT OWNER/MANAGER/USER.

14.6.3 Definition of ESA Phase II

Based on a properly executed Phase I report that concludes that advising a Phase II investigation is required, a Phase II must consist of:

1. The physical sampling of the site, using the recommendations of the Phase I report as a minimum guideline
2. A comprehensive written report detailing:
   - The rationale for the sampling that took place
   - The sampling protocols and procedures employed
   - An explanation of the analytical results
   - If necessary, a description of the recommended remedial action needed to restore the site to the appropriate condition for its intended use

14.6.4 Definition of ESA Phase III

Based on a properly executed Phase II report that concludes that Phase III (remediation) is required, Phase III must consist of:

1. The design and implementation of the remediation of the site
2. All necessary reports and permits to achieve cleanup of the site to the agreed-upon, site-specific standards
3. Final documented approval from the regulatory authorities that the cleanup has been completed and the proposed property use may proceed.

The degree to which an ESA is conducted will vary according to the environmental concerns identified for each real estate transaction and the applicable federal, state, and local laws. However, all ESAs must be comprehensive enough so that environmental risks are properly considered in any business decision involving the real estate transaction.

CERCLA requirements other than appropriate inquiry: The ESA Phase I practice does not address whether requirements in addition to appropriate inquiry have been met in order to complete qualification for CERCLA's innocent landowner defense; That is: "If the defendant establishes by a preponderance of the evidence that (a) he exercised due care with respect to the hazardous substance concerned, taking into consideration the characteristics of such hazardous substance, in light of all relevant facts and circumstances, *and* (b) he took precautions against foreseeable acts or omissions of any such third party and the consequences that could foreseeably result from such acts or omissions."

The innocent landowner defense is designed to protect landowners who have followed the proper procedures ensure that there has not, and will not, be improperly disposed of hazardous waste on the property. In the event that such waste contamination is found in the future, this defense can limit the monetary liability of the landowner for the cleanup. As the reader can see, the innocent landowner defense has two parts. The first part is the due diligence performed in ESA Phase I. The second part is protecting the parcel of land from improper use by third parties, such as fencing in a vacant lot to avoid illegal dumping.

14.6.5 Factors That Can Affect the Extent and Distribution of an ESA

In some cases, completion of Phase I may result in an ESA that is comprehensive enough to assess environmental risks adequately. For example, Phases II and III may not be necessary if available site information is adequate or if environmental risks are satisfactorily addressed by an indemnification.

If more than one phase is conducted, the phases occur in a sequential fashion, with information obtained from each phase used to define better the scope of work in the next phase. However, the schedule for completing the real estate transaction may require that several phases be conducted simultaneously.

The Phase I report is confidential and remains the property of the client. An ESA is confidential and privileged if it has been developed in anticipation of possible litigation. Accordingly, all parties involved in the performance and review of an ESA must take appropriate steps to protect the unauthorized disclosure of that confidential or privileged information.

14.7 OFF-SITE RESEARCH

Where there is the advantage of available time, the manager and project team preparing a site investigation must conduct a great deal of research, evaluation, and planning off-site.

(See Section 15 for required activities when the time advantage does not exist.) Any previous investigations of the site, including environmental site assessments and especially their sampling data, must be thoroughly examined and the results organized into usable planning information. This process is often referred to as data reduction.

Although there is no nationwide standard for writing a sampling plan, the appropriate regulatory agencies must be contacted to determine whether there is an established checklist of activities that must be addressed by the sampling plan. The above information will provide the basis for preparation of the control activities, including the health and safety system (see Section 16). As a participant, you need to be armed with an understanding of all the sections of this book.

Background information on site use and history, local geology, and groundwater use must be summarized. Interviews must be held with former and current site workers and community residents. Past uses of the site must be understood and documented. Regulatory agency reports and citations, prior site inspections, fire and police actions at the site, and newspaper reports that detail waste handling practices at the site, releases, and fires, although sometimes hard to come by, are particularly useful at this point.

CAUTION

THERE ARE MANY CASE HISTORIES OF INCOMPLETE OFF-SITE RE-SEARCH LEADING TO PROJECT HORROR STORIES. HISTORICAL DOC-UMENTATION AND NEIGHBORS' AND EMPLOYEES' RECOLLECTIONS WERE OVERLOOKED THAT COULD HAVE POINTED TO BURIED, AND SIGNIFICANT, HAZARDOUS MATERIALS DEPOSITS. THEY WERE THEN ONLY BROUGHT TO LIGHT LATE IN THE CONTROL PHASE, CAUSING ECONOMIC HARDSHIP AND REPLANNING WITH ALL KINDS OF DELAYS.

When a site, whether a Superfund or a brownfield site, requires remediation, the following steps must be taken. The following subsections are directed towards Superfund sites but might be equally applicable to brownfield sites.

14.8   SITE INVESTIGATION WORK PLAN

All sections of the Site Investigation work plan must be carefully executed. Any deviation can adversely affect the project planning. At best, the work will have to be repeated; at worst, the entire remediation system will have to be redesigned.

Once a site is determined to require remediation, a Site Investigation work plan must be designed and written to satisfy all pertinent local, state, tribal, and federal regulatory requirements. The rationale for the technical approach of the plan must be sound and the details of the investigation must be well organized. The main elements of project work are:

- The sampling and analysis plan
- The health and safety plan (HASP)
- The quality assurance/quality control (QA/QC) plan

14.8.1   Sampling and Analysis Plan

The plan must contain clear instructions for accomplishing plan goals to the HAZWOPER personnel who are assigned to conduct sampling and analysis. The plan also gives site owners, operators, and regulatory officials a clear understanding of what can be accomplished by carrying out the plan, including a reason for each element of the investigation. Further, there needs to be a basis for modification of the sampling strategy at any time that newly acquired data call for changes.

When completed, the sampling and analysis plan must provide for:

- Screening the site for potential contamination, including detailed step-by-step instructions on how and where on the site air monitoring and soil, surface water, groundwater, and bulk material sampling will be conducted
- Identifying areas of high-concentration hot zones (see Section 7.1) of contamination
- Characterizing the distribution of contamination
- Determining off-site migration and pathways of contaminant dispersion
- Determining site-specific parameters for inclusion in human health and ecological models
- Collecting information needed to evaluate the feasibility of various remedial alternatives

A variety of professionals, technicians, and services are usually called for by the site sampling plan. Analytical chemists and labs will determine the presence or absence of contaminants in samples and, if present, the concentrations. Professional geologists will define the ways in which the hydrogeological environment at the site will affect sampling. They can aid in the determination of migration pathways. Chemical and environmental engineers and analytical chemists can aid in designing the field-sampling program for the site.

At a minimum, the sampling plan will include unique sample numbers, the medium that is to be sampled (air, water, soil, and container contents), the rationale for sample location, and how and where each type of sample will be analyzed along with chain-of-custody forms for each sample. With a detailed site map to indicate sampling locations, the results will generate a clearer picture of the extent and types of contamination. See Section 12 for further details on sampling. Some states have specific laboratory requirements that must be addressed in the QA/QC plan. A site-specific HASP (see Section 16) is an OSHA requirement, and a HASP and QA/QC are required by some states in addition to the site-sampling plan.

14.8.2 Purpose of a Quality Assurance/Quality Control Plan (QA/QC)

○─π **Following a proper QA/QC plan will allow for forensic-quality data.**

A detailed QA/QC plan ensures that the sampling is performed in a manner consistent with the data quality objectives established for a site. The owners of the site, along with the engineers and regulators, will decide the sampling strategy for the site. As you will remember from Section 12, the value of forensic-quality results cannot be overstated.

With little or no background information, the decision might be to screen the entire site, that is, to take samples on a predetermined grid pattern on the whole site. This decision will also be based on regulatory requirements and the individual PRP's risk tolerance. The QA/QC plan will include:

- Proper sampling protocols
- Sampling quality objectives and guidance on how to maintain them
- Assignment of responsibility for maintaining quality for each component of the project

The components specified include:

- Field procedures (often EPA sampling guidelines)
- Laboratory protocols (often EPA analysis protocols)
- Sample identification

- Data validation
- Chain-of-custody tracking

The QA/QC plan also describes in detail the documentation necessary to ensure that the data collected are technically sound and legally defensible.

14.9   SITE CONTROL

⊙—🔑 **Site control means the physical control of access to the site by excluding un-qualified and unprotected personnel such as untrained personnel, onlookers, and children.**

Following investigation, your HAZWOPER responsibility might be to engage in control. Under HAZWOPER, this means the physical control of access to the site to prevent access by unqualified and unprotected personnel such as onlookers and children. Enclosing the site in chain link fencing would be an appropriate method of site control. Building warehouses to store material, hiring a security firm, and installing security cameras and an alarm system might also be part of a control plan. Control will end when the site has been remediated to the point that all responsible authorities are satisfied that a waste hazard no longer exists. This could take decades! Control is the longest-duration activity in most cases. It ranges from fencing off the site and its segregation into zones according to their approved activities to sampling and monitoring to prevention of hazardous material dissemination from the site.

An appropriate site control program must be developed prior to the implementation of cleanup operations. The site control program must be established during the planning stages of a hazardous waste operation. It will be modified as new information becomes available. The appropriate sequence for implementing site control measures will be determined on a site-specific basis.

The site control personnel, the project manager and staff, have the responsibility and must be granted the required authority for:

- Establishment of site boundaries and preparation of a detailed site map
- Physical security of the site, prohibiting unauthorized site access
- Establishment of work zones: (exclusion [hot] zone, contamination reduction [warm] zone, and support [cold] zone); to prevent unauthorized and unprotected personnel from entering controlled zones and direct site workers and visitors through a specific access control corridor
- Enforcement of mandatory use of the buddy system
- Enforcement of decontamination and work practices designed to reduce accidental spread of hazardous materials by preventing tracking of hazardous substances from the contaminated area by workers
- Confinement of work activities to the appropriate areas
- Establishment of an emergency action plan (EAP) for evacuation of all site personnel and visitors in the case of an emergency
- Establishment of universally understood communication systems
- Establishment of safe work practices
- Identification of and directions to the nearest medical facility along with contact information for police, fire, emergency response, ambulance, and other mutual aid agencies as necessary

14.10   THE SITE MAP

The purpose of the site map is to assist site personnel in planning and organizing investigation, control, and remediation activities. The initial site map must be developed during the initial phase of site investigation. This map must be periodically updated during the course of site operations to reflect:

- New information, such as information gained after initial site entry or from subsequent sampling and analysis activities
- Construction activities such as extraction, injection, and monitoring well and other remediation equipment installation
- Changes in site conditions, including changes resulting from accidents, ongoing site operations, hazards not previously identified, new materials introduced on-site, unauthorized entry or vandalism, and weather conditions

The site map must be developed prior to the initial site entry using information obtained during the preliminary evaluation. The map must include notations and depiction of:

- Prevailing wind direction
- Site drainage points
- All natural and man-made topographic features, including the location of buildings
- Containers, impoundments, pits, ponds, and tanks
- Location of specific work
- Any other relevant site features
- Locations of all potential hazards that were identified through interviews and records research
- The site's neighboring property monitoring results
- Observed and suspected hazards
- On-site and off-site air and soil sampling results
- Potential exposure pathways

14.10.1   The Three Work Zones

One of the basic elements of an effective site control program is the creation of work zones at the site. The concept of having three work zones was introduced in Section 7.1.1. See Figure 7.1. The purpose of establishing work zones is to:

- Reduce the accidental spread of hazardous substances by workers or equipment from the contaminated areas to the clean areas and off-site to workers' families and homes
- Confine work activities to the appropriate areas, thereby minimizing the likelihood of accidental exposures
- Facilitate the location and evacuation of personnel in case of an emergency
- Prevent unauthorized and unprotected personnel from entering controlled areas

When the work zones are being established at a site, the site map can provide a useful format for compiling relevant site data. In the absence of sampling results, site maps can provide essential information on potential and suspected hazards and potential exposure pathways.

Although a site may be divided into as many zones as necessary to ensure minimal employee exposure to hazardous substances, the three most frequently identified zones are the exclusion zone (or hot zone), the contamination reduction zone (or warm zone), and the support zone (or cold zone). Movement of personnel and equipment between these zones must be minimized and restricted to specific access control points along the contamination reduction corridor to prevent cross-contamination. In a larger remediation project, there may be many hot zones identified within a larger hot zone.

Note

The authors will use the common names of the three work zones and their dividing lines. The exclusion zone will be referred to as the hot zone. The contamination reduction zone will be referred to as the warm zone. The support zone will be referred to as the cold zone. The line dividing the hot zone from the warm zone will be referred to as the hotline. The line between the warm zone and the cold zone will be referred to as the cold line. The contamination reduction corridor has no universally accepted common name but is often referred to simply as the "CRC."

The Hot Zone.   The hot zone is the area in which contamination is either known or expected to occur and where the greatest potential for exposure exists. The outer boundary of the hot zone, called the hot line, separates the area of contamination from the warm zone. The hot line will initially be established by visually surveying the site and determining the extent of hazardous substances, discoloration, distressed vegetation, or any drainage, leachate, or spilled material present. Then, through atmospheric monitoring, the hot zone can be further delineated. HAZWOPER personnel must remember that conditions can change at a site in a matter of seconds (explosion) or minutes (fire, rapid material release, or weather changes). The hot line must be flexible to meet changing conditions. Other factors to consider in establishing the hot line include:

- Providing sufficient space to protect personnel outside the hot zone from potential fire or explosion
- Allowing an adequate area within which to conduct site operations
- Reducing the potential for contaminant migration

The hot line can be physically secured using fences or ropes. However, this makes an obstacle for HAZWOPER workers to cross in the case of a site emergency. The preferred method of demarcation is the use of 4-in. CAUTION or WARNING tape. Some site Safety officers (SSOs) prefer to use WARNING tape (black text on red tape) for the hot line and CAUTION tape (black text on yellow tape) for the cold line separating the warm zone from the cold zone.

During subsequent site operations, the boundary may be modified and adjusted as more information becomes available. The hot zone may also be subdivided into different areas of contamination based on the known or expected type and degree of hazards or the incompatibility of waste streams. If the hot zone is subdivided in this manner, additional demarcations are necessary.

Access to and from the hot zone must be restricted to access control points at the hot line and cold line. Access control points are used to regulate the flow of personnel, material, and equipment into and out of the contaminated area and verify that site control rules are being followed rigidly. Only properly, currently trained, and authorized workers are allowed access. All site personnel must be properly protected with the appropriate PPE and must go through decon each time they exit the hot zone.

Separate entrances and exits should be established to separate personnel and equipment movement into and out of the hot zone. This is usually one corridor separated by tape on stakes driven into the ground. This layout allows the access control personnel to keep track more easily of entries and exits.

All persons who enter the hot zone must wear the appropriate level of PPE for the degree and types of hazards present (see Section 6).

If the hot zone is subdivided, different levels of PPE might be appropriate. Each subdivision of the hot zone must be clearly marked to identify the hazards and the required level of PPE. The SSO has the responsibility for determining all workers' proper level of PPE.

The Warm Zone. The warm zone is the area in which decontamination procedures take place.

WARNING

NO WORKERS SHALL BE ALLOWED TO ENTER THE HOT ZONE UNTIL ALL DECON FACILITIES ARE UP AND OPERATIONAL. THE DECON OFFICER MUST GIVE THE GO-AHEAD FOR THE COMMENCEMENT OF WORK IN THE HOT ZONE. ALSO, READERS MUST REMEMBER THAT IF WORKERS ARE ENTERING THE HOT ZONE UNDER ACTUAL OR SUSPECTED IDLH CONDITIONS, THERE MUST BE ONE STANDBY RESCUER IN THE SAME LEVEL OF PROTECTION FOR EACH ENTRANT. STANDBY RESCUERS ARE STAGED IN THE COLD ZONE, READY TO ENTER THE HOT ZONE LITERALLY ON A MOMENT'S NOTICE. (SEE SECTION 7 FOR DETAILED DESCRIPTIONS OF THE VARIOUS DECONTAMINATION PROCESSES FOR PERSONNEL AND EQUIPMENT.)

While it is the transition area between the hot zone and the cold zone, the warm zone also acts as a buffer. The purpose of the warm zone is to reduce the possibility that the cold zone will become contaminated or affected by the site hazards.

The cold line marks the boundary between the warm zone and the cold zone and separates the clean areas of the site from those areas used to decontaminate workers and equipment. Access control points between the warm zone and the cold zone must be established to ensure that workers entering the warm zone are wearing the proper PPE and that workers exiting the warm zone to the cold zone remove or decontaminate all potentially contaminated PPE. All access from the cold zone access control points to the hot line access control points must be clearly marked. This controlled access corridor is called the contamination reduction corridor.

The Cold Zone. The cold zone is the uncontaminated area where workers will not be exposed to hazardous substances or dangerous conditions. Because the cold zone is free from contamination, personnel working within it may wear normal work clothes. Any potentially contaminated tools, heavy equipment such as backhoes and drilling rigs, and sampling equipment (outer containers) must remain inside the hot zone until they are no longer needed, at which point they will be either disposed of or decontaminated.

Designation of the cold zone will be based on all available site characterization data and will be located upwind from the hot zone. The cold zone will be in an area that is known to be free of elevated concentrations of uncontained hazardous substances. The cold zone will contain the staging area where newly characterized and containerized hazardous waste will await shipment to either a TSDF or an EPA regulated landfill.

All hazardous waste shipments MUST be accompanied by a Uniform Hazardous Waste Manifest.

14.11 SPECIAL SITE WORKER SAFETY

14.11.1 The Buddy System

The buddy system should be used at all times and is required by OSHA for initial entries where IDLH conditions are expected. It is a system whereby all

workers have another worker assigned to work closely with them and provide assistance when needed. Each worker is assigned to do that for another worker.

When carrying out activities in the hot zone, workers are required to use the buddy system to ensure that rapid assistance can be provided in the event of an emergency. The buddy system is an approach used to organize workgroups so that each worker is designated to be observed by at least one other worker. During initial site entry, use of a buddy system in which additional workers are assigned to provide safety backup is required. OSHA requires a minimum of four workers to be assigned to initial entries and to entries where the level of contaminants in the atmosphere is expected to be greater than the IDLH levels. Whenever workers are in the hot zone, OSHA requires corresponding standby workers for each worker, as well as separate decon personnel. The authors strongly recommend this approach until the site has been adequately characterized. The SSO may then make the determination to shrink the zones and possibly to lower the required level of PPE.

The site supervisor is responsible for enforcing the buddy system. The authors are aware of at least one life being lost where the buddy system was not followed.

As part of the buddy system, workers must remain in close proximity and maintain visual contact with each other to provide assistance in the event of an emergency. Should an emergency situation arise, workers must use communication signals agreed upon prior to entering the contaminated area. The responsibilities of workers utilizing the buddy system include:

- Providing the working partner with assistance
- Observing the working partner for signs of chemical exposure, heat exhaustion, or other unusual actions
- Periodically checking the integrity of the working partner's PPE
- Notifying the SSO or other site personnel if emergency assistance is needed

Workers must not rely entirely on the buddy system to ensure that help will be provided in the event of an emergency. To augment this system, workers in contaminated areas must remain either in line of sight or in direct communication contact with the SSO or site supervisor at all times. Hands-free radio communication for all site workers is strongly recommended.

14.11.2 Communication

○──π **Communication at a HAZWOPER site is of paramount importance. Whether carried out by audible or visual signals, or, more reliably, by radio, it is essential and required by OSHA.**

Communication systems must be established for both internal and external communication. Internal communication refers to communication among workers operating in the hot zone or warm zone or between the SSO and those workers. This communication must be constantly monitored by the SSO or designee at all times from the command post.

An internal communication system must be established using standard communication devices such as radio and audible or visual signals. Verbal communication can be difficult as a result of on-site background noise and the use of PPE. Therefore, commands and audio or visual signals must be developed prior to entering the hot zone. A secondary set of nonverbal signals must be established for use when communication devices fail or emergency situations occur.

External communication refers to communication between on-site and off-site personnel. An external communication system must be maintained in order to:

- Coordinate emergency response efforts with off-site responders
- Report progress or problems to off-site management
- Maintain contact with essential off-site personnel

The primary means of external communication are telephone and radio.

14.11.3 Safe Working Practices and Standard Operating Procedures (SOPs)

○━π **The use of site-specific SOPs is required to meet project quality control and remediation objectives.**

It has been proven that no matter how hard workers try to do their best, they cannot do a job properly if they do not understand the step-by-step instructions. Workers must be trained in and understand every facet of the task they are being asked to perform. Otherwise, confusion will rule the site. What is worse, without clear concise SOPs in emergencies, panic and injury can result. SOPs must be reviewed and updated weekly, as necessary, to reflect work site changes and worker's input.

As part of the site control plan and health and safety plan (HASP; see Section 16), procedures must be established to ensure worker safety. Worker safety procedures include preparation of the site for remediation activities, administrative controls, engineering controls, safe work practices, and SOPs. Worker safety procedures must be prepared in advance of conducting on-site operations and must be available for review by all assigned personnel at the site command post.

Engineering controls and safe work practices according to Section 6 must be implemented to reduce and maintain employee exposure levels at or below the permissible exposure limits (PELs) or other published exposure limits for those hazardous substances at the site. If engineering controls and safe work practices are insufficient to protect adequately against exposure, PPE must be used to protect employees against possible exposure to hazardous substances.

Safe work practices and SOPs help the worker eliminate guesswork. Workers, no matter how hard they try, can only work as safely and productively as they are trained to work. Well-thought out, detailed, and clearly presented SOPs can move any organization to the forefront of productivity, safety, and worker satisfaction.

14.11.4 Medical Assistance

As part of the site control program, the Project Manager must ensure that the identification and location of the nearest medical facilities are clearly posted. This facility is where site personnel can receive assistance in the event of an emergency. Information such as the names, phone numbers, addresses, and procedures for contacting the facilities must be maintained. This information must be posted conspicuously throughout the site, as well as near telephones or other external communication devices.

14.12 SITE REMEDIATION

○━π **Remediation is the process of restoring the air, soil, surface, and groundwater of a hazardous waste site to contamination concentrations that are acceptable to the regulators.**

Once a site has been identified as a Superfund, a brownfield site, or any other designated hazardous waste site and your organization is required to perform remediation in accordance with a voluntary compliance program with federal or state regulators, the remediation must begin. Remediation is the process of restoring the site air, soil, and water, to contamination concentrations that are acceptable to the regulators. Remediation rarely means restoring the site to pristine or virgin conditions. Some sites, as we explained before, will only be restored to site-specific standards agreed upon by the site owner and the regulator(s) prior to the start of work. These standards may be higher than published acceptable levels for groundwater or soil. These elevated levels are acceptable because the technology being used has been proven or is expected to capture the contaminants at the site without spread. A number of hurdles must be crossed before site cleanup can begin. Funding, regulatory, and the engineering issues of the site remedy are paramount.

When it comes to cleanup of a site, nothing may be assumed to be understood. Certainly there are standards for remediation at both the federal and state level. These standards are codified into law and usually must be adhered to strictly, depending on:

1. The intended reuse of the site
2. The type and extent of contamination
3. The location of the site
4. The technology intended for remediation
5. The ability of site owners, regulators, and the public to arrive at a consensus on the remediation approach
6. The availability of other land for development; the site may be able to be cleaned up to a different degree than other sites in that state or area

These are called *site-specific standards*. Working together, the regulators, the owner or potentially responsible parties (PRPs), the involved lawyers, and the local community can usually come together and hammer out an agreement acceptable to all.

14.12.1 Applicable or Relevant and Appropriate Requirements (ARARs)

Within the Superfund Amendments and Reauthorization Act (SARA) of 1986, Congress translated into law EPA's policy of using other environmental laws to guide response actions. SARA added CERCLA Section 121(d), which stipulates that "the remedial standard or level of control for each hazardous substance, pollutant, or contaminant must be at least that of any applicable or relevant and appropriate requirement (ARAR) under federal or state environmental law." For example, Clean Water Act (CWA) restrictions can be applicable to hazardous substances discharged into surface water from a Superfund site. Regulations codified in the National Contingency Plan (NCP) govern the identification of ARARs and require compliance with ARARs throughout the Superfund response process, including during certain removal actions.

Many states use this Superfund process for determining adequate cleanup standards of other sites, including brownfields, and other hazardous waste sites not included in federal Superfund sites.

In emergency situations, the responding party (EPA, another federal, state, or local agency, or PRP) will conduct a rapid hazardous waste-removal action to eliminate the threat. At sites where the threat is less immediate, the responding party will perform more extensive investigations to determine the appropriate remedial alternatives. In either case, the chosen management standards and cleanup levels must be protective, based on a site-specific risk assessment, and, for on-site response actions, consistent with state and federal ARARs identified for the site.

ARARs are used in conjunction with risk-based goals to govern Superfund response activities and establish cleanup goals. EPA uses ARARs as the starting point for determining adequate environmental protection. When ARARs are absent or not sufficiently protective, EPA uses data collected from the baseline risk assessment to determine cleanup levels. ARARs thus lend structure to the Superfund response process but do not supplant EPA's responsibility to reduce the risk posed by a Superfund site to an acceptable level.

Further guidance on ARARs can be found in the following documents: EPA Publication 540-R-98-020 and EPA Office of Solid Waste and Emergency Response (OSWER) Publication 9205.5-10A.

14.13   FUNDING

14.13.1   Who Pays for Remediation?

A current estimate of the cost to clean up all of the presently listed Superfund sites is in excess of 1 trillion dollars! Small Superfund sites cost between 1 million and 10 million dollars. Large sites can cost over 100 million dollars.

As you have seen, remediation funding for Superfund sites is, in part, public. Does that mean that you and I are paying for all of this work with our tax dollars? Actually, no. The federal government has moved away from simply using its extensive bank account of our money to clean up other people's messes. Through the use of PRP searches, the government is forcing the individuals and organizations responsible for the contamination to clean it up. Ideally, the EPA would like to expend its resources on research and development and emergency response exclusively.

Who pays for brownfields cleanup? That can pose an interesting problem. More times than not, the owner does not foot the bill. Brownfields are created because owners and operators of contaminating facilities simply abandon a distressed property. The property then goes up for a Sheriff's sale. If the state or local government does not receive a bid for the amount its is due in back taxes, the property usually continues to decay and become an increasing local problem.

Many states are avoiding collecting brownfields properties by allowing contaminated property transfers. Others will not do that. The authors feel that allowing property and liability transfer is an enlightened approach to the problem. Some states will require the buyer of the environmentally impacted property to post a bond for the estimated amount of the cleanup. If the buyer defaults, the state will be able to use the bond money to conduct the cleanup.

Cost Cap Insurance.   Another useful approach is the use of "cost cap" insurance. The site's controlling entity performs research into the cost of remediating the site to the satisfaction of the regulators. It then buys an insurance policy called a cost cap policy. For instance, the entity's engineering firm estimates the cleanup cost to be $1 million. The entity then buys a cost cap policy with a "cap" of $2 million. Then, if the remediation cost goes over $1 million, the insurance company pays up to the limit of the policy; the additional $1 million.

14.13.2   Joint and Several Liability

> **CERCLA's joint and several liability says that any and all parties who ever had control of or title to a property, or who can be proven to have contributed to the site contamination, are fully accountable for the contamination.**

CERCLA is a unique environmental regulation. It established, for the first time in environmental law, the idea of *joint and several liability with no time limit.* What this means is that if you hold legal title to a piece of land that is environmentally impacted and is impacting the surrounding properties, you must clean it up. Even if you acquired the land 100 years after it was contaminated and you do not even know how it became contaminated, it is your responsibility. Joint and several liability with no time limit means that the practices that took place at the site might have been legal at some time in the past, but in the interim those practices have been outlawed.

A case in point: the authors had a recent client that received all the holdings of an old metal fabricating company in a deal with another corporation. The old company had been out of business for decades. The acquiring company merely thought it was getting useful real estate. As it happened, of the five sites it acquired, all required remediation. The company had not performed any site assessments of the properties. The bill for the remediation will probably top $200 million. If you think this is an unusual circumstance, you are wrong!

14.13.3   Potentially Responsible Parties (PRPs)

> ⚷ **PRPs for a site include anyone who can be proven to have had an impact on the site. They can include owners, operators, and legal and illegal users.**

The authors were hired, hindsight always being 20-20, to work with investigators to perform PRP investigations. PRPs are anyone who can be linked to a site. They can include former owners, operators, illegal dumpers—anyone that can be brought to the table for negotiations. After several months of work, the client referred to in the above paragraph was still responsible for the lion's share of the remediation but had reduced its responsibility to around 88%, the other 12% being additional PRPs. It may not sound like much, but it could save the client over $24 million.

Some PRP searches have been going on for decades. The government and the primary PRP might or might not have started, or even finished, the remediation. The legal battles continue, however, until all persons and entities that can be tied to the site are found and, if possible, can be made to pay their share of the bill.

14.14   SITE SAMPLING AND MONITORING

14.14.1   Sampling and Monitoring

As discussed in Section 12, sampling and monitoring will be two common site tasks a HAZWOPER worker will perform. Learning how to use the instruments and interpret their data, along with logging that data, are extremely important duties. Accuracy is paramount. There are two broad categories of waste sites that the reader might be required to sample. *Uncontrolled sites* usually have no barriers prohibiting entry, the wastes are not containerized, or the containers are leaking directly onto the ground. Not all of the wastes have been located or characterized. *Controlled sites* are demarcated with fences, the wastes are contained in non-leaking drums or other packaging, and all wastes have been located.

Uncontrolled sites are considered the most hazardous to the HAZWOPER worker because the nature and extent of the contamination is unknown. Personal sampling, air, container, soil, surface, and groundwater sampling are often required on a continuous basis. PPE levels must be decided upon based on what waste exposures are found, and at what concentrations.

14.14.2   EPA-Defined Open Units

> ⚷ **The EPA considers open units to be any contaminated media that are not contained. Any contained contamination is considered a closed unit, even if the container is leaking.**

While open units might contain many types of wastes and come in a variety of shapes and sizes, they generally can be regarded as either waste piles or surface impoundments. Definitions of these two types of open units are given in 40 CFR Part 260.10.

- *Waste pile:* This can be any noncontainerized accumulation of solid, nonflowing hazardous waste that is held for treatment or storage and is not in a containment building.
- *Surface impoundment:* This can be a facility or part of a facility that is a natural topographic depression, man-made excavation, or diked area formed primarily of earthen materials, although it may be lined with man-made materials. It is designed to hold the accumulation of liquid wastes, or wastes containing free liquids, and one that is not an injection well. Examples are storage, settling, and aeration pits, ponds, and lagoons.

One of the distinguishing features between waste piles and surface impoundments is the state of the waste. Waste piles typically contain solid or nonflowing materials, whereas liquid wastes are usually contained in surface impoundments. The nature of the waste will also determine the mode of delivering the waste to the unit. Wastes are commonly pumped or gravity-fed into impoundments, while heavy equipment or trucks may be used to dump wastes in piles. Once the waste has been placed in an open unit, the state of the waste may be altered by environmental factors such as temperature and precipitation. Surface impoundments may contain several phases, such as floating solids, liquid phase(s), and sludges. Waste piles are usually restricted to solids and semisolids.

All of the potential solid and liquid phases contained in a waste unit should be considered in developing the sample design to meet the study's objective.

14.14.3 EPA-Defined Closed Units

There are a variety of designs, shapes, sizes, and functions of closed units. In addition to the challenges of the various designs and the safety requirements for sampling them, closed units are difficult to sample because they might contain liquid, solid, semisolid/sludge, or any combination of phases. Based on the study's design, it might be necessary to obtain a cross-sectional profile of the closed unit in an attempt to characterize the unit. The following are definitions of types of closed waste units described in 40 CFR 260.10:

- *Container:* any portable device in which waste is stored, transported, treated, disposed, or otherwise handled. Examples of containers are drums, overpacks, pails, IBCs, and roll-offs. Portable tanks, tank trucks, and tank cars vary in size and may range from simple to extremely complex designs. Depending on the unit's design, it may be convenient to consider some of these storage units as tanks for sampling purposes even though they meet the definition of a container.
- *Tank:* a stationary device, designed to contain an accumulation of waste, which is constructed primarily of non-earthen materials, which provide structural support.
- *Ancillary tank equipment:* any device including, but not limited to, such devices as piping, fittings, flanges, valves, and pumps that is used to distribute, meter, or control the flow of waste from its point of generation to a storage or treatment tank(s), between waste storage and treatment tanks to a point of disposal on-site, or to a point of disposal off-site.
- *Sump:* any pit or reservoir that meets the definition of a tank and those troughs or trenches connected to it that serve to collect liquid wastes. (Note: some outdoor sumps may be considered open units or surface impoundments.) A sump is designed to collect liquid at the lowest point of a system. A sump is not a drain!

Although any of the closed units may not be completely sealed and may be partially open to the environment, the unit needs to be treated as a closed unit for sampling purposes until a determination can be made. Once a closed unit is opened, a review of the proposed sampling procedures and level of protection can be performed to determine whether the personal protection equipment is suitable for the site conditions.

14.15 WASTE CONTAINERS: LOCATING, EXCAVATING, OPENING, AND SAMPLING

When first entering and working in an uncontrolled hazardous waste site, pay particular attention to where you step. Drums that have been buried and have contained liquids, especially liquids incompatible with extended outdoor storage in the drum, will usually be the first to fail. Ground moisture can cause corrosion from the outside of the drum, while the incompatible, corrosive, or reactive liquid can corrode the container from the inside.

WARNING

AS BURIED LIQUID DRUMS FAIL (CRACK OR DETERIORATE) AND THE CONTENTS LEAK OUT, THEY LEAVE A VOID IN THE EARTH. AS THE TOP AND SIDES OF THE DRUM ALSO DETERIORATE, A TRAP IS SET FOR THE UNWARY HAZWOPER WORKER. ONE OF THE AUTHORS HAS FALLEN INTO THESE VOIDS. IT IS QUITE A SHOCK TO WALK ALONG AND SUDDENLY FIND YOURSELF FOUR FEET BELOW GROUND. WHEN WORKING ON THE INITIAL SITE INSPECTION, AND INDEED, UNTIL THE SITE HAS BEEN SAFELY SECURED, WALK CAREFULLY.

The HAZWOPER regulations concentrate a great deal on drums and their management. A large part of hazardous waste storage over the years has been in 55-gallon steel drums. Figure 14.1 illustrates how badly storage might be managed.

One of HAZWOPER workers' most frequent, yet most potentially dangerous, jobs is the opening and sampling of unidentified, partially identified, or incorrectly identified drums. As the reader learned in Section 12, steel drums are just one of the types of containers that can contain liquids. Tanks, intermediate bulk containers (IBCs), plastic drums, bottles, and jars are also commonly used. Some of the procedures listed can also be safely used for container sampling applications. However, in this Section we will be concentrating on drums because they are the most common type of container encountered on waste sites.

FIGURE 14.1 "Valley of Drums": an enormous uncontrolled waste site. (*Courtesy of the USEPA.*)

All container contents must be sampled and characterized for disposal, consolidation, recycling, segregation, and classification purposes.

Depending on site conditions, equipment limitations, or limitations imposed by any of these procedures, the procedures may be varied by the site safety officer (SSO) as applicable.

14.15.1   Introduction to Drums: Construction and Usage

⚷ **The 55-gallon drum, also designated the 208 liter (L) drum, is the most commonly used container for hazardous waste.**

For about 40 years now the DOT has used specification-based packaging. That meant that as shippers, as long as the container we used met the specifications of the DOT for transporting our materials we were dealing with, we were within regulations. "Drums" usually brings to mind the 55-gallon (208-L) steel drum (Figures 14.2 and 14.3). More of these drums have been used over the last century than any other type of portable container to store, transport, and dispose of materials.

The standard steel drum was either the DOT-17H or the DOT-17E (Figure 14.6). Because these are the most commonly found drums at hazardous waste sites, they are the ones we will explain how to open, sample, and handle. Other drums that may be encountered are the polyethylene and high-density polyethylene (HDPE), usually used for foodstuffs, high-purity chemicals, and corrosives, among others. They have similar configurations. However, as you can see from Figures 14.4 and Figure 14.5 there are subtle differences in construction. As the readers will see later, those differences in type and materials of construction can help HAZWOPER workers predict how a certain type of drum might fail.

A standard steel bung-type (also called a tight-head) drum has a 2½-in. filling hole and a ⅝-in. vent hole in the top. These are called bungholes and are fitted with bung plugs. Bung-type drums have also been made with the fill opening on the side, halfway down, and with poly liners to allow them to carry corrosive liquids. Bung-type drums are *intended* for liquids only.

Open-mouth (also called open head) steel drums can come with no holes in the top, the same configuration as the bung type, or with two 2½-in. openings across the lid from each

FIGURE 14.2 Bung-type, also called tight-head, steel drum, 55-gallon (208-L) Can be polyethylene-lined, used for liquids.

FIGURE 14.3 Open-mouth, also called open-head, steel drum, 55-gallon (208-L) (gasket between top and drum). Can be polyethylene-lined, used for solids and sludges.

other. Open-mouth drums can be made with poly liners to allow them to carry corrosive solids. Open-mouth drums are *intended* for solids only.

Poly drums are found in the same configuration as the steel drums, except that they do not come with a side-mounted fill hole. Bung-type poly drums often carry corrosive liquids. Open-mouth poly drums often contain hazardous solid chemicals.

Drums come in many other sizes besides 55-gallon, including 5-, 10-, 20-, 30-, and 40-gallon sizes, as well as 75- and 95-gallon salvage drums or overpacks.

FIGURE 14.4 Bung-type, also called tight-head, polyethylene drum, 55-gallon (208-L). Often used for corrosive liquids.

FIGURE 14.5 Open-mouth, also called open-head, polyethylene drum, 55-gallon (208-L) (gasket between top and drum). Used for corrosive solids and sludges.

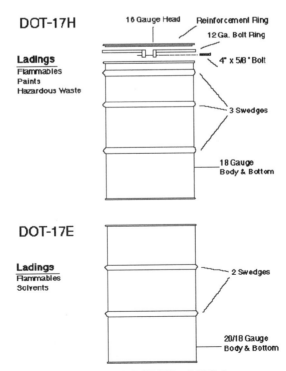

DOT-17H

16 Gauge Head   Reinforcement Ring

12 Ga. Bolt Ring

Ladings

Flammables
Paints
Hazardous Waste

4" x 5/8 " Bolt

3 Swedges

18 Gauge
Body & Bottom

DOT-17E

Ladings

Flammables
Solvents

2 Swedges

20/18 Gauge
Body & Bottom

FIGURE 14.6 Old-Style DOT-17H and 17-E drums.

Salvage Drums, Also Known as Overpacks. Salvage drums, which we will refer to by their more common name of overpacks, are drums that will easily hold a 55-gallon drum along with associated contaminated cleanup materials. They come in two basic designs. They always have open mouths.

The steel overpack is a larger version of the open-mouth drum. The drum to be over-packed must be lifted and placed into the overpack. The steel overpack meets DOT standards for shipping containers.

The polyethylene screw-on head drum is not designed as a primary container for shipment, that is, to be filled with material and shipped. Their sole purpose is to overpack a failing 55-gallon drum. This version, used more commonly in emergency response, has a gasketed screw-on head. For ease of loading, the drum head is screwed off and laid on the ground. The drum to be overpacked is "walked" onto the overpack lid. The overpack drum is then upended and screwed, over the "bad" drum, into the lid. The overpack is then upended for transport and disposal. This design allows one or two responders to safely package a drum without the use of lifting machinery such as a fork truck. The leaking drum will leak into the overpack, but the gasketed head makes it legal for transport.

Fiber-Pack Drums and Lab Packs. Drums can also be fabricated of rigid, wrapped paper. These are called fiberpacks. They are designed only for transporting dry solids and lab packs of specially packaged containers of solids and liquids in absorbent material.

When a HAZWOPER worker is confronted with a large number of small containers of chemicals, one question is how to ship them. Shipping each container individually would be costly, time-consuming, and wasteful of shipping resources. For that reason, the EPA and the DOT allow lab packs to be shipped. Lab packs are so named because they are frequently used to ship out-of-date and no longer used small containers from laboratories.

The small containers, usually glass bottles of compatible chemicals, are packed in a fiber pack drum. Each bottle is insulated from breakage from others and the wall of the fiber pack by vermiculite. The vermiculite also serves the purpose of absorbing any spilled material should one of the containers break in transportation. Lab packing should only be performed by a specialist with in-depth knowledge of chemical compatibility.

In the late 1990's, DOT issued final rules for Dockets HM-215B and HM-181H, respectively. For steel drum users, these primarily affected provisions on overpacks (salvage drums), minimum thickness, certain transitional provisions (we are allowed to use old-style drums until we run out of supply), and design type variations.

14.15.2 Required Drum Marking for Performance-Oriented Packaging (POP)

The new hazardous materials regulations are based on performance-oriented packaging (POP). A container can be manufactured in any fashion as long as the resulting package successfully conforms to the test provisions located in 49 CFR 178.600. The following items are not specified under the regulations: capacity, steel thickness, height or diameter requirement, and closures. The new drum markings are found at 49 CFR 178.503.

The new drum markings given as examples are for steel drums. For embossed metal drums, the letters "UN" may be applied in place of the United Nations symbol. Letter height must be a minimum of 12 mm for containers over 30 L, or 6 mm for containers under 30 L.

According to The Wiley Encyclopedia of Packaging Technology, Second Edition,* "a sample mark for an open-head steel drum of 1 millimeter thickness manufactured in 1994 by manufacturer M1234 and authorized to carry a Packing Group II or III solid with a gross mass of 300 kg (or less) is

---

The Wiley Encyclopedia of Packaging Technology, 2d ed., ed. A. L. Brody and K. S. Marsh, New York: John Wiley & Sons, 1997.

UN 1A2/Y300/S/94/USA/M1234 1.0

where UN = United Nations, 1 = drum, A = steel, 2 = open head, Y = Packing Group II or III, 300 = maximum gross mass in kg (net mass of solid plus mass of drum), S = solid, 94 = year of manufacture, USA = country of manufacture, M1234 = manufacturer's number of symbol, and 1.0 = thickness in millimeters.

A tighthead drum with a nominal 1.1-mm-thick head and bottom and 0.8-mm body manufactured in 1994 and authorized to carry a Packing Group II or III liquid with a specific gravity of 1.8 or less and with a product vapor pressure of 230 kPa (or less) at 55°C is marked

UN 1A1/Y1.8/230/94/USA/M1234 1.1/.8/1.1

where UN = United Nations, 1 = drum, A = steel, 1 = tight head, Y = Packing Group II or III, 1.8 = specific gravity (relative density of material to water), 230 = maximum hydro static pressure tested in kPa, 94 = year of manufacture, US = country of manufacture, M1234 = manufacturer's number of symbol, and 1.1/.8/1.1 = thickness of top, body and bottom in millimeters."

14.15.3 Intermediate Bulk Containers (IBCs)

One type of container that is coming into much wider use for both liquids and solids is the intermediate bulk container (IBC). As you can see from Figure 14.7, the IBC is designed to store or transport the most material in any given volume. Its cube shape is more economical of storage space than the cylinder (drum). IBCs are mentioned here for familiarization purposes. They are rarely found at hazardous waste sites. They are, however, used as new containers to ship consolidated, compatible solid or liquid hazardous wastes. (Figure 14.7 shows an IBC designed for liquids.)

Prior to sampling at a hazardous waste site, drums must be located, excavated (if necessary), inspected, staged, and opened. Only HAZWOPER-qualified personnel must be allowed to perform these tasks. Inspection is the observation and recording of visual qualities of each drum and any characteristics pertinent to the classification of the drum's contents. Staging involves the physical grouping of drums according to classifications established during the physical inspection. Opening of closed drums can be performed manually or remotely. Remote drum opening is recommended for worker safety. The most widely used method of sampling a drum involves the use of a glass thief (described in Section 14.15.10).

FIGURE 14.7 A 200-gallon intermediate bulk container (IBC) for use with liquids and solids (liquids version shown).

This method is quick, simple, relatively inexpensive, and requires no decontamination. The contents of a drum can be further characterized by performing various field tests.

14.15.4   Locating Containers

○—ㅠ   **All containers on a site must be located, even if they are empty. Locating buried containers can be difficult, but ground-penetrating radar (GPR) can guide location and excavation.**

Upon assignment to the initial site survey team, one of the HAZWOPER workers' prime responsibilities will be to locate containers of suspected hazardous material or waste. The workers will also have the task of locating unprotected and uncontrolled piles of materials and potentially contaminated surface impoundments of water. As site work progresses, as the reader will see in the later parts of this Section, workers will be tasked with finding the hidden materials: those that are contaminating site soil and groundwater.

If buried drums are suspected, geophysical investigation techniques such as ground-penetrating radar (GPR; see Section 12.6) can be used to determine the location and depth of drums. If drums are shallow in the ground, extra care must be taken that the excavating equipment does not drive over them and crush them. During excavation, the soil must be removed with great caution to minimize the potential for drum rupture. This will require a combination of delicate backhoe operation, nonsparking shovel work, and even hand excavation.

Until the contents are characterized, sampling personnel must assume that unlabeled drums contain hazardous materials. Labeled drums are frequently mislabeled, especially drums that are reused. Because a drum's label may not accurately describe its contents, extreme caution must be exercised when working with or around drums. Of course, weather and burial will deteriorate labels to illegibility in as little as six months.

If a drum containing a liquid cannot be moved without rupture, due to damage or deterioration, its contents must be transferred to a sound drum using an appropriate method of transfer based on the type of waste. Hand- and air-operated pumps are best for this application. In any case, preparations should be made to contain a spill should one occur.

If a drum of hazardous material is leaking, open, or deteriorated, it must be placed immediately in overpack containers.

The practice of tapping drums to determine their contents amount or type is neither safe nor effective and should not be used if the drums are visibly overpressurized or if shock-sensitive materials are suspected. Normally, a steel drum filled with liquid would give a different tone than one filled with a solid. A drum may have a distorted shape (head, bottom, or both may be bulging). A laser thermometer can be used detect the temperature of the drum from a distance, as high temperature often accompanies drum distortion. However, if the drum has been allowed sufficient time to cool, the distortions may be permanent. This instrument should not be relied upon alone, as drum temperature may not be an indication of overpressurization.

Drums that have been overpressurized, to the extent that the head is bulging several inches above the level of the chime (top or bottom rim of a steel drum), should not be moved until it is determined whether the drum is still pressurized. A number of devices have been developed for venting critically swollen drums. These devices will be discussed further in Subsection 14.15.9. Once the pressure has been relieved, the bung can be removed and the drum sampled.

Because there is potential for accidents to occur during handling, particularly initial handling, drums must only be handled if necessary. All personnel must be warned of the hazards prior to handling drums. Overpack drums and an adequate volume of absorbent material must be kept near areas where minor spills may occur. Where major spills might occur, a containment berm adequate to contain the entire volume of liquid in the drums must be constructed before any handling takes place. If drum contents spill, personnel trained in spill response must be used to isolate and contain the spill.

14.15.5 Container Excavation

> Always contact local utilities before doing any digging on the site. They will come to the site and locate and mark any buried utilities.

See Figure 14.8.

WARNING

IN NEARLY EVERY STATE, IT IS REQUIRED BY LAW TO CONTACT THE LOCAL UTILITIES COMPANIES (USUALLY 48 HOURS IN ADVANCE) PRIOR TO PERFORMING ANY EXCAVATIONS. THE COMPANIES WILL SEND TECHNICIANS TO THE SITE AND LOCATE ANY BURIED UTILITIES. INFORMATIONAL CONTACTS ARE AVAILABLE BY CALLING "DIG SAFELY" AT 888-258-0808. THEY ARE ALSO AVAILABLE ON THE WEB AT www.digsafely.com. THE SERVICE IS USUALLY FREE, OR PERFORMED FOR A MINIMAL FEE.

Shoring at Site Excavations. OSHA requires all site excavations created during initial site preparation or during hazardous waste operations to be shored or sloped as appropriate to prevent accidental collapse in accordance with Subpart P of 29 CFR Part 1926.

Other Excavation Safety Concerns. Engineering controls include the use of pressurized cabs or control booths on equipment. They also include the use of remotely operated material handling equipment. Site work practices such as removing all nonessential employees from potential exposure during opening of drums or other containers, wetting down dusty operations, and locating employees upwind of possible hazards must also be used.

If it is presumed that buried containers are on-site, then prior to beginning excavation activities, geophysical investigation techniques are required to approximate the location and depth of the containers. Steel drums and pipes are easily detected. Glass and plastic containers are more difficult to locate. However, ground-penetrating radar will locate nearly all kinds of buried objects.

FIGURE 14.8 What happens when you ignore precautionary information during excavation. (*Courtesy of USEPA.*)

In addition, it is important to ensure that all locations where excavation will occur are clear of utility lines, pipes, and poles (subsurface as well as above surface). The authors know of one crew that was excavating with a backhoe when the bucket struck a buried 13,200-volt power line. Fortunately, no one was seriously injured, but the explosive release of electrical current to ground blew the bucket off the backhoe.

Excavating and removing drums are generally accomplished with backhoes, sometimes using special attachments. Moving and handling drums, once they are excavated, is done with fork trucks and possibly skid loaders with special attachments. Once drums can be moved over smooth surfaces, drum trucks and drum dollies (Figures 14.15 and 14.16) are used as well. These devices will be discussed and illustrated later in Subsection 14.15.7. Drum excavation must be performed by an equipment operator who has experience in drum excavation. During excavation activities, drums must be approached in a manner that will avoid digging directly into them.

The soil around the drum should be excavated with nonsparking hand tools or other appropriate means. As the drums are exposed, a visual inspection must be made to determine the condition of the drums. Ambient air monitoring and other detection must be used to determine the presence of unsafe levels of any pre-determined or suspected contaminants. Based on this preliminary visual inspection, the appropriate safe method of drum excavation and handling can be determined.

14.15.6 On-Site Inspection and Hazard Categorization

Drum inspection, identification, initial sampling and analysis, and inventory must begin as soon as the first drum is excavated. Information such as location, date of removal, unique site drum identification number, overpack status, and any other identification marks should be recorded on a drum/tank sampling data sheet. Table 14.1 is a sample of an EPA-suggested on-site drum/tank sampling data sheet.

Appropriate procedures for handling drums depend on the condition of the drum first, and then the contents. Thus, prior to any handling, drums should be visually inspected to gain as much information as possible about their contents. Use this checklist to inspect drums for the following conditions:

❑ Drum condition, including visible signs of corrosion (stainless steel drums), rust (steel drums), bulging, punctures, damaged bungs, missing bung plugs, and leaking contents
❑ Drum type and construction
❑ Labels, symbols, words, or other markings on the drum indicating hazardous contents such as explosive, radioactive, toxic, flammable, or any other markings that might identify the contents or owner of the drum
❑ Signs that the drum is under pressure
❑ Shock sensitivity
❑ Air-monitoring results from air sampling conducted at the drum and surrounding soil, water, and liquids

The results of this inspection can be used to classify the hazards of the drum's contents into the following categories:

• Toxic
• Corrosive
• Flammable
• Explosive
• Shock-sensitive
• Radioactive
• Leaking or deteriorating

TABLE 14.1   An EPA-Suggested On Site Drum/Tank Sampling Data Sheet

Drum/Tank Sampling Data Sheet

Sampler Name(s):

Date:

Site Name:

Work Order Number:

Container Number/Sample Number:

SITE INFORMATION:

1. Terrain, drainage description: _____

_____

2. Weather conditions (from observation): _____
MET station on site: (Circle one) No Yes

CONTAINER INFORMATION:

1. Container type: Drum Tank Other: _____

2. Container dimensions: Shape: _____

Approximate size: _____

3. Label present: (Circle one) No Yes (description): _____

Other Markings: _____

_____

_____

4. Spill or leak present: (Circle one) No Yes Dimensions: _____

5. Container location: (Circle one) N/A See Map Other: _____

_____

SAMPLE INFORMATION: (Check and describe all applicable)

1. Description: _____ liquid _____ solid _____ powder or _____ crystals _____ sludge

2. Color: _____ Vapors: _____

Other: _____

_____

3. Local effects present: (damage—environmental, material _____

_____

FIELD MONITORING:

1. PID: _____ Background (clean zone)

_____ Probe used/Model used

_____ Reading from container opening

2. FID: _____ Background (clean zone)

_____ Reading from container opening

TABLE 14.1   An EPA-Suggested Drum/Tank Sampling Data Sheet (*Continued*)

---

3. Radiation Meter:

_____ Model used

_____ Background (clean zone)

_____ Reading from container opening

4. LEL/Oxygen Meter:

_____ Oxygen level from container opening

_____ LEL level from container opening

- Bulging
- Lab pack
- Empty

The Special Case of Bulging Drums

⊶ **Bulging drums must be approached and handled with care. Bulging drums have been overpressurized at some point. They may or may not be overpressurized when discovered.**

⊶ **Knowing the type and construction of an overpressurized drum will tell the worker how it might fail.**

⊶ **All drums that have been overpressurized eventually fail, due to internal pressure of less than 150 psi. Poly drum failures are the most unpredictable. There is no common pattern by which they rupture, such as at a seam.**

One of the greatest physical hazards to be encountered by the HAZWOPER worker is a bulging drum. When discovered, this should immediately send up a red flag. We know that the drum must have been overpressurized to cause such distortions. The question the worker must ask is, "Is it overpressurized now?" OSHA requires that drums and containers that appear to be under excessive internal pressure, as evidenced by bulging or swelling, shall not be moved until the cause for excess pressure is determined. Then appropriate containment procedures must be implemented to protect HAZWOPER employees from explosive relief of the drum. Further, the environment must be protected from any material that might be released from the drum upon depressurization and opening.

The Los Alamos National Laboratory (LANL) HAZMAT team has done a great deal of research in this area. Their research is presented in a video, *Bulging Drums—What Every Responder Should Know.* With the kind permission of the investigators, the following is a capsule of their research monograph "Pressure Effects and Deformation of Waste Containers." The researchers were LANL HAZMAT team members Michael D. Larrañaga, CFPS; David L. Volz; and Fred N. Bolton, PE, CIH.

Their tests revealed the following differences in the failure characteristics, due to high internal pressure, among drum types. Some of the drums contained water at various levels. The drums were pressurized with air through the bung hole.

- Fifty-five-gallon metal open-mouth (open head) drums:
  - The drums appeared to vent immediately adjacent to the nut-and-bolt fastener on the ring.
  - All drums tested self-vented pressures at or below 32 psig.

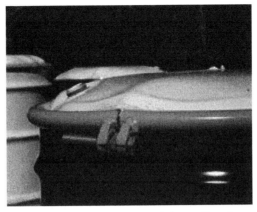

FIGURE 14.9 Bulging open-mouth drum self-venting at bolt ring joint seal. (*Courtesy of LANL; Photo by Michael D. Larrañaga, CFPS.*)

- Pinging (a periodic noise accompanying deformation) was noticeable between 15 and 20 psig. It increased dramatically immediately before drum failure.
- Body seams (top to bottom) experienced no visible distortion or apparent weakening.
- The 55-gallon metal open-mouth drums appeared to bulge at only the top and bottom ends (see Figure 14.9).

- Fifty-five-gallon metal bung-type (tight head) drums:
  - Ninety-five percent of the drums tested failed explosively.
  - Of the catastrophic failures, 68% failed at the bottom end, making the entire drum a projectile.
  - All drums tested failed at the top or bottom ends.
  - When drums were one-half or three-quarters filled with liquid, bottom failures appeared to be increasingly violent with increasing water levels up to three-quarters full.
  - Approximately 5 psig before failure, a significant amount of distortion of the drum chime (rim) occurred. (see Figure 14.10).

Statistical analyses of the test results indicate a probability that 99% of the failures will occur above 48.7 psig.

FIGURE 14.10 Bulging bung type drum with distorted chime. (*Courtesy of LANL; Photo by Michael D. Larrañaga, CFPS.*)

These observations show that bulging drums, especially the bung type, are extremely dangerous. Noticeable differences exist between closed- and open-head drums under pressure. However, both types of drums are inherently dangerous when pressurized and present many hazards. The testing provides a reasonable certainty that new bung type 55-gallon drums will fail above 48.7 psig and helps in determining a safe working envelope for working with these drums.

- Fifty-five-gallon plastic drums.
  - The five 55-gallon HDPE seamless drums tested failed explosively at pressures of 48, 48, 50, 30, and 58 psig.
  - Four of five failures occurred through the sides of the drums at no particular or identifying location; one failed at 30 psig out of the top end of the drum.

These are significant observations because they show the potential for seamless HDPE drums to fail out of the sides. Deformation was observed at the tops, bottoms, and sides of the drum.

The researchers concluded that deformation of a drum indicates that the drum has been subjected to internal pressure, *not that it is under pressure at the time of inspection.* The container's design makes it capable of violent rupture. Always approach deformed drums with caution. It was assumed that the experimental method of pressurization at increasing 5-psig intervals was representative of real pressurized drums in the field, although the speed and method of pressurization may be different. Also, it was assumed that the deformation observations correspond to real-world pressurized drum deformations in the field.

The results of this study indicate that significant differences exist among the failure characteristics of drum types and materials. The 55-gallon drum data were sufficient to support the development of a device for estimating internal pressures in 55-gallon metal drums. That is because of the similarity between pressure-versus-deformation curves. The safe discovery and mitigation of a pressurized drum can make the difference between a minor incident with limited and controlled consequences and a major one with the potential for loss of life and significant property damage. With the information contained in this report, Department of Energy personnel, private and municipal fire departments, HAZMAT teams, Explosive Ordinance Device (EOD) teams ("Bomb Squads"), and other emergency responders can make educated decisions during bulging drum incidents. Many observations were noted and should be used as indicators of danger when approaching bulging drums.

The researchers made the following training recommendations for HAZWOPER workers, based on the conclusions of this study. Learn the following:

- Indicators of drum pressurization
- Inherent hazards associated with drum incidents and chemical properties such as toxicity, flammability, corrosiveness, and reactivity
- Failure characteristics and differences between types of drums
- Pressurized drum-mitigation techniques such as remote venting, direct cooling (ice bath), or shooting with a projectile (water cannon, bow and arrow, or shotgun slug)

This research demonstrated the particular hazards of moving or even approaching a bulging drum. All drums eventually fail, all under 150 psi. Poly drums were the most unpredictable because there was no repeating pattern to their breaking point, such as a seam rupture. Almost all poly drums failed explosively, which could immerse a HAZWOPER worker in the contents (which are often highly corrosive)!

Steel drums were a different story. Liquid-containing bung-type drums all failed explosively, with a majority of them becoming projectiles. Open-mouth steel drums all self-vented at the meeting of the bolt ring.

Some of these drums may not be moved to a staging area safely. In that case, the area around the drum must be prepared for a spill using absorbents such as cat litter or absorbent socks or dikes. There should also be a new drum, a transfer pump for liquids, or an overpack standing by.

During drum excavation, the immediate work area atmosphere must be monitored.

14.15.7 Drum Staging

Prior to sampling, the drums must be staged to allow easy access. Ideally, the staging area will be located just far enough from the drum opening area and the characterized drum storage area to prevent a chain reaction if one drum should explode or catch fire when opened.

During staging, the drums should be physically separated into the following categories:

- Drums that contain liquids
- Drums that contain solids
- Drums that contain lab packs
- Drums that are empty

Note

If there is reasonable certainty of the contents of the drum (e.g.: an unopened drum with bung seals intact and manufacturer's label legible); the drum could be stored with the other materials in its class for shipment. This would avoid the staging step. Many sampling protocols will, however, require sampling of all containers.

This is done because the strategy for sampling and handling drums and other containers in each of these categories will be different. This may be achieved by visual inspection of the drum and its type and construction, labels, codes, and other identification. Solids and sludges are typically disposed of in open-mouth drums. Bung-type drums generally contain liquid. However, this is by no means always the case.

Where there is good reason to suspect that drums contain radioactive, explosive, or shock-sensitive materials, these drums should he staged in a separate, isolated area. Placement of explosives and shock sensitive materials in diked and fenced areas will minimize the hazard and the adverse effects of any premature detonation.

Where space allows, the drum opening area should be physically separated from the drum-removal and drum-staging operations. Drums are moved from the staging area to the drum opening area one at a time using forklift trucks equipped with drum grabbers or a barrel grappler. Drums will need to be restaged into compatible groups (see Section 3) after opening and sampling.

Staging of drums requires as much area as is available. To move drums from place to place requires special equipment and a delicate touch. Figures 14.11 to 14.16 illustrate some common and some state-of-the-art drum maneuvering equipment. Fifty-five-gallon drums can weigh anywhere from 20 to 45 pounds empty and up to 1,500 pounds full. Drums must never be rolled on their sides or swedges for transport purposes. This weakens the drum wall and can result in leaks or mixing of incompatible contents. Drums must only be rolled short, maneuvering distances, and only on their chimes. When at all possible, use a drum truck, drum dolly, or forklift.

All drums must be handled carefully, not only the ones that contain hazardous materials. A drum full of drinking water still presents a crushing hazard to the worker as it will weigh around 500 lb.

OSHA requires that drum or container staging areas shall be kept to the minimum number necessary to safely identify and classify materials and prepare them for transport and, that staging areas shall be provided with adequate access and egress routes.

FIGURE 14.11 Drum lifter for over-packing. (*Courtesy of Easy Lift Equipment Co., Inc.* [www.easylifteqpt.com].)

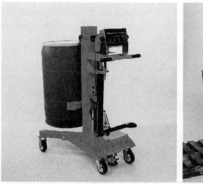 

FIGURE 14.12 EasyLift™ ERGO® drum transporters for use with poly or steel drums, with scale (L) or without (R). (*Courtesy of Easy Lift Equipment Co., Inc.* [www.easylifteqpt.com].)

FIGURE 14.13 EasyLift™ Eagle-Grip™ for use with poly, steel, or fiberpack drums, without (L) or with scale (R). (*Courtesy of Easy Lift Equipment Co., Inc.* [www.easylifteqpt.com].)

FIGURE 14.14 EasyLift™ Eagle-Grip™ 3 series for use with poly, steel, or fiberpack drums, carries two drums at once. (*Courtesy of Easy Lift Equipment Co., Inc.* [www.easylifteqpt.com].)

FIGURE 14.15 WESCO® 10 BT 1000-lb capacity drum truck. (note the sliding keeper on the shaft that holds the chime of drums from 5- to 85-gallon capacity). (*Courtesy of Wesco® Industrial Products, Inc.* [www.wescomfg.com].)

FIGURE 14.16 WESCO® drum dolly. (*Courtesy of Wesco® Industrial Products, Inc.* [www.wescomfg.com].)

During drum opening, sampling, and staging, the immediate work area atmosphere must be monitored.

14.15.8 Drum Opening

Although OSHA only requires unlabeled drums and drums labeled hazardous to be treated as hazardous waste, the authors recommend that all drums be considered hazardous until proven otherwise.

The following procedures shall be used when collecting samples from drums of unknown material:

Visually inspect all drums that are being considered for sampling for the following:

- Overpressurization causing bulging
- Crystals formed around the drum opening (OSHA requires the contents to be handled as a shock-sensitive waste until the contents are otherwise identified)
- Leaks, holes (stains on the side of the drum may show pinhole leaks)
- Labels, markings
- Composition and type (steel, polyethylene, or fiber pack, open-mouth or bung-type)
- Condition, age, rust
- Sampling accessibility

Drums showing evidence of pressurization and crystals must be further assessed to determine if remote drum opening is needed. If drums cannot be accessed for sampling, heavy equipment is necessary to stage drums for the sampling activities. Adequate time should be allowed for the drum contents to stabilize after a drum is handled. OSHA requires that fire-extinguishing equipment meeting the requirements of 29 CFR 1910, Subpart L, be on hand and ready for use to control incipient fires (incipient meaning before or at the beginning of the fire. Studies show that 95% of all fires can be stopped at the incipient stage). An ABC-type fire extinguisher is adequate for this purpose. Further, drums and containers must be opened in such a manner that excess interior pressure will be safely relieved. If pressure cannot be relieved from a remote location, OSHA requires that appropriate shielding be placed between the employee and the drums or containers to reduce the risk of employee injury.

Identify each drum that will be opened. Use Level B (see Section 6) protection for the following procedures.

Before opening, ground each metal drum that is not in direct contact with the earth (such as on a pallet) using grounding wires with clips and a grounding rod or metal structure. If a metal drum is in a metal salvage drum, both drums must be grounded.

Touch the drum opening equipment to the bung or lid and allow an electrical conductive path to form. Slowly remove the bung or drum ring and lid with spark-resistant tools (brass, bronze and manganese or beryllium).

Monitor any vapors from drums for explosive gases and toxic vapor with air monitoring instruments as the bung or drum lid is loosened and removed.

Depending on site conditions, HAZWOPER workers might be required to sample for one or more of the following:

- pH
- Toxics
- Halogen vapors
- Water reactive
- Flash point (flash point will require a small volume of sample for testing)

Note the state, quantity, phases, and color of the drum contents. Record all relevant results, observations, and information in a logbook or on a drum/tank sampling form. Review the screening results with any preexisting data to determine which drums will be sampled.

Drums in poor condition must be placed in salvage or overpack drums or their contents transferred to a drum in good condition.

During drum opening, the immediate work area atmosphere must be monitored.

14.15.9   Drum Opening Tools

Due to the large amount of wastes that are flammable and the resulting fire or explosion hazard, OSHA requires the use of nonsparking tools for opening flammable containers and unknowns.

FIGURE 14.17 WESCO® nonsparking drum wrench. (*Courtesy of Wesco® Industrial Products, Inc.* [www.wescomfg.com].)

Nonsparking Universal Bung Wrench

О—π **The most common drum-opening tool is the nonsparking universal bung wrench.**

Once a drum has been deemed safe to open, two common hand tools are used to open them prior to sampling: the nonsparking bung wrench and the drum deheader.

The most common method for opening drums manually, is the universal bung wrench (Figures 14.17 and 14.18). These wrenches have fittings made to remove nearly all commonly encountered bungs on steel and plastic drums. They are usually constructed of a nonsparking metal alloy such as brass, bronze and manganese or beryllium to reduce the likelihood of sparks.

WARNING

THE USE OF A NONSPARKING WRENCH DOES NOT COMPLETELY ELIMINATE THE POSSIBILITY OF A SPARK BEING PRODUCED.

The main disadvantage of the bung wrench is that the worker has to be standing directly beside the drum to use it. In most situations, this practice will be satisfactory. We recommend that all HAZWOPER personnel get in the habit of opening all containers SLOWLY. This will normally allow pressure to vent in a controlled fashion. Never open a drum with your face over the wrench or bung plug. When opening open-mouth drums, stand as far as practical from the bolt ring joint—this is where the drum will likely vent first.

FIGURE 14.18 Model 59SRM Morplug® spark resistant bronze wrench. (*Courtesy of Morse Manufacturing Company* [www.morsemfgco.com].)

We strongly recommend that this method never be used on unknowns (drums with unknown contents), especially where crystals have developed around the threads of the bung or on bulging drums (known affectionately in the HAZWOPER world as "rockers," for their propensity to rock from side to side on their bulging bottoms).

These drums must be opened remotely by using one of the devices designed for that purpose.

Drum Deheader.   One means by which a closed-head steel drum can be opened manually when a bung is not removable with a bung wrench is by using a drum deheader (Figure 14.19). Our illustration tool is constructed of non-sparking alloys and is designed to cut the lid of a drum off much like an old-fashioned can opener. Again, a major disadvantage is the worker's proximity to the drum. Another limitation of this device is that it can be attached only to bung-type drums. Drums with removable heads must be opened by other means. Remember, just because a drum has a removable head and is supposed to be used mainly for dry solids and sludges does not mean the drum was used properly. It may be full of liquid.

The portable powered model is shown in Figure 14.20.

To add an additional safety factor, remote control drum-opening stations can be used. There are electric powered drum deheaders and the pneumatic-powered drum openers.

The electric-powered drum deheader should be used when there is good assurance that the contents of the drum are not flammable, reactive, explosive, or shock sensitive. The unit seen in Figure 14.21 is excellent where there is an isolated drum opening area. The model illustrated is for bung-type liquid drums.

Remotely Operated Hydraulic Drum Opener.   The pneumatic drum opener is available in a number of configurations. The newest and possibly most valuable addition to this lineup was developed by Los Alamos National Laboratory (LANL) for use in the field by HAZMAT teams. It is powered by a standard SCBA cylinder. It can be easily attached to the top of any container. Using different-diameter spikes, it can be used for venting or for venting and sampling. Using a single 60-minute SCBA cylinder, this handy device can open up to 45 drums!

Remotely operated devices have been fabricated to open drums. One device uses hydraulic pressure to force a nonsparking spike through the top or head of a drum locked into an opening stand. It consists of a hydraulic pump that pressurizes hydraulic oil through a length of hydraulic line. Drums can also be pneumatically opened. The pneumatic device works the same as the hydraulic device, except it uses air pressure to operate the spike.

Hand Pick, Pickax, and Hand Spike.   These tools are usually constructed of brass or a nonsparking alloy with a sharpened point that can penetrate the drum lid or head when the tool is swung. These are not the commonly available hand chisels or spikes or pickaxes that

FIGURE 14.19   WESCO® Manufacturing patented nonsparking drum deheader. (*Courtesy of WESCO® Industrial Products, Inc.* [www.wescomfg.com].)

FIGURE 14.20 WESCO® Manufacturing powered drum deheader. (*Courtesy of WESCO® Industrial Products, Inc.* [www.wescomfg.com].)

FIGURE 14.21 WESCO® Manufacturing remote control drum deheading station. (*Courtesy of WESCO® Industrial Products, Inc.* [(www.wescomfg. com].)

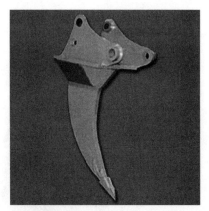

FIGURE 14.22   Generic backhoe spike.

you would purchase at the hardware store. These tools are generally nonsparking, mounted on longer poles with a pointed end.

CAUTION

THESE DEVICES ARE NO LONGER IN WIDESPREAD USE, DUE TO THEIR QUESTIONABLE SAFETY WHEN DEALING WITH FLAMMABLE, REACTIVE, SHOCK-SENSITIVE MATERIALS, AND OVERPRESSURIZED DRUMS.

Backhoe Spike.   A common means used to open drums remotely for sampling is a metal spike attached to a backhoe in place of the bucket. This method is very efficient and is often used in large-scale operations. The backhoe operator can be somewhat protected by using steel plate to armor the cab. This is an extremely versatile, mobile device, but it can be somewhat heavy-handed. Remotely operated drum-opening tools are the safest available means of drum opening. Remote drum opening is slow, but it provides a high degree of safety compared to manual methods of opening.

In the opening area, a single drum should be placed so as to allow sufficient space for backhoe maneuvering. Once staged, the drum can be quickly opened by punching a hole in the drum head or lid with the spike.

To prevent cross-contamination and adverse reaction from water-reactive or incompatible materials, the spike (Figure 14.22) should be decontaminated and dried after each drum is opened. Even though some splash or spray may occur when this method is used, the operator of the backhoe can be protected by a large shatter-resistant shield mounted in front of the operator's cage. This, combined with the normal personal protection gear, should be sufficient to protect the operator. Additional respiratory protection can be afforded by providing the operator with an onboard airline system.

14.15.10   Sampling of Hazardous Waste at the Site

The suspected type of waste, the wastes' state, and the type of analysis requested from the lab all play a role in how samples are taken, contained, preserved, and stored.

Select the appropriate sampling equipment based on the state of the material and the type of container. Sampling equipment must be made of nonreactive materials that will not alter the chemical or physical properties of the material that is to be sampled or react dangerously with the material.

Some samples, such as those that will be analyzed for VOCs, must be stored, cooled to 4°C, and protected from sunlight in order to minimize any potential reaction due to the light sensitivity of the sample. VOC sample bottles and jars are usually amber in color. They might also require the use of zero-headspace bottles. These are containers that have an inverted cone in the cap that displaces the liquid in the sample and allows for no air at the top of the sample (headspace). A small amount of spillage will occur with these types of containers.

Sample bottles for collection of waste liquids, sludges, or solids are typically wide-mouth amber glass jars with Teflon®-lined screw caps or Nalgene® bottles. Actual volume required for analysis should be determined in conjunction with the laboratory performing the analysis. Remember that when you are sampling for certain analytes, such as metals, the sample jar will come from the lab with some acid reagent inside. This is for fixing the sample. DO NOT POUR THIS LIQUID OUT OR SPILL IT ON YOUR HANDS OR CLOTHES. Its pH will be less than 2! The analytical lab will provide all sample containers, labels, chain-of-custody seals, and chain-of-custody forms. (See Figure 12.25.)

Waste sample handling procedures shall be as follows:

1. Label the sample container with the appropriate sample label and complete the appropriate field data sheet(s) or sample logs. Place sample container into two resealable plastic bags such as Zip-Loc® bags.

2. Place each bagged sample container into a shipping container that has been lined with plastic. Pack the container with enough noncombustible, absorbent cushioning material to minimize the possibility of containers breaking and to absorb any material that may leak.

3. Depending on the nature and quantity of the material to be shipped, different packaging may be required. The lab should be consulted prior to packing the samples.

4. Depending on the nature of the analyte, different packaging may be required. If the sample requires cooling, depending upon the distance to the lab, the samples might have to be shipped in a cooler packed in ice. Consult the analytical lab prior to packing the samples.

5. Complete a chain-of-custody record for each shipped sample container, place it into a resealable plastic bag, and affix it to the lid of the shipping container.

6. Secure and chain-of-custody seal the lid of the shipping container. Label the shipping container appropriately and arrange for the appropriate transportation mode consistent with the type of hazardous waste involved. The lab will usually pick up all samples at the site.

The following is a checklist of materials and equipment required for sampling:

❑ Personal protection equipment
❑ Wide-mouth amber glass or Nalgene® jars with Teflon® cap liner, approximately 500-mL volume or other appropriate types and sizes (such as zero-headspace 40 ml VOA (volatile organic analysis) vials) supplied by the lab
❑ Uniquely numbered sample identification labels with corresponding data sheets
❑ Drum/Tank Sampling Data Sheets and Field Test Data Sheets for Drum/Tank Sampling chain-of-custody records
❑ Decontamination materials
❑ Glass or plastic sample thief or coliwasa devices

❑ Coring device
❑ Stainless steel spatula or spoons
❑ Laser thermometer
❑ Drum overpacks
❑ Absorbent material for spills

Procedures for Drum Sampling Preparation.   Here is a checklist of arrangements that must be implemented prior to the excavation and sampling of drums.

❑ Determine the extent of the sampling effort, the sampling methods to be employed, and the types and amounts of equipment and supplies needed.
❑ Obtain necessary sampling *and* monitoring equipment.
❑ Decontaminate equipment and ensure that it is in working order.
❑ Prepare sampling schedule and coordinate with staff, clients, and regulatory agency, if appropriate.
❑ Perform a general site survey prior to site entry in accordance with the site-specific health and safety plan.
❑ Use stakes or flags to identify and mark all sampling locations. If required, the proposed locations may be adjusted based on site access, property boundaries, and surface obstructions.
❑ For field testing (characterization) of samples, a Field Test Data Sheet for Drum/Tank Sampling such as seen in Table 14.2 must be used.

All HAZWOPER workers must assume that unmarked drums contain hazardous materials until their contents have been categorized. Once a drum has been visually inspected and any immediate spill hazard has been eliminated by overpacking or transferring the drum's contents, the drum is affixed with an EPA hazardous waste label and transferred to a staging area. A description of each drum, its condition, any unusual markings, the location where it was buried or stored, and field monitoring information are recorded on a drum/tank sampling data sheet. *This data sheet becomes the principal record-keeping tool for tracking the drum on-site.*

After the drum has been opened, preliminary monitoring of headspace gases should be performed first with an LEL and oxygen meter. Afterwards, an organic vapor analyzer or other instruments should be used. If possible, these instruments should be intrinsically safe. In most cases it is impossible to observe the contents of these sealed or partially sealed drums. Since some layering or stratification is likely in any solution left undisturbed, a sample that represents the entire depth of the drum must be taken. When a previously sealed drum is sampled, a check should be made for the presence of a bottom sludge. This is easily accomplished by measuring the interior depth to apparent bottom, then comparing it to the known interior depth by measuring on the outside of the drum.

Drum Thief

O━π   **For sampling drums filled with liquid waste, the drum thief is the most common device used.**

The most widely used implement for sampling drums containing liquids is a tube commonly referred to as a drum thief (Figure 14.23). This tool is cost-effective, quick, and disposable. Thieves are typically 6 to 16 mm inside diameter and 42 in. long.

The drum thief is the ideal sampler to use to sample drums, transformers, and other liquid containers where small-volume, low-cost sampling is required. Each sampler is 42 in. in length and can be manufactured from chemical-resistant borosilicate glass, stainless steel,

TABLE 14.2   Field Test Data Sheet for Drum/Tank Sampling

Field Test Data Sheet for Drum/Tank Sampling

Sampler(s): _____ Date: _____

Site Name: _____

Work Order Number: _____

Container Number/Sample Number: _____

SAMPLE MONITORING INFORMATION:

1. PID: _____ Background (clean zone)

   _____ Probe used/Model used

   _____ Reading from sample

2. FID: _____ Background (clean zone)

   _____ Reading from sample

3. Radiation Meter: _____ Model used

   _____ Background (clean zone)

   _____ Reading from sample

4. LEL/Oxygen Meter:_____ Oxygen level (sample)

   _____ LEL level (sample)

SAMPLE DESCRIPTION: (Check or describe all appropriate)

_____ Liquid _____ Solid _____ Sludge _____ Color _____ Vapors

WATER REACTIVITY: (Check or describe all appropriate)

1. Add small amount of sample to water: _____ bubbles _____ color change to _____ vapor formation ___ heat _____ No Change

SPECIFIC GRAVITY TEST (compared to water):

1. Add small amount of sample to water: _____ sinks _____ floats

2. If liquid sample sinks, screen for chlorinated compounds. If liquid sample floats and appears to be oily, screen for PCBs (Chlor-N-Oil® kit).

Chlor-N-Oil® TEST KIT INFORMATION: (Check or describe all appropriate)

1. Test kit used for this sample: _____ Yes _____ No

2. Results: _____ PCB not present _____ PCB present, less than 50 ppm _____ PCB present, greater than 50 ppm _____ 100% PCB present

WATER SOLUBILITY TEST: (Check or describe all appropriate)

1. Add approximately one part sample to five parts water. You may need to stir and heat gently. **DO NOT HEAT IF WATER-REACTIVE!**

Results: _____ total _____ partial _____ no solubility

pH OF AQUEOUS SOLUTION:

1. Using 0–14 pH paper, check pH of water/sample solution: pH = _____.

TABLE 14.2   Field Test Data Sheet for Drum/Tank Sampling (*Continued*)

SPILL-FYTER® CHEMICAL CLASSIFIER STRIPS: (Check or describe all appropriate)

1. Acid/Base Risk: Color Change

_____ Strong acid (pH = 0) RED, _____ Moderately acidic (pH = 1–3) ORANGE, _____ Weak acid (pH = 5) YELLOW,

_____ Neutral (pH = 7) GREEN, _____ Moderately basic (pH = 9–11) Dark GREEN,

_____ Strong Base (pH = 13–14) Dark BLUE

2. Oxidizer Risk:

_____ Not Present WHITE, _____ Present BLUE, RED, OR ANY DIVERGENCE FROM WHITE

3. Fluoride Risk:

_____ Not Present PINK, _____ Present YELLOW

4. Petroleum Product, Organic Solvent Risk:

_____ Not Present LIGHT BLUE, _____ Present DARK BLUE

5. Halogens: Iodine, Bromine, and Chlorine Risk:

_____ Not Present PEACH, _____ Present WHITE OR YELLOW

SETAFLASH® IGNITABILITY TEST:

140°F Ignitable: _____ Non-Ignitable _____ °
160°F Ignitable: _____ Non-Ignitable _____ °

_____ Ignitable: _____ Non-Ignitable _____

_____ Ignitable: _____ Non-Ignitable _____

_____ Ignitable: _____ Non-Ignitable _____

_____ Ignitable: _____ Non-Ignitable _____

COMMENTS:

_____

_____

_____

_____

_____

_____

HAZCAT® KIT TESTS:

1. Test: _____

Outcome: _____

Comments: _____

_____

2. Test: _____

Outcome: _____

Comments: _____

_____

TABLE 14.2 Field Test Data Sheet for Drum/Tank Sampling (*Continued*)

3. Test: _____

Outcome: _____

Comments: _____

_____

4. Test: _____

Outcome: _____

Comments: _____

_____

5. Test: _____

Outcome: _____

Comments: _____

_____

HAZCAT® PESTICIDES KIT:

Present: _____ Not Present: _____

Comments: _____

_____

_____

FIGURE 14.23 Drum being sampled with a Conbar® 42-in. borosilicate glass drum thief. (*Courtesy of CONBAR® Environmental Products* [www.conbar.com].)

Teflon®, or HDPE plastic. For the most part, they are disposable. Decontaminating sample thieves generates another waste stream and if not performed properly may lead to sample contamination as well as chemical incompatibility problems.

Procedures for use:

- Remove the cover from the sample container.
- Insert the thief to the bottom of the drum or until a solid layer is encountered. About six inches of the thief will extend above the drum.
- Allow the liquid waste level in the drum to equilibrate in the thief.
- Cap the top of the thief with your thumb, ensuring liquid does not come into contact with your PPE if possible.
- Carefully remove the thief from the drum and insert the bottom end into the open sample container.
- Release your thumb from the top and allow the thief to drain until the container is approximately two-thirds full.
- Remove tube from the sample container.
- If disposable, insert the thief about halfway into the drum, break it, and place the pieces in the drum. If the thief is made of HDPE or Teflon®, place it in the drum at an angle and leave it in the drum.
- If reusable, place the thief in a container for decontamination.
- Cap the sample container tightly and clean off any spilled sample liquid. Place the sample container into a carrier.
- Replace the bung or otherwise re-seal the drum.
- Label the sample with a unique identifying number. Log all samples in the site logbook and on Drum/Tank Sampling Data Sheets.
- Perform hazard categorization analyses if included in the project scope.
- Transport the sample to the decontamination zone and package it for transport to the analytical laboratory, as necessary.
- Complete chain-of-custody records including chain-of-custody seals on sample container.

In many instances a drum containing waste material will have a sludge layer on the bottom. Slow insertion of the sample tube into this layer, then gradual withdrawal will allow the sludge to act as a bottom plug to maintain the fluid in the tube. The plug can be gently removed and placed into the sample container by the use of a stainless steel lab spoon.

Note

In some instances disposal of the tube by breaking it into the drum or otherwise disposing of it in the container may interfere with eventual plans for the removal of its contents. The use of this technique must be approved by the SSO, or other sampling methods must be adopted.

CAUTION

ALWAYS MAKE SURE TO RESEAL THE DRUM AFTER SAMPLING. OPEN CONTAINERS OF HAZARDOUS WASTE ARE SUBJECT TO HEAVY FINES FROM STATE AND FEDERAL AGENCIES. ALSO, ALWAYS WEAR PPE, INCLUDING CPC AND SAFETY GLASSES, WHEN SAMPLING.

The Composite Liquid Waste Sampler (Coliwasa).   The COLIWASA (Figure 14.24) is an extremely useful drum sampler designed to permit representative sampling of multiphase wastes from drums and other containerized wastes. One configuration consists of a 152 ×

FIGURE 14.24 A Conbar® Teflon® coli-wasa. (*Courtesy of CONBAR® Environmental Products* [www.conbar.com].)

4-cm inside diameter section of tubing with a neoprene stopper at one end attached by a rod running the length of the tube to a locking mechanism at the other end.

Manipulation of the locking mechanism opens and closes the sampler by raising and lowering the neoprene stopper.

The major drawbacks associated with using a Coliwasa concern decontamination and costs. The sampler is difficult to decontaminate in the field, and its high cost in relation to alternative procedures (drum thieves) make it an impractical throwaway item. It still has applications, however, especially in instances where a true representation of a multiphase waste is absolutely necessary.

Use the following checklist for properly using the coliwasa:

❑ Put the sampler in the open position by placing the stopper rod handle in the T-position and pushing the rod down until the handle sits against the sampler's locking block.

❑ SLOWLY lower the sampler into the liquid waste. Lower the sampler at a rate that permits the levels of the liquid inside and outside the Coliwasa to equilibrate. If the level of the liquid in the sample tube is lower than that outside the sampler, the sampling rate is too fast and will result in a nonrepresentative sample.

❑ When the sampler stopper hits the bottom of the waste container, push the sampler tube downward against the stopper to close the sampler. Lock the sampler in the closed position by turning the T-handle until it is upright and one end rests tightly on the locking block.

❑ Slowly withdraw the sample from the waste container with one hand while wiping the sampler tube with a disposable cloth or rag with the other hand.

❑ Carefully discharge the sample into the appropriate sample container by slowly pulling the lower end of the T-handle away from the locking block while the lower end of the sampler is positioned in a sample container.

❑ Cap the sample container tightly and label it. Place the sample container in a carrier.

❑ Replace the bung or otherwise seal the drum.

❑ Log all samples in the site log book and on Drum/Tank Sampling Data Sheets.

❑ Perform hazard categorization analyses if included in the project scope.

❑ Transport the sample to the decontamination zone and package it for transport to the analytical laboratory, as necessary.

❑ Complete chain-of-custody records including chain-of-custody seals.

Coring Device for Solids Sampling from Drums.   A coring device must be used to sample drum solids. Samples should be taken from different areas within the drum. Remember that solids are supposed to be stored in open-mouth drums. This is not always the case. If you sample a bung-type drum containing solids with a coring device, make sure to note this discrepancy on your sample sheet. This sampler consists of a series of extensions, a T-handle, and the coring device.

Use the following checklist for properly using the coring device.

❑ Assemble the sampling equipment.

❑ Remove the cover from the sample container.

❑ Insert the sampling device to the bottom of the drum. The extensions and the T-handle should extend above the drum.

❑ Rotate the sampling device to cut a core of material.

❑ Slowly withdraw the sampling device so that as much sample material as possible is retained within it.

❑ Transfer the sample to the appropriate sample container and label it. A stainless steel spoon or scoop may be used as necessary.

❑ Cap the sample container tightly and place it in a carrier.

❑ Replace the bung or otherwise seal the drum.

❑ Log all samples in the site logbook and on Drum/Tank Sampling Data Sheets.

❑ Perform hazard categorization analyses if included in the project scope.

❑ Transport the sample to the decontamination zone for additional field analysis; or package it for transport to the analytical laboratory, according to project protocol. Complete the chain-of-custody records including chain-of-custody seals.

14.15.11 Sampling Other Site Media

Equipment.   Selecting appropriate equipment to sample wastes is a challenging task due the uncertainty of the physical characteristics and nature of the wastes. It may be difficult to separate, homogenize, and containerize a waste due to its physical characteristics (viscosity or particle size). In addition, the physical characteristic of a waste may change with temperature, humidity, or pressure. Waste streams may vary depending on how and when a waste was generated, how and where it was stored or disposed of, and the conditions under which it was stored or disposed of. Also, the physical location of the wastes or the unit configuration may prevent the use of conventional sampling equipment.

Given the uncertainties that a waste may present, it is desirable to select sampling equipment that will facilitate the collection of samples while meeting the sampling protocol. It is important to not unintentionally bias the sample by excluding some of the sample population that is under consideration. However, due to the nature of some waste matrices or the physical constraints of the waste unit, it may be necessary to collect samples knowing that a portion of the desired media was omitted due to limitations of the equipment.

Any deviations from the sampling plan or difficulties encountered in the field concerning sample collection should be documented in a logbook, reviewed with the analytical data, and presented in the report. If the sampling event does not fall within the client's acceptable deviation, other methods will have to be developed and employed.

Other Waste Sampling Procedures

○━━π **Drums and containers are not the only types of material sampling a HAZWOPER worker might be required to perform. Large waste piles or pits, soil, tanks, surface impoundments or ponds, equipment that has been deconned, and monitoring wells may also have to be physically sampled.**

Waste sampling equipment must be made of nonreactive materials that will neither add to nor alter the chemical or physical properties of the material that is being sampled.

Waste Piles. Waste piles vary in size, shape, composition, and compactness and may vary in distribution of hazardous constituents and characteristics (strata). These variables will affect safety and access considerations. The number of samples, the type of sample(s), and the sample location(s) should be based on the site sampling plan. All equipment must be compatible with the waste and must have been cleaned to prevent any cross-contamination of the sample.

Decisions on how to sample waste piles must be made by the site's Program Manager or the Site Safety Officer. Their decision must be based on widely accepted sampling strategies and requirements from regulators. Be aware of "bridging" in waste piles that could lead to entrapment. If it can be avoided, do not climb on waste piles.

Surface Impoundments. Surface impoundments vary in size, shape, and waste content and may vary in distribution of hazardous constituents and characteristics such as layers depending on specific gravity. The number of samples, the type of sample(s), and the sample location(s) must be based on the sampling plan. Because of the potential danger of sampling waste units suspected of containing elevated levels of hazardous constituents, personnel must never attempt to sample surface impoundments used to manage potentially hazardous wastes from a boat. All sampling must be conducted from the banks, catwalks, or piers.

Stratified waste impoundments can be sampled using instruments discussed in Section 12.4 of this book. Liquid grab samples of homogenous wastes can be taken with a plastic beaker or other clean container attached to a pole.

Tanks. Sampling of tanks or vessels is considered hazardous due to the potential for them to contain large volumes of hazardous materials. Appropriate safety measures must be taken. Unlike drums, tanks may be compartmented or have complex designs. They may have multiple inlet, drain, and vent lines. They may have multiple access openings. Preliminary information about the tank's contents and configuration should be reviewed prior to the sampling operation to ensure the safety of sampling personnel and that the tank can be sampled adequately to determine contents. In addition to having discharge valves near the bottom of tanks and bulk storage units, most tanks have hatches at the top. It is desirable to collect samples from the top hatch because of the potential for the tank's contents to be stratified (such as water below the oil in a waste oil tank). Additionally, when you are sampling from the discharge valve, there is a possibility of the valve becoming stuck or broken in the open position. This situation could result in an uncontrolled release.

HAZWOPER personnel should not utilize valves on tanks or bulk storage devices unless they are operated by the owner or operator of the facility or a containment plan is in place should the valve stick open or break. If the HAZWOPER worker must sample from a tank discharge valve, the valving and piping arrangement of the particular tank must be clearly understood to ensure that all compartments are sampled.

Because of the many different types of designs and materials that may be encountered, Level B protection is required for the following procedures.

If sampling requires physical entry into the tank, ensure that all precautions for opening and inspecting a permit confined space are followed (see Section 10 for details). Use this checklist to sample tanks safely.

❑ Record information in a logbook concerning the tank, such as the type of tank, the tank capacity, markings, condition, and suspected contents.

❑ The samplers should inspect the ladder, stairs, catwalk, and any other parts of the tank that will be used to access the top hatch to ensure that they will support the samplers and their equipment.

❑ Before opening, ground each metal tank using grounding wires, clamps or clips, and a grounding rod or metal structure. To provide an adequate electrical bond, make sure to remove the paint on painted structures where grounding clips are attached.

❑ Any vents or pressure release valves should be opened slowly to allow the unit to vent to atmospheric pressure. Air monitoring for flammable gases and toxic vapors must be conducted during the venting, with the results recorded in a logbook. If dangerous concentrations of gases evolve from the vent or the pressure is too great, leave the area immediately.

❑ Touch the tank opening equipment to the bolts in the hatch lid and allow an electrical conductive path to form. Slowly remove bolts or open hatch with spark-resistant tools. If a pressure buildup is encountered or detected, cease opening activities and leave the area.

❑ Monitor the vapors from tanks for flammable gases and toxic vapors with air monitoring instruments.

❑ Note the state, quantity, number of phases, and color of the tank contents. Record all data, observations, and information in a logbook. Compare the screening results with any preexisting data to determine if the tank should be sampled.

❑ Select the appropriate sampling equipment based on the state of the material and the type of tank.

❑ Place sampling equipment and sample containers near tanks(s) to be sampled.

❑ To sample the tank for liquids, lower the bailer, sample bomb, Kabis® sampler (see Section 12), coliwasa, thief, or Teflon® or Tygon® tubing to the desired sampling depth. When sampling in areas where explosive or flammable atmospheres could occur, electric pumps are not permitted. Use an air-powered pump instead.

❑ Close the sampling device or create a vacuum and slowly remove the sampling device from the tank. Release the sample from the device into the sample container(s). Repeat the procedure until a sufficient sample volume is obtained.

❑ To sample the tank for solids or sludges use a hollow push tube or screw auger. Carefully remove the sample from the sampling device using a clean stainless steel spoon to place the sample into containers for analyses.

❑ Close the tank when sampling is complete.

❑ Label the sample container and log the sampling conditions.

❑ Segregate contaminated sampling equipment and sample containers contaminated by incompatible materials for decontamination. At a minimum, contaminated equipment should be cleaned with laboratory detergent and rinsed with tap water prior to returning it from the field.

Sampling of Contaminated Equipment or Structural Surfaces. Wipe samples can be taken on nonabsorbent surfaces such as concrete, metal, glass, and plastic. The wipe materials must be compatible with the solvent used and the analyses to be performed and should not come apart during use. The contract analytical lab will supply containers and advice on sampling media for wipe samples. Wipe samples shall not be collected for volatile organic compounds analysis. Due to their easy evaporation, the samples will not accurately reflect the level of contamination. VOC contamination detection will require the use of instrumentation (see Section 12).

Sampling personnel must be aware of hazards associated with the selected solvent and use appropriate PPE to prevent any skin contact or inhalation of these solvents. Further, outside gloves that touch the sample media must be discarded and replaced prior to each successive sample being taken, to prevent sample cross-contamination.

Typically, one square foot samples are taken. This can be accomplished in one of two ways. A template can be constructed of aluminum (or some other easily decontaminatable material) and placed on the surface. The alternative method is to use duct tape to mark a square foot area on the surface. The template has the advantage of easy use coupled with the disadvantage of requiring decontamination. The tape method requires no decontamination but is more laborious and possibly generates contaminated waste.

The prepared wipe, often simple "baby wipes," is removed from its container and used to wipe the entire area with firm strokes, using only one side of the wipe. The wipe is then placed into the sample container. This procedure is repeated until the area is free of visible contamination. Care should be taken to keep the sample container tightly sealed to prevent evaporation of the solvent. Samplers must also take care not to touch the used side of the wipe. Make sure to take the required number of field blanks (usually 1 per every 10 actual samples). A field blank, in this case, would be a baby wipe removed from its container and placed directly into a sample container.

Quality Assurance/Quality Control (QA/QC). All sampling, monitoring, data collection, and data validation will be in conformance with the site's QA/QC plan.

All data must be documented on chain-of-custody records, on Drum/Tank Sampling Data Sheets, on Field Test Data Sheets for Drum and Tank Sampling, or within site logbooks.

All instrumentation must be operated in accordance with operating instructions as supplied by the manufacturer, unless otherwise specified in the work plan. Equipment checkout and calibration activities must occur prior to sampling operation, and they must be documented. All instruments must also be postcalibrated after the event and the results documented.

Health and Safety. When working with potentially hazardous materials, HAZWOPER workers must follow OSHA U.S. EPA, and the organization's health and safety procedures. Section 16 provides much greater detail as to how to perform these tasks safely.

More specifically, the opening of closed containers is one of the most hazardous site activities. Maximum efforts should be made to ensure the safety of the sampling team. Proper protective equipment and a general awareness of the possible dangers will minimize the risk inherent to sampling operations. Employing proper drum opening techniques and equipment will also safeguard personnel. The use of remote sampling equipment whenever feasible is highly recommended.

Hazard Categorization

> **The goal of characterizing or categorizing the contents of drums is to obtain a quick preliminary assessment of any potential physical hazards that the drums' contents pose, such as toxicity, corrosivity, and flammability.**

The goal of characterizing or categorizing the contents of drums is to obtain a quick preliminary assessment of any potential physical hazard that the drums' contents pose, such as toxicity, corrosivity and flammability. You also want to characterize the types and levels of pollutants contained in the drums.

These activities generally involve rapid methods of analysis. The data obtained from these methods can be used to make decisions regarding drum staging or restaging, consolidation or compositing of the drum contents, and proper labeling and containerizing of drums for shipment off-site (see Table 14.2).

As a first step in obtaining these data, standard tests should be used to classify the drum contents into general categories such as toxics, reactives, water reactives, flammables, inorganic acids, organic acids, alkalis, heavy metals, pesticides, cyanides, inorganic oxidizers, and organic oxidizers. In some cases, further analyses should be conducted to identify the drum contents more precisely. For instance, most tests for reactivity and human health hazards must be conducted in a lab.

Several methods are available to perform these tests:

- pH of any aqueous solution (test strips or pH meter)
- The HazCat® chemical identification system
- The Chlor-N-Oil® test kit (for PCBs)
- Spill-fyter® chemical classifier strips
- Setaflash® closed-cup flash tester, for ignitability
- Water reactivity using a small amount of the waste (be careful with this test!)
- Specific gravity test (density compared to water)
- Water solubility test

The tests must be performed in accordance with the instructions. Use Field Test Data Sheets for Drum and Tank Sampling. Results of the tests must be documented on these data sheets.

In May of 2001, EPA finalized its Hazardous Waste Identification Rule (HWIR). The new regulation makes the following changes:

1. EPA has finalized the retention of the hazardous waste mixture and derived-from rules, a key part of the Resource Conservation and Recovery Act (RCRA) rules that define when wastes are regulated as hazardous wastes.

2. Under the mixture rule, a listed hazardous waste remains regulated as a hazardous waste when it is mixed with a nonhazardous waste.

3. Under the derived-from rule, waste generated from the treatment, storage, or disposal of a listed hazardous waste also remains regulated as a hazardous waste.

4. In addition, EPA has finalized two new exemptions that narrow the scope of these rules, tailoring them to match the risks posed by particular wastes:

   - An exemption for wastes listed solely for ignitability, corrosivity, and/or reactivity characteristics. Currently, mixtures of such wastes that are decharacterized are eligible for exemption. The revised rule extends this exemption to include all such wastes. These wastes are still subject to any applicable land disposal restriction requirements.
   - A conditional exemption for mixed wastes (that is, wastes that are both hazardous and radioactive) when they meet certain conditions and eligibility criteria.

In response to public comments, EPA has clarified that land disposal restrictions (LDRs) affect the exempted waste only where they are otherwise applicable under current regulation. (In other words, if a waste is not subject to LDRs under the current regulations, it will not be subject under the revised regulations). EPA has also moved the regulatory language for the conditional exemption for mixed waste to its own subsection (40 CFR 261.3(h)) in order to make it clearer.

Investigation-Derived Wastes (IDW).   IDWs are generated as a result of well installation, sampling and monitoring activities, and groundwater characterization. Routinely generated IDW consist of well water and soil cuttings (spoil) from well installation and development, monitoring, pump testing, and sampling activities. Personal protective equipment (PPE), disposable sampling equipment, field test kits, excess reagents, and sample bottles are also generated. Spills that occur during the depressurization of a 55-gallon drum would also be considered IDW.

How are these wastes handled? Under RCRA, EPA's guidance describes the allowable disposal of IDW within an area of contamination (AOC) as follows: "Storing IDW in a container within the AOC and then returning it to its source is allowable without meeting the specified Land Disposal Restrictions treatment standards. . . . Therefore, returning IDW that has been stored in containers within the AOC to its source does not constitute land disposal as long as containers are not managed in such a manner as to constitute a RCRA storage unit as defined in 40 CFR 260.10. In addition, sampling and direct replacement of waste within an AOC do not constitute land disposal."

That is for a Superfund site. In plain English, it says that if you sample soil or water and test it in place, you can pour it back. Now also, there is no federal requirement to manage well cuttings as hazardous waste.

For brownfields and other site investigations, regulations may apply to the storage and disposal of IDWs. If they are groundwater or surface water discharges, they may come under the CWA, for instance. Always determine IDW disposal procedures from regulators prior to disposal.

14.16 PACKAGING HAZARDOUS WASTE FOR SHIPMENT TO A TSDF

> All containers of hazardous waste stored on-site or shipped off-site must have properly filled in hazardous waste labels and must be included on a properly filled out uniform hazardous waste manifest.

One of the main tasks of a HAZWOPER worker will be preparing waste for shipment. One procedure that can be very helpful in reducing the number of containers that must be stored and shipped is consolidation or bulking. This practice is recommended. However, it is permitted only after a thorough characterization of the materials has been completed and the wastes are found to be compatible (see Section 4).

For transportation safety reasons, as well as economic ones, whenever possible, only full (not more than 2 in. from the top) drums shall be shipped. This eliminates most of the movement of the wastes in the drum. If solids are shipped and the drum is not full, the drum shall be topped-off with an inert material, such as cat litter, that is compatible with the waste.

Compatible liquids can often be consolidated into a tanker truck. This is very common when the material is being shipped to a recycling facility.

Liquids shall not be disposed of in a landfill. If liquid wastes can only be disposed of in this fashion, then they must first be solidified. Several products on the market are compatible with liquid wastes. Upon mixing, the liquid waste solidifies like thick gelatin.

Each employee involved in the packaging and shipment of hazardous materials must be adequately trained in the HAZWOPER regulations, the EPA, and the DOT regulations concerning hazardous materials, substances, and wastes. That reference material can be found in Sections 1 and 3 of this book. The DOT requires training every three years.

HAZWOPER workers involved in packaging hazardous waste for shipment must be able to determine the proper container type, labels required and information contained on those labels, and the proper completion of the Uniform Hazardous Waste Manifest. (See Figure 2.1 of this book.)

All hazardous materials or waste must be correctly identified on the shipping paper. The DOT requires that all materials be properly identified according to descriptions from their Hazardous Materials Table (49 CFR 172) (see www.hazmat.dot.gov/rules.html). Unless the actual material is identified, the material must be shipped in the hazard class in which it belongs. At a minimum, shipping papers or manifests must include:

1. Shipper's name and address
2. Consignee's name and address
3. Basic description: shipping name, hazard class, ID number, packing group—in that exact order
4. Weights and volume
5. Total quantity, type and kind of package

6. Emergency response information

7. Twenty-four-hour emergency response telephone number

8. Shipper's declaration, stating that everything meets DOT requirements

9. Page numbers

10. Title and signatures

The following information must accompany the shipping paper:

1. Basic description and chemical names

2. Immediate health hazards

3. Risks of fire or explosion

4. Immediate precautions to take in the event of an accident or incident

5. Immediate methods for handling spills or leaks

6. Preliminary first aid measures

This information can be provided by attaching an MSDS or ensuring that the transporter has a copy of the *Emergency Response Guidebook* (included in the Companion CD).

Many facilities use the Chemtrec® phone number as the emergency response contact.

CAUTION

THIS PHONE NUMBER IS ALLOWED ONLY TO BE USED IF THE OR-GANIZATION IS REGISTERED WITH CHEMTREC®. SEE www.chemtrec.org, WHERE YOU CAN DOWNLOAD THE APPLICATION FOR REGISTRATION WITH CHEMTREC®. THE FEE FOR ALL SERVICES IS $500.00 PER YEAR.

Managers must ensure that the phone number on the manifest will enable emergency responders to access someone immediately who has a good understanding of the hazardous material being shipped and emergency procedures. Further, they must ensure that all containers meet the DOT's performance-oriented packaging standards (POPs). Packaging must be embossed or labeled with the proper UN code according to what is being shipped. Packaging must be labeled with an approved hazard class label in the form of a minimum 4 $\times$ 4 in. DOT placard label. The contents label must include the material's proper shipping name and ID number and the shipper's and consignee's name and address.

Remember, any hazardous materials shipment weighing more than 1,000 pounds in aggregate weight may only be hauled by a driver with a CDL license that contains a hazardous materials endorsement. The vehicle must be placarded on all four sides identifying the hazardous materials being carried.

For more information on labeling and placarding, see Section 8.

14.17   *PROJECT DESIGN AND ENGINEERING*

After the site ownership has been determined, the site has been characterized as to the types and extent of contamination, and the funding for remediation is available, the project is ready for design and engineering. As you have read in other sections, there are literally thousands of hazardous materials that can contaminate the air, soil, surface water, or groundwater. Often, however, it is the hazardous constituents of materials not considered hazardous that must be cleaned up. Many petroleum products themselves are exempt from a majority of hazardous materials regulations. Once released to the environment, however, they constitute a large part of the remediation projects that must take place. The authors want to make the reader aware of some of the common technologies being used and some of the new or innovative technologies that are emerging. HAZWOPER workers might be employed in the construction and operation of any of these technologies.

Aside from cleaning up the contaminated site for reuse, it is most important that the remediation be designed and engineered to avoid any off-site migration of contaminants. We will also discuss technologies used for that purpose. Our examples concentrate on on-site or *in situ* remediation. When extremely large areas of land or water are contaminated, these are often the only feasible methods. Just the cost of excavating hundreds of thousands of cubic yards of soil or pumping and holding millions of gallons of water would sometimes be hundreds of times more than *in situ* treatment costs. And that does not even begin to include the actual treatment and disposal costs. Some off-site or *ex situ* methods will also be discussed.

HAZWOPER certified engineers will have input into the design and engineering phase of the project. HAZWOPER workers will be required for the construction phase, O&M, and routine monitoring.

14.18 COMMON TECHNOLOGY DESCRIPTIONS

As a 40-hour HAZWOPER certified worker, you may be involved in numerous tasks at a remediation site. If you have extra credentials, such as a forklift or backhoe operator's license, that may be your primary job. Sometimes the job can be as mundane as shoveling sludge or contaminated soils into drums for shipping off-site. At other times, HAZWOPER work may become quite complex.

Treatment technologies can be described as either active or passive. Active treatment includes pump-and-treat, excavation and solidification, and air sparging/soil vapor extraction. The active technologies require equipment and an outside energy source for their operation. Figure 14.25 illustrates the relative percentages of various treatment technologies that have been used in Superfund remedial actions. The figure also presents the relative usages of eight innovative treatment technologies. Hundreds of others not shown have been tried experimentally.

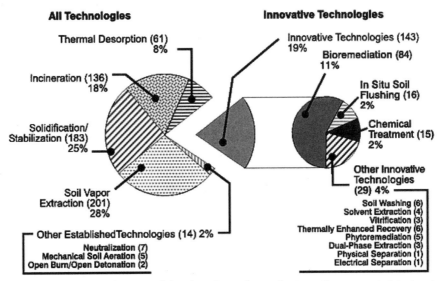

FIGURE 14.25 Superfund remedial actions: innovative applications of source control treatment technologies (FY 1982–FY 1999). (*Courtesy of the USEPA.*)

Passive technologies are those that either are constructed and allowed to take their course or rely upon natural attenuation. Passive remediation frequently includes some form of bio-remediation, phytoremediation, or permeable reactive barriers, among other options.

Years ago the only acceptable way to clean up a site was to remove all contaminated material. We quickly learned, however, that much of the expense of this type of remediation was in the packaging and transportation of the material to a treatment, storage, and disposal facility (TSDF). Then engineers and scientists started working on *in situ* remediation, which is now used in the vast majority of all cleanups taking place in the United States. The common types of *in situ* remediation include:

• Air sparging and soil vapor extraction
• Solidification and stabilization
• Incineration
• Thermal desorption
• Pump-and-treat

14.18.1   Air Sparging and Soil Vapor Extraction

Air sparging is a remedial technology that reduces concentrations of volatile constituents of petroleum products, solvents, and other typical VOCs that are absorbed and adsorbed in soils and dissolved in groundwater. This technology, also known as *in situ* air stripping and *in situ* volatilization, involves the injection, through air sparging wells, of contaminant-free air into the subsurface saturated zone, causing a phase transfer of hydrocarbons from a dissolved, absorbed, or adsorbed state to a vapor phase. The sparging wells are different from simple injection wells. Air injection wells simply blow air out of the bottom of the well casing. Sparging wells blow air out of a slotted length of the well casing in the saturated zone. This allows greater air contact with the contaminated media (soil or water). The air is then vented through other wells to equipment that removes the contaminants before discharging the clean air either to the atmosphere or back to the sparging wells.

To demonstrate air sparging, pour a glass of any carbonated beverage into a glass, place a straw into the liquid, and blow. Carbon dioxide will be liberated from the beverage along with the air that is blown into the liquid. This simulates using an air injection well for subsurface extraction of vaporizable component from water. If you were to put a great number of pinholes in the straw, up to the level of the liquid, put the end of the straw flat on the bottom of the glass, and then blow, you would see numerous small bubbles appear. You would also notice that the beverage became "flat" more quickly. This would simulate the advantage of the use of slotted wells for air sparging.

Air sparging is most often used together with soil vapor extraction (SVE), but it can also be used with other remedial technologies. When air sparging (AS) is combined with SVE, the SVE system creates a negative pressure in the unsaturated zone through a series of extraction wells to control the vapor plume migration. This combined system is called AS/SVE.

When used properly, air sparging has been found to be effective in reducing concentrations of volatile organic compounds (VOCs) found in petroleum product releases at underground storage tank (UST) sites. Air sparging is generally more applicable to the lighter gasoline constituents, such as benzene, ethylbenzene, toluene, and xylene (BTEX), because they readily transfer from the dissolved to the vapor phase. Air sparging is less applicable to diesel fuel and kerosene, the heavy-end distillates. Appropriate use of air sparging may require that it be combined with other remedial methods. For example, SVE or pump-and-treat is also valuable in feeding oxygen to aerobic microorganisms that feed on the contaminants. It can be used to remove contaminants from groundwater or from soil. An air sparging system can use either vertical or horizontal sparge wells. Well orientation is based on site-specific needs and conditions.

Air sparging should NOT be used by itself if the following site conditions exist.

- Where free product is present, such as a subsurface pool that is chiefly diesel fuel with only minor water content. Then air sparging can create groundwater mounding (actually building a hill of liquid), which could potentially cause free product to migrate and contamination to spread.
- Where there are nearby basements, sewers, or other subsurface spaces. Potentially dangerous vapor concentrations could accumulate in basements unless a highly effective vapor extraction system is used to control vapor migration.
- Where contaminated groundwater is located in a confined aquifer system. Air sparging cannot be used to treat groundwater in a confined aquifer because the injected air would be trapped by the saturated confining layer and could not escape to the unsaturated zone.

AS/SVE has also been used to create an 'air wall' on the periphery of a site to collect contaminants and avoid off-site migration.

Soil vapor extraction (SVE), also known as soil venting or vacuum extraction, is the most frequently used *in situ* remedial technology. It reduces concentrations of volatile constituents in released petroleum products adsorbed or absorbed on soils in the unsaturated (vadose) zone. (The vadose zone is between the earth's surface and the top of the groundwater and is a zone not saturated with water.)

SVE requires drilling extraction wells within the polluted area. These wells are drilled into the soil, but not the groundwater. Attached to the wells are pipes and blowers that create a vacuum. They draw air and vapors through the soil and up to the surface, where the vapors are captured or destroyed.

The number of air sparging and extraction wells can range from one to hundreds, depending on the size of the polluted area.

In SVE, a vacuum is applied to the extracting wells near the source of contamination in the soil. Volatile constituents of the contaminant mass evaporate and the vapors are drawn toward the extraction wells. The extracted air-vapor mixture is then treated before being released to the atmosphere. Common treatment methods include carbon adsorption and thermal oxidation. You will commonly see SVE systems at work at the periphery of landfills. They are attached to candle flares that burn off excess methane gas not collected in a landfill gas extraction system. This keeps fugitive methane from contaminating the atmosphere with odors and depleting the ozone layer. As it is burned off, the only byproducts (for pure methane) are CO_2 and water. The increased airflow through the subsurface can also stimulate aerobic biodegradation of some of the contaminants, especially those that are less volatile. Wells may be either vertical or horizontal. In areas of high groundwater levels, water table depression pumps may be required to offset the effect of upwelling or mounding induced by the vacuum. Water table depression pumps, like all liquid extraction pumps, have a design radius of influence (see below, Design Radius of Influence). These are wells such as municipal water supply wells. As they draw water out of the groundwater, the water level drops with proximity to the well in the form of a concentric flared cone. By drawing the groundwater in the area down by water table depression pumping, the SVE system can operate without affecting groundwater.

Effectiveness of SVE. This technology has been proven effective in reducing concentrations of volatile organic compounds (VOCs) and certain semivolatile organic compounds (SVOCs) found in petroleum products at UST sites. As indicated previously, SVE is generally more successful when applied to the lighter (more volatile) petroleum products such as gasoline. Diesel fuel, heating oils, and kerosene, which are less volatile than gasoline, are not readily removed by SVE, nor are lubricating oils and greases, which are nonvolatile. Because almost all petroleum products are biodegradable to a certain degree, these heavier petroleum products may be suitable for removal by bioventing. Injection of heated air, or other subsurface heating methods, also can be used to enhance the volatility of these heavier petroleum products because vapor pressure and rate of evaporation increase with temperature. However, energy requirements for volatility enhancement may be so large as to be economically prohibitive.

SVE is generally not appropriate for sites with a groundwater table located less than three feet below the land surface. Special considerations must be taken for sites with a groundwater table located less than 10 feet below the land surface because groundwater upwelling can occur within SVE wells under vacuum conditions, potentially occluding (clogging) well screens and reducing or eliminating vacuum-induced soil vapor flow.

SVE may also be appropriate near a building foundation to prevent vapor migration into the building. Here, the primary goal might be to control vapor migration and not necessarily to remediate soil. Here also a vacuum is applied to the contaminated soil matrix through extraction wells. This causes movement of vapors away from the foundation and toward these wells. The extracted vapors are then treated, as previously described.

Some of the factors that determine the effectiveness of SVE are:

- Permeability of the soil
- Soil structure and stratification
- Soil moisture
- Depth to groundwater

The permeability of the soil affects the rate of air and vapor movement through the soil; the higher the permeability of the soil, the faster the movement and (ideally) the greater the rate of vapor extraction.

Soil morphology (what the soil consists of) and stratification are important to SVE effectiveness because they can affect how and where soil vapors will flow within the soil matrix under extraction conditions. Structural characteristics such as clay lenses, layering of nonporous rock, and rock and hard earth fractures can result in preferentially unfavorable vapor flow behavior. This can lead to ineffective remediation or significantly extended remedial times.

High moisture content in soils can reduce soil permeability and, consequently, the effectiveness of SVE by restricting the flow of air through soil pores. In that case as air does flow, the SVE system is collecting chiefly water vapor. In coarse-grained soils (macropores), SVE works best because there is less air-vapor flow resistance. In medium-grained soils (mesopores), SVE's effectiveness begins to suffer. In fine-grained soils (micropores), SVE effectiveness suffers even more because of two factors. First, there is extremely high air-vapor flow resistance due to the tightly packed, fine soils. Second, in the soil there is a condition known as capillary fringe. This means that water raises through fine-grained packed soil due to capillary action. When SVE's vacuum is added, groundwater might be sucked up into the system. The result is that the SVE system is removing water vapor instead of the soil contaminants.

Design Radius of Influence (ROI).   This is the most important parameter to be considered in the design of an SVE system. The ROI is defined as the greatest distance from an extraction well at which a sufficient vacuum can be created and vapor flow induced, that is, in which local vacuum conditions are strong enough to enhance volatilization and extraction of the contaminants from the soil adequately. Extraction wells must be placed so that the overlap in their radii of influence completely covers the area of contamination.

Fluctuations in the groundwater table must also be considered when designing an SVE system. Significant seasonal or daily (tidal or precipitation-related) fluctuations might, at times, submerge some of the contaminated soil or a portion of the extraction well screen, making it unavailable for airflow. Avoiding this is most important for horizontal extraction wells, where the screen is parallel to the water table surface.

Surface Extraction Well Seals.   Surface seals might be included in an SVE system design to prevent surface water infiltration into the well. Seals can reduce the required air flow rates, reduce emissions of fugitive vapors, prevent vertical short-circuiting of air flow, and increase the design ROI. These results are accomplished because surface seals force air to be drawn from a greater distance from the extraction well. When surface seals are used, a higher vacuum might need to be applied to the extraction well, but it would be much more effective in vapor extraction.

Importance of Pilot SVE Studies.   Pilot studies are an extremely important part of the SVE design phase. Pilot studies can provide information on the concentration of volatile organic compounds (VOCs) that are likely to be extracted during the early stages of operation of the SVE system. A pilot test is recommended for evaluating SVE effectiveness and design parameters for any site, especially where SVE is expected to be only marginally to moderately effective. Pilot studies typically include short-term (1 to 30 days) extraction of soil vapors from a single extraction well, which may be an existing monitoring well at the site. However, longer pilot studies (up to six months) utilizing more than one extraction well may be appropriate for larger sites. Different extraction rates and wellhead vacuums are applied to the extraction wells to determine the optimal operating conditions.

Vapor concentrations are also measured at intervals during the pilot study to estimate initial vapor concentrations of a full-scale system. The vapor concentration, vapor extraction rate, and vacuum data are also used in the design process to select extraction and treatment equipment.

In some instances, it may be appropriate to evaluate the potential of SVE effectiveness using a computer screening model such as the EPA's HyperVentilate (see www.epa.gov/swerust1/cat/hyperven.html). HyperVentilate can be used to identify required site data, decide if SVE is appropriate at a site, evaluate air permeability tests, and estimate the minimum number of wells needed. It is not intended to be a detailed SVE predictive modeling or design tool.

14.18.2 Solidification and Stabilization

Solidification and stabilization are commonly occurring methods of treatment for hazardous waste solids at Superfund sites. They are ways of treating contaminated soils either *in situ* or *ex situ* so that the contaminants are bound in the media and cannot migrate. It involves adding binders to the soil that make the contaminants of concern stick to the soil and not migrate with any leachate.

Solidification refers to techniques that encapsulate the waste, forming a solid material. The treatment does not necessarily involve a chemical interaction between the contaminants and the solidifying additives. The most commonly used binder is cement.

Stabilization refers to techniques that chemically reduce the hazard potential of a waste by converting the contaminants into less soluble, less mobile, or less toxic forms. Stabilization usually involves mixing a contaminated medium, such as soil or sludge, with agents such as portland cement, lime, fly ash, cement kiln dust, or polymers to create a slurry, paste, or other semiliquid state. The mixture then is allowed time to cure into a solid form. The stabilization process also may include the addition of iron salts, silicates, clays, or pH adjustment agents. These can enhance the setting or curing time, increase the compressive strength of the stabilized waste, or reduce the leachability of contaminants. Other binders and reagents used include phosphates, lime, asphalt, sulfur, and polymers.

Solidification and stabilization are often used when the contaminant is considered *persistent,* such as heavy metals. Persistent contaminants are those that are not easily volatilized, or otherwise released from soil.

These treatments have the advantage of being quick and relatively cheap. Their main disadvantage is that the contamination, although stabilized, remains on the site.

Although widely used, these treatments do not offer a permanent solution to contamination at the site. Their choice may lead to deed and use restrictions for the property.

14.18.3 Incineration

Incineration, also called soil roasting, is commonly performed both *in situ* and *ex situ*. *Ex situ* costs might be more competitive at less than 5,000 cubic yards incinerated. At about 15,000 cubic yards or greater, *in situ* incineration is favored. What we mean by *in situ* is that a mobile incinerator is brought to the site. Contaminated soil must be excavated and fed through an incinerator that heats it to 1,600 to 2,200°F (870 to 1,200°C). Within this temperature range and with a residence time of several seconds, all petroleum hydrocarbons, and indeed most organic chemicals, including PCBs, are destroyed.

Cement calciners operating under these conditions have been used as incinerators for concentrated hazardous organic materials such as PCBs when cement product contamination does not occur.

The two main problems with incineration are that heating the soil to that temperature requires an extremely large amount of fuel and requires the excavation of the soil. Another disadvantage is that all organic matter that might be desirable in the treated soil is destroyed. An advantage is that heavily petroleum-contaminated soil can often add to the efficiency of the incinerator, the contamination becoming part of the fuel.

Incineration might not remove all heavy metals, with the exception of mercury and lead, from soils. When mercury and lead are present, the exhaust of the incinerator must be captured and these metals removed. There are also air-permitting requirements whenever an incinerator is in operation. Incineration has the advantage of destroying essentially all organic chemical contaminants and leaving clean, mineral soil.

Properly operating incinerators are those that provide at least 99.99% destruction and removal efficiency (DRE). Most exhaust streams will require scrubbing or filtration. Incineration is used about twice as often as thermal desorption.

14.18.4 Thermal Desorption

Thermal desorption temperatures range between 200°F and 1,000°F. Thermal desorption removes some contaminants from soil and other materials (like sludge and sediment) by using heat to *vaporize* the contaminants. These vapors can be collected with the same type of equipment as used in SVE. Particulates are separated from the vapors and disposed of safely. The clean soil is returned to the site. Thermal desorption is not the same as incineration, which uses heat to *destroy* the chemicals.

Thermal desorption uses equipment called a desorber to clean polluted soil. Soil is excavated and placed in the desorber. The desorber works like a large oven. When the soil becomes hot enough, vaporizable materials evaporate. To prepare the soil for the desorber, workers may need to crush it, dry it, blend it with sand, or remove debris. This allows the desorber to clean the soil more evenly and easily. Facility-size thermal desorption systems can clean over 20 tons of polluted soil per hour. The effort involved in cleaning up a site using thermal desorption depends on:

• The amount of polluted soil

• Moisture present in the soil (it could take most of the desorption energy just to evaporate water!)

• Debris present in the soil

• Type and amounts of contaminants present (fuel value, of great advantage in incineration, is of no advantage in thermal desorption)

Cleanup can take only a few weeks at small sites (several hundred cubic yards) of petroleum contamination. If the site is large (over 50,000 cubic yards) and the contaminant levels are high, cleanup can take years.

During each step of the process, workers must control dust from the soil and trap or destroy contaminants that are desorbed. Contaminant vapors are trapped in activated carbon or by cooling and then condensation in canisters. Alternatively, vapors can be incinerated or catalytically oxidized. Before returning the cleaned soil to the site, workers may spray it with water to cool it and control dust.

Thermal desorption has been used at many sites over the years. EPA requires exhaust gas testing to ensure that dust and gases are not released to the air in harmful amounts. Tests are also required of the soil to make sure it is clean before it is returned to the site. All equipment must meet federal, state, and local standards.

Thermal desorption works well at sites with dry soil and certain types of pollution, such as fuel oil, coal tar, chemicals that preserve wood, and solvents. Sometimes thermal desorption works well where some other cleanup methods cannot, such as at sites that have high concentrations of contaminants in the soil. Thermal desorption can be a faster cleanup method than SVE or microbiological destruction of contaminants. This is important if a polluted site needs to be cleaned up quickly so that it can be used for other purposes. The equipment for thermal desorption often costs less to build and operate than equipment for other cleanup methods using heat. Thermal desorption also tends to save the desirable organic matter in the soil that helps with growth of vegetation. The EPA has selected thermal desorption to clean up over 60 Superfund sites.

14.18.5 Pump-and-Treat for Groundwater Decontamination

A low-tech but often used process for cleaning up groundwater is the pump-and-treat method. Contaminated groundwater is simply pumped out of an extraction well, treated to remove the contaminants of concern using any of a variety of methods, and discharged. If the water is clean enough, it may be possible to return it to the groundwater via a recharge well. If not, it may have to be discharged to a publicly owned wastewater treatment works (POTW).

Pump-and-treat systems are used for almost any type of groundwater cleanup, regardless of contamination. They do, however, require a larger capital investment than some other technologies. As seen in Figure 14.26, they can also be used as a barrier to protect drinking

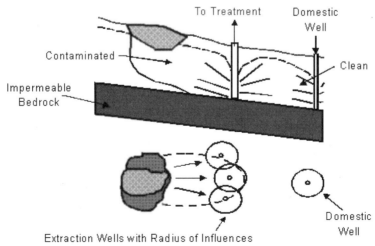

FIGURE 14.26 Elevation (top) and plan (bottom) views of a groundwater pump-and-treat layout. (*Courtesy of the Federal Remediation Technologies Roundtable.*)

water supplies. In that case, they may be an economical alternative to the construction of a slurry wall barrier to prevent contaminated groundwater from polluting a drinking water supply.

14.19 *SOME INNOVATIVE WASTE TREATMENT TECHNOLOGIES*

Note

As of the writing of this book, the EPA still considers these technologies innovative. The authors feel that a number of these technologies will soon be reclassified as "proven" or "common" technologies.

Some innovative technologies that have proven themselves are:

• Bioremediation
• Fracturing
• Groundwater circulating wells
• *In situ* Flushing
• *In situ* Oxidation
• Dual-phase extraction
• Natural attenuation
• Permeable reactive barriers
• Phytoremediation
• Vitrification
• Permeable reactive barriers
• Slurry walls

Forty-hour HAZWOPER certified personnel might be employed on any number of these types of remediation efforts. New approaches are also being designed and tested all the time. Most of them are simply improvements on proven technologies, but some are genuinely new. The EPA and state agencies will always insist that a new technology be tested and proven on a pilot scale before allowing the technology to become a site's mode of remedial action.

14.19.1 Bioremediation

Most of these innovative treatment technologies rely either upon microbiological reactions that render the contaminants harmless or convert them to less harmful materials or upon physical removal of the contaminants from the media (air, soil, or groundwater). Figure 14.27 illustrates the relative remediation usage of widely used bioremediation methods on superfund sites.

Bioremediation may occur through either aerobic (with oxygen) or anaerobic (without oxygen) processes. *Aerobic* treatment involves the conversion of chiefly organic chemical contaminants, in the presence of sufficient oxygen and nutrients, to oxygenated decomposition products, carbon dioxide, water, and microbial cell mass (biomass). *Anaerobic* treatment involves the metabolism of chiefly organic chemical contaminants, in the absence of oxygen. A major reaction product is methane. Nonoxygenated decomposition products, limited amounts of carbon dioxide, and trace amounts of hydrogen gas are produced. Under sulfate-reducing conditions, sulfate may be converted to sulfide or elemental sulfur. Under nitrate-reducing conditions, nitrogen gas (N_2) ultimately is produced.

Anaerobic bioremediation normally uses the addition to the contaminated media of nutrients, "bugs" (specially designed microbes that primarily eat only the contaminants of

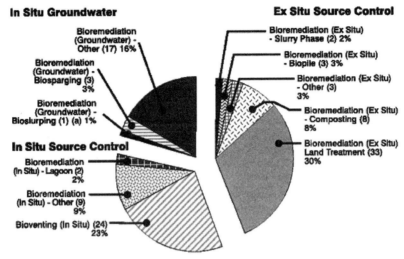

FIGURE 14.27 Superfund remedial actions: bioremediation methods for source control and *in situ* groundwater treatment (FY 1982–FY 1999). (*Courtesy of USEPA.*)

concern), and oxygen or oxygen-releasing compounds. Common methods for oxygen delivery include air or oxygen sparging, injection of a dilute solution of hydrogen peroxide, or use of oxygen release compounds (ORCs) such as magnesium oxides. Use of these additives is considered *enhanced bioremediation.*

Bioremediation can be conducted *in situ* or *ex situ.* The information about bioremediation presented here includes its use on soil, sediment, sludge, or other solid media, both *in situ* and *ex situ*, as well as on groundwater *in situ.* Examples of *in situ* processes include bioventing and *in situ* groundwater bioremediation. Bioventing systems deliver air from the atmosphere into the soil (as a source of oxygen) above the water table through injection wells placed in contaminated areas. *In situ* groundwater bioremediation involves engineering of subsurface conditions to induce or accelerate biodegradation of contaminants in an aquifer.

Examples of *ex situ* processes include slurry-phase land treatment and composting, sometimes collectively called "land farming." Slurry-phase treatment combines contaminated soil, water, and other additives under controlled conditions in bioreactors. This creates an optimum environment for microbial degradation. Composting involves mixing contaminant-laden waste with a bulking agent, such as straw or hay, to facilitate the delivery of optimum levels of air, water, and nutrients to the microorganisms.

A pioneer in bioremediation technology is WIK Associates, Inc. of New Castle, Delaware (www.wik.net). WIK has developed microbes that will completely metabolize most types of petroleum hydrocarbon contamination. WIK has conducted extensive field-testing of its Bugs+Plus™ system on numerous site remediation projects. The company has successfully remediated more than 50,000 tons of soil in the cleanup of oil and gasoline-related soil contamination. The technology has been used in both aboveground and underground applications with initial contamination concentrations as high as 80,000 mg/kg of total petroleum hydrocarbons (TPH). With Bugs+Plus™, soils are frequently remediated to nondetectable levels. Turnaround time is typically from four to six months for aboveground remediation. In a typical above ground remediation, the soil is land farmed. This process involves the excavation of the contaminated soil and subsequent depositing on a geomembrane to avoid leaching. Their Bugs+Plus™ microbiological treatment is then applied and tilled into the soil along with the nutrients and the entire site is covered with polyethylene to retain heat and moisture. Cleaned soil can then be backfilled into the excavation, eliminating costly transport fees. One requirement is a large enough site "footprint" for the project.

Longer cleanup times and the need for more rigorous engineering should be anticipated for *in situ* applications.

More than half of the bioremediation projects conducted at Superfund sites have been *in situ* projects. Land farming is the most common form of *ex situ* bioremediation, followed by composting.

14.19.2   Hydrofracturing and Pneumatic Fracturing for Enhancement of Subsurface Remediation

Injection of pressurized water or air through pipes into the ground can fracture low-permeability and overconsolidated sediments. The cracks that are formed are filled with porous media that serve as substrate for bioremediation or to improve pumping efficiency.

Hydrofracturing.   Hydrofracturing is a technology in which pressurized water is injected to increase the permeability of consolidated material or relatively impermeable unconsolidated material. Fractures promote more uniform delivery of treatment fluids and accelerated extraction of mobilized contaminants. Typical applications are linked with soil vapor extraction, *in situ* bioremediation, and pump-and-treat systems.

The fracturing process begins with the injection of water into a sealed borehole until the high pressure of the water exceeds the overburden pressure and a fracture is created. A slurry composed of a coarse-grained sand and guar gum gel or a similar substitute is then injected as the fracture grows away from the well. After pumping, the sand grains hold the fracture open while an enzyme additive breaks down the viscous fluid. The thinned fluid is pumped from the fracture, forming a permeable subsurface channel suitable for delivery or recovery of a vapor or liquid.

The hydraulic fracturing process can be used in conjunction with soil vapor extraction technology to enhance recovery. Hydraulically induced fractures are used to deliver fluids, substrates, and nutrients for *in situ* bioremediation applications.

Pneumatic Fracturing.   This treatment uses high-pressure air to perform the fracturing. With a direct push unit, such as the Geoprobe® discussed in Section 12, the ram is forced into the ground. At the desired depth, high-pressure air is released, fracturing the soil, and sometimes rock, under the site. The number of fracturing sites depends on the subsurface strata. As with well influence (ROI), discussed above, fracture site radii should be overlapped so that the whole site is permeable. The fractures or fissures created by the pneumatic fracturing process can then be used much as in hydrofracturing.

14.19.3   In-Well Air Stripping

Air is injected into the bottom of a double-screened well. Increased pressure at the bottom lifts the water in the well and forces it out the upper screen. Simultaneously, additional water is drawn in the lower screen. Once in the well, some of the VOCs in the contaminated groundwater are transferred from the dissolved phase to the vapor phase by air bubbles. The contaminated air rises in the well to the water surface, where vapors are drawn off and treated by a soil vapor extraction system. This SVE system, in addition to collecting the vapors from within the well, collects vapors from the surrounding vadose zone. The partially treated groundwater is never brought to the surface; it is forced into the unsaturated zone, and the process is repeated as water follows a hydraulic circulation pattern that allows continuous cycling of ground water. As groundwater circulates through the treatment system *in situ*, contaminant concentrations are gradually reduced.

Modifications to the basic in-well stripping process may involve additives injected into the stripping well to enhance biodegradation.

The duration of in-well air stripping is short to long term, the duration depending upon contaminant concentrations, Henry's law constants of the contaminants (for the distribution of contaminants between the vapor and liquid phases), the radius of influence, and site hydrogeology.

14.19.4 Circulating Wells

Circulating wells (CWs) provide a technique for subsurface remediation by creating a three-dimensional circulation pattern of the groundwater. Groundwater is drawn into a well through one screened section and is pumped through the well to a second screened section, where it is reintroduced to the aquifer. The flow direction through the well can be specified as either upward or downward to accommodate site-specific conditions. Because groundwater is not pumped above ground, pumping costs and permitting issues are reduced or eliminated. Also, the problems associated with storage and discharge are removed. In addition to ground-water treatment, CW systems can provide simultaneous vadose zone treatment in the form of bioventing or soil vapor extraction.

CW systems can provide treatment inside the well, in the aquifer, or a combination of both. For effective in-well treatment, the contaminants must be adequately soluble and mobile so that they can be transported by the circulating ground water. Because CW systems provide a wide range of treatment options, they provide some degree of flexibility to a remediation effort.

14.19.5 *In Situ* Soil Flushing

In situ soil flushing is the extraction of contaminants from the soil with water or other suitable aqueous solutions. Soil flushing is accomplished by passing the extraction fluid through soils using an injection or surface process. Extraction fluids must be recovered from the underlying aquifer and, when possible, are recycled.

14.19.6 Cosolvent Enhancement of Soil Flushing

Cosolvent flushing involves injecting a nonhazardous solvent mixture, a miscible organic solvent such as alcohol and water, into either vadose zone, saturated zone, or both, to extract organic contaminants. Cosolvent flushing can be applied to soils to dissolve either the source of contamination or the contaminant plume emanating from it. The cosolvent mixture is normally injected upgradient of the contaminated area, and the solvent with dissolved con-taminants is extracted downgradient and treated above ground.

Recovered groundwater and flushing fluids with the desorbed contaminants may need treatment to meet appropriate discharge standards prior to recycle or release to local, POTWs or receiving streams. To the maximum extent practical, recovered fluids should be reused in the flushing process. The separation of surfactants from recovered flushing fluid, for reuse in the process, is a major factor in the cost of soil flushing. Treatment of the recovered fluids results in process sludges and residual solids, such as spent carbon and spent ion exchange resin, which must be appropriately treated before disposal. Air emissions of volatile contam-inants from recovered flushing fluids must be collected and treated, as appropriate, to meet applicable regulatory standards. Residual flushing additives in the soil may be a concern and must be evaluated on a site-specific basis.

The duration of a soil-flushing process is generally short to medium term.

14.19.7 Chemical Oxidation of Soil and Groundwater Contaminants

Chemical oxidation involves a large family of methods for the destruction of organic chem-icals. We are interested in the destruction of those organics that are typically found as groundwater and soil pollutants—that is, petroleum compounds and their halogenated deriv-atives. The most common derivatives found are the chlorinated compounds, widely used as solvents. Chemical oxidation has been studied in laboratories and used in industry for over a century. For over a quarter of a century it has been used for destruction of hazardous wastes. The majority of this work was on *ex situ* wastes. However, we expect that over the next several decades steady progress will be made in the application of chemical oxidation to the destruction of *in situ* hazardous wastes.

Efforts have always been directed toward use of chemical oxidants that did not leave undesirable residuals after the waste destruction reaction. That naturally directed efforts toward oxygen and its compounds. Ozone and hydrogen peroxide destruction methods have been widely studied. Ozone must be generated where it is to be used. Hydrogen peroxide is commercially available in several strengths, including 50% by weight in water solution. This section focuses on the potential for *in situ* hydrogen peroxide use for destruction of hazardous organic compounds.

Chemical oxidation can be used for such common contaminants as trichloroethene (TCE), dichloroethene (DCE), methylene chloride, and benzene, toluene, ethylbenzene, and xylene (BTEX). It has also been used for the removal of dense nonaqueous phase liquids (DNAPLs) such as chlorinated solvents, and hydrocarbons.

Fenton's Reagent Used in Chemical Oxidation.   Fenton's reagent is iron-catalyzed hydrogen peroxide. Many metals have special oxygen transfer properties that enhance the oxidizing power of hydrogen peroxide (H_2O_2). The most common of these is iron. When used in the optimum manner, iron compounds together with hydrogen peroxide result in the generation of the highly reactive hydroxyl radicals ($\cdot OH$). The reactivity of this combination was first observed over 100 years ago by H. J. H. Fenton. Fenton's reagent has been used to destroy a variety of hazardous organic compounds, including petroleum products, solvents, phenols, formaldehyde, and complex wastes derived from dyestuffs, pesticides, wood preservatives, plastics additives, and rubber chemicals. The process may be applied to polluted wastewater, sludges, or contaminated soils.

Formation of the Powerful Hydroxyl Radical.   The iron for the Fenton's reagent, usually supplied as ferrous, reacts with hydrogen peroxide to produce the very powerful oxidant, the hydroxyl free radical.

$$Fe^{2+} + H_2O_2 \rightarrow Fe^{3+} + OH^- + \cdot OH$$

The biggest drawback currently to Fenton's reagent use seems to be the lack of understanding about how to control iron and peroxide addition to the contaminants to get the most rapid and complete destruction. However, the following conditions generally appear to be the most beneficial:

- Adjusting wastewater pH to 3–5
- Adding iron catalyst (as a solution of $FeSO_4$)
- Slowly adding the H_2O_2. If the pH is too high, the iron precipitates as ferric hydroxide $Fe(OH)_3$ and catalytically decomposes the H_2O_2 to the much less effective oxygen.

Typical iron-to-peroxide ratios are 1 part iron to between 5 and 25 parts hydrogen peroxide on a weight basis. Iron levels of less than 25–50 mg/L can require excessive reaction times. Fenton's reagent appears to have been most effective as a pretreatment tool, where chemical oxygen demands (CODs) are greater than 500 mg/L. In addition to free radical scavengers, the process is inhibited by iron chelants such as phosphates, EDTA, formaldehyde, and citric and oxalic acids. Because of the unique reactions of Fenton's reagent with various hazardous organics, it is recommended that the reaction always be characterized with the specific problem wastes in laboratory tests before proceeding to *in situ* or *ex situ* trials. The hydroxyl radical is one of the most powerful oxidants known, second only to the element fluorine in its reactivity.

Oxidation Conditions.   In the absence of iron, there is no evidence of hydroxyl radical formation by hydrogen peroxide itself. Hydrogen peroxide alone added to a phenolic waste results in almost no destructive oxidation of phenol. However, as $FeSO_4$ addition begins and the concentration of iron is increased, phenol destruction accelerates until a point is reached where further addition of iron becomes inefficient. An optimal dose range for iron catalysts is characteristic of Fenton's reagent, although the definition of the range varies between wastewaters. There are at least three factors that characterize an optimal dose of $FeSO_4$:

1. A minimal threshold concentration of 3–15 mg/L iron that allows the reaction to proceed within a reasonable period of time regardless of the concentration of the organic waste.

2. A constant ratio of iron to substrate above the minimal threshold, typically 1 part iron to 10–50 parts organic waste that yields the desired destruction.

3. Adding an additional amount of iron saturates any iron-chelating materials in the waste. That allows unsequestered iron to catalyze the formation of hydroxyl radicals.

Much still needs to be learned about how the powerful hydroxyl radicals oxidize organic wastes when pretreating a complex organic waste for toxicity reduction. As the hydrogen peroxide dose is increased, a steady reduction in COD may occur with little or no change in toxicity until a threshold is attained. Then, further addition of hydrogen peroxide results in a rapid decrease in wastewater toxicity due to the destruction of intermediate decomposition products that are toxic.

Fenton's reagent reactions are more pronounced at temperatures less than 20°C. Avoid temperatures above 40 to 50°C. In that range accelerated decomposition of hydrogen peroxide into oxygen and water occurs. As a practical matter, most Fenton's reagent use occurs at temperatures between 20 and 40°C.

The optimal pH for Fenton's reagent reactions occurs between pH 3 and pH 6. The drop in efficiency on the basic side is attributed to the transition of iron from hydrated ferrous ion to a colloidal ferric species. In the latter form, iron catalytically decomposes the hydrogen peroxide into oxygen and water without forming hydroxyl radicals.

An Application of Fenton's Reagent to Contaminated Soil.   The Geo-Cleanse® patented injection process is used to deliver hydrogen peroxide (H_2O_2) and trace quantities of ferrous sulfate ($FeSO_4$) and acid (to control pH) into a contaminated soil site. Fifty percent hydrogen peroxide is injected through multiple injectors.

The EPA reports that this process was effective in reducing contaminant concentrations in clays to below soil screening levels. Soil concentrations of up to 1,760 mg/kg of TCE have been reduced to below detection. Operating data indicate no adverse migration of organics to surrounding soils or groundwater.

14.19.8   Natural Attenuation (Monitored Natural Attenuation, MNA)

This process relies on natural processes to clean up or attenuate pollution in soil and groundwater. Natural attenuation occurs at most polluted sites. However, the right conditions must exist underground for sites to be cleaned properly. If not, cleanup will not be quick enough or complete enough. Scientists monitor or test these conditions to make sure natural attenuation is working. This is called monitored natural attenuation (MNA).

When the environment is polluted with chemicals, nature can work in three ways to clean it up:

1. Microbes that live in soil and groundwater use specific organic chemicals that they prefer for food. When they completely digest the chemicals, they discharge products of metabolism as carbon dioxide, water, and methane gas, depending upon conditions of oxygen availability.

2. Chemicals can adhere or sorb to soil, which holds them in place. This does not remove or destroy the chemicals, but it can keep them from polluting groundwater and leaving the site.

3. As pollution moves through soil and groundwater, it can mix with clean water. This reduces or dilutes the pollution, an unfavorable, but naturally occurring result.

The time it takes for MNA to clean up a site depends on several factors:

- Type and amounts of chemicals present
- Size and depth of the polluted area
- Type of soil and conditions present

These factors vary from site to site, but cleanup usually takes from years to decades. MNA is used when other methods will not work or are expected to take almost as long. Sometimes MNA is used as a final cleanup step after another method cleans up most of the pollution.

Some chemicals, such as oil and solvents, can evaporate, which means they change from liquids to gases within the soil. If these gases escape to the air at the ground surface, sunlight may help to destroy them.

MNA works best where the source of pollution has been removed. For instance, buried waste must be dug up and disposed of properly, or it can be removed using other available cleanup methods. After the source is removed, the natural processes rid the zone of the small amount of pollution that remains in the soil and groundwater. The soil and groundwater are monitored regularly to make sure they do become remediated.

MNA can be a safe process if used properly. No one has to excavate the contaminants, and nothing has to be added to the land or water. But MNA is not a do-nothing way to clean up sites. Regular monitoring is needed to make sure pollution does not leave the site. This ensures that people and the environment are protected during cleanup. Depending on the site, MNA may work just as well and almost as fast as other methods. Because MNA takes place underground, digging and construction are not needed. As a result, there is no waste to dispose of in landfills. This is less disruptive to the neighborhood and the environment. Also, it allows cleanup workers to avoid contact with the pollution. MNA requires less equipment and labor than most methods. Therefore, it can be a less expensive alternative. However, monitoring for many years can be costly. Also, the unavailability of the site for development for many years can be a strong drawback.

MNA is the only cleanup method being used at a few Superfund sites with groundwater pollution. At over 60 other sites with polluted groundwater, MNA is just one of the cleanup methods being used. MNA also is used for oil and gasoline spills from tanks.

14.19.9   Dual-Phase Extraction (DPE)

Dual-phase extraction is also known as multiphase extraction, vacuum-enhanced extraction, free product recovery, liquid-liquid extraction, or sometimes bioslurping. This technology uses a high-vacuum system to remove various combinations of contaminated ground water, separate-phase petroleum product, and hydrocarbon vapor from the subsurface. Extracted liquids and vapor are treated and collected for disposal or reinjected to the subsurface.

In DPE systems for liquid/vapor treatment, a high-vacuum system is utilized to remove liquid and gas from low-permeability formations. The vacuum extraction well includes a screened section in the zone of contaminated soils and groundwater. It removes contaminants from above and below the water table. The system lowers the water table around the well, exposing more of the formation. Contaminants in the newly created vadose zone are then accessible to vapor extraction. Once above ground, the extracted vapors or liquid-phase organics and groundwater are separated and treated. DPE for liquid/vapor treatment is generally combined with bioremediation, air sparging, or bioventing when the target contaminants include long-chain hydrocarbons. Use of dual-phase extraction with these technologies can shorten the cleanup time at a site. It also can be used with pump-and-treat technologies to recover groundwater in higher-yielding aquifers.

The DPE process for undissolved liquid-phase organics, also known as free product recovery, is used in cases where a fuel hydrocarbon layer more than 20 cm (eight inches) thick is floating on the water table. The free product is generally drawn up to the surface by a pumping system. Following recovery, it can be disposed of, reused directly in an operation not requiring high-purity materials, or purified prior to reuse. Systems may be designed to recover only product, mixed product and water, or separate streams of product and water.

The target contaminant groups for dual-phase extraction are VOCs, fuels, and low-density non-aqueous phase liquids (LNAPLs). Dual-phase vacuum extraction is more effective than SVE for heterogeneous clays and fine sandy soils. However, it is not recommended for lower-permeability formations due to the potential to leave isolated lenses of undissolved product in the formation.

Factors that may limit the applicability and effectiveness of the process include site geology, contaminant characteristics and their distribution. Combination with complementary technologies such as pump-and-treat may be required to remediate groundwater from high-yielding aquifers.

Dual-phase extraction requires both water treatment and vapor treatment.

14.19.10 Phytoremediation

Phytoremediation is a process that uses plants to remove, transfer, stabilize, and destroy contaminants in soil and sediment. The mechanisms of phytoremediation include enhanced rhizosphere biodegradation, phytoextraction (also called phytoaccumulation), phytodegradation, and phytostabilization. Figure 14.28 illustrates various phytoremediation actions.

Enhanced rhizosphere biodegradation takes place in the soil immediately surrounding plant roots. Natural substances released by plant roots supply nutrients to microorganisms, enhancing their biological activities. Plant roots also loosen the soil and then die, leaving paths for transport of water and aeration. This process tends to pull water to the surface zone and dry the lower saturated zones.

Phytoaccumulation is the uptake of contaminants by plant roots and the translocation and accumulation (phytoextraction) of contaminants into plant shoots and leaves.

Phytodegradation is the metabolism of contaminants within plant tissues. Plants produce enzymes, such as dehalogenase and oxygenase, that help catalyze degradation. Investigations

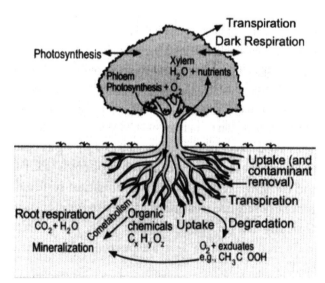

FIGURE 14.28 Various phytoremediation actions. (*Courtesy of USEPA.*)

are proceeding to determine if both aromatic and chlorinated aliphatic compounds are amenable to phytodegradation.

Phytostabilization is the phenomenon of production of chemical compounds by plants to immobilize contaminants at the interface of roots and soil.

Phytoremediation uses plants to clean up pollution in the environment. Plants can help clean up many kinds of pollution, including heavy metals, pesticides, explosives, and oil. The plants also help prevent wind, rain, and groundwater from carrying pollution, away from sites to other areas.

Phytoremediation works best at sites with low to medium amounts of pollution. Plants remove harmful chemicals from the ground when their roots take in water and nutrients from polluted soil, streams, and groundwater. Plants can clean up chemicals as deep as their roots can grow. Tree roots grow deeper than smaller plants, so they are used to reach pollution deeper in the ground. Once inside the plant, chemicals can be:

- Stored in the roots, stems or trunks, or leaves
- Changed into less harmful chemicals within the plant
- Changed into gases that are released into the air as the plant transpires (breathes)

Phytoremediation can occur even if the chemicals are not taken into the plant by the roots. For example, chemicals can stick or sorb to plant roots, or they can be changed into less harmful chemicals by microbes that live near plant roots. The plants are allowed to grow and take in or sorb chemicals. Afterward, they are harvested and destroyed, or recycled if metals stored in the plants can be reused. Usually, trees are left to grow and are not harvested. Plants grown for phytoremediation also can help keep harmful chemicals from moving from a polluted site to other areas. The plants limit the amount of chemicals that can be carried away by the wind or by rain that soaks into the soil or flows off the site.

Before phytoremediation begins, EPA studies whether plants grown to clean up pollution can be harmful to people. EPA tests the plants and air to make sure that the plants do not release harmful gases into the air.

Some insects and small animals may eat the plants used for phytoremediation. Scientists are studying these animals to see whether the plants can harm them. Scientists also are studying whether these animals pose harm to the larger animals that eat them. In general, as long as plants are not eaten, they are not harmful to people.

EPA uses phytoremediation because it takes advantage of natural plant processes. It requires less equipment and labor than other methods since plants do most of the work. Trees and plants can make a site more attractive as well. The site can be cleaned up without removing polluted soil or pumping polluted groundwater. This allows workers to avoid contact with harmful chemicals. Phytoremediation has been successfully tested in many locations and is being used at several Superfund sites.

The time it takes to clean up a site using phytoremediation depends on several factors:

- Type and number of plants being used
- Type and amounts of harmful chemicals present
- Size and depth of the polluted area
- Type of soil and conditions present

These factors vary from site to site. Plants may have to be replaced if they are destroyed by bad weather or animals. This adds time to the cleanup. Often it takes many years to clean up a site using phytoremediation.

14.19.11 Permeable Reactive Barriers (PRBs)

PRBs are a form of passive contamination control. See Figure 14.29. A PRB is placed down-gradient of the contamination source, such as a groundwater plume. Often used in conjunction with slurry walls, the PRB filters out the contaminants while allowing the

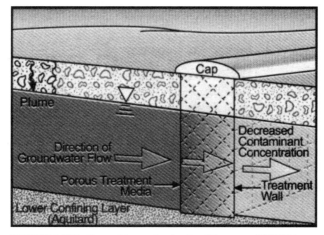

FIGURE 14.29 Permeable reactive barrier. (*Courtesy of USEPA.*)

groundwater to flow through. PRBs are filled with materials that will react in one or more ways with the contamination constituents. Some of the reactants are metal-based catalysts for degrading organic material, chelating agents for removing metals, reducing agents such as iron, bases such as calcium hydroxide and magnesium hydroxide. The most common weaker base is regular crushed limestone.

At some point these PRBs may become clogged, no longer reactive, or otherwise ineffective. At that point they will have to be excavated and properly disposed of as a new PRB takes their place.

PRBs are often used in conjunction with vertical engineered barriers (VEBs) (see next Subsection). The VEB funnels the plume toward the PRB. This is called a funnel-and-gate method.

14.19.12 Vertical Engineered Barriers (VEBs)

Although VEBs are not a cleanup technology *per se*, they are often used at remediation sites to keep contamination from migrating off-site while the remediation process takes place. The most common type of VEB is the slurry wall. It is low-tech, yet extremely effective on contamination plumes that are between ground level and about 30 ft below grade. A trench is excavated to below the contaminated area, ideally into bedrock. These subsurface barriers consist of vertically excavated trenches filled with slurry. The slurry, usually a mixture of bentonite and water, hydraulically shores the trench to prevent collapse and retards groundwater flow.

Slurry walls are typically placed at depths up to 30 m (100 ft) and are generally 0.6 to 1.2 m (2 to 4 ft) in thickness. Installation depths over 30 m (100 ft) are constructed using clamshell bucket excavation, but the cost per unit area of wall increases by about a factor of three. The most effective application of the slurry wall for site remediation or pollution control is to base (or key) the slurry wall 0.6 to 0.9 m (2 to 3 ft) into a low-permeability layer such as clay or bedrock. See Figure 14.30. Slurry walls are inert but are not a good choice when the contaminant is either very high or very low pH. The calcium-rich bentonite clay is easily degraded by extremes of pH.

Other types of VEBs include geosynthetic walls, grout walls, deep soil mixing, and sheet piles.

Geosynthetic walls are made by simply lining a trench with impermeable membrane. Grout walls are made by pumping grout into the ground at regular intervals to make an impermeable wall. This differs from a slurry wall in that there is no trenching required. Deep

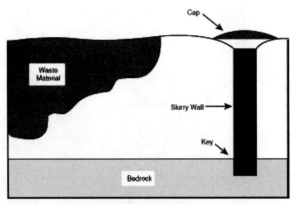

FIGURE 14.30   Elevation view of a slurry wall. (*Courtesy of the Federal Remediation Technologies Roundtable.*)

soil mixing uses augers to mix the soil with cement to form the barrier. Sheet piles are constructed by driving steel plates into the ground and overlapping them so that they form a wall.

14.20   TIERED APPROACH TO CORRECTIVE ACTION OBJECTIVES (TACO): ONE STATE'S ANSWER

Chicago, Illinois, is the home of brownfields remediation. At the top, geographically speaking, of the Rust Belt, Chicago had to do something about all of its real property that lay fallow due to real or perceived contamination. The state's response was the Tiered Approach to Corrective Action (TACO) under its Environmental Protection Act, 35 Illinois Administrative Code. This allowed sites, based upon their final use, to use a more streamlined approach to achieve their site-specific standards.*

14.20.1   Compliance with Remediation Objectives

Groundwater.   To determine if a TACO site meets the groundwater remediation objectives, the contaminant concentrations of discrete samples are compared to the applicable groundwater remediation objectives. A discrete sample is a sample collected from only one point. You will have achieved compliance if the analytical results from each sample do not exceed the applicable remediation objectives.

Soil.   When determining whether a TACO site meets the soil remediation objectives, you may choose among three methods of compliance, depending on the type of contaminant and the potential human receptors. You may reach compliance with soil remediation objectives in three ways:

---

*For a more complete explanation of TACO, refer to H. J. Rafson and R. N. Rafson, *Brownfields:* Redeveloping Environmentally Distressed Properties, New York: McGraw-Hill, 1999.

1. Compare the contaminant concentrations of discrete samples to the applicable soil remediation objectives.

2. Composite the soil samples by physically mixing the soil from more than one location prior to laboratory analysis. Next, compare the composite analytical result to the applicable soil remediation objectives.

3. Mathematically average individual analytical sample results, then compare the average to the applicable soil remediation objectives.

Whichever method or combination of methods you choose to use, you will have achieved compliance if the analytical results of the samples do not exceed the applicable remediation objective(s). There are, however, guidelines to follow when compositing and averaging.

Guidelines for TACO Soil Composting and Averaging. You can neither composite nor average soil samples for determining compliance with ingestion and inhalation objectives for the construction worker population. This is because construction workers are normally exposed to a specific location and not to the site as a whole. Construction worker exposure must be evaluated at all sites that depend on an industrial/commercial land use for remediation objectives. "Construction worker population" means people engaged on a temporary basis to perform work involving invasive construction activities such as earth-moving, building, and utility installation and repair during postremediation land use.

You may not composite soil samples if your contaminants of concern are volatile organic compounds (VOCs). Compositing of VOCs is prohibited because the compositing process itself provides a mechanism for the contaminants to escape into the atmosphere. Consequently, composite samples submitted for laboratory analysis underestimate the amount of VOCs actually present at the site.

How many samples do I need to collect at my site to determine compliance with the remediation objectives? The number of samples you collect depends on the specific program requirements under which the remediation is performed, such as LUST, Site Remediation Program, RCRA Closure and Corrective Action.

14.21 *TRAINING AIDS AND ADDITIONAL RESOURCES*

The authors recommend the following video for aiding readers to see real-life examples of what they have learned in this Section.

• *Bulging Drums—What Every Responder Should Know*

This video is available free of charge from the Los Alamos National Laboratory (http://www.lanl.gov/worldview).

14.22 *SUMMARY*

Section 14 has described the types of tasks to be performed by HAZWOPER personnel and the kinds of technology they will be applying to the remediation of Superfund and brownfield sites. HAZWOPER duties, responsibilities, and required authority were explained.

This HAZWOPER work deals with changing sites and widely varying hazardous materials and site conditions. Variability is an expected consequence of such employment.

We explained that Superfund sites must be remediated such that they are no longer a threat to the neighboring community. For brownfields, hazardous materials must be removed so that they can be developed from an unusable condition, or usable with serious limitations, to usable for immediate real estate development. Brownfields may contain wide varieties of wastes. Cleanup may take anywhere from days to decades to accomplish.

For an individual subject to HAZWOPER regulations, whether laborer, technician, engineer, or manager, there are three broad activities or phases you can become involved in and for which you must be prepared: site investigation, control, and remediation.

Before setting foot on a hazardous waste site, in addition to being 40-hour HAZWOPER trained and certified, you must have a minimum of information about site characteristics. That means you must be briefed about what has happened over the known history of the site. Further, you must know the current condition of the site.

You get this information from site investigation. When the site has been identified and under investigation for periods ranging from weeks to years, you will have the advantage of reviewing or learning the results of a large body of collected information. You can prepare yourself, or your team if you are the manager, with this information and then enter the site when ready. You will find this investigation activity exceptionally challenging when you are introduced to it during emergency response, as described in Section 15.

Site control concerns controlling access to the site to authorized HAZWOPER personnel. We described the fatal attraction an excavation activity has for children. If that pile or the hole that it came from is contaminated, so will the children become who play in it. Following investigation, your HAZWOPER responsibility is to engage in control. Under HAZWOPER, this means the control of the spread or dissemination of hazardous material as solid, liquid, gas, vapor, or fume from the site. Control will end when the site has been remediated to the point that all responsible authorities are satisfied that a waste hazard no longer exists. This could take decades!

Control is the longest-duration activity in most cases. Control includes fencing off the site and segregating it into zones according to their approved activities, sampling and monitoring, and prevention of hazardous material dissemination from the site. Control and remediation are done concurrently following site investigation.

Remediation includes the collection, accumulation, or the treatment, packaging, manifesting, and shipping of residual hazardous waste to an EPA approved hazardous waste landfill.

The hazards and safe management of drums found on Superfund and brownfield sites have been explained. We stressed careful management of drums from visual observation at a distance, through opening and analysis of contents, to final disposition.

Remediation methods that HAZWOPER personnel will be applying, a selection of those innovative methods predicted to be of growing use in the coming decades, have been summarized.

Remediation is completed when the responsible authorities are satisfied that no further hazardous materials remediation work needs to be done and agree to attest to that fact.

SECTION 15
EMERGENCY RESPONSE

OBJECTIVES

As a 40-hour HAZWOPER certified trainer, manager, or worker, you are required to have a working knowledge of emergency response. You must know exactly what you are and are not prepared to do in an emergency with this level of training. Essentially this is preparation for emergency response that is defensive in nature and primarily for response to routine, small-scale on-site spills. In this Section you will learn the job titles and functions of the various levels of emergency responders as well as the training requirements and the authors' view on experience requirements.

As HAZWOPER personnel read this Section, they will see that it is the culmination of everything in the preceding 14 Sections and in Section 16. Emergency responders really must have a good understanding of the preceding information. Although this Section will be giving the responder invaluable training references, it all must be digested and fully understood. That knowledge must be demonstrated by testing as well as practical response experience.

There are a number of kinds of emergency responders. This Section focuses on the duties and responsibilities of hazardous materials emergency responders. The authors emphasize that some levels of emergency response require additional, specialized training beyond the scope of this book. Although responder training, duties, and responsibilities are explained here in some detail, we do not represent this material as covering the in-depth training for all responder levels. We do, however, encourage all readers to engage in additional study and training so as to have a better overall understanding of the emergency response roles and duties in an incident situation. In this section we do cover all that an individual needs to know to be 40-hour HAZWOPER certified.

The authors have found that one part of emergency response that many people fail to understand is that responder organizations are not run in a democratic manner. At work, if you disagree with your supervisor, you may have some latitude and time to discuss the disagreement. Not so in emergency response. Emergency response teams are paramilitary organizations such as a fire or police department. Each of these has a command and control structure. That is, there is an overall Incident Commander (IC) who delegates authority to various other assigned division, group, or sector heads under the IC's command. Those heads delegate the work of the response to workers under their control.

Preplanning, training, annual retraining, and drilling exercises are essential to a well-organized team. Training and drilling can be on a small scale, such as a tabletop exercise. Exercises can be local, involving all members of mutual aid groups, or they can be regional, involving representatives from national response organizations.

Clarity and brevity in communication are vital at emergency response sites. State your information clearly and succinctly. Names of zones, teams, personnel, equipment, and others must be in plain English. Radio and telephone communications must refrain from code names and acronyms. Even the popular 10-code, such as 10-4, meaning "I understand," is not acceptable on the scene.

15.1   *WHO ARE HAZARDOUS MATERIALS EMERGENCY RESPONDERS AND TO WHAT TYPE OF EMERGENCIES ARE THEY REQUIRED TO RESPOND?*

"One in every five Americans lives within 3 miles of a site where EPA has acted to remove immediate threats to human health."—EPA's Emergency Response Team

O—π **Emergency responders are specially trained individuals who respond to spills and other releases of hazardous materials and wastes by stopping the release at its source, collecting the material, and disposing of it properly.**

O—π **Emergency responders protect the public, other response personnel, the environment, and property and equipment, in that order.**

Emergency responders, also called first responders, are those specially trained individuals who are called to the scene of an imminent or present hazardous materials emergency. That emergency may be any uncontrolled leak, spill, or other exposure that threatens people, the environment, or physical property. Responders are trained to evaluate the situation and then take appropriate actions to minimize exposure to the population, their fellow responders, the environment, and equipment and personal property, in that order. The emergency could be an explosion or fire that has released or threatens to release hazardous materials into the area. It could be a rail or road accident that has resulted in an uncontrolled leak or spill. The emergency response regulations apply to both emergency response and post-emergency cleanup of hazardous substance spills. The definition of hazardous substance used in these regulations is much broader than CERCLA's. They include all CERCLA hazardous substances, all RCRA hazardous wastes, and all DOT hazardous materials listed in 49 CFR Part 172. Therefore, most oils and oil spill responses are covered by these regulations.

Do not confuse HAZWOPER emergency responders with emergency medical technicians (EMTs) (paramedics), firefighters, wilderness, flood, missing persons, or collapsed building rescue teams. Although all of these courageous individuals are considered first responders, OSHA is clear in the HAZWOPER regulations that this HAZWOPER type of emergency responder is responding to hazardous materials incidents only. However, any or all of the above disciplines may be required to respond to an incident.

By its definition, a HAZWOPER emergency can involve any material and every kind of other hazard. It would be impossible to have equipment available to respond to ANY emergency. What can be done is to plan for foreseeable emergencies.

15.2   *WHAT TYPE OF TRAINING IS REQUIRED?*

O—π **Several government agencies and National organizations handle specific types of emergencies. Each requires a different kind of training, with some overlap.**

OSHA, EPA, FEMA (Federal Emergency Management Agency), US Coast Guard's NRC, USFA (United States Fire Administration), NFPA (National Fire Protection Association), and many states have different levels of training required before responders are allowed to work on an actual incident. All of these organizations and agencies training programs overlap to some degree.

Because this is an OSHA-oriented book, we will spend most of this section on the OSHA-required training. It must be understood, however, that a working knowledge of the other agencies' response capabilities is necessary in order to work effectively when an incident calls for a multijurisdictional response.

15.2.1 OSHA Training Levels

Think of OSHA emergency response (first responder) training as a series of building blocks or ranks with Awareness Level being the most junior and Incident Commander being the most senior level. As in a paramilitary organization, each rank builds on the experience and the response fundamentals that were learned in prior levels.

> **OSHA recognizes five increasing levels of emergency response (first responder) training; Awareness Level, first responder Operations Level, Hazardous Materials Technician, Hazardous Materials Specialist, and On-Scene Incident Commander.**

OSHA requires that employees who participate, or are expected to participate, in emergency response, shall be given training based on the duties and function to be performed. Further, they shall be trained prior to taking part in any actual emergency operation at an incident. There are also numerous individuals who are trained in many separate disciplines. In fact, in OSHA emergency response, so far as HAZWOPER workers are concerned, employees MUST be trained and competent at all the levels or ranks below them.

A common acronym for the duties common to all emergency responders is DECIDE, the National Fire Protection Association's (NFPA's) decision-making process for hazardous materials (HAZMAT) emergencies. The name comes from the six steps that can provide a framework for decision making in any HAZMAT emergency:

1. **D**etect the presence of hazardous materials and hazardous conditions.
2. **E**stimate potential human, environmental, and property harm that can occur without timely intervention.
3. **C**hoose response objectives.
4. **I**dentify action options.
5. **D**o the best option.
6. **E**valuate progress.

OSHA training is divided into the five categories below. Employees are only permitted to take actions that fall within their actual proven knowledge, experience, and training level.

1. First responder Awareness Level
2. First responder Operations Level
3. Hazardous Materials Technician
4. Hazardous Materials Specialist
5. On-scene Incident Commander

Let us look at each of these responder levels in detail for a thorough understanding of what responders are and are not expected to do at each level of competence.

CAUTION

DUE TO OVERLAPS IN TRAINING REQUIREMENTS, THERE HAS BEEN A GOOD DEAL OF DISAGREEMENT AMONG TRAINING ORGANIZATIONS AS TO WHAT LEVEL OF EMERGENCY RESPONSE IS TO BE COVERED IN 40-HOUR HAZWOPER TRAINING. THE AUTHORS BELIEVE THAT THE INFORMATION PRESENTED IN THIS SECTION, ALONG WITH THE OTHER SECTIONS, APPENDICES, AND ADDITIONAL TRAINING MATERIALS SUGGESTED IN THIS BOOK, IF UNDERSTOOD AND ASSIMILATED BY THE READER, WILL ADEQUATELY

PREPARE THE READER FOR ANY REASONABLE EXAM FOR FIRST RE-
SPONDER AT THE OPERATIONS LEVEL.

First Responder Awareness Level

O━━ㅈ **Awareness Level training is considered sufficient for a person to recognize a hazardous material release or potential release and notify authorities for assistance. The Awareness Level responder takes no active part in stopping the release or starting the cleanup.**

The Awareness Level first responder is trained to have an understanding of what hazards may occur at the job site. The site may be a chemical company, refinery, or any other type of industry or organization where hazardous materials or hazardous waste are stored, used, generated, produced, or otherwise present. At Awareness Level, HAZWOPER workers are trained to be just that—aware. They are specifically trained to take no further action beyond notifying the authorities of the release. Those authorities may be an organization's response unit, an organization's outside private response unit, the local fire company, or other mutual aid group.

OSHA recommends awareness level responders have sufficient training or sufficient experience to demonstrate competence objectively in the areas below:

1. Review and demonstrate competency in performing the applicable skills required under the HAZWOPER standard.
   - Understand what hazardous substances are, and the risks associated with them, in an incident.
   - Understand the potential outcomes associated with an emergency created when hazardous substances are present.
   - Be able to recognize the presence of hazardous substances in an emergency.
   - Be able to identify the hazardous substances, if possible.
   - Understand the role of the first responder awareness individual in the employer's Emergency Response Plan, including site security and control.
   - Be able to realize the need for additional resources and make appropriate notifications to the communication center.

2. Hands-on experience with the U.S. Department of Transportation's *Emergency Response Guidebook* (ERG) (on the Companion CD) and familiarization with OSHA Standard 29 CFR 1910.1201, *Retention of DOT Markings, Placards and Labels* (Section 8.2 of this book).

3. Review of the principles and practices for analyzing an incident to determine both the hazardous substances present and the basic hazard and response information for each hazardous substance present. (Use this Section, Section 16, and the HASP in Appendix K on the Companion CD.)

4. Review of procedures for implementing actions consistent with the local emergency response plan, the organization's standard operating procedures, and the current edition of DOT's ERG, including emergency notification procedures and follow-up communications. (Use this Section and Section 16.)

5. Review of the expected hazards, including fire and explosion hazards, confined space hazards, electrical hazards, powered equipment hazards, motor vehicle hazards, and walking-working surface hazards. (Use Sections 3, 5, and 11.)

6. Awareness and knowledge of the competencies for the first responder at the Awareness Level covered in the National Fire Protection Association's Standard No. 472, *Professional Competence of Responders to Hazardous Materials Incidents*.

First responders at the Awareness Level are specifically trained to gather only such information that is immediately available. They are *not* trained to approach incident areas for a closer look. They *are* trained to recognize hazards from a distance and contact the assigned communication center for the incident. They *are* trained to observe from and remain at a safe distance to inform the operations level or other higher-level responders of their observations.

OSHA requires at least a 16-hour course addressing the skills and competency requirements at the Awareness Level. The authors believe that the 16 hours of training can be based on material in this Section plus specific information about the duties and responsibilities of the trainee's own organization. The authors recommend that, in addition, Awareness Level responders be 40-hour HAZWOPER trained (56 hours total).

First Responder Operations Level

First responder Operations Level training is considered sufficient for personnel who are expected to take a defensive posture at a hazardous material release. They may take part in stopping the spread of the release but may not actually approach the release source.

At the Operations Level, trained workers will be the first to respond physically. Their response is, however, limited to a defensive posture. They are not trained to try to stop a release. Their function is to contain the release from a safe distance, keep it from spreading, and prevent exposures.

OSHA requires that first responders at the Operations Level have received at least eight hours of training or have had sufficient experience to objectively demonstrate competency in the areas below, in addition to the 16 hours listed for the Awareness Level.

1. Review and demonstration of competency in performing the applicable skills required under the HAZWOPER standard.
   - Know the basic hazard and risk assessment techniques. (Use the first 12 Sections of this book.)
   - Know how to select and use proper personal protective equipment provided to the first responder Operation Level. (Use Section 6 of this book.)
   - Understand basic hazardous materials terms. (Use Sections 3 and 4 of this book.)
   - Know how to perform basic control, containment, and confinement operations within the capabilities of the resources and personal protective equipment available with their unit. (To be learned as part of the 24 hours of training, in addition to the 40-hour HAZWOPER training.)
   - Know how to implement basic decontamination procedures. (Use Section 7.)
   - Understand the relevant standard operating procedures and termination procedures. (To be learned as part of the OSHA-required 24 hours training.)

2. Hands-on experience with the U.S. Department of Transportation's *Emergency Response Guidebook* (ERG), Manufacturer Material Safety Data Sheets, CHEMTREC®/CANUTEC®, shipper or manufacturer contacts, and other relevant sources of information addressing hazardous substance releases. Familiarization with OSHA Standard 29 CFR 1910.1201, *Retention of DOT Markings, Placards and Labels.* (Use Sections 8 and 9 and the ERG on the Companion CD.)

3. Review of the principles and practices for analyzing an incident to determine the hazardous substances present, the likely behavior of those hazardous substances and their containers, the types of hazardous substance transportation containers and vehicles, and the types and selection of the appropriate defensive strategy for containing the release. (Use Sections 3, 4, 14, and 15.)

4. Review of procedures for implementing continuing response actions consistent with the local Emergency Response Plan, the organization's standard operating procedures, and the current edition of DOT's ERG, including extended emergency notification procedures and follow-up communications. (To be learned as part of the 24-hour training; also use the ERG in the Companion CD.)

5. Review of the principles and practice for proper selection and use of personal protective equipment. (Use Section 6.)

6. Review of the principles and practice of personnel and equipment decontamination. (Use Section 7.)

7. Review of the expected hazards, including fire and explosions hazards, confined space hazards, electrical hazards, powered equipment hazards, motor vehicle hazards, and walking-working surface hazards. (Use Sections 3, 10, and 11.)

8. Awareness and knowledge of the competencies for the first responder at the Operations Level covered in the National Fire Protection Association's Standard No. 472, *Professional Competence of Responders to Hazardous Materials Incidents.*

OSHA requires at least a 24-hour course addressing the skills and competency requirements at the operations level. The authors believe that the 24 hours of training can be based on material in this section plus specific information about the duties and responsibilities of the trainee's own organization. The authors recommend that in addition to having completed awareness level requirements, Operations Level responders be 40-hour HAZWOPER trained (64 hours total).

First Responder Hazardous Materials Technician Level

○─π **Hazardous Materials Technicians are expected to actively stop the release of hazardous materials at an incident. They also have the duty to assist in the cleanup of release material.**

Hazardous Materials Technicians (HMTs) respond to releases or potential releases for the purpose of stopping the release. Their posture is more aggressive than that of the Operations Level. They will actually approach the point of release in order to close valves, plug, patch, or otherwise stop the release of a hazardous substance.

Although OSHA only partially specifies training duration for Hazardous Materials Technicians, OSHA does require first responders at the Hazardous Materials Technician level to have received at least 24 hours of training equivalent to that listed for the operations level. The authors believe that an additional 16 hours of Hazardous Material Technician level training is required to objectively demonstrate competency in the areas below. As base knowledge, the authors recommend the 40-hour HAZWOPER certification (total of 80 hours).

1. Review and demonstration of competency in performing the applicable skills required under the HAZWOPER standard (the 40-hour HAZWOPER certification).

- Know how to implement the employer's Emergency Response Plan. (Use this Section and Section 16)
- Know the classification, identification, and verification of known and unknown materials by using field survey instruments and equipment. (Use Section 12.)
- Be able to function within an assigned role in the Incident Command System (part of the 40 hours of technician level training).
- Know how to select and use proper, specialized, hazardous material personal protective equipment provided to the Hazardous Materials Technician. (Use Section 6.)
- Understand hazard and risk assessment techniques. (Use Sections 3 and 4.)
- Be able to perform advance control, containment, and confinement operations within the capabilities of the resources and personal protective equipment available with the unit. (To be learned as part of 40 hours of technician training, and Section 6.)
- Understand and implement decontamination procedures. (Use Section 7.)
- Understand termination procedures. (Use this section and technician training.)
- Understand basic chemical and toxicological terminology and behavior. (Use Sections 3, 4, and 5.)

2. Hands-on experience with written and electronic information relative to response decision making including but not limited to the U.S. Department of Transportation's *Emergency Response Guidebook* (ERG), manufacturers material safety data sheets, CHEMTREC®/

CANUTEC® (Transport Canada, *Transport of Dangerous Goods*), shipper or manufacturer contacts, computer databases and response models, and other relevant sources of information addressing hazardous substance releases. Familiarization with OSHA Standard 29 CFR 1910.1201, *Retention of DOT Markings, Placards and Labels.* (Practice using the ERG on the Companion CD; use Sections 8 and 9.)

3. Review of the principles and practices for analyzing an incident to determine the hazardous substances present, their physical and chemical properties, the likely behavior of the hazardous substance and its container, the types of hazardous substance transportation containers and vehicles involved in the release, the appropriate strategy for approaching release sites and containing the release. (Use Sections 3 and 4 and Section 14 on drum handling.)

4. Review of procedures for implementing continuing response actions consistent with the local emergency response plan, the organization's standard operating procedures, and the current edition of DOT's ERG, including extended emergency notification procedures and follow-up communications. (Use Section 16 and the 40 hours of technician training.)

5. Review of the principles and practice for proper selection and use of personal protective equipment. (Use Section 6.)

6. Review of the principles and practices of establishing exposure zones, proper decontamination, and medical surveillance stations and procedures. (Use Section 7.)

7. Review of the expected hazards, including fire and explosions hazards, confined space hazards, electrical hazards, powered equipment hazards, motor vehicle hazards, and walking-working surface hazards. (Use Sections 10 and 11.)

8. Awareness and knowledge of the competencies for the Hazardous Materials Technician covered in the National Fire Protection Association's Standard No. 472, *Professional Competence of Responders to Hazardous Materials Incidents.*

The authors' view is that attaining the above competencies will require the Hazardous Materials Technician to have completed 40-hour HAZWOPER training and 40 hours of additional hazardous materials technician training (80 hours total).

First Responder Hazardous Materials Specialist Level

○──π **Hazardous Materials Specialists are expected to give specific guidance to Hazardous Material Technicians relating to the specific material being released.**

Hazardous Materials Specialists (HMS) respond with, and provide support to, Hazardous Materials Technicians. Their duties parallel those of the Hazardous Materials Technician, but their training and knowledge is specifically focused on the various hazardous materials they may be called upon to contain.

Further, the Hazardous Materials Specialist acts as the site liaison with federal, state, local, and other government authorities in regard to site activities.

Although OSHA only partially specifies training duration for Hazardous Materials Specialists, OSHA does require first responders at the Hazardous Materials Specialist level to have received at least 24 hours of training equivalent to that listed for the operations level. The authors believe that an additional 16 hours of specialist training is required to be able to objectively demonstrate competency in the areas below. Additionally, the authors recommend 40-hour HAZWOPER certification.

1. Review and demonstration of competency in performing the applicable skills required under the HAZWOPER standard (40-hour HAZWOPER certification).
   - Know how to implement the local Emergency Response Plan. (Use Section 16 and the local plan.)
   - Understand classification, identification, and verification of known and unknown materials by using advanced survey instruments and equipment. (Use Section 12.)

- Know the state emergency response plan. (To be learned as part of the 40-hour specialist training.)
- Be able to select and use proper specialized hazardous material personal protective equipment provided to the hazardous materials specialist. (Use Section 6.)
- Understand in-depth hazard and risk techniques. (To be learned as part of the 40-hour specialist training.)
- Be able to perform specialized control, containment, and confinement operations within the capabilities of the resources and personal protective equipment available. (Use this Section.)
- Be able to determine and implement decontamination procedures. (Use Section 7.)
- Be able to develop a site safety and control plan. (Use Section 7.)
- Understand chemical, radiological, and toxicological terminology and behavior. (Use Sections 3 and 5.)

2. Hands-on experience with retrieval and use of written and electronic information relative to response decision making, including but not limited to the U.S. Department of Transportation's *Emergency Response Guidebook* (ERG), manufacturers Material Safety Data Sheets, CHEMTREC®/CANUTEC®, shipper or manufacturer contacts, computer databases and response models, and other relevant sources of information addressing hazardous substance releases. Familiarization with OSHA Standard 29 CFR 1910.1201, *Retention of DOT Markings, Placards and Labels*. (Learn to use the ERG on the Companion CD; also use Section 8.)

3. Review of the principles and practices for analyzing an incident to determine the hazardous substances present, their physical and chemical properties, and the likely behavior of the hazardous substance and its container, vessel, or vehicle. (Use Sections 3 and 4.)

4. Review of the principles and practices for identification of the types of hazardous substance transportation containers, vessels, and vehicles involved in the release as well as stationary containers and vessels; selecting and using the various types of equipment available for plugging or patching transportation containers, vessels, or vehicles; organizing and directing the use of multiple teams of Hazardous Material Technicians and selecting the appropriate strategy for approaching release sites and containing or stopping the release. (Use Sections 8 and 14.)

5. Review of procedures for implementing continuing response actions consistent with the local emergency response plan, the organization's standard operating procedures, including knowledge of the available public and private response resources, establishment of an incident command post, direction of hazardous material technician teams, and extended emergency notification procedures and follow-up communications. (To be learned as part of the 40-hour specialist training.)

6. Review of the principles and practice for proper selection and use of personal protective equipment. (Use Section 6.)

7. Review of the principles and practices of establishing exposure zones and proper decontamination, monitoring, and medical surveillance stations and procedures. (Use Section 7.)

8. Review of the expected hazards, including fire and explosions hazards, confined space hazards, electrical hazards, powered equipment hazards, motor vehicle hazards, and walking-working surface hazards. (Use Sections 4, 7 and 11.)

9. Awareness and knowledge of the competencies for the off-site specialist employee covered in the National Fire Protection Association's Standard No. 472, *Professional Competence of Responders to Hazardous Materials Incidents*.

The authors' view is that attaining the above competencies will require the Hazardous Materials Specialist to have completed 40-hour HAZWOPER and 40 hours of additional Hazardous Materials Specialist training (80 hours total).

First Responder Incident Commander Level

○─π   **The on-scene Incident Commander is expected to take overall responsibility for the hazardous material release control and cleanup. The Incident Commander must have authority over all personnel responding to the release incident.**

The apex of emergency response training is that required of the on-scene Incident Commander (IC). This individual must be granted the authority to control the incident scene. Ideally, this authority must be agreed upon by all potential responding organizations, prior to any incident occurring, during preplanning sessions. The Incident Commander directs all operations at the scene from the Command Post. If a facility relies entirely on a community mutual aid organization, by default the IC might be the fire chief or police chief.

We believe that an Incident Commander should first receive 40-hour HAZWOPER training. We further believe that the following training is necessary as an elementary basis for understanding the position: 24 hours of Operations Level, 16 hours of Hazardous Materials Specialist, 16 hours of Hazardous Materials Technician, and 40 hours of specialized Incident Command training. We firmly believe that this total of 136 hours of training is the minimum amount of training an inexperienced person needs to understand the Incident Command position. We further believe that, although certified as such, no person is truly ready for Incident Command until he or she has responded to at least 20 hazardous materials incidents at the lower levels of responsibility.

OSHA requires first responders at the On-Scene Incident Command level to have received at least 24 hours of training equivalent to that listed for the operations level. The authors believe that 24 hours of training is inadequate to be able to demonstrate competency objectively in the areas below. In the authors' opinion, Incident Commanders MUST have the training of all levels below them.

Incident Command may be conducted by one individual for the duration of the incident, or it may pass through several levels as the incident escalates or as more senior officials arrive. As an example, consider an awareness level worker coming upon a leaking valve at the back of a truck placarded as "UN/NA 1203 Flammable Liquid." The worker, the only responder on the scene for the time being, is the Incident Commander. The worker has not only the right, but the responsibility to:

- Call the facility's operations center.
- Stop all hot work in the area.
- Direct personnel and vehicle traffic away from the area.
- Aid the more advanced responders in locating the problem.
- Write up an incident report of what was seen and the response effort provided and retain it for personal reference. This will aid Awareness Level responders if they are questioned as to their actions during the incident at some future time.

By the same token, the IC may be the community's fire chief. As the incident escalates, that IC may be replaced several times by a local or state emergency response coordinator. Further, the position might be relinquished to an IC from Regional EPA, and if further assistance is needed, such as food, clothing, shelter, or airlift capabilities for mass casualties, the IC hat may be passed to an IC from the National EPA Emergency Response Team or from FEMA. As ICs relinquish command, they must brief the relieving IC on all relevant aspects of the incident. The relieved IC must remain available to the relieving IC for whatever period of time that relieving IC requires.

The IC duties and responsibilities require that the post be filled by the most senior qualified person on the scene. OSHA requires the IC to be fully capable of administering the site-specific Incident Command System (ICS). The authors recommend that the IC also be fully trained in the Unified Command System (UCS). Further, we recommend that all incident responses be managed using one UCS. In this scenario, leaders of the various local response or mutual aid organizations (fire, police, HAZMAT, medical, among others) all agree in advance on who will assume overall Incident Command.

The IC must be carefully chosen among the qualified applicants. All other first responders must be taught to direct the members of the press up the chain of command, but the buck stops with the IC. If the incident is at full throttle, requiring all of the IC's attention, the IC may have to delegate a special Public Information Officer (PIO). If the incident escalates to require other agencies, the IC may have to assign a Liaison Officer to coordinate between the various agencies and organizations.

From years of experience, the authors can state categorically that during any incident, the press might attempt to put the most catastrophic 'spin' on the situation. IC's we have trained must, at some point, go "in the fishbowl." This is their practice session for a real-life meeting with the press. In this exercise, other class students attempt, by the asking of pointed questions, to fluster the IC student. This is an eye-opening experience, for nearly all novices can be made to lose their cool. ICs who do so at an actual incident will be perceived as being unqualified or incapable of leading the command system. Angry, inappropriate, or humorous remarks have no place in a press conference.

CAUTION

1. NEVER MEET THE PRESS ON THEIR OWN GROUND. IF PRESS PEO-PLE SHOVE A CAMERA AND MICROPHONE IN YOUR FACE WHILE AN INFERNO IS RAGING IN THE BACKGROUND, DO NOT GIVE A LENGTHY INTERVIEW. DOING SO WILL GIVE THE IMPRESSION THAT YOU ARE NOT NEEDED TO DIRECT RESPONSE ACTIONS. DO NOT SMILE OR LAUGH, NO MATTER WHAT! ALL THE PUBLIC WILL EVER SEE IS YOU NOT REALIZING THE GRAVITY OF THE SITUA-TION AT BEST, OR ACTING AS AN INCOMPETENT AT WORST.

2. YOU MUST MAINTAIN CONTROL OF THE SITUATION. TELL THE PRESS THAT YOU WILL HAVE A STATEMENT FOR THEM LATER BUT THAT RIGHT NOW YOU ARE NEEDED FOR CRITICAL SUPER-VISION.

3. HOWEVER, DO NOT HIDE. AS SOON AS POSSIBLE, PERMIT THE PRESS, UNDER SECURITY CONTROL, AS CLOSE AS IS SAFE AND LET THEM TAKE THEIR PICTURES. THE PUBLIC HAS A RIGHT TO KNOW FACTS AS THEY BECOME CONFIRMED.

4. AS SOON AS POSSIBLE, ARRANGE FOR A PRESS CONFERENCE AWAY FROM THE SCENE, SUCH AS AT A LOCAL HOTEL. NEVER HOLD FORTH AT THE SCENE, IN THE COMMAND POST, OR ANY-WHERE WHERE YOU DO NOT CONTROL THE FLOW OF INFOR-MATION. OTHERWISE THE CONFERENCE COULD BECOME A DIS-TRACTION TO OTHER PERSONNEL GATHERING INFORMATION AND GIVING ORDERS.

5. STRESS TO ALL PERSONNEL THAT THEY SHOULD BE COURTEOUS TO THE PRESS BUT GIVE NO INFORMATION OTHER THAN TO DI-RECT THE PRESS TO THE PIO. PERSONNEL ARE TO UNDERSTAND THAT THEY ARE NOT HIDING ANYTHING, THEY SIMPLY DO NOT KNOW THE WHOLE PICTURE. ONLY THE IC OR THE DELEGATED PIO DOES.

Once again, OSHA only partially specifies training duration for Incident Commanders. OSHA does require first responders at the Incident Commander level to have received at least 24 hours of training equivalent to that listed for the Operations Level. The authors believe that the additional 112 hours of training described above is required to be able to objectively demonstrate competency in the areas below.

- Know and be able to implement the employer's Incident Command system.
- Know how to implement the employer's Emergency Response Plan.
- Know and understand the hazards and risks associated with employees working in chemical protective clothing.
- Know how to implement the local emergency response plan.
- Know of the state emergency response plan and of the Federal Regional Response Team.
- Know and understand the importance of decontamination procedures.

The IC is responsible for the entire incident response. Depending upon the size and nature of the incident, ICs will have to delegate much of their responsibilities to subordinates.

CAUTION

IF NOVICES ARE ONLY TAUGHT THE INCIDENT COMMAND SYSTEM, THE AUTHORS DO NOT BELIEVE THEM TO BE QUALIFIED FOR COMMAND. INCIDENT COMMAND TRAINING TAUGHT TO A NOVICE WILL REALLY MEAN VERY LITTLE. JUST AS COMPLETING A PAINT-BY-NUMBERS PICTURE DOES NOT MAKE YOU A VAN GOGH, SO THE IC TRAINING DOES NOT MAKE YOU AN IC. EXPERIENCE AT LOWER RESPONDER LEVELS AND OTHER MANAGEMENT, MILITARY, POLICE, OR FIREFIGHTING TRAINING OR EXPERIENCE IN EMERGENCY MANAGEMENT ARE REQUIRED TO PRODUCE AN IC WHO WILL MAKE THE CORRECT DECISIONS AND WHOM STAFF FOLLOW. AS GRADUATES FROM MILITARY ACADEMIES KNOW, WHEN YOU FIRST GO INTO BATTLE WITH NO PRACTICAL EXPERIENCE, YOU WILL LISTEN CAREFULLY TO THE ADVICE OF THE NONCOMMISSIONED OFFICER(S) UNDER YOU. IF AT ALL PRACTICAL, AN IC MUST BE AN INDIVIDUAL WHO HAS SERVED AT LEAST 20 TIMES AT A LOWER RANK IN EMERGENCY RESPONSES.

Incident Commander responsibilities also include, but are not limited to, the following duties and responsibilities:

- Announce to all workers and outside agencies that you are in command.
- Establish your Command Post (CP). Ideally, this will be in view of the incident (but always at a safe distance), taking into consideration the nature and size of the incident, weather, site topography, day or night, and wind direction and speed. The CP may be a police cruiser, a fire truck, or a trailer or other structure.
- Size up the situation.
- Control the communication setup. Require as much communication equipment as necessary.
- By whatever means available and to the extent possible, identify all hazardous substances or conditions present.
- Appoint a Site Safety Officer (SSO) and a Recorder at a minimum. The SSO will be responsible for the safety and health of the responders and community. The SSO must identify and evaluate hazards and provide direction with respect to the safety of operations for the emergency at hand. The SSO will ideally be an IC in training. The IC MUST have complete confidence in the SSO. The SSO must have the authority to alter, suspend, or terminate site activities when those activities involve an imminent danger condition. The SSO must immediately inform the IC of any actions to be taken to correct these hazards at the emergency scene. The Recorder will make a running log of every bit of information that is brought to the IC's attention. The recorder will also log all orders and requests for

assistance issued by the IC. This is imperative so that there is a complete record of the event, making the end-of-incident report that much easier to prepare.

- Direct use of engineering controls and use of any new technologies. This will allow for an acid test of new technologies and will also let the IC know whether the new technology was used successfully, should be modified, or should not be recommended for future use.
- Establish, through the SSO, maximum exposure limits and hazardous substance handling procedures.
- Ensure that the personal protective equipment worn is appropriate for the hazards to be encountered. At a minimum, all response workers exposed to inhalation hazards or potential inhalation hazards must wear positive-pressure SCBAs.
- Direct the SSO to use air monitoring to ensure safe levels of air contaminants before decreasing the level of respiratory protection of employees. Initially, have air monitoring results reviewed and summarized constantly.
- Limit the number of emergency response personnel at the emergency site, especially in those areas of potential or actual exposure to incident or site hazards. That limitation must be to those who are actively performing emergency operations.
- Direct your security team to keep all bystanders and press well away from the area. If the site is extremely large, this might require patrols of the site support (cold) zone border.
- Ensure use of the buddy system.
- Ensure that backup personnel are standing by with equipment ready to provide assistance or rescue. One backup person in the same level of protection is required for each responder in an IDLH atmosphere.
- Ensure that qualified basic life support personnel (EMTs) are standing by with medical equipment and transportation capability.
- Implement decontamination procedures in support of emergency operations.
- Demobilize the response. Ensure that written reports are sent to you by all supervisory personnel involved in the response action. Ideally, these reports will be submitted in a few days.
- Debrief all responders for lessons learned.
- Write up final end-of-incident report. Note that these reports may not have to be supplied to any other agencies but are essential for the IC's own records.

Due to the many additional specialized duties of Incident Commanders, the personal and legal responsibilities, and the necessity of their understanding the duties of the workers under them; the authors interpret the regulations as requiring the incident commander to have completed 40-hour HAZWOPER and 96 hours of additional Incident Commander training. This training must include adequate emphasis on each of the duties and responsibilities of those under the IC's command. As previously stated, the authors recommend a minimum participation in at least 20 incident responses before qualifying for Incident Command. Total: 136 hours.

Specialist Employees or Contractors

At the request of the IC, employees and contractors with special knowledge about the released material, material containers, special abatement procedures, environmental impact of the release, and so on; can be summoned to aid in the release abatement process.

These are employees or contractors of a participating organization who have in-depth knowledge of specific wastes, substances, chemicals, or materials, or specific types of containers, modes of transport, and the like. They might be Cargo Specialists, Engineers, Occupational Health Professionals, Certified Safety Professionals, Health Physicists, and the like. Their job is to provide technical advice or assistance to the IC at a hazardous materials release incident. To satisfy liability considerations, these individuals must be employees or contractors of one of the participating agencies, organizations, or facilities involved in the incident.

Along with maintenance of professional licenses or certificates, they must prove that they have received training or demonstrated competency in the area of their specialization annually.

Skilled Support Personnel

OSHA allows personnel not having 40-hour HAZWOPER certification or emergency response training, but having needed skills to abate the release, to work on an emergency response following an initial briefing and after being supplied with all protection provided to the authorized response personnel.

CAUTION
UNDER THE DIRECTION OF THE IC AND STAFF, OSHA ALLOWS THE USE OF WORKERS WITH CRITICAL SUPPORT SKILLS WHEN THEIR HELP IS NEEDED. THOSE SUPPORT WORKERS ARE EXEMPTED FROM TRAINING. THE AUTHORS STRONGLY RECOMMEND THEIR USE AS A LAST-DITCH EFFORT. EXHAUST ALL OTHER POSSIBILITIES OF USING EMERGENCY RESPONSE TRAINED PERSONNEL FIRST!

According to OSHA, personnel who are not necessarily employees of an organization involved in a response but who are skilled in the operation of needed equipment, such as mechanized earth moving or digging equipment or crane and hoisting equipment, may provide services. Characteristically, they are needed temporarily to perform immediate emergency support work that cannot reasonably be performed in a timely fashion by employees of the involved organizations. Even though they will or may be exposed to the hazards at an emergency response scene, OSHA does not require them to meet the training required of employees. However, OSHA requires that these personnel be given an initial briefing at the site prior to their participation in any emergency response. The initial briefing shall include instruction in the wearing of appropriate personal protective equipment, what chemical hazards are involved, and what duties are to be performed. All other appropriate safety and health precautions provided to the employer's (involved organizations') own employees shall be used to assure the safety and health of these personnel.

Although OSHA allows use of such untrained personnel, the authors strongly recommend that only hazardous materials response personnel be used in any capacity where the workers will be exposed to hazardous materials. For one major reason, untrained personnel do not understand the full implications of wearing PPE, respirators, and SCBA gear. They run the risk of collapse while working and thereby endangering other responders.

15.2.2 Spills and Other Releases

The most common types of emergency responses are spills of hazardous materials liquids. The responsibility of the emergency response team is to stop the spread of the spill, stop the flow from its source, and clean up the spill safely.

To protect personnel, the environment, and property, the responder must be familiar with the available equipment for use in containment, cleanup, and proper disposal of spilled materials.

Spills are the most common type of incident to which a HAZWOPER emergency responder will respond. Roadway tanker and railroad tank car spills are becoming frighteningly common. Hazardous waste cleanup sites often have containers that have deteriorated over time. They also routinely have buried containers. These containers often have liquid contents, and the containers become damaged while excavating, staging, opening, or sampling. Unfortunately, spills often occur in pumping liquids from one container to another. Spills can occur at pumps, flanges and gasket connections, and holes in hoses, and through inattention leading to overfilling storage tanks and other containers. Spills of solids are usually easier to clean up but can also pose significant risks, such as spills of pesticides and other poisons. Gases under pressure including liquefied gases, when accidentally released, are considered to be a release not a spill. There is usually no cleanup of a gas release because it cannot be contained. Releases of gases will require a defensive posture (unless there is a safely operable control valve), while spills of liquids and solids will require an offensive posture.

When responding to a spill, HAZWOPER emergency responders must be aware of the importance of:

- Acting as an IC if one is not present, or following directions of higher-level responders if one is present
- Identifying the hazards
- Notification of the proper authorities
- Securing the scene of the spill so that responders, the public, and the environment are not further exposed
- Documenting observations and actions taken.

Responders must then take the following steps immediately in the event of a spill or release:

- Evacuate and warn others as necessary
- If the spill is beyond the capacity of the spill kit at hand or emergency off-site help is needed, call the facility's emergency operations center
- Visually inspect the spill site to obtain enough information to describe the situation to contacted personnel or other responders. Be alert and keep clear if hazardous chemicals might be involved. At a minimum, gather the following information:
  1. Spill material (name of material if possible. If not, characteristics, such as flammable liquid, corrosive, toxic, or other)
  2. Quantity estimate, surface area covered and rate of flow
  3. Location and direction of spill flow
  4. People involved, including any injuries or exposures
  5. Any other relevant information

- At a safe distance from the spill, and out of the spill's path, meet and orient emergency service units.

If the spill is within your trained capacity to clean up, proceed to clean it up.

- Shut down equipment if you can do so safely.
- Safely isolate a leak in a line with valves, if possible, to reduce the total release.
- Safely upright a leaking drum or container that is leaking from the top. Roll a punctured container so the hole is facing up and chock it so it will not roll over.
- Apply sorbent material such as Oil Dri®, cat litter, absorbent towels, rolls, or booms.
- Plug or patch the leak with a rubber patching and plugging kit, wooden peg, patch putty, chemical tape, or a magnetic belly band.
- Block floor and storm drains using drain pads.
- Build a dike to prevent spread of material using soil, sand, clay, absorbent, or spill-control booms or 'socks.'

- Direct personnel away from the spill area.
- Make sure all contaminated cleanup media are properly containerized and labeled "Hazardous Waste" or "Nonhazardous Waste," as applicable.

On the same day of the spill, provide event information to your supervisor and the SSO of the involved facility to aid in further development of the site spill response plan and documentation of response actions. These bits of information will aid the SSO in the assessment of the need for further training or need for more or other equipment. Also, the spill kit must be refilled with equipment, containers, tools, and the like to be available for the next spill.

In addition to the above steps, other notification actions may be necessary. If the spill or release is greater than the Reportable Quantity (RQ) of a hazardous material, the spill or release must be reported to the National Response Center (NRC) within 15 minutes of the responsible parties, including a Project Manager, SSO, or Public Information Officer, becoming aware. It may also require notification to state and local authorities; however, this varies by location.

Note

The authors have heard of several incidents where the personnel at TSDF and hazardous waste sites have not contacted the National Response Center (NRC) when it was required. The reason is usually that local personnel believe they can remediate adequately and they do not want negative publicity, which is understandable. However, those individuals do not realize the tremendous resources that the NRC can bring into play if necessary to remediate the spill. NRC notification gives you a valuable source of information and technical and logistical support should you need it. If you do not need it, you simply file a report with the NRC documenting the event and your control and cleanup measures.

As shown in Figure 15.1, reusable adhesive mats are available to keep hazardous material leaks out of storm and sanitary drains. They are heavy enough so that they will not float away when the material collects at the drain.

As shown in Figure 15.2, dikes can be formed using overlapping socks. Pillows can be added for extra absorbency. These are made of material specially designed not only to absorb, but also to neutralize. Two different types, one for acids and one for bases, are available.

Dikes are useful not only for cleaning up spills and directing flows of released material, but as a preventive measure, such as when opening a corroded drum. They come in various

FIGURE 15.1 New Pig® Drainblocker® drain plug. (*Courtesy of New Pig® Corporation* [www. newpig.com].)

FIGURE 15.2 HAZMAT PIG® dikes and pillows. (*Courtesy of New Pig® Corporation* [www. newpig.com].)

forms, including as nonabsorbent materials that contain or divert only; as dikes or socks that absorb only; as dikes or socks that both absorb and neutralize, among others. They can have a number of advantages over loose absorbent, such as cat litter, including ease of cleanup.

Often the HAZWOPER worker will use spill kits that contain all the absorbents necessary, along with an overpack container to store and ship all the wastes, including the waste container. As shown in Figure 15.3, this kit contains all the absorbents necessary for confining up to 61 gallons of hazardous liquids. It is rated for Packing Groups I, II and III—certified for shipping by air, land, sea or rail. It can be moved using built-in forklift entries or an overpack dolly (shown here). It also comes with a 12-in. round DRAINBLOCKER® drain plug and a bottle of Shop Dri® floor sweep adsorbant.

FIGURE 15.3 New Pig® 95-gallon overpack spill kit. (*Courtesy of New Pig® Corporation* [www.newpig.com].)

15.3 INCIDENT COMMAND AND UNIFIED COMMAND STRUCTURE

15.3.1 The Five Functions of the Incident Command System (ICS) and the Three Levels of Responsibility

○━ㅠ The ICS is built around five major management functions: Command, Planning, Operations, Logistics, and Finance and Administration.

○━ㅠ The ICS is an elastic structure, from one person doing everything, to large organizations with many individual functional groups. The ICS can be used not only in hazardous material emergencies but indeed in any type of emergency situation.

The ICS is built around five major management functions: Command, Planning, Operations, Logistics, and Finance and Administration. As stated before, the ICS organization is largely paramilitary in nature. These five functions are applied to any incident whether large or small. The Incident Commander retains responsibility for these functions unless they are delegated to another individual. In some incidents or applications only a few of the organization's functional elements may be formally established or delegated to another individual. However, if there is a need to expand the organization, additional positions exist within the ICS framework to meet any need. The key personnel performing the following functions must designate qualified alternates.

- *Command* has overall responsibility for the incident and determines strategic or incident objectives. Command establishes priorities based on the nature of the incident, the resources available, and agency policy.
- *Planning* develops an Incident Action Plan (IAP) to accomplish the strategic objectives. Planning collects and evaluates information, and specifies needed resources and supplies.
- *Operations* develops the tactical organizations, their objectives, and the required tasks and directs all resources (personnel, material, equipment, and other resources) to carry out the Incident Action Plan.
- *Logistics* provides those resources and all other services needed to support the incident command system.
- *Finance and Administration* monitors costs related to the incident and provides accounting, procurement, time recording, cost analysis, and overall fiscal guidance.

○━ㅠ The ICS has three levels of responsibility: strategic, tactical, and task. The decisions, plans, and operations that are strategic affect the success or failure of the entire incident response. Those that are tactical affect the success or failure of individual operations within the incident response. Tasks are the detailed jobs that comprise individual operations.

There are three levels of responsibility under the IC.:

1. *Strategic:* assumes overall or (strategic) responsibility for the success or failure of the incident response. Develops the incident action plan (IAP), sets priorities, and assigns resources. Assigns specific objectives to the tactical level. The IC and staff work at this level. Assigns an alternate IC. Has an Aide (Recorder) who records every decision and command made along with all incoming information. Assigns a public information officer. In military terms, strategic decisions are those that win or lose the war.

WARNING

THE MOST COMMON ERROR THAT INCIDENT COMMANDERS MAKE IS TO GET INVOLVED AT THE HANDS-ON LEVEL. AL-

THOUGH MANY INCIDENT COMMANDERS WANT TO BE UP FRONT WITH THE ACTION, THEY MUST STAY AT THE COMMAND POST. SHOULD THEY BECOME INJURED OR OTHERWISE INCA-PACITATED, THE ENTIRE OPERATION WILL BE IN JEOPARDY. HOWEVER, THEY MUST ALWAYS HAVE AN ALTERNATE COMMANDER READY TO TAKE CHARGE.

2. *Tactical:* directs operations within the IAP. Several tactical officers might have required functions, such as site security and crowd control (police), firefighting (fire), medical (EMTs), hazardous materials (HAZMAT), food, drink, rest area (special services), and others as needed. In military terms, tactical decisions are those that win or lose the battle.

3. *Task:* the specific jobs performed by the responders to meet tactical objectives. In military terms, a task would be to take out an enemy mortar or other weapons site.

Incident command, as we said above, can be very simple or become extremely complex. When more resources are needed, the IC system will be called an expanded IC system. This will require further delegation to divisions, groups, or sectors. Whatever names are assigned to these other functional groups, the IC must establish them initially and then not deviate. Any necessary deviation from the IAP, must be immediately communicated to all workers. Avoid deviation unless absolutely necessary. It causes counterproductive confusion.

This IC system, and the scope of incident objectives, were originally developed by the USCG for oil spill response. However, the authors find it valid for any type of emergency response. As the Coast Guard puts it in their *Field Operations Guide,* 2000:

INCIDENT OBJECTIVES—Statements of guidance and direction necessary for the selection of appropriate strategies, and the tactical direction of resources. Incident objectives are based on realistic expectations of what can be accomplished when all allocated resources have been effectively deployed. Incident objectives must be achievable and measurable, yet flexible enough to allow for strategic and tactical alternatives.

15.3.2   The Unified Command Structure (UCS)

○─𝌆 **For a safe and effective response, it is essential that all parties agree in the preplanning stage on who will be in overall charge. The individual groups maintain integrity, but there can only be one overall leader and an agreed-upon structure. This is called the Unified Command Structure (UCS).**

Using the unified command structure (UCS) is essential to the success of the response to any large-scale incident. The UCS requires all potentially involved ICs to agree ahead of time on who will be in overall command. Once that most important step is agreed upon, the planning for how the unified groups will conduct themselves can begin.

The authors are aware of situations where the UCS was not used, leading to wasted time, less successful cleanups, and greater potential for personnel exposure. Operating within the UCS is not familiar to most people. Police are trained to follow their superiors. Firefighters will follow theirs. If these, along with other mutual aid organizations, are not brought in at the UCS planning stage, there will be confusion at the incident. All mutual aid organizations must order their personnel to follow the commands of the IC or of designated leaders who take their orders from the IC. The individual organizations retain their identity but will function as team members with others assigned to an ordered task.

The unified command structure (UCS) is shown in Figure 15.4. It can satisfy the basic needs for any incident involving hazardous materials. We have previously described the emergency responder levels: Awareness, Operations, Technician, and Specialist all report to the Operations Section Chief. Operations Section Chiefs will be trained Incident Commanders, although that is not their role in this large incident structure. The 'Unified Command' can be as simple as the Incident Commander and the assigned Recorder. It could be

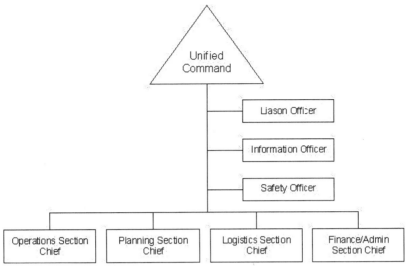

FIGURE 15.4   The Unified Command Structure (UCS). (*Courtesy of the U.S. Coast Guard.*)

as complex as a group of highly experienced individuals, with national and regional support, but still with one agreed-upon IC. For example, initially at the World Trade Center terrorist attack, the on-scene Incident Commander was The Fire Chief of the City of New York. As the magnitude of the incident unfolded the world observed the seamless assembly of a large scale Unified Command Structure with the Federal Emergency Management Agency taking the lead role. It was awesome to see the rapid assembly of local, national, and international fire, police, search and rescue, and military forces that were brought to bear in support of the heroic emergency responders already on the scene.

Using the checklist in Table 15.1 an individual preparing for incident command can get a better feel for the responsibilities of that position. Readers will note that almost all of the IC's responsibilities are delegated to section or group officers or chiefs and their staff. This structure, although designed for national-scale emergencies, can be helpful in any emergency. The structure is extremely flexible. One person may wear many hats. On the other end of the scale, each block can be expanded to include entire staffs to deal with intricate emergency details.

Tables 15.2 through 15.4 are responsibilities checklists for other key personnel.

Note

See Section 16 and Appendix K on the Companion CD for Health and Safety Plan details.

Common Responsibilities

○─π **Common responsibilities apply to all emergency response personnel.**

The following responsibilities apply to all ICS personnel:

- Receive assignment, notification, reporting location, reporting time, and travel instructions from your home agency.
- Upon arrival at the incident, check in at designated check-in locations. Check-in locations may be found at:

Incident command post

Base or camps, staging areas, helibases, division supervisors (for direct line assignments)

Agency representatives from assisting or cooperating agencies report to Liaison Officer at the Command Post after checking in.

- All radio communications to incident communications center shall be addressed "(Incident Name) Communications."
- Use clear text and ICS terminology (no "10-codes") in all radio transmissions.
- Receive briefing from immediate supervisor.

TABLE 15.1   UCS, IC Responsibilities Checklist

Responsibility	Completed
Mobilize, implement, and manage the incident specific unified command organization.	☐
Develop and prioritize incident objectives and strategies.	☐
Establish the Incident Command Post.	☐
Authorize the ordering, deploying, and demobilization of response resources.	☐
Keep higher authorities informed of incident status.	☐
Supervise and coordinate the command staff and the operations, planning, logistics, and finance sections.	☐
Oversee and authorize release of information to news media.	☐
Ensure public safety and the safety of all response operations.	☐
Ensure incident funding is available.	☐
Notify natural resource trustees and coordinate with NRDA[a] representatives.	☐
Coordinate incident investigation responsibilities.	☐
Seek appropriate legal counsel.	☐
Order the demobilization of the incident when appropriate.	☐

Source: Courtesy of the U.S. Coast Guard.
[a]CERCLA directed the Department of the Interior to prepare rules for Natural Resource Damage Assessments (NRDA) at hazardous waste sites and for emergency incidents involving CERCLA substances.

TABLE 15.2   UCS, Liaison Officer Responsibilities Checklist

Responsibility	Completed
Provide a point of contact for assisting and cooperating agencies.	☐
Identify agency representatives from each agency, including communications link and location.	☐
Maintain a list of assisting and cooperating interagency contacts.	☐
Assist in establishing, identifying, and coordinating interagency contacts.	☐
Identify contractors, emergency groups, and individuals that may be needed by the incident response and coordinate their assignment in assisting or cooperating roles.	☐
Coordinate the response to offers of the use of alternative response technology or other removal technology that may be needed by the incident response.	☐

Source: Courtesy of the U.S. Coast Guard.

- Acquire work materials.
- Organize, assign, and brief subordinates.
- Complete forms and reports required of the assigned position and send material through supervisor to documentation unit.
- Ensure continuity using relief briefings.
- Respond to demobilization orders.
- Brief subordinates regarding demobilization.

TABLE 15.3 UCS, Public Information Officer Responsibilities Checklist

Responsibility	Completed
Activate and manage the joint information center.	☐
Serve as primary media coordinator and point of contact.	☐
Develop presentations for media briefings and the press information package.	☐
Coordinate the schedule of press conferences, media releases, and other activities.	☐
Serve as media assistant to the unified command during media conferences.	☐
Draft all press releases and obtain unified command approval for media releases.	☐
Maintain current information summaries and/or displays on the incident and provide information on status of incident to assigned personnel.	☐
Arrange for tours and other interviews that may be required.	☐
Obtain media information that may be useful to the planning section or the unified command.	☐

Source: Courtesy of the U.S. Coast Guard.

TABLE 15.4 UCS, SSO Responsibilities Checklist

Responsibility	Completed
Identify and evaluate all safety and health hazards, including public health threats and responder safety concerns.	☐
Ensure the performance of preliminary and continuous site characterization and analysis, which shall include the identification of all actual or potential physical, biological, and chemical hazards known or expected to be present on-site.	☐
Participate in planning meetings to identify any health and safety concerns in the incident action plan.	☐
Review the incident action plan for safety implications.	☐
Investigate accidents that have occurred within incident areas.	☐
Ensure the preparation and implementation of the site-specific health and safety plan (HASP) in accordance with the area contingency plan (ACP) and state and federal OSHA regulations.	☐
Ensure site safety briefings are conducted each day, or more frequently.	☐
Advise state agencies, local agencies, and the medical community on all potential public health concerns.	☐
Determine levels of personal protective equipment required for all areas.	☐
Review and approve the medical plan.	☐
Assign assistants and manage the incident safety organizations.	☐

Source: Courtesy of the U.S. Coast Guard.

Section Chief Responsibilities.   Note that some emergency responders refer to section chiefs as "Unit Leaders." The four section chiefs lead operations, planning, logistics, and finance and administration. Responsibilities that must be accomplished by all section chiefs include (these responsibilities are not repeated in each section listing):

- Participate in incident planning meetings, as required.
- Determine current status of unit activities.
- Confirm dispatch and estimated time of arrival of staff and supplies.
- Assign specific duties to staff; supervise staff.
- Determine resource needs.
- Develop and implement accountability, safety, and security measures for personnel and resources.
- Supervise demobilization of unit, including storage of supplies.
- Provide supply unit leader with a list of supplies to be replenished.
- Maintain unit records, including unit activity log.

Expansion Capabilities of the ICS.   To demonstrate the expansion capabilities of the ICS, here is a breakdown of the individual titles, responsibilities, and actions from one of the sections reporting to the Incident Command staff as seen in Figure 15.4. The following is excerpted from the USCG *Field Operations Guide,* 2000.

Operations Section.   Of the four Section Chiefs, the most important to the responders below the IC level is the Operations Section Chief. The responders will all report to the Operations Section Chief or the Operations Section Chief's subordinate supervisors, for assignment tasks. The planning Section [not detailed here] provides strategic and tactical planning. The Logistics Section [not detailed here] ensures that required materials and equipment are available when needed. Finance and Administration [not detailed here] acquires and disburses funds to pay for services, materials, and equipment.

Operations Section Chief.   The Operations Section Chief, a member of the General Staff, is responsible for managing all operations directly applicable to the primary mission. The Operations Section Chief activates and supervises elements in accordance with the Incident Action Plan and directs its execution; activates and executes the Site Safety and Health Plan; directs the preparation of unit operational plans; requests or releases resources; makes expedient changes to the Incident Action Plans as necessary; and reports such to the Incident Commander. Branch Directors, Division and Group Supervisors report to the Operations Section Chief in carrying out incident tasks.

a. Review Common Responsibilities.

b. Develop operations portion of Incident Action Plan (IAP).

c. Brief and assign operations personnel in accordance with IAP.

d. Supervise execution of the IAP for Operations.

e. Request resources needed to implement Operation's tactics as part of the IAP development.

f. Ensure safe tactical operations.

g. Make, or approve, expedient changes to the IAP during the operational period, as necessary.

h. Approve suggested list of resources to be released from assigned status (not released from the incident).

i. Assemble and disassemble teams/task forces assigned to Operations section.

j. Report information about changes in the implementation of the IAP, special activities, events, and occurrences to Incident Commander as well as to Planning Section Chief and Public Information Officer.

k. Maintain Unit/Activity Log.

Staging Area Manager. The 'staging area', located in a support zone, is a space in which all personnel, materials, equipment, and other necessary resources are assembled and prepared for task work. Under the Operations Section Chief, the Staging Area Manager is responsible for managing all activities within the designated staging areas.

a. Review Common Responsibilities.

b. Implement pertinent sections of the IAP.

c. Establish and maintain boundaries of staging areas.

d. Post signs for identification and traffic control.

e. Establish check-in function, as appropriate.

f. Determine and request logistical support for personnel and equipment, as needed.

g. Advise Operations Section Chief of all changing situation and conditions on scene.

h. Respond to requests for resource assignments.

i. Respond to requests for information, as required.

j. Demobilize or reposition staging areas, as needed.

k. Maintain The Staging Activity Log.

Branch Director. The Branch Directors, when activated, are under the direction of the Operations Section Chief, and are responsible for implementing the portion of the IAP appropriate to the Branches.

a. Review Common Responsibilities.

b. Develop, with subordinates, alternatives for Branch control operations.

c. Attend planning meetings at the request of the Operations Section Chief.

d. Review Division/Group Assignment Lists for Divisions/Groups within Branch. Modify lists based on effectiveness of current operations.

e. Assign specific work tasks to Division/Group Supervisors.

f. Supervise Branch operations.

g. Resolve logistics problems reported by subordinates.

h. Report to Operations Section Chief when: IAP is to be modified; additional resources are needed; surplus resources are available; hazardous situations or significant events occur.

i. Approve accident and medical reports originating within the Branch.

j. Maintaining the Branch Activity Log.

Division/Group Supervisor. The Division and/or Group Supervisor reports to the Operations Section Chief or Branch Director, when activated. The supervisor is responsible for implementing the assigned portion of the IAP, assigning resources within the division/group, and reporting progress of control operations and status of resources within the division/group.

a. Review Common Responsibilities.

b. Implement IAP for division/group.

c. Provide available IAP to team/task force leaders.

d. Identify geographic areas or functions assigned to the divisions and groups.

e. Review division/group assignments and incident activities with subordinates and assign tasks.

f. Keep Incident Communications and/or Resources Unit advised of all changes in status of resources assigned to the division and/or group.

g. Coordinate activities with other divisions.

h. Determine need for assistance on assigned tasks.

i. Submit situation and resources status information to Branch Director or Operations Section Chief.

j. Report special occurrences or events such as accidents or sickness to the immediate supervisor.

k. Resolve logistics problems within the division/group.

l. Participate in developing Branch plans for the next operational period.

m. Maintaining the Division or Group Activity Log.

Task Force Leader. The Task Force Leader reports to a Division or Group Supervisor and is responsible for performing tactical assignments of the Task Force. The leader reports work progress, resources status, and other important information to a division or group supervisor, and maintains work records on assigned personnel.

a. Review Common Responsibilities.

b. Monitor work progress and make changes, when necessary.

c. Coordinate activities with other Task Forces, and single resources.

d. Submit situation and resource status information to the Division or Group Supervisor.

e. Maintain The Task Force Activity Log.

Note

The above breakdown of titles, responsibilities, and tasks demonstrates the flexibility and expandability of an ICS. An ICS of the magnitude described here would be appropriate for national- or regional-scale emergency response, such as the terrorist attacks beginning in 2001, with a large release amount (asbestos particulates) and a large affected population. The more common ICS the reader will work under will usually be a much smaller operation. Written summaries of information and orders received and actions taken must be kept of any emergency response action.

15.4 PLANNING THE SUCCESSFUL EMERGENCY RESPONSE

Planning for an emergency response incident is as important as responding to an incident. Planning should be methodical and include all potential participants.

There are recognized building blocks leading to the accomplishment of a successful emergency response. OSHA defines them in the following broad categories. They are the establishment of:

- Pre-emergency planning
- Personnel roles and lines of authority and communication
- Emergency recognition and prevention
- Safe distances and places of refuge
- Site security and control
- Evacuation routes, procedures, and accountability
- Decontamination procedures not covered by the site safety and health plan
- Emergency medical treatment and first aid
- Emergency equipment and procedures for handling emergency incidents

15.4.1 Preemergency Planning

In order for any response to be successful, all of the responders must be specifically trained in their roles. Refresher training is required annually. At this stage, all organizations that are reasonably expected to take part in an actual incident must be involved.

The organization to which the response teams will report is the host. Planning sessions are the time to hammer out details.

- What are the natures of incidents that may occur?
  - What chemicals or other hazardous materials are present in the area of service?
  - Where are they located?
  - What are the names and contact numbers of the personnel usually in charge of those materials?
- What are the area site layouts?
  - How many entrances are there to each site?
  - What is the nature of cach site? Is it a hazardous waste site or a chemical factory, for instance?
- Are there personnel from each site that are expected to aid in the response? Are they trained to the level required?
- If no employees are expected to respond to any emergency, are there evacuation routes and emergency muster destinations?
- What outside support agencies are there within a reasonable response time radius?
- Is the organization equipped to handle any emergency that might likely occur at each site? If not, which is the next-closest team that can handle the response?

As you can clearly see, preplanning is an awesome task. It should involve everyone expected to participate in any response. See more detail of Emergency Response Plans in Section 16 of this book.

15.4.2 Personnel Roles and Lines of Authority and Communication

It is essential that all persons at the site understand their individual roles in an emergency. Section 15.2 above goes into roles and lines of authority in detail.

No employee must be asked, required, or ordered to perform any task for which he or she has not been trained and proven competent. There is no room for heroes at a hazardous material incident. The lone exception is described in Section 15.2.1 under Skilled Support Personnel. We reiterate: use these personnel only after the qualifications of any trained personnel have been exhausted.

15.4.3 Emergency Recognition and Prevention

⊙━━🔑 **Under HAZWOPER, OSHA recognizes three types of releases that can potentially occur: (1) incidental releases, (2) releases that may be incidental or require an emergency response depending on the circumstances, and (3) releases that require an emergency response regardless of the circumstances.**

In order to comply with the HAZWOPER regulations, OSHA divides releases of hazardous materials in the workplace into three categories. It is essential that you, as a HAZWOPER worker, supervisor, or trainer, understand these categories and the actions that are required to control them.

Potential releases of hazardous substances in the workplace can be categorized into three distinct groups in terms of the planning provisions of 1910.120(q):

1. Releases that are clearly incidental regardless of the circumstances
2. Releases that might be incidental or might require an emergency response depending on the circumstances
3. Releases that clearly require an emergency response regardless of the circumstances

Incidental Releases. An incidental release is a release of hazardous substance that does not pose a significant safety or health hazard to employees in the immediate vicinity or to the employee cleaning it up, nor does it have the potential to become an emergency within a short time frame. Incidental releases are limited in quantity, exposure potential, or toxicity and present only minor safety or health hazards to employees in the immediate work area or those assigned to clean them up.

If the hazardous substances in the work area are always stored in very small quantities, such as a laboratory that handles amounts in five-gallon sizes down to milligrams, they may not pose a significant safety and health threat. The risks of having a release that escalates into an emergency are minimal. In this setting incidental releases will generally be the norm and employees will be trained to protect themselves in handling incidental releases following the training requirements of the Hazard Communication standard (29 CFR 1910.1200).

For example, a tanker truck is receiving a load of hazardous materials at a tanker truck loading station. At the time of an accidental spill, the product can be contained by employees in the immediate vicinity and cleaned up utilizing absorbent without posing a threat to the safety and health of employees. HAZWOPER employees may respond to such incidental releases.

This situation describes an incidental spill under the HAZWOPER standard. An incidental spill poses an insignificant threat to health or safety and may be safely cleaned up by employees who are familiar with the hazards of the chemicals with which they are working.

Releases That May Be Incidental or Require an Emergency Response Depending on the Circumstances. The properties of hazardous substances, such as toxicity, volatility, flammability, explosiveness, reactivity, and corrosiveness, as well as the particular circumstances of the release itself, such as quantity, confined space considerations, ventilation, and other factors, will have an impact on what employees can handle safely and what procedures will be followed.

Additionally, there are other factors that might mitigate the hazards associated with a release and its remediation. They include the knowledge of the employee in the immediate work area, the response and personal protective equipment (PPE) at hand, and the preestablished standard operating procedures for responding to releases of hazardous substances. There are usually some engineering control measures that will mitigate the release that employees can activate to assist them in controlling and stopping the release.

These considerations (properties of the hazardous substance, the circumstances of the release, and the mitigating factors in the work area) combine to define the distinction between incidental releases and releases that require an emergency response. The distinction is facility-specific and is a function of the emergency response plan.

For example: A spill of trichloroethylene in a facility that manufactures trichloroethylene may not require an emergency response because of the advanced knowledge of the personnel in the immediate vicinity and equipment available to absorb and clean up the spill. However, the same spill inside a dry cleaner's, with personnel that have had only the basic hazard communication training on trichloroethylene, may require an emergency response by more highly trained personnel. The dry cleaner's emergency response plan in this case would call for evacuation except in cases of the most minor spills. Evacuation and emergency response would be necessary for only much larger spills at the chemical manufacturing facility.

Releases That Require an Emergency Response Regardless of the Circumstances. Some releases of hazardous substances pose a sufficient threat to health and safety that, by their very nature, they require an emergency response, regardless of the circumstances surrounding the release or the mitigating factors. An employer must determine the potential for anticipated emergencies in a reasonably predictable worst-case scenario and plan response procedures accordingly.

Take, for example, a truck driver transporting hazardous materials. At the time of an accidental release in the parking lot of a truck stop, the product cannot be contained by the driver in the immediate vicinity and cleaned up using the truck's pail of absorbent. Because of the larger problem, the truck driver evacuates the area and calls for outside help, as instructed by the employer.

In this instance, if in the event of a spill of a hazardous substance an employer instructs all employees to evacuate the danger area, then the employer may not be required to train those employees under the HAZWOPER standard. However, the ability to decide whether a spill is an incidental spill or one requiring an emergency response requires training. Also, any employees who are expected to become actively involved in an emergency response due to a release of a hazardous material are covered by the standard and must be trained accordingly.

Note

OSHA has limited jurisdiction for over-the-road vehicle operation. In the instance of spills occurring while the material is on the vehicle or otherwise in transportation, OSHA's HAZWOPER standard does not cover the operator *per se*. It does, however, cover emergency response personnel who respond to the incident. If the driver of the vehicle is expected to become involved in an emergency response, then the driver is an emergency responder and must be trained. The driver is covered under DOT hazardous materials regulations.

When you are trying to make a determination as to whether or not an emergency response is required, the following conditions must be considered:

- A medical emergency exists
- The release requires evacuation of employees or the public from an area
- The release poses, or has the potential to pose, conditions that are immediately dangerous to life and health (IDLH)
- The release poses a serious threat of fire or explosion (exceeds or has the potential to exceed the lower explosive limit or lower flammable limit)
- The release requires immediate attention because of imminent danger
- The release may cause high levels of exposure to toxic substances
- There is uncertainty that the employees in the work area can handle the severity of the hazard with the PPE and equipment that has been provided and the exposure limit could easily be exceeded
- The situation is unclear, or data are lacking on important factors

15.4.4   Safe Distances and Places of Refuge

OSHA requires that employees required to evacuate due to an emergency be provided with safe places of refuge.

OSHA requires that employers required to have emergency response plans include safe distances and places of refuge from any reasonably foreseeable incident.

Safe distance is the number of feet, meters, kilometers, or miles away from the incident where a worker would be safe without any PPE. These distances depend upon the type of release or incident, wind direction, site topography, and whether the incident occurs during the day or at night. See Section 8.4 of this book for further information in the Table of Initial Isolation Distances.

In certain incidents, especially those that involve gases, vapors, or mists, when the best (safest) course is to hunker down and ride it out; then evacuation must take place to a place of refuge. However, evacuation could be unsafe for unprotected personnel. For example, a large chemical plant uses chlorine gas in their process. The gas is stored in a 20,000-pound above ground tank. A line from the tank ruptures and all of the contents rapidly escape. In the place of refuge, personnel might be told to turn off ventilation systems and window air conditioners, close all windows and place rags at the bottom of doors. This is called "sheltering in place."

15.4.5 Site Security and Control

Any type of incident unless it is in an extremely remote area, is likely to draw a crowd. Think of TV scenes with reporters covering fires. People at the scene are likely to hamper operations as well as become victims of the release or fire. For example, they may walk through puddles of hazardous material and spread the contamination, leading to a larger and longer cleanup.

At a hazardous waste remediation site or a TSDF, site security measures such as controlled access and fencing must already be in place. In an emergency response where the threat travels past the site's boundaries, additional help will be required to keep people out of that area.

If the site of the incident is at our prior example site, the chemical plant, the situation may be different. Yes, the site has guards at the gates and fences, but perhaps it is in the middle of a neighborhood. What then? That same release of chlorine gas may affect hundreds or even thousands of people, depending on conditions.

In that case, the police and the media may be the best allies. They can spread the word to stay inside. The police can patrol the area, ensuring compliance until the all-clear message is given.

15.4.6 Evacuation Routes, Procedures, and Accountability

○─π **All employees are responsible for knowing the most direct emergency evacuation route from where they are working.**

When an incident or release at a site is imminent or has occurred, all workers must understand what to do when the word to evacuate or the evacuation alarm is sounded. The authors suggest redundant (including both audible and visual) evacuation notification. That is, two or more ways of notifying personnel to evacuate, such as flags, strobe lights, alarms, two-way radios, or phones.

One of the very first pieces of equipment installed at a hazardous waste cleanup site or factory is a wind sock. We have all seen them. One or more socks are placed so as to be visible from any part of the site. We must always know which way the wind is blowing. In whatever direction the wind sock is pointing, that is the direction the wind is blowing toward. If at all possible, personnel should evacuate into the wind. The exception to this would be if a release is upwind of the worker. In that case, the workers should evacuate from the nearest exit and walk around the site until he or she is upwind of the release.

Simply evacuating is not enough, however. There must be a primary and backup muster area. This is the area where evacuees gather at safe distance for a head count. It is essential that you make it to the muster area. Simply evacuating the area is not enough. The authors have seen cases where rescue teams have had to do a hard target search (meaning that they had to look everywhere), putting their lives on the line, for employees who had been evacuated but had gone to lunch! Needless to say, those employees were severely disciplined.

15.4.7   Decontamination Procedures Not Covered by the Site Safety and Health Plan

As touched on in Section 7 of this book, in a site emergency, employees may be required to undergo an abbreviated decontamination procedure. In the case of a large-scale or uncontrollable emergency, such as fire or explosion, the employees may have to evacuate without any decontamination. At a hazardous waste site, that may mean that the employee tracks contamination out of the hot zone, through the warm zone, and into the cold zone.

To prepare for such an event, it is advisable to have a mobile decontamination kit. This kit could be as simple as a blow-up kiddie pool, a five-gallon jug of decon solution and a five-gallon jug of fresh water, and some towels. It may not be pretty, but it can get the job done. All of the worker's PPE should then be put into plastic bags for later complete decontamination and the employee should be sent to the showers.

The decon kit will also have to be cleaned, as well as any contamination tracked outside the hot zone. At least the worker will be safe.

15.4.8   Emergency Medical Treatment and First Aid

At a HAZWOPER site there will likely be accidents and injuries. One job of the SSO is to determine which of these incidents fall under the category of first aid and which require emergency medical treatment.

As we all know, first aid includes small cuts and abrasions, small burns, and the like. These incidents can most likely be responded to with a first aid kit. However, any injury or illness beyond this basic scope requires medical professionals. Due to the extremely high level of physical stress put on responders' bodies, it is also highly recommended that all workers know CPR. They can be adequately trained in one day, with a yearly refresher. Remember, CPR rarely revives a person. It is simply used to maintain blood oxygenation and flow to the brain until the professionals arrive.

For all incidents beyond the Band-Aid® level, seek medical treatment. Preplanning is paramount, as with all aspects of emergency response. The outside aid agency the company or organization calls must be made aware of the nature of the hazards it will be required to deal with. It is extremely helpful to bring the paramedics in at the planning stage. They may have to be specially trained to wear chemical protective clothing, for instance, in order to evacuate victims. Often the firefighters evacuate the victim and the paramedics perform only the medical duties.

15.4.9   Emergency Equipment and Procedures for Handling Emergency Incidents

Professional HAZMAT emergency response teams are normally better outfitted and trained than the emergency response team at the site. We recommend that an organization include the local HAZMAT team in any incident planning.

At the HAZWOPER work site, there will rarely be a fully outfitted emergency response team. Usually, the work site will have first aid capabilities, eyewash, and emergency showers only. Medical emergency response will be left to the professionals. In other cases, such as when responding to an environmental release, the responders will come equipped with what they need and stay at the site until the threat of medical emergency has ended. Emergency response crews must outfit their response vehicles with the equipment that they expect to use based on experience. See Tables 15.5 through 15.9 and Figures 15.5 through 15.8. They must consider the manufacturing and industrial facilities in the area, types and volumes of traffic on roads in their response area, terrain, weather patterns, and availability of assisting response from other agencies. Most often, emergency response crews will be part of the local fire department because the fire department already has the organization and skills that are needed in an emergency response incident such as:

- *A command and control structure:* Fire departments have people trained in giving and taking orders. A fire department is not a democracy. It is a paramilitary organization where members do what they are told because their lives and the lives of others are at stake. There is no time for lengthy discussions of alternatives.
- *Trained firefighters:* Many HAZMAT emergencies result in or are the result of a fire.
- *A well-defined communications structure:* When you dial 911 in most areas of this country, you get police, fire, and ambulance. The firefighting crew is already notified in the case of any non-police emergency.
- *A large inventory of specialized equipment:* Firefighters are accustomed to operating and maintaining a large variety of equipment. They always maintain that equipment in operable condition and are masters at stowing tremendous amounts of gear in a limited space.
- *Emergency medical technicians (EMTs):* Although private ambulance firms are available almost everywhere, nearly every fire department has an EMT staff. An EMT staff is always needed at a HAZMAT emergency.
- *A trained public affairs officer (PAO):* This is one area that is often overlooked in preparing for emergency response. Area residents can become quite anxious if accurate information is not disseminated in a timely fashion. It might be necessary to evacuate an area. If not managed properly, that can cause injuries. A fire department has immediate access to all media outlets and can get the correct information out fast.

There are also qualified private HAZMAT crews and crews from state and federal authorities.

The following tables and figures describe and show an example of what the authors consider to be a well-supplied emergency response vehicle.

Compartments are designated D for driver's side or O for officer's side and numbered consecutively from front to rear. Thus, compartments D1 and D2 are the first and second compartments immediately to the rear of the driver's door.

TABLE 15.5 An Example of a Fully Stocked HAZMAT Response Vehicle

Driver's side (cab) item description	Quantity
Radio	1
Cellphone	1
Laptop computer	1
Wireless connection for laptop	1
Printer (for receiving hard copy data)	1
GPS unit	1
Unit keys	1
Binoculars	1
ERG (latest)	1
NFA Guide to HAZMAT Emergency Responders	1
Local county map book	1
Adjacent counties map book	1
Hearing protectors (muffs)	2
Fuel card	1
Fuel logbook	1
Two-million candlepower rechargeable spotlight	1
Emergency triangle flare kit	1

FIGURE 15.5 An example of a fully stocked HAZMAT response vehicle (driver's side forward).

TABLE 15.6 An Example of a Fully Stocked HAZMAT Response Vehicle (Driver's Side Forward Contents)

Driver's side (D1) item description	Quantity	Driver's side (D2) item description	Quantity
Safety belts (top shelf)	4	Responder® Level B encapsulation suits	6
Webbing straps (top shelf)	8	Barricade® Level B encapsulation suits	6
Small carabiners (top shelf)	2	Tychem® Level B coveralls	8
Medium carabiners (top shelf)	8	Teflon® Level B two-piece suits	4
Large carabineers (top shelf)	2	Butyl booties	8
Dupont Teflon® Level A suits	4	Latex booties	24
Kappler Responder® Level A suits	4	PVC boots	16
Cooling vests	4	Neoprene boots	16
		PVA gloves	20
		PVC gloves	20
		Butyl gloves	20
		Neoprene gloves	20
		Viton gloves	20
		Nitrile gloves	20
		Silver Shield® gloves	50
		Nitrile disposable gloves	50
		Vinyl disposable gloves	50
		Leather palm work gloves	20
		Black dot work gloves	20
		Sweatbands	25
		Chemical splash goggles	6
		Dust and mist masks	24
		Disposable earplugs	25

FIGURE 15.6   An example of a fully stocked HAZMAT response vehicle (driver's side aft).

TABLE 15.7   An Example of a Fully Stocked HAZMAT Response Vehicle (Driver's Side Aft Contents)

Driver's side (D3) item description	Quantity	Driver's side (D4) item description	Quantity
Responder® flash protection	4	Eight-five-gallon metal overpack drum	1
Turnout gear	2	Fifty-five-gallon metal overpack drum	1
Rain suits	6	Twenty-gallon poly lab packs (inside 85-gallon drum)	2
Scott® SCBA communication sets	6	Eight-gallon metal lab pack (inside 55-gallon drum)	1
Spare small SCBA facepieces	1	Five-gallon plastic buckets (inside 55-gallon drum)	2
Spare large SCBA facepieces	2	Hand truck	1
Spare extra-large SCBA facepieces	1	Propane flare tripod and hose	1
Nomex® hoods	4		
SCBA cylinder covers	4		
Duct tape	4		
Kappler Chem-Tape®	2		
Kneepads	12		
Velcro wrist and ankle straps	24		
Antifog spray	1		
Scott® communication interface cables	6		
APR® T-bars	4		
Organic vapor APR cartridges	8		
Scott® voicemitter™ covers	8		
CPC wristcups	14		
Scott® one-hour SCBA	6		
Scott® one-hour spare cylinders	6		
Nomex® coveralls	6		
SCBA facepieces, large	6		
Cairns® 660 helmets	2		
Cairns® HP-1 helmets	4		
Command board easel	1		

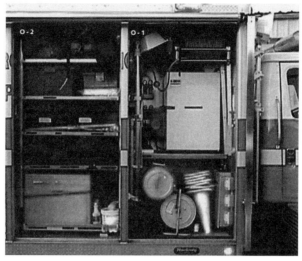

FIGURE 15.7 An example of a fully stocked HAZMAT response vehicle (officer's side forward).

TABLE 15.8 An Example of a Fully Stocked HAZMAT Response Vehicle (Officer's Side Forward Contents)

Officer's side (O1) item description	Quantity	Officer's side (O2) item description	Quantity
Electric pigtails	7	EMT comprehensive first aid kit	1
Freezer	1	Breathing oxygen kit	1
Cool-vest gel packs	36	Spare breathing oxygen cylinder	1
Electric cord reels (125 ft)	2	Scale	1
Electric cord reels (220 V)	1	Spare clothing	8
Fifteen-hundred-watt quartz halogen tripod lights	1	Decon kit	1
Awning rods	3	Tarps	2
Five-gallon water cooler	1	Containment bags	8
Shore line cord	1	Pipe leak diverter	1
Explosion-proof lights and chargers	3	New Pig® pipe plugging kit	1
ICS boards	3	New Pig® pipe patching kit	1
Traffic cones	10	New Pig® drum and tank leak kit	1
Twenty-five-kilowatt PTO generator	1	Drum rolls	2
		Sampling kit	1
		Drum belly patches	2
		Drain plug kit	1
		Inflatable plug kit with pump	1
		Plug and dike container	1
		Nonsparking tool kit	1
		Toolbox with assorted tools	1
		Vetter patch and lance kit	1
		Air drill and chisel kit	1
		Lid lock dome clamps	6
		Assorted plug kit	1
		Assorted patching materials kit	1
		Liquid leak detector bottle	1
		Five-gallon caustic neutralizer	1
		Five-gallon acid neutralizer	1
		Five-gallon lime	2
		Five-gallon soda ash	2
		Nonstatic scoop	1
		Plastic scoop	1

FIGURE 15.8   An example of a fully stocked HAZMAT response vehicle (Officer's side aft).

TABLE 15.9   An Example of a Fully Stocked HAZMAT Response Vehicle (Officer's Side Aft Contents)

Officer's side (O3) item description	Quantity
PMI® rope ½-in. green	1 600-ft
PMI® rope ½-in. white/orange	1 200-ft
PMI® rope ½-in. white	1 600-ft
Extra-large carabiners	22
Large carabiners	23
Straps	2
Roof roller	2
Medium anchor plates	2
Large anchor plates	1
Rope protection wrap	1
Edge pads	2
Miscellaneous rope pads	3
Miscellaneous ½-in. rope	3
L-U-tracks	2
Straight rack	2
Anchor straps	1
Backboard straps	1
Webbing	300 ft
Prusik cord	100 m
Daisy chain	1
Wrist harness	1
Rescue figure 8	4
Etrier	1
Handled ascender	1
Rescender	5
Single pulleys	4
Double pulleys	4
Prusik minding	5

TABLE 15.9 An Example of a Fully Stocked
HAZMAT Response Vehicle (Officer's Side Aft
Contents) (*Continued*)

Officer's side (O4) item description	Quantity
Gloves	24
Load-releasing straps	3
Stretcher spider	1
Y-Lifting bar	1
Duty twin runners	2
Small-body harness	2
Medium-body harness	3
Large-body harness	2
Extra-large-body harness	2
Petzl® helmets	4
Mars kits	1
Sked stretcher	1
Stokes® straps	4
Lock-out/tag-out kit	1
Chemical-absorbent pillows	12
Chemical-absorbent pads	50
Chemical-absorbent booms	4
Chemical-absorbent mini-booms	8
Five-pound bags chemical absorbent pulp	2
Command board easel	1
Plug rug	2
Tripod	1
Tube hoist	1
Backboard	1
Stokes® basket	1
Zumro® inflatable shelter	1
Inflation kit	1
Chlorine A kit	1
Chlorine B kit	1
Chlorine C kit	1
Pallet puller	1
Drum upender	1
Gas shut-off tool	1
Plastic roll	1
Sampling rods	6
Pick-head ax	1
Flat-head ax	1
Hooligan bar	1
Bolt cutters	1
Grounding rods	2
Six-foot lentil	1
Poly street brooms	2
Street brooms	2
Chemical squeegees	2
Chemical hand pump	1
Petroleum hand pump	1
Nonsparking shovels	4
Tables	2
Nonsparking drum sling	1
Nylon® drum strap	1

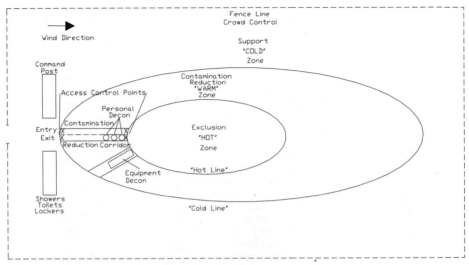

FIGURE 15.9   The three zones for emergency response.

15.4.10   Location and Layout of Work Zones

OSHA, NIOSH, EPA, and USCG recommend dividing the incident site into three zones, establishing access control points, and delineating a contamination reduction corridor. Figure 15.9 is a diagram of the recommended zones. Although this model is based on incident or emergency response, it holds true for long-term environmental cleanup projects as well. This has been explained in detail in Section 7 of this book.

Note

Based on the results of site air monitoring and site soil, water, and hazardous material sampling, the size of the work zones may shrink. There are some very good reasons for this: there will be less area to control, there will be shorter distances for the workers to walk to or escape from the hot zone, there will be less spread of contamination through tracking, and communication with workers will be easier.

At an emergency response, preparation time is limited to minutes or maybe a few hours. The site-investigation information could still be relatively complete if the incident is in an organization's own facility. However, if the site is unfamiliar territory to the responders, they must approach it with extreme caution. Investigation time will be severely limited.

Responders must be prepared for the most hazardous conditions and then, as information is gathered, modify actions and PPE protection accordingly. Then, in coordination with the incident commander's directions, they must act to preserve responders' and bystanders' safety. Following that, their responsibility is in the establishment and operation of the most immediate steps in the control phase. They are the short-term tasks that stop major release of hazardous materials from the site.

15.5   COOPERATIVE EXERCISES

Exercises of your organizations' emergency response capabilities are essential. These exercises can be scaled to fit the needs of the organization and the surrounding community.

There are a number of response elements that the practice or tabletop exercises that your organization conducts can emphasize to the participants. All participants will have the opportunity to demonstrate that they know their role on the team during an incident. Organization members will see how well they can use their emergency response plan. Further, all participants will learn how effective the overall emergency response planning process has been. At the exercise stage, hopefully all of the deficiencies in training, planning, and logistics will be uncovered.

OSHA does not define the various types of exercises that should be practiced to ensure readiness for an actual emergency. Several agencies do define the types of exercise that they recommend. FEMA categorizes exercises as tabletop, functional, and full-scale. EPA categorizes exercises as either tabletop or field. USCG uses a functional-type exercise called OSC/RRT and a field-type exercise known as OSC/local. Private sector organizations might classify their exercise types differently from the public sector types.

The authors recommend that HAZWOPER workers, who are responsible for emergency response, carry out exercises with their entire site, company, or organization at least annually. Tabletop exercises can often be as instructional as full-scale operations. Both types of incident exercise start out with a scenario and then work through the steps necessary to respond to the incident.

Preplanning for the exercise is as important as preplanning for an actual incident. Always include any mutual aid group that your organization will require in an actual emergency. Even if your organization has virtually complete emergency response capabilities on-site, incidents could occur that would require outside aid. Let us take a look at the two common types of emergency preparedness exercises. Prior to any exercise, all participants must be trained adequately in the role they will play. See Section 15.2.1 for details. After everyone is satisfactorily trained, planning can begin for an exercise.

The workers who are expected to respond, along with all supporting personnel, should be part of the planning stage. This is especially important if there is newer management directing a more experienced workforce. Valuable lessons can be learned from the foot soldiers.

As an example, one of the authors' training sessions was at a top-down organization. None of the actual workers was in on the planning stages. As soon as we started the exercise, a maintenance mechanic on the response team told us that the fire hydrants in the response area of the plant (an integral part of the company's response plan) were disconnected due to road repairs. Aside from being a safety violation, this provided a great lesson learned about preplanning.

15.5.1   Tabletop Exercises

> **Tabletop exercises are one of the primary modes of answering the question "What if . . . ?" They are inexpensive, relatively easy to conduct, do not require large amounts of equipment, and provide valuable data regarding your organization's abilities and further training needs.**

A tabletop exercise is an excellent economical way to plan and have the participants execute (verbally and in writing) their roles.

The tabletop starts with all response participants assembled, all trained in their roles. If practical, a scale model of the site available is excellent for better visualization. If this cannot be arranged, a site map or aerial photographs can be substituted. The exercise begins as the emergency response-trained observer starts a stopwatch and tells a randomly chosen worker the incident scenario. The scenario is announced to the incident commander by that randomly chosen worker calling the organizations operations center, which at an emergency response, remediation site or TSDF may be either an office or the command post.

The IC then declares himself or herself in charge and starts using the ICS to gather information and apply resources. Depending upon the resources at hand and the scope of the emergency, the IC will call in the outside mutual aid organizations, such as fire, police, and ambulance. The observer does not prompt the participants. The observer may answer questions pertaining to observable variables such as current weather, weather forecast, time of day, and wind direction. To make the tabletop exercise realistic, times of actions must be calculated. That is to say, if the IC tells an operations level responder to go to the site and give a report of the situation, sufficient time must be allotted for donning the proper PPE, gathering and field calibrating instruments, taking readings, and reporting to the IC.

Many participants forget that the buddy system is required and act out their task individually. Here is a good time to throw a monkey wrench into the works. The observer might say, for instance, "You have forgotten to take your buddy with you. You have been overcome and cannot call for help." (Question for the IC: "What do you do now?")

It is important for observers to remember that there *are* right and wrong answers in these exercises. Participants should not be mocked or put down but encouraged by the fact that incorrect decisions at the exercise stage are learning points. Incorrect decisions at an actual incident may be fatal!

The exercise then runs through the scenario. Workers who are not immediately involved with the response are evacuated or kept at safe distance. Supervisors perform head counts to check for missing personnel. The public, if endangered, must be warned what actions to take for their safety.

Depending on the type of organization and the size of the response area, these exercises can be anywhere from difficult to extremely complex. They are never time-wasters!

One of the authors performed refresher tabletop exercises at a chemical company that had a 10,000-gallon chlorine tank and a 20,000-gallon ammonia tank on the premises. The scenario was that the chlorine tank had a catastrophic failure, there were two workers down in the area, and the plant had to deal safely with about 100 workers. Only 20 workers were trained in emergency response. It was assumed that the tanks were filled to capacity. It was also assumed that the entire contents were released into the atmosphere. Not one trainee knew that the EPA reportable quantity (RQ) for chlorine was 10 lb and that a spill greater than the RQ required notification of the National Response Center.

The command post chosen was the control room of the plant. The control room had been chosen because it was centrally located and, due to the hazards of the plant processes, had backup power and an adequate air filtration system. It was a typical bunker. The problem with this choice was that the IC and staff could not see any of the emergency site due to the room's location. All information came from outside sources. (After the exercise, the company agreed that it was a good idea to install additional video monitors to those already in place to provide a more complete view of the exterior of the plant.)

This exercise had representatives from the fire and police department present to handle fires, crowd control, and medical emergencies. As an example of lack of Unified Command, the facility manager insisted on being the IC, even though he had been brought to that facility only about six months earlier. Neither the fire nor the police departments thought that this individual was qualified to be an IC. They managed to argue about who would be the IC for almost an hour! As it worked out, the fire chief of the local department became the agreed-upon IC as his department had a fully equipped and trained hazardous materials team and all of the responders had years of experience.

Several other deficiencies were also noted. The facility manager had recently purchased incident modeling software and was in the process of learning how to use it. The fire department had different software that they were proficient in using. The participants then spent a good bit of time choosing which modeling software would be used.

A number of key people were absent the day of the exercise. Their places were taken by trained replacements who had no experience. This caused further confusion. Also, some plant personnel who were not trained for response were in plant operator positions. When they were told to evacuate, they did so. However, they could not perform plant shutdown procedures safely. That was obviously a major deficiency. (These plant operators were later all fully trained and now exercise with the other trained responders.) Take all of the above, and add in the probability of a curious neighboring metropolitan population, and the reader will begin to appreciate how complex these exercises can become.

Although the exercise seemed to represent a highly hazardous incident, the response solution was fairly simple. The plant was on the coast. The incident occurred during a time when wind at the site was blowing towards the sea. The command solution was to notify all personnel to shelter in place and wait for the all clear.

15.5.2 Full-Scale Exercises

🔑 **Full-scale exercises are the 'graduate school' of emergency response training. They are as close to real life as many emergency response trainees may ever get. A site is chosen, a scenario of an incident is prepared, and local individuals take roles as the "injured." Response to and cleanup of the site are video recorded for later review and lessons learned.**

Once all of the problems have been worked out at the tabletop level, an organization can 'graduate' to a full-scale exercise. This can be performed at the site or plant level or at the local level, bringing in outside agencies.

If conducted at the site or plant level, the company should make it easy on itself and not train during some crucial process time. Is this cheating? You may think so, but a plant that takes hours or days for shutdown usually takes hours or days for startup. That takes the exercise from a one-day event to a one-week event.

Full-scale exercises differ from tabletops because they are real-time incidents with real people. These should always be tried at plant scale before local or regional involvement. Allow for growth and learning before you expand.

Full-scale exercises will require the entire site's cooperation. Some operators may be exempted from the exercise, but they should be included in the next exercise. It is best to get the most out of any exercise, so plan to have some casualties that must be treated. Have enough PPE available to allow the maximum amount of personnel to suit up. This is an excellent time to try out emergency alarms, monitoring instruments, evacuation, sheltering, and muster locations. If there are any glitches in your communication system, they can be brought to light during this exercise. If you know that you really cannot react to a full-blown emergency with the assets at hand, be honest. Call the fire and police in on the exercise. (Make sure you have arranged it with them ahead of time!) After the HAZWOPER workers have cut their teeth on a site-wide exercise, you can move on to local or regional exercises.

Even though this is an exercise, the authors recommend that EMTs be on-site and prepared to handle any true accidents or injuries, such as heat-related disorders, that might occur.

Local and regional exercises should be held annually. Before you and your organization become involved, try it as a spectator. If you are interested in participating, contact your local emergency planning committee (LEPC), FEMA, USCG, or the EPA's Office of Solid Waste and Emergency Response (OSWER). Most full-scale local or regional exercises would gladly include your organization in their plans.

For this type of event, the guiding agency (usually FEMA or the USCG) enlists a large number of people (high school students, for instance) as casualties. This gives the area's ambulance and hospital personnel a realistic view of their capabilities. It also shows all the participants where improvement is necessary and where the combined effect of multiple releases, or releases coupled with a natural emergency such as a hurricane, will multiply the challenges. At this scale your organization will also learn the limitations of the region's agencies and the almost unlimited capabilities of the national agencies.

Although neither a HAZWOPER incident nor a natural emergency, the Oklahoma City bombing was an excellent case study in how local and national assets work together to respond quickly and efficiently to a huge disaster.

15.6 COMPUTER MODELING PROGRAMS

🔑 **Computer modeling is an important source of information about what is likely to occur at an incident. It can be used for training purposes, but it is most valuable in predicting actual dispersion of releases into the atmosphere.**

There are numerous private-label computer modeling software programs. The choice is yours. As a starting point, the authors recommend the suite of programs named CAMEO, which is freely available in downloadable or CD version from the EPA website. CAMEO, which stands for computer-aided management of emergency operations, is an integrated set of software modules jointly developed and sponsored by the National Oceanic and Atmospheric Administration (NOAA), the Office of Response and Restoration, and the U.S. Environmental Protection Agency, Chemical Emergency Preparedness and Prevention Office (EPA CEPPO). As readers will no doubt know, NOAA is the nation's weather expert. CAMEO is designed to help first responders and emergency planners plan for, and quickly respond to, chemical accidents. HAZWOPER emergencies require rapid actions by firefighters, police, and other emergency personnel. These actions are often hampered by a lack of accurate information about the substances spilled and the safe actions to be taken to protect responders and the public. CAMEO is intended to be a solution to this problem. CAMEO is available for both Apple Macintosh and Windows-compatible computers. It includes:

- A database of hazardous chemicals
- MARPLOT, an electronic mapping program
- ALOHA, a computer model that predicts the movement of chemical vapors and gases in the atmosphere
- The Chemical Reactivity Worksheet, which predicts potential reactivity between two or more chemicals if they are mixed together

CAMEO's chemical database contains response recommendations for about 6,000 chemicals. It also contains 80,000 chemical synonyms and identification numbers. You can quickly search CAMEO to identify unknown substances during an incident. Once a chemical is identified, CAMEO provides firefighting and spill response recommendations, physical properties, health hazards, and first aid guidance.

The ALOHA (Aerial Locations of Hazardous Atmospheres) air-dispersion model predicts the downwind dispersion of a hazardous gas or vapor cloud. Graphical outputs include estimates of the cloud footprint (representing the area where hazardous gas concentrations may reach a level of concern) and chemical concentration over time at locations of particular concern. Inputs include the rate and duration of release of the chemical to the atmosphere. ALOHA is a computer program that uses information you input, along with physical property data from its extensive chemical library, to predict how a hazardous gas or vapor cloud might disperse in the atmosphere after an accidental chemical release. ALOHA can predict rates of chemical release from broken gas pipes, leaking tanks, and evaporating puddles and can model the dispersion of both neutrally buoyant and heavier-than-air gases. (Lighter-than-air gases being of less concern to the HAZWOPER Responder.)

ALOHA can display a footprint plot of the area downwind of a release where concentrations may exceed a user-set threshold level. It also displays plots of source strength (release rate), concentration, and dose over time. ALOHA accepts weather data transmitted from portable monitoring stations and can plot footprints on electronic maps displayed in a companion mapping application, MARPLOT, as shown in Figure 15.10.

MARPLOT is a general-purpose mapping application jointly developed by NOAA and EPA that runs on both Macintosh and PC computers with Windows®. It is designed to be easy to use and fast and consume as little disk and memory space as possible so that you

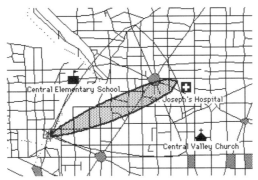

FIGURE 15.10 ALOHA's representation of a gas cloud footprint using MARPLOT to show the projected affected locations. (*Courtesy of USEPA.*)

can create, view, and modify maps quickly and easily. It also allows you to link objects on your computer maps to data in other programs, including CAMEO.

Map data for MARPLOT come from a variety of sources. All of the TIGER/Line data from the Bureau of the Census (roads, water bodies, railroads, parks, and so on) is available in MARPLOT format and can be downloaded for free. Maps are also available on the LandView IV DVD, which also contains EPA-regulated sites, demographic data (for number of people affected), geographic boundaries (states, counties, cities, congressional districts, and so on), Geographic Names Information System (GNIS) features, and selected federal lands from the USGS National Atlas.

MARPLOT can display an ALOHA footprint on an electronic map. You can add information to MARPLOT maps, such as the locations of facilities storing hazardous materials and populations of special concern (such as hospitals, day care centers, and schools). You can keep more information about these locations (such as addresses, hours of opening, and phone numbers for emergency contacts) in other CAMEO databases. Using MARPLOT's search feature, you can quickly access and view this information to assess the degree of hazard posed by an incident and decide how to respond to it.

MARPLOT maps are presented by the Interagency LandView Team: the U.S. Environmental Protection Agency, the U.S. Census Bureau, the U.S. Geological Survey, and the National Oceanic and Atmospheric Administration.

15.7 THE UNITED STATES FIRE ADMINISTRATION HAZARDOUS MATERIALS GUIDE FOR FIRST RESPONDERS

O━π **The USFA Guide is a most valuable response resource. It is an excellent companion to the ERG.**

The following two subsections are adapted from the FEMA/USFA description on the USFA website, courtesy of the USFA.

15.7.1 How to Use the Specific Material Guide (Chemical-Specific) Pages

The Specific Material Guides (RED TAB) provide detailed response information for 430 materials. The materials are selected based on the probability of their being encountered and the degree of the hazard they pose to the First Responder and the surrounding community. An example of a specific material guide (hydrogen cyanide) is shown in Figure 15.11.

HYDROGEN CYANIDE
(STABILIZED)
UN 1051

POISON 6

FLAMMABLE 3

Shipping Name: Hydrogen cyanide, stabilized with less than 3 percent water
Other Names: AC   Hydrocyanic acid solution
               HCN   Prussic acid
               Hydrocyanic acid

WARNING! • POISON! BREATHING THE VAPORS OR SKIN CONTACT CAN KILL YOU!
- Firefighting gear (including SCBA) provides NO protection. If exposure occurs, remove and isolate gear immediately and thoroughly decontaminate personnel
- EXTREMELY FLAMMABLE!

Hazards:
- Odor is not a reliable indicator of the presence of toxic amounts of vapor
- May react with itself without warning with explosive violence
- Container may BLEVE or explode when exposed to fire
- Vapors may travel long distances to ignition sources and flashback
- Vapors in confined areas (e.g., tanks, sewers, buildings) may explode when exposed to fire
- Vapors are slightly lighter than air but will collect and stay in low areas
- Combustion products are less toxic than the material itself

Awareness and Operational Level Training Response:
- DO NOT ATTEMPT RESCUE!
- Stay upwind and uphill
- Determine the extent of the problem
- BACK OFF! - Isolate a wide area around the release or fire, deny entry and call for expert help
- Remove all ignition sources
- For container exposed to fire evacuate the area in all directions because of the risk of BLEVE or explosion
- Evacuate the immediate area and downwind for a large release
- Notify local health and fire officials and pollution control agencies
- If material or contaminated runoff enters waterways, notify downstream users of potentially contaminated water

Description:
- Colorless liquid that boils at 78° F
- Sweet odor like bitter almonds; many people cannot smell it
- Dissolves slowly in water but is soluble in water
- Extremely flammable
- Vapors are slightly lighter than air but will collect and stay in low areas
- Transported in red and white candy striped containers
- Produces large amounts of vapor
- Freezes at 8° F

Operational Level Training Response:
RELEASE, NO FIRE:
- Stop the release if it can be done safely from a distance
- Use large amounts of water well away from the material to disperse vapors - contain runoff
- Ventilate confined area if it can be done without placing personnel at risk
FIRE:
- If material is on fire and conditions permit, DO NOT EXTINGUISH; combustion products are less toxic then the original material. Cool exposures using unattended monitors.
- Specially trained personnel operating from a safe distance can fight fires using alcohol resistant (AFFF) foam or dry chemical if available in sufficient amounts. Under favorable conditions, experienced crews can use coordinated fog streams to sweep the flames off the surface of the burning liquid. Do not direct straight streams into the liquid.
- Cool exposed containers with large quantities of water from unattended equipment or remove intact containers if it can be done safely
- If cooling streams are ineffective (unvented container distorts, bulges or shows any other signs of expanding), withdraw immediately to a secure location

First Aid:
- DO NOT ATTEMPT RESCUE!
- The contaminated victim poses a health risk to the responder
- Decontaminate the victim from a safe distance with a stream of water; have the victim remove clothing if possible; provide Basic Life Support/CPR as needed
- Decontaminate the victim as follows:
  - Inhalation - remove the victim to fresh air and give oxygen if available
  - Skin - remove and isolate contaminated clothing (including shoes) and wash skin with soap and large volumes of water for 15 minutes
  - Eye - rinse eyes with large volumes of water or saline for 15 minutes
  - Swallowed - do not make the victim vomit
- Victims should be examined by a physician as soon as possible
- Toxic effects may be delayed
- Do NOT perform direct mouth to mouth resuscitation; use a bag/mask apparatus
- Note to physician: can produce cyanide toxicity; if symptoms indicate, initial treatment includes the cyanide antidote kit

CAS: 74-90-8

FIGURE 5.11   Sample page from USFA *Hazardous Materials Guide For First Responders.* (*Courtesy of FEMA/USFA.*)

The common chemical name for the material is at the top center of each page. Underneath the chemical name is the UN/NA number (see Section 8.3.3 of this book). Any chemical that does not have a UN/NA number either does not pose a hazard or is not shipped in any significant quantity. The proper shipping name (required on manifests) of the material appears directly under the chemical name. Beneath the shipping name are other names for the chemical.

In the upper right-hand corner of the page are the proper DOT placard(s) for the material when transported within the United States. A complete list of current DOT placards is included in the Guide. For a few materials, no specific DOT placard designation has been made in 49 CFR and this area is left blank on the page.

In the upper left corner of the page is the NFPA 704M® label, providing summary information on acute health, fire, and reactivity hazards plus any special concerns, such as water reactivity, that apply to the material. An explanation of the number designations used in the NFPA label taken from NFPA 704M® is found in Section 8.3.1 of this book.

The label is commonly found on storage containers or posted at fixed facilities. When it is posted on fixed facilities, each designation represents the worst hazard in that category within the building or facility. The label is not found on materials in transport.

NFPA 704M® designations were taken, when available, from the 1994 editions of NFPA 49 and NFPA 325. If the material was not rated in these references, values were determined and assigned from published data, where available. Alternatively, they were based on reasonable estimation from data published on structurally similar materials, using the definitions for these designations from NFPA 49. If the information in the label is from the NFPA 49 or NFPA 325, the designation "NFPA" appears along the right edge of the label. If there is no such designation, the content of the NFPA 704M® label was determined by the Guide's authors.

Below the list of synonyms may appear a section in red entitled "WARNING." This section is vitally important. It provides crucial information about hazards that are immediately life-threatening to the First Responder. A WARNING indicates a very dangerous material because of the health risk or because of the extreme fire, explosion, or reactivity risk(s). Most materials will not have a WARNING section. This does not mean that they are not dangerous and cannot injure or kill, only that they are not likely to do so if they are handled properly.

Below the WARNING section is a section entitled "HAZARDS." This section describes the physical, chemical, or toxic properties of the material that create risks for the First Responder. This section includes such things as explosion hazard, flammability risk, and acute health hazards. Hazards are arranged in the approximate order of their importance to the first responder.

Next to the HAZARDS is a section entitled "DESCRIPTION." This section describes what the material looks and smells like, along with some important information about the physical properties of the material, such as whether it floats or sinks in water or whether it is heavier or lighter than air (if it is a gas or vapor). The information in this section may be useful in verifying the identity of a hazardous material and anticipating some of its actions.

Note that these WARNING and HAZARDS sections are similar to what can be found on the MSDS. However, the next paragraph describes information not found on an MSDS.

In the middle of the page are sections entitled "AWARENESS" and "OPERATIONAL LEVEL TRAINING RESPONSE." These sections list the appropriate actions for the first responder trained to each of these two levels of expertise. Not all of the statements listed may be appropriate for every situation, but actions should not be more aggressive than those listed. The most important recommendations are given first. Remember that these are initial recommendations for the First Responder. They may be modified by the on-scene HAZWOPER Incident Commander. Awareness Level response actions are all defensive in nature. Operational Level response actions are divided into two general situations, those involving releases of material without an accompanying fire and those where a fire is involved, whether or not the material itself is burning. Operational Level responders must remember that actions listed under Awareness and the first part of Operational Level response should be completed *before* the second part of Operational Level response actions is begun.

At the bottom of the page is a section in green entitled "FIRST AID." These recommendations are to be used in caring for victims who are out of the hot zone (see Section 7 of this book). Rescue of victims from within a hot zone should only be performed by trained personnel protected by appropriate PPE. This is not generally a First Responder action.

Removal of hazardous material from the skin, eyes, or clothing of a victim (decontamination) is usually the most important first aid action that can be initiated. It is best performed only by appropriately trained and equipped individuals. However, rapid removal of the material might make the difference between a minor injury and a serious injury. The details of specific decontamination techniques are beyond the scope of this book. In general, using large quantities of water to rinse off materials is almost always the first choice for decontamination in the field. Materials that are so toxic that first aid should not be performed on contaminated victims because of the risk of serious injury to the responder are clearly labeled. There are very few antidotes for treating victims exposed to chemicals, and these are listed in this section for the benefit of hospital personnel who may care for these victims. Other first aid information is also provided.

Finally, at the middle of the page below the first aid section is a CHEMICAL ABSTRACTS SERVICE REGISTRY NUMBER (CAS: ___-__-_). This is a specific identifying number given to each chemical by the Chemical Abstract Service. Mixtures are usually not assigned CAS numbers. While not commonly used in shipping, the CAS number may be found on containers and material safety data sheets (MSDS) and is used by many more detailed references as an indexing number. It is provided as another positive identifier and to allow quick reference to other databases.

15.7.2   How to Use the Materials Summary Response Table

The Materials Summary Response Table (YELLOW TAB) provides summary information on 1422 additional materials. These materials are less likely to be encountered by the First Responder. This table is arranged in alphabetical order using the most common chemical name of the material. For each material the UN number and DOT placard designation are provided, if available. NFPA designations are provided for all materials. In the case where NFPA designations were not available from NFPA 49 or 325, values were assigned by the Guide authors using NFPA 49 definitions along with available data on the material or on structurally similar materials. NFPA 704M® designations taken from NFPA sources are shown in green, while those assigned by the Guide authors are in black.

15.8   NECESSARY REFERENCES

O—π **The four essential references for the Emergency Responder are the DOT's** *Emergency Response Guidebook* **(ERG) (on the Companion CD to this book); the USFA's** *Hazardous Materials Guide for First Responders,* **NFPA 471,** *Recommended Practice for Responding to Hazardous Materials Incidents,* **and the USCG** *Field Operations Guide for Hazardous Materials Spills* **(FOG).**

There are at least four references with which the reader should be comfortable and familiar. The *Emergency Response Guidebook* (ERG) 2000 was discussed in Section 8.4. This is the first reference with which even the Awareness Level First Responder must be familiar. A complete, searchable version is on the Companion CD to this Book.

The second reference is the United States Fire Administration (USFA) *Hazardous Materials Guide for First Responders*. Although written for the broad audience of professional and volunteer firefighters, it contains a wealth of quick reference information for the HAZWOPER first responder at the operations level and above.

The third reference is NFPA 471. *Recommended Practice for Responding to Hazardous Materials Incidents* (current edition). Although written for firefighters, it is also useful for HAZWOPER First Responders.

While the ERG and USFA HAZMAT guide are both field references, the *Recommended Practice for Responding to Hazardous Materials Incidents* and the U.S. Coast Guard *Field Operations Guide for Hazardous Materials Spills,* Field Operations Guide ICS-OS-420-1 are important self-help guidance for the serious student or the seasoned professional.

These references are relied upon by most of the response organizations in the country. Further references can be found in Appendix E on the Companion CD.

15.9   TRAINING AIDS AND ADDITIONAL RESOURCES

The authors recommend the following videos for aiding readers to see real-life examples of what they have learned in this Section.

- *HAZMAT Incident Management: The Eight Step Process®:*
  - *1 Site Management and Control*
  - *2 Identifying the Problem*
  - *3 Hazard and Risk Evaluation*
  - *4 Protective Clothing and Equipment*
  - *5 Information Management and Resource Coordination*
  - *6 Implementing the Response Objectives*
  - *7 Decontamination*
  - *8 Terminating the Incident*
  - *I. C. S.—The Incident Command System*

These videos are available from the Emergency Film Group (www.efilmgroup.com).

- *Sounding the Alarm: Awareness Level Training*
- *Spill Response: A Refresher Session*
- *Don't Panic: Responding to a Hazardous Materials Incident*

These videos are available from Safe Expectations (formerly BNA Communications, Inc.) (www.safeexpectations.com).

15.10   SUMMARY

Forty-hour HAZWOPER-certified individuals have learned that their training is roughly equivalent to Operations Level emergency response. At this level, you now know what type of response you are expected to provide at an incident and what response you are not expected to provide. Knowing what you may and may not do are equally important. This Section has given the 40-hour HAZWOPER-certified individual worker the additional information needed to pursue higher levels of emergency response training.

HAZWOPER personnel reading through this Section will have realized that it is the culmination of the preceding 14 sections. The emergency responder really must know the preceding keys and explanatory information. Although this Section has given the responder several invaluable references, these must be digested and fully understood and the knowledge gained must be demonstrated by classroom examination as well as practical performance.

The authors continued to emphasize in this Section that higher levels of emergency response require additional, specialized training beyond the scope of this book. We have explained that additional training in detail. The authors encourage all readers to take the additional training to have a better overall understanding of Emergency Responder roles and duties in an incident situation.

If you learned only one thing from this Section (and we trust that you learned much more), it is that emergency response according to the Incident Command structure is a systematic, paramilitary process. Because time is critical, responders will be expected to take orders from superiors without question. However, you, as a responder, should be armed with enough knowledge to know if following any commands will put you or others in jeopardy.

The reader has learned that preplanning, annual retraining, and frequent drilling exercises are essential to a well-organized team. Training and drilling should be conducted with all expected participants, including all outside mutual aid groups. Readers also have learned that there is a wealth of response technical information, planning assistance, and aid available from government agencies. There are even entire organizations that can perform the response function for your organization.

SECTION 16
HEALTH AND SAFETY SYSTEM

OBJECTIVE

Our major objective in Section 16 is to make readers aware of the organizational structure and content that OSHA requires for the maintenance of employee safety and health where HAZWOPER work is ongoing. Reading through mountains of OSHA, EPA, and other documentation, you will find this structure sometimes called a program and sometimes called a system or even a plan. We choose to call this structure a system because of the systematic approach all organizations must take to make it work.

You will also find this same initiative called both safety and health and health and safety. Our best guess is that OSHA originally tended to put the word "safety" first because it appears first in their title. Then, over the years, the health and safety plan (HASP) became widely recognized by OSHA, EPA, NIOSH, and most organizations doing Superfund and brownfields redevelopment work. The HASP became viewed as so necessary (and possibly sufficient) preparatory to site remediation work that the wording has gradually developed in that order. Make no mistake. Both concerns, health and safety, go hand in hand. We trust all readers will agree with that after reading Section 5, "Toxicology." If you want to be healthy, you had better work safely!

We interpret the OSHA requirement in this critical area to be a systematic, organization-wide, management-sponsored and adopted, unified system with employee buy-in, promoting employee health and safety, one that has a guiding policy that all members of an organization can readily understand and that both management and employees will want to buy into and can easily remember. That policy requires programs for its execution. The programs must have approved and workable plans to spell out the details. All plans, obviously, must support the policy. Workers must be trained to understand the system, and company policy must require compliance as a condition for employment.

Elsewhere you, as a manager or trainer, might have read explanations of what is required of such a system. Our objective is to allay any concerns or fears those explanations might have caused, due first to the complexity of the explanations, and second to the apparent magnitude of your compliance obligations.

The fact is that OSHA wants you, the organization owner or employer, to have a straightforward, workable, and usable system, one that demonstrates management commitment to health and safety of employees and places strong emphasis on continuous improvement. Then it will succeed much like the quality initiatives Dr. Edwards Deming championed. If you have done any work on ISO-9000, ISO-14000, or other quality or environmental management programs, you will notice some similarities in the management style sought by their proponents. OSHA's regulators do not want you to create a system that ends up consisting of dusty stacks of program and planning documentation. If that happens, as it can, the programs and plans will be infrequently consulted, little used, and not updated or auditable.

Another objective of this Section is to emphasize that any well-managed organization that has already taught and put into practice the health and safety elements of the previous sections of this book might have most of the system requirements in operation. That organization just needs to show that it has policy, programs, and plans that meet reasonable requirements, documented and in routine use by qualified personnel.

However, we must point out that OSHA, EPA, and a number of other concerned agencies have conducted audits of operating HAZWOPER sites for several years. The auditors have found a number of critical deficiencies in the actual day-to-day operations with respect to the HASP. Our objective in response to those findings is to make readers aware of the common failings that were reported. Then we want to present details of health and safety planning, management, and operation that will help avoid those pitfalls.

The site-specific HASP is the prime document auditors and inspectors will want to see upon their arrival at a HAZWOPER site. Then they will want to determine how successful the HASP is in addressing site-specific issues during ongoing site work.

If your organization hasn't done enough along these lines to date, and your facility is covered by the OSHA Acts (see Sections 1 and 2 of this book), your first objective is to understand the salient points of the first 12 Sections of this book. Then you must develop a policy that the employer (or management from the top down) will wholeheartedly buy into. An example of this organization-wide cooperation is the OSHA Voluntary Protection Program (VPP) explained in Section 1.3 of this book.

Of all of the programs and plans that comprise a health and safety system, the most widely known is the site-specific HASP. OSHA requires health and safety system information for a HAZWOPER facility to be embodied in written documents. To assist those needing to prepare a HASP, we include as Appendix K on the Companion CD what we consider to be an all-inclusive model of a HASP, one that should cover the requirements for any site-specific conditions the reader might have. Adapted from one used to cover work at the largest ever Superfund site, it is many times larger than most of these plans need be. If you have a small organization, your HASP might only have a few pages, with references to other company health and safety documents prepared for other purposes.

In addition to the HASP, the programs and plans in the health and safety system must ensure that there is employee understanding of site-specific workplace hazards and the many ways of protecting themselves and others. These programs and plans include emergency action and emergency response, the health effects of toxic materials and then appreciation of the need for regular medical exams, the use of personal protective equipment (PPE), and decontamination following exposure.

To satisfy these requirements, many organizations already have applicable policies, programs, and plans that have been prepared for other purposes. However, OSHA, along with numerous other organizations in joint audits of HAZWOPER site operations, has maintained that lack of site specificity has been the chief deficiency of most plans.

We explain that, in addition to having a policy, programs, and plans, the system must have an organization of key personnel to make it work effectively. Those personnel must have management-approved roles, responsibilities, and authority. That organization can be simple or complex. In a small company, one well-qualified individual could fill several key roles. The president or owner could take on the duties and responsibilities of project manager (PM), site safety officer (SSO), and site supervision and engage an occupational health professional as a contract service. A much larger company could have dozens of individuals filling those roles.

Consider Section 16 as a wrap-up of the health and safety information in all the previous book sections. This is particularly true of earlier book sections dealing with material hazards, toxicology, PPE, decon, confined spaces, and other workplace hazards. All of those topics bear directly upon workers' health and safety preservation.

A site HASP describes the actual observed and the potential hazards of the work site. The HASP will state all company policies. They can be administrative safety controls, engineering controls, and work practices required to minimize site hazards. The most important factor in reducing workplace injuries is implementation of the plan. This requires management's commitment to provide adequate resources for training, accountability, self-audits, and employee involvement.

Lack of an up-to-date, site-specific HASP is a serious infraction of OSHA regulations. If facility management does not know what to do to protect its workers, how are the supervisors and workers going to know what to do?

The terms *worker* and *employee* are used interchangeably. Remember that all workers are required to be employees of a company. Contract workers are employees of a contracting company. Both the company assigning work and the contracting company must have their employees meet the same required health and safety standards.

16.1   *WHAT IS A HEALTH AND SAFETY SYSTEM?*

> **A Health and Safety System is a systematic, organization-wide, management sponsored and adopted, unified 'system' with employee buy-in, promoting employee health and safety.**

Every employer should have an operating system that protects the health and safety of the company's workers, both employees and contractors. Where hazardous materials or their wastes are involved, OSHA requires the employer to have such a system. The authors interpret the OSHA requirement in this critical area to be a systematic, organization-wide, management-sponsored and adopted, unified system with employee buy-in, promoting employee health and safety.

16.1.1   Deficiencies That OSHA Auditors Have Found In Health and Safety Systems

Over a period of years starting in the 1990s, OSHA, in cooperation with other interested agencies, dispatched a team of auditors to examine the overall health and safety systems in operation at a number of HAZWOPER sites. They found widespread awareness of OSHA's HAZWOPER standard. Most contractors and subcontractors had prepared written HASPs. Many, however, used boilerplate language, gathered from past experiences, and generic plans that lacked essential site-specific details. Also, implementation and evaluation of written plans often were lacking.

Sufficient Authority for Change Not Granted to Key Personnel.   At several sites visited by the audit team, contract specifications bound contractors to specific safety and health procedures. Ones that were later found by OSHA and the contract-issuing agency (usually the EPA or the Army Corps of Engineers, managing the project for EPA) to be unsuited to site hazards. For example, at one site, contract requirements limited sampling to contaminants specified in the contract even though evidence of the presence of other hazardous substances was found during site operations.

At another site, the contract required the use of full-face air-purifying respirators despite the fact that sampling results collected during site operations did not reveal airborne levels of site contaminants. As explained in Section 6 of this book, new exposures or unanticipated levels of exposure, either high or low, may necessitate procedural changes in sampling, personal protective equipment (PPE), or medical monitoring for effective hazard control. In such cases, the latitude to change PPE requirements or otherwise reduce control measures can be as important to worker protection and program effectiveness as the authority to increase control levels. With PPE, remember from Section 6 that more is not always better! Site safety officers (SSOs) must consider the physical strain additional PPE places on the worker.

It follows that contracts must provide the agreed-upon site safety officer (SSO) with the authority to establish and modify site safety and health procedures throughout site operations based on existing and anticipated hazards. Conditions can change rapidly or unexpectedly at hazardous waste sites. Modifications to safety and health procedures might be necessary at any time to ensure employee protection. The SSO must be sufficiently knowledgeable to interpret and use site data to make safety and health decisions. The SSO's authority to make such decisions must be clearly established in the HASP.

Inappropriate PPE Requirements and Key Personnel Failures.   At several audited sites, the contractor's original written plan called for inappropriately stringent PPE requirements. Supervisors enforced them even though subsequent sampling results showed that lower levels of PPE could be used. In some cases, SSOs appeared to lack the experience and professional judgment necessary to make the indicated changes. At one site, the SSO was unavailable to oversee site procedures, having been allowed to go on vacation without a qualified replacement—an obvious failure in project management.

O—π   **OSHA's severest criticism of jobsites was the lack of a comprehensive, site-specific Health and Safety Plan (HASP).**

HASPs Found Not to Be Site-Specific.   The auditors also found that although boilerplate documents may be a useful starting point for site-specific plans and may ensure that each HAZWOPER standard paragraph is addressed, the final HASP must identify the specific site hazards that are present. Further, specific administrative and engineering controls must be in place to protect against those hazards.

Despite easily recognizeable site-specific variations in hazards, many of the HASPs reviewed by the audit team appeared to be boilerplate documents. Rather than dealing with the specific hazards, tasks, and safety and health procedures applicable to the site, the plans often contained generic guidance on safety and health topics. Sections on decontamination, spill containment, confined space, medical surveillance, and heat stress commonly reflected this approach. Detail also was lacking in job hazard analyses; the plans overlooked many site tasks and their associated hazards, such as equipment maintenance activities.

O—π   **As readers will recall from Section 7, 'Decon'; the decon procedures must be up and running before HAZWOPER work can begin. OSHA found numerous sites where the equipment was at the site, but was not erected with specific procedures and personnel in place.**

Decontamination Planning and Execution Deficiencies.   OSHA's decontamination zone requirements (Section 7 of this book) are intended to contain and reduce hazardous exposures. When establishing and maintaining work zones, there must be procedures to take into account changing site operations or weather conditions.

The audit team frequently found that work zones were configured more for ease of operation than for hazard isolation. For example, at one site, trailers were designated as clean areas to accommodate the project's administrative needs, despite the fact the trailers were adjacent to a contaminated area. Sample results indicated contamination had spread from the contaminated area into the trailers. On the same site, all roads were designated as clean areas, allowing unimpeded movement of site equipment, even though much of that equipment moved into and between exclusion zones without undergoing decontamination.

Note

Authors of this book have audited several sites where decontamination equipment was on-site but not set up and operational, as required by OSHA, prior to commencement of work. In fact, site operations continued for several days as the decon equipment remained in storage at the site.

On most audited sites, the use of a contamination reduction zone (CRZ), providing a buffer between contaminated and clean zones was rare except for personnel decontamination areas. OSHA's work zone requirements, as we explained in Section 7, are intended to contain and reduce hazardous exposures. When work zones are established and maintained, there must be procedures to take into account changing site operations or weather conditions. We have emphasized that where chemical hazards can be carried from one part of the site to another via dust particles, clothing, equipment, or water runoff, and where heavy equipment must move between zones and on-site roads, the use of a CRZ is essential to decontaminate equipment and personnel. It prevents the transfer or tracking of contamination to clean areas.

Lack of Corrective Action Procedures and HASP Updating. Some audited sites documented safety and health deficiencies but failed to document corrective action or update the site HASP. At one site, the daily safety log contained several references to heat stress difficulties and subcontractor noncompliance with PPE requirements. However, no other documentation was available to indicate whether these issues had been addressed.

Emergency Action Plan and Emergency Response Plan Deficiencies. Thermal treatment to desorb contaminants from soil is a versatile and cost-effective technology. It also presents potentially serious hazards, including fire, explosion, and contact with high-voltage circuits. Emergency action and response routinely must be coordinated with local responders and rehearsed on-site. Although site emergency response often requires the assistance of local fire, ambulance, and medical personnel, it is not uncommon to find local responders unfamiliar with site hazards or with their respective roles in emergencies.

At one rural site that lacked 911 service and relied on volunteer dispatch, fire, and ambulance services, the fire chief had not been contacted about the site's 30,000-gallon LPG (liquefied petroleum gas) tank and expressed concern about his crew's ability to respond to a site fire or explosion that might involve it.

At another site, the local hospital had not been provided a copy of the site's emergency response plan as required by the local emergency planning committee and had no knowledge of site activities. The volunteer ambulance service for that same site also was unfamiliar with site hazards and with the contamination likely to be encountered when transporting or providing aid to potential victims. In addition, workers at several sites expressed skepticism that the emergency alarm system could be heard above the noise of operating equipment. Further, they were not convinced that evacuation routes would be accessible in the event of an actual emergency. The authors recommend frequent emergency evacuation drills even on projects of short duration. Section 15 of this book emphasizes that ALL local responders must be included in response planning and drills.

16.2   CORE ELEMENTS OF THE SYSTEM

The core elements of the Health and Safety System are the health and safety policy, management and employee 'buy-in' to the policy, and core programs and plans of the health and safety system.

First and foremost, an organization must have a health and safety policy. There must be buy-in to that policy by everyone from top management to all worker levels. Then there must be programs, plans, and practices to spell out the working details of the policy. Many times one organization's program is another's plan or practice. That is not a problem. What is critically important is that all employees develop a rich understanding of what the policy is meant to do by the details in the programs, plans, and practices.

As we develop the core elements, readers might recognize that our description of the important characteristics of policy and its embodiment in programs and plans resembles other current initiatives. Ones that affect organizational success and continued operation are:

- OSHA's Process Safety Management of Highly Hazardous Chemicals and Process Hazard Analysis, 29 CFR 1910.119
- EPAs Risk Management Plans (RMP), 40 CFR 68
- ISO 14000 for Environmental Management
- ISO 9000-2000 registration and compliance, showing dedication to quality products and services

16.2.1   Health and Safety Policy

O—π   **For any 'system' to actually work day-by-day it must have the full backing of an organizational 'policy'. Workers must be trained to understand the system and company policy must require compliance as a condition for employment.**

For any system actually to work day by day, it must have the full backing of an organizational policy. Workers must be trained to understand the system, and company policy must require compliance as a condition for employment.

Most organizations have a limited number of policies. That is because policies are top-level statements. They are authorized and signed by top management. Policies tell what all employees in the organization, top management included, shall (that is an order!) keep foremost in mind as they do their daily work and planning. A policy should be brief. It may only have a few sentences. It must be a guiding statement that all members of the organization can readily understand. It must be a statement of actions that both management and employees will want to buy into and can easily remember. Organizations may also have practices. These are much more numerous than policies and are far narrower in scope. They are also more frequently changed and updated with changing working conditions.

Any policy requires programs for its execution. The programs must have approved and workable plans and practices to spell out the details. All plans and practices, obviously, must support the policy. Policies don't change according to the whim or management style of changing managers. The organizational culture becomes embodied in its policies. An example of a practice flowing from a policy is the empowerment of an employee to stop an ongoing process when the employee detects a safety hazard.

The policy we are now leading you toward is a health and safety policy. You may well ask, Does any of this really work? The answer is yes. The authors have seen safety policies in effect over periods of decades at two major corporations. The results have been demonstrable in the reductions of accidents and injuries and the maintenance of high health standards relative to all industry.

What might your health and safety policy be? It might be as simple as this:

"Our employees are committed to making health and safety their first priority in planning and performing their daily work. That means that before beginning any job or task they shall be prepared to use the practices and protective measures that will preserve the health and safety of all persons involved. Those practices and protective measures shall be a part of the living documentation found in the programs, plans, and practices of our organization's Health and Safety System. Working in strict accordance with the rules and regulations of that system shall be a condition of employment."

16.2.2   Management and Employee Buy-in to the Policy

Without management's total agreement that a policy is vital to the organization's success, it won't work. As Dr. Deming said many times, only management can establish policies and systems. His understanding was based upon a lifetime of producing astonishing results in working to improve the operations of organizations around the world. If management does not prepare and then fully endorse the policies and systems of an organization, there is little the employees can do to make them effective. However, if the employees don't buy in to the policies and systems, there will be constant friction between employees and management and the effectiveness of the policies and systems will be very low. The unavoidable conclusion is that both parties must be wholeheartedly in agreement with both policies and systems.

16.2.3   Core Programs and Plans of the Health and Safety System

O—π   **The core programs and plans for an effective Health and Safety System are: the Health And Safety Plan (HASP), Employee Training Program, Medical**

Surveillance Program, Emergency Action Plan (EAP), Emergency Response Plan (ERP), and Record Keeping And Documentation Updates Program.

OSHA has specified a number of core programs and plans that are critical to the success of a health and safety system. They are listed here with brief summaries. Section 16.4 presents details of these plans and programs.

Health and Safety Plan (HASP). The HASP is a site-specific document that provides for the protection of HAZWOPER personnel. This includes assignment of key personnel and training in assigned duties and responsibilities. The HASP is a written plan that must give complete guidance and instructions for activities during routine operations and during emergency response operations. Companies that will not be conducting emergency response operations may omit that part of the HASP, delegating it in writing to another organization by agreement.

Employee Training Program. All workers must be trained and that training documented. Employees must have a full understanding of their rights and responsibilities under all the core programs and plans.

All other workers at a site, facility, or during response to an incident, managers and staff included, are responsible for site safety. Although the law is clear that supervisors have the ultimate responsibility for those they supervise, the workers must share in this responsibility. For this reason, training and training documentation have a key role in the overall safety system.

Medical Surveillance Program. The medical surveillance program is a regulatory requirement. All workers, including employees and contractors, are members of this program. This is a program of examinations and job-site monitoring to ensure the health of all workers. This program consists of preemployment physicals, routine periodic physicals and monitoring, and postemployment physicals. It also includes detailed record keeping of exposures and the results of any medical tests and routine monitoring.

Emergency Action Plan (EAP). The EAP is a written plan that tells workers precisely what they must do when a site emergency is declared (29 CFR 38). In particular, it tells them the types of site alarms (bells, sirens, flashing lights) and their meanings, safe evacuation routes, and assembly areas. Emergencies include fires, storms, floods, hazardous materials releases, and any other type of emergencies. This is the only emergency plan if site employees are not trained in, or expected to provide, emergency response; as described in Section 15.

Emergency Response Plan (ERP). The project manager must develop and implement an emergency response plan (ERP) in accordance with requirements of 29 CFR 1910.120(l) if site workers are expected to perform emergency response at that site. If the site emergency escalates beyond the site workers' ability to control it, the responding workers must follow the EAP and evacuate the area.

Record-Keeping and Documentation Updates Program. Record keeping has several important aspects. Being able to find records easily for management or OSHA inspection is of prime importance. Records retention takes some thought and planning. It requires a retention program.

Documentation updating needs review at least every year. Each program and plan must be reviewed for its current applicability.

Next we will describe the duties, responsibilities, and authority of the key employees involved in making these programs and plans work. Then starting with Section 16.4 we will describe the elements of these six programs and plans.

16.3   DUTIES AND RESPONSIBILITIES OF KEY SYSTEM EMPLOYEES

O—π   The Project Manager (PM), The Site Safety Officer (SSO), Occupational Health Professional (OHP), Technical Staff, Engineers and Scientists, Site Supervision, and Technicians, Operators, Skilled Trades Workers, and Laborers are key personnel in the system.

16.3.1   Employed by Your Organization or Contract Employees?

Whether employed by your organization or by an outside contractor, all employees are obligated to follow the rules and regulations that are part of the programs and plans of the health and safety system. The outside contractor must commit to instructing contract employees to follow all the rules and regulations that organization employees must follow. Contractors must participate in all site drills and exercises that might impact on their part of the site.

16.3.2   Project Manager (PM)

Organization management, as part of its overall delegation of authority, assigns to the project manager the authority for control of all operations at a specific site or sites. That includes the operation of the health and safety system. The site safety officer, occupational health professional, technical staff, and site supervision all report directly to the project manager.

16.3.3   Site Safety Officer (SSO)

Under the overall direction of the project manager, the SSO's sole duty shall be the management of the health and safety system. All other key personnel shall defer to the instructions of the SSO in matters of health and safety. The SSO must be a "competent person" as defined by OSHA. That is, the SSO shall have the authority to cease or alter any site operation upon deeming it necessary for the health and safety of any worker. The SSO must have detailed knowledge of all operations that are transpiring or will transpire at the site. Current certification as a certified industrial hygienist (CIH) or as a certified safety professional (CSP) are two recognized credentials for this position.

16.3.4   Occupational Health Professional (OHP)

Tracking employee health and making recommendations for improvement are the tasks of the OHP. This individual shall have, by virtue of training and experience, the ability to conduct medical examinations and screening tests. The OHP shall also be capable of analyzing the results of these exams and tests and make recommendations based on those analyses. Current Board certification as licensed occupational health professional certified in occupational medicine by the American Board of Preventive Medicine is an excellent credential for this position. Due to the cost of having such a specialist on staff, many facilities use a contractor for this position. The OHP reports to the project manager and is an advisor to the SSO as needed.

16.3.5   Technical Staff: Engineers and Scientists

The technical staff required at any site may vary quite a bit. Again, depending upon the site and the project's complexity, one or many people might be involved. In a small operation, the project manager might also serve in this capacity. Otherwise, all such personnel report to the project manager.

16.3.6   Site Supervision

The site supervisor or supervisors, under the direction of the project manager, have the authority to direct the project work of technicians, operators, skilled trades workers, and laborers. Supervision shall seek the advice of the SSO and technical staff personnel about any details of project work that are outside supervision's area of expertise.

16.3.7   Technicians, Operators, Skilled Trades Workers, and Laborers

These are the personnel that perform the hands-on work of the project described in Sections 13, 14, and 15 of this book. They report directly to supervision.

16.4   HEALTH AND SAFETY PLAN (HASP)

> **Possibly the most important element of the system is the site-specific HASP. The HASP has all the information necessary to safely deal with any potential health or safety concern.**

In the 1986 amendments to the Comprehensive Environmental Response, Compensation, and Liability Act (CERCLA), Congress assigned the task of modifying the National Contingency Plan (NCP) (40 CFR 300). The task was to modify the NCP to provide for protection of health and safety of employees involved in response actions. That task was assigned to the Administrators of the EPA and OSHA, the Secretary of the DOT, and the Director of NIOSH. To satisfy this directive, standards requiring the development of a site-specific health and safety plan (HASP), were established by OSHA under the HAZWOPER standard. Those standards were incorporated into the NCP (40 CFR 300.150). Additionally, the NCP requires compliance with standards and regulations of OSHA's Construction Safety (29 CFR 1926) and General Industry Standards (29 CFR 1910), where applicable.

Brief History of E-HASP, A Joint OSHA and EPA Effort.   In the past several years, EPA, OSHA, and a number of other concerned organizations have jointly pursued a program to conduct health and safety field audits of hazardous waste site operations. During these audits and during standard OSHA inspections, numerous deficiencies in written HASPs have been found. These deficiencies generally indicate a lack of knowledge and/or understanding of the required content of the HASP. The most consistent deficiency is a lack of site-specificity in the plans.

In the early 1990s, EPA's Environmental Response Team (ERT) published a DOS-based software package that generates a model HASP. This software is downloadable from the Environmental Response Team's website. It contains some automatic fill options and simple decision logic but at this date seems limited in its programmed features and in its capacity to generate a site-specific HASP. Based on field evidence, both agencies are interested in updating the original EPA software. This interest is understandable in light of OSHA's introduction, in November 1998, of the proposed Safety and Health Program Rule (29 CFR 1900.1).

This rule would require employers to establish a workplace safety and health program to ensure compliance with OSHA standards and the general duty clause. Although a large number of employers already have such programs in force, a large number of other employers would have to start from scratch. This proposed rule is meeting with great resistance from some legislators as well as business interests, while labor organizations are hailing the proposed rule as a high-water mark for OSHA.

The updated OSHA/EPA E-HASP Guide uses modern (Windows-based) software, more site-specific text, and expanded decision logic to assist users in determining the appropriate controls of health and safety hazards for their sites. The chemical database that is linked to the software must also be expanded and updated. The electronic, interactive E-HASP Guide will be a useful tool for health and safety professionals to provide model language that is acceptable to OSHA in preparing a site's HASP. These same professionals can also draw on the chemical database and embedded decision logic to assist them in identifying the hazards associated with site-specific contaminants and in choosing effective site controls for worker protection.

Independent of passage of OSHA's proposed rule, we see E-HASP, in its fully operating version, as a potential aid for those organizations without a detailed, site-specific HASP. Already, a number of companies are offering private-label versions purported to supply the necessary policies, audits, training records, checklists, and forms. Any such program can, however, be manipulated so that individuals could generate an approved HASP while not really knowing much about the subject.

The model HASP provided in Appendix K on the Companion CD to this manual, however, is exhaustive. We believe that with deletions that are not applicable to a specific site, and supplemented by the reference material in this book, and site-specific details, it can lead to a HASP that exceeds all requirements in the HAZWOPER standard.

16.4.1 Site-Specific HASP Requirements

While all of the following sections must be included in the HASP, a site may determine that a portion of a section does not apply, for example cold temperature extremes for a tropical climate. If a portion of a section is not applicable, it need not be included, but an explanation of nonapplicability must be provided.

Introduction.   The site-specific HASP requires an introduction. Its main purpose is to describe the site the HASP must cover and the HASP's applicability to operations. In developing this description, the preparer (an individual with sufficient site, organization, and hazardous waste background; usually a CSP, CIH, or other qualified individual), must include:

- A brief site description and plot plan

- Background information (e.g., site history, prior site activities)

- Areas of known site contamination

- A capsule explanation of previous and ongoing site characterization

- Site operations to be performed

- References to supporting documents, such as the EAP and ERP

16.4.2 Key Personnel

○━π **All key personnel must know their roles and how to safely perform their jobs.**

The HAZWOPER standard does not require a listing of all key personnel in the site-specific HASP. However, due to the importance of this list of individuals to the overall safety and health effort at a hazardous waste site, a listing of all key personnel is strongly recommended by the authors.

Key personnel responsibilities must be assigned, and appropriate authority delegated, by organization management for any hazardous waste site. All key personnel must have assigned alternates.

At a minimum, the key personnel section to be included in the site-specific HASP should identify the:

- Project manager (PM)
- Site safety officer (SSO)
- Occupational health professional (OHP)
- Technical staff, engineers, and scientists (emergency response coordinator, decontamination station officer, rescue team personnel, security officer, and command post supervisor)
- Site supervision
- Technicians, operators, skilled trades workers, and laborers

16.4.3 Hazard Analysis and Assessment

> **A Hazard Analysis of each task, along with a full assessment of any potential safety or health hazards is the first task to be performed.**

The HAZWOPER standard does not give specific guidance as to the methodology to be used to meet the requirements of 29 CFR 1910.120(b)(4)(ii)(A) regarding hazard analysis and assessment. However, the following should be understood and addressed in any complete plan.

Hazard assessment is a way of identifying hazards that a worker will or might face when doing an assigned job. In manufacturing, chemical processing, and other industrial operations, where large enough quantities of covered materials are involved, OSHA's Process Safety Management and Process Hazards Analysis regulations must be followed. At a minimum, the hazard assessment must include the following steps:

- Identification of an operation or job to be assessed
- Break down of the job or operation into tasks
- Identification of the hazards associated with each task
- Determination of the necessary administrative and engineering controls for the hazards
- Installation and approved operation of the controls

Of course, more detailed hazard assessment procedures are also acceptable and encouraged. More specific information on the hazard assessment process is contained in Section 3 of this book.

16.4.4 Training

> **Documented training of all site personnel in their roles and responsibilities is required.**

In 29 CFR 1910.120(e), various kinds and levels of training are mandated, depending on the task to be performed and the authority and responsibility of the worker. The following information, at a minimum, must be explained in the training:

- Key personnel responsible for site safety and health
- Safety, health, and other hazards present on-site
- Use of personal protective equipment (PPE)

- Safe work practices
- Safe use of on-site engineering controls and equipment
- Medical surveillance program requirements, including signs and symptoms of overexposure
- Site decontamination procedures
- Site ERP and EAP
- Confined space entry procedures
- Site spill containment program and procedures

Additionally, 29 CFR 1910.120 (b)(4)(iii) specifies that a preentry briefing be given to each site worker, manager, and supervisor and any other individual allowed access to the site by a competent person. Documentation of these briefings must be maintained at the site command post.

Note

Once the site is considered a hazardous waste site, the authors agree that the minimum level of training should be 40 hours for workers and 48 hours for supervisors. Although OSHA allows 24-hour training for some limited site exposure, such as monitoring well sampling, the authors feel that all site entrants should receive the full 40-hour training.

16.4.5   Personal Protective Equipment (PPE)

O—π   **After a Hazard Analysis is performed and all available engineering and administrative controls are in place, the proper PPE for the task can be selected.**

Careful selection and use of PPE is essential to protect the health and safety of workers. The purpose of PPE is to shield or isolate workers from the chemical, physical, radiological, and biological hazards that may be encountered at the site and to prevent the contamination of workers' clothes, automobiles, homes, and families. The PPE program contained in the site-specific HASP must, at a minimum, address:

- PPE selection based on site hazards
- PPE use and limitations
- Task duration
- Maintenance and storage
- Decontamination and disposal
- Training and proper fitting
- Donning and doffing procedures
- Inspection procedures prior to, during, and after use
- Effectiveness evaluation procedures
- Limitations due to temperature extremes, as well as other appropriate medical and physical concerns.

Additional information on PPE is contained in Section 6 of this book.

16.4.6   Temperature Extremes

Limitations due to temperature extremes often result in the necessity to modify work/rest cycles or work hours or otherwise reduce the time employees must spend in chemical protective clothing (CPC). Section 6 of this book describes hazards relating to temperature extremes; it provides guidance on how to make work schedule determinations and evaluate the potential for temperature-related disorders or conditions. The temperature extreme program must, at a minimum, address:

- Identification of potential hazards early in the planning phase of the development and operation of required contingency plans
- Proper monitoring of worker physiology
- Early implementation of preventive measures and standard operating procedures (SOPs) prior to commencement of operations so that sound worker practices are developed and followed
- Proper initial training of workers to recognize the symptoms of temperature extreme-related disorders or conditions in themselves and their fellow workers
- Implementation of a buddy system
- Proper acclimatization of all workers to new or changing work conditions

Additional information on temperature extremes can be found in Section 10.

16.4.7 Medical Surveillance

Medical surveillance of workers at hazardous waste sites is necessary to protect the health of the worker, establish fitness for duty and ability to wear required PPE, and ensure documentation of exposure to hazardous materials. The elements of the medical surveillance program contained in the site-specific HASP must, at a minimum, address:

- Employees covered by the program
- Frequency of medical exams and consultations
- Content of medical exams and consultations
- Information provided to the occupational health professional
- Occupational health professional's written opinion and the worker's rights to review that information
- Record-keeping requirements

More specific information regarding medical surveillance is contained in Section 16.6.

16.4.8 Exposure Monitoring and Air Sampling

The monitoring component of the site-specific HASP must be based on all chemical, physical, and other workplace hazards identified in the site characterization. At a minimum, it must address:

- Sampling strategy and schedule for personal OSHA monitoring (samples to be taken in the workers' breathing zone to measure personal exposures)
- Ambient air monitoring for determination of level of protection
- Liquids and soil sampling for determination of level of protection
- Off-site air, soil, surfacewater and groundwater sampling and monitoring for determining off-site migration and effectiveness of cleanup
- A list of instrumentation and equipment to be used
- Pre- and post-use calibration schedules and maintenance of instruments and equipment

- Applicable quality assurance and quality control (QA/QC) procedures and analytical methods
- Sampling and monitoring record keeping and procedures for granting worker access to, and explanation of, results.

More specific information on exposure monitoring and air sampling is contained in Section 12.

16.4.9   Site Control

The site control planning and execution function is to restrict access to the site to authorized personnel. Also, the site control measures program contained in the site-specific HASP must, at a minimum, include:

- A site map showing work zones
- Physical barriers, such as fencing, to be used to control access to the site by curious bystanders
- Definition and use of the buddy system
- Site communication procedures, including emergency procedures
- Safe work practices and SOPs
- Location of nearest medical assistance

Additional information on site control is contained in Sections 7 and 12. Other workplace hazards are covered in Section 11.

16.4.10   Decontamination

No specific decontamination requirements are stated in 29 CFR 1910.120(k). The authors recommend the methodology presented in the U.S. Environmental Protection Agency (EPA) document titled *Standard Operating Safety Guide* (*SOSG*). The SOSG establishes the decontamination layout and required procedures based on the level of PPE used at the site.

With the acceptance of the above recommendation, the decontamination elements contained in the site-specific HASP must, at a minimum, include:

- Training
- Location and layout of decontamination stations and areas
- Decontamination methods
- Required decontamination equipment
- SOPs to minimize worker contact with contaminants during decontamination
- SOPs for decontamination personnel
- Procedures for collection, storage, and disposal of clothing, equipment, and any other materials that have not been completely decontaminated

Additional information on decontamination is contained in Section 7.

16.4.11   Emergency Response Plan

○━━π   **An ERP must be developed if any site workers are expected to respond to emergencies at that site.**

The project manager must develop and implement an emergency response plan (ERP) in accordance with requirements of 29 CFR 1910.120(l) if site workers are expected to respond to emergencies at that site. The ERP to be included in the site-specific HASP must, at a minimum, address:

- Pre-emergency planning
- Personnel roles, responsibilities, and lines of communication
- Emergency recognition, preparedness drills, and follow-up procedures
- Safe distances and places of refuge
- Site security and control
- Evacuation routes and procedures
- Emergency (abbreviated) decontamination procedures
- Emergency medical treatment and first aid
- Emergency alerting and response procedures
- PPE and emergency equipment
- Site topography and layout
- Incident reporting procedures
- List of local emergency response contacts
- Potential worst-case weather by season
- Critique of emergency response and prevention procedures

Additional information about the ERP is contained in Appendix K of the Companion CD.

16.4.12 Emergency Action Plan

An EAP must be developed if site workers are NOT expected to respond to emergencies at that site.

If employees are expected to evacuate the site and not participate in emergency response activities, the site must have an emergency action plan (EAP) in accordance with requirements of 29 CFR 1910.38(a). The EAP to be included in the site-specific HASP must, at a minimum, address:

- Emergency escape procedures and route assignments
- Procedures to be followed by personnel who stay behind to conduct critical operations before they evacuate
- Procedures to account for all employees after evacuation
- Rescue and medical duties for assigned personnel
- Names and phone numbers of personnel and organizations to be contacted for additional aid resources
- Description of the alarm procedures used to alert personnel of emergency and evacuation situations
- EAP training requirements and methods to evaluate employee knowledge of the plan
- Procedures and frequency for EAP drills, review, and update of the plan

Additional information about the emergency action plan is contained in Appendix K on the Companion CD.

16.4.13 Confined Space Entry

The confined space entry procedures for the HASP are derived from 29 CFR 1910.146 and American National Standards Institute (ANSI) Recommendation Z117.1-1989, among other references. The confined space entry program portion of the site-specific HASP must, at a minimum, address:

- Personnel duties and responsibilities
- Identification, posting, and evaluation of confined spaces on-site
- Hazard controls (engineering, administrative and PPE)
- Entry permit contents, requirements, and approval
- Entry procedures
- Lock-out/tag-out requirements and procedures
- Additional safeguards and emergency procedures
- Training requirements

Additional information about confined space entry is contained in Section 10.

16.4.14 Spill Containment

The spill containment program provides procedures to contain and isolate the entire volume of a hazardous waste spill and minimizes worker exposure to hazardous waste spills.

The spill containment program to be included in the site-specific HASP must, at a minimum, address:

- Initial spill actions and response
- Organization of the response team
- Spill cleanup procedures
- Postincident review and evaluation

Additional information on spill containment may be found in Section 13.

16.5 EMPLOYEE TRAINING PROGRAM

Readers will know, by this time, that every Section of this book, and every program and plan required by OSHA, stresses employee training in the essential subject matter. Further, there must be a method of examination to demonstrate that the employee has not only been trained, but has learned that essential subject matter. Therefore there must be initial health and safety training sessions for every new employee, giving the new member the scope of the entire health and safety system. Then there must be routine training sessions, say a half-hour per week, on a continuing basis. Where do you get this training material? The Key Facts from Sections 3 through 11 of this book are the starting point. They must be amplified by the addition of your own critical site-specific key facts.

16.6 MEDICAL SURVEILLANCE PROGRAM

The medical surveillance requirements have been derived from several sources, including the *Occupational Safety and Health Guidance Manual for Hazardous Waste Site Activities* (also referred to as the "Four Agency Document", NIOSH/OSHA/USCG/EPA, 1985); 29 CFR 1910.120; and generally accepted work practices.

16.6.1 The Mission of Medical Surveillance

The six-part mission of the health and safety system's medical surveillance is to acquire a medical log of each exposed worker's health before, during, and at completion of employment. This subsection will show the type of in-depth detail required in all of the programs and plans required in the health and safety system.

Documentation of Workers in the Medical Surveillance Program. The medical surveillance program records must define the physical health of all workers. Documentation must include the purpose of all medical examinations and tests as well as the reasons for conducting such tests. All medical records, including results of personal monitoring, must be maintained for a period of at least 30 years postemployment. There are a number of private companies that specialize in retention of privileged medical information over long periods of time for a fee.

The sequence of events associated with the medical surveillance process, depending upon the worker's job description, is shown in Figure 16.1.

Baseline or Pre-Employment Physical Examination. This examination establishes the baseline medical condition of the worker. This is otherwise known as a "fitness-for-duty" exam. Depending upon the nature of the workers' assignments, this exam could exclude them from hazardous waste operations. This exam will determine the worker's ability to operate effectively and safely while wearing different types of protective equipment. For instance, a worker who is physically unable to wear a respirator due to chronic asthma or emphysema might be declared unfit for duty. Other conditions, such as high blood lead levels or evidence of scarring of the lung, might lead to an unfit-for-duty diagnosis for lead or asbestos abatement work, for instance.

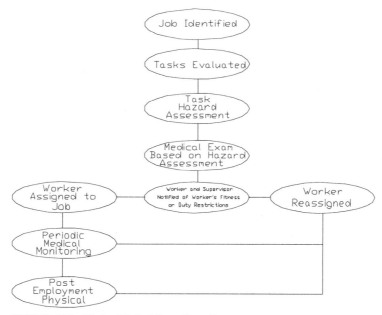

FIGURE 16.1 Worker Medical Surveillance Process

Periodic Monitoring (Screening).   This monitors the health of workers on an established schedule. Yearly respirator physicals are a good example. The purpose of periodic monitoring is to determine whether workers are developing early signs of toxic effects from their contact with site materials. It can also provide indications of the workers' adherence to PPE requirements and work habits, among other factors. Pulmonary function tests and blood tests for lead are two examples.

During these exams, the OHP is required to document any work-related and non-work-related exposures the worker has experienced.

Due to the importance of a HAZWOPER worker's alertness to the myriad hazards at the site, screening might include tests for controlled substances. This monitoring and the associated questioning will also uncover non-work-related exposures that should be considered in further assignments. Tobacco, alcohol, and prescription drug use are three examples.

Examination after Illness or Injury.   This "fitness-for-duty" exam is designed to detect any reasons for temporary or even permanent duty restrictions brought about by an illness or injury. When workers are given a clean bill of health, they are considered fit for full duty.

Termination Examination (Exit Physical).   Just as the preemployment physical gives the worker's health baseline, the termination exam gives a snapshot of the worker's health at separation. This exam is critical so that the organization can be aware of termination health status with regard to any future health-related legal claims.

Maintenance of Medical Records.   Medical records shall be maintained for 30 years after employment. These records must remain confidential and may only be released to a third party with the written consent of the worker. For the protection of the company, documentation of employee exposure and medical conditions shall be maintained as a part of the employee's personnel records. The authors recommend that these records be maintained indefinitely due to changing regulations and the long latency period for some occupational illnesses.

Employees must be granted access to their own records under 29 CFR 1910.20 "Access to Employee Exposure and Medical Records."

16.6.2   Preparing Baseline Medical Information

Identification of Covered Workers.   The project manager must prepare a list of workers to be covered for hazardous waste duties. As necessary background information for determining baseline examination characteristics, the occupational health professional must be provided with the following assistance and information by the SSO prior to an employee examination:

- All data related to expected, or known, covered worker exposure levels to hazardous substances
- A description of personal protective equipment (PPE) expected to be worn by the employee
- A description of the duties expected to be performed by the employee
- Available information from previous medical surveillance examinations
- Updated medical and occupational history

Contractor Responsibilities.   Each contractor must also implement a records retention program with respect to medical records and medical surveillance records of the contractor's employees. In addition, the contractor's occupational health professional(s) must document that they have a copy of the Occupational Safety and Health Administration regulations, 29 CFR 1910.120, "Hazardous Waste Operations and Emergency Response," and 29 CFR 1910.20, "Access to Employee Exposure and Medical Records."

16.6.3 Medical Examination Content

○─ℼ **Medical examination content will be decided by the SSO and the OHP.**

Medical examinations, whether baseline, periodic, or at termination, must include a medical and work history. Such examinations must place special emphasis on symptoms related to exposure to hazardous substances and their health effects. Emphasis must also be placed on fitness for duty when conducting site tasks. The content of the medical examinations must be based on applicable laws and regulations and known or potential exposure to contaminants.

The greatest credibility is expected to be given to examinations where the content is determined by a licensed occupational health professional certified in occupational medicine by the American Board of Preventive Medicine. However, at a minimum, the occupational health professional making the determination must be knowledgeable and experienced in occupational medicine screening and surveillance. If no occupational health professional is on the staff of the employer, the content of the examination is determined by the SSO in concert with a contract occupational health professional. An example of a matrix of medical examination by job task is shown in Table 16.1.

Initial Examination. Each HAZWOPER worker (employee or contract worker) must receive a baseline or preassignment medical examination. This examination must be based on an activity hazard assessment prior to the worker being assigned to a hazardous or potentially hazardous activity (e.g., exposure to toxic materials, repetitive motion, heat/cold stress). The examination must include, at a minimum, the items listed below.

Baseline Medical Examination Parameters. The baseline medical examination parameters are determined by the SSO in coordination with the occupational health professional after review of the assessment of exposure hazards. However, at a minimum, the following checklist will guide the SSO as to what must be included:

❏ A complete medical and work history (OSHA "Ten Minute Questionnaire" included as Appendix H on the Companion CD)
❏ Physical examination
❏ Pulmonary function test (if worker will be expected to wear an air purifying respirator)
❏ Eye examination
❏ Audiogram
❏ Urinalysis
❏ Blood chemistry
❏ EKG (as appropriate)
❏ Heavy metal screen (as appropriate)
❏ Radiological bioassay (as appropriate)
❏ Evaluation of stresses related to repetitive motion

The authors recommend development of a table of hazardous substances expected at the work site. This shall include the target organs affected, the potential health effects, and the medical monitoring to be performed. It shall be prepared by the project manager and SSO. An example is provided in Table 16.2. (The information in the table must be consistent with information in the hazard assessment.)

Periodic Medical Monitoring. Employees working on hazardous waste sites, which may include chemical, physical, and radiological hazards, must be provided with medical examinations every 12 months unless the occupational health professional believes a shorter or longer duration is appropriate or required. The periodic medical monitoring must be completed every 24 months at a minimum. The content of the examination is:

TABLE 16.1 Example of Periodic Examination Based on Job Task

	Medical and work history	Physical examination	Pulmonary function test	X-ray	EKG	Eye examination	Hearing test	Urinalysis
Project manager	X	X	A	A	A	A	A	A
Data collector	X	X	X	A	X	X	X	X
Supervisor	X	X	A	A	A	A	A	A
Heavy equipment operator	X	X	X	A	X	X	X	X
Truck Driver	X	X	X	A	X	X	X	X
Front-end loader or backhoe driver	X	X	X	A	X	X	X	X
Laborer	X	X	X	A	X	X	X	X
Other	A	A	A	A	A	A	A	A

X—Recommended
A—As appropriate

	Blood chemistry	Heavy metals	Bioassay	Other (as determined by occupational health professional)
Project manager	A	A	A	
Data collector	X	X	A	
Supervisor	A	A	A	
Heavy equipment operator	X	X	A	
Truck driver	X	X	A	
Front-end loader or backhoe driver	X	X	A	
Laborer	X	X	A	
Other	A	A	A	

X—Recommended
A—As appropriate

TABLE 16.2 Hazardous Substances Expected to be Encountered,
the Target Organ, Potential Health Effects, and Recommended Medical Monitoring

Hazardous substance	Target organ	Potential health effects	Medical monitoring
Hydrocarbons (some common hydrocarbons will be identified here)			
Toluene	CNS, skin	CNS depression, dermatitis	History of physical exam focusing on nervous system and changes in skin
Trichloroethene	Liver, CNS, kidneys, respiratory, skin	Liver disease and kidney injury, dermatitis, CNS depression, cancer, ventricular arrhythmias	History of preexisting liver disease or decreased lung functions, measurement of liver enzymes and liver function, urine screen, physical exam focusing on nervous system, skin, and respiratory system
Heavy metals (some common RCRA-regulated heavy metals will be listed here)			
Lead	Kidney, blood, GI tract impairment	Renal system, CNS dysfunction	Urine screen, measurement of kidney function where relevant, CBC, history and physical exam focusing on CNS
Herbicides (specific herbicides will be identified here)			
2,4-D	Skin, CNS	Chloracne, peripheral neuropathy	History and physical exam focusing on skin nervous system, urinalysis

- Based on applicable laws and regulations
- Determined by the occupational health professional
- Designed to detect changes from the baseline and previous examinations
- Designed to identify physiological changes

Worker site-specific exposure data, based on the parameters in the previous subsection, and a hazard assessment must be provided to the examining occupational health professional.
Examination after Illness or Injury. Follow-up examinations must be provided as soon as possible to a worker should any of the following situations occur:

- Notification to, or recognition by, the supervision, the project manager, the SSO, or the occupational health professional that the worker has developed signs or symptoms indicating sensitivity or overexposure
- A possible exposure to a worker above the permissible exposure limit or other guiding, published exposure limit
- Lost time illness by a worker of three working days or more
- Any OSHA recordable injury to the worker

In the case of worker injury or illness, the SSO or designated alternate is responsible for notifying the project manager of the incident and of any substance involved. Then an examination must be carried out by the occupational health professional, who must determine the scope of the examination. The worker shall not return to work until the occupational health professional certifies that the worker is fit to return to work, any activity restrictions are identified, and documentation of fitness for duty is provided.

Termination Examination.   The employer must provide a termination medical examination when an employee is terminated or reassigned to an area or activity where the employee is not exposed to hazardous substances. The occupational health professional must determine the termination examination content. If termination occurs within six months of a periodic examination, the occupational health professional may determine that an additional examination is not necessary. Documentation of the decision not to provide a termination examination, and its basis, must be provided in the medical file for the employee.

All HAZWOPER personnel must be included in the medical surveillance program. Also included are the following personnel:

- All employees who are exposed to hazardous substances or health hazards above published exposure limits (e.g., OSHA PELs, ACGIH TLVs®, NIOSH RELs), without regard to the use of respirators, for 30 days or more a year
- All employees who wear a respirator for 30 days (or fractions of days) or more a year or as required by 29 CFR 1910.134
- All employees required to wear an air-purifying respirator
- All employees who are injured, become ill, or develop signs or symptoms due to possible overexposure involving hazardous substances or health hazards from an emergency response or hazardous waste operation

16.6.4   Maintenance and Availability of Medical Records

The employees must be notified of recommended limitations on their assigned work. The occupational health professional must provide a written opinion in the records indicating that the employee has been informed of the results of all exams and of any medical conditions that require further examination or treatment. In addition, the following specific information must be maintained:

- Name and social security number of employee
- Occupational health professional's written opinion, recommended limitations, and results of examinations
- Employee medical complaints related to exposure to hazardous substances
- Information provided to the occupational health professional from the employer
- Engineering controls, work practices, and PPE that are in use and in operation for employee protection

Personnel medical records and exposure monitoring records must be maintained according to the requirements of OSHA and must be maintained for 30 years postemployment. Employees must be given access to their medical records within a maximum of 30 days of their request in accordance with 29 CFR 1910.20. The employer must hold the employee medical records in confidence to the extent permitted by law.

16.7   *EMERGENCY ACTION PLAN (EAP)*

○━π   **An EAP must have the following sections at a minimum: EAP goals and required procedures, emergency escape route assignment, procedures for critical operations personnel, procedures to account for all employees, rescue and med-**

ical duties, reporting fires and other emergencies, EAP contact personnel, emergency evacuation alarm system, EAP training requirements, fire prevention plan, fire prevention housekeeping, fire prevention training, fire prevention maintenance, and procedures for the review and update of the EAP.

Not all sites have the organizational capability for complete response to emergencies. Most sites will, in fact, evacuate employees and only perform such activities as emergency shutdown, first aid, and CPR. With this in mind, OSHA states that it is not necessary for these sites to prepare an emergency response plan. Instead, these sites must prepare an emergency action plan (EAP), as described in Section 16.4.12 and Appendix K on the Companion CD and described further below.

16.7.1 EAP Goals and Required Procedures

The EAP must describe those actions to be taken to provide employee protection in case of fire, hazardous waste or other hazardous material releases, floods, and other emergencies. The EAP must be totally integrated and coordinated with the site's emergency response plan. This EAP checklist will aid you, the preparer, in ensuring that you are including the minimum elements required. The following elements must be included:

❑ Emergency escape procedures and emergency escape route assignments

❑ Procedures to ensure that all contractors on site coordinate their EAPs with the facility EAP to prevent conflicts and confusion

❑ Procedures to be followed by personnel who stay behind to conduct critical operations (shutdown procedures) before they evacuate

❑ Procedures to account for all employees after emergency evacuation has been completed

❑ Rescue and medical duties (first aid, CPR, etc.) for those individuals who are to perform them

❑ Methods for reporting fires and other emergencies

❑ Names and phone numbers of personnel and organizations to be contacted for further aid should it be required

❑ Alarm system(s) to be used to alert personnel to the emergency/evacuation

❑ Training each employee must receive in order to carry out the requirements of the EAP effectively, and the methods for evaluating employee knowledge of the plan

❑ Fire prevention plan

❑ Procedures for the review and update of the EAP (scheduled rehearsals)

16.7.2 Emergency Escape Route Assignment

This section of the EAP must contain the information necessary for the safe, orderly evacuation of site employees. The contents of this section must focus on the procedures for evacuation and the establishment of emergency escape routes. A minimum of one primary and one alternate route to be used for the evacuation of personnel must be established. The following checklist may be used in establishing safe emergency evacuation routes. See Section 7 for further explanation of zones.

❑ Routes must be directed from the exclusion zone through an upwind contamination reduction zone to a support zone, and from the support zone to an off-site location should conditions require a general site evacuation

❑ Direct evacuation routes should be predominately upwind from the exclusion zone. In some cases, as at very large sites, some exits may need to be placed downwind.

❑ Workers must be informed during training that they are not safe until they have reached the designated safety area.

❑ The accessibility of potential routes must be considered. Obstructions such as locked gates, trenches, pits, drums, tanks, and other barriers must be considered, as well as the additional time and equipment needed to maneuver around or through them.

❑ Two or more routes, separate from each other, must be planned that lead to safe areas. Ensure that routes do not overlap or intersect.

❑ Mark all evacuation routes as "safe" or "not safe" on a daily basis, depending on wind direction and other conditions at the site.

The SSO must consider mobility limitations for personnel wearing PPE and other emergency equipment, including:

❑ Developing procedures to use ladders, planking, and other materials to traverse hazardous terrain (ditches, heights, trenches)

❑ Checking clearances of access ports, crawlspaces, hatches, manholes, and/or tunnels to ensure that personnel wearing protective equipment can enter and exit safely through them

❑ Establishing a routine for ensuring that all evacuation routes are kept clear and marked for immediate recognition

In the event of an emergency that necessitates an evacuation of the site, the employer must:

❑ Provide audio and/or visual evacuation alarm notification supplemented by the use of portable radios

❑ Train all personnel to evacuate upwind of any activities along established safe, well marked, evacuation routes and then proceed to the predetermined assembly location so that all personnel can be accounted for

❑ Ensure that personnel are trained to proceed to the closest exit with their buddies, and then continue to the safe area associated with the evacuation route

❑ Train personnel to remain in that area until the reentry signal is sounded or supervision provides further instructions

❑ Provide a map that shows evacuation routes from the site to an immediate safe area

❑ In the event of a major incident, indicate further assembly areas and safe distances on the map

16.7.3 Procedures for Critical Operations Personnel

Procedures and instructions must be developed to clearly identify critical operations and actions required of those personnel performing critical operations during site evacuation. Examples are shutting down and deenergizing equipment that may contribute to the emergency. All critical operations personnel shall be equipped with radios to account to the SSO for their delayed evacuation and estimated time of evacuation. This checklist of procedures and instructions includes:

❑ A listing of all critical operations, the personnel assigned to perform them, and the procedure for updating the listing

❑ A description of any additional training requirements for designated personnel

❑ A step-by-step procedure (SOP) to complete the critical tasks

❑ Estimated time required for the employee to complete the assigned critical tasks

❑ Procedures for their delayed evacuation (if different from normal evacuation procedures)

❑ Procedures for reporting to the designated safe area

❑ Procedures for emergency communications during the conduct of critical tasks and their delayed evacuation

❑ A description of procedures for evacuating prior to completion of critical shutdown

16.7.4 Procedures to Account for all Employees

Procedures must be developed that account for all personnel immediately following an evacuation. This is of paramount importance for the safety of the workers and the safety of the rescue team. The authors suggest the use of muster points. These are specific areas, denoted by signage or painting on concrete or asphalt, where the employees meet for a head count during emergency evacuations of the site. Few agonies equal sending a Rescue Team into danger to rescue a worker who has been improperly unaccounted for at the muster point. These procedures must:

• Explain how to account for, and report, the number of personnel evacuated

• Provide a means for notifying the SSO and emergency personnel when employees are missing

• Include procedures for accounting for personnel performing critical operations

• Provide a means of supplying the names and work locations of all personnel at the site (for example, a sign-in/sign-out log at the site's main entry point)

16.7.5 Rescue and Medical Duties

Personnel may need to perform lifesaving CPR or emergency first aid prior to the arrival of the local emergency personnel (fire, medical, Rescue Teams). This section of the EAP must describe the actions to be taken by site personnel with first aid and CPR responsibilities. At a minimum, this section must contain the following information:

• Procedures for notifying emergency personnel

• A current listing of all qualified, currently certified personnel with first aid and CPR duties

• Training, and refresher training, frequency requirements for personnel with first aid and CPR duties

• Description of conditions under which employees should attempt to perform first aid or CPR

• Description of medical and rescue duties taken until the arrival of the emergency medical technicians (EMTs)

• Procedures for reporting the incident to the appropriate personnel

16.7.6 Reporting Fires and Other Emergencies

All personnel at the site must know how to report emergencies. This section of the EAP must contain the information needed for personnel to report fires and other emergencies at the site so that competent aid will be received promptly. At a minimum, this section must contain:

• A description and location of communications equipment available at the site for emergency reporting

• Procedures for the use of site communications equipment

- Procedures for reporting the emergency
- Complete listing of emergency telephone numbers, radio frequencies, emergency signals, and related contact information. For instance, the universal visual 'sign' for requesting immediate first aid is the crossing of one's arms repeatedly across the chest

16.7.7 EAP Contact Personnel

A list of primary and alternate key personnel must be developed that identifies individuals and organizations with the expertise to explain and provide technical information on the use of the EAP. Name, title, and telephone number must be listed for these personnel or organizations and applicable alternates. This list must be continually updated and made available to all employees and visitors. Additionally, the list must include the names and contact information for site safety personnel (PM, SSO, and OHP) and their alternates.

16.7.8 Emergency Evacuation Alarm System

This section must contain a complete description of all alarms and signals (and related backup systems) to notify personnel of an emergency. All applicable alarms and signals (evacuation or other emergency), as well as the actions to be taken in the event the alarm is sounded, must be fully explained. All alarm systems must take into account the limitations imposed on the workers by the required PPE and site activities. Alarms must be redundant. There must be audible (siren, bullhorn) as well as visual (strobe light, flag) alarms. Alarm systems and backup systems must be in compliance with 29 CFR 1910.165.

16.7.9 EAP Training Requirements

Each employee at the site must be properly trained in all aspects of the EAP. Critical operations personnel and those personnel with first aid and CPR responsibilities must receive the additional training necessary to carry out their duties effectively. This training must be conducted and documented in accordance with the training section of the site-specific HASP. The following items in this section of the EAP must be performed and documented:

- Designate and train a sufficient number of persons to assist in the safe and orderly emergency evacuation of employees.
- Identify critical operations and designate and train a sufficient number of persons to conduct those operations.
- Train all personnel working at the site in evacuation and other required emergency procedures, as applicable.
- Provide a listing of all required EAP training to be completed prior to beginning work at the site.
- Coordinate unique training requirements with the SSO and provide a description of all required training not covered by the training section of the HASP.

All personnel must receive required training prior to beginning work at the site. The methods for evaluating personnel competency in carrying out the requirements of the EAP must be reviewed with each individual covered by the EAP at the following times:

- Initially, when the plan is developed
- Upon worker assignment
- Whenever the employee's responsibilities or designated actions under the plan change
- Whenever the plan is changed

16.7.10 Fire Prevention Plan

This section must contain procedures that reduce the vulnerability of the workplace to fire. Fire prevention plan requirements are specified by OSHA in 29 CFR 1910.38 and include discussions of housekeeping, training, and maintenance. The plan must be updated as hazards change. This fire prevention plan satisfies OSHA's requirement for a fire protection plan.

In addition to the categories listed above, the plan must include:

- A list of the major workplace fire hazards
- Names or titles of personnel responsible for the control of workplace hazards
- A list of types of fire protection equipment or systems and the hazards they control
- A list of workers trained and authorized to fight fires using all available equipment ('fire party')
- Pre-fire planning in coordination with the local emergency response services to familiarize them with workplace process hazards

WARNING

THERE MUST BE NO SMOKING, MATCHES, OR LIGHTERS ALLOWED IN THE CONTROLLED AREA AT ANY TIME. HOT WORK (WELDING, GRINDING, AND BURNING) MUST ONLY BE PERMITTED AFTER A THOROUGH REVIEW OF THE ACTIVITY AND THE SITE BY THE SSO. THE SSO MUST SIGN OFF ON A HOT WORK PERMIT AND ASSIGN A FIREWATCH FOR EACH INSTANCE OF HOT WORK. IN THE CASE OF WELDING OR BURNING, A COOL-DOWN PERIOD MUST BE ESTABLISHED. DURING THIS PERIOD, THE FIREWATCH MUST REMAIN AT THE SITE UNTIL ANY RISK OF THE HOT WORK IGNITING COMBUSTIBLES IS ELIMINATED. THE AUTHORS' SUGGESTED COOL-DOWN PERIOD IS 30 MINUTES.

16.7.11 Housekeeping

Procedures must be developed to control accumulations of material and residues so that they are not the source of a fire emergency. Fire prevention housekeeping plans include:

- Proper handling, storage, and control procedures for flammable and combustible waste materials
- Listing of potential ignition sources (welding, grinding, burning, internal combustion engines, and the like) and their control procedures
- Housekeeping procedures that maintain the means of egress free of obstructions (keeping aisles and doors unblocked, for instance)

16.7.12 Training

Site workers must be informed of the fire hazards associated with the materials and processes to which they are exposed. They must be trained in response procedures for fires. Workers who are required to respond to small fires with extinguishers must be trained in the proper selection, use, and limitations of the available extinguishers and that training must be documented. A fire is considered out of control (beyond the incipient stage) as soon as the contents of a portable fire extinguisher are exhausted and the fire continues. In the case of firefighting using hose stations, the fire is considered out of control if the use of the hose is not abating the fire, the fire increases in magnitude, or there is an explosion.

16.7.13 Maintenance

Proper maintenance, inspection, and testing of fire protection equipment and systems are key to eliminating or controlling fire development. Equipment must be maintained according to manufacturers' specifications. In addition, National Fire Protection Association NFPA 25, *Inspection, Testing and Maintenance of Water-Based Fire Protection System,* and other NFPA codes covering the particular equipment or device must be consulted and applied. Hand-held fire extinguishers must be inspected and recharged (if necessary) annually. A weather-resistant tag recording that inspection must be affixed to each extinguisher. Monthly checks of each extinguisher are strongly recommended. Whether maintenance is performed in-house or by contractors, the individuals performing the work must be properly trained. Names and titles of personnel responsible for maintenance must be kept on file. Maintenance, inspection, and testing procedures apply to:

- Equipment installed to detect fuel leaks and control heating
- Portable extinguishers, automatic sprinkler systems, and fixed extinguishing systems (sprinkler heads, sprinkler control valves, fire pumps)
- Detection systems for smoke, carbon monoxide, heat, or flame
- Fire alarm systems
- Emergency backup systems and the equipment they support

16.7.14 Procedures for the Review and Update of the EAP

This section must contain the procedures to review and update the EAP. At a minimum, this section must describe procedures for:

- The periodic (at least annually) review and, if necessary, update of the plan
- Training employees on the latest changes to the plan
- Coordination and integration of the latest version of the EAP with the HASP, site emergency response organizations, and the ERP (if applicable).

16.8 EMERGENCY RESPONSE PLAN (ERP)

The site-specific emergency response plan must be designed as a separate section of the HASP and be compatible and integrated with the disaster, fire, and emergency response plans of local, state, and federal agencies. The purpose of the ERP is to protect workers in emergency situations resulting from the release of all types of hazardous wastes and substances, including extremely hazardous substances, CERCLA hazardous substances, RCRA hazardous wastes, and any substance listed by the DOT as a hazardous material. The requirements for an ERP at hazardous waste sites are codified in 29 CFR 1910.120. The ERP must be developed and implemented prior to beginning site operations. Hazardous waste site operations must not begin until the ERP is in place.

CAUTION

ANY ORGANIZATION THAT STORES OVER THE THRESHOLD QUANTITIES (TQ) OF THE LISTED HAZARDOUS SUBSTANCES FALLING UNDER THE EPA'S RELEASE MANAGEMENT PLAN (RMP) REGULATION MUST ALREADY HAVE AN ERP. IN MOST CASES, THE ERP UNDER RMP WILL FULFILL THE REQUIREMENTS OF THIS SECTION.

Sites with RCRA-permitted treatment, storage, and disposal facilities for hazardous waste, having the required contingency plan meeting the requirements of their permit, would not need to duplicate the same planning elements. Those items of the ERP that are properly addressed in the RMP contingency plan may be integrated to the ERP.

The objective of this section is to describe the minimum required elements of the ERP, which are as follows:

- Preemergency planning
- Personnel roles, lines of authority, and communication
- Emergency recognition and prevention
- Safe distances and places of refuge
- Site security and control
- Evacuation routes and procedures
- Decontamination procedures
- Emergency medical treatment/first aid
- Emergency alerting and response procedures
- PPE and emergency equipment
- Procedures for reporting incidents to local, state, and federal governmental agencies
- Critique of response and follow-up

Elements identified above may require data that has already been created and documented in other sections of the HASP (site characterizations, hazard assessments, maps, and transportation routes). Copies of this documentation must be incorporated into the ERP.

16.9 HEALTH AND SAFETY SYSTEM RECORD-KEEPING AND DOCUMENTATION UPDATING PROGRAM

Record-keeping has several important aspects. Being able to find records easily for management or OSHA inspection is of prime importance. Records retention takes some thought and planning. It requires a retention program. You will need to retain records for all those in current employment. You are required to maintain the records of those employees who separate from the organization for a period of 30 years postemployment.

Documentation updating needs review at least every year. Reviews shall be conducted at earlier intervals following accidents, emergencies, and near misses. Each program and plan must be reviewed for its current applicability. Place a review update form in the file. At each scheduled review, a knowledgeable key person must make the review and either make changes for management approval and get that approval or signify that no changes were required. Then sign and date the form. Whenever there are incidents or near misses, there are reasons for documenting what has happened. Then there is a need for a "lessons learned" addition to the documentation. These should not wait for scheduled updates. They are required immediately. They could also be part of weekly health and safety reviews, or sooner, if there is an imminent danger.

16.10 SUMMARY

We trust that readers will agree with the authors' insistence upon sound organizational structure for the maintenance of employee health and safety where HAZWOPER work is ongoing. We have explained our choice in calling this structure a system because of the systematic approach all organizations must take to make it work.

The HASP (health and safety plan), over the years, has become the most widely recognized plan by OSHA, EPA, NIOSH, and organizations doing Superfund and brownfields redevelopment work. By now the HASP has become accepted as necessary preparation prior to any HAZWOPER site work.

Importance of Management Sponsorship and Employee Buy-in

The authors have presented their interpretation of the OSHA requirements for an effective health and safety system. It must be a systematic, organization-wide, management-sponsored and adopted, unified system that includes hearty employee acceptance, or buy-in. The system cannot operate without a guiding policy. All members of an organization must find it easy to understand and remember. Approved programs and workable plans must naturally follow to make the policy a living statement.

OSHA management wants you, the managers, trainers, and other key personnel at a HAZWOPER site, to have a workable and usable system. A system that demonstrates management commitment to health and safety of employees and one that places strong emphasis on continuous improvement. Don't forget the amazing success of the quality initiatives that Dr. Edwards Deming championed. You will notice some similarities in the management style sought by quality management consultants and OSHA.

Requirement for a Living Document

As we stated earlier, OSHA does not want you, as an owner or manager, to create a system that consists of stacks of program documentation and plans that are infrequently consulted, little used, and not updated or even auditable. We want to emphasize that any well-managed organization that has already taught and put into practice the health and safety elements of the previous sections of this book already has most of the system requirements in operation. You just need to show that you have policy, programs, and site-specific plans documented and in routine use by site personnel.

We have provided guidelines in this Section for policy, programs, and plans that can make an effective system. We provide Appendix K in downloadable form on the Companion CD as a model site-specific HASP. By "site-specific" we mean that it contains explanations of every issue that almost any site must address. Don't be daunted by it. If you are a HASP preparer, every one you prepare will be far smaller than the Appendix K example. You just have to eliminate inapplicable portions and add site-specific information.

Overcoming HASP Deficiencies Found in Previous Audits

However, we have pointed out that OSHA has been a party to audits of operating HAZWOPER sites for several years. Auditors have found some recurring deficiencies in actual day-to-day operations that were following HASPs. We have presented details of health and safety planning, management, and operation intended to avoid those deficiencies. You have been alerted that the site-specific HASP is the prime document auditors and inspectors will want to see, and observe in operation, upon their arrival at a HAZWOPER site. We have explained the HASP features auditors are seeking.

If you have done little along these lines to date and your facility is covered by the OSHA Acts (see Sections 1and 2 of this book), your first objective is to understand the salient points of the first 12 Sections of this book. Then you must develop a policy that the employer (or management from top down) will wholeheartedly buy into.

If you have a small organization, your HASP might only have a few pages, with references to other company health and safety documents prepared for other purposes.

In addition to the HASP, the programs and plans in the health and safety system must ensure that there is employee understanding of: the workplace hazards of materials and wastes, the many ways of protecting themselves and others in the workplace, emergency action and emergency response, the health effects of toxic materials; and then appreciation of the need for regular medical exams, the use of personal protective equipment, and decontamination following exposure.

To satisfy these requirements, many organizations already have applicable policies, programs and plans that have been prepared for other purposes. However, OSHA, along with numerous other organizations in joint audits of HAZWOPER site operations, has maintained that lack of site specificity has been the chief deficiency of most plans.

We explain that, in addition to having a policy, programs, and plans, the system must have an organization of key personnel to make it work effectively. Those personnel must have management-approved roles, responsibilities, and authority. OSHA has found common examples where the required delegation of authority was absent. Site health and safety performance suffered.

HASP Must Match the Organization's Needs

We showed that a HASP can be simple or complex. In a small company, key personnel roles could look quite different from those in a large organization and site. The president or owner of a small organization would have the authority and, if qualified, could take on the duties and responsibilities of project manager, site safety officer, site supervision, and technical advisors. The owner could engage an occupational health professional and other key specialists as contract services. On the other hand, a much larger company, at a much larger and more complicated site, could have dozens of individuals working in those functions.

We have advised that readers consider Section 16 as a wrap-up of the health and safety information of all the previous book sections. This is particularly true of earlier Sections dealing with material hazards, toxicology, PPE, decon, confined spaces, and other workplace hazards.

Keep the System Documentation up to Date

OSHA requires a health and safety system for a HAZWOPER facility to be a written document, kept up to date and readily available for inspection. We have recommended that it have the following elements: policy, programs that describe how the policy will be carried out, site health and safety plan, an emergency action plan (EAP), and, if workers are trained and equipped to respond to an emergency, an emergency response plan (ERP).

A site health and safety plan describes the actual observed and the potential hazards of the work site. The HASP will state all company policies (administrative controls), engineering controls, and work practices required to minimize those hazards. The most important factor in reducing workplace injuries is implementing the plan. Implementation requires management's commitment to provide adequate resources for training, accountability, self-audits, and employee involvement.

Lack of an up-to-date, site-specific health and safety plan is a serious infraction of OSHA regulations. If the organization does not know specifically what to do to protect its workers, how are the supervisors and workers going to know what to do?

Workers must be trained to understand the system and company policy must require compliance as a condition for employment.

The terms *worker* and *employee* are used interchangeably. Remember that all workers are required to be employees of some organization. Contract workers are employees of a contracting company. Both the organization assigning work and the contracting company must have their employees meet, and be protected by, the same required health and safety standards.

APPENDIX A
GLOSSARY AND ACRONYMS

A.1 GLOSSARY

Note

The definitions given here for words and phrases are written to assist HAZ-WOPER personnel in their work with hazardous materials and wastes. Included are terms used by OSHA, EPA, DOT, FEMA, USCG, the National Response Center (NRC), the U.S. Fire Administration, and others.

abandoned well   A well which is permanently not being used or that is in a state of such disrepair that it cannot be used for its intended purpose.

abatement   The actions taken to reduce the amount or degree of the hazard, intensity of the release, or threatened release of a hazardous material.

absolute pressure (psia)   The total pressure within a vessel, pipe, or other container, recorded as pounds per square inch, absolute. It includes the atmospheric pressure plus the gauge pressure. The pressure we read on a gauge is gauge pressure, psig. The gauge reads zero when on the shelf and not installed for measurement. The gauge measurement never includes the atmospheric pressure. The atmospheric pressure at sea level is about 14.7 psia. In a total vacuum, the pressure would be 0.0 psia.

absorbed dose   In exposure assessment, this is the amount of a substance that penetrates an exposed organism's absorption barriers (for example, skin, lung tissue, gastrointestinal tract) through physical or biological processes. The term is synonymous with *internal dose.*

absorption   To take up and hold a gas in a liquid or solid; or take up a liquid in a solid. Water absorbs (or dissolves) carbon dioxide to make soda water. A sponge absorbs water. Do not confuse a*b*sorption with a*d*sorption. (See **adsorption.**)

absorption material   Material used to soak up liquid hazardous materials, such as Oil Dri®, disposable towels and rags, and specialized products for absorbing one material and not another, such as specialty wipes and pads from New Pig™ that will absorb oil but not water.

acceptable entry conditions   The conditions that must exist in a permit space to allow entry and ensure that employees involved with a permit-required confined space entry can safely enter into, and work within, the space without additional PPE, such as respirators.

access control point   An opening in the physical boundary between work zones that is used to regulate movement of personnel or equipment between the zones of a hazardous waste site.

accident   An unexpected event that generally results in injury, loss of property, or disruption of service.

accident site   The location of an unexpected occurrence, failure or loss, either at a facility or along a transportation route, resulting in a release of hazardous materials.

acclimatization   The process whereby an employee gradually adjusts to work under extreme conditions such as high or low temperatures, high relative humidity, and high or low altitude; the physiological and behavioral adjustments of an organism to changes in its environment.

acid   Any chemical material (liquid, gas, or solid) that undergoes dissociation in water with the formation of hydrogen ions. Acids give a sour taste and may cause severe burns. They turn litmus paper red and have pH values of 0 to about 6.9. Acids will neutralize bases or alkaline media and will react with bases to form salts. "Vinegar" is the common name for dilute acetic acid produced by natural fermentation.

acid deposition   A complex chemical and atmospheric phenomenon that occurs when emissions of sulfur and nitrogen compounds and other substances are transformed by chemical processes in the atmosphere. Often it occurs far from the original sources, and the transformed substances are deposited on earth in either wet or dry form. The wet forms, popularly called "acid rain," can fall to earth as rain, snow, or fog. The dry forms are acidic gases or particulates.

acid neutralizing capacity   Measure of the ability of a base to resist changes in pH.

acidic   The condition of a material when it contains a sufficient amount of acid to lower the pH below 7.0.

ACGIH   American Conference of Governmental Industrial Hygienists, an organization of professionals in governmental agencies or educational institutions engaged in occupational safety and health programs. ACGIH develops and publishes recommended occupational exposure limits for chemical substances and physical agents (www.acgih.org).

acrid   Irritating and bitter. Sulfur oxides (SOx) cause an acrid odor.

action level   (1) A quantitatively measured level of a chemical, biological, or radiological agent in the atmosphere, at which actions (air monitoring, medical surveillance, respirator training, starting engineering controls) must be taken. This is required to prepare workers for potentially dangerous exposure levels and prepare the workplace to reduce the potential for worker exposure. Usually expressed as a vapor, fiber, or particle concentration in air; ppm, ppb, $\mu g/m^3$, or mg/m^3 for an eight-hour time-weighted average (See **TWA.**) (2) In the Superfund program, the existence of a contaminant concentration in the environment high enough to warrant action or trigger a response under SARA and the National Oil and Hazardous Substances Contingency Plan. The term is also used in other regulatory programs.

activated carbon   A highly adsorbent form of carbon used to remove odors and toxic substances from liquid or gaseous emissions. It is used to remove dissolved organic matter, chlorine, and other contaminants from drinking water. It is also used in motor vehicle evaporative control systems.

activated charcoal   A form of activated carbon from other sources. Charcoal is an amorphous form of carbon, formed by heating in the absence of oxygen, wood, nutshells, animal bones, and other carbonaceous materials. Charcoal becomes activated by heating it with steam, while excluding oxygen, to 800–900°C. During this treatment, a porous, submicroscopic internal structure is formed that gives it an extensive internal surface area. Activated charcoal is commonly used as a gas or vapor adsorbent in air-purifying respirators and as a solid sorbent in air sampling.

activated sludge   Product that results when primary effluent in wastewater treatment is mixed with microbiological-laden sludge and then agitated and aerated to promote biological treatment. This speeds the breakdown of organic matter in raw sewage undergoing secondary waste treatment.

active ingredient   Ingredient of a product that actually does what the product is designed to do. The remaining ingredients may be inert but have other useful properties.

activity plans   Written procedures in a school's asbestos-management plan that detail the steps a **Local Education Authority (LEA)** will follow in performing the initial and additional cleaning, operation, and maintenance program tasks; periodic surveillance; and reinspection required by the Asbestos Hazard Emergency Response Act (AHERA).

acute effect   Adverse health effect on a human or animal that has severe symptoms developing rapidly and coming quickly to a crisis. Usually associated with high doses over short periods. Also see **chronic effect.**

acute exposure   A dose (usually a large amount of a chemical, biological, or radiological agent) that is delivered to a receptor (usually a person) in a single event or over a short period of time.

acute lethality   Death of animals immediately or within 14 days after a single dose of, or exposure to, a toxic substance.

acute toxicity   The ability of a substance to cause severe biological harm or death soon after a single exposure or dose. Also, any poisonous effect resulting from a single short-term exposure to a toxic substance.

add-on control device   An air pollution control device such as carbon adsorber or incinerator that reduces the pollution in an exhaust gas. The control device usually does not affect the process being controlled and thus is add-on technology, as opposed to a scheme to control pollution through altering the basic process itself.

administered dose   In exposure assessment, the amount of a substance given to a test subject (human or animal) to determine dose-response relationships. Since exposure to chemicals is usually inadvertent, this quantity is often called "potential dose."

administrative controls   Procedures enforced by management necessary to ensure safe operation of a facility. These include hazard assessments, record keeping, adjustments of work/rest cycles, and disciplinary actions, among others.

administrative order   A legal document signed by an OSHA or EPA official directing an individual, business, or other entity to take corrective action or refrain from an activity. It describes the violations and actions to be taken and can be enforced in court. Such orders may be issued, for example, as a result of an administrative complaint whereby the respondent is ordered to pay a penalty for violations of a statute.

administrative order on consent   A legal agreement signed by an EPA official and an individual, business, or other entity through which the violator agrees to pay for correction of violations, take the required corrective or cleanup actions, or refrain from an activity. It describes the actions to be taken, may be subject to a comment period, applies to civil actions, and can be enforced in court.

administrative record   All documents that EPA considered or relied on in selecting the response action at a Superfund site, culminating in the Record of Decision for Remedial Action or an Action Memorandum for Removal Actions.

adsorption   To take up and hold, by a strong force, gas or liquid molecules on the surfaces of a solid. Activated carbon is an adsorbent media in organic vapor respirator cartridges.

aerated lagoon   A holding or treatment pond in which air is injected or vigorously mixed with the water to cause it to dissolve in the water. This increased oxygen concentration speeds up the natural process of biological decomposition of organic waste by stimulating the growth and activity of microorganisms.

aeration   A process that promotes biological degradation of organic matter in water by increasing the oxygen content. The process may be passive (as when wastewater is exposed to ambient air), or active (as when a turbine or sparging device introduces the air).

aeration tank   A vessel in which aeration takes place.

aerobic   Life processes that require the presence of oxygen (See **Anaerobic.**)

aerobic treatment   Processes by which microbiological organisms decompose complex organic compounds in the presence of oxygen and use the liberated energy for reproduction and growth. (Such processes include extended aeration, trickling filtration, and rotating biological contactors.)

aerosol   Fine mechanical dispersoids of liquids creating particles small enough to remain airborne much longer than larger size particles.

affected landfill   Under the Clean Air Act, municipal solid waste landfills that meet criteria for capacity, age, and emissions rates set by the EPA. They are required to collect and combust their gas emissions. The chief combustion component is methane.

AFFF ("A Triple F")   Aqueous Film Forming Foam, an additive to water used to extinguish burning fuels by smothering.

afterburner   In incinerator technology, a burner and chamber located downstream of the initial incinerator. The incinerator exhaust passes through the higher-temperature afterburner flame in order to destroy the most resistant materials, such as PCBs.

agent   Any substance, force, radiation, organism, or influence that affects the body. Effects may be beneficial or injurious.

Agency for Toxic Substances and Disease Registry (ATSDR)   Charged under Superfund to assess the presence and nature of health hazards at specific Superfund sites. It acts to help prevent or reduce further exposure and the illnesses that result from such exposures and to expand the knowledge base about health effects from exposure to hazardous substances.

Agency for Toxic Substances and Disease Registry's Hazardous Substance Release/ Health Effects Database (ATSDR-HazDat)   The scientific and administrative database that provides access to information on both the release of hazardous substances from Superfund sites or emergency events and the effects of hazardous substances on the health of human populations.

agricultural chemicals   Chemicals such as pesticides, herbicides, fungicides, insecticides, and fertilizers used in agricultural applications to control pests and disease or control/promote growth. The toxicity of agricultural chemicals is classified from I (most toxic) to IV (least toxic). There are specific OSHA regulations governing worker safety in the presence of agricultural chemicals. DOT also has special regulations regarding shipping (for instance: not with foodstuffs).

agricultural pollution   Farming wastes, including runoff and leaching of pesticides and fertilizers; erosion and dust from plowing; improper disposal of animal manure and carcasses; crop residues; and debris.

air   The composition of dry air is nitrogen, 78.1%; oxygen, 20.9%; carbon dioxide and rare gases, 0.1%; and argon, 0.9%. Water vapor (humidity) varies.

air bill   The shipping paper prepared from a bill of lading that accompanies each parcel of an air shipment.

airborne contaminants   Fine solids, liquids, or gaseous materials that are mixed in the air and that flow with air movements.

air changes per hour (ACH)   The number of times the total volume of air in a space is theoretically replaced in one hour. OSHA requires a minimum of four air changes per hour in the workplace.

air distribution system   The system used to deliver breathing air to a respirator or fully encapsulating suit.

air-line respirator   A respirator that is connected to a compressed, breathable air source by a hose of small diameter. The air is delivered continuously (positive pressure) or intermittently (pressure demand) in a sufficient volume to meet the wearer's breathing requirements. One type of atmosphere supplying respirator (ASR).

air permeability   Permeability of soil with respect to air. Important in the design of soil-gas surveys. Measured in centimeters per second.

airborne particulates   Total suspended particulate matter found in the atmosphere as solid particles or liquid droplets. Chemical composition of particulates varies widely, depending on location, work being conducted, and time of year. Sources of airborne particulates include dust, emissions from industrial processes, combustion products from the burning of wood and coal, combustion products associated with motor vehicle or other internal combustion engine exhausts, and reactions of contaminants in the atmosphere.

air-purifying respirator (APR)   A respirator that uses absorption, adsorption, or chemical reaction to remove specific gases, mists, dusts, fibers, and vapors from the air and trap them in the respirator's cartridge. Also, a respirator that uses densely woven fibers in a filter to remove particulate matter from the air. An air-purifying respirator must be used only when there is sufficient oxygen to sustain life. Further, the specific air contaminant levels must be below the concentration removal limits of the device for those specific contaminants. (See **maximum use level** and **maximum use concentration.)**

air pollutant Any substance in air that could, in high enough concentration, harm man, other animals, vegetation, or material. Pollutants may include almost any natural or artificial composition of airborne matter. They may be in the form of solid particles, liquid droplets, gases, or a combination thereof. Generally, they fall into two main groups: (1) those emitted directly from identifiable sources and (2) those produced in the air by interaction between two or more primary pollutants or by reaction with normal atmospheric constituents, with or without photoactivation. Exclusive of pollen, fog, and dust, which are of natural origin, about 100 contaminants have been identified. Air pollutants are often grouped in categories for ease in classification; some of the categories are solids, sulfur compounds, volatile organic chemicals, particulate matter, nitrogen compounds, oxygen compounds, halogen compounds, radioactive compounds, and odors.

air pollution The presence of contaminants or pollutant substances in the air that interfere with human health or welfare or produce other harmful environmental effects.

air pollution control device Any mechanism or equipment that cleans emissions generated by a source (e.g., an incinerator, industrial smokestack, or automobile exhaust system) by removing pollutants that would otherwise be released to the atmosphere.

air quality standards The level of pollutants prescribed by regulations that are not to be exceeded during a given time in a defined area.

air quality criteria The levels of pollution and lengths of exposure above which adverse health and welfare effects may occur.

air-reactive materials Materials that can ignite at normal temperatures when exposed to air in the absence of an ignition source. Finely divided metals poured through air are an example.

air sparging Injecting air into a liquid to produce fine bubbles (increased surface area), to increase decomposition rate of contaminants, or to volatilize VOCs. Air is sparged into an aquifer to strip or flush volatile contaminants from the ground water. The contaminants can then be captured by a vapor extraction system.

air stripping A treatment system that removes volatile organic compounds (VOCs) from contaminated groundwater or surface water by forcing a rising air stream through falling water and causing VOCs to leave with exhaust air. A type of air pollution control device.

air toxics Any air pollutant for which a National Ambient Air Quality Standard (NAAQS) does not exist (i.e., excluding ozone, carbon monoxide, PM-2.5, sulfur dioxide, nitrogen oxide) that may reasonably be anticipated to cause cancer; respiratory, cardiovascular, or developmental effects; reproductive dysfunctions; neurological disorders; inheritable gene mutations; or other serious or irreversible chronic or acute health effects in humans.

air surveillance Use of air monitoring and air sampling to identify and quantify airborne contaminants on- and off-site. It includes monitoring changes in air contaminants that occur over the lifetime of work at a site.

algal blooms Sudden spurts of algal growth, which can affect water quality adversely. They indicate nutrient enrichment of surface water and potentially hazardous changes in local water chemistry. Algal blooms can be responsible for fish kills due to algae lowering oxygen concentrations to below levels required for fish respiration.

algaecide A substance or chemical used specifically to kill or control algae.

aliquot A measured portion of a sample taken for analysis. One or more aliquots make up a sample.

alkali Common alkalis form soluble soaps with fatty acids. Alkalis are also referred to as bases. Alkalis can cause severe burns to the skin. Alkalis turn litmus paper blue and have pH values in solution from above 7 (neutral) to 14. The fat from cooked meat can be heated with caustic soda to form common clothes-washing soap, lye soap.

alkaline   The condition of water or soil that contains a sufficient amount of alkali substance to raise the pH above 7.0.

alkalinity   The capacity of bases to neutralize acids. An example is the capacity of lime added to an acid spill to neutralize the acidity.

allergic reaction   Abnormal physiological response to a chemical stimuli by a person sensitive to that particular allergen.

allergic respiratory reaction   Labored breathing, coughing, or gasping caused by inhaling a particular substance. An example is sneezing and tiredness caused by common hay fever.

allergic skin reaction   Reddening, swelling, or itching of the skin following contact with a substance to which a person has become sensitized due to previous skin contact or natural body conditions. Some people have an allergic skin reaction to common yard grass, poison ivy, or poison oak.

alopecia   Loss of hair.

alpha (α) radiation   Large-particle radiation that can be blocked with a sheet of paper. The least hazardous radioactive radiation to human health.

alternate method   Any method of sampling and analyzing for an air or water pollutant that is not a reference or equivalent method but that has been demonstrated in specific cases, to EPA's satisfaction, to produce results adequate for compliance monitoring.

alternative compliance   A policy that allows facilities to choose among methods for achieving emission reduction or risk reduction instead of command-and-control regulations that specify standards and how to meet them. Use of a theoretical emissions bubble over a facility to cap the amount of pollution emitted while allowing the company to choose where and how (within the facility) it complies; emissions trading, for instance.

alternative fuels   Substitutes for traditional liquid, petroleum-derived motor vehicle fuels such as gasoline and diesel. It includes mixtures of alcohol-based fuels with gasoline, methanol, ethanol, compressed natural gas, and others.

ambient air   Any unconfined portion of the atmosphere: open air, surrounding air.

ambient air quality standards   (See **criteria pollutants** and **National Ambient Air Quality Standards.**)

ambient measurement   A measurement of the concentration of a substance or pollutant within the immediate environs of an organism (such as OSHA monitoring); taken to relate it to the limits of safe exposure.

ambient temperature   Temperature of the surrounding air or other medium. For air, usually considered between 65°F and 75°F.

amenorrhea   Absence of menstruation.

anaerobic   A microbiological process that occurs in the absence of oxygen.

anaerobic decomposition (also called anaerobic digestion)   Reduction of the net energy level and change in chemical composition of organic matter caused by microorganisms in an oxygen-free environment. A common product of anaerobic digestion is methane.

analgesia   Loss of sensitivity to pain. An analgesic lessens pain. Aspirin is a common analgesic.

anesthetic   A substance that, when given to a human or animals, causes a total or partial loss of sensation. Overexposure to anesthetics can cause impaired judgment, dizziness, drowsiness, headache, unconsciousness, and even death. Chloroform was the first anesthetic discovered. Modern anesthetics for dental and other surgery are more effective and have fewer side effects.

anhydride   A compound derived from another compound by removing elements composing water (hydrogen and oxygen).

anhydrous   Containing essentially no water. A substance in which no water molecules are present as solvent, solute, or hydrate or as water of crystallization.

annulus   The space between two concentric tubes or casings or between the casing and the borehole wall in a well.

anorexia   Loss of appetite.

anosmia   Loss of the sense of smell.

anoxia   The detrimental effects, including dizziness, unconsciousness, and others, due to lack of oxygen in inspired air.

ANSI   American National Standards Institute. A privately funded organization that identifies industrial and public national consensus standards and coordinates their development. ANSI approves, through testing, HAZMAT PPE, among other things. The ANSI website is www.ansi.org.

antidote   Remedy to relieve, prevent, or counteract the effects of a poison.

anuria   Absence or defective excretion of urine.

apnea   Breathing temporarily stopped.

appearance   Physical state of a material, such as solid, liquid, gaseous; or color, texture, or size.

applicable or relevant and appropriate requirements (ARARs)   Any state or federal statute that pertains to protection of human life and the environment in addressing specific conditions or use of a particular cleanup technology at a Superfund site.

aquatic toxicity (AQTX)   Adverse effects on marine life that result from their being exposed to a toxic substance.

aqueous   Water-based solution or suspension. Frequently, a gaseous compound dissolved in water. Eye drops and nose drops are water-based solutions. Orange juice is a water-based solution and pulp suspension combined.

aqueous solubility   The maximum concentration of a substance that will dissolve in pure water at a reference temperature and pressure.

aquifer   A water-bearing formation of permeable rock, sand, or gravel, sandwiched between two impermeable layers of rock, capable of yielding water to a well or spring.

area of review   In the Underground Injection Control (UIC) program, the area surrounding an injection well that is reviewed during the permitting process to determine if flow between aquifers will be induced by the injection operation.

area source   Any source of air pollution that is released over a relatively small area but that cannot be classified as a point source. Such sources may include vehicles and other small engines, small businesses and household activities, or biogenic sources such as a forest that releases hydrocarbons.

aromatics   A type of organic compound containing one or more benzene rings in its structure. Benzene (one benzene ring), naphthalene (two benzene rings joined), and anthracene (three benzene rings joined) are examples. The common gasoline components BTEX (benzene, ethylbenzene, toluene, and xylene) are all aromatic hydrocarbons. Benzene is a listed carcinogen.

arsenicals   Pesticides and other compounds containing arsenic.

artesian (aquifer or well)   Water held under pressure in porous rock or soil confined by impermeable geological formations. An artesian well produces water without requiring any mechanical pumping.

asbestos-containing waste materials (ACWM)   Mill tailings or any waste that contains commercial asbestos and is generated by a source covered by the Clean Air Act under the National Emission Standard for Hazardous Air Pollutants (NESHAPS).

asbestos   The common name for a group of naturally occurring silicate minerals that separate into thin but strong fibers. There are six asbestos minerals that have been used commercially: chrysotile, amosite, crocidolite, anthophyllite, tremolite, and actinolite, the most common being chrysotile. Asbestos is a known carcinogen.

asbestosis   Chronic scarring of the lungs, caused by inhaling airborne asbestos fibers. The disease makes breathing progressively more difficult and can be fatal. The onset is believed to take between 15 and 30 years from the first exposure.

aseptic   The condition of being sterile or free from contamination. The opposite of **septic.**

ash   The mineral content of materials remaining after complete combustion.

asphyxia   Lack of oxygen and interference with the oxygenation of the blood. Can lead to unconsciousness.

asphyxiant   Any gas or vapor that either chemically, biologically, or physically displaces the breathable atmospheric oxygen, or otherwise makes it impossible for a person to breathe, which causes unconsciousness or death by suffocation. Most *simple asphyxiants* are harmful to the body only when they become so concentrated that they reduce oxygen in air (normally 20.9%) to levels below 19.5%. An example is nitrogen. Some chemicals such as carbon dioxide function as *chemical asphyxiants* by reducing the blood's ability to carry oxygen. (See **reducing agent.**) Asphyxiation is a major potential hazard of working in confined spaces.

aspiration hazard   Danger of drawing material into the lungs leading to an inflammatory response.

assessment   In HAZWOPER work, the process of determining the nature and degree of hazard of a hazardous material or hazardous materials incident.

assigned protection factor (APF)   A measure of efficiency assigned to various respirators by OSHA and NIOSH. Note that OSHA and NIOSH assign different values to the same type of respirator. (See **protection factor.**)

asthma   Disease characterized by recurrent attacks of dyspnea, wheezing, and perhaps coughing caused by spasmodic contraction of the bronchiole in the lungs.

ASTM Committee F23   Made up of producers and users of Chemical Protective Clothing (CPC). The Committee develops test methods and voluntary standards assessing the performance of protective clothing against occupational hazards.

ASTM Standard F739   Standard test method for resistance of protective clothing materials to permeation.

ASTM Standard F1001   Standard chemical test battery used to compare material resistance to 21 chemical families.

ASTM Standard F1052   Standard practice for pressure testing of Level A totally encapsulating chemical protective suits. The standard allows a 20% pressure drop to occur during the test.

asymptomatic   Neither causing nor exhibiting symptoms.

assay   A test for a specific chemical, microbe, or effect.

atmosphere-supplying respirator (ASR)   A respirator that provides breathing air from a source independent of the surrounding atmosphere. There are two types: air-line respirators and self-contained breathing apparatus (SCBAs).

atmospheric pressure   The pressure exerted in all directions by the atmosphere. At sea level, mean atmospheric pressure is 29.92 in. of mercury, 14.7 psia, or 407 in. of water gage.

attainment area   An area considered to have air quality as good as or better than the national ambient air quality standards as defined in the Clean Air Act. An area may be an attainment area for one pollutant and a nonattainment area for others.

attenuation   The process by which a compound is reduced in concentration over time, through absorption, adsorption, degradation, dilution, and/or transformation. It can also be the decrease with distance of sight caused by dispersion of light by particulate pollution.

ataxia   Loss of muscular coordination.

atmosphere (atm)   A unit of pressure measurement. One standard atmosphere (atm) = 14.7 psia. The average sea level atmospheric pressure, say at Atlantic City, NJ, is about 14.7 psia. The atmospheric pressure in Denver, CO, is closer to 12.2 psia.

atrophy   Wasting or diminution in the size of tissue, organs, or the entire body caused by lack of use.

attendant   The individual stationed outside a confined space who monitors the authorized entrants, controls access into the confined space, and is alert to any hazards that may arise. The attendant's most important tasks are to evacuate the space or call for aid in case of emergency.

autoignition temperature   Minimum temperature to which a substance must be heated without application of flame or spark to cause the substance to ignite.

background level   (1) The concentration of a substance in an environmental medium (air, water, or soil) that occurs naturally or is not the result of human activities. (2) In exposure assessment, the concentration of a substance in a defined control area, during a fixed period of time before, during, or after a data-gathering operation.

backwashing   Reversing the flow of water back through the filter media to remove entrapped solids.

bactericide   A substance used to kill bacteria. Hypochlorite and chlorine solutions are common bactericides.

baffle   A flat board or plate, deflector, guide, or similar device constructed or placed in flowing water or slurry systems to cause more uniform flow velocities to absorb energy and to divert, guide, or agitate liquids. Used in ventilation systems, baffles disrupt preferential flow patterns to allow more complete air exchange.

bailer   (1) A pipe with a valve at the lower end, used to remove slurry from the bottom or side of a well as it is being drilled. (2) A long pipe or stick with a container on the end or a container connected to a length of rope or measuring tape. Used to collect groundwater samples from wells, open boreholes, or bodies of water.

BAL   British Anti-Lewisite. A name for the drug dimecaprol, a treatment for toxic inhalations. BAL is a **chelating agent** used to remove toxic materials from the body. Lewisite is a wartime or riot-control gas that causes vomiting and other symptoms.

barrier coating(s)   A layer of a material that obstructs or prevents passage of liquid or vapor material through a surface that is to be protected; e.g., grout, caulk, or various sealing compounds; sometimes used with polyurethane membranes to prevent corrosion or oxidation of metal surfaces, chemical impacts on various materials, or, for example, to prevent radon infiltration into a home through walls, cracks, or joints.

barrier clothing   Clothing that protects the wearer against such hazards as particulates, liquids, gases, vapors, solids, biological agents, radiation, fire, or heat. (See **chemical protective clothing.**)

base   Substances that (usually) liberate OH anions when dissolved in water. Bases react with acids to form salts and water. Bases have a pH greater than 7, turn litmus paper blue, and may be corrosive to human tissue. Another name for **alkalis** or **caustics.**

benign   A growth that is not recurrent or not tending to progress. Not malignant. Not cancerous.

beryllium   A metal hazardous to human health when inhaled as an airborne pollutant. It can be discharged by machine shops, ceramic plants, and foundries.

best available control technology (BACT)   An emission limitation based on the maximum degree of emission reduction (considering energy, environmental, and economic impacts) achievable through application of production processes and available methods, systems, and techniques. BACT does not permit emissions in excess of those allowed under any applicable Clean Air Act provisions. Use of the BACT concept is allowable on a case-by-case basis for major new or modified emissions sources in attainment areas and applies to each regulated pollutant.

best demonstrated available technology (BDAT)   As identified by EPA, the most effective commercially available means of treating specific types of hazardous waste. The BDATs may change with advances in treatment technologies.

best management practice (BMP)   Methods that have been determined to be the most effective, practical means of preventing or reducing pollution from nonpoint sources.

beta (β) radiation   Smaller than alpha rays, beta radiation is made up of electrons. Beta rays can be reduced or stopped by a layer of clothing or by a few millimeters of a substance such as aluminum.

bioaccumulants   Substances that increase in concentration in living organisms as they take in contaminated air, water, or food because the substances are very slowly metabolized or excreted. Refers to the process whereby certain substances such as pesticides or heavy metals move up the food chain. They work their way into rivers or lakes and are eaten by aquatic organisms such as fish, which in turn are eaten by large birds, animals, or humans. The substances become concentrated in tissues or internal organs as they move up the chain. Lead and mercury are bioaccumulated: lead in the bones and teeth; mercury in the fatty tissues.

bioassay   A test to determine the relative strength of a substance by comparing its effect on a test organism with that of a standard preparation.

bioavailabiliity   Capability of being absorbed and ready to interact in organism metabolism.

biodegradable   Organic material's tendency to decompose as a result of attack by microorganisms.

biological hazards or biohazards   Infectious agents that present a risk to living organisms either through infection or through physical hazard exposure, such as hospital wastes, poison ivy and poison oak, bee stings, and snake bites.

biological monitoring   Periodic examination of body substances, such as blood, urine, or tissue, to determine the extent of hazardous material absorption as compared with background exposure. The term is also used for the medical surveillance of HAZMAT workers at a site due to the extreme conditions of wearing so much PPE. Perhaps better referred to as *physiological monitoring;* response workers are monitored for blood pressure, pulse, body weight (for dehydration) and body temperature and must reach an established baseline before suiting up and returning to the work area.

biological oxygen demand   Also called *biochemical oxygen demand.* BOD_5 is the standard way of describing how much oxygen dissolved in water is consumed by oxidation of biological materials during the stated period of time (during five days). The unit indicates the pounds of oxygen consumed by each 100 lb of biological materials during the time stated. If BOD is followed by "theor.," it indicates the pounds of oxygen theoretically required to completely oxidize 100 lb of the biological materials.

A useful measure of the concentration of biologically degradable material present in a waste stream.

biological oxidation   Decomposition of complex organic materials by microorganisms. Occurs in self-purification of water bodies and in activated sludge wastewater treatment.

biological treatment   A treatment technology that uses microbiological organisms to consume organic waste.

biomass   All of the living material in a given area; often refers to vegetation.

biome   Entire community of living organisms in a single major ecological area.

biomonitoring   (1) The use of living organisms to test the suitability of effluents for discharge into receiving waters and to test the quality of such waters downstream from the discharge. (2) Analysis of blood, urine, tissues, etc., to measure chemical exposure in humans.

biopsy   Removal and examination of tissue from the living body.

bioremediation   Use of living organisms to clean up oil spills or remove other pollutants from soil, water, or wastewater; use of organisms such as nonharmful insects to remove agricultural pests or counteract diseases of trees, plants, and garden soil.

biosphere   The portion of Earth and its atmosphere that can support life.

biostabilizer   A machine that converts solid waste into compost by grinding and aeration.

biota   The animal and plant life of a given region.

biotransformation   Conversion of a substance into other compounds by organisms; includes biodegradation.

blackwater (sewage)   Water that contains animal, human, or food waste. As opposed to **graywater,** which is wastewater from sinks and dishwashers.

blank   Usually a particulate air sample container that is opened, captures an ambient air sample, and then is immediately closed at the site. Blanks are analyzed with the other samples of contaminated air. Blank counts are deducted from the true sample counts. Also called *field blanks.*

blanking or blinding   The absolute closure of a pipe, line, or duct by the fastening of a solid plate (such as a spectacle blank/blind or a skillet blank/blind) that completely covers the bore and is capable of withstanding the maximum fluid or gas pressure in the pipe, line, or duct with no leakage beyond the plate.

blood-borne pathogen   Any microorganism that is present in human blood that can cause disease in humans. OSHA 29 CFR 1910.1030 is the regulation covering the protection of workers from blood-borne pathogens.

bloom   A proliferation of algae or higher aquatic plants in a body of water; often related to pollution, especially when pollutants accelerate growth; such as nitrogen, phosphorus, and potassium from fertilizers.

BOD$_5$ The amount of dissolved oxygen consumed in five days by biological processes breaking down organic matter in wastewater.

body burden   Total amount of a toxic material that a person has ingested, inhaled, absorbed, or had injected into the body, from all sources over time.

bodily enter   See **confined-space entry.**

boiling liquid expanding vapor explosion (BLEVE)   Condition in which contained flammable liquids are excessively heated, resulting in the violent rupture of a container and the rapid vaporization, and subsequent explosion, of the material. The possibility of a BLEVE increases with the volatility of the material.

boiling point (BP)   Temperature at which a liquid changes to its vapor state at a given pressure. Flammable materials with low boiling points generally present special fire hazards.

BOM, or BuMINES   Bureau of Mines, U.S. Department of Interior.

bonding   Safety practice where two containers, or a delivery nozzle and a container, are connected by clamps and adequate conducting wire. This completed circuit equalizes electrical potential between the containers and helps prevent static sparks that could ignite flammable materials during pouring, loading, or unloading. The friction of liquids traveling through hoses and falling through air builds up static electricity.

boom   (1) A floating device used to contain oil on a body of water. (2) A piece of equipment used to apply pesticides, fertilizers, or nutrients from a tractor or truck.

booties   Worn over work boots or CPC suits, these are disposable items that help to minimize contamination, protect CPC from wear, and accelerate decon.

borehole   Hole made with drilling equipment.

brackish   Mixed fresh and saltwater.

bradycardia   A slow heartbeat with pulse rate below 60/minute.

breakthrough time   The time it takes a chemical, from first contact with a CPC material, to be detected on the other side of that material.

British Thermal Unit (BTU)   Quantity of heat required to raise the temperature of 1 lb of water 1°F at 39.2°F, its temperature of maximum density.

bronchitis   Inflammation of the bronchial tubes in the lungs.

brownfields   Abandoned, idled, or underutilized industrial or commercial facilities or sites where expansion or redevelopment is complicated by real or perceived environmental contamination. They can be in urban, suburban, or rural areas. EPA's Brownfields Initiative helps communities mitigate potential health risks and restore the economic viability of such areas or properties.

buddy system   System of organizing employees into work groups in such a manner that each employee of the work group is designated to be observed by at least one other employee in the work group. The purpose of the buddy system is to provide rapid assistance to employees in the event of an emergency.

buffer   Substance that reduces the change in hydrogen ion concentration (pH) that otherwise would be produced by adding acids or bases to a solution.

buffer strips   Strips of grass or other erosion-resisting vegetation between or below cultivated strips or fields; especially when bounded by water. Besides erosion, these strips of land help filter out excess fertilizers so that they do not enter the surface water. Excess fertilizer in surface water causes algal growth, which consumes oxygen; with the consequence of loss of marine life.

bulk container   A large cargo container, such as a tank truck or tank rail car, used for transporting dry solid or semisolid materials.

bulk density   The weight per unit volume of a solid particulate material as it is normally packed, with voids between particulates containing air. Usually expressed as lb/ft^3 or g/cm^3.

BUNA™   Trademark for synthetic rubber based on butadiene polymers such as Buna-N (butadiene-acrylonitrile copolymer) or Buna-S (butadiene-styrene copolymer).

bung   The name for the plugs that cover the two openings in the top or the one opening in the side of a liquid drum (bung-type or tight-head drum).

butyl rubber   Synthetic rubber copolymers of isobutylene and isoprene with good resistance to acids and oxidizing agents.

byproduct   Material, other than the principal product, generated as a consequence of an industrial process or as a breakdown product in a living system.

°C   See **degrees Celsius (centigrade).**

cadmium (Cd)   A heavy metal (RCRA-regulated) having an atomic weight of 112.4. At one time a common constituent of rust-preventative paints, now commonly found in nickel-cadmium batteries.

calorie   Standard unit of heat. A calorie is the amount of heat required to raise 1 g of water 1°C.

cap   A layer of clay, usually covered by an impermeable membrane, installed over the top of a closed landfill to prevent entry of rainwater and minimize leachate while containing methane and noxious odors.

capacity assurance plan   A statewide plan that supports a state's ability to manage the hazardous waste generated within its boundaries over a 20-year period.

capillary action   Movement of liquids through very small passages. A towel partially immersed in water will become totally wet due to capillary action. This is partially how moisture rises to the top of tall trees.

carbon adsorber   An add-on control device that uses activated carbon to adsorb volatile organic compounds and other materials from a gas stream. The VOCs are later recovered from the carbon by heating it, a process called *desorption.*

carbon adsorption   (1) Referring to air filtration, such as in respirator cartridges, it is the strong attraction between some vapors and gases and activated carbon. (2) A treatment system that removes contaminants from ground water or surface water by forcing it through tanks containing activated carbon treated to trap the contaminants.

Note

Activated (pure, dry, with a large surface area) carbon can both a*b*sorb and a*d*sorb.

carbon dioxide (CO₂) Heavy, colorless gas produced by combustion and decomposition of organic substances and as a byproduct of chemical processes. It is part of the vapor we exhale. It will not burn and is relatively nontoxic and unreactive. It can cause oxygen-deficient environments in large concentrations. It is useful as a fire-extinguishing agent to block oxygen and smother fire.

carbon monoxide (CO) Colorless, odorless, flammable, and very toxic gas produced by the incomplete combustion of carbon compounds and as a byproduct of many chemical processes. A chemical asphyxiant, it reduces the blood's ability to carry oxygen. (See **carboxyhemoglobin.**)

carbon tetrachloride (CCl₄) Compound consisting of one carbon atom and four chlorine atoms, once widely used as a industrial raw material, as a solvent, and in the production of CFCs. Its use as a solvent ended when it was discovered to be carcinogenic. When involved in a fire, it can decompose to poisonous and corrosive phosgene gas.

carboxyhemoglobin Hemoglobin in which the iron is bound to carbon monoxide (CO) instead of oxygen. Hemoglobin (red blood cells) have approximately a 100 times greater affinity for CO than oxygen.

carcinogen Substance or agent capable of causing or producing cancer in mammals. Human carcinogens are classified by the International Agency for Research on Cancer (IARC) as follows:

- *Group 1: The agent is carcinogenic to humans.* This category is used only when there is sufficient evidence of carcinogenicity in humans.

- *Group 2A: The agent is probably carcinogenic to humans.* This category is used when there is limited evidence of carcinogenicity in humans and sufficient evidence of carcinogenicity in experimental animals. Exceptionally, an agent may be classified into this category solely on the basis of limited evidence of carcinogenicity in humans or of sufficient evidence of carcinogenicity in experimental animals strengthened by supporting evidence from other relevant data.

- *Group 2B: The agent is possibly carcinogenic to humans.* This category is generally used for an agent for which there is the absence of sufficient evidence in experimental animals. It may also be used when there is inadequate evidence of carcinogenicity in humans or when human data are nonexistent but there is sufficient evidence of carcinogenicity in experimental animals. In some instances, an agent for which there is inadequate evidence or no data in humans but limited evidence of carcinogenicity in experimental animals together with supporting evidence from other relevant data may be placed in this group.

- *Group 3: The agent is not classifiable as to its carcinogenicity to humans.* Agents are placed in this category when they do not fall into any other group.

- *Group 4: The agent is probably not carcinogenic to humans.* This category is used for an agent for which there is evidence suggesting lack of carcinogenicity in humans together with evidence suggesting lack of carcinogenicity in experimental animals. In some circumstances, agents for which there is inadequate evidence of or no data on carcinogenicity in humans but evidence suggesting lack of carcinogenicity in experimental animals, consistently and strongly supported by a broad range of other relevant data, may be classified in this group.

- It is listed as a carcinogen or potential carcinogen in the Annual Report on Carcinogens published by the National Toxicology Program (NTP) (latest edition); or

- It is regulated by OSHA as a carcinogen.

carcinoma Malignant (growing) tumor or cancer; a new growth made up of epithelial cells tending to grow rapidly, infiltrate other cells, and give rise to metastasis (spreading).

cargo manifest (shipping papers) A document that describes all of the contents being carried by the transporting vehicle or vessel.

CAS (Chemical Abstracts Service) number An assigned number used to identify a chemical. CAS stands for Chemical Abstracts Service, an organization that indexes information published in *Chemical Abstracts* by the American Chemical Society and that provides index guides with information about particular substances may be located in the abstracts. Sequentially assigned CAS numbers identify specific chemicals, except when followed by an asterisk(*), which signifies a compound (often naturally occurring) of variable composition. The numbers have no chemical significance. The CAS number is a concise, unique means of material identification. As of August 2000, the CAS registry had indexed over 25,000,000 substances. The CAS website is www.cas.org.

cask   A thick-walled container (usually lead) used to transport radioactive material. Also called a *coffin*.

catalyst   A substance that changes the rate of a chemical reaction (makes it go faster or slower) without being consumed. Catalysts can cause reactions to occur that otherwise would not occur or cause reactions to occur at lower temperatures and pressures than otherwise would be required. The commercial production of ammonia depends upon reacting nitrogen and hydrogen in the presence of a catalyst at the proper elevated temperature and pressure. Without the catalyst, that method of ammonia production would not be practical.

catalytic converter   An air pollution-abatement device that removes carbon monoxide, hydrocarbons, and nitrogen oxides pollutants from motor vehicle engine exhaust. The converter oxidizes the first two to carbon dioxide and water. It reduces the nitrogen oxides to nitrogen and oxygen.

catalytic incinerator   An air pollution control device that oxidizes volatile organic compounds (VOCs) by using a catalyst to promote the combustion process. Catalytic incinerators require much lower temperatures than conventional thermal incinerators, thus saving fuel and other costs.

cataract   Loss of transparency of the crystalline lens of the eye or its capsule.

categorical exclusion   A class of actions that either individually or cumulatively would not have a significant effect on the human environment and therefore would not require preparation of an environmental assessment or environmental impact statement under the National Environmental Policy Act (NEPA).

cathodic protection   A technique to prevent corrosion of a metal surface by making the protection device the cathode of an electrochemical cell.

caustic   See **alkali.**

cavitation   The formation and collapse of gas pockets or bubbles on the blade of a pump impeller or the gate of a valve; collapse of these pockets or bubbles drives water with such force that it can cause pitting of the gate or valve surface. It will also cause a liquid centrifugal pump to cavitate; that is, to stop pumping liquids.

CC   Closed cup. Identifies one of the methods used to measure flash points of flammable liquids (for instance, Setaflash®).

cc, cm³   Cubic centimeter. Metric unit of volume. Practically equivalent to a milliliter (mL).

ceiling (C)   Maximum allowable human exposure limit for airborne substances; not to be exceeded even instantaneously. Also called *ceiling limit.*

cells   (1) In hazardous waste disposal, holes where waste is dumped, compacted, and covered with layers of dirt on a daily basis. (2) The smallest structural part of living matter capable of functioning as an independent unit.

cementitious   Densely packed and nonfibrous friable materials.

central collection point   Location at which a generator of regulated wastes consolidates wastes that were originally generated at various locations (*satellite collection points*) in the facility. The wastes are gathered together for treatment on-site or for transportation elsewhere for treatment or disposal.

Centers for Disease Control (CDC)   This Department of Health and Human Services agency includes the National Center for Environmental Health and the National Institute for Occupational Safety and Health (NIOSH).

centipoise   Unit of the measure of viscosity equal to 1/100 Poise. Viscosity of water at 20°C is approximately 1 centipoise (1 cP).

centimeter, cm   1/100 m. A cm is approximately 0.4 in.

certified industrial hygienist (CIH)   A professional who has met the training and proficiency requirements in the field of industrial hygiene and is so certified by the American Board of Industrial Hygiene.

CFC   Chlorofluorocarbon. R-12 Freon® is an example. An ozone-depleting chemical.

CFR   Code of Federal Regulations. This document codifies all the rules made by the executive departments and agencies of the federal government. It is divided into 50 volumes, known as titles, that represent broad areas subject to federal regulation. Title 29 of the CFR (referenced as 29 CFR) lists all OSHA regulations.

challenge agent   The chemical liquid, vapor, solid, mixture, solution, or microbe against which a protective suit is tested. The degree to which these agents affect the CPC materials is the basis for the particular CPC material's compatibility chart.

characteristic wastes   Defined by the EPA as hazardous waste if they exhibit one or more of the following characteristics: ignitability, corrosivity, reactivity, or toxicity.

characterization of ecological effects   Part of ecological risk assessment that evaluates ability of a stressor to cause adverse effects under given circumstances.

chelating agent   See **BAL.** Chemical compounds capable of forming multiple chemical bonds to a metal ion. They are used to treat metal poisoning by increasing excretion. Chelating agents for lead poisoning are BAL, calcium disodium versenate, cuprimine, and chemet. Prophylactic chelation, the routine use of chelating agents to keep body burdens of toxic metals low despite exposures, is illegal.

chemical   A product of a chemical company. Some elements are chemicals, such as oxygen and nitrogen (considered chemicals by the American Chemical Society). Most chemicals are compounds (such as sulfuric acid) or mixtures (most solvents are not one pure compound). A chemical is also a substance or material in the definitions by various agencies.

chemical cartridge respirator   A respirator that uses various chemical substances to purify inhaled air of certain gases and vapors. This type of respirator is effective for concentrations no more than 10 times the TLV® of the contaminant, if the contaminant has warning properties (odor or irritation) below the TLV®.

chemical compound   A distinct and pure substance formed by the union of two or more elements in definite proportion by weight.

chemical element   A fundamental substance composed of only one kind of atom; the simplest form of matter. All known chemical elements are shown on the latest *periodic chart*. There are less than 100 stable chemical elements. Examples are hydrogen, oxygen, calcium, sulfur, iron, and lead.

Chemical Emergency Preparedness and Prevention Office of the EPA (CEPPO)   Provides leadership, advocacy, and assistance to prevent and prepare for chemical emergencies, respond to environmental crises, and inform the public about chemical hazards in their community. To protect human health and the environment, CEPPO develops, implements, and coordinates regulatory and nonregulatory programs. The CEPPO website is www.epa.gov/swercepp/.

chemical family   Group of elements (for instance the halogens: chlorine, fluorine, bromine, iodine) or compounds (for instance, the hydrocarbons methane, ethane, propane; containing only hydrogen and carbon) given a common, general name.

chemical formula   Gives the number and kinds of atoms that comprise a molecule of a chemical. A structural formula shows the bonding arrangements between atoms in a molecule. Both types of formulas are on the information sheet that comes with pharmaceuticals, for example.

chemical hazard   The exposure of workers to any regulated or nonregulated hazardous materials with the potential for causing harm to people, the environment, or property when released.

chemical hygiene plan (CHP)   Per 29 CFR 1910.1450, OSHA standard, *Occupational Exposures to Hazardous Chemicals in Laboratories*, effective 5/1/90. A written plan that includes specific work practices, standard operating procedures, equipment, engineering controls, and policies to ensure that employees are protected from hazardous exposure levels to all potentially hazardous chemicals in use in their work areas. The OSHA standard provides for training, employee access to information, medical consultations, examinations, hazard identification procedures, respirator use, and record-keeping practices.

chemical name   Scientific designation, by the International Union of Pure and Applied Chemistry, of a name that clearly identifies a chemical. Especially useful for hazard evaluation purposes.

chemical oxygen demand (COD)   Measure of the quantity of oxidizable components present in wastewater, chiefly hydrocarbons, but also nitrogen- and sulfur-containing materials. The consumed oxygen is only a measure of the chemically oxidizable components using the specific test method. COD must be determined when dealing with waste streams that are sent to POTWs.

chemical pneumonitis   Inflammation of the lungs caused by accumulation of fluids due to chemical irritation.

chemical protective clothing (CPC)   Any item of clothing used to protect any part of the body from contact with a hazardous material.

chemical reactivity   Ability or tendency of a material to take part in a chemical reaction. Undesirable and dangerous effects such as heat, explosions, or the production of noxious substances can result.

chemical resistance   How well a material protects against chemical exposure.

chemiluminescence   Emission of light during a chemical reaction at about ambient temperature. Examples are the oxidation of decaying wood containing certain bacteria, the luciferin on fireflies, and the chemical lights used by campers and the military. Chemical lights are started by breaking a glass tube of one chemical inside a plastic tube of another chemical.

CHEMNET   The HAZMAT industry's mutual aid emergency response network. It was established in 1985 to provide timely emergency response and technical assistance at the scene of HAZMAT transportation incidents. The network is composed of emergency response teams from participating chemical companies and teams provided by commercial, for-hire contractors under contract to the American Chemistry Council. Participation in the CHEMNET program provides shippers of hazardous materials with limited emergency response resources to access a nationwide network of emergency response teams. These teams are available to respond to an incident involving the company's products. The network is available 24 hours a day and can only be activated through the CHEMTREC Emergency Center.

CHEMTREC®   Twenty-four-hour toll-free telephone number (800-424-9300), intended primarily for use by those who respond to chemical transportation emergencies. Established by the Chemical Manufacturer's Association (CMA) it is now operated by the CMA's successor, the American Chemistry Council.

chime   On steel 55-gallon (208-L) bung-type drums, the top and bottom rims of the drum. On open-mouth drums, the bottom rim of the drum. The chime is used for stabilizing the drum as it is lifted and handled.

chloracne   Acne-like skin eruption caused by excessive contact with certain chlorine-containing substances. PCBs are an example.

chlorinated hydrocarbons   Chemicals containing chlorine, carbon, and hydrogen. These include a class of persistent, broad-spectrum insecticides that linger in the environment and accumulate in the food chain. Among them are DDT, aldrin, dieldrin, heptachlor, chlordane, lindane, endrin, mirex, hexachloride, and toxaphene. Other examples include trichloroethylene (TCE) and perchloroethylene (Perc), used as an industrial solvents.

chlorinated solvent   An organic solvent containing chlorine atoms (e.g., methylene chloride and 1,1,1-trichloromethane). Chlorinated solvents have been used as propellants in aerosol spray containers, in highway paint, and in dry cleaning fluids.

chlorination   The application of chlorine to drinking water, sewage, or industrial waste to disinfect or to oxidize undesirable compounds.

chlorinator   A device that adds chlorine, in gas or liquid form, to drinking water or wastewater to kill microorganisms.

chlorine-contact chamber   That part of a water treatment plant in which effluent is disinfected by chlorine.

chlorofluorocarbons (CFCs)   Compounds containing chlorine, fluorine, and carbon. A family of inert, nontoxic, and easily liquefied chemicals used in refrigeration, air conditioning, packaging, and insulation and as solvents and aerosol propellants. Because CFCs are not destroyed in the lower atmosphere, they drift into the upper atmosphere, where their chlorine components destroy ozone.

chromium   A metallic element of atomic weight 52. The metal is thought to be nontoxic. However, it is a RCRA-regulated heavy metal largely because of the toxicity of hexavalent Cr^{+6} compounds, more so than the trivalent Cr^{+3} compounds. Toxicity is in the form of irritation and diseases of the respiratory tract, including cancer. Chromium compounds were a common paint additive.

chronic effect   An adverse effect on a human or animal body, with symptoms that develop slowly over a long period of time. (See **acute**.)

chronic exposure   Low doses repeatedly delivered to a receptor (human or animal) over a long period of time.

chronic toxicity   Adverse effects resulting from repeated doses of, or exposures to, a material over a relatively prolonged period of time. Ordinarily used to denote effects noted in experimental animals.

circle of influence   The circular outer edge of a depression produced in the water table by the pumping of liquid from a well.

clarification   Clearing action that occurs during wastewater treatment when solids settle out. This is often aided by centrifugal action and chemically-induced coagulation (by the addition of flocculants) in wastewater.

Class I area   Under the Clean Air Act, a Class I area is one in which visibility is protected more stringently than under the national ambient air quality standards; includes national parks, wilderness areas, monuments, and other areas of special national and cultural significance.

Class I substance   One of several groups of chemicals with an ozone depletion potential of 0.2 or higher, including CFCs, Halons®, carbon tetrachloride, and methyl chloroform (listed in the Clean Air Act), and HBFCs and ethyl bromide (added by EPA regulations).

Class II substance   A substance with an ozone depletion potential of less than 0.2. All HCFCs are currently included in this classification.

Class A fire extinguishers   Will put out fires in **ordinary combustibles**, such as wood and paper. The numerical rating for this class of fire extinguisher refers to the amount of water the fire extinguisher holds. Class A, B, C is now the common standard.

Class A, B, C fire extinguishers   Can be used on ordinary combustibles, flammable liquids, and electrical fires. These are the generic "all-round" fire extinguishers in most common use today.

Class B fire extinguishers   Should be used on fires involving **flammable liquids**, such as grease, gasoline, and oil. The numerical rating for this class of fire extinguisher states the approximate number of square feet of a flammable liquid fire that a nonexpert person can be expected to extinguish.

Class C fire extinguishers   Suitable for use on **electrically energized fires**. This class of fire extinguishers does not have a numerical rating. The presence of the letter C indicates that the extinguishing agent is non-conductive.

Class D fire extinguishers   Designed for use on **flammable metals** and often specific for the type of metal in question. There is no picture designator for Class D extinguishers. These extinguishers generally have no rating, nor are they given a multipurpose rating for use on other types of fires.

Class A explosive   A material or device that presents a maximum hazard through detonation (DOT definition).

Class B explosive   A material or device that presents a flammable hazard and functions by deflagration (DOT definition).

Class C explosive   A material or device that contains restricted quantities of either Class A or Class B explosives or both, but presents a minimum hazard (DOT definition).

Class A poison   A poisonous gas or liquid of such nature that a very small amount of the gas, or vapor of the liquid, is dangerous to life (DOT definition).

Class B poison   A substance that is known to be so toxic to human life that it affords a severe health hazard during transportation (DOT definition).

Clean Air Act (CAA)   Restricts the types and amounts of pollutants that may be released into the air and requires permits for large, and sometimes small, polluters. Superfund cleanup responses must comply with CAA requirements, and the substances that are listed as hazardous air pollutants under section 112 of the CAA are considered to be Comprehensive Environmental Response, Compensation, and Liability Act (CERCLA) hazardous substances.

clean fuels   Blends or substitutes for gasoline fuels, including compressed natural gas, methanol, ethanol, and liquefied petroleum gas.

cleanup   Actions taken to deal with a release or threat of release of a hazardous substance that could affect humans or the environment. The term *cleanup* is sometimes used interchangeably with the terms *remedial action, removal action, response action,* and *corrective action.*

cleanup operation   An operation in which hazardous substances are removed, contained, incinerated, neutralized, stabilized, cleared up, or in any other manner processed or handled with the ultimate goal of making the site safer for people or the environment.

Clean Water Act (CWA)   The primary federal law that protects our nation's waters, including lakes, rivers, aquifers, and coastal areas. Includes the permitting system for waterways, the National Pollutant Discharge Elimination System (NPDES).

closed-loop recycling   Reclaiming or reusing wastewater for nonpotable purposes in an enclosed process.

closure   The procedure a landfill operator must follow when a landfill reaches its legal capacity for solid waste. It includes ceasing acceptance of solid waste and placing a cap on the landfill site. It may also include an operation and maintenance period that can extend for decades.

CNS   central nervous system, containing the brain and spinal cord.

coagulation   A treatment process that causes clumping of particles in wastewater to settle out impurities, often induced by chemicals such as lime, alum, and iron salts.

cold line   At hazardous waste sites and at emergency response sites, the line that separates the contamination reduction zone (warm zone) from the support (cold) zone.

cold zone   The uncontaminated area outside of the warm zone. Also known as the *support zone.*

colloids   Very small, finely divided particles that do not dissolve and remain dispersed in a liquid due to their extremely small size and electrical charge.

contaminant/contamination   An undesired and nonbeneficial substance.

combined sewer overflows   Discharge of a mixture of stormwater and sewage to surface waters when the flow capacity of a sewer system is exceeded during rainstorms.

combined sewers   A sewer system that carries both sewage and stormwater runoff. Normally, its entire flow goes to a waste treatment plant, but during a heavy storm the volume of water may be so great as to cause overflows of untreated mixtures of storm water and sewage into receiving waters. Stormwater runoff may also carry toxic chemicals from industrial areas or streets into the sewer system.

combustible liquids   29 CFR 1910.106 defined by OSHA as those having a flash point at or above 37.8°C (100°F). They do not ignite as easily as flammable liquids; however, they can be ignited and must be handled with caution. The DOT and NFPA define them in the following categories:

• *Class II combustible liquid*   Fl.P. (flash point) at or above 100°F and below 140°F.

• *Class IIIA combustible liquid*   Fl.P. at or above 140°F and below 200°F

• *Class IIIB combustible liquid*   Fl.P. at or above 200°F

combustion   The chemical process of rapid oxidation accompanied by the generation of heat.

command post   The control center for an emergency response or other cleanup. The base for the Incident Commander and technical staff. The prime source of all information about an incident.

common name   Designation for material other than a chemical name, such as code, trade, brand, or generic name.

compaction   Reduction of the bulk of solid waste by crushing, rolling, tamping, and other applications of pressure.

compatibility   A condition in which two or more materials can be in contact for an extended time without some reaction occurring. The opposite of **incompatibility.**

compliance monitoring   Collection and evaluation of data, including self-monitoring reports, and verification to show whether pollutant concentrations and loads contained in permitted discharges are in compliance with the limits and conditions specified in the permit.

compliance schedule   A negotiated agreement between a pollution source and a government agency that specifies dates and procedures by which a source will reduce emissions and thereby comply with a regulation.

compost   The relatively stable humus material that is produced from a composting process in which bacteria in soil, mixed with garbage and degradable trash, break down the mixture into organic fertilizer. Homeowners can also compost leaves, grass clippings, and other vegetable wastes in their backyards to help reduce the amount of waste going to landfills.

composting   The natural biological decomposition of organic material in the presence of air to form a humus-like material. Controlled methods of composting include mechanical mixing and aerating or placing the compost in open-air piles and mixing or turning it periodically to improve aeration.

composting facilities   (1) An off-site facility where the organic component of municipal solid waste is decomposed under controlled conditions. (2) An aerobic process in which organic materials are ground or shredded and then decomposed to humus in windrow piles or in mechanical digesters, drums, or similar enclosures.

composite   An item constructed of different materials. CPC is usually a composite of materials with the characteristics of strength and chemical protection.

composite sample   A number of soil or water samples taken from distinctly different areas of a remediation site, mixed together and analyzed as one sample.

compound   In chemical terminology, a pure substance or chemical. Water, trichloroethylene, and a PCB are chemical compounds.

Comprehensive Environmental Response, Compensation, and Liability Act CERCLA)   The Superfund Law, Public Law PL 96-510, found at 40 CFR 300. The EPA has jurisdiction.

The Comprehensive Environmental Response, Compensation, and Liability Act (CERCLA) Information System (CERCLIS)   An EPA database of information about Superfund sites. This information is intended for EPA employees to use for management of the Superfund program.

compressed gas   A gas contained under pressure. As opposed to a gas dissolved in a liquid (acetylene in acetone) or a gas liquefied by refrigeration (cryogenic liquid nitrogen). Also, any material or mixture having in the container absolute pressure exceeding 40 psia at 70°F (21°C) or having an absolute pressure exceeding 104 psia at 130°F (54°C) (DOT definition). The air in an SCBA cylinder is a compressed gas.

compressed natural gas (CNG)   An alternative fuel for motor vehicles; considered one of the cleanest because of low hydrocarbon emissions. Its combustion exhausts are relatively non-ozone producing. However, vehicles fueled with CNG, if not catalytically converted, could emit a significant quantity of nitrogen oxides from combustion of the nitrogen in the air.)

concentration   Relative amount of a substance when combined or mixed with other substances. The *volume percent concentrations* in dry atmospheric air, from sea level to any altitude on earth, are approximately 21% oxygen, 78% nitrogen, and 1% argon. Further, volume percent concentration means that in 100 cu ft of dry air there are 21 cu ft of oxygen. *Weight percent concentration:* as an example, 3% hydrogen peroxide in water, the common peroxide bought at a drug store, is 3 lb of hydrogen peroxide in 100 lb of peroxide solution.

condensation   The change of state from gaseous or vapor to liquid, usually by means of a drop in temperature. Gases are not easily condensed; vapors are.

conditions to avoid   Conditions encountered during handling or storage of a material that could cause it to become unstable.

conditionally exempt small quantity generator (or conditionally exempt generators)   Generators of less than 220 lb (100 kg) of hazardous waste per calendar month are known as CESQGs. They are exempt from most regulation; they are required merely to determine whether their waste is hazardous, notify appropriate state or local agencies, and ship it by an authorized transporter to a permitted facility for proper disposal.

conductance   A rapid method of estimating the dissolved solids content of water. Conductivity is a measure of the ability of a solution to carry an electrical current, used to detect the amount of salts (electrolyte) in water. It is the inverse of resistance. A copper wire is an efficient conductor of electrical current.

cone penetrometer testing (CPT)   A direct push (DP) system used to measure lithology based on soil penetration resistance. Sensors in the tip of the cone of the DP rod measure tip resistance and side-wall friction, transmitting electrical signals to digital processing equipment on the ground surface.

confined space   See **permit-required confined space.** Any area that a person can physically enter that has one or more of the following: limited openings for entry and exit that would make escape or rescue difficult in an emergency. May have a lack of ventilation. Not intended or designated for continuous human occupancy. Examples include sewers, culverts, tanks, silos, pits, mixers, process piping, condensers, and kilns.

confined space entry   The action by which a person passes through an opening into a confined space. Entry includes ensuing work activities in that space and is considered to have occurred as soon as any part of the entrant's body breaks the plane of an opening into the space.

confinement   Methods used to limit the physical area or size of a released material. (Examples: dams, dikes, and absorbent.) This term is used to define material already released. (See **containment.**)

conjunctivitis   Inflammation of conjunctiva, the delicate membrane that lines the eyelid and covers the eyeball.

consent decree   A legal document, approved by a judge, that formalizes an agreement reached between EPA and potentially responsible parties (PRPs) through which PRPs will conduct all or part of a cleanup action at a Superfund site, cease or correct actions or processes that are polluting the environment, or otherwise comply with EPA initiated regulatory enforcement actions to resolve the contamination at the Superfund site involved. The consent decree describes the actions PRPs will take and may be subject to a public comment period.

conservation easement   Easement restricting a landowner to land uses that that are compatible with long-term conservation and environmental values.

conservation   Preserving and renewing, when possible, human and natural resources. The use, protection, and improvement of natural resources according to principles that will ensure their highest ecological or social benefits.

consignee   The person or entity that is to receive a shipment.

constituent(s) of concern   Specific chemicals that are identified for evaluation in the site assessment process.

construction and demolition waste   Waste building materials, dredging materials, tree stumps, and rubble resulting from construction, remodeling, repair, and demolition of homes, commercial buildings, and other structures and pavements. May contain lead, asbestos, or other hazardous substances.

construction ban   If, under the Clean Air Act, EPA disapproves of an area's planning requirements for correcting nonattainment, EPA can ban the construction or modification of any major stationary source of the pollutant for which the area is in nonattainment.

Consumer Product Safety Commission (CPSC)   16 CFR 1000–end (Part 1500, '*Hazardous Substances and Articles; Administration and Enforcement Regulations*). Regulates consumer products, conducts testing, and issues recalls and advisories about unsafe or potentially unsafe products.

container   Any bag, barrel, bottle, box, can, cylinder, drum, reaction vessel, storage tank, or the like that can contain a hazardous chemical. Under the Hazard Communication Standard, pipes or piping systems and engines, fuel tanks, or other operating systems in a vehicle are not considered to be containers. Also, an article of transport equipment that is:

1. Of a permanent character and strong enough for repeated use

2. Specifically designed to facilitate the carriage of goods by one or more modes of transport without intermediate reloading

3. Fitted with devices permitting its ready handling, particularly its transfer from one mode to another (DOT definition)

containment   Control methods used to keep the material in its container. Examples include plugging and patching and, in the case of a 55-gallon drum, simply rolling it over until the hole faces up (See **confinement.**)

contaminant/contamination   An unwanted and nonbeneficial substance or containing such substance.

contamination reduction corridor   At a hazardous waste site or emergency response site, the area that controls access into and out of the exclusion (hot) zone and where personnel decontamination activities take place.

contamination reduction zone   The buffer zone between the exclusion zone and support zone. The contamination reduction corridor runs through the contamination reduction zone.

contingency plan   A document setting out an organized, planned, and coordinated course of action to be followed in case of a fire, explosion, or other accident that releases toxic chemicals, hazardous waste, or radioactive materials that threaten human health or the environment.

continuous discharge   A routine release to the environment that occurs without interruption, except for infrequent shutdowns such as for maintenance or process changes.

Contract Laboratory Program (CLP)   National network of EPA personnel, commercial laboratories, and support contractors whose fundamental mission is to provide data of known and documented quality. The CLP supports EPA's Superfund effort under the 1980 Comprehensive Environmental Response, Compensation, and Liability Act (CERCLA) and the 1986 Superfund Amendments and Reauthorization Act (SARA).

control technique guidelines (CTG)   EPA documents designed to assist state and local pollution authorities to achieve and maintain air quality standards for certain sources (e.g., organic emissions from solvent metal cleaning known as degreasing) through **reasonably available control technology** (RACT).

controlled reaction   A chemical reaction under temperature and pressure conditions maintained within safe limits to produce a desired product or process.

convulsions   Seizures manifested by discontinuous, involuntary, skeletal-muscular contractions, either brief contractions repeated at short intervals or longer ones interrupted by intervals of muscular relaxation.

cornea   Transparent structure of the external layer of the eyeball.

corrective action   The process of remediating or cleaning up a spill, or contaminants released into the environment, to levels satisfactory to regulators. Also, action taken by the emergency response Incident Commander to control and contain a hazardous materials emergency.

corrosion   The oxidation, destruction, dissolution, and wearing away of metal caused by chemical reactions such as between water and iron pipes, acids contacting a metal surface, and contact between two different metals.

corrosion rate   For metals, usually expressed as loss of a sample surface in thousandths of an inch (mils) per year, with results accompanied by the temperature, humidity, corrosive test substance, and any other conditions under which the corrosion rate was measured.

corrosive   Liquid or solid that causes visible destruction or irreversible alterations in skin tissue at the site of contact. Or, in the case of leakage from its packaging, any liquid that exhibits a severe corrosion rate on steel, aluminum, or other packaging material. Acids and alkalies are typical corrosives.

cost/benefit analysis   A quantitative evaluation of the costs that are estimated to be incurred by implementing an environmental regulation versus the overall benefits to society of the proposed action.

cost-effective alternative   An alternative control or corrective method identified after analysis as being the best available in terms of reliability, performance, and cost. Although costs are one important consideration, regulatory and compliance analysis does not require EPA to choose the least expensive alternative. For example, when selecting or approving a method for cleaning up a Superfund site, the Agency balances costs with the long-term effectiveness of the methods proposed and the potential danger posed by the site.

cost recovery   In waste cleanup activities, a legal process by which potentially responsible parties who contributed to contamination at a Superfund site can be required to reimburse the trust fund for money spent during any cleanup actions by the federal government.

cost sharing   In environmental efforts, a publicly financed program through which society, as a beneficiary of environmental protection, shares part of the cost of pollution control with those who must actually install the controls. In Superfund, for example, the government may pay part of the cost of a cleanup action with those responsible for the pollution paying the major share.

cover material   Soil used to cover compacted solid waste in a sanitary landfill.

cradle-to-grave or manifest system   A procedure in which hazardous materials are identified and followed as they are produced, treated, transported, and disposed of by a series of permanent, linkable, descriptive documents (e.g., manifests). Commonly referred to as the cradle-to-grave system.

criteria   Descriptive factors taken into account by EPA in setting standards for various pollutants. These factors are used to determine limits on allowable concentration levels and to limit the number of violations per year. When issued by EPA, the criteria provide guidance to the states on how to establish their standards.

criteria pollutants   The 1970 amendments to the Clean Air Act required EPA to set National Ambient Air Quality Standards (NAAQSs) for certain pollutants known to be hazardous to human health. EPA has identified and set standards to protect human health and welfare for six pollutants: ozone, carbon monoxide, total suspended particulates, sulfur dioxide, lead, and nitrogen oxides. The term *criteria pollutants* derives from the requirement that EPA must describe the characteristics and potential health and welfare effects of these pollutants. It is on the basis of these criteria that standards are set or revised.

critical pressure   The pressure required to liquefy a gas at its critical temperature. At any higher temperature, the gas cannot be liquefied at any pressure.

critical temperature   The temperature above which a gas cannot be liquefied by pressure. The critical pressure is that pressure required to liquefy a gas at its critical temperature.

cross-contamination   In sampling and monitoring, occurs when unsterilized equipment or containers are used. In HAZWOPER work, when personnel or materials are moved across zones without undergoing decon procedures. The movement of underground contaminants from one level or area to another due to invasive subsurface activities.

cryogenic   Relating to extremely low temperature, as in 'refrigerated' gases. Generally referring to temperatures below $-250°F$. You will see the sign "Refrigerated Nitrogen" on liquid nitrogen transport vehicles. The temperature of liquid nitrogen, not pressurized, is about $-320°F$. Severe (cryogenic) skin burns can result from unprotected exposure to cryogenic liquids (See **frostbite**).

Cu ft, ft³   Cubic foot or feet; cu ft is preferred.

Cu m, m³   Cubic meter or meters; m³ is preferred.

cumulative ecological risk assessment   Consideration of the total ecological risk from multiple stressors to a given ecological zone (ecozone).

cumulative exposure   The sum of exposures of an organism to a pollutant over a period of time.

curettage   Cleansing of a diseased surface.

cutaneous   Pertaining to or affecting the skin.

cyanosis   Dark purplish coloration of skin and mucous membrane caused by deficient oxygenation of the blood. A symptom of CO_2 overexposure.

cyclone collector   A device that uses centrifugal force to remove dust or ash particles from polluted air or exhaust gases.

dangerously reactive material   Material that can react with atmospheric water, and/or air, producing hazardous conditions. Sodium and phosphorus are examples.

data quality objectives (DQOs)   Qualitative and quantitative statements of the overall level of uncertainty that a decision-maker will accept in results or decisions based on environmental data. They provide the statistical framework for planning and managing environmental data operations consistent with user's needs.

dBA   Decibels on the A scale. A method of rating exposure to noise. Constant noise in the working environment of greater than 85 dBA requires hearing protection.

DDT   The first chlorinated hydrocarbon insecticide widely used. Chemical name dichlorodiphenyltrichloroethane. It has a half-life of 15 years and can collect in fatty tissues of humans and animals. EPA banned registration and interstate sale of DDT for virtually all but emergency uses in the United States in 1972 because of its persistence in the environment and accumulation in the food chain.

deadmen   (1) Anchors drilled or cemented into the ground to provide additional reactive mass for direct push sampling rigs. (2) The concrete blocks to which USTs are secured to prevent 'floating.' (3) Switches that must be engaged by operators to run certain types of machinery, such as cranes. If the operator releases the dead-man switch, the machine will immediately stop operating.

decant   To draw off the upper layer of liquid after the heaviest material (a solid or another liquid) has settled.

decay products   Degraded radioactive materials, often referred to as *daughters*; radon decay products (radon daughters) of most concern from a public health standpoint are polonium-214 and polonium-218. Radon is considered a carcinogen.

dechlorination   Removal of chlorine from a substance. Dechlorination of PCBs yields the relatively innocuous biphenyl.

DECIDE   The NFPA's decision-making process for hazardous materials (HAZMAT) emergencies. There are six steps that can provide a framework for decision-making in any HAZMAT emergency:

1. **D**etect the presence of hazardous materials and hazardous conditions.
2. **E**stimate potential human, environmental, and property harm that can occur without timely intervention.
3. **C**hoose response objectives.
4. **I**dentify action options.
5. **D**o the best option.
6. **E**valuate progress.

decomposition   Breakdown of a chemical, material, or substance into parts, elements or simpler compounds. Hydrogen peroxide naturally decomposes to form oxygen and water. Also, the breakdown of organic material by microorganisms, changing the chemical composition and physical appearance of materials.

decontamination   The physical or chemical process of removing or neutralizing hazardous substances from the bodies of employees and their equipment.

deep-well injection   Deposition of raw or treated, filtered hazardous waste by pumping it into deep wells, where it is contained in the pores of permeable subsurface rock. The wells can be over 1000 ft deep.

defatting   Removal of natural oils from the skin by fat-dissolving solvents or other chemicals. Common dishwashing liquid defats the skin with overuse.

deflocculating agent   A material added to a suspension to prevent settling.

defluoridation   The removal of excess fluoride in drinking water to prevent the staining of teeth.

defoliant   An herbicide that removes leaves from trees and growing plants.

degasification   A water treatment that removes dissolved gases from the water.

degradation   The loss of valuable properties of a material. Change in chemical composition. This includes shrinkage, swelling, becoming brittle, and changes of color.

degree-day   A rough measure used to estimate the amount of heating required in a given area. It is defined as the difference between the mean daily temperature and 65°F. Degree-days are also calculated to estimate cooling requirements.

degree of hazard   A relative measure of how much harm can be caused by a hazardous substance.

degrees celsius (centigrade)   The temperature on a scale in which the freezing point of water is 0°C and the boiling point is 100°C. To convert to degrees fahrenheit, use the following formula: $°F = (°C \times 1.8) + 32$.

degrees fahrenheit   The temperature on a scale in which the boiling point of water is 212°F and the freezing point is 32°F. To convert to degrees celsius, use the following formula: $°C = (°F - 32) \div 1.8$.

delegated state   A state (or other governmental entity, such as a tribal government) that has received authority to administer an environmental regulatory program in lieu of a federal counterpart. As used in connection with NPDES, UIC, and PWS programs, the term does not connote any transfer of federal authority to a state.

deliquescent   Water-soluble salts (usually powdered) absorb moisture from air and soften or dissolve as a result. Common table salt will take on water at relative humidity greater than about 80%.

delist   Use of the petition process to have a facility's toxic waste designation rescinded.

demand-side waste management Prices whereby consumers use purchasing decisions to communicate to product manufacturers that they prefer environmentally sound products packaged with the least amount of waste, made from recycled or recyclable materials and containing no hazardous substances.

demineralization A treatment process that removes dissolved minerals from water.

demulcent Material capable of soothing or protecting inflamed, irritated mucous membranes. Cough lozenges act as a demulcent for the mouth and throat membranes.

denitrification The biological reduction of nitrate to nitrogen gas by denitrifying bacteria in soil or water.

dense nonaqueous phase liquid (DNAPL) Nonaqueous phase liquids such as chlorinated hydrocarbon solvents or petroleum fractions with a specific gravity greater than 1.0 that will sink through a water column until they reach a confining layer. Because they are at the bottom of aquifers instead of floating on the water table, typical monitoring wells may not indicate their presence.

density Ratio of weight-to-volume of a material, usually in grams per cubic centimeter or pounds per cubic foot. For example, lead is much more dense than aluminum.

depressant A substance that reduces the body's functional activity, such as the ability to keep warm, or an instinctive desire, such as appetite. Beverage alcohol is a depressant.

dermal Used on, applied to, or otherwise in contact with the skin.

dermal absorption/penetration Process by which a chemical penetrates the skin and enters the body as an internal dose.

dermal toxicity Ratings corresponding to the following definitions are derived from data obtained from the test methods, as described by the Consumer Products Safety Commission in 16 CFR 1500.4 and categories of toxicity as described in 16 CFR 1500.3.

- *Non-Toxic:* The probable lethal dose of undiluted product to 50% of the test animals determined from dermal toxicity studies (LD_{50}) is greater than 2 g per kilogram of body weight.
- *Toxic:* The probable lethal dose of undiluted product to 50% of the test animals determined from dermal toxicity studies (LD_{50}) is greater than 200 mg and less than or equal to 2 g per kilogram of body weight.
- *Highly Toxic:* The probable lethal dose of undiluted product to 50% of the test animals determined from dermal toxicity studies (LD_{50}) is less than or equal to 200 mg per kilogram of body weight.

dermatitis Inflammation of the skin from any cause.

dermatosis A broader term than dermatitis; it includes any cutaneous abnormality, thus encompassing folliculitis, acne, pigmentary changes, and nodules and tumors.

desalination (1) Removing salts from ocean or brackish water by use of various technologies. (2) Removal of salts from soil by artificial means, usually leaching.

desiccant A substance that removes moisture (water vapor) from the air to maintain a dry atmosphere in a container or other closed system. Silica gel is a common desiccant.

designated representative Any individual or organization to which an employee gives written authorization to exercise such employee's rights under the Hazard Communication Standard.

designated area An area of, or device within, a work area to be used for work with select carcinogens, reproductive toxins, and other materials that have a high probability of causing acute toxicity. Also, a specific area; for example: "Smoking allowed only in designated areas." An administrative control intended to minimize the potential for employee exposure to hazardous chemicals.

designated pollutant An air pollutant which is neither a criteria nor hazardous pollutant, as described in the Clean Air Act, but for which new source performance standards exist.

The Clean Air Act does require states to control these pollutants, which include acid mist, total reduced sulfur (TRS), and fluorides.

design capacity   The average daily flow that a treatment plant or other facility is designed to accommodate.

designed for continuous worker occupancy   Intended as a place of regular work and supplied with ventilation and other conditions necessary to support healthy life.

designer bugs   Popular term for microbes developed through biotechnology that can degrade specific toxic chemicals at their source in toxic waste dumps, soil, groundwater, or other water bodies.

design value   The monitored reading used by EPA to determine an area's air quality status; e.g., for ozone, the fourth-highest reading measured over the most recent three years is the design value.

destruction and removal efficiency (DRE)   A percentage that represents the number of molecules of a compound removed or destroyed in an incinerator relative to the number of molecules entering the system (e.g., a DRE of 99.99% means that 9,999 molecules are destroyed for every 10,000 that enter; 99.99% is known as "four nines." For some pollutants, the RCRA removal requirement may be as stringent as six nines.)

desulfurization   Removal of sulfur from fossil fuels to reduce pollution from sulfur oxides when the fuels are used in combustion.

detectable leak rate   The smallest leak (from a storage tank), expressed in terms of gallons or liters per hour, that a test can reliably discern with a stated probability of detection.

detection limit   The lowest concentration of a chemical that can reliably be distinguished by instrumental analysis from a zero concentration.

detergent   Synthetic surface-active or washing agent (a substitute for soap) that helps to cleanse materials of contaminants. Some can contain phosphate and other compounds that kill useful bacteria and encourage algae growth when they are in wastewater that reaches receiving waters.

dewater   (1) Remove or separate a portion of the water in a sludge or slurry so it can be handled and disposed of as a solid. A dewatered sludge can still contain a high percentage of water. (2) Remove or drain the water from a tank or trench.

diaphoresis   Perspiration, especially profuse.

diatomaceous earth (DE)   A chalk-like material (fossilized diatoms) used as a filtering material. DE is commonly used in regenerative swimming pool filters. Also used as an active ingredient in some powdered pesticides, and as the nitroglycerine absorbent in dynamite.

diazinon   An insecticide. In 1986, EPA banned its use on open areas such as sod farms and golf courses because it posed a danger to migratory birds. The ban did not apply to agricultural, home lawn, or commercial establishment uses.

dibenzofurans   A group of organic compounds, some of which are toxic, sometimes present in incinerator exhausts.

diffused air   A type of aeration that forces oxygen into wastewater or sewage by pumping air through perforated pipes inside a holding tank. Similar to spaging.

diffusion   The movement of suspended or dissolved materials from a more concentrated to a less concentrated area. The process tends to distribute the particles or molecules more uniformly.

digester   In wastewater treatment, a closed tank accomplishing a treatment step; in solid-waste conversion, a unit in which bacterial action is induced and accelerated in order to break down organic matter and establish the proper carbon-to-nitrogen ratio.

digestion   The biochemical decomposition of organic matter, resulting in partial gasification, liquefaction, and mineralization of pollutants.

dike   A barrier constructed to control or confine hazardous substances and prevent them from entering sewers, ditches, streams, or other water bodies.

diluent   Any liquid or solid material used to reduce the concentration of an active ingredient.

dilution ratio   The relationship between the volume rate-of-flow of water in a stream and the volume rate-of-flow of an incoming stream. It affects the ability of the receiving stream to assimilate waste.

dilution ventilation   Fresh airflow designed to dilute indoor air contaminants to acceptable levels.

dioxin   Any of a family of compounds known chemically as dibenzo-*p*-dioxins. Concern about them arises from their potential toxicity as contaminants in combustion and incineration emissions and commercial products. Tests on laboratory animals indicate that they are some of the more toxic anthropogenic (man-made) compounds. A family of highly toxic, chlorine-containing compounds that can often be found in incineration gases apparently dependent upon incineration temperature. In the past, found in water as a byproduct of the pulp and paper industry.

direct discharger   A municipal or industrial facility that introduces a waste stream to the environment through a defined conveyance or system such as outlet pipes; a point source.

direct push   Technology used for performing subsurface investigations by driving, pushing, and/or vibrating small-diameter hollow steel rods into the ground. Also known as direct drive, drive point, or push technology. A Geoprobe® is an example.

direct-reading instrument   A device that measures and displays the value being sought, without further calculation. For example, the concentration of a contaminant in air or water.

discharge   Flow of surface water in a stream or canal or the outflow of groundwater from a flowing artesian well, ditch, or spring. Can also apply to discharge of liquid effluent from a facility or to chemical emissions into the air through designated venting mechanisms.

disinfectant   A chemical or physical process that kills pathogenic microorganisms in water or air or on surfaces. Chlorine and hypochlorites are often used to disinfect sewage treatment effluent, water supplies, wells, and swimming pools.

dispersant   A chemical agent used to break up concentrations of organic material in water bodies, such as spilled oil.

disposal facilities   Repositories for solid waste, including landfills and combustors intended for permanent containment or destruction of waste materials. Excludes transfer stations and composting facilities.

disposal   Final placement or destruction of toxic, radioactive, or other wastes; surplus or banned pesticides or other chemicals; polluted soils; and drums containing hazardous materials from removal actions or accidental releases. Disposal may be accomplished through use of approved secure landfills, surface impoundments, land farming, deep-well injection, ocean dumping, or incineration.

dissolved oxygen (DO)   The oxygen freely available in water, vital to fish and other aquatic life and for the prevention of odors. DO levels are considered a most important indicator of a water body's ability to support desirable aquatic life. Secondary and advanced waste treatment are generally designed to ensure adequate DO in waste-receiving waters.

dissolved solids   Disintegrated organic and inorganic material in water. Excessive amounts make water unfit to drink or useful in industrial processes. Periodic blowdown, or discharge to waste treatment, is used in the circulatory cooling tower water to lower the content of dissolved solids before they start settling out in the system.

distributor   A business, other than a chemical manufacturer or importer, that supplies hazardous chemicals or other materials to other distributors or to employers.

doffing   Taking off protective clothing or other PPE.

donning   Putting on protective clothing or other PPE.

dose   An amount of a substance that a worker (as in an incident) or an animal (as in a toxicology test) receives such as by breathing or absorption through the skin; sometimes called an exposure.

dose-response   Shifts in toxicological responses of an individual (such as alterations in severity) or populations (such as alterations in incidence of a response) that are related to changes in the dose of any given substance. A material is said to have a dose-response relationship if the greater the dose, the greater the chance of disease or illness. Tobacco is such a material.

dose-response relationship   Correlation between the amount of exposure to an agent or toxic chemical and the resulting effect on the body.

dosimeter   An instrument to measure dosage; many so-called dosimeters actually measure exposure rather than dosage. Dosimetry is the process or technology of measuring or estimating dosage.

DOT reportable quantity (RQs)   The quantity of a substance specified in a U.S. Department of Transportation regulation that triggers labeling, packaging, and other requirements related to shipping such substances.

double block and bleed   The closure of a line, duct, or pipe by closing and locking or tagging two in-line valves and opening and locking or tagging a drain or vent valve in the line between the two closed valves.

downgradient   The direction in which groundwater flows; similar to downstream for surface water.

draft permit   A preliminary permit drafted and published by EPA; subject to public review and comment before final action on the application.

drawdown   (1) The drop in the water table or level of water in the ground when water is being pumped from a well. (2) The amount of water used from a tank or reservoir. (3) The drop in the water level of a tank or reservoir.

dredging   Removal of sediment from the bottom of water bodies. This can disturb the ecosystem and cause silting that kills aquatic life. Dredging of contaminated muds can expose biota to heavy metals and other toxics. Dredging activities may be subject to regulation under Section 404 of the Clean Water Act.

drive point profiler   An exposed groundwater DP system used to collect multiple depth-discrete groundwater samples. Ports in the tip of the probe connect to an internal stainless steel or Teflon® tube that extends to the surface. Samples are collected via suction or airlift methods. Deionized water is pumped down through the ports to prevent plugging and cross-contamination while driving the tool to the next sampling depth.

dry chemical (fire extinguisher)   Powdered fire extinguishing agent, usually composed of sodium bicarbonate or potassium bicarbonate.

dual entry   An OSHA term describing when workers from two or more companies enter and/or work in the same permit-required confined space.

dual-phase extraction   Active withdrawal of both liquid and gas phases, or aqueous and non-aqueous phase liquids from a well, usually involving the use of a vacuum pump.

dump   A site used to dispose of solid waste without environmental controls.

duplicate   A second aliquot or sample that is treated the same as the original sample in order to determine the precision of the analytical method.

dust   Solid particles generated by handling, crushing, grinding, rapid impact, detonation, and decrepitation of organic or inorganic materials, such as rock, ore, metal, coal, wood, and grain. Dusts do not tend to flocculate, except under electrostatic forces; they do not diffuse in air but settle under the influence of gravity. Most dusts are an inhalation, fire, and dust explosion hazard.

dustfall jar   An open container used to collect very large particles from the air for measurement and analysis.

dysplasia   An abnormality of development.

dyspnea   Sense of difficulty in breathing; shortness of breath; difficult or labored breathing.

dysuria   Difficult or painful urination.

ecological exposure   Exposure of a nonhuman organism to a stressor.

ecological impact   The effect that a man-caused or natural activity has on living organisms and their nonliving (abiotic) environment.

ecological risk assessment   The application of a formal framework, analytical process, or model to estimate the effects of human actions(s) on a natural resource and interpret the significance of those effects in light of the uncertainties identified in each component of the assessment process. Such analysis includes initial hazard identification, exposure and dose-response assessments, and risk characterization.

ecology   The relationship of living things to one another and their environment, or the study of such relationships.

ecosystem   The interacting system of a biological community and its nonliving environmental surroundings.

ecotox thresholds (ETs)   Sufficient amounts of media-specific contaminant concentrations that indicate further site investigation is needed. Superfund site managers use ETs as screening tools to efficiently identify contaminants that may pose an ecological threat. Also used to focus further site activities on those contaminants and the media in which they are found.

edema   Abnormal accumulation of clear, watery fluid in body tissue.

effective concentration (EC$_{50}$)   Concentration of a material in water, a single dose of which is expected to cause a biological effect on 50% of a group of test animals.

effluent   Wastewater—treated or untreated—that flows out of a treatment plant, sewer, or industrial outfall. Generally refers to wastes discharged into surface waters.

ejector   A device used to disperse a measured rate of addition of a chemical treatment solution into water being treated.

elastomers   Man-made materials that have useful stretching, compression, and chemical-resistance properties. Examples are butyl, neoprene, and Viton®. They have become common substitutes for rubber in many applications.

electromagnetic geophysical methods   Ways to measure subsurface conductivity via low-frequency electromagnetic induction.

electrostatic precipitator (ESP)   A device that removes particles from a gas stream (smoke) after combustion occurs. The ESP imparts an electrical charge to the particles, causing them to adhere to oppositely charged metal plates inside the precipitator. Automatic, periodic rapping on the plates causes the particles to fall into a hopper for disposal.

electromotive force (EMF)   Also referred to as voltage.

CAUTION

ALTHOUGH MUCH SCIENTIFIC RESEARCH HAS BEEN DONE, THERE IS NO EVIDENCE THAT POWER LINES PRODUCE HARMFUL RADIATION. THE RESEARCH IS ONGOING INTO THIS AREA, ALSO REFERRED TO AS *ELECTRICAL AND MAGNETIC FIELDS.*

electrolyte   Ionic solutions (containing positive and negative ions) are the electrolyte that conduct electric current in solution moving between two electrodes. Sulfuric (battery) acid, containing H^+ and SO_4^{2-} ions, is an example.

electron volt (eV)   Unit of energy equal to the energy acquired by an electron accelerating through a potential difference of 1 volt. (See **ionization potential.**)

elongation   Stretching without breaking. One physical property used to compare CPC fabrics. Elastomers will tend to stretch; nonelastomers will tend to tear or crack.

embolism   Obstruction of a blood vessel by a transported clot, a mass of bacteria, or other material.

embryo   Organism in the early stages of development before birth.

embryotoxin   Material harmful to a developing embryo at a concentration that has no detectable adverse effect on the pregnant female.

emergency   A sudden and unexpected event calling for immediate action.

emergency (chemical)   Situation created by an accidental release or spill of hazardous chemicals that poses a threat to the safety of workers, residents, the environment, or property.

Emergency Planning and Community Right-to-Know Act (EPCRA)   Also known as Superfund Amendments and Reauthorization Act (SARA) Title III, provides an infrastructure at the state and local levels to plan for chemical emergencies. Facilities that store, use, or release certain chemicals may be subject to various reporting requirements under EPCRA. Reported information is then made publicly available so that interested parties may become informed about potentially dangerous chemicals in their community.

emergency removal action   Action or actions undertaken in a time-critical situation to prevent, minimize, or mitigate a release that poses an immediate and/or significant threat to human health or welfare, or to the environment (See also **removal action**).

emergency response   A response effort by employees from outside the immediate release area or by other designated responders (e.g., mutual aid groups, local fire departments, etc.). The response is to an occurrence, which results, or is likely to result, in an uncontrolled release of a hazardous substance. Responses to incidental releases of hazardous substances where the substance can be absorbed, neutralized, or otherwise controlled at the time of release by employees in the immediate release area or by maintenance personnel are not considered to be emergency responses within the scope of the OSHA standard. Responses to releases of hazardous substances where there is no potential safety or health hazard (e.g., fire, explosion, or chemical exposure) are not considered to be emergency responses. Example: A nonvolatile hazardous liquid leaks from a storage tank into a containment dike.

Emergency Response Notification System (ERNS)   Database used to store information on notifications of oil discharges and hazardous substance releases. ERNS is now part of the National Response Center.

emetic   Agent that induces vomiting.

emissions   Gaseous, vapor, and particulate discharges into the atmosphere from smokestacks, other vents, and surface areas of commercial or industrial facilities; from residential chimneys; and from motor vehicle, locomotive, or aircraft exhausts.

emission standard   The maximum amount of air polluting discharge legally allowed from a single source, mobile or stationary.

emissions trading   The creation of surplus emission reductions at certain stacks, vents, or similar emissions sources and the use of this surplus to meet or redefine pollution requirements applicable to other emissions sources. This allows one source to increase emissions when another source reduces them, maintaining an overall constant emission level. Facilities that reduce emissions substantially may bank their credits or sell them to other facilities or industries.

emphysema   Irreversibly diseased lung condition in which the alveolar walls have lost their elasticity, resulting in an excessive reduction in the lungs' capacity.

employee, HAZMAT   A worker who may be exposed to hazardous chemicals under normal operating conditions or in foreseeable emergencies.

employer, HAZMAT   A person engaged in a business where chemicals are either used, distributed, or produced for use or distribution, including a contractor or subcontractor.

emulsifier   A chemical that aids in suspending one liquid in another. Usually an organic chemical in an aqueous solution.

encapsulation   The treatment of asbestos-containing material with a liquid that covers the surface with a protective coating or embeds fibers in an adhesive matrix to prevent their release into the air. Also used in lead abatement.

enclosure   Putting an airtight, impermeable, permanent barrier around asbestos-containing materials to prevent the release of asbestos fibers into the air. Also used in lead abatement.

endangerment assessment   A study to determine the nature and extent of contamination at a site on the National Priorities List and the risks posed to public health or the environment. EPA or the state conducts the study when a legal action is to be taken to direct potentially responsible parties to clean up a site or pay for it. An endangerment assessment supplements a remedial investigation.

endothermic   A chemical reaction that absorbs heat from its surroundings.

energy recovery   Obtaining energy from waste through a variety of processes (e.g., combustion, 'trash-to-steam').

enforcement   DOT, EPA, OSHA, state, or local legal actions to obtain compliance with transportation, environmental, or worker safety laws, rules, regulations, or agreements and the obtaining of penalties or criminal sanctions for violations. Enforcement procedures may vary depending on the requirements of different laws and related implementing regulations. In some situations, if investigations by agencies uncover willful violations, criminal trials and penalties are sought.

engineered controls   Method of managing environmental and health risks by placing a barrier between the contamination and the rest of the site, thus limiting exposure pathways.

engineering controls   Systems that reduce potential hazards by isolating the worker from the hazard or removing the hazard from the work environment. Examples are ventilation, isolation, containment, filtration, encapsulation, and enclosure.

engulfment   The surrounding and effective capture of a person by a liquid or finely divided (flowable) solid substance that can be aspirated to cause death by filling or plugging the respiratory system or can exert enough force on the body to cause death by strangulation, constriction, or crushing.

entry permit (permit)   The written or printed document that is provided by the employer to allow and control entry into a permit-required confined space and that contains the information specified in paragraph (f) of 29 CFR 1910.146.

entry supervisor   The person (such as the employer, foreman, or crew chief) responsible for determining if acceptable entry conditions are present at a permit space where entry is planned. (This duty is frequently delegated to an Inspector.) Also responsible for authorizing entry and overseeing entry operations and terminating entry, as required.

Note

An entry supervisor also may serve as an attendant or as an authorized entrant, as long as that person is trained and equipped as required by this section for each role he or she fills. Also, the duties of entry supervisor may be passed from one individual to another during the course of an entry operation.

enrichment   The addition of nutrients (e.g., nitrogen, phosphorus, carbon compounds) from sewage effluent or agricultural runoff to surface water; greatly increases the growth potential (provides enrichment) for algae and other aquatic plants.

entrain   To pick up gas, vapor, or particulates in a flowing gas, liquid, or vapor stream.

environment   The sum of all external conditions affecting the life, development, and survival of an organism.

environmental assessment   The measurement or prediction of the concentration, transport, dispersion, and final fate of a released hazardous substance in the environment.

environmental emergencies   Incidents involving the release (or potential release) of hazardous materials into the environment that require immediate action. Example: A forklift driver has accidentally sheared the drain valve off the bottom of a tank containing volatile hazardous material.

environmental fate   The destiny of a chemical or biological pollutant after release into the environment.

environmental hazard   A condition capable of posing an unreasonable risk to air, water, or soil quality and to plants or wildlife.

environmental impact statement   A document required of federal agencies by the National Environmental Policy Act for major projects or legislative proposals significantly affecting the environment. A tool for decision making, it describes the positive and negative effects of the undertaking and cites alternative actions.

environmental indicator A measurement, statistic, or value that provides a proximate gauge or evidence of the effects of environmental management programs or of the state or condition of the environment.

environmental justice (EJ) The fair treatment and meaningful involvement of all people, regardless of race, color, national origin, or income, in the development, implementation, and enforcement of environmental laws, regulations, and policies. The opposite of Environmental Racism.

environmental lien A charge, security, or encumbrance on a property's title to secure payment of cost or debt arising from response actions, cleanup, or other remediation of hazardous substances or petroleum products.

environmental medium A major environmental category that surrounds or contacts humans, animals, plants, and other organisms (e.g., surface water, groundwater, soil, or air) and in which chemicals or pollutants are entrained or through which chemicals or pollutants move.

Environmental Response Team (ERT) Established under Section 311 of the Clean Water Act to provide on-site expertise as required by the National Contingency Plan (NCP) section on Special Forces; a team of EPA experts located in Edison, NJ, and Cincinnati, OH, who can provide around-the-clock technical assistance to EPA regional offices and states during all types of hazardous waste site emergencies and spills of hazardous substances.

environmental sample A sample that is considered to contain no contaminants, or low concentrations of contaminants, as compared to a hazardous sample.

EPA ID number The unique code assigned to each generator, transporter, and treatment, storage, or disposal facility by regulating agencies to facilitate identification and tracking of chemicals or hazardous waste.

epidemiology Science that deals with the study of the distribution of disease or other health-related states and events in human populations, as related to age, sex, occupation, ethnicity, and economic status in order to identify and alleviate health problems and promote better health.

epiphora Excessive flow of tears.

epistaxis Nosebleed.

ergonomics Study of human characteristics for the appropriate design of living and work environments.

erosion The wearing away of land surface by wind or water, intensified by land-clearing practices related to farming, residential or industrial development, road building, or logging.

erythema Abnormally red skin from capillary congestion.

estuary Region of interaction between rivers and near-shore ocean waters, where tidal action and river flow mix fresh and salt water. Such areas include bays, mouths of rivers, salt marshes, and lagoons. These brackish water ecosystems shelter and feed marine life, birds, and wildlife.

etiologic agent A living microorganism (germ) that may cause human disease.

etiology All of the factors that contribute to the cause of a disease or an abnormal condition.

evaporation The conversion of a material from liquid state to vapor state below the material's boiling point.

evaporation ponds Areas where sewage sludge is dumped and dried.

evaporation rate Relative rate at which a particular material will vaporize when compared to the rate of vaporization of a standard material. Evaporation rate can be useful in evaluating the health and fire hazards of a material. Many MSDS preparers use *n*-butyl acetate (BuAc) as the standard and assign it an evaporation rate of 1.0. Some example evaporation rates are:

- *Fast greater than 3.0,* examples: MEK = 3.8, acetone = 5.6, hexane = 8.3
- *Medium 0.8–3.0,* examples: 95% ethyl alcohol = 1.4, naphtha = 1.4
- *Slow less than 0.8,* examples: xylene = 0.6, isobutyl alcohol = 0.6, water = 0.3, mineral spirits = 0.1

exclusion zone   The controlled area, located on the site, where contamination is either known or expected to occur and where the greatest potential for exposure exists. Also known as the *hot zone.*

exothermic   A chemical reaction that gives off heat. Sulfuric acid dissolved in water produces a warmer solution.

explosive   Material, such as nitroglycerine, that produces a sudden, almost instantaneous release of gas and heat when subjected to abrupt shock, spark, or high temperature, or other form of initiation.

explosive limits   See **upper/lower explosive limit** and **upper/lower flammable limit.** For all intents and purposes explosive limits and flammable limits are practically the same.

exposure   State of being open and vulnerable to a hazardous material by inhalation, ingestion, skin contact, absorption, or injection; includes potential (accidental or possible) exposure.

exposure assessment   Identifying the pathways by which toxins may reach individuals, estimating how much of a toxin an individual is likely to be exposed to, and estimating the number likely to be exposed.

exposure limits   Concentration in air of a material that is thought to be acceptable and causing no health hazard for human exposure by the rating organization, under specific exposure conditions. The most widely accepted exposure limits are OSHA permissible exposure limits (PELs) (the only federally enforceable limits), NIOSH recommended exposure limits (RELs), and ACGIH threshold limit values (TLV®s).

extinguishing media   Fire extinguisher or extinguishing method appropriate for use on specific material.

exposure pathway   The path from sources of pollutants via air, soil, water, or food to man and other living species or environmental settings.

extraction procedure (EP toxic)   Determining toxicity by a procedure that simulates leaching; if a certain concentration of a toxic substance can be leached from a waste, that waste is considered hazardous, i.e., EP toxic.

extraction well   A well used to remove contaminated groundwater or air containing contaminant vapors for collection and treatment.

extremely hazardous substances   Any of over 400 chemicals identified by EPA as toxic and listed under SARA Title III. The list is subject to periodic revision.

eye irritation   Ratings corresponding to the following definitions are derived from data obtained from test methods described by the Consumer Products Safety Commission in 16 CFR 1500.42 and graded pursuant to the Draize Scale for scoring ocular lesions and temporal reversibility criteria as set forth in NAS Publication 1138.

1. *Practically Nonirritating:* The undiluted product, when instilled into the eyes of rabbits produces no noticeable irritation, or slight transient conjunctiva irritation (average Draize score 0.00–15.0).
2. *Slightly Irritating:* The undiluted product, when instilled into the eyes of rabbits, produces slight to moderate conjunctiva irritation, slight corneal involvement, and/or slight iritis (average Draize score 15.1–25.0).
3. *Moderately Irritating:* The undiluted product, when instilled into the eyes of rabbits, produces moderate corneal involvement with or without severe iritis (average Draize score range 25.1–50.0). The effects clear within 21 days.
4. *Severely Irritating (or Corrosive):* The undiluted product, when instilled into the eyes of rabbits, produces severe corneal involvement with or without severe iritis (average Draize score range 50.1–110.0). The effects persist for 21 days or more.

°F   See **degrees Fahrenheit.**

facility

1. Any building, structure, installation, equipment, pipe or pipeline (including any pipe leading into a sewer or publicly owned treatment works), well, pit, pond, lagoon, impoundment, ditch, storage container, motor vehicle, rolling stock, or aircraft; or

2. Any site or area where a hazardous substance has been deposited, stored, disposed of, or placed, or otherwise come to be located; but does not include any consumer product in consumer use or any water-borne vessel.

facilities plans   Plans and studies related to the construction of treatment works necessary to comply with the Clean Water Act or RCRA. A facilities plan investigates needs and provides information on the cost-effectiveness of alternatives, a recommended plan, an environmental assessment of the recommendations, and descriptions of the treatment works, costs, and a completion schedule.

facility emergency coordinator   Representative of a facility covered by environmental law (e.g., a chemical plant) who participates in the emergency reporting process with the local emergency planning committee (LEPC).

fasciculation   Muscular twitching.

feasibility study   (1) Analysis of the practicability of a proposal; e.g., a description and analysis of potential cleanup alternatives for a site such as one on the National Priorities List. The feasibility study usually recommends selection of a cost-effective alternative. It usually starts as soon as the remedial investigation is underway; together, they are commonly referred to as the RI/FS. (2) A small-scale investigation of a problem to ascertain whether a proposed research approach is likely to provide useful data.

fecal coliform bacteria   Bacteria found in the intestinal tracts of mammals. Their presence in water or sludge is an indicator of pollution and possible contamination by pathogens.

federal register   Publication of U.S. government documents officially promulgated under the law; documents whose validity depends upon such publication. It is published on each day following a government working day. It is, in effect, the daily supplement to the Code of Federal Regulations (CFR).

Federal Insecticide, Fungicide, and Rodenticide Act (FIFRA)   Requires EPA registration for all pesticides sold in the United States. It also regulates the manufacture and use of pesticides and allows EPA to restrict or prohibit use of particularly harmful pesticides.

ferrous metals   Magnetic metals derived from iron or steel; products made from ferrous metals include appliances, furniture, containers, and packaging like steel drums and barrels. Recycled products include tin/steel cans, strapping, and metals from appliances processed into new products.

fiber   A form of matter. Can be crystalline or amorphous (noncrystalline), with a high ratio of length to diameter. An asbestos fiber is defined as having a minimum length-to-width ratio of 3:1, for example.

fibrosis   Formation of fibrous tissue, as a reparative or reactive process following particulates deposition in the lungs, in excess of amounts normally present in healthy lung tissue walls. This reduces the oxygen and carbon dioxide exchange efficiency.

fill   Man-made deposits of natural soils or rock products and waste materials.

filter   The part of a respirator that reduces the flow of particulates, gases, or vapors to the breather. More commonly referred to as a *cartridge*.

filter strip   Strip or area of vegetation used for removing sediment, organic matter, and other pollutants from runoff and wastewater.

filtration   A process used in all wastewater treatment. It is under the control of qualified operators. It removes solid (particulate) matter from water. *Secondary filtration* uses porous media such as sand or a man-made filter. It removes particles that can contain pathogens. *Primary filtration* removes rags, debris, and the large wastes before the filtrate goes to secondary filtration.

financial assurance for closure   Documentation or proof that an owner or operator of a facility, such as a landfill or other waste repository, is capable of paying the projected costs of closing the facility and monitoring it afterwards as provided in RCRA regulations.

finding of no significant impact (FNSI)   A document prepared by a federal agency showing why a proposed action would not have a significant impact on the environment and thus would not require preparation of an environmental impact statement. An FNSI is based on the results of an environmental assessment.

fines   Finely crushed or powdered material or fibers; especially those considerably smaller than the average in a mix of various sizes.

fire diamond (NFPA® 704M)   A diamond-shaped symbol, divided into four parts, designed by the NFPA to give an easily visible and understandable numerical rating for the labeled material's degree of hazard: health (blue), flammability (red), reactivity (yellow), and specific (white). 4 is the highest hazard rating, while 0 is the lowest hazard rating.

fire point   Lowest temperature at which liquid will produce sufficient vapor in air to flash near its surface and continue to burn, vs. **flash point** (See below).

first draw   The water that comes out when a tap is first opened, likely to have the highest level of lead contamination from plumbing materials containing lead.

first responders   The first persons to arrive on the scene of a hazardous materials incident. Usually officials from a local emergency service, firefighters, or police. In HAZWOPER they are defined in ascending order of training as: awareness level, operations level, hazardous materials technician, hazardous materials specialist, and incident commander.

fix a sample   A sample is fixed in the field by adding chemicals that prevent water quality indicators of interest in the sample from changing before laboratory measurements are made.

fixed-location monitoring   Sampling of an environmental or ambient medium for pollutant concentration at one location continuously or repeatedly.

flame resistant   Material that resists burning, melting, or other degradation when exposed to heat or flame.

flame retardant   Material that is treated to extinguish itself upon ignition from a flame source within three seconds of removal of the flame source.

flammable aerosol   Product packaged in an aerosol container that can release a flammable material. Many commercially available hairsprays and deodorants, for example.

flammable compressed gas   Any flammable material or mixture in a container having a pressure exceeding 40 psi at 100°F (37.78°C).

flammable gas   Any material that is a gas at ambient temperature (20°C/68°F) and pressure (101.3 kPa/14.7 psi) forms a flammable mixture with air at a concentration of 13% by volume or less; or has a flammable range of one atmosphere with air, of at least 12% regardless of the lower limit (DOT).

flammable limits   Flammables have a minimum concentration below which propagation of flame does not occur on contact with a source of ignition. This is known as the lower flammable limit (LFL) or lower explosive limit (LEL). There is also a maximum concentration of vapor or gas in air above which propagation of flame does not occur. This is known as the upper flammable limit (UFL) or upper explosive limit (UEL). These units are expressed in percent of gas or vapor in air by volume.

flammable liquid   29 CFR 1910.106. Defined by OSHA, as a liquid with a flash point, in air, below 100°F. Flammable liquids are subdivided as follows:

- *Class 1A:* Flash point below 73°F and boiling point below 100°F. Example: pentane.
- *Class 1B:* Flash point below 73°F and boiling point at or above 100°F. Example: acetone, benzene, toluene, methanol.
- *Class 1C:* Flash point at or above 73°F and below 100°F. Example: p-Xylene DOT defines flammable liquids as those with flash points of less than 140°F.

flammable range   The difference between the lower and upper explosive limits, expressed in terms of percentage of vapor or gas in air by volume. Also often referred to as the *explosive range*.

flammable solid   Solid that will ignite readily and continue to burn or is liable to cause fires under ordinary conditions or during transportation through friction or retained heat from manufacturing or processing, and that burns so vigorously and persistently as to create a serious transportation hazard. Includes flares or solid propellants that are not explosive.

flare   A control device that burns hazardous materials to prevent their release into the environment; may operate continuously or intermittently, usually on top of a stack. Flares are frequently used at landfills to destroy gases being emitted by the landfill.

flash back   Occurs when a trail of flammable material is ignited by a distant spark or ignition source. The flame then travels along the trail of the material back to its source.

flash fire   A rapid burning of flammable gases or vapors characterized by high temperature, short duration, and a considerable shock wave.

flash point (Fl.P.)   Temperature at which a liquid will give off enough flammable vapor to ignite. The minimum temperature at which a liquid gives off vapor within a test vessel in sufficient concentration to form an ignitable mixture with air near the surface of the liquid. Two tests are used, open cup and closed cup. There are different test methods. Their resulting flash points may vary for the same material, depending on the method used, so the test method is indicated when the flash point is given.

floc   A clump of solids formed in sewage by biological or chemical action.

flocculation   Process by which clumps of solids in water or sewage agglomerate through biological or chemical action so they can be separated from water or sewage by filtration or settling.

floodplain   The flat or nearly flat land along a river or stream or in a tidal area that is covered by water during a flood. When used in conjunction with a number of years (e.g. 100 year flood plain), the area expected to experience flooding on an average of once per that amount of time.

flue gas   The gases exiting a chimney or stack after combustion in the system burner. It can include nitrogen oxides, carbon oxides, water vapor, sulfur oxides, particles, and many chemical pollutants.

flue gas desulfurization   A technology that employs a sorbent, usually lime or limestone, to remove sulfur oxides (chiefly sulfur dioxide) from the gases produced by burning fossil fuels. Flue gas desulfurization is current state-of-the art technology for major sulfur oxides SO_X emitters, like power plants.

fluidized   A mass of solid particles that is made to flow like a liquid by injection of water or gas is said to have been fluidized. In water treatment, a bed of filter media is periodically fluidized by backwashing water through the filter to clean its surface.

fluorinated ethylenepropylene (F.E.P.)   Teflon®, used to provide extra chemical resistance for CPC.

fluorocarbons (FCs)   Any of a number of organic compounds analogous to hydrocarbons in which one or more hydrogen atoms are replaced by fluorine used as refrigerants replacing CFCs. FCs containing chlorine are called chlorofluorocarbons (CFCs). They are believed to be contributing to the destruction of the ozone layer in the stratosphere, thereby allowing more harmful solar radiation to reach the Earth's surface.

flush   (1) To open a cold-water tap to clear out all the water that may have been sitting for a long time in the pipes. In new homes, to flush a system means to send large volumes of water gushing through the unused pipes to remove loose particles of solder and flux. (2) To force large amounts of water through a system to clean out piping or tubing and storage or process tanks.

flux   (1) A flowing or flow. (2) A substance used to help metals fuse together, such as soldering.

fly ash Noncombustible residual particles expelled by flue gas.

foam See **AFFF.** Firefighting material consisting of small bubbles of air, water, and concentrating agents. Foam will put out a fire by blanketing it, excluding air and blocking the escape of volatile vapor.

fog Visible suspension of fine liquid droplets in a gas (e.g. water in air).

food chain A sequence of organisms, each of which uses the next-lower member of the sequence as a food source.

forbidden A material that is prohibited from being offered or accepted for transportation.

foreseeable emergency Potential occurrence such as equipment failure, rupture of containers, or failure of control equipment that could result in an uncontrolled release of a hazardous substance or material.

formaldehyde (CH_2O) A colorless, pungent, and irritating gas, used chiefly as a disinfectant and preservative and in synthesizing other compounds, such as resins.

formula The scientific expression of the chemical composition of a material. Examples are water, H_2O; sulfuric acid, H_2SO_4; and sulfur dioxide, SO_2.

fossil fuel Fuel, chiefly hydrocarbons, derived from ancient organic remains, e.g., peat, coal, crude oil, and natural gas.

fracture A break in a rock formation due to structural stresses, e.g., faults, shears, joints, and planes of fracture cleavage.

free product A liquid petroleum hydrocarbon in the pure or concentrated form, not mixed with, or dissolved in, water. (See **nonaqueous phase liquid.**)

freeboard (1) Vertical distance from the normal water surface to the top of a confining wall. (2) Vertical distance from the sand surface to the underside of a trough in a sand filter. (3) Vertical distance from the normal water surface to the top of the hull of a boat.

freezing point Temperature at which a material changes its physical state from liquid to solid. The same as **melting point** for pure compounds. This information is important because a frozen material may burst its container or the hazards it presents could change.

friable asbestos Any material containing more than 1% asbestos and that can be crumbled or reduced to powder by hand pressure. May include previously nonfriable material that becomes broken or damaged by mechanical force.

friable Capable of being crumbled, pulverized, or reduced to powder by hand pressure.

frostbite Damage to tissue (e.g., skin) from exposure to an extremely cold atmosphere or contact with extremely cold liquids or solids. Skin contact with cryogenic liquids such as liquid nitrogen, for example, causes frostbite immediately.

fugitive emission Gas, liquid, solid, vapor, fume, mist, fog, or dust that accidentally escapes from process equipment or a product. Pumps with leaking seals are a typical cause of fugitive emissions.

fully encapsulating suit (FES) Also called totally encapsulating suit (TES); fully protective gear that keeps gases, vapor, liquid, and solids from any contact with skin and prevents them from being inhaled or ingested.

fume Airborne suspension consisting of minute solid particles (usually metals) arising from the heating of a solid. This heating is often accompanied by a chemical reaction where the particles react with oxygen in the air to form an oxide. For example, metal fume emitted during welding. The particles are usually less than 1 micron in diameter.

fumigant A pesticide vaporized to kill pests. Used in buildings and greenhouses.

fungicide A chemical that controls or inhibits fungus growth.

future liability Refers to potentially responsible parties' obligations to pay for additional response activities beyond those specified in the record of decision or consent decree.

g Gram or grams. Metric unit of weight. Approximately 454 g equal 1 lb.

gamma (γ) radiation The most hazardous radiation to humans. Blocking of gamma radiation can be achieved using several feet of thickness of concrete or a few inches of lead.

gangrene Death of tissue combined with putrefaction.

gas Formless fluid that occupies the space of its enclosure. A state of matter in which the material has very low density and viscosity, can expand and contract greatly in response to changes in temperature and pressure, easily diffuses into other gases, and readily and uniformly distributes itself throughout any container. A gas can be changed to the liquid state at its condensing or cooling temperature, or to the solid state at its freezing or melting point, at sufficiently lower temperature. Examples include sulfur dioxide, ozone, and carbon monoxide. Can settle to the bottom or raise to the top of an enclosure when mixed with other materials of higher or lower density. Oxygen, nitrogen, and the mixture air are gases that liquefy only in the vicinity of $-300°F$. Humidity (water) in the air is a vapor. It is easily condensed to a liquid as dew on cool evenings and returns to the vapor state in the heat of the next day.

gas chromatograph/mass spectrometer (CG/MS) Instrument that identifies the molecular composition and concentrations of various chemicals in air, gaseous mixtures, water, and soil samples.

gasification Conversion of solid material such as coal into a gas for use as a fuel.

gas-tight suit A Level A vapor-protective suit and gas-tight suit as defined by OSHA in HAZWOPER. The suit is acceptable for use if it can maintain 4 in. (water column) of positive air pressure, with less than a 20% drop in pressure after four minutes in a standard inflation test. (See **ASTM Standard F1052**.)

gastric lavage Washing out of the stomach using a tube and fluids.

gastritis Irritation of lining of the stomach. Symptoms are stomach pains, vomiting, or diarrhea.

gastroenteritis Inflammation of the stomach and intestines.

gastrointestinal tract Stomach and intestines as a functional unit.

gauge pressure Gas, vapor, or liquid pressure measured with an instrument that does not include atmospheric pressure. The common American unit is psig. The metric units are kPag. "Gage" is a common engineering spelling of "gauge."

general duty clause (OSHA) States that each employer must "furnish to each of his employees employment and a place of employment which are free from recognized hazards that are causing or likely to cause death or serious physical harm to his employees."

general permit A permit applicable to a class or category of pollutant dischargers.

general reporting facility A facility having one or more hazardous chemicals above the 10,000-lb threshold for planning quantities. Such facilities must file MSDS and emergency inventory information with the SERC, LEPC, and local fire departments.

general ventilation Removal of contaminated air from the entire work area and replacing it with clean air, as opposed to local ventilation, which is removal of contaminated air from the immediate work area. There are two types of general ventilation: *exhaust* ventilation and *supply* ventilation. Supply ventilation blows fresh air into the workplace and should only be used with great caution. Exhaust ventilation exhausts atmospheric contaminants from the workplace and replaces the contaminated air with fresh air. OSHA requires four air changes per hour in the workplace; a confined space, for example.

generator

1. Any person or business that produces hazardous waste or first causes hazardous waste to become subject to RCRA regulations. Generators include small or large businesses, manufacturing plants, and other facilities. Generators are subject to specific hazardous waste regulations. Small quantity generators, less than 1,000 kg/month, large quantity generators, greater than 1,000 kg/month. (See **CESQG**.)

2. A fuel-powered, mechanical means of producing electricity on the work site, usually portable.

generic name Designation or identification of a chemical, pharmaceutical, or other substance by other than its accepted chemical or trademarked name.

genetic   Pertaining to or carried by genes. Hereditary.

Geographic Information System (GIS)   Electronically manages geographically referenced data. Through GIS, such data can be displayed, assembled, stored, and manipulated and frequently displayed in a map format. Examples of the use of GIS include county boundaries, land use, and pollution-monitoring locations.

geological log   A detailed description of all underground features (depth, thickness, type of formation) discovered during the drilling of a well.

geophysical log   A record of the structure and composition of the earth encountered when drilling a well or similar type of test hole or boring.

germicide   Any compound that kills disease-causing microorganisms.

gingivitis   Inflammation of the gums.

global warming   An increase in the near surface temperature of the Earth. Global warming has occurred in the distant past as the result of natural influences, but the term is most often used to refer to the warming predicted to occur as a result of increased emissions of greenhouse gases. Scientists generally agree that the Earth's surface has warmed by about 1°F in the past 140 years. The Intergovernmental Panel on Climate Change (IPCC) recently concluded that increased concentrations of greenhouse gases are causing an increase in the Earth's surface temperature and that increased concentrations of sulfate aerosols have led to relative cooling in some regions, generally over and downwind of heavily industrialized areas.

glovebag   A polyethylene or polyvinyl chloride bag-like enclosure affixed around an asbestos-containing source (most often thermal system insulation) permitting the material to be removed while minimizing release of airborne fibers to the surrounding atmosphere.

glovebox   A unit used for filling and opening containers of extremely hazardous materials such as radioactive and biohazard wastes. The box is kept under a vacuum and that exhausted air is cleaned so as to not be hazardous to the worker. The worker manipulates the hazardous contents through gloves built through the wall of the glovebox and cannot, during normal operations, come into contact with the material.

gram   Metric unit of weight. One U.S. ounce is 28.4 grams and 1 pound is about 454 grams.

gram/kilogram (g/kg)   Expression of dose used in oral and dermal toxicology testing to indicate the grams of substance dosed per kilogram of animal or human body weight; g/kg preferred.

granular activated carbon (GAC) treatment   Also referred to as *carbon adsorption;* a filtering system often used in small water systems and individual homes to remove organics. Also used by municipal water treatment plants. GAC can be highly effective in lowering elevated levels of radon, chlorine, organics, and many other materials in air or water.

gray water   Domestic wastewater composed of wash water from kitchen, bathroom, and laundry sinks, tubs, and washers. It does not include sewage.

greenhouse effect   The warming of the Earth's atmosphere, attributed to a buildup of carbon dioxide and other gases in a layer of the atmosphere. This layer allows the sun's rays to heat the Earth. However, it traps the infrared radiation from the earth to space, thereby preventing a counterbalancing loss of heat.

greenhouse gas   A gas, such as carbon dioxide or methane, that contributes to the layer above the earth that reduces the earth's heat loss by radiation to space.

ground cover   Plants grown to keep soil from eroding.

ground water under the direct influence (UDI) of surface water   Any water beneath the surface of the ground with: (1) significant occurrence of insects or other microorganisms, algae, or large-diameter pathogens; (2) significant and relatively rapid shifts in water characteristics such as turbidity, temperature, conductivity, or pH that closely correlate to climatological or surface water conditions. Direct influence is determined for individual sources in accordance with criteria established by a state.

grounding   Safety practice to conduct electrical charge by a conducting cable attached from equipment to an earth ground, thereby preventing ignition of a material by discharge sparks.

ground-penetrating radar (GPR)   A geophysical method that uses high-frequency electromagnetic waves to obtain subsurface information.

groundwater   Water found in the saturated portions of geologic formations beneath the surface of land or water. The supply of fresh water found beneath the Earth's surface, usually in aquifers, which supplies wells and springs. Because groundwater is a major source of drinking water, there is growing concern over contamination from leaching agricultural chemicals, industrial pollutants, and leaking underground storage tanks.

groundwater discharge   Groundwater entering near coastal waters that has been contaminated by landfill leachate, deep well injection of hazardous wastes, septic tanks, and other contaminating releases.

half-life   (1) The time required for a pollutant to lose one-half of its original concentration. For example, the biochemical half-life of DDT in the environment is 15 years. (2) The time required for half of the atoms of a sample of radioactive element to undergo self-transmutation or decay (half-life of radium is 1620 years). (3) The time required for the elimination of half a total dose from the body.

halogens   A chemical family that includes fluorine, chlorine, bromine, and iodine.

Halon®   Trifluorobromomethane (Halon® 1301). A compound with a long atmospheric lifetime that breaks down in the stratosphere causing depletion of ozone. Banned from production in the mid-1990s, Halon® was formerly used in fire protection, especially in areas, such as computer and electrical rooms where water cannot be used. Halon® can still be encountered in fire-suppression systems. It is being aggressively captured and recycled, just as CFCs are.

hazard   A circumstance or condition that can do harm. Hazards are categorized into two broad categories, material and physical. Material hazards include toxic, corrosive and chemical, flammable and explosive, and radiological.

hazard assessment   The determination of the lack of safety or degree of risk based on all parts of an exposure situation, including the characteristics of the material(s) to which one is exposed and the conditions that determine degree of exposure.

hazard classes   A series of nine descriptive terms established by the UN Committee of Experts to categorize the hazardous nature of chemical, physical, and biological materials. These hazard classes and divisions are used by the DOT for purposes of manifesting, labeling, and placarding shipments of hazardous materials. Some companies will also use these classes and divisions for identifying stored materials and storage facilities. These categories are:

Class 1. Explosives

   Division 1.1 Explosives with a mass explosion hazard

   Division 1.2 Explosives with a projection hazard

   Division 1.3 Explosives with predominantly a fire hazard

   Division 1.4 Explosives with no significant blast hazard

   Division 1.5 Very insensitive explosives; blasting agents

   Division 1.6 Extremely insensitive detonating articles

Class 2. Gases

   Division 2.1 Flammable gases

   Division 2.2 Nonflammable, nontoxic compressed gases

   Division 2.3 Gases toxic by inhalation

Class 3. Flammable liquids and combustible liquids

Class 4. Flammable solids; spontaneously combustible materials; and dangerous when wet materials

Division 4.1 Flammable solids

Division 4.2 Spontaneously combustible materials

Division 4.3 Dangerous when wet materials

Class 5. Oxidizers and organic peroxides

Division 5.1 Oxidizers

Division 5.2 Organic peroxides

Class 6. Toxic materials and infectious substances

Division 6.1 Toxic materials

Division 6.2 Infectious substances

Class 7. Radioactive materials

Class 8. Corrosive materials

Class 9. Miscellaneous dangerous goods

Hazard Communication Standard   The OSHA regulation (29 CFR 1910.1200) that requires chemical manufacturers, suppliers, and importers to assess the hazards of the chemicals that they make, supply, or import and to inform employers, customers, and workers of these hazards through MSDS information.

hazard evaluation   The quantification (stating a numerical value) of the actual or potential impact, or risk, that a hazardous substance poses to public health and the environment.

Hazard Ranking System (HRS)   The principal screening tool used by EPA to evaluate risks to public health and the environment associated with abandoned or uncontrolled hazardous waste sites. The HRS calculates a score based on the potential of hazardous substances spreading from the site through the air, surface water, or groundwater and on other factors such as density and proximity of human population. This score is the primary factor in deciding if a site should be on the National Priorities List (NPL) and, if so, what ranking it should have compared to other sites on the list.

hazardous   Capable of posing an unreasonable risk to health and safety (DOT definition).

hazardous atmosphere   An atmosphere that may expose employees to the risk of death, incapacitation, impairment of ability to self-rescue (that is, escape unaided from a permit space), injury, or acute illness from one or more of the following causes:

1. Atmospheric oxygen concentration below 19.5% or above 23.5%

2. Flammable gas, vapor, or mist in excess of 10% of its lower flammable limit (LFL)

3. Atmospheric concentration of any substance for which a dose or a permissible exposure limit is published in Subpart G, Occupational Health and Environmental Control, or in Subpart Z, Toxic and Hazardous Substances, of 29 CFR and that could result in employee exposure in excess of its dose or permissible exposure limit

4. Airborne combustible dust at a concentration that meets or exceeds its LFL

Note

This concentration may be approximated as a condition in which the dust obscures vision at a distance of 5 ft (1.52 m) or less.

Note

An atmospheric concentration of any substance that is not capable of causing death, incapacitation, impairment of ability to self-rescue, injury, or acute illness due to its health effects is not covered by this provision.

5. Any other atmospheric condition that is immediately dangerous to life or health.

Note

For air contaminants for which OSHA has not determined a dose or permissible exposure limit, other sources of information, such as material safety data sheets that comply with the Hazard Communication Standard, Section 1910.1200 of 29 CFR, published information, and internal documents can provide guidance in establishing acceptable atmospheric conditions.

hazardous air pollutants (HAPs)   Air pollutants that are not covered by ambient air quality standards but that, as defined in the Clean Air Act, may present a threat of adverse human health effects or adverse environmental effects. Such pollutants include asbestos, beryllium, mercury, benzene, coke oven emissions, radionuclides, and vinyl chloride.

hazardous chemical   1. A chemical that can cause chronic or acute harm upon exposure 2. A chemical listed in a specific EPA, OSHA, FDA, USDA, or DOT regulation. 3. An EPA designation for any hazardous material requiring an MSDS under OSHA's Hazard Communication Standard. Such substances are capable of producing fires and explosions or adverse health effects like cancer and dermatitis. Hazardous chemicals are distinct from hazardous waste.

hazardous decomposition   Breaking down or separation of a substance into its constituent parts, elements, or simpler compounds accompanied by the release of heat, gas, or hazardous materials.

hazardous ingredients   Hazardous substances that make up a mixture.

hazardous material   A substance or material that has been determined by the Secretary of Transportation to be capable of posing an unreasonable risk to health, safety, and property when transported in commerce and that has been so designated (DOT definition).

hazardous material emergency   An uncontrolled or unexpected release of a hazardous material.

Hazardous Materials Identification System (HMIS®)   A system developed by the National Paint and Coatings Association (NPCA) to provide information on health, flammability, and reactivity hazards that are encountered in materials in the workplace. A number is assigned to a material indicating the degree of hazard, from 0 for the least up to 4 for the most severe. Letters are used to designate personal protective equipment. Similar but not identical to NFPA 704M®.

Hazardous Material Response Team (HAZMAT Team)   An organized group of employees, designated by the employer, who are expected to perform work to handle and control actual or potential leaks or spills of hazardous substances requiring possible close approach to the substance. The team members perform responses to releases or potential releases of hazardous substances for the purpose of control or stabilization of the incident. A HAZMAT team is not a fire brigade, nor is a typical fire brigade a HAZMAT team. A HAZMAT team, however, may be a properly-trained separate component of a fire brigade or fire department.

hazardous polymerization   Said to occur when an unwanted or unexpected polymerization reaction occurs that could result in releases of a large amount of heat or an accidental release of the material or injury to workers. (See **polymerization**.)

hazardous sample   A sample that is considered to contain high concentrations of hazardous material.

hazardous substance   Any substance designated or listed under 1 through 7 of this definition, exposure to which results or may result in adverse effects on the health or safety of employees:

1. Any substance defined under Section 101(14) of CERCLA

2. Any toxic pollutant listed under Section 307(a) of the Federal Water Pollution Control Act

3. Any hazardous air pollutant listed under Section 112 of the Clean Air Act

4. Any imminently hazardous chemical substance or mixture with respect to which the Administrator has taken action pursuant to Section 7 of the Toxic Substances Control Act

5. Any biologic agent and other disease causing agent that after release into the environment and upon exposure, ingestion, inhalation, or assimilation into any person, either directly from the environment or indirectly by ingestion through food chains, will or may reasonably be anticipated to cause death, disease, behavioral abnormalities, cancer, genetic mutation, physiological malfunctions (including malfunctions in reproduction), or physical deformations in such persons or their offspring

6. A material and its mixtures or solutions that is identified by the letter "E" in column (1) of the Hazardous Materials Table, 49 CFR 172.101, when offered for transportation in one package or in one transport vehicle if not packaged, and when the quantity of the material therein equals or exceeds the Reportable Quantity (RQ)

7. Any hazardous waste having the characteristics identified under or listed pursuant to Section 3001 of the Solid Waste Disposal Act (but not including any waste the regulation of which under the Solid Waste Disposal Act has been suspended by Act of Congress)

Note

The term does not include petroleum including crude oil or any fraction thereof, which is not otherwise specifically listed or designated as a hazardous substance under the definitions above. Also, the term does not include natural gas, natural gas liquids, liquefied natural gas, and synthetic gas usable for fuel (or mixtures of natural gas and such synthetic gas). (An OSHA definition.)

hazardous substance research centers (HSRC)   These oversee basic and applied research, technology transfer, and training involving problems relating to hazardous substance management. These activities are conducted regionally by five multiuniversity centers that focus on different aspects of hazardous substance management.

hazardous waste   A solid waste that is:

[A] A waste or combination of wastes that is **LISTED** in 40 CFR 261.31 (EPA).

[B] A **CHARACTERISTIC** waste as defined in 40 CFR

261.21 Ignitable Waste . . . . (I) (Flash Point less than 60°C or 140°F)

261.22 Corrosive Waste . . . . (C) (pH less than 2 or greater than 12.5)

261.23 Reactive Waste (R) (Normally unstable, reacts violently or forms toxic gases when in contact with water, is a high explosive, etc.)

261.24 Toxicity Characteristic Waste . . . . (E) Exhibits the characteristic of toxicity using the Toxicity Characteristic Leaching Procedure (TCLP)

Note

Some states will require the Extraction Procedure Toxicity Test (EP Toxicity).

[C] Those substances defined as hazardous wastes in 49 CFR 171.8 (DOT).

Note

For HAZWOPER purposes we define hazardous wastes as wastes that meet EPA's definition for solid waste and possess the characteristics of ignitability, corrosivity, reactivity, or toxicity (as defined by RCRA). Also included are wastes on an EPA list of hazardous wastes or on the DOT Hazardous Materials Tables.

Hazardous Waste Identification Number   Identification number assigned by the EPA, per RCRA law, to identify and track hazardous wastes. Consists of a letter and three digits. For example, D001 is the ID number for waste characterized as flammable.

hazardous waste landfill   An EPA-regulated excavated or engineered site where hazardous waste is deposited and covered.

hazardous waste minimization   Reducing the amount or toxicity of waste produced by a generator by either source reduction or environmentally sound recycling.

hazardous waste operation   Any operation conducted within the scope of the HAZ-WOPER standard 29 CFR 1910.120.

hazardous waste site, or site   Any facility or location within the scope of the HAZWOPER standard at which hazardous waste operations take place.

hazards analysis   Procedures used to (1) identify potential sources of release of hazardous materials from fixed facilities or transportation accidents; (2) determine the vulnerability of a geographical area to a release of hazardous materials; and (3) compare hazards to determine which present greater or lesser risks to a community.

hazards identification   Providing information on which facilities have extremely hazardous substances, what those chemicals are, how much there is at each facility, how the chemicals are stored, and whether they are used at high temperatures.

headspace   The vapor mixture trapped above a solid or liquid in a sealed vessel.

health hazard   Chemical material, mixture of chemical materials, pathogen, or radioactive substance for which there is statistically significant evidence, based on at least one study conducted in accordance with established scientific principles, that acute or chronic health effects may occur in exposed personnel.

   The term includes chemicals that are carcinogens, toxic or highly toxic agents, reproductive toxins, irritants, corrosives, sensitizers, hepatotoxins, nephrotoxins, neurotoxins, agents that act on the hematopoietic system, agents that damage the lungs, skin, eyes, or mucous membranes, and radioactive materials. It also includes stress due to temperature extremes. Further definition of the terms used above can be found in Appendix A to 29 CFR 1910.1200 (the OSHA Hazard Communication Standard, HAZCOMM).

heavy metals   Metallic elements with high atomic weights, for example: mercury (200.1), cadmium (112.4), and lead (207.2). Can damage living things at low concentrations and tend to accumulate in the food chain; the RCRA-regulated metals. Those atomic weights are high compared to aluminum (27) or iron (55.8).

hematopoietic system   The blood-forming mechanism of the human body.

hematuria   Presence of blood in the urine.

hemolysis   Separation of the hemoglobin from red blood corpuscles.

Henry's Law   Constant frequently used to estimate the environmental movement and fate of organic chemicals in water. Chemical Properties Handbook (McGraw-Hill, 1999) lists this constant in units of atm/mol fraction as 11,515 for ethylene, 1630 for carbon tetrachloride, and 308 for benzene at ambient conditions.

hepatic   Pertaining to the liver.

hepatotoxin   A substance that causes injury to the liver.

herbicide   Usually a manufactured chemical designed to control or destroy plants, weeds, or grasses. (See **defoliant**.)

high-intensity discharge   Generic term for the electrical activity in mercury vapor, metal halide, and high-pressure sodium lamps and their fixtures.

high-density polyethylene (HDPE)   Material used to make plastic drums and other containers that is resistant to a number of chemicals. HDPE is readily recycled. Incomplete incineration can produce noxious vapors.

high-risk community   Community located within the vicinity of numerous sites of facilities or other potential sources of environmental exposure and health hazards that may result in high levels of exposure to contaminants or pollutants.

highly toxic   A chemical in any of the following three categories:

1. A chemical with a median lethal dose (LD_{50}) of 50 mg or less per kilogram of body weight when administered orally to albino rats between 200 and 300 g each.

2. A chemical with a median lethal dose (LD_{50}) of 200 mg or less per kilogram of body weight when administered by continuous contact for 24 hours (or less if death occurs within 24 hours) with the bare skin of albino rabbits weighing between 2 and 3 kg each.

3. A chemical that has a median lethal concentration (LC_{50}) in air of 200 parts per million by volume or less of gas or vapor, or 2 mg per liter or less of mist, fume, or dust when administered by continuous inhalation for 1 hour (or less if death occurs within 1 hour) to albino rats weighing between 200 and 300 g each.

holding pond   Pond or reservoir, usually made of earth, built to store polluted runoff.

holding time   The maximum amount of time a sample may be stored before analysis.

hollow-stem auger drilling   Conventional drilling method that uses augurs to penetrate the soil and provide samples via the hollow stem. As the augers are rotated, soil cuttings are conveyed to the ground surface via auger spirals.

homogeneous area   In accordance with the Asbestos Hazard and Emergency Response Act (AHERA) definition, an area of surfacing materials, thermal surface insulation, or miscellaneous material that is uniform in color and texture.

homogenous waste   Waste that appears to be uniform in color, consistency, and specific gravity.

host   (1) In genetics, the organism, typically a bacterium, into which a gene from another organism is transplanted. (2) In medicine, an animal infected or parasitized by another organism.

hot line   The outer boundary of the exclusion (hot) zone. It separates the area of highest contamination from the contamination reduction (warm) zone. It provides an adequate area in which to conduct site operations taking into account potential contaminant migration.

hot work   Work that produces arcs, sparks, flames, heat (greater than 400°F), or other sources of ignition.

hot work permit   The employer's written authorization to perform operations (for example, grinding, riveting, welding, cutting, burning, and heating) capable of providing a source of ignition.

household hazardous waste (HHW)   Items such as paints, stains, oven cleaner, motor oil, batteries, solvents, pesticides, and other materials or products containing volatile chemicals that can catch fire, react, or explode or that are corrosive or toxic. They are commonly disposed of in the trash by households. While these items are not federally regulated as hazardous waste, they contain hazardous constituents. HHW refers to items such as these that can be disposed of in MSW landfills but are often collected by communities and managed as hazardous waste.

HVAC   Heating, ventilation, and air conditioning.

hydraulic gradient   In general, the direction of groundwater flow measured as changes in the depth of the water table.

hydrogen sulfide (H_2S)   Gas emitted during decomposition of materials containing sulfur. It has the odor of rotten eggs and is both flammable and toxic. It does not have a PEL due to the rapid fatiguing of the olfactory senses. That is, even though the concentration of H_2S may be increasing, the worker will detect its odor less and less. As soon as the odor of rotten eggs is detected, leave the area at once. It has an OSHA 10-minute ceiling level of 50 ppm and an IDLH level of 100 ppm.

hydrocarbon   Organic compound composed only of carbon and hydrogen. Petroleum, natural gas, and coal are the main sources of hydrocarbons for industry. Common hydrocarbon products may contain other elements or compounds as impurities.

hydrogeological cycle   The natural process recycling water from the atmosphere down to (and through) the earth and back to the atmosphere.

hydrogeology   The geology of groundwater, with particular emphasis on the chemistry and movement of water.

hydrologic cycle   Movement or exchange of water between the atmosphere and earth.

hydrology   The science dealing with the properties, distribution, and circulation of water.

hydrophilic   Materials that easily dissolve in water, or materials that absorb and retain water, causing them to swell and frequently to gel. Detergent molecules have both hydrophilic and hydrophobic (or oleophilic) character, allowing them to remove oily materials and dissolve them in water.

hydrophobic   Materials not soluble in, or having a strong aversion for, water.

hygroscopic   A material that readily absorbs available moisture from the atmosphere. Table salt and sugar are examples.

hyperemia   Congestion of blood in a body part.

hypergolic   Self-igniting upon contact of its components without a spark or external aid. Some liquid rocket propellants are hypergolic when mixed, such as hydrazine (strong reducing agent) and hydrogen peroxide (strong oxidizer).

hypersensitivity diseases   Diseases characterized by allergic responses to pollutants; diseases most clearly associated with indoor air quality are asthma, rhinitis, and pneumonic hypersensitivity. Hypersensitivity can be acquired from repeated exposure to a chemical.

hypocalcemia   Calcium deficiency of the blood.

hypoxia   Insufficient oxygen, especially applied to body cells.

hypoxic waters   Waters with dissolved oxygen concentrations of less than 2 ppm, the level generally accepted as the minimum required for most marine life to survive and reproduce.

ignition source   Anything that provides heat, spark, or flame sufficient to cause combustion or explosion.

ignition temperature   Lowest temperature at which a combustible material, when heated, will catch fire in air and will continue to burn independently of the source of heat. The ignition temperature of paper is about 454°F.

immediate use   Means that the hazardous chemical will be under the control of and used only by the person who transfers it from a labeled container. Further, it will be used up within the work shift in which it is transferred. For example, pouring paint from a labeled drum into a small bucket for convenience. Hazardous materials contained for immediate use are not required to be labeled; however, labeling is strongly advised.

immediately dangerous to life or health (IDLH)   An atmospheric concentration of any toxic, corrosive, or asphyxiant substance; also, the presence of reactive materials under conditions that pose an immediate threat to life or would interfere with an individual's ability to escape from a dangerous atmosphere without a respirator. This includes oxygen deficiency, which means that the concentration of oxygen, by volume, in the atmosphere is below 19.5%. A supplied air respirator must be provided in that case.

For the purposes of HAZWOPER, also included in this definition are other safety hazards such as rusted or broken ladders and staging, exposed electrical busses, or uncontrolled filling lines that could cause drowning or engulfment.

Note

Rapid drowning can occur in low-density liquids. It has occurred in actively aerated wastewater. Swimming in aerated water can be extremely difficult or nearly impossible.

imminent hazard   One that would likely result in unreasonable adverse effects on humans or the environment; or risk unreasonable hazard to an endangered species during the time required for a pesticide registration cancellation proceeding.

imminent threat   A high probability that exposure is occurring.

immiscibility   The inability of two or more liquids to dissolve readily into one another, such as oil and water.

importer   First business with employees within the customs territory of United States that receives hazardous chemicals produced in other countries for the purpose of supplying them to distributors or employers within the United States.

impermeable   Not easily penetrated. The property of a material or soil that does not allow, or allows only with great difficulty, the movement or passage of water or other liquids.

impervious   Material that does not allow a given concentration of another substance to pass through, or penetrate, it within a proposed use time period.

impoundment   A body of water or sludge confined by a dam, dike, floodgate, or other barrier.

inches of mercury column   A unit used in measuring pressures. One inch of mercury column equals a pressure of 1.66 kPa (0.491 psi). The common unit of measurement of barometric pressure in the United States.

inches of water column   A unit used in measuring pressures. One inch of water column equals a pressure of 0.25 kPa (0.036 psi). The common unit of measurement of the pressure of flowing air in heating and air conditioning ducts due to the accuracy of measurement at lower pressures.

incident   The release or potential release of a hazardous substance or material into the environment.

incident characterization   The process of identifying the substance involved in an incident, determining exposure pathways, and projecting the effect they will have on people, property, wildlife and plants, and the disruption of services.

incident command system (ICS)   The organizational arrangement wherein one person, often the fire chief of the impacted district, is in charge of an integrated, comprehensive emergency response organization and the emergency incident site, backed by an emergency operations center staff with resources, information, and advice.

incident command post   A facility located at a safe distance from an emergency site, where the Incident Commander, key staff, and technical representatives can make decisions and deploy emergency manpower and equipment.

incident evaluation   The process of assessing the impact that released, or potentially released, substances pose to public health and the environment.

incineration   A treatment technology involving destruction of waste by controlled burning at high temperatures (above 1500°F). For example, a remediation method used for burning sludge to remove the water and reduce the remaining residues to a safe, nonburnable ash that can be disposed of safely on land, in some waters, or in underground locations. Incineration below 1500°F frequently releases noxious gases and vapors.

incinerator   A furnace for burning waste under controlled conditions.

incompatible   Materials that could cause dangerous reactions from direct contact with one another.

incompatible waste   A waste unsuitable for mixing with another waste or material because it may react to cause a hazard.

indicator   (1) In biology, any biological entity or processes, or community whose characteristics show the presence of specific environmental conditions. (2) In chemistry, a substance that shows a visible change, usually of color, at a desired point in a chemical reaction. (3) A device that indicates the result of a measurement; e.g., a pressure gauge or a moveable scale.

industrial pollution prevention (P^2)   Combination of industrial source reduction and toxic chemical use substitution.

industrial process waste   Residues produced during manufacturing operations.

industrial sludge   Semiliquid residue or slurry remaining from treatment of industrial water and wastewater.

industrial source reduction   Practices that reduce the amount of any hazardous substance, pollutant, or contaminant entering any waste stream or otherwise released into the environment. Also reduces the threat to public health and the environment associated with such releases. Term includes equipment or technology modifications, substitution of raw materials, and improvements in housekeeping, maintenance, training, or inventory control.

industrial waste   Unwanted materials from an industrial operation; may be liquid, sludge, solid, or hazardous waste.

inert   A characteristic of a material meaning that it will not react with other materials. For example, helium and argon are inert gases. Nitrogen is relatively inert.

inerting   The displacement of the atmosphere in a confined space by a noncombustible gas (such as nitrogen) to such an extent that the resulting atmosphere is noncombustible.

Note

This procedure produces an IDLH oxygen-deficient atmosphere.

inert ingredients   Anything other than the active ingredient in a product; not having active properties.

infectious agent   Any organism, such as a pathogenic virus, parasite, or bacterium, that is capable of invading body tissues, multiplying, and causing disease.

infectious waste   Hazardous waste capable of causing infections in humans, including contaminated animal waste; human blood and blood products; isolation waste, pathological waste; and discarded sharps (needles, scalpels, or broken medical instruments). (See **biohazard.**)

infiltration   (1) The penetration of water through the ground surface into subsurface soil or the penetration of water from the soil into sewer or other pipes through defective joints, connections, or manhole walls. (2) The technique of applying large volumes of wastewater to land to penetrate the surface and percolate through the underlying soil.

infiltration rate   The quantity of water that can enter the soil in a specified time interval.

inflammable   Capable of being easily set on fire and continuing to burn, especially violently. Same as **flammable.**

inflammation   Series of reactions produced in body tissue by an irritant, injury, or infection. Characterized by swelling and redness caused by an influx of blood and fluids.

influent   Water, wastewater, or other liquid flowing into a reservoir, basin, or treatment plant. (See **effluent.**)

information   Knowledge acquired concerning the conditions or circumstances particular to an incident.

infoterra   International environmental referral and research network made up of 177 countries coordinated by the United Nations Environment Programme (UNEP). The services offered by UNEP-Infoterra/USA include responding to requests from the international community for environmental information through document delivery, database searching, bibliographic products, purchasing information, and referrals to experts. In addition, UNEP-Infoterra/USA assists U.S. residents in identifying sources of international environmental information.

ingestion   Swallowing, or taking in of a substance through the mouth.

inhalation   Breathing in of a substance in the form of a gas, vapor, fume, mist, or dust.

inhalation toxicity   Ratings corresponding to the following definitions are derived from the test methods and categories of toxicity described in 16 CFR 1500.3.

1. *Nontoxic* The probable lethal concentration of the undiluted product to 50% of the test animals (LC_{50}) is greater than 200 mg per liter by volume when inhaled continuously for one hour or less.

2. *Toxic* The probable lethal concentration of the undiluted product to 50% of the test animals (LC_{50}) is greater than 2 mg and less than or equal to 200 mg per liter by volume when inhaled continuously for one hour or less.

3. *Highly Toxic* The probable lethal concentration of the undiluted product to 50% of the test animals (LC_{50}) is less than or equal to 2 milligrams per liter by volume when inhaled continuously for one hour or less.

inhibitor   A substance (chemical) added to another substance to prevent an unwanted chemical change from occurring. Rust inhibitors are added to vehicle antifreeze liquids.

injection   One of the four routes of entry into the human body. Injection is direct contact with the bloodstream, such as, through a needle stick or a contaminated nail or shard of glass.

injection well   A well into which fluids are injected for purposes such as waste disposal or improving the recovery of crude oil.

injection zone   A geological formation receiving fluids through a well.

innovative treatment technologies   New or inventive methods to treat hazardous waste effectively and reduce risks to human health and the environment. Technologies whose routine use is inhibited by lack of data on performance and cost.

inorganic materials   Materials derived from other than vegetable or animal sources; they generally do not contain carbon atoms. Examples are sand, salt, water, and metals. Exceptions are the inorganic carbonates, such as baking soda (sodium bicarbonate).

insecticide   A pesticide compound specifically used to kill or prevent the growth of insects.

in situ   In its original place; unmoved; unexcavated; remaining at the site or in the subsurface.

in situ **flushing**   Introduction of large volumes of water, at times supplemented with cleaning compounds, into soil, waste, or groundwater to flush hazardous contaminants from a site.

in situ **oxidation**   Technology that oxidizes contaminants dissolved in groundwater, converting them to nontoxic compounds.

in situ **stripping**   Treatment system that removes or strips volatile organic compounds from contaminated ground or surface water by forcing an airstream through the water and causing the compounds to evaporate.

insoluble   A material that is incapable of being dissolved in a given liquid. Wax is considered insoluble in water.

inspection   An examination to determine if certain criteria have been met.

institutional controls   Statutory authority for resource management, public health and safety, and environmental protection. In cases where institutional jurisdiction stems from resource management concerns, such as water rights, agreements, and land use, institutional control includes permits or authorizations required to proceed.

interface   The common boundary between two substances such as water and a solid, water and a gas, or two immiscible liquids such as water and oil.

intermediate bulk container (IBC)   A rigid or flexible container larger than a 55-gallon drum but smaller that a truck or rail bulk container. Designed to contain between 100 and 1,000 gallons, IBCs can be used to store and transport dry or liquid materials.

interstitial fibrosis   Scarring of the lungs.

interstitial monitoring   The continuous surveillance of the space between the walls of a double-walled underground storage tank.

intelligence   Information obtained from existing records or documentation, placards, labels, signs, special configuration of containers, inspection, visual observations, technical records, eyewitnesses, and others.

International Agency for Research on Cancer (IARC)   Part of the World Health Organization. (See **carcinogen.**)

inventory (TSCA)   Inventory of chemicals produced pursuant to Section 8(b) of the Toxic Substances Control Act.

inversion   A layer of warm air that prevents the rise of cooler air below it and thereby traps pollutants beneath it; can cause an air pollution episode; usually occurs during morning and evening. (See **lapse.**)

investigation   On-site and off-site survey(s) conducted to provide intelligence and qualitative and quantitative assessments of hazards associated with a site.

in vitro   Testing or action outside an organism (e.g., inside a test tube or culture (Petri) dish).

in vivo   Testing or action inside an organism.

ion   An electrically charged atom or group of atoms. In salt water, there are Na^+ and Cl^- ions.

ion exchange treatment   A common water-softening method used in areas that have naturally hard water supplies. Hard water contains high concentrations of limestone-type elements (calcium and magnesium). Ion exchange exchanges soluble salt-forming sodium ions for insoluble salt-forming calcium and magnesium ions in the water. The water-softener ion exchange resin is recharged with common salt (NaCl).

ionization potential (IP)   The energy required to ionize an atom or molecule. The energy is usually given in terms of electron volts (eV). Different materials have different IPs so they can be detected and identified. The most common field instrument for this monitoring is the photoionization detector, PID.

ionizing radiation   Radiation that can strip electrons from atoms; e.g., alpha, beta, and gamma radiation.

iridal   Pertaining to the iris of the eye.

iridocyclitis   Inflammation of both the iris and the ciliary body of the eye.

IRIS   EPA's Integrated Risk Information System, an electronic database containing the Agency's latest descriptive and quantitative regulatory information on chemical constituents.

irradiation   Exposure to radiation of wavelengths shorter than those of visible light (gamma, X-ray, or ultraviolet) for medical purposes, to sterilize milk or other foodstuffs, or to induce polymerization of monomers or vulcanization of rubber.

irrigation   Applying water or wastewater to land areas to supply the water and nutrient needs of plants.

irritant   Substances that, by contact in sufficient concentration for a sufficient period of time, will cause an inflammatory response or reaction of the eye, skin, or respiratory system.

isolation   The process by which a permit-required confined space is removed from service and completely protected against the release of energy and material into the space. This includes the following practices: blanking or blinding; misaligning or removing sections of lines, pipes, or ducts; a double block and bleed system; lock-out and/or tag-out of all sources of energy; or blocking or disconnecting all mechanical linkages.

isomers   Compounds that have same molecular weight and atomic composition but differ in molecular structure. Iso-octane and normal octane (*n*-octane) are isomers. However, iso-octane has an motor fuel octane rating of 100. Iso-octane has a branched-chain structure that assists combustion in an auto engine. Normal octane (*n*-octane) has a straight-chain structure. Normal octane's motor fuel rating is lower.

isotope   A variation of an element that has the same atomic number of protons but a different weight because of the number of neutrons. Various isotopes of the same element may have different radioactive behaviors. Some are highly unstable.

jaundice   Yellowish discoloration of tissue, whites of the eyes, and bodily fluids with bile pigment caused by any of several pathological conditions that interrupt the liver's normal production and discharge of bile.

joint and several liability   Under CERCLA, this legal concept relates to the liability for Superfund site cleanup and other costs on the part of more than one potentially responsible party (i.e., if there were several owners or users of a site that became contaminated over the years, they could each be considered potentially liable for cleaning up the entire site).

ketone   An organic compound containing a $C=O$ group, such as methyl ethyl ketone (MEK) or acetone.

ketosis   Condition marked by excessive production or accumulation of ketone compounds in the human body caused by disturbed carbohydrate metabolism.

key personnel   Those personnel or organizations considered to be essential to ensure the safe operation of a facility, site, project, operation, or task.

kinetic energy   Energy possessed by a moving object or water body.

kilogram (kg)   Metric unit of weight, about 2.2 pounds. kg preferred.

label   Any written, printed, or graphic sign or symbol displayed on or affixed to containers of hazardous chemicals. It should contain the identity of the material, appropriate hazard warnings, and name and address of the chemical manufacturer, importer, or other responsible party.

laboratory scale (activity)   Generally considered as chemical work using the smallest practical amounts of materials to test reactions or processes. The work involves containers of substances used for reactions and transfers that are designed for easy and safe handling by one person. Workplaces that produce commercial quantities of materials are excluded from the definition of laboratory. The general classifications of scales in increasing size are laboratory, pilot plant, and production scale.

lacrimation   Secretion and discharge of tears.

lacrimator   Material that produces tears; tear gas, oleoresin capsicum (pepper spray). Habañero peppers have 100,000 to 200,000 Scoville heat units, while pepper spray can have 2,000,000 Scoville heat units.

lagoon   (1) A shallow pond where sunlight, bacterial action, and oxygen work to purify wastewater; also used for storage of wastewater or spent nuclear fuel rods. (2) Shallow body of water, often separated from the sea by coral reefs or sandbars.

land application   Discharge of wastewater onto the ground for treatment or reuse.

land ban   Phasing out of land disposal of most untreated hazardous wastes, as mandated by the 1984 RCRA amendments.

Land Disposal Restrictions (LDRs)   These rules require that hazardous wastes be treated before they are land disposed to destroy or immobilize hazardous constituents that might otherwise migrate into soil and groundwater.

land farming (of waste)   A disposal process in which hazardous waste deposited on or in the soil is degraded by either naturally occurring or man-made microorganisms.

landfill   (1) Sanitary landfills are disposal sites for nonhazardous solid wastes spread in layers, compacted to the smallest practical volume, and covered by material applied at the end of each operating day. Also called *municipal landfills*. (2) Hazardous waste landfills are disposal sites for hazardous waste, selected and designed to minimize the chance of release of hazardous substances into the environment. They are specially designed disposal units for disposal of hazardous solid waste. Modern landfills generally have double synthetic liners to prevent releases. Active landfill areas are covered daily. They are specially covered (capped) and maintained when the landfill is no longer used. Their placement, design, construction, and operation are regulated by the EPA. Also called *secure chemical landfills*.

lapse   The opposite of inversion. Occurs on clear, sunny days.

large quantity generator   Generator that produces more than 2,200 pounds (1,000 kg) of hazardous waste per calendar month (about five full 55-gallon drums) are considered to be LQGs. They must follow certain regulations.

laser-induced fluorescence (LIF)   A method for measuring the relative amount of oily contamination in soil using a subsurface probe.

latency period   Time that elapses between exposure and the first manifestations of disease or illness. Asbestosis is believed to have a latency period of 15 to 30 years, for example.

lavage   Washing of a hollow organ, such as the stomach, using a tube and fluids.

LC$_{50}$ Lethal concentration 50, median lethal concentration. The concentration of a material in air that on the basis of laboratory tests (respiratory route) is expected to kill 50% of a group of test animals when administered as a single exposure in a specific time period, usually one hour. LC$_{50}$ is expressed as parts of material per million parts of air, by volume (ppmv), for gases and vapors; as micrograms of material per liter of air, or milligrams of material per cubic meter of air (mg/m^3) for dusts and mists, as well as for gases and vapors.

LC_{LO}   Lethal concentration low. The lowest concentration of a substance in air reported to have caused death in humans or animals. The reported concentrations may be entered for periods of exposure that are less than 24 hours (acute) or greater than 24 hours (subacute and chronic).

LD_{50}   Lethal dose 50. The single dose of a substance that causes the death of 50% of an animal population from exposure to the substance by any route other than inhalation. LD_{50} is usually expressed as milligrams or grams of material per kilogram of animal weight (mg/kg or g/kg). The animal species and means of administering the dose (oral, intravenous, etc.) should also be stated.

LD_{LO}   Lethal dose low. The lowest dose of a substance introduced by any route other than inhalation reported to have caused death in humans or animals.

leachate   Water that collects contaminants as it trickles through wastes, pesticides, or fertilizers. Leaching may occur in farming areas, feedlots, and landfills and may result in hazardous substances entering surface water, groundwater, or soil.

leachate collection system   A system that gathers leachate and pumps it to the surface for treatment.

leaching   The process by which soluble constituents are dissolved and filtered through the soil by a percolating fluid.

lead (Pb)   A heavy metal whose compounds are hazardous to health if breathed or swallowed. Their use in gasoline, paints, and plumbing compounds has been sharply restricted or eliminated by federal laws and regulations due to the serious health hazards they present, especially in young children.

lesion   Abnormal change, injury, or damage to tissue or to an organ.

leukemia   Progressive, malignant disease of the blood-forming organs.

Level A suit   The highest available level of respiratory, skin, splash, and eye protection. Requires a fully encapsulating vapor protective suit with supplied breathing air. A Level A hazardous material has a high vapor pressure, is toxic through skin absorption, or is carcinogenic.

Level B suit   The same level of respiratory protection as Level A but less skin protection. Suits are usually fully encapsulating but not gas-tight. Level B chemicals are not vapors or gases or other materials with skin toxicity or carcinogenic effects.

Level C suit   No skin, splash, or eye protection. Used with APRs. For minimal respiratory protection.

Level D suit   No skin or respiratory protection. Normal work uniform. Used in areas where there are no respiratory, skin, or splash hazards.

lifetime average daily dose   Figure for estimating excess lifetime cancer risk.

lifetime exposure   Total amount of exposure to a substance that a human would receive in a lifetime (usually assumed to be 70 years).

light nonaqueous phase liquid (LNAPL)   A nonaqueous phase liquid with a specific gravity less than 1.0. Because the specific gravity of water is 1.0, most LNAPLs float on top of the water table. Most common petroleum hydrocarbon fuels and lubricating oils are LNAPLs.

limited or restricted means for entry or exit   Any space where an occupant must crawl, climb, twist, be constrained in a narrow opening, follow a lengthy path, or otherwise exert unusual effort to enter or leave, or where the entrance may become sealed or secured against opening from inside.

limited quantity   With the exception of poison B materials, the maximum amount of a hazardous material for which there is a specific labeling and packaging exception.

limit of detection (LOD)   The amount, expressed in units of concentration that describes the lowest concentration level of an analyte that an analyst can determine (using a certain method) to be different from an analytical blank.

line breaking   The intentional opening of a pipe, line, or duct that is or has been carrying flammable, corrosive, or toxic material, an inert gas, or any fluid or other material at a volume, pressure, or temperature capable of causing injury. If lines cannot be broken and capped or plugged, see **blanking or blinding.**

liner   (1) A relatively impermeable barrier designed to keep leachate inside a landfill. Liner materials include plastic sheeting and dense clay. (2) An insert or sleeve for sewer pipes to prevent leakage or infiltration.

lipid granuloma   Mass of chronically inflamed tissue that is usually infective.

lipid pneumonia   Chronic condition caused by the aspiration of oily substances into the lungs.

lipid solubility   The maximum concentration of a chemical that will dissolve in fatty substances. Lipid soluble substances are insoluble in water. They will very selectively disperse through the environment via uptake in living tissue. DDT has lipid solubility.

liquefaction   Changing a gas or vapor to a liquid by cooling. Gas liquefaction requires temperatures below about $-160°F$.

liquid injection incinerator   Commonly used system that relies on high pressure to prepare liquid wastes for incineration by breaking them up into tiny droplets using spray nozzles to allow easier combustion.

listed wastes   Specific wastes determined by EPA to be hazardous and published in EPA lists. These lists are organized into five categories: source-specific wastes, nonspecific source wastes, commercial chemical products, toxic, and highly toxic wastes. Wastes listed as hazardous under RCRA but that have not been subjected to the toxic characteristics listing process because the dangers they present are considered self-evident.

liter (L)   A metric measure of volume; one quart equals about 0.9 L. One liter equals about 1,000 cc or 1,000 mL.

lithology   Mineralogy, grain size, texture, and other physical properties of granular soil, sediment, or rock.

Local Education Authority (LEA)   In the asbestos program, an educational authority at the local level that exists primarily to operate schools or to contract for educational services, including primary and secondary public and private schools. A single unaffiliated school can be considered an LEA for AHERA purposes.

local effects   Toxic or irritation effects that occur at the site of contact with a chemical or substance.

Local Emergency Planning Committee (LEPC)   A committee appointed by the state emergency response commission, as required by SARA Title III, to formulate a comprehensive emergency plan for its jurisdiction.

local exhaust ventilation   Drawing off of contaminated air directly from its source. Welders and painters often use local exhaust ventilation.

low-density polyethylene (LDPE)   Plastic material used for both rigid containers and plastic film applications.

lower explosive limit (LEL)   The lower limit of explosibility (and flammability for practical purposes) of a gas or vapor at ordinary ambient temperatures expressed in percent of the gas or vapor in air by volume. This limit is assumed constant for temperatures up to 120°C (250°F). Above this, it should be decreased by a factor of 0.7 because explosibility increases with higher temperatures.

lower flammable limit (LFL)   Lowest concentration (lowest percentage of the substance in air) that will burn (or for practical purposes, explode) when an ignition source (heat, spark, electric arc, or flame) is present.

low-level radioactive waste (LLRW)   Wastes less hazardous than most of those associated with a nuclear reactor; generated by hospitals, research laboratories, and certain industries. The Department of Energy, Nuclear Regulatory Commission, and EPA share responsibilities for managing them.

lower detection limit   The smallest signal above background noise an instrument can reliably detect. (See **limit of detection.**)

lowest acceptable daily dose   The largest quantity of a chemical that will not cause a toxic effect, as determined by animal studies.

lowest achievable emission rate   Under the Clean Air Act, the rate of emissions that reflects (1) the most stringent emission limitation in the implementation plan of any state for such source unless the owner or operator demonstrates such limitations are not achievable; or (2) the most stringent emissions limitation achieved in practice, whichever is more stringent. A proposed new or modified source may not emit pollutants in excess of existing new source standards.

lowest observed adverse effect level (LOAEL)   The lowest level of a stressor that causes statistically and biologically significant differences in test samples as compared to other samples subjected to no stressor.

macropores   Secondary soil features such as root holes, rocky soil, or desiccation cracks that can create significant conduits for movement of NAPL and dissolved contaminants or vapor-phase contaminants. The largest-sized soil pores.

magnetic separation   Use of magnets to separate ferrous materials from mixed municipal waste streams.

malaise   Feeling of general discomfort, distress, or uneasiness.

malignant   As applied to a tumor, cancerous and capable of undergoing metastasis, or invasion of surrounding tissue.

major modification   Defines modifications of major stationary sources of emissions with respect to Prevention of Significant Deterioration and New Source Review under the Clean Air Act.

major spill   The uncontrolled release of a hazardous substance into the environment to such a degree that operations personnel cannot control or contain the spill. Spill control and cleanup require mobilization of emergency response personnel.

major stationary sources   Term used to determine the applicability of Prevention of Significant Deterioration and new source regulations. In a nonattainment area, any stationary pollutant source with potential to emit more than 100 tons per year is considered a major stationary source. In PSD areas, the cutoff level may be either 100 or 250 tons, depending upon the source.

management plan   Under the Asbestos Hazard Emergency Response Act (AHERA), a document that each local education authority is required to prepare if it elects to leave asbestos in place. It describes all activities planned and undertaken by a school to comply with AHERA regulations, including triannual building inspections to identify asbestos-containing materials, response actions, and operations and maintenance programs to minimize the risk of exposure.

manifest (Uniform Hazardous Waste Manifest)   A multicopy shipping form used to identify the type and quantity of waste, the generator, the transporter, and the TSDF to which the waste is being shipped. The manifest includes copies for all participants in the waste shipment chain.

manifest system   Tracking of hazardous waste from cradle to grave (generation through disposal) with accompanying documents known as manifests.

manufacturer's formulation   A list of substances or component parts as described by the maker of a coating, pesticide, or other product containing chemicals or other substances. (See **material safety data sheets.**)

mapp gas   methylacetylene propadiene (stabilized). A common replacement gas for acetylene. Used in welding and burning.

marsh   A type of wetland that does not accumulate appreciable peat deposits and is dominated by herbaceous vegetation. Marshes may be either fresh- or saltwater, tidal or nontidal.

mass An amount of material. On earth, mass and weight can be considered the same. However, the same mass (a person's body) will have a different weight on the moon than on the earth because the moon's gravity is less than the earth's.

material category In the asbestos program, broad classification of materials into thermal systems insulation (TSI), surfacing material, and miscellaneous material.

material safety data sheet (MSDS) A compilation of information required under the OSHA Hazard Communication Standard on the identity of hazardous chemicals, health and physical hazards, exposure limits, and precautions. Section 311 of SARA requires facilities to submit MSDSs under certain circumstances. MSDSs must be supplied by the manufacturer or importer upon request.

material type For asbestos, classification of suspect material by its specific use or application; e.g., pipe insulation, fireproofing, and floor tile.

maximum available control technology (MACT) The emission standard for sources of air pollution requiring the maximum reduction of hazardous emissions, taking cost and feasibility into account. Under the Clean Air Act Amendments of 1990, the MACT must not be less than the average emission level achieved by controls on the best-performing 12% of existing sources, by category of industrial and utility sources.

maximum contaminant level (MCL) The maximum permissible level of a contaminant in water delivered to any user of a public system. MCLs are enforceable standards.

maximum contaminant level goal (MCLG) Under the Safe Drinking Water Act, a nonenforceable concentration of a drinking water contaminant, set at the level at which no known or anticipated adverse effects on human health occur and that allows an adequate safety margin. The MCLG is usually the starting point for determining the regulated maximum contaminant level.

maximum use concentration (MUC) The concentration of contaminants in the worksite that must not be exceeded. It is assumed that if the APR is used above the MUC, the wearer will be exposed to potentially damaging health effects, despite wearing the respirator. These MUC designations are listed on the side of the cartridge or canister. The MUC is assigned by the manufacturer for each NIOSH-approved cartridge or canister.

maximum use limit (MUL) The maximum amount of protection provided by a specific respirator. The MUL is calculated by multiplying the respirator's **assigned protection factor (APF)** by the **permissible exposure limit (PEL)** for the contaminant. (APF $\times$ PEL = MUL.)

mechanical aeration Use of mechanical energy to inject air into water to cause a waste stream to dissolve oxygen at a greatly increased rate over natural means.

mechanical separation Using mechanical means to separate waste into various components.

mechanical turbulence Random irregularities of airflow caused by buildings or other nonthermal processes. Also turbulence caused in liquids by mechanical agitators or mixers.

media Specific environments—air, water, or soil—that are the subject of regulatory concern and activities.

medical surveillance A periodic comprehensive review of a worker's health status; acceptable elements of such surveillance program are listed in the OSHA HAZWOPER standard.

medical waste Any solid waste generated in the diagnosis, treatment, or immunization of human beings or animals, in research pertaining thereto, or in the production or testing of biologicals, excluding hazardous waste identified or listed under 40 CFR Part 261 or any household waste as defined in 40 CFR Subsection 261.4 (b)(1).

melting point Temperature at which a solid substance changes to a liquid state. For mixtures, a melting range may be given.

meniscus The curved top of a column of liquid where it touches a solid surface. Easily visible in a small tube.

mercury (Hg)   A heavy metal (atomic weight of 200.6) that is toxic if breathed, swallowed, or absorbed through the skin. Organomercury compounds are highly toxic.

mesopores   Medium-sized soil pores. Liquids, vapors, and gases can travel fairly freely through this stratum. They will not travel as freely as through macropores.

metabolism   Chemical and physical processes whereby the body functions.

metastasis   Transmission of a disease from one part of the body to another. Transfer of the causal agent (cell or microorganism) of a disease from a primary focus to a distant one through the blood or lymphatic vessels. Also, spread of malignancy from site of primary cancer to secondary sites.

meter   A metric unit of length, equal to about 39 inches. Also, an instrument that measures something, such as an oxygen meter, which measures the percentage of oxygen in air.

methane   A colorless, nonpoisonous, flammable gas created by anaerobic decomposition of organic compounds. A major component of natural gas used in the home.

methanol   An alcohol that can be used as an alternative fuel or as a gasoline additive. It is less volatile than gasoline; when blended with gasoline it lowers the carbon monoxide emissions but increases hydrocarbon emissions. Used as pure fuel, its emissions are less ozone-forming than those from gasoline. Poisonous to humans and animals if ingested.

methemoglobinemia   Presence of methemoglobin in the bloodstream caused by the reaction of materials with the hemoglobin in red blood cells that reduces their oxygen-carrying capacity.

microbial growth   The amplification or multiplication of microorganisms such as bacteria, algae, diatoms, plankton, and fungi.

miscellaneous ACM   Interior asbestos-containing building material or structural components, members, or fixtures, such as floor and ceiling tiles; does not include surfacing materials or thermal system insulation.

microcurie (μCi)   A measure of radioactivity of a substance based on activity. Activity is defined as the number of disintegrations (also known as transformations) that occur per unit time. One μCi is equal to 3.7×10^4 transformations/second.

microgram (μg)   One-millionth of a gram. 10^{-6} grams; μg preferred.

micrometer (μm)   One-millionth (10^{-6}) of a meter; micron preferred.

micron (micrometer, μm)   A unit of length equal to one millionth of a meter, approximately 1/25,000 of an inch. Its symbol is the Greek letter mu, written μ. Mu also signifies one-millionth of any unit, as μg equals microgram, or one one-millionth of a gram. Micrometer is preferred.

micropores   The smallest-diameter routes for water, vapor, or gases to travel through soil. Usually found in densely packed fine soils. These pores are actually so small that they will contribute to capillary action of water and other liquids. This soil morphology is the least amenable to soil vapor extraction because of the reduced airflow capabilities.

mil   1/1,000 in., 0.001 in.

millimeter (mm)   1/1000 of a meter.

milligram (mg)   A unit of weight in the metric system. One thousand milligrams equals one gram. 1/1000 of a gram or 10^{-3} grams.

milligrams per cubic meter (mg/m³)   Unit used to measure air concentrations of dusts, gases, mists, and fumes. Used in measurement of air pollutants, milligrams of pollutant per cubic meter of air.

milligrams per kilogram (mg/kg)   The same as ppmw, parts-per-million, by weight. A dosage measurement used in toxicology testing to indicate a dose administered per kg of body weight.

milliliter (mL)   A metric unit used to measure volume. 1/1000 of a liter. One milliliter also equals approximately one cubic centimeter.

millimeter of mercury (mmHg)   The unit of pressure equal to the pressure exerted by a column of liquid mercury one millimeter high at a standard temperature. Units used to measure **vapor pressure** and **atmospheric pressure.** A measure of pressure as the height, in millimeters, that a mercury column with a vacuum at it's top, will rise above a reservoir of mercury in a U-tube, with the other end open to the pressure to be measured. Standard, sea-level atmospheric pressure is 760 mmHg. U.S. weather reports give atmospheric pressure in inches of mercury, usually in the 29 to 30 in. of mercury range.

minimization   A comprehensive program to minimize or eliminate wastes, usually applied to wastes at their point of origin.

minor source   New emissions sources or modifications to existing emissions sources that do not exceed NAAQS emission levels.

miscible   The extent to which liquids or gases can be mixed or blended to form one phase. Alcohols are generally miscible (soluble) in water. Oils are generally immiscible (not miscible, or not soluble) in water.

mist   Slowly falling liquid droplets in the air generated by condensation from the vapor to the liquid state. Also formed by mechanically breaking up liquid by splashing or atomizing. The droplets are larger than fog droplets that remain suspended in air longer. Liquid particles measuring 40 to 500 micrometers are formed by condensation of vapor. By comparison, fog particles are smaller than 40 micrometers in diameter.

mitigation   Actions taken to prevent or reduce the severity of threats to human health and the environment.

mixture   Heterogeneous association of materials that cannot be represented by a chemical formula. Sand and salt mixtures are spread on highways to melt ice. A mixture does not undergo chemical change as a result of interaction among the mixed materials.

mobile incinerator systems   Hazardous waste incinerators that can be transported from one site to another.

moisture content   The fraction of weight lost from a water-containing solid material (such as soil or sludge), upon heating to a constant weight at a temperature that does not cause the solid to decompose. Moisture content is expressed as the percent of water in the moist (pre-dried) solid. For a fully saturated medium, moisture content is an indicator of the porosity of soil.

molecule   The smallest division of a compound, material, or chemical that still retains or exhibits its characteristic properties.

molar solution   A solution made up of one mole (measured in grams) of a compound in enough solvent to make a thousand milliliters of the solution.

mole   In chemistry, the same concept as the mol in engineering; both pronounced the same. They are fundamental units for expressing the weight of compounds. In chemistry, one gram-molecular-weight (or gram-mole) of a compound is the molecular weight of the compound in grams. In U.S. chemical engineering the pound-mol (pound-molecular-weight) is defined as the molecular weight of a compound in pounds. One pound-mol of sulfuric acid weighs 98 lb.

Note

Moles (or mols) are useful in analyzing reactions. For example, how many grams (or pounds) of hydrogen, when reacted with how many grams (or pounds) of oxygen, will produce 36 grams (or pounds) of water (H_2O)?   We write: $2H_2 + O_2 \rightarrow 2H_2O$. H_2 has a molecular weight of 2. O_2 has a molecular weight of 32 and H_2O a molecular weight of 18. So 4 grams (or pounds) of hydrogen reacted with 32 grams (or pounds) of oxygen will give 36 grams (or pounds) of water.

molecular weight (MW)   The sum of the atomic weights of the atoms in a molecule. The hydrogen atom has an atomic weight of 1. The oxygen atom, an atomic weight of 16. Thus, the molecular weight of water, H_2O, is 18.

molten salt reactor   A thermal treatment unit that rapidly heats waste in a heat-conducting fluid bath of a carbonate salt.

monitoring   The process of sampling and measuring certain environmental parameters on a real-time basis for spatial and time variations. For example, air monitoring may be conducted with direct-reading instruments to indicate relative changes in air contaminant concentrations at various times or during various operations.

monitoring well   (1) Well used to obtain water quality samples or measure groundwater levels. (2) Well drilled at a hazardous waste management facility or Superfund site to collect groundwater samples for the purpose of physical, chemical, or biological analysis to determine the amounts, types, and distribution of contaminants in the groundwater beneath the site.

moratorium   During the negotiation process, a period of 60 to 90 days during which EPA and potentially responsible parties may reach settlement but no site response activities may be conducted.

morbidity   Rate of disease incidence.

mortality   Death rate.

mppcf   Millions of particles per cubic foot of air, based on impinger samples counted by light-field techniques (OSHA). Used in determining respiratory protection and in measuring dust concentrations to determine explosion hazards.

mucous membrane   The mucus-secreting lining that lines the hollow organs of the body, notably the nose, mouth, stomach, intestines, bronchial tubes, and urinary tract.

multimedia approach   Joint approach to several environmental media, such as air, water, and land.

multiple chemical sensitivity   A diagnostic label for people who suffer multisystem illnesses as a result of contact with, or proximity to, a variety of airborne agents and other substances.

municipal solid waste (MSW)   Discarded material, such as common garbage or refuse generated by industries, commercial and institutional facilities, and homes.

municipal discharge   Discharge of effluent from wastewater treatment plants which receive waste water from households, commercial establishments, and industries in the coastal drainage basin. Combined sewer/separate storm overflows are included in this category.

municipal sewage   Wastes (mostly liquid) originating from a community; may be composed of domestic wastewaters and/or industrial discharges.

municipal sludge   Semiliquid residue remaining from the treatment of municipal water and wastewater.

mutagen   Substance or agent capable of altering the genetic material (mutating) a living cell.

_n-_   Normal; used as a prefix in chemical names signifying a straight-chain structure. The other commonplace prefix is iso-, signifying a nonstraight, or branched-chain, structure.

N   Normal, a measurement of concentration in chemical analysis. A 1 N (one normal) solution contains one gram-equivalent weight of a substance in 1 L of solution.

nano-   Prefix meaning one-billionth.

narcosis   Stupor or unconsciousness produced by narcotics or other materials. Nitrogen narcosis or "Rapture of the Deep" is caused by high pressure forcing nitrogen into the blood.

National Ambient Air Quality Standards (NAAQS)   Standards established by EPA that apply for outdoor air throughout the country.

National Contingency Plan (NCP)   Policies and procedures for responding to both oil spills and hazardous substance releases. This national response capability plan promotes the overall coordination among a hierarchy of responders and contingency plans. The federal regulation that guides determination of the sites to be corrected under both the Superfund program and the program to prevent or control spills into surface waters or elsewhere.

National Emissions Standards for Hazardous Air Pollutants (NESHAPS)   Emissions standards set by EPA for an air pollutant not covered by NAAQS that may cause an increase in fatalities or in serious, irreversible, or incapacitating illness. Primary standards are designed to protect human health, secondary standards to protect public welfare (e.g., building facades, visibility, crops, and domestic animals).

National Environmental Performance Partnership Agreements   System that allows states to assume greater responsibility for environmental programs based on their relative ability to execute them.

National Estuary Program   A program established under the Clean Water Act Amendments of 1987 to develop and implement conservation and management plans for protecting estuaries and restoring and maintaining their chemical, physical, and biological integrity, as well as controlling point and nonpoint pollution sources.

National Fire Protection Association (NFPA)   Sets standards that aid in HAZWOPER work, most notably the 704M® labeling system. The National Fire Protection Association is a voluntary membership organization whose aim is to promote and improve fire protection and prevention. The NFPA publishes 16 volumes of codes known as the National Fire Codes.

National Institute for Occupational Safety and Health (NIOSH)   Part of the U.S. Department of Health and Human Services, Public Health Service, Centers for Disease Control and Prevention. Known for their promulgation of worker exposure levels called Recommended Exposure Limits (RELs) and publication of the *NIOSH Pocket Guide to Chemical Hazards*. NIOSH continues, as a counterpart to OSHA, to conduct research on health and safety concerns, test and certify respirators, and train occupational health and safety professionals.

National Institute of Environmental Health Sciences (NIEHS)   Reduces the burden of human illness and dysfunction from environmental causes through multidisciplinary biomedical research programs; prevention and intervention efforts; and communication strategies that encompass training, education, technology transfer, and community outreach.

National Municipal Plan   A policy created by EPA and the states in 1984 to bring all publicly owned treatment works (POTWs) into compliance with Clean Water Act requirements.

National Oil and Hazardous Substances Contingency Plan/National Contingency Plan (NOHSCP/NCP)   (See **National Contingency Plan.**)

National Pollutant Discharge Elimination System (NPDES)   A provision of the Clean Water Act that prohibits discharge of pollutants into waters of the United States unless a special permit is issued by EPA, a state, or, where delegated, a tribal government on an Indian reservation.

National Priorities List (NPL)   EPA's list of the most serious uncontrolled or abandoned hazardous waste sites identified for possible long-term remedial action under Superfund. The list is based primarily on the score a site receives from the Hazard Ranking System. EPA is required to update the NPL at least once a year. A site must be on the NPL to receive money from the Trust Fund for remedial action.

natural resource damages (NRD)   Defined as injury to, destruction of, or loss of natural resources. The measure of damages under the Comprehensive Environmental Response, Compensation, and Liability Act (CERCLA) and the Oil Pollution Act (OPA) is the cost of restoring damaged natural resources to their normal condition, compensation for the interim loss of damaged resources pending recovery, and the reasonable costs of a damage assessment.

National Response Center   The federal operations center that receives notifications of all releases of oil and hazardous substances into the environment; open 24 hours a day, it is operated by the U.S. Coast Guard, which evaluates all reports and notifies the appropriate agency for response action.

National Response Team (NRT)   Response planning and coordination is accomplished at the federal level through the National Response Team (NRT), an interagency group co-chaired by the EPA and the U.S. Coast Guard with representatives from 16 federal agencies. Although the NRT does not respond directly to incidents, it is responsible for three major activities related to managing responses: (1) distributing information; (2) planning for emergencies; and (3) training for emergencies. The NRT also supports the Regional Response Teams (RRTs).

National Secondary Drinking Water Regulations   Commonly referred to as NSDWRs.

National Toxicology Program (NTP)   The NTP publishes an Annual Report on carcinogens.

nausea   Tendency to vomit, a feeling of sickness at the stomach.

necrosis   Death of plant or animal cells or tissues. In plants, necrosis can discolor stems or leaves or kill a plant entirely. In animals, necrosis can turn skin black or white.

negotiations (under Superfund)   After potentially responsible parties are identified for a site, EPA coordinates with them to reach a settlement that will result in the PRP paying for or conducting the cleanup under EPA supervision. If negotiations fail, EPA can order the PRP to conduct the cleanup or EPA can pay for the cleanup using Superfund monies and then sue to recover the costs.

neoplasm   New or abnormal tissue growth that is uncontrollable and progressive.

nephrotoxic   Poisonous to the kidney.

neoprene   An elastomeric, rubbery synthetic material with moderate chemical resistance.

neuritis   Inflammation of the nerves.

neutralization   The act of changing the pH of a chemical to (or close to) pH 7 (neutral). (Example: neutralizing an acid spill with soda ash.)

neutralize   To render material less corrosive, that is, to change the pH of an acidic or alkaline material to the neutral level of approximately pH 7.

New Source   Any stationary source built or modified after publication of final or proposed regulations that prescribe a given standard of performance.

New Source Performance Standards (NSPS)   Uniform national EPA air emission and water effluent standards that limit the amount of pollution allowed from new sources or from modified existing sources.

New Source Review (NSR)   A Clean Air Act requirement that state implementation plans must include a permit review that applies to the construction and operation of new and modified stationary sources in nonattainment areas to ensure attainment of national ambient air quality standards.

nitrate   A compound containing nitrogen as NO_3^-. Nitrate particles can exist in the atmosphere. Nitrates dissolved in drinking water in excess of 10 mg/L can have harmful effects on humans and animals, and can cause severe illness in infants. Nitrates are common plant nutrients and inorganic fertilizers. Nitrates are found in septic systems, animal feed lots, agricultural fertilizers, manure, industrial wastewaters, sanitary landfills, and garbage dumps. Excess nitrates in water bodies cause unfavorable aquatic growths.

nitric oxide (NO)   A gas formed by combustion of nitrogen from air under high temperature and high pressure in an internal combustion engine. Nitrogen dioxide (NO_2) is formed along with NO.

nitrification   The process whereby ammonia in wastewater is oxidized to nitrites and then to nitrates by bacterial or chemical reactions.

nitrite   A compound containing nitrogen as (NO_2^-) (1) An intermediate in the process of nitrification. (2) Nitrite is used in food preservation as sodium nitrite.

nitrogen dioxide (NO_2)   The result of nitric oxide combining with oxygen. In the atmosphere it becomes a major component of photochemical smog.

nitrogen oxides (NO_x)   A general formula for oxides of nitrogen, chiefly nitrogen oxide (NO) and nitrogen dioxide (NO_2). They react with moisture to produce nitric acid (HNO_3). A corrosive irritant to tissue, causing congestion and pulmonary edema. Symptoms of acute exposure can develop immediately or over about a day. Chronic exposure to low levels can cause irritation, cough, headache, and tooth erosion. Exposure to between 5 and 50 ppm of NO_2 can cause slowly evolving pulmonary edema.

NO_x mixtures are commonly produced by combustion processes, including operation of motor vehicle engines. Smog, as has occurred in Los Angeles and other cities, can contain enough NO_X to be visible as a faint reddish tint in a fog or cloud, due to the red color of NO_2. Smog causes a stinging sensation in the eyes.

nitrogenous wastes Animal or vegetable residues that contain significant amounts of nitrogen.

nitrophenols Synthetic organopesticides containing carbon, hydrogen, nitrogen, and oxygen.

noble metal Chemically inactive metal such as gold; does not corrode easily.

no further remedial action planned (NFRAP) Database contains information on sites that have been removed from the inventory of Superfund sites. Archive status indicates that to the best of the EPA's knowledge, Superfund has completed its assessment of a site and has determined that no further steps will be taken to list that site on the National Priorities List (NPL).

nonaqueous phase liquid (NAPL) Contaminants that remain undiluted with water as the original bulk liquid in the subsurface, for example spilled oil. (See **free product.**)

Nonattainment Area Area that does not meet one or more of the National Ambient Air Quality Standards for the criteria pollutants designated in the Clean Air Act.

nonbinding allocations of responsibility (NBAR) A process for EPA to propose a way for potentially responsible parties to allocate costs among themselves.

nonferrous metals Nonmagnetic metals such as aluminum, lead, and copper. Products made all or in part from such metals include containers, packaging, appliances, furniture, electronic equipment, and aluminum foil. Metals not containing iron.

nonflammable Incapable of being easily ignited or burned when lighted, usually in an air atmosphere. Also, a DOT hazard class for any compressed gas other than a flammable one. Nitrogen and helium are nonflammable. Hydrogen is highly flammable in air.

nonfriable asbestos-containing materials Any material containing more than one percent asbestos (as determined by polarized light microscopy) that, when dry, cannot be crumbled, pulverized, or reduced to powder by hand pressure.

nonhazardous industrial waste Wastes and wastewaters from manufacturing processes regulated under Subtitle D that are not considered to be MSW, hazardous waste, or other wastes under Subtitle C and D. (EPA)

nonionizing electromagnetic radiation (1) Radiation that does not change the structure of atoms but does heat tissue and may cause harmful biological effects. (2) Microwaves, radio waves, and low-frequency electromagnetic fields from high-voltage transmission lines.

nonpermit confined space Any confined space that does not contain or, have the potential to contain, any atmospheric hazards or any physical hazards capable of causing death or serious physical harm.

nonpoint sources Diffuse pollution sources (i.e., without a single point of origin or not introduced into a receiving stream from a specific outlet). The pollutants are generally carried off the land by stormwater. Common nonpoint sources are agriculture, forestry, urban, mining, construction, dams, channels, land disposal, saltwater intrusion, and city streets.

nonpotable Water that is unsafe or unpalatable to drink because it contains pollutants, contaminants, or infective agents.

nontoxic The product will not injure a person *if it is used properly.* Non-toxic cleaners will not injure a person if used as cleaners, but may injure a person if ingested.

Non-toxic items, such as crayons that are used by children and are traditionally misused (eaten) are certified to contain no harmful chemicals or toxins that would result in injury if the product were misused. ASTM D-4236 is a labeling practice that ensures proper testing has been performed and the product, when used according to directions, creates no health hazards and is nontoxic.

normal solution (chemistry) A solution made by dissolving 1 g-equivalent weight of a substance in sufficient distilled water to make 1 L of solution. The weight of a gram equivalent depends on how the substance ionizes in solution. A gram equivalent of hydrochloric acid (HCl) and a mole of HCl are the same. A gram equivalent of sulfuric acid (H_2SO_4) is only half the weight of a mole of H_2SO_4. That is because each mole of H_2SO_4 releases two equivalents of hydrogen ions.

notice of deficiency An EPA request to a facility owner or operator requesting additional information before a preliminary decision on a permit application can be made.

notice of intent to deny Notification by EPA of its preliminary intent to deny a permit application.

noxious Objectionable to most humans, possibly because of odors, minor irritations, minor changes in taste. Not necessarily toxic or even harmful. The mercaptans added at very low levels to commercial cooking and heating gas produce a noxious odor. It is a safety mechanism so that gas leaks can be detected at extremely low levels by the odor of the mercaptans.

NRR Noise reduction rating. A value used in rating hearing protection devices.

nuclide An atom characterized by the number of protons, neutrons, and energy in the nucleus.

nuisance particulates Have a long history of little adverse effect on the lungs and do not produce significant organic disease or toxic effect when exposures are kept under reasonable control. They can, however, cause an annoying sneezing reaction.

nutrient Any substance assimilated by living organisms that promotes growth. The term is generally applied to nitrogen and phosphorus in wastewater, but is also applied to other essential and trace elements.

nutrient pollution Contamination of water resources by excessive inputs of nutrients. In surface waters, excess algal production is a major concern.

nystagmus Spastic, involuntary motion of the eyeballs.

Occupational Safety and Health Administrations (OSHA) OSHA's goals are to save lives, prevent injuries, and protect the health of America's workers. OSHA and its state partners have approximately 2100 inspectors, plus complaint discrimination investigators, engineers, physicians, educators, standards writers, and other technical and support personnel in more than 200 offices throughout the country. This staff establishes protective standards, enforces those standards, and reaches out to employers and employees through technical assistance and consultation programs. OSHA's website is www.osha.gov.

odor A description of what a human smells of a substance. We describe an odor using our sense of smell.

odor threshold Lowest concentration of a substance's vapor, in air, that can be detected with the nose and sense of smell.

Office of Emergency and Remedial Response (OERR) Manages the Superfund program.

Office of Emergency and Remedial Response Geographic Information System (OERRGIS) Work Group Coordinates and shares information on GIS projects related to the Superfund and Oil Programs within the OERR and works with EPA regional offices on GIS-related issues.

Office of Enforcement and Compliance Assurance (OECA) Works in partnership with EPA regional offices and state, tribal, and other federal agencies to ensure compliance with the nation's environmental laws. By employing an integrated approach of compliance assistance, compliance incentives, and innovative civil and criminal enforcement, OECA and its partners seek to maximize compliance and reduce threats to public health and the environment.

Office of Pesticide Programs (OPP) Regulates the use of all pesticides in the United States and establishes maximum levels for pesticide residues in food, thereby safeguarding the nation's food supply.

Office of Pollution Prevention and Toxics (OPPT)   Focuses on promoting pollution prevention efforts for controlling industrial pollution; safer chemicals through a combination of regulatory and voluntary efforts; risk reduction to minimize exposure to existing substances such as lead, asbestos, dioxin, and polychlorinated biphenyls; and public understanding of risks by providing understandable, accessible, and complete information on chemical risks to the broadest audience possible.

Office of Prevention, Pesticides, and Toxic Substances (OPPTS)   Oversees the Office of Pesticides Programs (OPP) and the Office of Pollution Prevention and Toxics (OPPT). OPPTS promotes pollution prevention and the public's right to know about chemical risks. Some of OPPTS's top priorities include dealing with emerging issues like endocrine disrupters and lead poisoning prevention.

Office of Research and Development (ORD)   Conducts leading-edge research and fosters the sound use of science and technology to help fulfill EPA's mission to protect human health and safeguard the environment. As a part of that effort, ORD performs research and development to identify, understand, and solve current and future environmental problems; provides responsive technical support to EPA's mission; integrates the work of ORD's scientific partners (other agencies, nations, private sector organizations, and academia); and provides leadership in addressing emerging environmental issues and in advancing the science and technology of risk assessment and risk management.

Office of Sustainable Ecosystems and Communities (OSEC)   Helps implement integrated, geographic approaches to environmental protection with an emphasis on ecological integrity, economic sustainability, and quality of life—known as Community Based Environmental Protection (CBEP). OSEC develops and supports demonstration projects, tools, and policies that sustain CBEP activities.

Office of Science Coordination and Policy (OSCP)   Provides coordination, leadership, peer review, and synthesis of science policy within the Office of Prevention, Pesticides, and Toxic Substances. OSCP program areas include biotechnology, endocrine disrupters, and the Federal Insecticide, Fungicide, and Rodenticide Act (FIFRA) Scientific Advisory Panel (SAP).

Office of Solid Waste (OSW)   Operates under authority of the Resource Conservation and Recovery Act (RCRA). OSW protects human health and the environment by ensuring responsible national management of hazardous and nonhazardous waste.

Office of Solid Waste and Emergency Response (OSWER)   Develops guidelines and standards for the land disposal of hazardous wastes and guidelines for management of underground storage tanks. OSWER also implements a program to respond to abandoned and active hazardous waste sites and accidental releases, including some oil spills, and encourages the use of innovative technologies for contaminated soil and groundwater.

off-site facility   A hazardous waste treatment, storage, or disposal area that is located away from the generating site.

Oil Pollution Act (OPA)   Created largely in response to rising public concern following the Exxon Valdez incident to improve the nation's ability to prevent and respond to oil spills. OPA established provisions that expand the federal government's authority and provide the money and resources necessary to respond to oil spills. The OPA also created the national Oil Spill Liability Trust Fund, which can provide up to 1 billion dollars per spill incident.

oil spill   An accidental or intentional discharge of oil that reaches bodies of water. Can be controlled by chemical dispersion, combustion, mechanical containment, absorption, or adsorption. Spills from tanks and pipelines can also occur away from water bodies, contaminating the soil, getting into sewer systems, and threatening underground water sources.

olfactory   Relating to the sense of smell.

oliguria   Scanty or low volume of urine.

on-site facility   A hazardous waste treatment, storage, or disposal area that is located on the generating site.

opacity   The amount of light obscured by particulate pollution in the air; clear window glass has zero opacity, a brick wall is 100% opaque. Measurements of opacity are indicators of changes in performance of particulate control systems.

opaque   Impervious to light rays. Milk and wood are opaque. Motor oil is *translucent,* meaning light can pass through it, but you can't see through it. Clear glass is *transparent,* meaning you can see through it.

open burning   Uncontrolled fires in an open dump.

open dump   An uncovered site used for disposal of waste without environmental controls.

open transfer   Any transfer that at any time involves contact of a moving fluid with the atmosphere, air, or oxygen. Open transfer of flammable liquids, especially Class IA liquids, is dangerous due to the release of flammable vapors into the work area. Since there is a risk of fire or explosion if an ignition source is present, these transfers should be done only in a hood.

operable unit   Term for each of a number of separate activities undertaken as part of a Superfund site cleanup. A typical operable unit would be removal of drums and tanks from the surface of a site.

operation and maintenance (O&M)   1. Activities to protect the integrity of a site's cleanup plan. O&M measures are initiated by a state after cleanup objectives have been reached, and the site is determined to be operational and functional (O&F) based on state and federal agreement. 2. All kinds of equipment and facilities should have approved, O&M procedures.

operating conditions   Conditions specified in a RCRA permit that dictate how an incinerator must operate as it burns different waste types. A trial burn is used to identify operating conditions needed to meet specified performance standards.

oral   Used in, or taken into the body through, the mouth. (See **ingestion**.)

oral toxicity   Ratings corresponding to the following definitions are derived from data obtained from the test methods and categories of toxicity as described in 16 CFR 1500.3.

1. *Nontoxic:*   The probable lethal dose of undiluted product to 50% of the test animals determined from ingestion studies (LD_{50}) is greater than 5 g per kilogram of body weight.

2. *Toxic:*   The probable lethal dose of undiluted product to 50% of the test animals determined from ingestion studies (LD_{50}) is greater than 50 mg and less than or equal to 5 g per kilogram of body weight.

3. *Highly toxic:*   The probable lethal dose of undiluted product to 50% of the test animals determined from ingestion studies (LD_{50}) is less than or equal to 50 mg per kilogram of body weight.

organic chemicals   With only minor exceptions, they are compounds composed of carbon, along with hydrogen, oxygen, and other elements and having chain or ring structures. The hydrocarbons contain only hydrogen and carbon. Benzene is an aromatic (ring structure) organic compound. Gasoline is a complex mixture of organic compounds, almost totally hydrocarbons, having both ring and chain structures. Note that there are some silicones, considered organics, that contain only silicon, oxygen, and hydrogen. Other silicones do contain carbon atoms.

organic matter   Carbonaceous material contained in plant or animal matter and originating from domestic or industrial sources.

organism   Any form of animal or plant life.

organophosphates   Pesticides that contain phosphorus; usually short-lived in the environment, but some can be toxic to humans when first applied.

organophyllic   A substance that easily combines with organic compounds.

organotins   Organic chemical compounds containing the metal tin, used in antifouling paints to protect the hulls of boats and ships, buoys, and pilings from marine organisms such as barnacles. These compounds are extremely toxic to humans also.

original generation point   Where regulated hazardous material first becomes waste.

osmosis   The passage of a solvent from a weak solution to a more concentrated solution across a semipermeable membrane. The membrane allows passage of the solvent, usually water, but not the dissolved solids.

outfall   The place where effluent is discharged into receiving waters.

Outreach and Special Projects Staff (OSPS)   Coordinates and implements for the Office of Solid Waste and Emergency Response (OSWER) the agency's principles and new initiatives, such as Brownfields, Environmental Justice (EJ), and the Tribal initiatives. Through its unique cross-program perspective, OSPS involves all stakeholders and seeks to leverage OSWER resources through partnerships with EPA headquarters and regions, public and private organizations, and the general public.

overexposure   Exposure to a hazardous material beyond the allowable exposure levels.

oxidant   An element, chemical, or material that acts as an oxidizer in a chemical reaction. A collective term for some of the primary constituents of photochemical smog. The most common oxidant is oxygen in the atmosphere. Chlorine is also an oxidant.

oxidation   Reaction of a substance with oxygen or another oxidizer (or oxidizing agent), such as chlorine.

oxidation pond   A man-made (anthropogenic) body of water, open to the atmosphere, in which waste is consumed by bacteria, used most frequently with other waste-treatment processes; a sewage lagoon.

oxidation-reduction potential   The electric potential available for the transfer of electrons from one compound or element (the oxidant) to another compound (the reductant); used as a qualitative measure of the power to cause oxidation of contaminants in water treatment systems.

oxide pox   Dermatitis caused by contact with oxides under poor personal hygienic conditions.

oxidizer   See **oxidation.** Typical strong oxidizers are the peroxides, perchlorates, chlorates, chlorine, and fluorine. Presence of an oxidizer near any fuels increases the fire hazard greatly.

oxidizing agent   Same as an oxidant. An element, chemical, or substance that brings about an oxidation reaction. (See **oxidation.**)

oxygenated solvent   An organic solvent containing oxygen as part of the molecular structure. Alcohols and ketones are oxygenated compounds often used as paint solvents.

oxygen deficient atmosphere   An atmosphere containing less than 19.5% oxygen by volume. That concentration of oxygen by volume below which atmosphere supplying respiratory protection must be provided.

oxygen-enriched atmosphere   An atmosphere containing more than 23.5% oxygen by volume. Oxygen enriched atmospheres can cause normally noncombustible materials to become explosively flammable.

ozone (O_3)   A very powerful but unstable oxidizer, it must be generated on-site. It cannot be stored. Ozone releases the powerful oxygen free radical to accomplish oxidation. The ozone layer in the upper atmosphere is what protects the earth from dangerous levels of solar radiation.

Ozone is found in two layers of the atmosphere, the stratosphere and the troposphere. In the stratosphere (the atmospheric layer 7 to 10 miles or more above the earth's surface), ozone is a natural form of oxygen that provides a protective layer shielding the earth from ultraviolet radiation. In the troposphere (the layer extending up 7 to 10 miles from the earth's surface), ozone is a chemical oxidant and major component of photochemical smog. It can seriously impair the respiratory system and is one of the most widespread of all the criteria pollutants for which the Clean Air Act required EPA to set standards. Ozone in the troposphere is produced through complex chemical reactions of nitrogen oxides, which are among the primary pollutants emitted by combustion sources; hydrocarbons, released into the atmosphere through the combustion, handling, and processing of petroleum products; and sunlight.

ozonation/ozonator   Application of ozone to water for disinfection or for taste and odor control. The ozonator is an electrical device that converts oxygen to ozone.

ozone depletion   Destruction of the stratospheric ozone layer, which shields the earth from ultraviolet radiation harmful to life. This destruction of ozone is caused by the breakdown of certain chlorine- and bromine-containing compounds (chlorofluorocarbons and Halon®s), which break down when they reach the stratosphere and then catalytically destroy ozone molecules.

ozone depletion potential (ODP)   The ratio of the impact on ozone, and the ozone layer, of a chemical, compared to the impact of a similar mass of CFC-11. The ODP of CFC-11 is defined to be 1.0. Other CFCs and HCFCs have ODPs that range from 0.01 to 1.0. The Halons® have ODPs ranging up to 10. HFCs, having no chlorine or bromine have zero ODP.

ozone hole   A thinning break in the stratospheric ozone layer. Designation of amount of such depletion as an ozone hole is made when the detected amount of depletion exceeds 50%. Seasonal ozone holes have been observed over both the Antarctic and Arctic regions, part of Canada, and the extreme northeastern United States.

ozone layer   The protective layer in the atmosphere, about 15 miles above the ground, that absorbs some of the sun's ultraviolet rays, thereby reducing the amount of potentially harmful radiation that reaches the earth's surface.

packaging   The assembly of one or more containers and any other components necessary to ensure minimum compliance with a program's storage and shipment packaging requirements. Also, the containers involved.

palpitation   Irregular, rapid heartbeat.

pandemic   A widespread epidemic throughout an area, nation, or the world.

parameter   A variable, measurable property whose value is a determinant of the characteristics of a system; e.g., temperature, pressure, and density are parameters of the atmosphere.

parathesia   Sensation of pricking, tingling, or creeping on the skin that has no objective cause.

particulate   A suspension of fine solid or liquid particles in air, such as dust, fog, fume, mist, smoke or sprays. Generally, anything that is not a fiber (which has a minimum length-to-width ratio of 3:1). An environmentally important component of solid and liquid combustion reactions. Coal-burning power plants and cement plants can be large generators of particulates.

Particulates Not Otherwise Regulated (PNOR)   Dust is a potential health hazard. OSHA has established a PEL for dust, referred to as Particulates Not Otherwise Regulated. The PELs for PNOR are as follows: 15 mg/m³ for total dust and 5 mg/m³ for the respirable fraction. You must perform a separate evaluation for dust exposure using the PEL for PNOR.

parts per million (ppmv or ppmw)   *ppmv* is the unit for measuring the concentration of a gas or vapor in air. It tells the volume of the gas or vapor in a million volumes of air; *ppmw* is used to indicate the concentration of particulates in air, such as pounds of particulates per million pounds of air.

The unit *ppmw* is also used to report the level of contaminants in drinking water, wastewater, and even solids. One ppmw is the same as 1 mg/L (one milligram per liter) of water. One liter of water weighs approximately 1,000 g. Some more toxic materials are measured in concentrations of parts per billion (ppb) and parts per trillion (ppt).

passive treatment walls   Technology in which a chemical reaction takes place when contaminated groundwater comes in contact with a barrier such as limestone or a wall containing iron filings.

pathogens   Microorganisms (bacteria, viruses, or parasites) that can cause disease in humans, animals, and plants.

pathway   The physical course a chemical or pollutant takes from its source to the exposed organism. (See **vector.**)

pathways of dispersion The medium (water, groundwater, soil, and air) through which a material is transported into and through the environment.

penetration Visible passing of a chemical in liquid form through CPC fabric or through a tear, seam, or zipper.

percent saturation The amount of a substance that is dissolved in a solution compared to the amount that could be dissolved in it.

percent volatile Percent volatile by volume is the percentage of a liquid or solid (by volume) that will evaporate when exposed to the atmosphere at an ambient temperature of 70°F (unless some other temperature is specified). Examples: butane, gasoline, and paint thinner (mineral spirits) are 100% volatile; their individual evaporation rates vary, but in time each will evaporate completely.

percolation (1) The movement of water downward and radially through subsurface soil layers, usually continuing downward to ground water. Can also involve upward movement of water. (2) Slow seepage of water through a filter.

performance bond Cash or securities deposited before a landfill operating permit is issued, held to ensure that all requirements for operating and subsequently closing the landfill are faithfully performed. The money is returned to the owner after proper closure of the landfill is completed. If contamination or other problems appear at any time during operation, or upon closure, and are not addressed, the owner must forfeit all or part of the bond that is then used to cover clean-up costs.

performance data (for incinerators) Information collected, during a trial burn, on concentrations of designated organic compounds and pollutants found in incinerator emissions. Data analysis must show that the incinerator meets performance standards under operating conditions specified in the RCRA permit.

performance standards (1) Regulatory requirements limiting the concentrations of designated organic compounds, particulate matter, and hydrogen chloride in emissions from incinerators. (2) Operating standards established by EPA for various permitted pollution control systems, asbestos inspections, and various program operations and maintenance requirements.

permeability The ability of gases, vapors, or liquids to pass through soil, plastic sheeting, fabrics, or other materials in a specified direction.

permeation The passage of a chemical through a protective clothing material on a molecular level. Gases and vapors may pass through the material, leaving no detectable physical signs of damage to the clothing. It is the highest level of testing available for CPC fabrics.

permeation rate The rate at which permeation occurs. Commonly expressed as micrograms per square centimeter (of fabric) per minute ($\mu g/cm^2/min$).

permissible exposure limit (PEL) Legally enforced exposure limit for a substance established by OSHA. The PEL indicates the permissible concentration of air contaminants to which nearly all workers may be repeatedly exposed 8 hours a day, 40 hours a week, over a working lifetime (40 years), without adverse effects. These materials are listed in Table Z-1, Z-2, or Z-3 of OSHA Regulations found at 29 CFR 1910.1000, *Air Contaminants*. (See Appendix G on the Companion CD.)

permit EPA definition: an official license that specifically allows a facility to treat, store, or dispose of hazardous waste and outlines the precautions that must be taken to manage the waste in a manner that adequately protects human health and the environment. Owners or operators of hazardous waste TSDFs must obtain a permit in order to operate. NPDES issues permits allowing certain amounts of pollutants to be discharged into the air and water with accurate monitoring.

permit The form, completed by the Inspector, containing all appropriate information required for safe entry into a permit-required confined space. (See **entry permit** for OSHA definition.)

permit-required confined space (permit space) See **confined space.** Any confined space that has one or more of the following characteristics:

1. Contains or has a potential to contain a hazardous atmosphere
2. Contains a material that has the potential for engulfing an entrant
3. Has an internal configuration such that an entrant could be trapped or asphyxiated by inwardly converging walls or by a floor that slopes downward and tapers to a smaller cross-section
4. Contains any other recognized serious safety or health hazard.

permit system   The employer's written procedure for preparing and issuing permits for entry and for returning the permit space to service following termination of entry.

persistent chemical   A substance that resists biodegradation and/or chemical transformation when released into the environment. Persistent chemicals tend to accumulate on land, in the air, and in the water. Notable examples are the PCBs.

personal hygiene   Precautionary measures, such as washing face and hands and showering, taken to maintain good health when exposed to harmful materials.

personal air samples   Air samples taken with a pump that is directly attached to the worker with the collecting filter and cassette placed in the worker's breathing zone (required under OSHA asbestos standards and EPA worker protection rule).

personal protective equipment (PPE)   Equipment used to shield or isolate individuals from the chemical, physical, and biological hazards that may be encountered at a hazardous waste site. PPE can protect the respiratory system, skin, eyes, face, hands, feet, head, body, and hearing. Clothing and equipment worn by HAZMAT workers cleaning up Superfund sites, pesticide mixers, loaders and applicators and reentry workers, and emergency responders.

pest   An insect, rodent, nematode, or other form of terrestrial or aquatic animal life that is injurious to health or the environment.

pesticide   Substances or mixtures thereof intended for preventing, destroying, repelling, or mitigating any pest.

petroleum   Crude oil or any fraction thereof that is liquid under normal conditions of temperature and pressure. The term includes petroleum-based substances including a complex blend of hydrocarbons derived from crude oil through the processes of separation, conversion, upgrading, and finishing, such as motor fuel, jet oil, lubricants, petroleum solvents, and asphalt.

petroleum distillate   Complex mixture of hydrocarbons, liquid at normal ambient conditions, separated from crude oil and other refinery process streams by distillation. The lower-boiling temperature fractions from crude oil.

pH   "Potenz hydrogen"; a scale of 0 to 14 representing the relative acidity or alkalinity of an aqueous solution. Pure water has pH of about 7. Substances in aqueous solution will ionize to various extents giving different concentrations of H+ and OH− ions.

phenols   Organic compounds that are byproducts of petroleum refining, tanning, and textile, dye, and resin manufacturing. They contain carbon, hydrogen, and oxygen. Low concentrations cause taste and odor problems in water; higher concentrations can kill aquatic life and injure humans. The pure compound phenol was the first disinfectant discovered.

phlegm   Thick mucus from the respiratory passage.

phosphates   Inorganic and organic chemical compounds containing phosphorus as $(PO_4)^{3-}$.

phosphogypsum piles (stacks)   Principal byproduct generated in production of phosphoric acid from phosphate rock. These piles may generate radioactive radon gas.

phosphorus   An essential element in our food and nutrients for microorganisms and plant life. Too high a concentration of phosphates can contribute to the eutrophication (excess nutrients in a water body) of lakes and other water bodies. Increased phosphorus levels result from discharge of phosphorus-containing materials, mostly fertilizers, into surface waters.

photochemical oxidants   Oxidizing chemicals formed by the action of sunlight on atmospheric pollutants.

photochemical smog   Air pollution caused by chemical reactions of photochemical oxidation on various pollutants emitted from internal combustion engines, fossil fuel power plants, incinerators, and other sources.

photoionization detector (PID)   An instrument that uses ultraviolet light as a means of ionizing a sample of chiefly volatile organic compounds in air using a specific ionization potential. Especially useful in air monitoring surveys and extremely sensitive to organic compounds, such as solvents and hydrocarbon fuels. The current generated is a relative measure of the concentration of all materials having an ionization potential that is lower than the instrument's specific ionization potential.

photophobia   Intolerance to light.

photosynthesis   The manufacture by plants of carbohydrates and oxygen from carbon dioxide in the atmosphere mediated by chlorophyll in the presence of sunlight.

physical hazard   Other than material hazards. They include entrapment, engulfment, and release of hazardous energy. Also includes unguarded holes, broken or missing ladders, and incomplete staging.

physical state   Condition or state of a material (solid, liquid, or gas) at a given temperature and pressure. Note that vapor is not a separate state, but an important portion of the gas state. (See **vapor.**)

phytoremediation   Using plant life as a low-cost remediation option for sites with widely dispersed contamination at low concentrations. (See **phytotreatment.**)

phytotoxic   Harmful to plants.

phytotreatment   The cultivation of specialized plants that absorb specific contaminants from the soil through their roots or from the atmosphere into their foliage. This reduces the concentration of contaminants in those media, but incorporates them into biomasses that may be released back into the environment when the plant dies or is harvested.

pico   Prefix meaning one-trillionth.

picocurie (pCi)   (See **microcurie.**) One picocurie equals 3.7×10^7 transformations per second.

picocuries per liter (pCi/L)   Unit of measure for levels of radon gas; Becquerels per cubic meter is the metric equivalent.

pilot tests   Testing a cleanup technology under actual site conditions to identify potential problems prior to full-scale implementation.

planned removal (non-time-critical removal)   The removal of released hazardous substances that pose a threat or potential threat to human health or welfare or to the environment from a site within a nonimmediate time period. Under CERCLA: Actions intended to minimize increases in exposure such that time and cost commitments are limited.

plume   (1) A visible or measurable discharge of a contaminant from a given point of origin. Can be visible or thermal in water, or visible in the air as, for example, a plume of smoke. (2) The zone of radiation leaking from a damaged reactor. (3) Area downwind within which a release could be dangerous for those exposed to leaking fumes.

plutonium   A radioactive metallic element of atomic weight 244.

PM-10 and PM-2.5   PM 10 is measure of particles in the atmosphere with a diameter of less than 10 or equal to a nominal 10 micrometers. PM-2.5 is a measure of smaller particles, 2.5 micrometer or less in the air. PM-10 has been the pollutant particulate level standard against which EPA has been measuring Clean Air Act compliance. On the basis of newer scientific findings, the Agency is considering regulations that will make PM-2.5 the new standard.

pneumoconiosis   Respiratory tract and lung condition caused by inhalation and retention of irritant mineral or metallic particles. A medical X-ray examination can detect changes, which include fibrosis.

point source   A stationary location or fixed facility from which pollutants are discharged; any single identifiable source of pollution; e.g., a pipe, ditch, ship, ore pit, factory smokestack.

poison   Any substance that is injurious to health and may lead to death when relatively small amounts are either taken internally or applied externally.

poison, Class A   DOT term for an extremely dangerous poison such as a poisonous gas or liquid of such a nature that a very small amount of the gas or vapor of the liquid mixed with air is dangerous to life.

poison, Class B   DOT term for liquid, solid, paste, or semisolid substances other than class A poisons or irritating materials known, or presumed by animal tests, to be so toxic to man to be a health hazard during transportation.

poison control (or information) center   Provides medical information on a 24-hour basis for accidents involving ingestion of potentially poisonous materials. Usually listed in the front part of the Yellow Pages phone book.

pollutant   A substance or mixture of substances that, after release into the environment and upon the exposure of any organism or vegetation, will, or may reasonably be anticipated to, cause adverse effects in such organisms, their offspring, or vegetation.

pollutant transport   The mechanisms by which a substance may migrate outside the immediate location of the release or discharge of the substance. For example, pollution of groundwater by hazardous waste leachate migrating from a landfill.

pollution prevention (P²)   (1) Identifying areas, processes, and activities that create excessive waste products or pollutants in order to reduce or prevent them through, alteration, or eliminating a process. Such activities, consistent with the Pollution Prevention Act of 1990, are conducted across all EPA programs and can involve cooperative efforts with such agencies as the Departments of Agriculture and Energy. (2) EPA has initiated a number of voluntary programs in which industrial or commercial partners join with EPA in promoting activities that conserve energy, conserve and protect water supply, reduce emissions or find ways of utilizing them as energy resources, and reduce the waste stream. Among these are:

- *Agstar,* to reduce methane emissions through manure management
- *Climate Wise,* to lower industrial greenhouse-gas emissions and energy costs
- *Coalbed Methane Outreach,* to boost methane recovery at coal mines
- *Design for the Environment,* to foster including environmental considerations in product design and processes
- *Energy Star* programs, to promote energy efficiency in commercial and residential buildings, office equipment, transformers, computers, office equipment, and home appliances
- *Environmental Accounting,* to help businesses identify environmental costs and factor them into management decision-making
- *Green Chemistry,* to promote and recognize cost-effective breakthroughs in chemistry that prevent pollution
- *Green Lights,* to spread the use of energy-efficient lighting technologies.
- *Indoor Environments,* to reduce risks from indoor-air pollution
- *Landfill Methane Outreach,* to develop landfill gas-to-energy projects
- *Natural Gas Star,* to reduce methane emissions from the natural gas industry
- *Ruminant Livestock Methane,* to reduce methane emissions from ruminant livestock
- *Transportation Partners,* to reduce carbon dioxide emissions from the transportation sector
- *Voluntary Aluminum Industrial Partnership,* to reduce perfluorocarbon emissions from the primary aluminum industry
- *WAVE,* to promote efficient water use in the logging industry
- *Wastewi$e,* to reduce business-generated solid waste through prevention, reuse, and recycling

polychlorinated biphenyls (PCBs)   Family of pathogenic and teratogenic compounds that accumulate in body tissues. They were commonly used for several decades, up until about the 1980s. Until about the 1970s they were not believed to be hazardous. They were very popular as transformer dielectrics (materials that tend not to conduct electricity, the opposite of a conductor) and as heat transfer media because of their extreme resistance to combustion and decomposition.

polyelectrolytes   Synthetic chemicals that help solids to coagulate or clump together during wastewater treatment. This aids downstream filtration.

polymer   Long chains of molecules bonded having structural units of the original monomer molecules. Common polymers are Nylon®, PVC, epoxy, polyethylene, and polypropylene.

polymerization   A chemical reaction in which small molecules (monomers) combine to form larger molecules (polymers). A hazardous polymerization is one that can release large amounts of heat. The heat can cause fires, explosions, or pressure increases that can burst containers. Materials that can polymerize usually contain inhibitors that can delay the reaction.

The commonly used epoxies are polymer adhesives and repair materials. When you mix the monomer and the activator (usually labeled Part A and Part B), the mixture gives off heat as it forms the epoxy polymer.

poly vinyl chloride (PVC)   A durable thermoplastic material offering good protection against acids. Widely used in plastic piping.

porosity   Degree to which soil, gravel, sediment, or rock is permeated with pores or cavities through which water or air can move.

postclosure   The time period following the shutdown of a waste management or manufacturing facility; for monitoring purposes, often considered to be 30 years.

post emergency response   Means that portion of an emergency response performed after the immediate threat of a release has been stabilized or eliminated and cleanup of the site has begun. If postemergency response is performed by an employer's own employees who were part of the initial emergency response, it is considered to be part of the initial response and not postemergency response. However, if a group of an employer's own employees separate from the group providing initial response performs the cleanup operation, then the separate group of employees will be considered to be performing postemergency response.

potable water   Water that is safe for drinking and cooking.

potentially responsible parties (PRPs)   Individuals, companies, or any other party that is potentially liable for payment of Superfund cleanup costs.

POTW   Publicly owned treatment works.

pounds per square inch (psi)   In the metric system, vessel pressure is rated in kPa. For technical accuracy, pressure measurements in the British system must be expressed as psig (pounds per square inch gauge) or psia (pounds per square inch absolute). That is, to learn the absolute pressure of a gas, liquid, or vapor inside a vessel, take the gauge pressure plus ambient atmospheric pressure. At sea level that would be the gauge reading in psig plus approximately 14.7 psi.

pour point   Temperature at which a liquid ceases, or begins, to flow, or at which it congeals.

PO_x   A general term for the several oxides of phosphorus.

precipitate   Solid particles separating from a solution due to greater than 100% solubility concentration in the solvent caused by cooling of the solution. Also, particles separating from a solution due to a chemical reaction that formed them at a concentration greater than their 100% solubility in the solution at that temperature.

precipitation treatment   Removal of hazardous solids from liquid waste by causing a precipitation reaction.

precipitator   Pollution control device that collects particles from a gas stream. An electrostatic precipitator charges the particles, then attracts them to oppositely charged plates.

precordial   In front of the heart, stomach.

preliminary assessment   The process of collecting and reviewing available information about a known or suspected waste site or release.

preliminary assessment and site inspection (PA/SI) EPA uses PA/SI to evaluate the potential for a release of hazardous substances from a site. Information collected during the PA and SI is used to calculate a hazardous ranking system (HRS) score. Sites with an HRS score of 28.50 or greater are eligible for listing on the National Priorities List (NPL) and require the preparation of an HRS scoring package.

pretreatment Processes used to reduce, eliminate, or alter the nature of wastewater pollutants from nondomestic sources before they are discharged into publicly owned treatment works (POTWs).

primacy Having the primary responsibility for administering and enforcing regulations.

primary skin irritant Noncorrosive substance that produces severe skin irritation.

primary waste treatment First steps in wastewater treatment; screens and sedimentation tanks are used to remove most materials that float or will settle. Primary treatment removes about 30% of carbonaceous biochemical oxygen demand from domestic sewage.

principal organic hazardous constituents (POHCs) Hazardous compounds monitored during an incinerator's trial burn, selected for high concentration in the waste feed and difficulty of combustion.

process wastewater Any water, not part of the final product but used in processing, that comes into contact with any raw material, product, byproduct, or waste.

produce To manufacture, process, formulate, or repackage.

prohibited condition Any condition in a permit space that is not allowed by the permit during the period when entry is authorized.

prohibited work Any work process that is specifically forbidden by a confined space entry permit. Examples are hot work (welding, burning, and grinding) and painting.

project manager The individual who has authority to direct all site activities. The project manager has authority to direct site response and ensures overall management of projects.

proposed plan A plan for a site cleanup that is available to the public for comment.

proprietary information Any information a company chooses to keep private. It must not be unlawful to keep that information private. A company may choose to divulge proprietary information to another party for business purposes, while requiring that the other party not divulge it any further than stated in a written agreement.

proprietary ingredient Substance used in a product that is kept a secret by the user company. Its use is not protected by patents. Employees are generally required to promise not to reveal its identification outside the company using the ingredient. A proprietary ingredient need not be listed on an MSDS if doing so would materially harm the owners market share as a producer of the material. However, complete instructions on how to properly use, store, cleanup, and dispose of the product, part of which is a proprietary ingredient, MUST be listed in the MSDS. Proprietary ingredients must be disclosed to response personnel and medical personnel in emergencies.

prostration Physical exhaustion, incapacitation.

protection factor (PF) Measure of efficiency of a given respirator. For a given respirator the PF is equal to the concentration of the contaminant outside the facepiece divided by the concentration inside the facepiece.

proteins Complex nitrogenous organic compounds of high molecular weight made of amino acids; essential for growth and repair of human and animal tissue. Many, but not all, proteins are enzymes.

proteinuria Presence of protein in the urine.

protocol A series of formal steps for conducting a test.

public comment period The time allowed for the public to express its views and concerns regarding an action by EPA (e.g., a *Federal Register* notice of proposed rule-making, a public notice of a draft permit, or a notice of intent to deny).

public hearing   A formal meeting wherein EPA officials hear the public's views and concerns about an EPA action or proposal. EPA is required to consider such comments when evaluating its actions. Public hearings must be held upon request during the public comment period.

public notice   (1) Notification by EPA informing the public of Agency actions such as the issuance of a draft permit or scheduling of a hearing. EPA is required to ensure proper public notice, including publication in newspapers and broadcast over radio and television stations. (2) In the safe drinking water program, water suppliers are required to publish and broadcast notices when pollution problems are discovered.

psychotropic   Acting on the mind.

published exposure level   The exposure limits published by OSHA, ACGIH, NIOSH, or other recognized organizations.

pulmonary edema   Fluid in the lungs.

pumping test   A test conducted to determine aquifer or well characteristics.

purge   The process of removing the contents of a space, container, or other vessel by emptying the space of its contents. Vessels and piping are given a nitrogen purge to clear flammables from a system before performing hot work. In the case of confined spaces, the term means removing the hazardous atmosphere from a confined space and replacing it with fresh air or nitrogen. (See **inerting.**) Removing stagnant air or water from a sampling zone or equipment prior to sample collection, as in purging a well.

putrefaction   Biological decomposition of organic matter; associated with anaerobic conditions.

pyrolysis   A form of solid chemical or material decomposition, and vaporization of liquids and gases, produced by heating in the absence of air. Charcoal is produced from wood by pyrolysis. If air were present, burning or combustion would occur.

pyrophoric   Class of materials that ignite spontaneously in air below 130°F. Occasionally friction will ignite them. Finely divided materials such as dusts or powders are more pyrophoric than larger granules. Normally noncombustible materials such as iron become pyrophoric when finely divided and poured through air.

qualified person   Person with specific training, knowledge, and experience in the area for which the person has the responsibility and the authority to control. A certified industrial hygienist may be a qualified person for preparing a site health and safety plan.

qualitative   Relating to, or determining, the kind of a substance or the kind of an analysis, statement, or opinion. "Pure water," "good and bad," "hard and soft," and "hazardous" are qualitative statements.

quality assurance/quality control (QA/QC)   A system of procedures, checks, audits, and corrective actions being followed to ensure that all kinds of efforts, including research, design, performance testing, environmental monitoring and sampling, and other technical and reporting activities, are of the highest achievable quality.

quantitative   Providing a specific numerical value, or range of values, in an analysis, statement, or labeling. Examples: "This oil contains 50 ppmw of PCBs." "Your grade in the exam is 93." "These workers have between 5 and 10 years of experience with hazardous materials."

quench tank   Water-filled tank used to cool incinerator residues or hot materials during industrial processes.

rad (radiation absorbed dose)   Unit used to measure a quantity called absorbed dose. Relates to the amount of energy actually absorbed in some material and is used for any type of radiation and any material.

radiation   Transmission of energy though space or any medium. Also known as radiant energy.

radio frequency radiation   See **nonionizing electromagnetic radiation.**

radioactive decay   Spontaneous breakup of an atom with emission of charged particles and radiation; also known as radioactive disintegration and radioactivity.

radioactive substances   Substances that spontaneously undergo radioactive decay.

radioisotopes   Chemical variants of radioactive elements with potentially carcinogenic, teratogenic, and mutagenic effects on the human body. However, in carefully controlled medical treatments they are used to destroy cancer cells.

radionuclide   An artificially produced radioactive element. Examples are plutonium, americium, and curium. Can have a long life as soil or water pollutant.

radius of influence (ROI)   (1) The radial distance from the center of a wellbore to the point where there is no lowering of the water table or potentiometric surface (the edge of the cone of depression). (2) the radial distance from an extraction well that has adequate air flow for effective removal of contaminants when a vacuum is applied to the extraction well.

radon   A colorless, naturally occurring, radioactive and therefore toxic inert gas formed by radioactive decay of radium atoms deep in the earth. The heaviest member (atomic weight 222) of the inert gas family that begins with helium.

radon daughters/radon progeny   Short-lived radioactive decay products of radon that decay into longer-lived lead isotopes that can attach themselves to airborne dust and other particles and, if inhaled, damage the linings of the lungs.

radon decay products   A term used to refer collectively to the immediate products of the radon decay chain. These include Po-218, Pb-214, Bi-214, and Po-214, which have an average combined half-life of about 30 minutes.

reaction   Chemical transformation or change. A decomposition or polymerization of a substance. The interaction of two or more substances to form a new substance or substances that may have radically different properties from the interacting substances. The reaction of metallic sodium with poisonous, greenish-yellow, chlorine gas produces white table salt (sodium chloride).

reactive material   A chemical substance or mixture that will vigorously polymerize, decompose, condense, or become self-reactive due to shock, pressure, or temperature. Includes explosive materials, organic peroxides, pressure-generating materials, and water-reactive materials.

reactivity   The tendency of a substance to undergo a chemical reaction or change that may result in dangerous side effects, such as an explosion, burning, and corrosive or toxic emissions.

reagent   A substance used in a chemical reaction to produce another substance; or in analysis to detect the composition of a substance.

real-time monitoring   Monitoring and measuring environmental contaminants as personnel perform their work or a remediation process operates. Site safety officers use these data for day-to-day decision-making about worker health, PPE requirements, and the environment.

reasonably available control technology (RACT).   Control technology that is reasonably available and both technologically and economically feasible. Usually applied to existing sources in nonattainment areas; in most cases is less stringent than new source performance standards.

reasonably available control measures (RACM)   A broadly defined term referring to technological and other measures for pollution control.

receptor   Ecological entity exposed to a **stressor.**

recharge area   A land area in which water reaches the zone of saturation from surface infiltration, e.g., where rainwater soaks through the earth to reach an aquifer.

recharge rate   The quantity of water per unit of time that replenishes or refills an aquifer.

recharge   The process by which water is added to a zone of saturation, usually by percolation from the soil surface; e.g., the recharge of an aquifer.

recommended exposure limit (REL) (NIOSH)   The highest allowable airborne concentration that is not expected to injure a worker. Expressed as a ceiling limit or as a time weighted average, usually for an 8-hour work shift.

record of decision (ROD)   Public document that explains which cleanup alternatives will be used to clean up a Superfund site. The ROD for sites listed on the NPL is created from information generated during the remedial investigation/feasibility study.

RCRAInfo   RCRAInfo is EPA's comprehensive information system, providing access to data supporting the Resource Conservation and Recovery Act (RCRA) of 1976 and the Hazardous and Solid Waste Amendments (HSWA) of 1984. RCRAInfo replaces the data recording and reporting abilities of the Resource Conservation and Recovery Information System (RCRIS) and the Biennial Reporting System (BRS).

recycling   The series of activities by which discarded materials are converted into raw materials and used in the production of new products.

reduction   In chemical reactions, the addition of hydrogen, removal of oxygen, or addition of electrons to an element or compound.

reducing agent   An element or compound that (1) combines with oxygen or (2) loses electrons to the reaction during a reduction reaction. Hydrogen is a common industrial reducing agent.

Regional Response Team (RRT)   Representatives of federal, local, and state agencies who may assist in coordination of activities at the request of the on-scene coordinator before and during a significant pollution incident such as an oil spill, major chemical release, or Superfund response.

Registry of Toxic Effects of Chemical Substances (RTECS)   Published by NIOSH. Presents basic toxicity data on thousands of materials. The objective is to identify all known toxic substances and reference original studies.

regulatory agency   The agency in charge of promulgating regulations and enforcing those regulations. OSHA is charged with worker safety. DOT is charged with transportation safety. The EPA is responsible for implementing, monitoring, and enforcing environmental regulations. OSHA, DOT, and EPA can all have approved state programs that deal with that state's regulations.

regulated asbestos-containing material (RACM)   Friable asbestos material or nonfriable ACM that will be or has been subjected to sanding, grinding, cutting, or abrading or has crumbled or been pulverized or reduced to powder in the course of demolition or renovation operations.

regulated medical waste   Under the Medical Waste Tracking Act of 1988, any solid waste generated in the diagnosis, treatment, or immunization of human beings or animals, in research pertaining thereto, or in the production or testing of biologicals. Included are cultures and stocks of infectious agents; human blood and blood products; human pathological body wastes from surgery and autopsy; contaminated animal carcasses from medical research; waste from patients with communicable diseases; and all used sharp implements, such as needles and scalpels, and certain unused sharps. (See **biohazard.**)

regulated material   A substance or material that is subject to regulations set forth by OSHA, the EPA, DOT, or any other federal agency.

release   Any spilling, leaking, pumping, pouring, emitting, emptying, discharging, injecting, escaping, leaching, dumping, or disposing of hazardous or toxic chemical or extremely hazardous substance into the environment.

rem (Roentgen equivalent in man)   Unit of radiation dose in man called "equivalent dose." This relates the absorbed dose (rad) in human tissue to the effective biological damage of the radiation. Not all radiation has the same biological effect, even for the same amount of absorbed dose. Equivalent dose is often expressed in terms of thousandths of a rem, or millirem (mrem). To determine equivalent dose (rem), multiply absorbed dose (rad) by a quality factor (Q) that is unique to the type of radiation.

Remedial Action (RA)   The actual construction or implementation phase of a Superfund site cleanup that follows remedial design.

remedial actions   As defined in the National Contingency Plan, responses to releases on a National Priority List that are consistent with a treatment-oriented remedy that is protective of human health and the environment and permanently and significantly reduces toxicity, mobility, or volume of hazardous substances.

Remedial Design (RD)   The phase in Superfund site cleanup when the technical specifications for cleanup remedies and technologies are decided. Remedial action (RA) follows the remedial design phase and involves the actual construction or implementation phase of Superfund site cleanup. The RD/RA is based on the specifications described in the record of decision (ROD).

Remedial Investigation/Feasibility Study (RI/FS)   Performed at the sites with actual known or potential environmental problems. The RI serves as the mechanism for collecting data, while the FS is the mechanism for developing, screening, and evaluating alternative remedial actions. The RI and FS are conducted concurrently. Data collected in the RI influence the development of remedial alternatives in the FS, which in turn affect the data needs and scope of treatability studies and additional field investigations.

remedial project manager (RPM)   The EPA or state official responsible for overseeing on-site remedial action.

remedial response   Long-term action that stops or substantially reduces a release or threat of a release of hazardous substances that is serious but not an immediate threat to public health.

remediation   Cleanup or other methods used to remove or contain a toxic spill or hazardous materials from a Superfund site or brownfields site.

removal action   Short-term immediate actions taken to address releases of hazardous substances that require expedited response.

renal   Pertaining to the kidney.

representative sample   A portion of gaseous, liquid, or solid material that is as nearly identical in content and consistency as possible to that in the larger body of material being sampled.

reportable quantity (RQ)   The amount (five categories: 1, 10, 100, 1,000, or 5,000 lb) of a CERCLA or EPCRA hazardous substance that, when released into the environment, requires notification to the appropriate government agency. The federal agency is the EPA's National Response Center (NRC). State and local authorities may also need to be notified. RQs come under CERCLA, EPCRA, and the CWA.

reproductive health hazard   Any agent that has a harmful effect on the adult male or female reproductive system or the developing fetus or child.

rescue service   In confined spaces work, the personnel or department designated to rescue employees from permit spaces.

residual   Amount of a pollutant remaining in the environment after a natural or technological process has taken place; e.g., the sludge remaining after initial wastewater treatment, or particulates remaining in air after it passes through a scrubbing or other process.

residual risk   The extent of health risk from air pollutants remaining after application of the Maximum Achievable Control Technology (MACT).

Resource Conservation and Recovery Act (RCRA)   This law encourages environmentally sound methods for disposal of household, municipal, commercial, and industrial waste. Its primary goals are to protect human health and the environment from the potential hazards of waste disposal and to conserve energy and natural resources, reduce the amount of waste generated, and ensure that wastes are managed in an environmentally sound manner. RCRA is divided into sections called Subtitles.

respirable size particulates   Particulates in the size range that permits them to penetrate deep into the lungs upon inhalation.

respirator (approved)   A device that has met the requirements of 30 CFR Part 11, is designed to protect the wearer from inhalation of harmful atmospheres, and has been approved by the National Institute for Occupational Safety and Health (NIOSH).

respiratory system Breathing system, consisting of (in descending order) the nose, mouth, nasal passages, nasal pharynx, pharynx, larynx, trachea, bronchi, bronchioles, air sacs (alveoli) of the lungs, and muscles of respiration.

respiratory protection Devices that will protect the wearer's respiratory system from overexposure by inhalation of airborne contaminants or where there is an insufficient oxygen supply. Respiratory protection is used when a worker must work in an area where he or she might be exposed to contaminant concentrations in excess of the PEL, or there is insufficient oxygen.

response actions or response operations Actions taken to recognize, evaluate, and control an incident.

responsible party Someone who can provide additional information on the characteristics of a hazardous material and also recommend appropriate emergency procedures.

restoration Measures taken to return a site to previolation prerelease conditions.

retrieval system The equipment (including a retrieval line, chest or full-body harness, wristlets, if appropriate, and a lifting device or anchor) used for nonentry rescue of persons from permit spaces. The retrieval system must be available, meaning that the entrant's retrieval line is attached to a mechanical device so that rescue can begin as soon as the rescuer becomes aware that rescue is necessary.

reverse osmosis Treatment process used in water systems that reverses the normal process of osmosis by sufficiently pressurizing a water solution to force it through a semipermeable membrane. Reverse osmosis removes most drinking water contaminants. Also used in wastewater treatment. Large-scale reverse osmosis plants are in use worldwide.

reversible effect An effect that is not permanent; especially adverse effects which diminish when exposure to a toxic chemical stops.

risk The recognition that there is a probability that harm will occur during an activity.

risk assessment The use of a factual information base to define the health effects of exposure of individuals or populations to hazardous materials and situations.

risk characterization The last phase of the risk assessment process that estimates the potential for adverse health or ecological effects to occur from exposure to a stressor and evaluates the uncertainty involved.

risk factor Characteristics (e.g., race, sex, age, obesity) or variables (e.g., smoking, occupational exposure level) associated with increased probability of a toxic effect.

risk management The process of weighing policy alternatives and selecting the most appropriate regulatory action. Integrating the results of risk assessment with engineering data and with social and economic concerns to reach an action decision.

risk management plans (RMP) The Clean Air Act (CAA) requires covered facilities (referred to as stationary sources) to develop RMPs to prevent accidental releases of dangerous chemicals. Covered stationary sources are those that have certain regulated substances present in excess of applicable thresholds (threshold quantities (TQs)). These plans must also determine "worst case" scenarios of releases to the community.

rodenticide A chemical or agent used to destroy rats or other rodent pests or to prevent them from damaging food, crops, materials, and structures.

roentgen (R) Unit used to measure a radiation exposure. This can only be used to describe an amount of gamma and X-rays, and only in air. The main advantage of this unit is that it is easy to measure directly, but it is limited because it is only for deposition in air, and only for gamma and X-rays.

rotary kiln incinerator Incinerator with a rotating combustion chamber that keeps waste tumbling and moving while exposed to incineration flame, thereby allowing all of it to be exposed to the flame temperature and to achieve more complete combustion.

routes of exposure; routes of entry The manner in which a contaminant can enter the body. The four major routes are inhalation, ingestion, skin absorption, and injection.

sacrificial anode An easily corroded material deliberately installed in a vessel, pipe, or other water supply system component to allow the material to be consumed by (sacrifice it to) corrosion while other components of the system remain relatively corrosion-free. The sacrificial anode in a hot water heater is constructed of magnesium and zinc.

Saint Andrew's cross (X) Used in packaging for transport; it means harmful, stow away from foodstuffs.

safe Condition of exposure under which there is a practical certainty that no harm will result to exposed individuals.

Safe Drinking Water Act (SDWA) Protects the quality of drinking water in the United States. This law focuses on all waters actually or potentially designed for drinking use, whether from aboveground or underground sources.

safety Freedom from man, equipment, material, and environmental interactions that can result in injury, illness, or loss.

safety and health plan; health and safety plan (HASP) Written, site-specific safety criteria that establish requirements for protecting the health and safety of site workers during all activities conducted at an incident or project site.

salinity The percentage of salt in water.

salts The reaction products formed when acids react with bases. The most common example is sodium chloride (table salt). Minerals that water picks up as it passes through the air, over and under the ground, or from households and industry.

sampling The collection of a representative portion of media. Examples are the collection of a part of a hazardous waste stream, a water sample from a contaminated stream, or collection of air, gases, or vapors for off-site analysis.

sampling frequency The interval between the collection of successive samples.

sarcoma A tumor that is often malignant.

SARA Superfund Amendments and Reauthorization Act. Title III of SARA is known as the Emergency Planning and Community Right-to-Know Act (EPCRA) of 1986. A revision and extension of CERCLA, SARA is intended to encourage and support local and state emergency planning efforts. It provides citizens and local governments with information about potential chemical hazards in their communities. SARA calls for facilities that store hazardous materials to provide officials and citizens with data on the types (flammables, corrosives, etc.); amounts on hand (daily, yearly); and their specific locations. Facilities are to prepare and submit inventory lists, MSDSs, and Tier 1 and 2 inventory forms.

saturated zone The area below the water table where all open spaces are filled with water under pressure equal to or greater than that of the atmosphere.

saturation The condition of a liquid when it has taken into solution the maximum possible quantity of a given substance at a given temperature and pressure.

sclerae Tough, white, fibrous covering of the eyeball.

screening risk assessment Risk assessment performed with few data and many assumptions to identify exposures that should be evaluated more carefully for potential risk.

scrubber Air pollution control device, usually in the form of a tall, cylindrical vessel that uses a countercurrent, falling spray of liquid or falling granular material to trap pollutants in an upward-flowing stream of gaseous emissions.

secondary treatment The second step in most publicly owned waste treatment systems in which bacteria consume the organic parts of the waste. It is accomplished by bringing together waste, bacteria, and oxygen in trickling filters or in the activated sludge process. This treatment removes floating and settleable solids and about 90% of the oxygen-demanding substances and suspended solids. Chemical and other treatment can follow as tertiary treatment.

sedimentation Letting solids settle out of wastewater by gravity during treatment.

sedimentation tanks Wastewater tanks in which floating wastes are skimmed off and settled solids are removed for disposal.

sediments   Soil, sand, and minerals washed from land into water, usually after rain. They accumulate in reservoirs, rivers, and harbors, destroying fish and wildlife habitat and clouding the water so that sunlight cannot reach aquatic plants. Careless farming, mining, and building activities will expose sediment materials, allowing them to wash off the land after rainfall.

semivolatile organic compounds (SVOCs)   Organic compounds that volatilize slowly at standard conditions (20°C and 1 atmosphere of pressure).

sensitization   State of immune-response reaction in which further exposure elicits an immune or allergic response. A condition in which a person previously exposed to a certain material is more sensitive upon further contact.

sensitizer   Substance that, on first exposure, causes little or no reaction in man or test animals but on repeated exposure may cause a marked response not necessarily limited to the contact site.

settling tank   A holding area for wastewater, where heavier particles sink to the bottom for removal and disposal.

severe   A relative term used to describe the degree to which hazardous material releases or exposures can cause adverse effects to human health and the environment.

sharps   Hypodermic needles, syringes (with or without the attached needle), Pasteur pipettes, scalpel blades, blood vials, needles with attached tubing, and culture dishes used in animal or human patient care or treatment, or in medical, research, or industrial laboratories. Also included are other types of broken or unbroken glassware that were in contact with infectious agents, such as used slides and cover slips, and unused hypodermic and suture needles, syringes, and scalpel blades.

short-circuiting   In confined spaces ventilation, when exhausted stale air is drawn back into the space, instead of clean, fresh air.

short-term exposure limit (STEL)   ACGIH-recommended exposure limit. Maximum concentration to which workers can be exposed for a short period of time (15 minutes) for only four times throughout the day with at least one hour between exposures.

siderosis   Pneumoconiosis caused by the inhalation of iron particles. Also, tissue pigmentation caused by contact with iron.

sign   (1) Abnormality in the body indicating poisoning or disease that is observable by another person. (2) A posted command, warning, or direction.

signal words   Distinctive words that serve to alert the reader to the existence and relative degree of a hazard. Signal words are limited to:

1. *Danger:*   Materials that are highly toxic; corrosive to living tissue; extremely flammable; radioactive; or are suspected human carcinogens.

2. *Warning:*   Materials that are moderately toxic, have severe skin irritation potential, cause allergic skin reactions, or are flammable.

3. *Caution:*   Materials that have a low order of toxicity; produce only slight to moderate skin irritation, or are combustible.

significant potential source of contamination   A facility or activity that stores, uses, or produces compounds with potential for significant contaminating impact if released into the source water of a public water supply.

significant violations   Violations by point source dischargers of sufficient magnitude or duration to be a regulatory priority (EPA definition).

silicosis   Condition of massive fibrosis of the lungs causing shortness of breath because of prolonged inhalation of silica dusts.

site   An area or place within the jurisdiction of the EPA and/or a state.

site assessment program   A means of evaluating hazardous waste sites through preliminary assessments and site inspections to develop a hazard ranking system score.

site inspection   The collection of information from a Superfund or brownfield site to determine the extent and severity of hazards posed by the site. It follows and is more extensive than a preliminary assessment. The purpose at Superfund sites is to gather the information necessary to score the site, using the hazard ranking system, and to determine if it presents an immediate threat requiring prompt (emergency) removal.

site safety plan   A crucial element in all removal actions, it includes information on equipment being used, precautions to be taken, and steps to take in the event of an on-site emergency. (See **safety and health plan; health and safety plan.**)

site safety and health supervisor or officer (SSO)   The individual located on a hazardous waste site that is responsible to the site owner or manager. The SSO has the authority and knowledge necessary to implement the site health and safety plan (HASP) and verify compliance with applicable safety and health requirements.

"Skin"   A notation (sometimes used with PEL or TLV® exposure data) that indicates that the stated substance may be absorbed by the skin, mucous membranes, and eyes, either airborne or by direct contact, and that this additional exposure must be considered part of the total exposure to avoid exceeding the PEL or TLV® for that substance.

skin irritation   Ratings corresponding to the following definitions are derived from data obtained from the test methods as described in the 16 CFR 1500.41 (CPSC) and/or National Academy of Sciences (NAS) publication 1138 and categories of toxicity as described in 16 CFR 1500.3.

1. **Practically Nonirritating:**   The undiluted product causes no noticeable irritation or causes slight inflammation (edema and erythema skin reaction values of 0 to 1) of intact or abraded skin of rabbits during the study period. Primary irritation index of 0 to 1.9.

2. **Moderately irritating:**   The undiluted product causes well-defined inflammation (edema and erythema skin reaction values of 2) during the study period. Primary irritation index of 2 to 4.9.

3. **Primary skin irritant:**   The undiluted product cause moderate to severe inflammation (edema and erythema skin reaction values of 3 or 4) of the intact or abraded skin of rabbits during the study period. Primary irritation index of 5 or more.

4. **Corrosive:**   The undiluted product causes visible destruction or irreversible alterations of the tissue structure at the site of contact on intact or abraded skin of rabbits during the study period.

sludge   A semisolid residue from any of a number of air or water treatment processes; can be a hazardous waste.

slurry   A pourable mixture of solid and liquid.

small quantity generator   Generator of less than 2,200 lb (1,000 kg) of hazardous waste per calendar month. Regulated to a lesser degree than an LQG.

smelter   A facility that melts or fuses metal ore, often with an accompanying chemical change, to separate its metal content. Uncontrolled emissions can cause pollution. Smelting is the process involved.

smog   Air pollution typically associated with reactions between oxidants and combustion products. When strong enough, it leaves a stinging sensation in the eyes and can cause breathing difficulty, especially in those people with respiratory diseases, such as emphysema and asthma.

smoke   Dry particles and droplets generated during combustion or incineration. They are combined with, and suspended in, the vapors and gases that are also formed.

soil gas   Gaseous elements and compounds in the small spaces between particles of the earth and soil. Such gases can be moved or driven out under pressure.

soil pores   The void spaces between soil particles. No matter how tightly (densely) the soil particles are packed, there will always be void spaces between them. In clean, virgin soil below the water table, the pore spaces are filled with water, and above the water table the pore spaces are filled with varied amounts of air, possibly organic gases such as methane, CO_2, and hydrogen sulfide, as well as water. (See **macropores, mesopores,** and **micropores.**)

solidification and stabilization   Removing of wastewater from a waste or changing it chemically to make it less permeable and less susceptible to transport by water.

solid waste   Discarded material, such as garbage, refuse, and sludge (including solids, semi-solids, liquids, or contained gaseous material). By EPA definition, all hazardous waste must also be solid waste.

solubility in water   The maximum percentage of a material (by weight) that will dissolve in water at a specified temperature and ambient pressure. Solubility is strongly dependent on temperature and only slightly dependent on pressure. The solubility of almost all materials increases with temperature. A few materials have negative solubility in water over certain temperature ranges. Instead of having greater solubility at higher temperature, they have lower solubility. Sodium sulfate is more soluble in 40°C water than in boiling water. Essentially, all nitrates are soluble.

solution   A single liquid phase consisting of solute(s) in a solvent or solvents. Dissolved substances are called solutes. The solute(s) will be in molecular or ionic form in the solvent. Seawater is mostly a solution of salt in water, forming sodium and chloride ions. If water is evaporated from seawater, at some point the solubility limit of salt will be exceeded and salt will begin to crystallize and precipitate from solution.

solvent   A liquid in which other substances can be dissolved. Water and alcohols are the most common solvents.

sorbent   (1) A material that removes toxic gases and vapors from air inhaled through a canister or cartridge. (2) Material used to collect gases and vapors during air sampling.

soot   Fine particles, usually black, formed by combustion. They consist chiefly of carbon. Soot produces the blackening around fireplaces and chimney stacks.

sorption   The action of soaking up, absorbing, adsorbing, or attracting substances; process used in many pollution control systems and responses to spills.

source area   The location of liquid hydrocarbons or other material releases, or the zone of highest soil or groundwater concentrations, or both, of the released material of concern.

source reduction   This refers to the design, manufacture, purchase, or use of materials or equipment to reduce the amount or toxicity of materials being released from a point or area source before they enter the waste stream.

SO_x   Oxides of sulfur where x equals the number of oxygen atoms, commonly sulfur dioxide (SO_2) and sulfur trioxide (SO_3). They are typical pollutants from coal-burning power plants. When mixed with rain (H_2O), they form sulfurous and sulfuric acids and are a cause of acid rain.

sparge or sparging   Injection of air below the water table to strip dissolved volatile organic compounds from the water or to oxygenate groundwater to facilitate aerobic biodegradation of organic compounds.

spasm   Involuntary, convulsive muscular contraction.

specific chemical identity   Chemical name, CAS number, or other information that reveals the precise chemical designation of the substance.

specific conductance   Rapid method of estimating the dissolved solids content of a water supply by testing its ability to carry an electrical current.

specific gravity (Sp. Gr.)   The ratio of the mass of a unit volume of a substance to the mass of the same volume of a standard substance at a standard temperature. Water at 4°C (39.2°F) is the standard usually referred to for liquids; for gases, dry air (at the same temperature and pressure as the gas) is often taken as the standard substance. Weight of a material compared to the weight of an equal volume of water. An expression of the relative density of a material. It will tell whether a material will float or sink when spilled in water.

spill   The uncontrolled release of a hazardous substance into the environment that onsite personnel are capable of containing. (See **major spill.**)

Spill Prevention, Containment, and Countermeasures Plan (SPCCP)   Plan covering actions to be taken before, during, and after the release of hazardous substances as defined in the Clean Water Act.

spoil   Dirt or rock removed from its original location, destroying the composition of the soil in the process, as in well drilling, strip-mining, dredging, or construction.

stability   Ability of a material to remain unchanged. For MSDS purposes, a material is stable if it remains in the same form under expected and reasonable conditions of storage or use. Conditions that may cause instability (dangerous change) are stated. Examples are temperatures above 150°F and shock from dropping.

stabilization   Conversion of the active organic matter in sludge into inert, harmless material.

stakeholder   Any organization, governmental entity, or individual that has a stake (significant interest) in, or may be impacted by, a given approach to environmental regulation, pollution prevention, energy conservation, and similar activities.

standard   A standard is a defining document or a set procedure that provides rules, guidelines or characteristics for a product or a test procedure. It standardizes how the product will be made or how the test will be conducted, for instance. Standards are then approved by recognized bodies, such as ANSI, ASME, or the NFPA, and then adopted by regulators such as OSHA, EPA, or DOT.

standard temperature and pressure   Typically defined as 20°C or 68°F and one atmosphere of pressure.

state authorization   The process by which states are given authority to run their own OSHA program or RCRA program instead of OSHA or EPA.

State Emergency Response Commission (SERC)   Commission appointed by each state governor according to the requirements of SARA Title III. The SERCs designate emergency planning districts, appoint local emergency planning committees, and supervise and coordinate their activities.

state environmental goals and indication project   Program to assist state environmental agencies by providing technical and financial assistance in the development of environmental goals and indicators.

State Implementation Plans (SIP)   EPA-approved state plans for the establishment, regulation, and enforcement of air pollution standards.

stationary source   A fixed-site producer of pollution, mainly power plants and other facilities using industrial combustion processes.

sterilizer   One of three groups of antimicrobials registered by EPA for public health uses. EPA considers an antimicrobial to be a sterilizer when it destroys or eliminates all forms of bacteria, viruses, and fungi and their spores. Because spores are considered the most difficult form of microorganism to destroy, EPA considers the term *sporicide* to be synonymous with *sterilizer*.

stomatitis   Inflammation of the mucous membrane of the mouth.

storage   The temporary placement of a hazardous material in a location, which provides protection to personnel, the material, and the environment.

stratification   A condition often found in confined spaces where gases seek their own level based on their density relative to air much as immiscible liquids do based on specific gravity. It is the reason that all levels of a space must be tested. For example: methane gas will have a tendency to float to the top of a space, especially if undisturbed, much the same way hydrogen sulfide will sink to the bottom of a stratified atmosphere. Separating into layers.

stressors   Physical, chemical, or biological entities that can induce adverse effects on ecosystems or human health.

stupor   Partial or nearly complete unconsciousness.

subcutaneous   Beneath the skin.

sublime   Change from the solid to the vapor phase without passing through the liquid phase. Dry ice, frozen carbon dioxide (CO_2), does not melt. At -78.4°C and any higher temperature, at ambient pressures, it sublimes from solid to gas.

Subtitle C   Section of RCRA establishing a regulatory framework for managing the generation, storage, treatment, and disposal of certain wastes defined as hazardous wastes.

Subtitle D   This section of RCRA establishes a system for managing solid waste, including both garbage/trash and nonhazardous industrial waste.

Subtitle I   Section of RCRA regulating toxic substances and petroleum products stored in underground storage tanks, such as at commercial gas stations.

sump   A pit or tank, at the lowest local level, that catches liquid runoff for drainage or disposal.

Superfund   The program operated under the legislative authority of CERCLA and SARA that funds and carries out EPA solid waste emergency and long-term removal and remedial activities. These activities include establishing the National Priorities List, investigating sites for inclusion on the list, determining their priority, and conducting and/or supervising cleanup and other remedial actions.

support zone   In hazardous material site remediation or emergency response, uncontaminated area where workers are unlikely to be exposed to hazardous substances or dangerous conditions. The cold zone.

surface impoundments   Lined natural or synthetic depressions or diked areas that can be used to treat, store, or dispose of waste.

surfacing ACM   Material sprayed or troweled onto structural members (beams, columns, or decking) for fire protection; or on ceilings or walls for fireproofing, acoustical or decorative purposes. Includes textured plaster and other textured wall and ceiling surfaces.

surfactant   (surface active agent)   A material that lowers the surface tension of water. A detergent compound that promotes lathering and solution or dispersion of oily materials in water.

suspended solids   Solid materials that float on the surface of, or are suspended in, sewage or other liquids. Removal treatment involves skimming from the surface, filtration, or addition of materials that cause settling. One treatment, dissolved air filtration (DAF), uses flocculants and sparged air to float light suspended solids to the surface and agglomerate heavier solids for settling.

swedge   On steel 55-gallon (208-L) drums, the protruding, integral rings around the drum. On bung-type drums, one is located approximately one-third of the way down the side of the drum, the other approximately two-thirds of the way down the side of the drum. Open-mouth drums can have three swedges. Swedges are usually, but not always, used to aid in lifting the drum.

synergism   An interacting effect of two or more chemicals that results in an effect greater than the sum of their separate effects.

synergistic effect   A synergistic effect occurs when any two chemicals acting together produce an effect that is greater than the simple sum of their effects when acting alone.

synonym   Another name or names by which a material is known. "Caustic soda" is a synonym for sodium hydroxide.

synthetic organic chemicals (SOCs)   Man-made organic chemicals. Petrochemicals and most pharmaceutical "wonder" drugs are typical examples. Some SOCs are volatile. They are classed as VOCs that can become troublesome pollutants.

systemic   Spread throughout the body, affecting all body systems and organs, not localized in one spot or area.

systemic effects   Acute or chronic adverse health effects that occur in parts of the body removed from the site of exposure to the material.

tachycardia   Excessively rapid heartbeat, with a pulse rate above 100 beats per minute while at rest.

tachypnea   Increased rate of respiration.

target organ toxin   Toxic substance that attacks a specific organ of the body.

Technical Assistance to Brownfields Communities (TAB)   Program, part of EPA's Brownfields Initiative, that helps communities clean and redevelop properties that have been damaged or undervalued by environmental contamination. The purpose of these efforts is to create better jobs, increase the local tax base, improve neighborhood environments, and enhance the overall quality of life.

Technical Assistance Grant (TAG)   Provides money for activities that help a community participate in the decision-making process of site-specific cleanup strategies at eligible Superfund sites.

Technology Innovation Office (TIO)   Acts as an advocate for new technologies, working to increase the application of innovative treatment technologies to contaminated waste sites, soils, and groundwater. As a part of this effort, TIO has worked with many partners inside EPA, in other federal agencies, and in the private sector to improve the nation's understanding of cleanup technologies and reduce the impediments to their widespread use.

teratogen   Substance or agent to which exposure of a pregnant female can result in malformation in the fetus.

tertiary treatment   Advanced cleaning of wastewater that goes beyond the secondary or biological stage, destroying or removing hazardous materials, removing nutrients such as phosphorus, nitrogen, and most BOD and suspended solids. Sometimes referred to as "polishing."

testing   The process by which the atmospheric hazards that may confront entrants of a permit space are identified and evaluated. Testing includes specifying the tests that are to be performed in the permit space. Confined space testing must follow this order: O_2, LEL, and then toxics. (OSHA requirement.)

Note

Testing enables employers both to devise and implement adequate control measures for the protection of authorized entrants and to determine if acceptable entry conditions are present immediately prior to and during entry.

thermal treatment   Use of elevated temperatures up to several hundred degrees to drive vapors from hazardous waste using temperatures greater than 1,500°F is considered **thermal destruction.**

thermoplastics   Heat-welded or heat-laminated fabrics such as PVC or chlorinated polyethylene (CPE). Plastics that soften or melt, rather than decompose, under moderate heating to a few hundred degrees F.

threshold   The lowest dose or exposure to a chemical at which a specific effect is observed.

Threshold Limit Value (TLV®) (ACGIH)   Airborne concentration of a material to which nearly all persons can be exposed day after day, without adverse effects. TLV®s are expressed in three ways:

1. **TLV®-C:**   Ceiling limit, concentration that should not be exceeded even instantaneously.

2. **TLV®-STEL:**   Short-term exposure limit, maximum concentration for a continuous 15-minute exposure period.

3. **TLV®-TWA:**   Time-weighted average, concentration for a normal eight-hour work day or 40-hour work week.

threshold planning quantity (TPQ)   Per 40 CFR 302, the amount of material at a facility that requires emergency planning and notification per CERCLA. A quantity designated for each chemical on the list of extremely hazardous substances that triggers notification by facilities to the state emergency response commission that such facilities are subject to emergency planning requirements under SARA Title III. Also referred to as TQ

time-weighted average (TWA)   Refers to concentrations of airborne toxic materials that have been weighted for a certain time duration, usually eight hours. In air sampling, the average air concentration of contaminants during a given period.

tinnitus   Ringing sound in the ears.

torr Abbreviation for Torricelli; measurement of pressure equal to 1 mm Hg. 760 torr equals 1 atmosphere (760 mm Hg).

total dissolved solids (TDS) Common measurement of water. When TDS is determined by summing the results of separate analyses for all major ions, it is analogous to salinity.

total petroleum hydrocarbons (TPH) Measure of the concentration or mass of petroleum hydrocarbon constituents present in a given amount of soil or water. The word total is a misnomer; few, if any, of the procedures for quantifying hydrocarbons can measure all of them in a given sample. Volatile ones are usually lost in the process and nonquantified and nonpetroleum hydrocarbons sometimes appear in the analysis.

total suspended solids (TSS) Measure of the suspended, or undissolved, solids in wastewater, effluent, or water bodies. (See **suspended solids.**)

total suspended particles (TSP) Method of monitoring airborne particulate matter by total weight.

ToxFAQs™ Series of fact sheets about hazardous substances developed by the Agency for Toxic Substances and Disease Registry's (ATSDR) Division of Toxicology. The fact sheets are a guide to the most frequently asked questions (FAQs) about exposure to hazardous substances found around hazardous waste sites and the effects of exposure on human health (www.atsdr.cdc.gov/toxfaq.html).

toxic chemical Any chemical listed in EPA rules under Toxic Chemicals Subject to Section 313 of the Emergency Planning and Community Right-to-Know Act of 1986.

toxic chemical release form Information form required of facilities that manufacture, process, or use (in quantities above a specific amount) chemicals listed under SARA Title III.

toxic concentration The concentration at which a substance produces a toxic effect.

toxic dose The dose level at which a substance produces a toxic effect.

toxicity The degree to which a substance or mixture of substances can harm humans or animals. Acute toxicity involves harmful effects in an organism through a single or short-term exposure. Chronic toxicity is the ability of a substance or mixture of substances to cause harmful effects over an extended period, usually upon repeated or continuous exposure sometimes lasting for the entire life of the exposed organism. Subchronic toxicity is the ability of the substance to cause effects for more than one year but less than the lifetime of the exposed organism. The ability of a substance to produce injury once it reaches a susceptible site in or on the human body, other organisms, or vegetation (phytotoxins).

Toxicity Characteristics Leaching Procedure (TCLP) The TCLP test, EPA test method SW-846-1311, is used to determine toxicity of a material prior to disposal. If it is found to be nontoxic, the material may be disposed of in a sanitary landfill. If the test proves toxicity, the waste must be disposed of in an EPA-approved hazardous waste landfill.

toxicology The study of the nature, effects, and detection of toxic materials in living organisms. Also, the study of substances that are otherwise harmless but prove toxic under particular conditions.

toxic pollutants Materials that cause death, disease, or birth defects in organisms that ingest or absorb them. The quantities and exposures necessary to cause these effects can vary widely.

Toxics Release Inventory (TRI) Database contains information concerning waste management activities and the release of toxic chemicals by facilities that manufacture, process, or otherwise use such materials. Citizens, businesses, and governments can then use this information to work together to protect the quality of their land, air, and water.

toxic substance Chemical or material:

1. For which research has provided sufficient evidence that exposure to it will cause an acute or chronic health hazard, and

2. That is listed in the RTECS manual, provided that the chemical or material:

- Causes harm at any dose level;
- Causes cancer or reproductive effects in animals at any dose level;
- Has a median lethal dose level of less than 500 mg per kilogram of body weight when administered orally to rats;
- Has a median lethal dose level of less than 1,000 mg per kilogram of body weight when administered by continuous contact to the bare skin of albino rabbits; or
- Has a median lethal concentration in air of less than 2,000 ppm by volume of gas or vapor, or less than 20 mg per liter of mist, fume, or dust when administered to albino rats.

Toxic Substances Control Act (TSCA) Public Law PL 94-469, found in 40 CFR 700-799. EPA has jurisdiction. Effective January 1, 1977. Controls the exposure to and use of raw industrial chemicals not subject to other laws. Chemicals are to be evaluated prior to use and can be controlled based on risk. The act provides for a listing of all chemicals that are to be evaluated prior to manufacture or use in the United States.

toxic waste A waste that can produce injury if inhaled, swallowed, or absorbed through the skin.

trade name Trademark name or commercial trade name for a material, given by the manufacturer.

trade secret Any confidential formula, pattern, process, device, information, or compilation of information used in an employer's business that gives the employer an opportunity to obtain an advantage over competitors. Trade secrets are not patented, otherwise they would be public information. Trade secrets are also **proprietary information.** Trade secret ingredients need not be disclosed on an MSDS. They must, however, be disclosed to medical professionals who are administering medical treatment to an exposed individual.

transpiration The process by which water vapor is lost to the atmosphere from living plants. The term can also be applied to the quantity of water thus dissipated.

transporter Hazardous waste transporters pick up properly packaged and labeled hazardous waste from generators, together with a Uniform Hazardous Waste Manifest, and transport it to designated facilities that treat, store, recycle, or dispose of the waste. Transporters are subject to specific hazardous waste regulations by both EPA and DOT.

treatability studies Tests of potential cleanup technologies conducted in a laboratory.

treatment (1) Any method, technique, or process designed to remove material or pollutants from air, drinking water supplies, contaminated water bodies, solid waste, waste streams, effluents, and air emissions. (2) Methods used to change the biological character or composition of any regulated medical waste so as to reduce substantially or eliminate its potential for causing disease.

treatment, storage, and disposal facility (TSDF) Facility that receives hazardous waste from generators or other facilities for treatment, storage or disposal of waste.

trial burn An incinerator test in which emissions are monitored for the presence of specific organic compounds, particulates, sulfur and nitrogen oxides, and halogen-containing compounds such as hydrogen chloride (hydrochloric acid).

trichloroethylene (TCE) A stable, low-boiling point colorless liquid, toxic if inhaled. Used as a solvent or metal degreasing agent and in other industrial applications. A VOC.

turbidimeter A device that measures the **turbidity** caused by suspended solids in a liquid.

turbidity A cloudy condition in water due to suspended solids.

ultraviolet rays Radiation from the sun that can be useful or potentially harmful. UV rays from one part of the spectrum (UV-A) enhance plant life. UV rays from other parts of the spectrum (UV-B) can cause skin cancer or other tissue damage. The ozone layer in the atmosphere partly shields us from ultraviolet rays reaching the earth's surface.

uncontrolled hazardous waste site An area identified as an uncontrolled hazardous waste site by a governmental body, whether federal, state, local, or other, where an accumulation of hazardous substances creates a threat to the health and safety of individuals or the environment or both. Some sites are found on public lands, such as those created by former municipal, county, or state landfills where illegal or poorly managed waste disposal has taken place. Other sites are found on private property, often belonging to generators or former generators of hazardous substance wastes. Examples of such sites include, but are not limited to, surface impoundments, landfills, dumps, and tank or drum farms. Normal operations at TSD sites are not covered by this definition.

underground injection control (UIC) wells　(Class I UICs.) Steel and concrete-encased shafts penetrating thousands of feet underground into which hazardous wastes are deposited by force and under pressure. Diluted liquid industrial wastes are injected through wells thousands of feet into geologic formations that serve as environmental protection barriers for as long as the waste remains hazardous. Those permitted to use deep well disposal for hazardous material must file petitions demonstrating that the waste will not migrate to the environment in a hazardous form for at least 10,000 years.

underground storage tank system (UST)　Tank and any underground piping connected to the tank that has at least 10% of its combined volume underground. Under the Resource Conservation and Recovery Act (RCRA), EPA has established regulatory programs to prevent, detect, and clean up releases from USTs containing petroleum or hazardous substances. Almost all gasoline stations have USTs.

unit　(1) Any type of quantity that qualifies as a physical measurement. The gallon is a unit of liquid measurement. (2) This term can also refer to tanks, containers, incinerators, surface impoundments, containment buildings, and waste piles. (Hazardous waste unit.)

unsaturated zone　The area above the water table where soil pores are not fully saturated, although some water may be present. Also called the 'vadose zone'.

unstable　Tending toward decomposition or other unwanted chemical change during normal handling or storage.

upper detection limit　The largest concentration that an instrument can reliably detect.

upper explosive limit (UEL)　Highest concentration (expressed in percent vapor or gas in the air by volume) of a substance that will burn or explode when an ignition source (heat, electric arc, or flame) is present. In most cases the UEL and UFL are so close as to be practically considered the same.

upper flammability limit (UFL)　Highest concentration (highest percentage of the substance in air) that will burn when an ignition source (heat, electric arc, or flame) is present.

urea-formaldehyde foam insulation (UFFI)　A material once used to conserve energy by sealing crawlspaces, attics, etc.; no longer used because emissions of formaldehyde were found to be a health hazard.

use　To package, handle, react, or transfer.

utricaria　Nettle rash; hives; elevated, itching white patches.

vadose zone　The zone between land surface and the water table within which the moisture content is less than saturation (except in the capillary fringe). Soil pore space also typically contains air or other gases. The capillary fringe, is included in the vadose zone because it is saturated due to capillary action, even though it is saturated with groundwater.

vapor　A temperature range within the gas state of a material where it can be condensed to a liquid or returned to a vapor under temperature changes usually achievable under atmospheric conditions. (Also see **gas**.)

vapor capture system　Any combination of hoods and ventilation system that captures or contains vapors so they may be directed to an abatement or recovery device.

vapor density　Weight of vapor or gas compared to the weight of an equal volume of air; an expression of the relative density of the vapor or gas.

vapor dispersion　The movement of vapor clouds in air due to wind, thermal action, gravity spreading, and mixing.

vapor plumes　Flue gases and vapors visible because they contain water or other liquid droplets.

vapor pressure Pressure exerted by a saturated vapor above its liquid in a closed container at a given temperature. Vapor pressure is strongly influenced by temperature but practically not influenced by contained pressure. If a vapor is kept in confinement over its liquid so that the vapor can accumulate above the liquid (the temperature being held constant), the vapor pressure approaches a fixed limit called the maximum (or saturated) vapor pressure, dependent only on the temperature and the liquid. Vapor pressure is the key to evaporation rate. The higher the vapor pressure of a material, the faster it will evaporate. Important facts to remember:

- Vapor pressure of a substance at 100°F will always be higher than the vapor pressure of the substance at 60°F.

- Vapor pressures are usually reported on MSDSs in mmHg. 760 mmHg is equivalent to 14.7 psia (1 atm). The vapor pressure of water at room temperature is about 16 mmHg. The vapor pressure of acetone is about 150 mmHg.

- The lower the boiling point of a substance, the higher its vapor pressure.

vapors The 'gaseous' form of substances that are normally in the solid or liquid state (within about 150°C of room temperature and at near atmospheric pressures). The vapor can be changed back to the solid or liquid state either by increasing the pressure without increasing the temperature or by decreasing the temperature alone. Vapors diffuse slowly or rapidly through most barriers. Evaporation is the process by which a liquid is changed into the vapor state. Solvents with low boiling points will volatilize (evaporate) readily. Examples include benzene, methyl alcohol, and toluene.

variance Government permission for a delay or exception in the application of a given law, ordinance, or regulation.

vector (1) An organism, often an insect or rodent, that carries disease. (2) Plasmids, viruses, or bacteria used to transport genes into a host cell. A gene is placed in the vector; the vector then infects the bacterium.

ventilation Circulating fresh air to replace contaminated air.

ventilation rate The rate at which indoor air enters and stale air leaves a building. Expressed as the number of changes of outdoor air per unit of time (air changes per hour, ACH, or the rate at which a volume of outdoor air enters in cubic feet per minute, CFM).

ventilation/suction The act of admitting fresh air into a space in order to replace stale or contaminated air; achieved by blowing air into the space. Similarly, suction represents the admission of fresh air into an interior space by lowering the pressure outside of the space at the suction side of the blower, thereby drawing the contaminated air outward.

vertigo Feeling of revolving in space; dizziness, giddiness. Usually associated with being in a high place, as on scaffolding.

vinyl chloride A chemical compound. The monomer used in producing PVC plastics. Vinyl chloride is a known carcinogen.

violations Other-Than-Serious, Serious, Willful, Criminal Willful, Repeated (OSHA types in ascending order. From $7,000.00 to $70,000.00 fines and more including imprisonment.)

viscosity Tendency of a fluid to resist flow. Viscosity is not necessarily related to density. Vehicle motor oils are rated (the SAE number, or range) according to their ability to maintain a working viscosity, or protective lubrication, over selected temperature ranges.

Viton® A copolymer of vinylidene fluoride and hexaflouropropylene. A common elastomeric CPC material offering fair to good chemical protection.

VOA Volatile Organic Analysis. Measurement of the volatile organics present in a water sample; that is, those that will volatilize at room temperature. To prevent their loss in sampling, zero headspace sample containers are used.

volatile Any substance that evaporates readily.

volatile liquids Liquids that easily vaporize or evaporate at room temperature and atmospheric pressure.

volatile organic compounds (VOCs) Formerly used widely in solvents, cleaning agents, paint and other coatings; because they can be chosen to evaporate with controlled rapidity and provide good solvent and ingredient suspension properties. Their use is now being regulated to reduce the VOC (pollutant) load in the atmosphere.

volatility A measure of how quickly a substance vaporizes at ordinary (ambient) temperatures. The higher the vapor pressure of a substance, the higher its volatility. Gasoline is more volatile than water.

volume reduction Processing waste materials to decrease the amount of space they occupy, usually by compacting, shredding, incineration, or composting.

waste (1) Unwanted materials as residue from a manufacturing process or any other use where recycling or reuse are impractical. (2) Refuse from places of human or animal habitation.

waste characterization Identification of specified chemical, radiological, and microbiological constituents and physical characteristics of a waste material.

waste minimization Measures or techniques that reduce the amount of wastes generated during industrial production processes; the term is also applied to recycling and other efforts to reduce the amount of waste going into production process waste streams.

waste piles Noncontainerized, lined or unlined accumulations of solid, nonflowing waste.

waste prevention See **source reduction.**

waste stream A flow of unwanted material. The total flow of solid waste from homes, businesses, institutions, and manufacturing plants that is recycled, burned, or disposed of in landfills, or segments thereof such as the residential waste stream or the recyclable waste stream.

waste treatment lagoon Impoundment made by excavation or earth fill for biological treatment of wastewater.

waste treatment stream The continuous movement of waste from generator to treater and disposer.

waste-heat recovery Recovering heat discharged by one process to provide heat needed by a second process.

waste-to-energy facility/municipal waste combustor Facility where recovered municipal solid waste is converted into a usable form of energy, usually via combustion to provide steam to run turbines to generate electricity.

wastewater The spent or used water from a home, community, farm, or industry that contains dissolved or suspended matter.

wastewater infrastructure Plan or network for the collection, treatment, and disposal of sewage in a community. The level of treatment will depend on the size of the community, the type of discharge, and/or the designated use of the receiving water.

wastewater operations and maintenance Actions taken after construction to ensure that facilities constructed to treat wastewater will be operated, maintained, and managed to reach prescribed effluent levels in an optimum manner.

wastewater treatment plant A facility containing a series of tanks, screens, filters, and other processes by which pollutants are removed from water. If owned by a municipality, known as a Publicly Owned Treatment Works (POTW).

water pollution The presence in water of enough harmful or objectionable material to damage the water's quality.

water-reactive Material that reacts with water to release a gas that is either flammable or presents a health hazard.

W.C. (water column) A measure of gauge pressure, typically in inches of water column as a difference of water level in the two legs of a U-tube. One leg of the tube is attached to the pressure being read and the other end is open to the atmosphere. Used to measure relatively low pressures. It is a measure used to rate the air pressure produced by fans and blowers. One inch of W.C. is equal to only 0.0361 psi.

water quality standards   State-adopted and EPA-approved ambient standards for water bodies. The standards prescribe the use of the water body and establish the water quality criteria that must be met to protect designated uses.

water solubility   The maximum possible concentration of a chemical compound soluble in water at a specified temperature and pressure. If a substance is water-soluble, it can very readily disperse through the environment.

water table   The depth below ground to the surface of groundwater.

way bill   The shipping paper prepared from a bill of lading that accompanies each parcel of a rail shipment. An **air bill** accompanies shipments by air.

weir   (1) A V-shape or rectangle cut in a plate placed in an open channel to measure the flow rate of water by the height the water raises in the cut. (2) A wall or obstruction used to control flow from the discharge end of settling tanks and clarifiers to ensure a uniform flow rate of clarified water and avoid short-circuiting.

well   A bored, drilled, or driven shaft or a dug hole whose depth is greater than the largest surface dimension and whose purpose is to reach underground water supplies, gas, or oil or to store or dispose of fluid waste below ground.

well field   Area containing one or more wells that produce usable amounts of water, gas, or oil.

well injection   The subsurface emplacement of fluids into a well.

well monitoring   Measurement by on-site instruments or laboratory methods of well water quality.

well plug   A watertight, gastight seal installed in a bore hole or well to prevent flow of surface fluids into the well or to access the well. Used in decommissioning wells that will no longer be used.

well point   A hollow vertical tube, rod, or pipe terminating in a perforated pointed shoe and fitted with a fine-mesh screen.

wellhead protection area   A protected surface and subsurface zone surrounding a well or well field supplying a public water system to keep contaminants from reaching the well water.

wetlands   An area that is saturated by surface water or groundwater with vegetation adapted for life under those soil conditions, such as swamps, bogs, fens, marshes, and estuaries.

wettability   The relative degree to which a fluid will spread into or coat a solid surface in the presence of other immiscible fluids.

wettable powder   Dry formulation that must be mixed with water or other liquid before it is applied.

work area   A room or other defined space in a workplace where hazardous materials are produced, used, stored, or packaged and where employees are present.

workplace   An establishment at one geographical location containing one or more work areas.

work plan   Written directives, including standard operating procedures (SOPs), that specifically describe all work activities that are to take place at a work site.

xenobiota   Any biotum displaced from its normal habitat; a chemical foreign to a biological system.

X-ray   A type of electromechanical radiation used to make images of the internal structures of the body or other materials or structures of limited density; for example, X-ray inspection of welds on steel.

X-ray fluorescence (XRF)   A process that uses low-energy X-ray tube-emitted x-rays. X-ray quanta knock out lower shell electrons in sample atoms. This is called excitation. An electron from a higher shell fills the excited electron vacancy. The energy difference between shells is emitted as an X-ray fluorescence photon. The energy of the emitted radiation photon is characteristic to each individual element in the periodic table. This method does not make the sample radioactive and leaves it unchanged. Used extensively in the investigation of lead-based paint, it is also used to analyze for various metals in soils at hazardous waste sites.

yellow-boy Iron oxide flocculant (clumps of solids in waste or water); usually observed as orange-yellow deposits in surface streams with excess iron content. (See **floc, flocculation.**)

yield The quantity of water (expressed as a rate of flow or total quantity per year) that can be collected for a given use from surface or groundwater sources.

zero air Atmospheric air purified to contain less than 0.1 ppm total hydrocarbons.

zinc fume fever Caused by inhalation of zinc oxide fume characterized by flu-like symptoms, a metallic taste in the mouth, coughing, weakness, fatigue, muscular pain, and nausea, followed by fever and chills.

Z list OSHA's Toxic and Hazardous Substances Tables Z-1, Z-2, and Z-3 of air contaminants, found in 29 CFR 1910.1000. These tables record PELs, TWAs, and ceiling concentrations for the materials listed. Any material found in these tables is considered to be hazardous.

A.2 ACRONYMS

AA Atomic absorption

A&I Alternative and innovative (wastewater treatment system)

AAEE American Academy of Environmental Engineers

AAR Association of American Railroads

AAS Atomic absorption spectroscopy

ACGIH American Conference of Governmental Industrial Hygienists

ACBM Asbestos-containing building materials

ACFM Actual cubic feet per minute

ACM Asbestos-containing materials

ACS American Chemical Society

ACWA American Clean Water Association

ACWM Asbestos-containing waste material

AEA Atomic Energy Act

AEE Alliance for Environmental Education

AERE Association of Environmental and Resource Economists

AES Auger electron spectrometry

AFA American Forestry Association

AFCEE Air Force Center for Environmental Excellence

AHERA Asbestos Hazard Emergency Response Act

AHU Air handling unit

AIHA American Industrial Hygiene Association

AIHC American Industrial Health Council

AIP Auto ignition point

ALARA As low as reasonably achievable

ALJ Administrative law judge

ANPR Advance notice of proposed rulemaking

ANSI   American National Standards Institute

APCA   Air Pollution Control Association

APHA   American Public Health Association

API   American Petroleum Institute, an organization of the petroleum industry

APR   Air-purifying respirator

ARAR   Applicable or relevant and appropriate standards, limitations, criteria, and requirements

ARCHIE   Automated Resource for Chemical Hazard Incident Evaluation

ASR   Atmosphere-supplying respirator

ASRL   Atmospheric Sciences Research Laboratory

ASTM   American Society for Testing and Materials

ATERIS   Air Toxics Exposure and Risk Information System

ATSDR   Agency for Toxic Substances and Disease Registry, U.S. Department of Health and Human Services, NIOSH

ATSDR-HazDat   Agency for Toxic Substances and Disease Registry's Hazardous Substance Release/Health Effects Database

AWO   American Waterway Operators

AWRA   American Water Resources Association

BACM   Best available control measures

BACT   Best available control technology

BADT   Best available demonstrated technology

BAF   Bioaccumulation factor

BaP   Benzo(a)pyrene

BAT   Best available technology

BATEA   Best available treatment economically achievable

BCT   Best control technology

BCPCT   Best conventional pollutant control technology

BDAT   Best demonstrated achievable technology

BDCT   Best demonstrated control technology

BDT   Best demonstrated technology

BEJ   Best engineering judgment; best expert judgment

BIOPLUME   Computer modeling software to predict the maximum extent of existing plumes

BMP   Best management practice(s)

BMR   Baseline monitoring report

BOD   Biological oxygen demand (also sometimes referred to as *biochemical oxygen demand*)

BOF   Basic oxygen furnace

BP   Boiling point

BSI   British Standards Institute

BTEX   Benzene, toluene, ethylbenzene, xylene; common components of gasoline

CAA   Clean Air Act; compliance assurance agreement

CAAA   Clean Air Act Amendments

CADET   Computer Aided Design for Exercise Training

CAER   Community Awareness and Emergency Response

CAG   Carcinogen Assessment Group

CAMEO   Computer Aided Management of Emergency Operations

CAO   Corrective action order

CAP   Corrective action plan; criteria air pollutant

CAR   Corrective action report

CAS   Chemical Abstract Service

CCHW   Citizens Clearinghouse for Hazardous Wastes

CCID   Confidential Chemicals Identification System

CDC   Centers for Disease Control and Prevention, Atlanta, GA; part of the US Department of Health and Human Services, Public Health Service

CDD   Chlorinated dibenzo-*p*-dioxin

CDF   Chlorinated dibenzofuran

CECATS   CSB Existing Chemicals Assessment Tracking System

CEI   Compliance evaluation inspection

CEM   Continuous emission monitoring

CEPP   Chemical emergency preparedness and prevention

CEPPO   The EPA's Chemical Emergency Preparedness and Prevention Office

CERCLA   Comprehensive Environmental Response, Compensation, and Liability Act (1980)

CERCLIS   The Comprehensive Environmental Response, Compensation, and Liability Act (CERCLA) Information System

CESQG   Conditionally exempt small quantity generator

CEQ   Council on Environmental Quality

CFC   Chlorofluorocarbons

CFM   Chlorofluoromethanes; cubic feet per minute

CFR   Code of Federal Regulations

CGA   Compressed Gas Association

CGI   Combustible gas indicator

CHEMNET   Chemical Industry Emergency Mutual Aid Network

CHEMTREC   Chemical Transportation Emergency Center, a 24-hour-a-day service (1-800-424-9300) provided by the American Chemistry Council for chemical emergency questions and guidance

CHIP   Capability and Hazard Identification Program; chemical hazard information profiles

CHMM   Certified Hazardous Materials Manager

CHRS   Chemical Hazards Response Information System of the U.S. Coast Guard

CIS   Chemical Information System

CLEANS   Clinical Laboratory for Evaluation and Assessment of Toxic Substances

CLEVER   Clinical Laboratory for Evaluation and Validation of Epidemiologic Research

CLIPS   Chemical List Index and Processing System

CLP   Contract Laboratory Program

CLP   EPA's Contract Laboratory Program

CM   Corrective measure

CMA   Chemical Manufacturers' Association (now the American Chemistry Council)

CNG   Compressed natural gas

CNS   Central nervous system

COD   Chemical oxygen demand

COCO   Contractor-owned/contractor-operated

CPDA   Chemical Producers and Distributors Association

CPM   Counts per minute

CRP   Community relations plan

CSB   Chemical Safety Board

CSEPP   Chemical Stockpile Emergency Preparedness Program

CSIN   Chemical Substances Information Network

CSMA   Chemical Specialties Manufacturers Association

CSP   Certified Safety Professional

CTARC   Chemical Testing and Assessment Research Commission

CWA   Clean Water Act (also known as FWPCA)

DDT   Dichlorodiphenyltrichloroethane

DECON   Decontamination

DFM   Diesel fuel, marine

DHHS   Department of Health and Human Services

DNAPL   Dense nonaqueous phase liquid

DO   Dissolved oxygen

DOC   Department of Commerce

DOD   Department of Defense

DOE   Department of Energy

DOI   Department of Interior

DOL   Department of Labor (OSHA and MSHA come under the DOL)

DOT   Department of Transportation

DPM   Disintegrations per minute (used in radiation technology)

DRE   Destruction and removal efficiency

DRI   Direct-reading instruments

EA   Endangerment Assessment; Enforcement Agreement; Environmental Action; Environmental Assessment; Environmental Audit

EAA   Environmental Assessment Association

EAG   Exposure assessment group

EBS   Emergency Broadcast System

ECD   Electron capture detector

ECL   Environmental chemical laboratory

EDB   Ethylene dibromide

EDC   Ethylene dichloride

EDRS   Enforcement Document Retrieval System

EDTA   Ethylene diamine tetraacetic acid

EECs   Estimated environmental concentrations

EERU   Environmental Emergency Response Unit

EESL   Environmental ecological and support laboratory

EETFC   Environmental Effects, Transport, and Fate Committee

EIA   Environmental impact assessment

EIL   Environmental impairment liability

EIR   Endangerment information report; environmental impact report

EIS   Environmental impact statement; environmental inventory system

EIS/AS   Emissions inventory system/area source

EIS/PS   Emissions inventory system/point source

EJ   Environmental justice

EL   Exposure level

EM   Electromagnetic conductivity

EMAP   Environmental Mapping and Assessment Program

EMF   Electromotive force (commonly referred to as voltage); electrical and magnetic fields

EMS   Emergency medical service

EMT   Emergency medical technician

EOC   Emergency operations (Command Post) center

EOD   Explosive ordinance disposal (bomb squad)

EOP   Emergency operations plan

EOS   Emergency operations simulation

EPA   Environmental Protection Agency

EPCA   Energy Policy and Conservation Act

EPACT   Environmental Policy Act

EPACASR   EPA Chemical Activities Status Report

EPCRA   Emergency Planning and Community Right-to-Know Act

EPTC   Extraction Procedure Toxicity Characteristic (EP toxicity)

ERCS   Emergency Response Cleanup Services, under EPA contract

ERNS   EPA's Emergency Response Notification System

ERT   EPA's Emergency Response Team

ET   EPA's ecotox thresholds

eV   Electron volt (see **ionization potential**)

FATES   FIFRA and TSCA Enforcement System

FEFx   Forced expiratory flow

FEMA   Federal Emergency Management Agency

FEPCA   Federal Environmental Pesticide Control Act; enacted as amendments to FIFRA

FEV   Forced expiratory volume

FFDCA   Federal Food, Drug, and Cosmetic Act

FID   Flame ionization detector

FIFRA   Federal Insecticide, Fungicide, and Rodenticide Act

FINDS   Facility Index System

FI   Field investigation team

Fl.P.   Flash point

FM   Factory Mutual Insurance Company; tests and certifies instruments and equipment used in hazardous locations

FML   Flexible membrane liner

FOIA   Freedom of Information Act

FONSI   Finding of no significant impact

FPD   Flame photometric detector

FR   Federal Register

FS   Feasibility study

FURS   Federal Underground Injection Control Reporting System

FUV   Far ultraviolet light detector

FWPCA   Federal Water Pollution and Control Act (also known as CWA)

GAC   Granular activated carbon

GC/MS   Gas chromatograph/mass spectrometer

GIS   Geographic Information System

GLC   Gas liquid chromatography

GPR   Ground-penetrating radar

GWM   Groundwater monitoring

HAP   Hazardous air pollutant

HAPPS   Hazardous Air Pollutant Prioritization System

HATREMS   Hazardous and Trace Emissions System

HAZCOMM   OSHA's Hazard Communication Standard (29 CFR 1910.1200)

HAZMAT   Hazardous materials; Hazardous materials team

HAZOP   Hazard and operability study

HAZWOPER   Hazardous Waste Operations and Emergency Response

HBFC   Hydrobromofluorocarbon

HC   Hazardous constituents; hydrocarbon

HCCPD   Hexachlorocyclo-pentadiene

HCFC   Hydrochlorofluorocarbon

HDPE   High-density polyethylene

HEPA   High-efficiency particulate air; an air filter capable of filtering 99.97% of particles greater than 0.03 μm in diameter, from the air; equivalent of P100 filter for APRs

HFC   Hydrofluorocarbon

HHW   Household hazardous waste

HI-VOL   High-volume sampler

HM-EEM   FEMA's Hazardous Materials Exercise Evaluation Methodology

HMIS   Hazardous Materials Information System (both Paint and Coatings Council and D.O.D.)

HMIX   Hazardous Materials Information Exchange

HMTA   Hazardous Materials Transportation Act

HMTR   Hazardous Materials Transportation Regulations

HPLC   High-performance liquid chromatography

HRS   EPA's Hazard Ranking System

HSRC   EPA's Hazardous Substance Research Centers

HSWA   Hazardous and Solid Waste Amendments

HVAC   Heating, ventilation, and air-conditioning

HW   Hazardous waste

HWDMS   Hazardous Waste Data Management System

HWM   Hazardous waste management

IARC   International Agency for Research on Cancer

IAQ   Indoor air quality

ICS   Incident Command System

ICWM   Institute for Chemical Waste Management

IDLH   Immediately dangerous to life and health

IEMIS   Integrated Emergency Management Information System

IEMS   Integrated Emergency Management System

IHMM   Institute of Hazardous Materials Managers

IP   Ionization potential (see **electron volt**)

IPCC   Intergovernmental Panel on Climate Change

IR   Infrared radiation

IRIS   EPA's Integrated Risk Information System

ISEA   International Safety Equipment Association

ITC   Innovative Technology Council

IUPAC   International Union of Pure and Applied Chemistry, the organization that is responsible for the periodic table of the elements and the naming of chemicals

IWS   Ionizing wet scrubber

LAER   Lowest achievable emission rate

LC$_{50}$ Lethal concentration, usually dispersed in respirable air, for 50% of the test population; LC$_{50}$ (rat) refers to a rat as the subject

LD$_{50}$ Lethal dose, administered to a subject through injection or ingestion, for 50% of the test population. LD$_{50}$ (rat) refers to a rat as the subject

LD$_{L0}$ The lowest dosage of a toxic substance that kills test organisms

LDR   Land disposal restrictions

LEL   Lower explosive limit

LEPC   Local Emergency Planning Committee

LERC   Local Emergency Response Committee

LFG   Landfill gas

LFL   Lower flammable limit

LNAPL   Light nonaqueous phase liquid

LOD   Limit of detection

LQG   Large quantity generator

MATC   Maximum acceptable toxic concentration

MCL   Maximum contaminant level

MCS   Multiple chemical sensitivity

MDL   Method detection limit

mg/L   Milligrams per liter

mg/m^3   Milligrams per cubic meter

μCi   Microcurie

μg/L   Micrograms per liter

μg/m^3   Micrograms per cubic meter

mppcf   Million particles per cubic foot of air (a measure of respirable dust)

mR/hr   Milliroentgens per hour (a measure of radioactive exposure)

MSDS   Material safety data sheet

MSHA   Mine Safety and Health Administration (part of the DOL)

MSW   Municipal solid waste

MTBE   Methyl tertiary butyl ether

MUC   Maximum use concentration (as defined by a respirator and filter manufacturer)

MUL   Maximum use limits = **PEL × PF**

MUTA   Mutagenicity

NAA   Nonattainment area

NAAQS   National Ambient Air Quality Standard

NACD   National Association of Chemical Distributors

NAPL   Nonaqueous phase liquid

NAS   National Academy of Sciences

NBR Nitrile-butadiene rubber

NCI/NTP National Cancer Institute/National Toxicology Program

NCP National contingency plan

NEC National Electrical Code

NESHAP National Emission Standard for Hazardous Air Pollutants

NFPA National Fire Protection Association

NFRAP No further remedial action planned

NIEHS National Institute of Environmental Health Sciences

NIOSH National Institute for Occupational Safety and Health

NO Nitric oxide

NOAA National Oceanographic and Atmospheric Administration

n.o.s. Not otherwise specified (DOT)

N_2O Nitrous oxide

NO_x Nitrogen oxides

NORM Naturally occurring radioactive material

NPDES National Pollutant Discharge Elimination System, EPA's permitting system for the controlled release of certain pollutants to the environment

NPL National Priorities List, sites destined to become Superfund sites

NRC Nuclear Regulatory Commission

NRC EPA's National Response Center (800-424-8802)

NRD Natural resource damages

NRT EPA's National Response Team

NSEC National System for Emergency Coordination

NSEP National System for Emergency Preparedness

NSPS New Source Performance Standards

NSR New Source Review

NTIS National Technical Information Service

NTP National Toxicology Program

NTTC National Tank Truck Carriers

O_3 Ozone

O&M Operation and maintenance

OECA EPA's Office of Enforcement and Compliance Assurance

OERR EPA's Office of Emergency and Remedial Response

OERRGIS EPA's Office of Emergency and Remedial Response Geographic Information System

OHMTADS EPA's Oil and Hazardous Materials Technical Assistance Data System

OPA Oil Pollution Act

OPP EPA's Office of Pesticide Programs

OPPT EPA's Office of Pollution Prevention and Toxics

OPPTS EPA's Office of Prevention, Pesticides and Toxic Substance

ORD EPA's Office of Research and Development

ORM Other regulated material; various specific classes other than the nine DOT hazard classes—ORM-A through ORM-E

OSC On-scene coordinator

OSCP EPA's Office of Science Coordination and Policy

OSEC EPA's Office of Sustainable Ecosystems and Communities
OSHA Occupational Safety and Health Administration
OSPS EPA's Outreach and Special Projects Staff
OSW EPA's Office of Solid Waste
OSWER Office of Solid Waste and Emergency Response
OVA Organic vapor analyzer (portable organic vapor monitoring device)
P² Pollution Prevention
PAHs Polyaromatic hydrocarbons
PA/SI EPA's Preliminary Assessment and Site Inspection
Pb Lead
PCBs Polychlorinated biphenyls
PCE Perchloroethylene
PCM Phase contrast microscopy
pCi Picocurie
PEL Permissible exposure limit (OSHA)
PF Protection factor
PID Photoionization detector (measures and identifies materials according to their ionization potential)
PIO Public information officer
PHC Principal hazardous constituent
PLIRRA Pollution Liability Insurance and Risk Retention Act
PLM Polarized light microscopy
PM$_{2.5}$ Particulate matter smaller than 2.5 microns in diameter
PM$_{10}$ Particulate matter (nominally 10 microns and less)
PM$_{15}$ Particulate matter (nominally 15 microns and less)
PNA Polynuclear aromatic hydrocarbons
PNOR Particulates not otherwise regulated (PEL for dusts not on OSHA's Z-lists)
POHC Principal organic hazardous constituent
POP Persistent organic pollutant; performance-oriented packaging
POTW Publicly owned treatment works
ppb Parts per billion
PPE Personal protective equipment
ppm Parts per million
ppt Parts per trillion
PRPs Potentially responsible parties
psia Pounds per square inch, absolute
psig Pounds per square inch, gauge (above atmospheric pressure)
PSM Process Safety Management; point source monitoring
PTFE Polytetrafluoroethylene (Teflon®)
PVA Poly vinyl alcohol
PVC Poly vinyl chloride
QA/QC Quality assurance/quality control
QAO Quality Assurance Officer
QAPP Quality assurance program plan
QL Quantification limit

RA   Remedial action; risk analysis; risk assessment

RACM   Reasonably achievable control measures

RACT   Reasonably achievable control technology

RAD   Radiation adsorbed dose (unit of measurement of radiation absorbed by humans)

RAIS   Risk Assessment Information System (Oak Ridge National Laboratory)

RAP   Remedial accomplishment plan; response action plan

RBC   Red blood cell

RCRA   Resource Conservation and Recovery Act of 1976 (PL 94-580)

RCRIS   Resource Conservation and Recovery Information System

RD/RA   Remedial Design/Remedial Action

R&D   Research and development

REL   Recommended Exposure Limit (NIOSH)

REM   (Roentgen equivalent man)

REP   Radiological emergency preparedness

RFI   Remedial field investigation

RI/FS   Remedial Investigation/Feasibility Study

RMCL   Recommended maximum contaminant level

RMP   Risk Management Programs

ROD   Record of decision

RODS   Records of decision system

RRC   Regional response center

RRP   Regional response plan

RRT   Regional Response Team

RSPA   Research and Special Programs Administration (DOT)

RTECS   Registry of Toxic Effects of Chemical Substances at NIOSH

RQ   Reportable quantity

SARA   Superfund Amendments and Reauthorization Act of 1986

SBR   Styrene-butadiene rubber

SCBA   Self-contained breathing apparatus

SCFM   Standard cubic feet per minute

SDWA   Safe Drinking Water Act

SDWIS   Safe Drinking Water Information System

SERC   State Emergency Response Commission

SETAC   Society for Environmental Toxicology and Chemistry

SI   International System of Units; site inspection

SIC   Standard Industrial Classification

SMOA   Superfund Memorandum of Agreement

SO$_2$ Sulfur dioxide

SOPs   Standard operating procedures

SRAP   Superfund Remedial Accomplishment Plan

STEL   Short-term exposure limit

STP   Sewage treatment plant; standard temperature and pressure

SQG   Small quantity generator

SWDA   Solid Waste Disposal Act

TAB   Technical Assistance to Brownfields Communities

TAG   Technical assistance grant

TAMS   Toxic Air Monitoring System

TAPDS   Toxic Air Pollutant Data System

TCD   Thermal conductivity detector

TCDD   Dioxin (tetrachlorodibenzo-*p*-dioxin)

TCDF   Tetrachlorodibenzofurans

TCE   Trichloroethylene

TCLP   Toxicity characteristic leachate procedure

TD$_{50}$ A numerical description of carcinogenic potency

THC   Total hydrocarbons

TIO   Technology Innovation Office

TLV®s   Threshold limit values (ACGIH)

TLV®-C   TLV-ceiling

TLV®-STEL   TLV-short-term exposure limit

TLV®-TWA   TLV-time-weighted average

TMDL   Total maximum daily limit; total maximum daily load

TNT   Trinitrotoluene

TOC   Total organic carbon

TPQ   Threshold planning quantity (also referred to as TQs)

TRANSCAER   Transportation Community Awareness and Emergency Response

TRI   Toxics Release Inventory

TSCA   Toxic Substances Control Act

TSCATS   TSCA Test Submissions Database

TSDF   Treatment, storage, and disposal facility

TSI   Thermal system insulation

TSP   Total suspended particulates

TSS   Total suspended (nonfilterable) solids

TTO   Total toxic organics

TVOC   Total volatile organic compounds

TWA   Time-weighted average

UEL   Upper explosive limit

UFL   Upper flammable limit

UNEP   United Nations Environment Programme

USCG   United States Coast Guard

USGS   United States Geological Survey

UST   Underground storage tank

UTM   Universal Transverse Mercator

UV   Ultraviolet light

UVA, UVB, UVC   Ultraviolet radiation bands (A, B, C)

VCP   Voluntary cleanup program

VOC   Volatile organic compounds

VP   Vapor pressure

WAP   Waste analysis plan

WB   Wet bulb

WHP   Wellhead protection program

WHPA   Wellhead protection area

XRF   X-ray fluorescence

ZHE   Zero headspace extractor

SI Prefixes and Symbols (SI = Système International d'Unités [International System of Units])

exa (E) = 10^{18}

peta (P) = 10^{15}

tera (T) = 10^{12} = trillion

giga (G) = 10^9 = billion

mega (M) = 10^6 = million

kilo (k) = 10^3 = thousand

milli (m) = 10^{-3} = thousandth

micro (μ) = 10^{-6} = millionth

nano (n) = 10^{-9} = billionth

pico (p) = 10^{-12} = trillionth

femto (f) = 10^{-15}

atto (a) = 10^{-18}

INDEX

ABOUT THE AUTHORS

All of the authors are Members of the American Institute of Chemical Engineers.

E. Ellsworth Hackman, III, Ph.D., P.E. has been President of NST/Engineers, Inc. since 1973; NST/Engineers is a Delaware firm engaged in safety, environmental, chemical, and quality engineering and training. Dr. Hackman is a licensed Professional Engineer in Delaware and Pennsylvania, and certified by the National Council of Examiners for Engineering and Surveying. He has a Ph.D. in Physical Chemistry (combustion research) from the University of Delaware, an M.S. in Chemical Engineering from the University of Pennsylvania, and a B.S. in Chemistry from Juniata College. He is the author of *Toxic Organic Chemicals: Destruction and Waste Treatment*. Dr. Hackman's experience includes PCBs waste treatment process evaluation; air pollution control system design; ISO 9000 quality systems development; and chemical plant start-up, safety, and operating instructions preparation. His hazardous materials experience includes work with propellants, explosives, biological and chemical agents, cryogenics, acids and acid gases, cyanides, and VOCs. Dr. Hackman has also led over 250 sessions of a popular corporate, in-house, continuing education program (Air Products University); and numerous seminars on the selection of processes for hazardous waste treatment.

Christian L. Hackman is a Certified Environmental Manager whose 25 years of hands-on experience in hazardous waste management includes work with the U.S. Navy, DuPont Engineering Test Center, and NST/Engineers, Inc. He has trained over 1,000 men and women in HAZWOPER, Emergency Response, Incident Command System, Confined Spaces, Personal Protective Equipment, Lead and Asbestos Abatement, and numerous other employee safety courses. His project experience includes site remediation design and implementation, on-site safety audits and inspections, Environmental Site Assessment, determination of long-term remediation costs for "cost-cap" insurance, and air and wastewater permitting projects. Mr. Hackman has, for the last four years, been the Chairperson of the New Castle County (Delaware) Chamber of Commerce—Environmental and Engineering Forum.

Matthew E. Hackman is a Professional Engineer in Alabama, Connecticut, Delaware, Illinois, Massachusetts, New Jersey, Rhode Island, and Texas. He is also Certified by the National Council of Examiners for Engineering and Surveying. Mr. Hackman received his B.S. in Chemical Engineering from the University of Delaware, an M.S. in Chemical Engineering from Princeton, and an MBA from Northeastern University. He is a Certified Hazardous Materials Manager (Senior Level) and past president of the New England ACHMM chapter. He has been responsible for the design and implementation of numerous remediation projects. Air and Wastewater permitting are a specialty. He is a Licensed Site Professional (LSP) in Massachusetts and a Licensed Environmental Professional (LEP) in Connecticut. Formerly the Technical Director for a Part B TSDF, he is currently the LSP for two multimillion dollar HUD Demonstration Disposition multifamily housing renovation and remediation projects in Boston, Massachusetts.

SOFTWARE AND INFORMATION LICENSE

The software and information on this diskette (collectively referred to as the "Product") are the property of The McGraw-Hill Companies, Inc. ("McGraw-Hill") and are protected by both United States copyright law and international copyright treaty provision. You must treat this Product just like a book, except that you may copy it into a computer to be used and you may make archival copies of the Products for the sole purpose of backing up our software and protecting your investment from loss.

By saying "just like a book," McGraw-Hill means, for example, that the Product may be used by any number of people and may be freely moved from one computer location to another, so long as there is no possibility of the Product (or any part of the Product) being used at one location or on one computer while it is being used at another. Just as a book cannot be read by two different people in two different places at the same time, neither can the Product be used by two different people in two different places at the same time (unless, of course, McGraw-Hill's rights are being violated).

McGraw-Hill reserves the right to alter or modify the contents of the Product at any time.

This agreement is effective until terminated. The Agreement will terminate automatically without notice if you fail to comply with any provisions of this Agreement. In the event of termination by reason of your breach, you will destroy or erase all copies of the Product installed on any computer system or made for backup purposes and shall expunge the Product from your data storage facilities.

LIMITED WARRANTY

McGraw-Hill warrants the physical diskette(s) enclosed herein to be free of defects in materials and workmanship for a period of sixty days from the purchase date. If McGraw-Hill receives written notification within the warranty period of defects in materials or workmanship, and such notification is determined by McGraw-Hill to be correct, McGraw-Hill will replace the defective diskette(s). Send request to:

Customer Service
McGraw-Hill
Gahanna Industrial Park
860 Taylor Station Road
Blacklick, OH 43004-9615

The entire and exclusive liability and remedy for breach of this Limited Warranty shall be limited to replacement of defective diskette(s) and shall not include or extend to any claim for or right to cover any other damages, including but not limited to, loss or profit, data, or use of the software, or special, incidental, or consequential damages or other similar claims, even if McGraw-Hill has been specifically advised as to the possibility of such damages. In no event will McGraw-Hill's liability for any damages to you or any other person ever exceed the lower of suggested list price or actual price paid for the license to use the Product, regardless of any form of the claim.

THE McGRAW-HILL COMPANIES, INC. SPECIFICALLY DISCLAIMS ALL OTHER WARRANTIES, EXPRESS OR IMPLIED, INCLUDING BUT NOT LIMITED TO, ANY IMPLIED WARRANTY OF MERCHANTABILITY OR FITNESS FOR A PARTICULAR PURPOSE. Specifically McGraw-Hill makes no representation or warranty that the Product is fit for any particular purpose and any implied warranty of merchantability is limited to the sixty day duration of the Limited Warranty covering the physical diskette(s) only (and not the software or information) and is otherwise expressly and specifically disclaimed.

This Limited Warranty gives you specific legal rights; you may have others which may vary from state to state. Some states do not allow the exclusion of incidental or consequential damages, or the limitation on how long an implied warranty lasts, so some of the above may not apply to you.

This Agreement constitutes the entire agreement between the parties relating to use of the Product. The terms of any purchase order shall have no effect on the terms of this Agreement. Failure of McGraw-Hill to insist at any time on strict compliance with this Agreement shall not constitute a waiver of any rights under this Agreement. This Agreement shall be construed and governed in accordance with the laws of New York. If any provision of this Agreement is held to be contrary to law, that provision will be enforced to the maximum extent permissible and the remaining provisions will remain in force and effect.